PLURIPOTENT STEM CELLS

Edited by **Deepa Bhartiya**
and **Nibedita Lenka**

Pluripotent Stem Cells

http://dx.doi.org/10.5772/45917
Edited by Deepa Bhartiya and Nibedita Lenka

Contributors

Jijun Hao, Guofeng Han, Hongtao Wang, Kun-Hsiung Lee, Herman S. Cheung, Kiminobu Sugaya, Jacob Kimmel, Antonio Liras, Aline S Gaban, Cristina Segovia, Mari Pekkanen-Mattila, Kristiina Rajala, Katriina Aalto-Setälä, Takuro Horii, Izuho Hatada, Laura Sperling, Patricia Miang Lon Ng, Cheng-I Wang, Walden Ai, Ying Shi, Shohreh Mashayekhan, Andrew Laslett, Juan Carlos Polanco, Guanghu Wang, Erhard Bieberich, Antonio Lo Nigro, Catherine M. Verfaillie, Minoru Tomizawa, Derrick E. Rancourt, Mehdi Shafa, Perla Cota-Elizondo, Roxana Nat, Andreas Eigentler, Georg Dechant, Irina Kerkis, Nelson Foresto Lizier, Cristiane Wenceslau, Alexandre Kerkis, Thomas Lufkin, Joel Sng, Hideyuki Kobayashi, Charles Hong, Hidetoshi Masumoto, Jun K Yamashita, Virt, Andreas M. Weiss, Michael Breitenbach, Candace Kerr, Michael Wassler, Deepa Bhartiya

Published by InTech

Janeza Trdine 9, 51000 Rijeka, Croatia

Publishing Process Manager Sandra Bakic
Typesetting InTech DTP Team
Cover InTech Design Team

Additional hard copies can be obtained from orders@intechopen.com

Pluripotent Stem Cells, Edited by Deepa Bhartiya and Nibedita Lenka
p. cm.
ISBN 978-953-51-1192-4

For EU product safety concerns:
IN TECH d.o.o., Prolaz Marije Krucifikse Kozulić 3, 51000 Rijeka, Croatia,
info@intechopen.com or visit our website at intechopen.com.

Meet the editors

Deepa Bhartiya is currently head of Stem Cell Biology Department at National Institute for Research in Reproductive Health, Mumbai, India. Her group has derived two well characterized human embryonic stem cell lines: KIND1 and KIND2. Pre-clinical evaluation of safety, efficacy and feasibility of pancreatic and tripotent cardiac progenitors obtained by directed differentiation are ongoing in animal models. Besides the group, she is also interested in a unique population of pluripotent stem cells which exist in adult body tissues termed very small embryonic-like stem cells (VSELs). VSELs have been studied in cord blood, bone marrow and mammalian gonads resulting in several publications.

Nibedita Lenka serves as a senior faculty scientist at National Centre for Cell Science, Pune, India. The prime focus of her group pertains to exploring the guiding cues underlying the cell fate decision machinery using pluripotent embryonic stem cells as a model system. Besides, her group is indulged in understanding the controversial phenomenon of transdifferentiation from mesenchymal stem cells derived from sources ranging bone marrow, umbilical cord blood and cord tissues. The ongoing investigations aim at understanding the mechanistic basis of stem cells maintenance, the temporal action and cross-talk among factors contributing to mesodermal and neuroectodermal specifications and subsequent differentiation into functional cardiomyocytes and neurons with special focus on dopaminergic neuronal subtypes respectively, and exploring their therapeutic efficacy by using rodent models.

Contents

Preface

Pluripotency is basically defined as the potential to give rise to all three germ layer derivatives reflecting to 200 odd cell types present in the body, and except the extra-embryonic cell types. It could be either (i) inherent as seen in case of embryonic stem (ES) cells derived from the inner cell mass of blastocyst stage embryo, embryonic/primordial germ (EG/PG) cells derived from developing gonads, embronal carcinoma (EC) cells or (ii) acquired. The latter category includes the cells from adult tissues reprogrammed to attain the pluripotent state and functioning similar to ES cells. This could be achieved either by somatic cell nuclear transfer (SCNT) or by incorporating a set of transcription factors specific to ES cells into somatic cells and their transient activation leading to induction of pluripotency in them, the latter being designated as induced pluruipotent stem cells (iPSCs) pioneered by Dr. Shinya Yamanaka (the recipient of Nobel Prize in Physiology / Medicine for the year 2012 along with Sir Dr. John Gordon, the father of reprogramming phenomenon).

Undoubtedly the recent progress in Stem Cells research field has opened up a wider horizon with various interesting avenues for explorations in basic Biology and Development and on successful broad spectral implication of stem cells in cell replacement therapy, gene therapy, live stock improvement and tissue engineering as well as in pharmaceutical industries. However, a priori requirement is to have the mechanistic understanding and address the safety vs. the efficacy issue including the ethical concerns. In an effort in this line, the book, "Pluripotent Stem Cells" is a compendium addressing the aforesaid aspects. Indeed wide gamut of topics has been covered in the book under various sections with contributions from experts in the stated field concerning diverse aspects of pluripotent stem cells. We hope that the book would add up to the existing knowledge on pluripotent stem cells.

Dr. Deepa Bhartiya
National Institute for Research in Reproductive Health (NIRRH),
India

Dr. Nibedita Lenka
National Centre for Cell Science,
Pune University Campus,
Ganeshkhind, Pune, India

Section 1

Pluripotent Stem Cells: The Genesis and Means

Chapter 1

An Overview of Pluripotent Stem Cells

Deepa Bhartiya, Punam Nagvenkar,
Kalpana Sriraman and Ambreen Shaikh

Additional information is available at the end of the chapter

http://dx.doi.org/10.5772/55130

1. Introduction

This book is entitled **Pluripotent Stem Cells** (PSCs) and various contributors have written on different aspects of the PSCs. But I will fail as an editor of this book if I do not bring to the reader's attention the all the sources of PSCs (Figure 1).

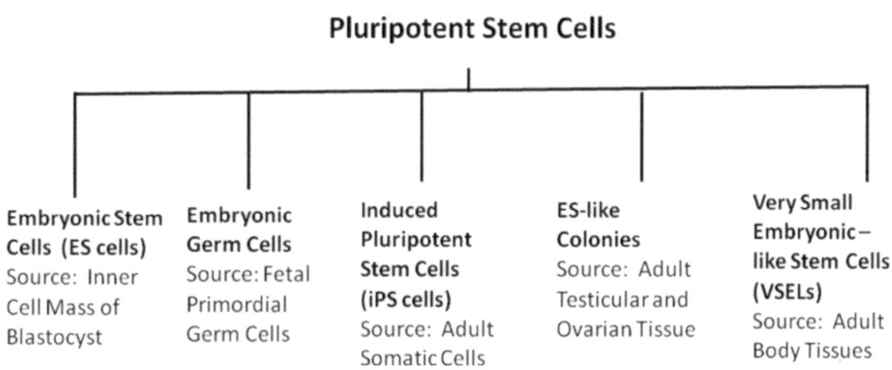

Figure 1. Potential sources for pluripotent stem cells

Professor Thomson and Prof Gearhart published landmark papers in 1998 wherein they published derivation of PSCs from inner cell mass of spare human blastocyst [1] and from early fetal germ cells [2] respectively. Recently Professor Yamanaka was awarded the Nobel prize for medicine for establishing protocols to reprogram somatic cells to embryonic state with the help of 4 factors [3, 4]. Besides this there are several papers which have reported derivation of ES-like colonies from adult testicular biopsies in both mice [5, 6] and men [7-10]. Similarly Gong et al [11] reported ES-like culture using ovarian tissue. There is a huge body of literature suggesting that mesenchymal stem cells (MSCs) have pluripotent characteristics and can transdifferentiate [12]. We have recently published that adult gonads [13, 14] umbilical cord blood/tissue, bone marrow [15] etc. harbor a sub-population of similar kind of pluripotent stem cells termed very small embryonic-like stem cells (VSELs). We also developed a case for VSELs which may be resulting in ES-like colonies rather than de-differentiation of spermatogonial stem cells into pluripotent state [16]. Moreover, the VSELs have confused the field of MSCs, since they are always present as a sub-population amongst MSCs but have remained unnoticed and the pluripotent properties were conferred incorrectly on to the MSCs. VSELs are not widely accepted at present, but have been shown to have promising application towards regenerative medicine.

Thus the aim of the present chapter is to update the readers with the recent advances with embryonic stem cells, induced pluripotent stem cells and VSELs which have been implicated with maximum potential for use in cell-based therapies.

2. Embryonic stem cells

Embryonic stem (ES) cells, as the name suggests, are derived from embryos, more specifically from the inner cell mass (ICM) of the blastocyst. ES cells are characterized by two hallmark properties viz., self-renewal - ability to proliferate indefinitely and pluripotency - capacity to give rise to cells of all the three embryonic germ lineages such as ectoderm, mesoderm and endoderm. They possess a high nucleo-cytoplasmic ratio and telomerase activity. ES cells display high activity of endogenous alkaline phosphatase and express several nuclear and cell-surface markers of pluripotency. They tend to cluster together when cultured in suspension on a non-adherent surface to form 3D aggregates known as embryoid bodies that may be simple or cystic. Moreover, they produce teratomas on injection in immune deficient (SCID) mice, are clonogenic and are capable of producing chimeras when injected into blastocysts in the mouse model.

3. Mouse ES cells

ES cells were first derived from ICM of mouse blastocyst stage embryos [17, 18]. Besides ICM of blastocyst mouse ES (mES) cells have also been derived from cleavage stage embryos and even from biopsied individual blastomeres of two- to eight-cell stage embryos [19- 21]. In general, mES cells can be cultured on a layer of mitotically inactive mouse embryonic fibroblasts (MEF) in the presence of serum and leukaemia inhibitory factor (LIF). The cytokine LIF sustains the self-renewing and pluripotency features of mES cells. LIF, a soluble glycoprotein of interleukin (IL)-6 family of cytokines acts via binding to heterodimers of the LIF-receptor and the signal transducer gp130 resulting in activation of STAT3 signaling [22-24]. In absence of serum, LIF is incapable of maintaining pluripotency of mES cells; however, in combination with bone morphogenetic protein-4 (BMP4) prevents differentiation of mES cells [25]. BMP4 induces expression of *Id* (Inhibitor of differentiation) genes via the Smad pathway. Overexpression of *Id* indeed allows proliferation of mES cells in the presence of LIF and without need of BMP4 or serum.

4. Human ES cells

A breakthrough occurred with the derivation of human ES (hES) cells in 1998 [1]. Since the first report on derivation of hES cell lines at least 1071 hES cell lines have been derived worldwide [26]. Besides spare human blastocysts, hES cell lines have also been derived from morula stage embryos [27], abnormally developing and arrested embryos [28], single blastomeres of 8-cell stage embryos [29] and 4-cell stage embryos [30, 31]. Mitotically inactivated feeder cells and serum containing medium along with basic fibroblast growth factor (bFGF) are generally used to maintain hES cells. LIF and its related cytokines fail to support hES cells in serum-containing media that supports mES cells despite the existence of a functional LIF/STAT3 signaling pathway in hES cells [1, 32, 33]. In contrast to mES cells, FGF and TGF/Activin/Nodal signaling are essential for the self-renewal of hES cells [34]. Although, elements of the BMP pathway exist in hES cells [35], but unlike mES cells, BMPs added to hES cells in conditions that would otherwise support self-renewal, cause rapid differentiation [36]. Recent studies have revealed multiple interactions between the FGF, TGFβ, and BMP pathways in hES cells. Activin induces bFGF expression [37], and bFGF induces Tgfβ1/TGFβ1 and Grem1/GREM1 (a BMP antagonist) expression and inhibits Bmp4/BMP4 expression in both fibroblast feeders and in hES cells [38].

Although similar in their characteristics such as expression of Oct-4, Nanog, alkaline phosphatase activity, formation of embryoid bodies, teratoma formation, some potential differences exist between mES cells and hES cells. In contrast to mES cells which show expression of SSEA-1, hES cells express SSEA-3/4, TRA-1-60/81. Further, the average population doubling time for hES cells is longer compared to mES cells (30-35 hr vs. 12-15 hr).

5. *In vitro* culture and differentiation of hES cells

Although hES cell lines were first derived on MEF feeder layers, continuous efforts towards developing xeno-free culture system has resulted in establishment of human feeders derived from fallopian tube epithelium [39], fetal foreskin, muscle [40, 41], or amniotic epithelium [42]. Attempts have been made to derive new hES cell lines in more defined conditions including serum-free or feeder-free conditions in the presence of extracellular matrices such as matrigel and fibronectin [43-45]. Crook et al [46] derived six clinical-grade hES cell lines using GMP-grade human feeder grown in a medium with GMP-quality FBS and propagated the cell lines using a GMP formulation of Knockout Serum Replacement (KO-SR). Although not xeno-free, the cell lines meet clinical quality. Sidhu et al [47] reported the derivation of hES cell line in culture using human-derived collagen coated plates and KO-SR to maintain human feeder fibroblasts. A fully defined xeno-free medium (RegES), capable of supporting the expansion of hES cell lines, induced pluripotent stem (iPS) cells and adipose stem cells has been described [48]. Recently, Wang et al [49] have developed a xeno-free and feeder-cell-free culture system for propagating hES cells and hiPS cells using human plasma and human placenta extracts.

Human ES cells have the ability to form 200 odd cell types in our body. Essentially, ES cells can be differentiated spontaneously by embryoid body formation or by directed differentiation using a cocktail of growth factors. Several growth factors have been shown to direct differentiation of ES cells namely activin-A and transforming growth factor (TGF-β1) mainly induce mesodermal cells; retinoic acid (RA), epidermal growth factor (EGF), BMP-4, and bFGF activate ectodermal and mesodermal cells; β nerve growth factor (NGF) and hepatocyte growth factor (HGF) differentiate all three embryonic germ layers [50-53]. Directed differentiation is a more controlled process involving stage specific sequential addition of growth inducers and inhibitors which are known to effect key pathways. For e.g. activin A and BMP4 are two such growth factors which have been used widely for cardiogenic differentiation. Various studies have shown that hES cells can be differentiated into neuronal [54], hematopoietic [55], endothelial [56], muscle [57], cardiac [58, 59] pancreatic [60, 61], hepatic [62] lineages. Although hES stem cell lines are similar with respect to self-renewal and expression of pluripotency markers, published literature however suggests that they exhibit differences in their differentiation ability under identical culture conditions [63, 64].

6. Potential use of ES cells

The remarkable features of hES cells has served as an important breakthrough for basic research and has great potential for regenerative medicine. ES cells may act as key research tools for understanding the complex events that occur during embryonic development which may explain the causes of birth defects. They are ideal candidates for studying apoptosis in early stage of embryo, mechanism of differentiation, mutagenesis, immune rejection and aging. Human ES cells and their derivatives may be used for testing therapeutic drug efficacy and toxicity. They also have wide applications in tissue engineering. Following their culture on polymer scaffold, it has been reported to coax stem cells to form tissues with characteristics of developing human cartilage, liver, neurons and blood vessels.

Despite being associated with the risk of inducing teratomas and immune rejection, the vital potential application of hES cells is the generation of cells and tissues that could be used for cell-based therapies. Human ES cells directed to differentiate into specific cell types offer the possibility of a renewable source of replacement cells and tissues to treat a myriad of diseases and disabilities including Parkinson's and Alzheimer's diseases, spinal cord injury, burns, heart failure and diabetes etc. The first FDA-approved phase-1 clinical trial for safety began with Geron's (Menlo Park, CA, USA) GRNOPC1 derived oligodendrocyte progenitor cells to treat complete thoracic-level spinal cord injury [65]. The trial was initially stalled for occurrence of microscopic cysts in animal transplants but was later approved [66, 67]. However, in November 2011 Geron dropped out of stem cell research for financial reasons and said that they would continue to monitor existing patients, and were attempting to find a partner that could continue their research. The recent success of a prospective clinical study of Advanced Cell Technology (CA and MA, USA) to establish the safety and tolerability of subretinal transplantation of hES cell-derived retinal pigment epithelium (RPE) in patients with Stargardt's Macular Dystrophy (SMD) and Dry age-related Macular Degeneration (Dry AMD) represents an important step towards therapeutic use of hES cells [68]. Although long-term follow up is essential and eye is an immune-privileged site; it is still encouraging to note that there are no associated signs of hyperproliferation, tumorigenicity, ectopic tissue formation, or immune- rejection after 4 months of transplantation.

7. Induced pluripotent stem cells

A major progress in the stem cell field was generation of induced pluripotent stem (iPS) cells by the reprogramming of somatic cells to an embryonic stem cell state using a cocktail of transcription factors. In 2006, Takahashi and Yamanaka reprogrammed mouse fibroblasts through retroviral transduction with 24 candidate genes [3]. The pool of genes was gradually reduced to four transcription factors, Oct4, Sox2, c-Myc, and Klf4. The results were rapidly confirmed by various researchers [69-71]. Soon the technology was successfully applied to generate iPS cells from human fibroblasts [4, 72, 73]. Concurrently, another group identified Oct4, Sox2, Nanog, and Lin28 to be sufficient to reprogram human cells, with Oct4 and Sox2 appearing essential and the other two factors either strongly (Nanog) or modestly (Lin28) influencing the efficiency of reprogramming [74].

The different ways for generation of mouse and human iPS cells using various reprogramming factors has been well summarized by Maherali and Hochedlinger [75] and Kiskinis and Eggan [76]. The choice of a gene delivery system is a key aspect for generation of iPS cells and has been very well reviewed by Oh et al [77]. Many researchers have reported use of integrating viral vectors such as retroviral [4, 73, 78] and lentiviral vectors [74, 79], non-integrating viral vectors such as adenoviral [80] and Sendaiviral vectors [81], nonviral methods such as plasmid DNA [82], piggyBac transposons [83, 84], recombinant proteins [85, 86], mRNAs [87] and small molecules such as valproic acid [88]. Moreover, derivation of iPS cells from patients suffering from the neurodegenerative disease amyotrophic lateral sclerosis (ALS) [89] as well as patients with other diseases, including juvenile onset type 1 diabetes mellitus, Parkinson disease (PD) [90], and spinal muscular atrophy (SMA) [91] has been reported.

8. Advantages and disadvantages of iPS cells

As a potential application in cell based therapy, one of the major advantages of iPS cells is the avoidance of immune rejection, since they are derived from a patient's own cells, as well as ethical issues associated with the use of human embryos. Furthermore, iPS cells are similar to ES cells in many aspects, including cell morphology, expression of pluripotency markers, long telomeres and capability to form embryoid bodies, teratoma, and viable chimeras [92, 93]. Apart from use in cell-based therapy, iPS cells derived from patients with disease can serve as an effective model to understand the mechanisms of diseases.

However, use of iPS cells have several drawbacks and are mostly related to current reprogramming methods. Viral vectors employed for gene delivery has led to the integration of multiple viruses into iPS cell genomes, resulting in tumorigenesis due to genetic abnormalities in the cells. Moreover, the efficiency of reprogramming of human iPS cells from fibroblasts is very low, approximately less than 0.02% [94]. The use of Myc gene as a reprogramming factor and/or the reactivation of a silenced Myc gene might cause iPS cells to become cancer cells [95].

Recently, three studies published in Nature showed that the reprogramming process and the subsequent culture of iPS cells *in vitro* can induce genetic and epigenetic abnormalities in these cells. Gore et al [96] found on an average of five point mutations in each of the iPS cell line analyzed, with the majority of the mutations being non-synonymous, nonsense or splice variants, and were enriched in genes mutated or having causative effects in cancers. Hussein and colleagues [97] showed that copy number variations (CNVs) occurred at a high rate during the process of reprogramming leading to genetic mosaicism in early-passage iPSCs. Analysis of the CG methylation patterns by Lister et al [98] identified numerous differentially methylated CG regions (CG-DMRs) between iPS cells and ES cells. The presence of a core set of CG-DMRs in every iPS cell line suggests hotspots of failed epigenomic reprogramming. These studies raise concerns over the implications of such aberrations for future applications of iPS cells. A much more in-depth research is necessary to understand about the reprogramming process and the biological consequences of these genomic and epigenomic changes needs to be investigated.

9. Very small embryonic like stem cells

The ethical and other technical issues concerning the use of ES cells in regenerative medicine have led to search for alternative stem cells with therapeutic potential. In this regard adult stem cells can potentially provide a therapeutic alternative to ES or iPS cells. Though adult stem cells are known to be tissue specific and can only differentiate into cells of their tissues of origin, nevertheless several studies have reported that adult stem cells can differentiate in to cells of completely different lineage. The process is termed as adult stem cell plasticity. Wagers and Weissman proposed few potential mechanisms and explanations for the observed adult stem cell plasticity [99]. The potential mechanisms include trans-differentiation or de-differentiation of stem cells, presence of multiple different stem cells in a tissue, presence of pluripotent stem cells in addition to adult stem cells and cell fusion of stem cell with cell of different lineage. However, several lines of evidence support existence of pluripotent stem cells in adult tissues that can differentiate into all three lineages explaining adult plasticity the best. Many investigators have reported presence of pluripotent stem cells in adult tissues and were defined either as mesenchymal stem cells (MSCs) [100], multipotent adult progenitor cells (MAPCs) [101], marrow isolated adult multilineage inducible cells (MIAMI) [102], multipotent adult stem cells (MASCs) [103], very small embryonic like stem cells (VSELs) [104]. Although these cells may represent an overlapping type of stem cells, the most characterized among these cells to the single cell level is VSELs and they have been isolated and identified in several adult body organs.

VSELs are defined as epiblast derived stem cells, which are deposited early during organogenesis and may serve as source of tissue commited stem cells. Pluripotent VSELs ($Oct4^+$, $SSEA1^+$, $Sca1^+$, Lin^-, $CD45^-$) were first reported by Ratajczak and group in various adult mice tissues [105]; highest numbers being in brain, kidneys, muscles, pancreas and bone marrow [106]. These are diploid cells with high telomerase activity, express other pluripotent (Rex-1, Nanog, SSEA and Klf-4) and germ cell (Mvh, Stella, Fragilis, Nobox and Hdac-6) markers and decrease in numbers with age [107]. An important evidence for pluripotency of VSELs is hypomethylated status of OCT-4 promoter and its association with transcription promoting histones [108] as well as presence of bivalent domains [109]. Like embryonic stem cells they do not express MHC class I and HLA-DR antigens and are also negative for mesenchymal stem cell markers like CD90, CD105, CD29. They are very small in size (3-5 um in mice), have a large nucleo-cytoplasmic ratio, and open chromatin structure for OCT-4 and Nanog promoter [107]. OCT-4 expression at mRNA and protein level in VSELs has been confirmed using sequence specific primers. VSELs have the ability to differentiate into three germ layers *in vitro*, however unlike ES cells, VSELs neither complement during blastocyst development nor form teratomas in immuno-deficient mice [110]. Attempts have been made to propagate them on feeder layers, but they do not self-renew as easily as the established embryonic stem cell lines possibly because of altered methylation status of some developmentally crucial genes. Similar VSELs have also been isolated from human umbilical cord blood, mobilized peripheral blood, and adult bone marrow by flow cytometry as $CD133^+$, lin^-, $CD45^-$ [104] and also by differential centrifugation method [15, 111].

VSELs are descendants of epiblast stage pluripotent stem cells. They get deposited in various body organs including the gonads in early stages of development, as a quiescent stem cell population which possibly serves as a back up to the tissue committed stem cells (TCSCs) [112]. These two populations of stem cells (VSELs and TCSCs) together are responsible in bringing about tissue renewal, homeostasis and regeneration after injury throughout life and decrease in number with age. The co-existence of two stem cell populations (the more primitive being quiescent and the progenitor being more rapidly dividing) has been recently proposed [113, 114]. VSELs are the DNA label-retaining (e.g. BrdU), quiescent stem cells with lower metabolic state whereas the tissue committed stem cells actively divide and do not retain DNA label over time. They are highly mobile, respond to the SDF-1 gradient and enter into circulation in case of any injury to bring about regeneration and homeostasis. They are also considered as a missing link to support the germ-line hypothesis of cancer development [115, 116].

VSELs in Umbilical Cord Blood (UCB): A population of human cells similar to murine bone marrow derived VSELs was first reported by Kucia et al in umbilical cord blood [117]. These UCB derived VSELs (Lin-/CD45-/CD133+) ranged between 6-8 um in size, possess large nuclei and express nuclear embryonic transcription factors OCT-4, Nanog and cell surface SSEA-4. The strategy of isolation of VSELs from cord blood is hampered by their small size as they get discarded along with debris. Recently our studies reported that VSELs settle along with RBCs and are not enriched in interphase layer of MNCs obtained after ficoll separation of cord blood [15].These VSELs expressed pluripotent markers OCT-4,primitive marker CD133 along with primordial germ cell marker stella and fragilis indicating their epiblast origin. Our studies have also shown the presence of VSELs in the discarded fractions of bone marrow and cord blood obtained after processing [15].

VSELs in adult mammalian gonads: Initial studies by Ratajczak group have shown that mouse testis harbor VSELs [106]. Our group has identified presence of VSELs in testis of human and mice as well as in ovaries of human, sheep, monkey, rabbit and mice [13]. These VSELs are localized in the basal layer of cells adjacent to the basement membrane in seminiferous tubules [13] and were found interspersed with the ovarian surface epithelial cells [14]. The main approach in identifying the VSELs in adult mammalian gonads involves studying differential expression of a pluripotent marker OCT-4. OCT-4 is an octamer binding transcription factor required for maintaining pluripotency of cell. Published literature on OCT-4 in somatic stem cells has confused stem cell researchers [118-120] because of the presence of several pseudo-genes and alternatively spliced transcripts [118, 121]. Thus a careful designing of primers for RT-PCR analysis and proper selection of antibodies becomes essential to detect specific transcripts. Also a careful selection of OCT-4 antibodies is essential to detect pluripotent stem cells [119]. We used a polyclonal OCT-4 antibody that enabled the simultaneous identification of VSELs with nuclear OCT-4 and tissue committed stem cells with cytoplasmic OCT-4. In addition, careful selection of primers for OCT-4A and total OCT-4 for Q-PCR studies has helped us generate interesting results [13-15, 122].

VSELs in Testis: We have documented that adult human testis harbors a population of pluripotent VSELs (with nuclear OCT-4A) which are more primitive to A_{dark} Spermatogonial Stem Cell (SSC) (with cytoplasmic OCT-4B). The VSELs possibly give rise to A_{dark} SSCs, which in turn undergo clonal expansion as evident by the presence of cytoplasmic bridges between the rapidly dividing cells [13]. OCT-4 is not immuno-localized in more differentiated male germ cells. Based on this study a new hierarchy of testicular cells was proposed with all testicular cells originating from VSELs and not from SSCs as generally believed. Similarly presence of VSELs distinct from SSCs was also identified in mouse testicular tissue.

VSELs in Ovaries: The long- held dogma in female biology is that women and other mammalian females are born with fixed and non-renewing pool of germ cells, which are enclosed in structures called follicles. Their number decrease with age due to ovulation or atresia and their exhaustion lead to menopause. However in last 8 years several investigators with access to modern molecular techniques have convincingly demonstrated that adult mammalian ovaries harbor stem cells and undergo postnatal oogeneisis and thus have challenged the central dogma. Presence of PSCs in adult ovary has been demonstrated by many groups [11, 14, 123, 124]. Our group has identified two distinct types of stem cells in ovarian surface epithelium (OSE) of human and other mammalian species [14, 122]. The two stem cells are VSELs that express OCT-4 in nucleus, which are pluripotent and slightly larger progenitor committed cells (termed Ovarian Germ Stem Cells-OGSCs) that express OCT-4 cytoplasmically. This is very similar to reported presence of VSELs and Spermatogonial stem cells in adult mammalian testis as mentioned earlier. We have recently reviewed various publications on ovarian stem cells and explained the results in the context of VSEL biology [122]. Readers are encouraged to read the review for more details.

Based on our studies in ovarian stem cells and other literature, we have proposed a model for oogenesis and follicular assembly in adult mammalian ovaries [122]. According to the model, VSELs with nuclear Oct-4 that are located in the OSE undergo asymmetric cell division and give rise to cells with cytoplasmic Oct-4 (OGSCs, which intensely stain with Haematoxylin). The OGSCs undergo further proliferation, meiosis and differentiation to assemble into primordial follicles in the OSE. The granulosa cells are formed by the epithelial cells through epithelial mesenchymal transition. As the follicles grow and further mature they shift into the ovarian medulla.

10. Clinical potential of VSELs

The clinical potential of VSELs, isolated from cord blood or bone marrow by flow cytometry, is just beginning to emerge. In various disease models like myocardial infarct [125,126], stroke [127], skin burn injury [128], neural regeneration [129] etc. these cells get mobilized into circulation within 24 hours. For myocardial regeneration, the VSELs are very efficient to improve LV ejection fraction and attenuation of myocardial hypertrophy [126]. As they become scarce with age, regeneration becomes inefficient resulting in age-related disease manifestations.

The identification of VSELs in gonads has far reaching implications in reproductive health issues. Understanding the biology of VSELs in gonads may help explain the mechanisms of different pathologies of gonads and may pave for new treatments for infertility. However, application of VSELs to improve reproductive health needs to be researched and established. We have recently studied the differential effect of busulphan on the relatively quiescent VSELs versus rapidly dividing germ cells in adult mice gonads (unpublished results). The VSELs were found to be resistant to the treatment, however were unable to differentiate probably due to the altered niche of VSELs due to treatment. Ratajczak group recently reported that VSELs in mouse bone marrow are resistant to total body irradiation [130]. They observed that there was increase in proliferation of VSELs post treatment, although were unable to reconstitute the bone marrow. These studies open up newer and exciting avenues for fertility preservation in cancer survivors who are rendered infertile by various cancer treatments.

11. Advantages of VSELs over ES cells

VSELs can be derived easily from autologus source and do not form teratoma easily [131]. Thus both the major concerns associated with ES cells of immune-rejection and risk of teratoma formation is taken care of.

12. Advantages of VSELs over iPS cells

There is no need for reprogramming somatic cells (which may harbor mutations) to embryonic state when pluripotent ES-like stem cells can be harvested from adult tissues. They may also be superior to iPS cells since they are derived from a very quiescent stem cell population and are thus 'young' cells with long telomeres that could be isolated from an aged body, in contrast to iPS cells which are derived from terminally differentiated somatic skin fibroblasts (with shortened telomeres) that tend to accumulate DNA mutations over time. In addition, VSELs do not have epigenetic issues associated with iPS cells. Unlike iPS cells, there is no requirement of viral vectors and hence risk of transformation of VSELs into cancer cells is avoided.

13. Future perspectives

Embryonic stem cells are considered to be 'magic bullets' having a great potential for cell-based therapy, however future clinical use of ES cells are still plagued by ethical issues. Hence there is urgent need to expand research in derivation and culture of pluripotent stem cells from alternate sources. Induced pluripotent stem cells though believed to be ideal candidates need to be exploited further to realize their clinical potential. Considering the potential advantages of VSELs over ES and iPS cells, the need for research to harness potentials of VSELs is high. Currently the availability of large number of VSELs for effective use in clinical applications is limited. Research is progressing towards expansion of VSELs in culture and is still in nascent stages. Also, many key questions have to be answered before realizing the full potential of stem cells.

Author details

Deepa Bhartiya*, Punam Nagvenkar, Kalpana Sriraman and Ambreen Shaikh

*Address all correspondence to: deepa.bhartiya@yahoo.in; bhartiyad@nirrh.res.in

Stem Cell Biology Department, National Institute for Research in Reproductive Health, Parel, Mumbai, India

References

[1] Thomson JA, Itskovitz-Eldor J, Shapiro SS, Waknitz MA, Swiergiel JJ, Marshall VS, Jones JM (1998) Embryonic Stem Cell Lines Derived from Human Blastocysts. Science 282:1145-1147.

[2] Shamblott MJ, Axelman J, Wang S, Bugg EM, Littlefield JW, Donovan PJ, Blumenthal PD, Huggins GR, Gearhart JD (1998) Derivation of Pluripotent Stem Cells from Cultured Human Primordial Germ Cells. Proc Natl Acad Sci U S A. 95:13726-13731.

[3] Takahashi K, Yamanaka S (2006) Induction of Pluripotent Stem Cells from Mouse Embryonic and Adult Fibroblast Cultures by Defined Factors. Cell 126: 663–676.

[4] Takahashi K, Tanabe K, Ohnuki M, Narita M, Ichisaka T, Tomoda K, Yamanaka S (2007) Induction of Pluripotent Stem Cells from Adult Human Fibroblasts by Defined Factors. Cell 131: 861–872.

[5] Kanatsu-Shinohara M, Inoue K, Lee J, Yoshimoto M, Ogonuki N, Miki H, Baba S, Kato T, Kazuki Y, Toyokuni S, Toyoshima M, Niwa O, Oshimura M, Heike T, Nakahata T, Ishino F, Ogura A, Shinohara T (2004) Generation of pluripotent stem cells from neonatal mouse testis. Cell 119: 1001-1012.

[6] Guan K, Nayernia K, Maier LS, Wagner S, Dressel R, Lee JH, Nolte J, Wolf F, Li M, Engel W, Hasenfuss G (2006) Pluripotency of spermatogonial stem cells from adult mouse testis. Nature 440: 1199-1203.

[7] Conrad S, Renninger M, Hennenlotter J, Wiesner T, Just L, Bonin M, Aicher W, Buhring HJ, Mattheus U, Mack A, Wagner HJ, Minger S, Matzkies M, Reppel M, Hescheler J, Sievert KD, Stenzl A, Skutella T (2008) Generation of pluripotent stem cells from adult human testis. Nature 456: 344-349.

[8] Golestaneh N, Kokkinaki M, Pant D, Jiang J, DeStefano D, Fernandez-Bueno C, Rone JD, Haddad BR, Gallicano GI, Dym M (2009) Pluripotent stem cells derived from adult human testes. Stem Cells Dev 18: 1115-1126.

[9] Kossack N, Meneses J, Shefi S, Nguyen HN, Chavez S, Nicholas C, Gromoll J, Turek PJ, Reijo-Pera RA (2009) Isolation and characterization of pluripotent human spermatogonial stem cell-derived cells. Stem Cells 27: 138-149.

[10] Mizrak SC, Chikhovskaya JV, Sadri-Ardekani H, van Daalen S, Korver CM, Hovingh SE, Roepers-Gajadien HL, Raya A, Fluiter K, de Reijke TM, de la Rosette JJ, Knegt AC, Belmonte JC, van der Veen F, de Rooij DG, Repping S, van Pelt AM (2009) Embryonic stem cell-like cells derived from adult human testis. Human Reprod 25: 158-167.

[11] Gong SP, Lee ST, Lee EJ, Kim DY, Lee G, Chi SG, Ryu BK, Lee CH, Yum KE, Lee HJ, Han JY, Tilly JL, Lim JM (2010) Embryonic Stem Cell-Like Cells Established by Culture of Adult Ovarian Cells in Mice. Fertil Steril 93:2594-2601.

[12] Phinney DG, Prockop DJ (2007) Concise Review: Mesenchymal Stem/Multipotent Stromal Cells: The State of Transdifferentiation and Modes of Tissue Repair--Current Views. Stem Cells 25:2896-902.

[13] Bhartiya D, Kasiviswanathan S, Unni SK, Pethe P, Dhabalia JV, Patwardhan S, Tongaonkar HB (2010) Newer Insights into Premeiotic Development of Germ Cells in

Adult Human Testis using Oct-4 as a Stem Cell Marker. J Histochem Cytochem 58:1093-1106.

[14] Parte S, Bhartiya D, Telang J, Daithankar V, Salvi V, Zaveri K, Hinduja I (2011) Detection, Characterization, and Spontaneous Differentiation *In Vitro* of Very Small Embryonic-Like Putative Stem Cells in Adult Mammalian Ovary. Stem Cells Dev 20:1451-64.

[15] Bhartiya D, Shaikh A, Nagvenkar P, Kasiviswanathan S, Pethe P, Pawani H, Mohanty S, Rao SG, Zaveri K, Hinduja I (2012) Very Small Embryonic-Like Stem Cells with Maximum Regenerative Potential get Discarded during Cord Blood Banking and Bone Marrow Processing for Autologous Stem Cell Therapy. Stem Cells Dev 21:1-6.

[16] Bhartiya D, Kasiviswanathan S, Shaikh A (2012) Cellular Origin of Testis derived Pluripotent Stem Cells: A Case for VSELs. Stem Cells Dev 21:670-674.

[17] Evans M, Kaufman M (1981) Establishment in Culture of Pluripotent Cells from Mouse Embryos. Nature 292:154–156.

[18] Martin GR (1981) Isolation of a Pluripotent Cell Line from Early Mouse Embryos Cultured in Medium Conditioned by Teratocarcinoma Stem Cells. Proc Natl Acad Sci USA 78:7634–7638.

[19] Chung Y, Klimanskaya I, Becker S, Marh J, Lu SJ, Johnson J, Meisner L, Lanza R (2006) Embryonic and Extraembryonic Stem Cell Lines Derived from Single Mouse Blastomeres. Nature 439: 216–219.

[20] Wakayama S, Hikichi T, Suetsugu R, Sakaide Y, Bui HT, Mizutani E, Wakayama T (2007) Efficient Establishment of Mouse Embryonic Stem Cell Lines from Single Blastomeres and Polar Bodies. Stem Cells 25: 986–993.

[21] Lorthongpanich C, Yang SH, Piotrowska-Nitsche K, Parnpai R, Chan AW (2008) Development of Single Mouse Blastomeres into Blastocysts, Outgrowths and the Establishment of Embryonic Stem Cells. Reproduction 135:805–813.

[22] Smith AG, Heath JK, Donaldson DD, Wong GG, Moreau J, Stahl M, Rogers D (1988) Inhibition of Pluripotential Embryonic Stem Cell Differentiation by Purified Polypeptides. Nature 336:688–690.

[23] Williams RL, Hilton DJ, Pease S, Willson TA, Stewart CL, Gearing DP, Wagner EF, Metcalf D, Nicola NA, Gough NM (1988) Myeloid Leukaemia Inhibitory Factor Maintains the Developmental Potential of Embryonic Stem Cells. Nature 336:684–687.

[24] Yoshida K, Chambers I, Nichols J, Smith A, Saito M, Yasukawa K, Shoyab M, Taga T, Kishimoto T (1994) Maintenance of the Pluripotential Phenotype of Embryonic Stem Cells Through Direct Activation of gp130 Signalling Pathways. Mech Dev 45:163–171.

[25] Ying QL, Nichols J, Chambers I, Smith A (2003) BMP Induction of Id Proteins Suppresses Differentiation and Sustains Embryonic Stem Cell Self-Renewal in Collaboration With STAT3. Cell 115:281–292.

[26] Loser P, Schirm J, Guhr A, Wobus AM, Kurtz A (2010) Human Embryonic Stem Cell Lines And Their Use In International Research. Stem Cells 28:240-246.

[27] Strelchenko N, Verlinsky O, Kukharenko V, Verlisnky Y (2004) Morula-derived Embryonic Stem Cells. Reprod. Biomed. Online 9: 623-629.

[28] Zhang X, Stojkovic P, Przyborski S, Cooke M, Armstrong L, Lako M, Stojkovic M (2006) Derivation of Human Embryonic Stem Cells From Developing And Arrested Embryos. Stem Cells 24:2669-2676.

[29] Klimanskaya I, Chung Y, Becker S, Lu SJ, Lanza R (2007) Derivation of Human Embryonic Stem Cells from Single Blastomeres. Nat Protoc 2: 1963-1972.

[30] Feki A, Hovatta O, Jaconi M (2008) Derivation of Human Embryonic Stem Cell Lines from Single Cells of 4-Cell Stage Embryos: Be Aware of the Risks. Hum Reprod 23: 2874.

[31] Geens M, Mateizel I, Sermon K, De Rycke M, Spits C, Cauffman G, Devroey P, Tournaye H, Liebaers I, Van de Velde H (2009) Human Embryonic Stem Cell Lines Derived from Single Blastomeres of Two 4-Cell Stage Embryos. Hum Reprod 24: 2709-2717.

[32] Daheron L, Opitz SL, Zaehres H, Lensch MW, Andrews PW, Itskovitz-Eldor J, Daley GQ (2004) LIF/STAT3 Signaling Fails to Maintain Self-Renewal of Human Embryonic Stem Cells. Stem Cells 22: 770–778.

[33] Humphrey RK, Beattie GM, Lopez AD, Bucay N, King CC, Firpo MT, Rose-John S, Hayek A (2004) Maintenance of Pluripotency in Human Embryonic Stem Cells is STAT3 Independent. Stem Cells 22:522–530.

[34] Vallier L, Alexander M, Pedersen RA (2005) Activin/Nodal and FGF Pathways Cooperate to Maintain Pluripotency of Human Embryonic Stem Cells. J Cell Sci 118:4495–4509.

[35] Rho JY, Yu K, Han, JS, Chae JI, Koo DB, Yoon HS, Moon SY, Lee KK, Han YM (2006) Transcriptional Profiling of the Developmentally Important Signalling Pathways in Human Embryonic Stem Cells. Hum Reprod 21:405–412.

[36] Xu RH, Chen X, Li DS, Li R, Addicks GC, Glennon C, Zwaka TP Thomson JA (2002) BMP4 Initiates Human Embryonic Stem Cell Differentiation to Trophoblast. Nat Biotechnol 20: 1261–1264.

[37] Xiao L, Yuan X, Sharkis SJ (2006) Activin A Maintains Self-Renewal and Regulates Fibroblast Growth Factor, Wnt, and Bone Morphogenic Protein Pathways in Human Embryonic Stem Cells. Stem Cells 24: 1476–1486.

[38] Greber B, Lehrach H, Adjaye J (2007) Fibroblast Growth Factor 2 Modulates Transforming Growth Factor β signaling in Mouse Embryonic Fibroblasts and Human ESCs (hESCs) to Support hESC Self-renewal. Stem Cells 25: 455–464.

[39] Bongso A, Fong CY, Ng SC, Ratnam SS (1994) Isolation and culture of inner cell mass cells from human blastocysts. Hum Reprod 9:2110–2117.

[40] Amit M, Margulets V, Segev H, Shariki K, Laevsky I, Coleman R, Itskovitz-Eldor J (2003) Human Feeder Layers for Human Embryonic Stem Cells. Biol Reprod 68:2150–2156.

[41] Richards M, Fong CY, Chan WK, Wong PC, Bongso A (2002) Human Feeders Support Prolonged Undifferentiated Growth of Human Inner Cell Masses and Embryonic Stem Cells. Nat Biotechnol 20:933–936.

[42] Miyamoto K, Hayashi K, Suzuki T, Ichihara S, Yamada T, Kano Y, Yamabe T, Ito Y (2004) Human placenta feeder layers support undifferentiated growth of primate embryonic stem cells. Stem Cells 22:433–440.

[43] Genbacev O, Krtolica A, Zdravkovic T, Brunette E, Powell S, Nath A, Caceres E, McMaster M, McDonagh S, Li Y, Mandalam R, Lebkowski J, Fisher SJ (2005) Serum-free derivation of human embryonic stem cell lines on human placental fibroblast feeders. Fertil Steril 83:1517-1529.

[44] Klimanskaya I, Chung Y, Meisner L, Johnson J, West MD, Lanza R (2005) Human Embryonic Stem Cells Derived Without Feeder Cells. Lancet 365:1636-1641.

[45] Ludwig TE, Levenstein ME, Jones JM, Berggren WT, Mitchen ER, Frane JL, Crandall LJ, Daigh CA, Conard KR, Piekarczyk MS, Llanas RA, Thomson JA (2006) Derivation of Human Embryonic Stem Cells in Defined Conditions. Nat Biotechnol 24:185-187.

[46] Crook JM, Peura TT, Kravets L, Bosman AG, Buzzard JJ, Horne R, Hentze H, Dunn NR, Zweigerdt R, Chua F, Upshall A, Colman A (2007) The Generation of Six Clinical-Grade Human Embryonic Stem Cell Lines. Cell Stem Cell 1: 490–494.

[47] Sidhu KS, Ryan JP, Tuch BE (2008). Derivation of a New hESC Line, Endeavour-1 and its Clonal Propagation. Stem Cells Dev 17:41-52.

[48] Rajala K, Lindroos B, Hussein SM, Lappalainen RS, Pekkanen-Mattila M, Inzunza J, Rozell B, Miettinen S, Narkilahti S, Kerkelä E, Aalto-Setälä K, Otonkoski T, Suuronen R, Hovatta O, Skottman H (2010) A Defined and Xeno-Free Culture Method Enabling the Establishment of Clinical Grade Human Embryonic, Induced Pluripotent and Adipose Stem Cells. PLoS One 5:e10246.

[49] Wang Q, Mou X, Cao H, Meng Q, Ma Y, Han P, Jiang J, Zhang H, Ma Y (2012) A Novel Xeno-Free and Feeder-Cell-Free System for Human Pluripotent Stem Cell Culture. Protein Cell 3:51-59.

[50] Slager HG, Van Inzen W, Freund E, Van der Eijnden-Van Raaij AJ, Mummery CL (1993) Transforming Growth Factor-beta in the Early Mouse Embryo: Implications for the Regulation of Muscle Formation and Implantation. Dev Genet 14: 212-224.

[51] Rohwedel J, Maltsev V, Bober E, Arnold HH, Hescheler J, Wobus AM (1994) Muscle Cell Differentiation of Embryonic Stem Cells Reflects Myogenesis in vivo: Developmentally Regulated Expression of Myogenic Determination Genes and Functional Expression of Ionic Currents. Dev Biol 164: 87-101.

[52] Bain G, Kitchens D, Yao M, Huettner JE, Gottlieb DI (1995) Embryonic Stem Cells Express Neuronal Properties in vitro. Dev Biol 168: 342-357.

[53] Schuldiner M, Yanuka O, Itskovitz-Eldor J, Melton DA, Benvenisty N (2000). Effects of Eight Growth Factors on the Differentiation of Cells Derived from Human Embryonic Stem Cells. PNAS 97: 11307-11312.

[54] Kirkeby A, Grealish S, Wolf DA, Nelander J, Wood J, Lundblad M, Lindvall O, Parmar M (2012) Generation of Regionally Specified Neural Progenitors and Functional Neurons from Human Embryonic Stem Cells Under Defined Conditions. Cell Rep 1:703-714.

[55] Chang KH, Bonig H, Papayannopoulou T (2011) Generation and Characterization of Erythroid Cells from Human Embryonic Stem Cells and Induced Pluripotent Stem Cells: An Overview. Stem Cells Int 2011:791604.

[56] Li Z, Hu S, Ghosh Z, Han Z, Wu JC (2011) Functional Characterization and Expression Profiling of Human Induced Pluripotent Stem Cell- and Embryonic Stem Cell-Derived Endothelial Cells. Stem Cells Dev 20:1701-1710.

[57] Barberi T, Bradbury M, Dincer Z, Panagiotakos G, Socci ND, Studer L (2007) Derivation of Engraftable Skeletal Myoblasts from Human Embryonic Stem Cells. Nat Med 13: 642-648.

[58] Laflamme MA, Chen KY, Naumova AV, Muskheli V, Fugate JA, Dupras SK, Reinecke H, Xu C, Hassanipour M, Police S, O'Sullivan C, Collins L, Chen Y, Minami E, Gill EA, Ueno S, Yuan C, Gold J, Murry CE (2007) Cardiomyocytes Derived from Human Embryonic Stem Cells in Pro-Survival Factors Enhance Function of Infarcted Rat Hearts. Nat Biotechnol 25:1015–1024.

[59] Yang L, Soonpaa MH, Adler ED, Roepke TK, Kattman SJ, Kennedy M, Henckaerts E, Bonham K, Abbott GW, Linden RM, Field LJ., Keller GM (2008) Human Cardiovascular Progenitor Cells Develop from a KDR^{+} Embryonic-Stem-Cell-Derived Population. Nature 453: 524-528.

[60] Jiang J, Au M, Lu K, Eshpeter A, Korbutt G, Fisk G, Majumdar AS (2007) Generation of Insulin Producing Islet-Like from Human Embryonic Stem Cells. Stem Cells 25:1940-1953.

[61] Kroon E, Martinson LA, Kadoya K, Bang AG, Kelly OG, Eliazer S, Young H, Richardson M, Smart NG, Cunningham J, Agulnick AD, D'Amour KA, Carpenter MK,

Baetge EE (2008) Pancreatic Endoderm Derived from Human Embryonic Stem Cells Generates Glucose-Responsive Insulin-Secreting Cells in vivo. Nat Biotechnol 26:443–452.

[62] Touboul T, Hannan NR, Corbineau S, Martinez A, Martinet C, Branchereau S, Mainot S, Strick-Marchand H, Pedersen R, Di Santo J, Weber A, Vallier L (2010) Generation of Functional Hepatocytes from Human Embryonic Stem Cells Under Chemically Defined Conditions that Recapitulate Liver Development. Hepatology 51:1754-1765.

[63] Osafune K, Caron L, Borowiak M, Martinez RJ, Fitz-Gerald CS, Sato Y, Cowan CA, Chien KR, Melton DA (2008) Marked Differences in Differentiation Propensity among Human Embryonic Stem Cell Lines. Nat Biotechnol 26:313-315.

[64] Tavakoli T, Xu X, Derby E, Serebryakova Y, Reid Y, Rao MS, Mattson MP, Ma W (2009) Self-renewal and Differentiation Capabilities are Variable Between Human Embryonic Stem Cell Lines I3, I6 and BG01V. BMC Cell Biol 10: 44.

[65] Mayor S (2010) First Patient Enters Trial to Test Stem Cells in Spinal Injury. BMJ 341:c5724.

[66] Alper J (2009) Geron Gets Green Light for Human Trial of ES-cell Derived Product. Nat Biotechnol 27:213-4.

[67] De Francesco L (2009) Fits and start for Geron. Nat Biotechnol 27:877.

[68] Schwartz SD, Hubschman JP, Heilwell G, Franco-Cardenas V, Pan CK, Ostrick RM, Mickunas E, Gay R, Klimanskaya I, Lanza R (2012) Embryonic Stem Cell Trials for Macular Degeneration: A Preliminary Report. Lancet 379:713-20.

[69] Maherali N, Sridharan R, Xie W, Utikal J, Eminli S, Arnold K, Stadtfeld M, Yachechko R, Tchieu J, Jaenisch R, Plath K, Hochedlinger K (2007) Directly Reprogrammed Fibroblasts Show Global Epigenetic Remodeling and Widespread Tissue Contribution. Cell Stem Cell 1:55–70.

[70] Okita K, Ichisaka T, Yamanaka S (2007) Generation of Germline-Competent Induced Pluripotent Stem Cells. Nature 448: 313–317.

[71] Wernig M, Meissner A, Foreman R, Brambrink T, Ku M, Hochedlinger K, Bernstein BE, Jaenisch R (2007). In vitro Reprogramming of Fibroblasts into a Pluripotent ES-Cell Like State. Nature 448: 318–324.

[72] Lowry WE, Richter L, Yachechko R, Pyle AD, Tchieu J, Sridharan R, Clark AT, Plath K (2008) Generation of Human Induced Pluripotent Stem Cells from Dermal Fibroblasts. Proc Natl Acad Sci 105: 2883–2888.

[73] Park IH, Zhao R, West JA, Yabuuchi A, Huo H, Ince TA, Lerou PH, Lensch MW, Daley GQ (2008) Reprogramming of Human Somatic Cells to Pluripotency with Defined Factors. Nature 451: 141–146.

[74] Yu J, Vodyanik MA, Smuga-Otto K, Antosiewicz-Bourget J, Frane JL, Tian S, Nie J, Jonsdottir GA, Ruotti V, Stewart R, Slukvin II, Thomson JA (2007) Induced Pluripotent Stem Cell Lines Derived from Human Somatic Cells. Science 318: 1917–1920.

[75] Maherali N, Hochedlinger K (2008) Guidelines and Techniques for the Generation of Induced Pluripotent Stem Cells. Cell Stem Cell 3:595–605.

[76] Kiskinis E, Eggan K (2010) Progress toward the clinical application of patient-specific pluripotent stem cells. J Clin Invest 120:51-59.

[77] Oh SI, Lee CK, Cho KJ, Lee KO, Cho SG, Hong S (2012) Technological Progress in Generation of Induced Pluripotent Stem Cells for Clinical Applications. Scientific World Journal 2012:417809.

[78] Aasen T, Raya A, Barrero MJ, Garreta E, Consiglio A, Gonzalez F, Vassena R, Bilic J, Pekarik V, Tiscornia G, Edel M, Boue S, Izpisua Belmonte JC (2008) Efficient and Rapid Generation of Induced Pluripotent Stem Cells from Human Keratinocytes. *Nat* Biotechnol 26:1276–1284.

[79] Sommer CA, Stadtfeld M, Murphy GJ, Hochedlinger K, Kotton DN, Mostoslavsky G (2009) Induced Pluripotent Stem Cell Generation Using a Single Lentiviral Stem Cell Cassette. Stem Cells 27:543-9.

[80] Stadtfeld M, Nagaya M, Utikal J, Weir G, Hochedlinger K (2008) Induced Pluripotent Stem Cells Generated Without Viral Integration. Science 322:945–949.

[81] Fusaki N, Ban H, Nishiyama A, Saeki K, Hasegawa M (2009) Efficient Induction of Transgene-Free Human Pluripotent Stem Cells Using a Vector Based on Sendai Virus, an RNA Virus that Does Not Integrate Into the Host Genome. Proceedings of the Japan Academy Series B 85: 348–362.

[82] Okita K, Nakagawa M, Hyenjong H, Ichisaka T, Yamanaka S (2008) Generation of Mouse Induced Pluripotent Stem Cells Without Viral Vectors. Science 322:949–953.

[83] Woltjen K, Michael IP, Mohseni P, Desai R, Mileikovsky M, Hämäläinen R, Cowling R, Wang W, Liu P, Gertsenstein M, Kaji K, Sung HK, Nagy A (2009) piggyBac Transposition Reprograms Fibroblasts To Induced Pluripotent Stem Cells. Nature 458:766–770.

[84] Yusa K, Rad R, Takeda J, Bradley A (2009) Generation of Transgene-Free Induced Pluripotent Mouse Stem Cells by the piggyBac Transposon. Nature Methods 6: 363–369.

[85] Kim D, Kim CH, Moon JI, Chung YG, Chang MY, Han BS, Ko S, Yang E, Cha KY, Lanza R, Kim KS (2009) Generation of Human Induced Pluripotent Stem Cells by Direct Delivery of Reprogramming Proteins. Cell Stem Cell 4:472–476.

[86] Zhou H, Wu S, Joo JY, Zhu S, Han DW, Lin T, Trauger S, Bien G, Yao S, Zhu Y, Siuzdak G, Scholer HR, Duan L, Ding S (2009) Generation of Induced Pluripotent Stem Cells Using Recombinant Proteins. Cell Stem Cell 4: 381–384.

[87] Warren L, Manos PD, Ahfeldt T, Loh YH, Li H, Lau F, Ebina W, Mandal PK, Smith ZD, Meissner A, Daley GQ, Brack AS, Collins JJ, Cowan C, Schlaeger TM, Rossi DJ (2010) Highly Efficient Reprogramming to Pluripotency and Directed Differentiation of Human Cells with Synthetic Modified mRNA. Cell Stem Cell 7: 618–630.

[88] Huangfu D, Maehr R, Guo W, Eijkelenboom A, Snitow M, Chen AE, Melton DA (2008) Induction of Pluripotent Stem Cells by Defined Factors is Greatly Improved by Small-Molecule Compounds. Nature Biotechnol 26: 795–797.

[89] Dimos JT, Rodolfa KT, Niakan KK, Weisenthal LM, Mitsumoto H, Chung W, Croft GF, Saphier G, Leibel R, Goland R, Wichterle H, Henderson CE, Eggan K (2008) Induced Pluripotent Stem Cells Generated from Patients with ALS can be Differentiated into Motor Neurons Science 321:1218–1221.

[90] Park IH, Arora N, Huo H, Maherali N, Ahfeldt T, Shimamura A, Lensch MW, Cowan C, Hochedlinger K, Daley GQ (2008) Disease-specific induced pluripotent stem cells. Cell 134:877–886.

[91] Ebert AD, Yu J, Rose FF Jr, Mattis VB, Lorson CL, Thomson JA, Svendsen CN (2009). Induced Pluripotent Stem Cells from a Spinal Muscular Atrophy Patient Nature 457:277–280.

[92] Amabile G, Meissner A (2009) Induced pluripotent stem cells: current progress and potential for regenerative medicine. Trends Mol Med 15:59–68.

[93] Marion RM, Strati K, Li H, Tejera A, Schoeftner S, Ortega S, Serrano M, Blasco MA (2009) Telomeres Acquire Embryonic Stem Cell Characteristics in Induced Pluripotent Stem Cells. Cell Stem Cell 4:141–154.

[94] Takahashi K, Okita K, Nakagawa M, Yamanaka S (2007) Induction of Pluripotent Stem Cells from Fibroblast Cultures Nature Protocols 2: 3081–3089.

[95] Takahashi K, Ichisaka T, Yamanaka S (2006) Identification of Genes Involved in Tumor-Like Properties of Embryonic Stem Cells. Methods in Molecular Biology 329: 449–458.

[96] Gore A, Li Z, Fung HL, Young JE, Agarwal S, Antosiewicz-Bourget J, Canto I, Giorgetti A, Israel MA, Kiskinis E, Lee JH, Loh YH, Manos PD, Montserrat N, Panopoulos AD, Ruiz S, Wilbert ML, Yu J, Kirkness EF, Izpisua Belmonte JC, Rossi DJ, Thomson JA, Eggan K, Daley GQ, Goldstein LS, Zhang K (2011) Somatic Coding Mutations in Human Induced Pluripotent Stem Cells. Nature 471:63-37.

[97] Hussein SM, Batada NN, Vuoristo S, Ching RW, Autio R, Narva E, Ng S, Sourour M, Hamalainen R, Olsson C, Lundin K, Mikkola M, Trokovic R, Peitz M, Brustle O, Ba-

zett-Jones DP, Alitalo K, Lahesmaa R, Nagy A, Otonkoski T (2011) Copy Number Variation and Selection During Reprogramming to Pluripotency. Nature 471:58-62.

[98] Lister R, Pelizzola M, Kida YS, Hawkins RD, Nery JR, Hon G, Antosiewicz-Bourget J, O'Malley R, Castanon R, Klugman S, Downes M, Yu R, Stewart R, Ren B, Thomson JA, Evans RM, Ecker JR. Hotspots of Aberrant Epigenomic Reprogramming in Human Induced Pluripotent Stem Cells. Nature 471:68-73.

[99] Wagers AJ, Weissman IL (2004) Plasticity of Adult Stem Cells. Cell 116:639–648.

[100] Dominici, M., Le Blanc, K., Mueller, I., Slaper-Cortenbach, I., Marini, F., Krause, D., Deans, R., Keating, A., Prockop, D. J., Horwitz, E. (2006). Minimal criteria for defining multipotent mesenchymal stromal cells. The International Society for Cellular Therapy position statement. Cytotherapy 8:315–317.

[101] Jiang Y, Jahagirdar BN, Reinhardt RL, Schwartz RE, Keene CD, Ortiz-Gonzalez XR, Reyes M, Lenvik T, Lund T, Blackstad M, Du J, Aldrich S, Lisberg A, Low WC, Largaespada DA, Verfaillie CM (2002) Pluripotency of Mesenchymal Stem Cells Derived from Adult Marrow. Nature 418:41–49.

[102] D'Ippolito G, Diabira S, Howard GA, Menei P, Roos BA, Schiller PC (2004) Marrow-Isolated Adult Multilineage Inducible (MIAMI) Cells, A Unique Population of Postnatal Young and Old Human Cells with Extensive Expansion and Differentiation Potential. J Cell Sci 117:2971–2981.

[103] Beltrami AP, Cesselli, D, Bergamin N, Marcon P, Rigo S, Puppato E, D'Aurizio F, Verardo R, Piazza S, Pignatelli A, Poz A, Baccarani U, Damiani D, Fanin R, Mariuzzi L, Finato N, Masolini P, Burelli S, Belluzzi O, Schneider C, Beltrami CA (2007) Multipotent Cells Can Be Generated In Vitro from Several Adult Human Organs (Heart, Liver And Bone Marrow). Blood 110:3438–3446.

[104] Kucia M, Reca R, Campbell FR, Zuba-Surma E, Majka M, Ratajczak J, Ratajczak MZ (2006) A Population of Very Small Embryonic-Like (VSEL) CXCR4(+)SSEA-1(+)Oct-4+ Stem Cells Identified in Adult Bone Marrow. Leukemia 20:857-869.

[105] Ratajczak MZ, Kucia M, Majka M, Reca R, Ratajczak J (2004) Heterogeneous Populations of Bone Marrow Stem Cells--Are We Spotting on the Same Cells from the Different Angles? Folia Histochem Cytobio 42:139-146.

[106] Zuba-Surma EK, Kucia M, Wu W, Klich I, Lillard JW Jr, Ratajczak J, Ratajczak MZ. (2008). Very Small Embryonic-Like Stem Cells Are Present in Adult Murine Organs: Imagestream-based Morphological Analysis and Distribution Studies. Cytometry A 73A:1116-1127.

[107] Zuba-Surma EK, Wu W, Ratajczak J, Kucia M, Ratajczak MZ (2009) Very Small Embryonic-Like Stem Cells in Adult Tissues-Potential Implications for Aging. Mech Ageing Dev 130:58-66.

[108] Shin DM, Zuba-Surma EK, Wu W, Ratajczak J, Wysoczynski M, Ratajczak MZ, Kucia M (2009) Novel Epigenetic Mechanisms that Control Pluripotency and Quiescence of Adult Bone Marrow Derived Oct4(+) Very Small Embryonic-Like Stem Cells. Leukemia 23:2042-2051.

[109] Shin DM, Liu R, Wu W, Waigel SJ, Zacharias W, Ratajczak MZ, Kucia M (2012) Global Gene Expression Analysis of Very Small Embryonic-Like Stem Cells Reveals That the Ezh2-Dependent Bivalent Domain Mechanism Contributes To Their Pluripotent State. Stem Cells Dev 21:1639-52.

[110] Ratajczak MZ, Machalinski B, Wojakowski W, Ratajczak J, Kucia M (2007) A Hypothesis for An Embryonic Origin of Pluripotent Oct-4(+) Stem Cells in Adult Bone Marrow and Other Tissues. Leukemia 21:860-867.

[111] Bhartiya D (2012) Pluripotent Very Small Embryonic-Like Stem Cells Get Discarded During Cord Blood and Bone Marrow Processing. Stem Cells Dev 21:2563-2564.

[112] Ratajczak MZ, Zuba-Surma EK, Shin DM, Ratajczak J, Kucia M (2008) Very Small Embryonic-Like (VSEL) Stem Cells in Adult Organs and Their Potential Role in Rejuvenation of Tissues and Longevity. ExpGerontol 43:1009-1017.

[113] Li L, Clevers H (2010) Coexistence of Quiescent and Active Adult Stem Cells in Mammals. Science 327:542-5.

[114] De Rosa L, De Luca M (2012) Cell Biology Dormant and Skin Stem Cells. Nature 489:215-217.

[115] Ratajczak MZ, Shin DM, Kucia M (2009) Very Small Embryonic/Epiblast-Like Stem Cells: A Missing Link to Support the Germ Line Hypothesis of Cancer Development? Am J Path 174:1985-1992.

[116] Ratajczak MZ, Shin DM, Liu R, Marlicz W, Tarnowski M, Ratajczak J, Kucia M. (2010) Epiblast/Germ Line Hypothesis of Cancer Development Revisited: Lesson from the Presence of Oct-4(1) Cells In Adult Tissues. Stem Cell Rev 6:307-316.

[117] Kucia M, Halasa M, Wysoczynski M, Baskiewicz-Masiuk M, Moldenhawer S, Zuba-Surma E, Czajka R, Wojakowski W, Machalinski B and Ratajczak MZ (2007) Morphological and Molecular Characterization of Novel Population of cxcr4+ SSEA-4+ Oct-4+ Very Small Embryonic-Like Cells Purified From Human Cord Blood – Preliminary Report. Leukemia 21:297–303

[118] Liedtke S, Enczmann J, Waclawczyk S, Wernet P, Kögler G (2007) Oct4 and Its Pseudogenes Confuse Stem Cell Research. Cell Stem Cell 1:364-366.

[119] Liedtke S, Stephan M, Kogler G (2008) Oct4 Expression Revisited: Potential Pitfalls for Data Misinterpretation in Stem Cell Research. J Biol Chem 389:845-850.

[120] Wang X, Dai J (2010) Isoforms Of OCT4 Contribute to the Confusing Diversity in Stem Cell Biology. Stem Cells 28:885-893.

[121] Takeda J, Seino S, Bell GI (1992) Human Oct3 Gene Family: cDNA Sequences, Alternative Splicing, Gene Organization, Chromosomal Location, and Expression at Low Levels in Adult Tissues. Nucleic Acids Research 20:4613-4620.

[122] Bhartiya D, Sriraman K, Parte S (2012) Stem Cell Interaction with Somatic Niche may hold the Key to Fertility Restoration in Cancer Patients. Obstet and Gynec Intl 2012: 921082.

[123] Virant-Klun I, Rozman P, Cvjeticanin B, Vrtacnik-Bokal E, Novakovic S, Rülicke T, Dovc P, Meden-Vrtovec H (2009) Parthenogenetic Embryo-Like Structures in The Human Ovarian Surface Epithelium Cell Culture in Postmenopausal Women with No Naturally Present Follicles and Oocytes. Stem Cells Dev 18:137-149.

[124] Bukovsky A, Svetlikova M, Caudle MR (2005) Oogenesis in Cultures Derived from Adult Human Ovaries. Reprod Biol Endocrinol. 3:17.

[125] Zuba-Surma EK, Wojakowski W, Ratajczak MZ, Dawn B (2011) Very Small Embryonic-Like Stem Cells: Biology and Therapeutic Potential for Heart Repair. Antiox Redox Signal 15:1821-1834.

[126] Wojakowski W, Kucia M, Zuba-Surma E, Jadczyk T, Książek B, Ratajczak MZ, Tendera M (2010) Very Small Embryonic-Like Stem Cells in Cardiovascular Repair. Pharmacology & Therapeutics 129:21-28.

[127] Paczkowska E, Kucia M, Koziarska D, Halasa M, Safranow K, Masiuk M, Karbicka A, Nowik M, Nowacki P, Ratajczak MZ, Machalinski B (2009) Clinical Evidence that Very Small Embryonic-Like Stem Cells are Mobilized into Peripheral Blood in Patients after Stroke. Stroke. 40(4):1237-1244.

[128] Drukała J, Paczkowska E, Kucia M, Młyńska E, Krajewski A, Machaliński B, Madeja Z, Ratajczak MZ (2012) Stem Cells, Including A Population of Very Small Embryonic-Like Stem Cells, Are Mobilized into Peripheral Blood in Patients After Skin Burn Injury. Stem Cell Review 8:184-194.

[129] Ratajczak J, Zuba-Surma E, Paczkowska E, Kucia M, Nowacki P, Ratajczak MZ (2011) Stem Cells for Neural Regeneration--A Potential Application of Very Small Embryonic-Like Stem Cells. Journal of Physiology and Pharmacology 62:3-12.

[130] Ratajczak J, Wysoczynski M, Zuba-Surma E, Wan W, Kucia M, Yoder MC, Ratajczak MZ (2011). Adult Murine Bone Marrow-Derived Very Small Embryonic-Like Stem Cells Differentiate into the Hematopoietic Lineage after Coculture over OP9 Stromal Cells. Exp Hematol 39:225-237.

[131] Ratajczak MZ, Liu R, Marlicz W, Blogowski W, Starzynska T, Wojakowski W, Zuba-Surma E (2011) Identification of Very Small Embryonic/Epiblast-Like Stem Cells (Vsels) Circulating in Peripheral Blood During Organ/Tissue Injuries. Methods Cell Biol 103:31- 54.

Chapter 2

Pluripotent Adult Stem Cells: A Potential Revolution in Regenerative Medicine and Tissue Engineering

Tsz Kin Ng, Daniel Pelaez, Veronica R. Fortino,
Jordan Greenberg and Herman S. Cheung

Additional information is available at the end of the chapter

http://dx.doi.org/10.5772/54366

1. Introduction

Stem cells are undifferentiated cells defined by their abilities to self-renew and differentiate into mature cells. Stem cells found in fully developed tissues are defined as adult stem cells. The function of adult stem cells is the maintenance of adult tissue specificity by homeostatic cell replacement and tissue regeneration (Wagers and Weissman, 2004). Adult stem cells are presumed quiescent within adult tissues, but divide infrequently to generate a stem cell clone and a transiently-amplifying cell. The transiently-amplifying cells will undergo a limited number of cell divisions before terminal differentiation into mature functional tissue cells. The existence of adult stem cells has been reported in multiple organs; these include: brain, heart, skin, intestine, testis, muscle and blood, among others. This chapter focuses on four adult stem cell populations: hematopoietic, mesenchymal, periodontal ligament-derived, and spermatogonial (Table 1).

Hematopoietic stem cells are the most characterized adult stem cell population. They function to generate all cell lineages found in mature blood (erythroid, myeloid and lymphoid) and to sustain blood production during the entire life of an animal (Kondo et al., 2003). Adult bone marrow, umbilical cord blood and mobilized peripheral blood are sources of hematopoietic stem cells for transplantation in many blood-related diseases. Hematopoietic stem cells can be characterized by positive selection of CD34, CD45, and CD133 markers and negative selection of CD31, CD105 and CD146 markers (Tárnok et al., 2010).

Mesenchymal stem cells, also called marrow stromal cells, are another well-studied adult stem cell population. Mesenchymal stem cells were originally identified in the bone marrow, but have since been found in other systems such as adipose tissue, umbilical cord and menstrual blood (Ding et al., 2011). Mesenchymal stem cells differentiate into osteocytes, chondrocytes and adipocytes (Arita et al., 2011; Pittenger et al., 1999). Human mesenchymal stem cells can be characterized by the positive expression of CD29, CD44, CD73, CD90, CD105, CD146 and STRO-1, and the negative expression of CD31, CD34, CD45, CD49f and CD133 (Mödder et al., 2012; Tárnok et al., 2010).

Adult stem cells	Feasible sources	Characterization
Hematopoietic stem cells	Bone marrow, umbilical cord blood, mobilized peripheral blood	(+): CD34, CD45, CD133
		(-): CD31, CD105, CD146
Mesenchymal stem cells	Bone marrow, adipose tissue, umbilical cord, menstrual blood	(+): CD29, CD44, CD73, CD90, CD105, CD146, STRO-1
		(-): CD31, CD34, CD45, CD49f, CD133
Periodontal ligament-derived stem cells	Periodontal ligament	Mesenchymal stem cell markers: CD29, CD44, CD73, CD90, CD105, CD146, STRO-1
		Neural crest cell markers: p75, nestin, Slug, SOX10
Spermatogonial stem cells	Testis	(+): CD9, CD49f and GPR125

Table 1. Feasible sources and characterization of adult stem cells

Periodontal ligament, derived from the cranial neural crest, is a soft connective tissue embedded between the tooth root and the alveolar bone socket, supporting the teeth *in situ* and preserving tissue homeostasis. The periodontal ligament contains stem cell populations that can differentiate into cementum-forming cells or bone-forming cells (Seo et al., 2004). Periodontal ligament-derived stem cells are heterogeneous, composed of mesenchymal stem cells and putative neural crest cells. Therefore, human periodontal ligament-derived stem cell populations have been characterized not only by mesenchymal stem cell markers, but also by neural crest cell markers, such as p75, nestin, Slug and SOX10 (Huang et al., 2009; Mrozik et al., 2010).

Testicular spermatogonial stem cells are the germ-line cells for spermatogenesis, an ongoing process throughout the lifespan of the male animals. They are unipotent in nature and continuously generate differentiating daughter cells for subsequent production of spermatozoa (Fagoonee et al., 2011). Human spermatogonial stem cells can be purified by antibodies against cell surface markers CD9, CD49f and GPR125 (Conrad et al., 2008).

2. Pluripotent stem cells

Pluripotency refers to the ability of cells to self-renew and differentiate into all 3 germ layers (ectoderm, endoderm and mesoderm). Pluripotent stem cells are the origin of all somatic and germ-line cells in the developing embryo. The first pluripotent cells were derived in 1976 from a type of germ-line tumor known as a teratocarcinoma (Hogan, 1976). Embryonic stem cells, derived from the inner cell mass of a blastocyst prior to gastrulation, are still considered the gold standard for pluripotent stem cells. Even though adult cells are terminally differentiated, pluripotency has also been conferred to these cells in past studies, by the technique of somatic cell nuclear transfer (Perry, 2005), parthenogenesis of unfertilized eggs (Brevini et al., 2008), and reprogramming by cell fusion (Pralong et al., 2006). Research into adult cell pluripotency was slow to progress until a major breakthrough in 2006 brought with it the technique of "induced pluripotent stem cells". In this process adult skin fibroblasts were induced into a pluripotent state by the forced expression of key transcription factors (OCT4, SOX2, KLF4 and c-MYC; Takahashi et al., 2007) or (OCT4, SOX2, NANOG and LIN28; Yu et al., 2007). Despite the low reprogramming efficiency, this has become a convenient method for generating new pluripotent stem cell lines for research from differentiated adult cells.

Adult stem cells are thought to be tissue-specific and only able to differentiate into progeny cells of their tissues of origin. An increasing number of studies, however, report that adult stem cells are capable of giving rise to cells of an entirely distinct lineage. The concept of adult stem cell plasticity might be explained by 5 potential mechanisms: cell fusion, trans-differentiation, de-differentiation, heterogeneous stem cell populations, or pluripotency (Wagers and Weissman, 2004). Cell-cell fusion occurs at a low frequency, but is implicated in the transplantation of bone marrow cells to liver hepatocytes, cardiomyocytes and Purkinje neurons (Alvarez-Dolado et al., 2003). In cell fusion events, the stem cells acquire the mature phenotype of the tissue they are embedded within and can be easily mistaken for correct differentiation of the transplanted cells. Trans-differentiation is a direct lineage conversion by the activation of a dormant differentiation program to alter the lineage specificity of the cell. De-differentiation is another lineage conversion phenomenon in which a tissue-specific cell spontaneously de-differentiates into a more basal multipotent cell and re-differentiates to a new lineage. While the heterogeneity of the stem cell population employed can account for some of the apparent trans-differentiation and de-differentiation events observed *in vivo*, it is worth discussing as a separate factor in the resulting multi-lineage tissues, which are often seen after transplantation. The characterization of homogeneous stem cell populations that contribute to the regeneration of one cell type remains an active field of study for most cellular therapy applications. Lastly, pluripotent stem cells are present in adult tissues as minute sub-populations in certain stem cell niches. Such a population has already been identified and reported in bone marrow derived mesenchymal stem cells (Jiang et al., 2002). In addition, pluripotent stem cells in adult tissues can also arise from remnants of the migrating neural crest. The neural crest is a transient embryonic structure that affords various organs with cells which could undergo a more stochastic type of differentiation than other embryonic progenitor cells (Slack, 2008). Neural crest cells are pluripotent and may retain some of their characteristics after their migration and engraftment into their terminal sites.

3. Isolation of pluripotent adult stem cells

The expression of embryonic stem cell markers in some adult stem cells suggest a sub-population of pluripotent cells in these niches (Table 2). The common embryonic stem cell makers, such as OCT4, SOX2, NANOG, KLF4, LIN28, SSEA-1, SSEA-3, SSEA-4, TRA-1-60 and TRA-1-81, are all expressed in hematopoietic stem cells (Wang et al., 2010; Zhao et al., 2006; Zulli et al., 2008) and mesenchymal stem cells (Anjos-Afonso and Bonnet, 2007; Jaramillo-Ferrada et al., 2012; Riekstina et al., 2009; Sung et al., 2010). Similarly, expressions of most of these markers, except for LIN28, have been reported in periodontal ligament-derived stem cells, a tissue arising from the migrating cranial neural crest (Huang et al., 2009; Kawanabe et al., 2010). Previous studies show that spermatogonial stem cells also express most of the embryonic stem cell markers, except SSEA-3 and TRA-1-60 (Izadyar et al., 2008; Izadyar et al., 2011; Kanatsu-Shinohara et al., 2008; Panda et al., 2011; Zheng et al., 2009). These findings suggest that pluripotent stem cells exist as sub-populations in adult stem cell reservoirs.

Embryonic stem cell marker	HSC	MSC	PDLSC	SSC
SOX2	+	+	+	+
OCT4	+	+	+	+
NANOG	+	+	+	+
KLF4	+	+	+	+
LIN28	+	+		+
SSEA-1	+	+	+	+
SSEA-3	+	+	+	
SSEA-4	+	+	+	+
TRA-1-60	+	+	+	
TRA-1-81	+	+	+	+

HSC: hematopoietic stem cells; MSC: mesenchymal stem cells; PDLSC; periodontal ligament-derived stem cells; SSC: spermatogonial stem cells;

Table 2. Embryonic stem cell marker expression in different adult stem cell populations

The existence of cells with a defined pluripotency-associated phenotypic expression within adult tissues enables researchers to isolate and purify a homogeneous subpopulation of adult pluripotent stem cells. In fact, with the use of magnetic affinity cell sorting, adult human mesenchymal stem cells, shown to differentiate into endodermal, ectodermal and mesodermal cells, were isolated by antibody against SSEA-3 (Kuroda et al., 2010). Similarly, stem cells exhibiting the potential to generate specialized cells of the three embryonic germ layers can be isolated by positive SSEA-4 expression from human periodontal ligament (Kawanabe et al., 2010). Furthermore, human spermatogonial stem cells, sharing cellular and molecular similarities with human embryonic stem cells, can be purified by α_6 integrin (CD49f) antibody (Conrad et al., 2008). Moreover, a human hematopoietic stem cell subpopulation, highly efficient in generating long-term multi-lineage grafts, can also be isolated by the same α_6 integrin expression (Notta et al., 2011). In addition, stem cells from granulocyte colony-stimulating factor-mobilized human peripheral blood can divide indefinitely without reaching replicative senescence and differentiate into multiple lineages (Cesselli et al., 2009).

Recently, a cell surfaceome map of mouse embryonic stem cells and induced pluripotent stem cells was reported (Gundry et al., 2012). Previously unidentified cellular surface markers, such as CD31, CD49f, CD123 and CD326, indicated a purified population of pluripotent stem cells. Further analyses should be performed to determine the expression of these markers in different adult stem cell populations. Their presence in adult stem cell populations could facilitate the purification of homogeneous pluripotent stem cells within an otherwise heterogeneous pool of regenerative adult cells.

4. Characterization of pluripotent adult stem cells

The standard tests for pluripotency are teratoma and chimera formation assays. Teratomas can be formed when pluripotent stem cells are injected into immunodeficient animals; they consist of foci with derivatives of ectodermal, mesodermal and endodermal embryonic germs layers (Wobus et al., 1984). Chimeras can be generated when pluripotent stem cells are microinjected into mouse blastocysts and are induced to differentiate into multiple cell types during normal developmental processes (Becker et al., 1984). Teratoma formation assays can be used to test for the pluripotency of human stem cells, whereas both teratoma and chimera formation can test for the pluripotency of mouse stem cells. Spermatogonial stem cells isolated from human testis by positive expression of CD49f are able to form teratomas when injected into immunodeficient mice (Conrad et al., 2008). Mesenchymal stem cells isolated from murine bone marrow contribute to most of the somatic cell types (chimerism ranged between 0.1% and 45%) when they are singly injected into an early mouse blastocyst (Jiang et al., 2002). Moreover, human hematopoietic stem cells isolated by CD49f cell surface marker display multi-lineage chimerism when transplanted into the NOD-*scid*-*IL2Rgc*$^{-/-}$ mice (Notta et al., 2011). However, human bone marrow-derived mesenchymal stem cells purified by the SSEA-3 cell surface marker do not form teratomas in immunodeficient mouse testes even though cells positive for human ectodermal, endodermal and mesodermal lineage markers were detected within the injected mouse testes (Kuroda et al., 2010). Conversely, pluripotency assays of human periodontal ligament-derived stem cells isolated by SSEA-4 cell surface marker expression have not yet been reported (Kawanabe et al., 2010).

Although most of the adult stem cells are unable to form teratomas in immunodeficient mice, can they still be defined as pluripotent stem cells? Considering this apparent inability as well as the variability in teratoma formation efficiency even when using a known pluripotent stem cell line, a teratoma assay might not be a suitable assay for pluripotency of adult stem cells. Instead, *in vitro* and *in vivo* differentiation into cells of the 3 embryonic germ layers along with chimera formation in xeno-transplanted mice can be applied for testing adult stem cell potency. The conventional concept of development involves a hierarchical structure of cellular commitment extending outward from embryonic and pluripotent, to adult terminally differentiated tissues. However, recent ideas propose that all or most tissues in the postnatal body are continuously turning over and contain a pluripotent stem cell reservoir (Slack, 2008). These pluripotent stem cell populations are able to differentiate into multiple cell types depending on their microenvironmental cues. Therefore, the stem cell status should be defined by plasticity (Zipori, 2005). Pluripotency refers to the ability of cells to differentiate into any cell type of the 3 germ layers (ectoderm, endoderm and mesoderm), whereas multipotency refers to the ability of cells to differentiate only into a closely related family of cells (Ilic and Polak, 2011). All of the previously described adult stem cells (hematopoietic, mesenchymal, periodontal ligament-derived, and spermatogonial) could differentiate into specialized cells of the three germ layers: neurons (ectodermal lineage), adipocytes, cardiomyocytes, osteoblasts, and chondrocytes (mesoderm lineage), and hepatocytes and insulin-producing cells (endodermal lineage) (Conrad et al., 2008; Jiang et al., 2002; Kuroda et al., 2010; Kawanabe et al., 2010; Notta et al., 2011). Therefore, these adult stem cells could also be defined as pluripotent stem cells.

5. Advantages of pluripotent adult stem cells over embryonic stem cells and induced pluripotent stem cells

Human embryonic stem cells come from the inner cell mass of human blastocysts. Therefore, embryonic stem cells used for cell therapy are allogenic; the transplanted donor cells do not originate from the recipient. This raises a concern about the immunogenic response of the host, and the need for immune-suppressive therapy concurrent with embryonic stem cell transplantation (Charron et al., 2009). Moreover, embryonic stem cell-based therapy has been hampered by the moral, legal and ethical dilemma surrounding the use of human embryos for derivation of the stem cell lines (Zarzeczny and Caulfield, 2009). Furthermore, as the gold standard of pluripotent stem cells, embryonic stem cells have the potential to form teratomas in the host. Tumorigenic potential can be reduced by differentiating the embryonic stem cells into lineage-specific progenitor cells or mature tissue cells prior to transplantation (Schwartz et al., 2012). In order to better control standards of good manufacturing practices and reduce variability as much as possible, the *in vitro* manipulation of embryonic stem cells should be minimized as recommend by the Food and Drug Administration (Lysaght and Campbell, 2011). Furthermore, tumorigenic potential remains a concern if the entirety of the embryonic stem cell population does not completely differentiate into fully mature cells.

Differentiated adult cells used for the generation of the induced pluripotent stem cells can be collected from the recipient body, avoiding the contentious need for a human embryo. This also circumvents the problem of immune rejection. There are technical hurdles, however, concerning generation of induced pluripotent stem cells (Hayden, 2011). Firstly, the delivery of reprogramming factors (OCT4, SOX2, NANOG, LIN28, KLF4 and c-MYC) relies on the use of viral vectors for delivery (Takahashi et al., 2007). Retroviral sequences could integrate into the DNA of the host cells, potentially disrupting the gene structure as well as resulting in an aberrant phenotypic expression. Ultimately this could result in pathological mutations and cancer formation. Alternative methods such as direct protein or small molecule delivery have been adopted, although the reprogramming efficiency of these techniques is lower than with viral vectors (Kim et al., 2009; Shi et al., 2008). Secondly, two of the reprogramming factors, *c-MYC* and *KLF4*, are proto-oncogenes, which raise the concern of cancer formation further. Omitting *c-MYC* would lower the reprogramming efficiency, whereas silencing *c-MYC* could lead to its reactivation. Moreover, reprogramming can induce other genomic changes, such as DNA mutations (Gore et al., 2011), copy number variations (Hussein et al., 2011) and chromosomal aberrations (Mayshar et al., 2010). Genomic instability could have unpredictable and undesirable effects on the reprogrammed cells. Furthermore, induced pluripotent stem cells carry their epigenetic signatures from the original differentiated adult cells (Lister et al., 2011). The reprogrammed cells, therefore, unlike embryonic stem cells, may not develop into some cell types. In addition, induced pluripotent stem cells can still cause immune reactions when transplanted allogeneically.

The sources of adult stem cells are multiple and feasibly obtained from various adult tissues, such as bone marrow, blood, adipose tissue, teeth and testes (Table 1). These adult stem cells can be collected from the human body at anytime throughout life. This makes them readily available and does not raise the moral and ethical issues involved with the attainment of embryonic stem cells. Moreover, pluripotent adult stem cells can easily be isolated and purified by cell surface markers, such as CD49f, SSEA-3 and SSEA4 (Conrad et al., 2008; Kuroda et al., 2010; Kawanabe et al., 2010; Notta et al., 2011). The pluripotent status of these adult stem cells is naturally acquired and does not require reprogramming by the introduction of pluripotent transcriptional factors, thus eliminating the use of viral vectors and the chance of aberrant chromosomal changes. Furthermore, transplantation of mesenchymal stem cells and periodontal ligament-derived stem cells can be autogenic or allogeneic. Immuno-suppression is not necessary since mesenchymal stem cells have strong immunomodulatory properties against alloreactivity of T lymphocytes and dendritic cells (Chen et al., 2011). Similarly, mesenchymal stem cells and periodontal ligament-derived stem cells inhibit the proliferation of peripheral blood mononuclear cells (Wada et al., 2009). Spermatogonial stem cells, however, are killed by cytotoxic T lymphocytes after transplantation (Dressel et al., 2009), whereas allogeneic hematopoietic stem cell transplantation induces graft-vs-host disease (Strober et al., 2011). Therefore, transplantation of spermatogonial stem cells and hematopoietic stem cells should only be autogenic, without the application of immunosuppressive drugs. Similar to embryonic stem cells and induced pluripotent stem cells, pluripotent adult stem cells can differentiate into specialized cells of the three germ layers. Except for spermatogonial stem cells (Conrad et al., 2008), teratoma formation was not found in pluripotent hematopoietic stem cells, mesenchymal stem cells and periodontal ligament-derived stem cells (Kuroda et al., 2010; Kawanabe et al., 2010; Notta et al., 2011). This suggests a reduction in the probabilities of tumor formation post-transplantation, and the elimination of the need to manipulate the cells into mature tissue prior to transplantation. In addition, transplanted stem cell-induced regeneration may not be due to stem cell differentiation per se (Johnson et al., 2010; Williams and Hare, 2011). Instead, a paracrine effect has been hypothesized in which the adult stem cells secrete cytokines, chemokines, or protective proteins (Bai et al., 2012; Bráz et al., 2012) that nourish the host tissue cells and facilitate the healing process. This special feature has not yet been reported with the use of embryonic stem cells or induced pluripotent stem cells in a clinical setting.

6. Potential applications of pluripotent adult stem cells

Stem cell clinical trials have advanced rapidly for a broad spectrum of diseases, such as diabetes, neurodegeneration, immune diseases, heart disease, and bone disease. In 2011, there were 123 clinical trials using mesenchymal stem cells (Trounson et al., 2011). It is predicted that stem cell therapy will eventually become the treatment of choice in regenerative medicine, especially the use of adult stem cells. As stem cell products become more wide-spread and maintained under various conditions, the need for global standardization and regulation of processes will become necessary for the viable application of these products in a clinical setting. The Food and Drug Administration regulates interstate commerce in human cells and tissue-based products under the Public Health Service Act and the Code of Federal Regulations for Food and Drugs (Lysaght and Campbell, 2011). Human cells and tissue-based products are defined as "articles containing or consisting of human cells or tissues that are intended for implantation, transplantation, infusion, or transfer into a human recipient" (Lysaght and Campbell, 2011). Human cells and tissue-based products must be: (1) minimally manipulated, (2) intended only for homologous use, (3) not combined with another article (except for water, or sterilization, preservation, or storage agents), and (4) either: (a) have no systemic or metabolic effect, or (b) be for autologous use, allogeneic use in first- or second-degree blood relative, or reproductive use.

Pluripotent adult stem cells fall under the criteria for human cells and tissue-based products as stated by the Food and Drug Administration. Unlike induced pluripotent stem cells, pluripotent adult stem cells can be minimally manipulated as their pluripotent state occurs naturally. Unlike embryonic stem cells, pluripotent adult stem cells are suited for autologous use. Similar to embryonic stem cells and induced pluripotent stem cells, pluripotent adult stem cells are able to differentiate into specialized cells of the three germ layers. In addition, embryonic stem cells and induced pluripotent stem cells have the potential to form teratomas (an unfavorable side-effect in clinical applications) although a recent study suggests that the teratoma-forming cells could be removed by the antibody against SSEA-5 (Tang et al., 2011). In contrast, most pluripotent adult stem cells do not form teratomas *in vivo*, eliminating the need for preemptive differentiation of pluripotent adult stem cells into mature specialized cells.

If stem cell-aided regeneration is not due to stem cell differentiation to replace damaged cells (Johnson et al., 2010; Williams and Hare, 2011), pluripotent adult stem cells are favorable over embryonic stem cells and induced pluripotent stem cells. The secretion of cytokines, chemokines, and/or protective proteins from the adult stem cells could nourish the host tissue and facilitate the healing process (Bai et al., 2012; Bráz et al., 2012).

7. Summary

Adult stem cells are found all over the body. They can be conveniently obtained from different accessible tissues: bone marrow, blood, adipose tissue, teeth and testes. Pluripotent adult stem cells, which reside as a subpopulation within adult stem cells, can be easily isolated by pluripotent cell surface markers, such as SSEA-3, SSEA-4 and CD49f. Moreover, pluripotent adult stem cells can be characterized by their ability to differentiate into cells of 3 germ layers (ectoderm, mesoderm and endoderm) as well as by the chimera formation in xeno-transplanted mice. Pluripotent adult stem cells are better than embryonic stem cells and induced pluripotent stem cells as they are an autologous source, require minimal manipulation and do not have the ability to form teratomas. In addition, they are more appropriate to be used as a clinical product for therapeutic treatments, as a cellular replacement or secretory protein reservoir. However, there are uncertainties that still remain unanswered. Which stem cell types are optimal for regenerative medicine? What is the optimal cell number for transplantation? Should the cells be preemptively differentiated or used as is? Further research is needed to understand the mechanisms of stem cells in regenerating damaged tissues after transplantation.

Author details

Tsz Kin Ng[1], Daniel Pelaez[1], Veronica R. Fortino[2], Jordan Greenberg[2] and Herman S. Cheung[1,2*]

*Address all correspondence to: hcheung@med.maimi.edu

1 Geriatric Research, Education and Clinical Center, Miami Veterans Affairs Medical Center, Miami, FL, USA

2 Department of Biomedical Engineering, College of Engineering, University of Miami, Coral Gables, FL, USA

References

[1] Alvarez-dolado, M. R, Pardal, J. M, Garcia-verdugo, J. R, Fike, H. O, Lee, K, Pfeffer, C, Lois, S. J, & Morrison, A. Alvarez-Buylla. (2003). Fusion of bone-marrow-derived cells with Purkinje neurons, cardiomyocytes and hepatocytes. Nature, 425, 968-973.

[2] Anjos-afonso, F, & Bonnet, D. (2007). Nonhematopoietic/endothelial SSEA-1+ cells define the most primitive progenitors in the adult murine bone marrow mesenchymal compartment. Blood., 109, 1298-1306.

[3] Arita, N. A, Pelaez, D, & Cheung, H. S. (2011). Activation of the extracellular signal-regulated kinases 1 and 2 (ERK1/2) is needed for the TGFβ-induced chondrogenic and osteogenic differentiation of mesenchymal stem cells. *Biochem Biophys Res Commun.*, 405, 564-569.

[4] Bai, L, Lennon, D. P, Caplan, A. I, Dechant, A, Hecker, J, Kranso, J, Zaremba, A, & Miller, R. H. (2012). Hepatocyte growth factor mediates mesenchymal stem cell-induced recovery in multiple sclerosis models. Nat Neurosci. *In press*.

[5] Becker, K, Wobus, A. M, Conrad, U, & Schöneich, J. (1984). Injection of murine embryonal carcinoma cells and embryo-derived pluripotential cells into mouse blastocysts. *Cell Differ.*, 15, 195-202.

[6] Bráz, J. M, Sharif-naeini, R, Vogt, D, Kriegstein, A, Alvarez-buylla, A, Rubenstein, J. L, & Basbaum, A. I. (2012). Forebrain GABAergic neuron precursors integrate into adult spinal cord and reduce injury-induced neuropathic pain. *Bráz, J.M., R. Sharif-Naeini, D. Vogt, A. Kriegstein, A. Alvarez-Buylla, J.L. Rubenstein, and A.I. Basbaum. 2012. Forebrain GABAergic neuron precursors integrate into adult spinal cord and reduce injury-induced neuropathic pain. Neuron.*., 74, 663-675.

[7] Brevini, T. A, Pennarossa, G, Antonini, S, & Gandolfi, F. (2008). Parthenogenesis as an approach to pluripotency: advantages and limitations involved. *Stem Cell Rev.*, 4127-135.

[8] Cesselli, D, Beltrami, A. P, Rigo, S, Bergamin, N, Aurizio, F. D, Verardo, R, Piazza, S, Klaric, E, Fanin, R, Toffoletto, B, Marzinotto, S, Mariuzzi, L, Finato, N, Pandolfi, M, Leri, A, Schneider, C, Beltrami, C. A, & Anversa, P. (2009). Multipotent progenitor cells are present in human peripheral blood. *Circ Res.*, 104, 1225-1234.

[9] Charron, D, Suberbielle-boissel, C, & Al-daccak, R. (2009). Immunogenicity and allogenicity: a challenge of stem cell therapy. *J Cardiovasc Transl Res.*, 2, 130-138.

[10] Chen, P. M, Yen, M. L, Liu, K. J, Sytwu, H. K, & Yen, B. L. (2011). Immunomodulatory properties of human adult and fetal multipotent mesenchymal stem cells. *J Biomed Sci.* 18:49.

[11] Conrad, S, Renninger, M, Hennenlotter, J, Wiesner, T, Just, L, Bonin, M, Aicher, W, Bühring, H. J, Mattheus, U, Mack, A, Wagner, H. J, Minger, S, Matzkies, M, Reppel, M, Hescheler, J, Sievert, K. D, Stenzl, A, & Skutella, T. (2008). Generation of pluripotent stem cells from adult human testis. *Nature*, 456, 344-349.

[12] Ding, D. C, Shyu, W. C, & Lin, S. Z. (2011). Mesenchymal stem cells. *Cell Transplant.*, 20, 5-14.

[13] Dressel, R, Guan, K, Nolte, J, Elsner, L, Monecke, S, Nayernia, K, Hasenfuss, G, & Engel, W. (2009). Multipotent adult germ-line stem cells, like other pluripotent stem

cells, can be killed by cytotoxic T lymphocytes despite low expression of major histocompatibility complex class I molecules. *Biol Direct*. 4:31.

[14] Fagoonee, S, Pellicano, R, Silengo, L, & Altruda, F. (2011). Potential applications of germline cell-derived pluripotent stem cells in organ regeneration. *Organogenesis*, 7, 116-122.

[15] Gore, A, Li, Z, Fung, H. L, Young, J. E, Agarwal, S, Antosiewicz-bourget, J, Canto, I, Giorgetti, A, Israel, M. A, Kiskinis, E, Lee, J. H, Loh, Y. H, Manos, P. D, Montserrat, N, Panopoulos, A. D, Ruiz, S, Wilbert, M. L, Yu, J, Kirkness, E. F, Izpisua, J. C, Belmonte, D. J, Rossi, J. A, Thomson, K, Eggan, G. Q, & Daley, L. S. Goldstein, and K. Zhang. (2011). Somatic coding mutations in human induced pluripotent stem cells. *Nature*, 471, 63-67.

[16] Gundry, R. L, Riordon, D. R, Tarasova, Y, Chuppa, S, Bhattacharya, S, Juhasz, O, Wiedemeier, O, Milanovich, S, Noto, F. K, Tchernyshyov, I, Raginski, K, Bauschfluck, D, Tae, H. J, Marshall, S, Duncan, S. A, Wollscheid, B, Wersto, R. P, Rao, S, Van Eyk, J. E, & Boheler, K. R. (2012). A Cell Surfaceome Map for Immunophenotyping and Sorting Pluripotent Stem Cells. *Mol Cell Proteomics*. In press.

[17] Hayden, E. C. (2011). Stem cells: The growing pains of pluripotency. *Nature*, 473, 272-274.

[18] Hogan, B. L. (1976). Changes in the behaviour of teratocarcinoma cells cultivated in vitro. *Nature*, 263, 136-137.

[19] Hussein, S. M, Batada, N. N, Vuoristo, S, Ching, R. W, Autio, R, Närvä, E, Ng, S, Sourour, M, Hämäläinen, R, Olsson, C, Lundin, K, Mikkola, M, Trokovic, R, Peitz, M, Brüstle, O, Bazett-jones, D. P, Alitalo, K, Lahesmaa, R, Nagy, A, & Otonkoski, T. (2011). Copy number variation and selection during reprogramming to pluripotency. *Nature*, 471, 58-62.

[20] Ilic, D, & Polak, J. M. (2011). Stem cells in regenerative medicine: introduction. *Br Med Bull.*, 98, 117-126.

[21] Izadyar, F, Pau, F, Marh, J, Slepko, N, Wang, T, Gonzalez, R, Ramos, T, Howerton, K, Sayre, C, & Silva, F. (2008). Generation of multipotent cell lines from a distinct population of male germ line stem cells. *Reproduction.*, 135, 771-784.

[22] Izadyar, F, Wong, J, Maki, C, Pacchiarotti, J, Ramos, T, Howerton, K, Yuen, C, Greilach, S, Zhao, H. H, Chow, M, Chow, Y. C, Rao, J, Barritt, J, Bar-chama, N, & Copperman, A. (2011). Identification and characterization of repopulating spermatogonial stem cells from the adult human testis. *Hum Reprod.*, 26, 1296-1306.

[23] Jaramillo-ferrada, P. A, Wolvetang, E. J, & Cooper-white, J. J. (2012). Differential mesengenic potential and expression of stem cell-fate modulators in mesenchymal stromal cells from human-term placenta and bone marrow. *J Cell Physiol.*, 227, 3234-3242.

[24] Jiang, Y, Jahagirdar, B. N, Reinhardt, R. L, Schwartz, R. E, Keene, C. D, Ortiz-gonzalez, X. R, Reyes, M, Lenvik, T, Lund, T, Blackstad, M, Du, J, Aldrich, S, Lisberg, A, Low, W. C, Largaespada, D. A, & Verfaillie, C. M. (2002). Pluripotency of mesenchymal stem cells derived from adult marrow. *Nature*, 418, 41-49.

[25] Johnson, T. V, Bull, N. D, Hunt, D. P, Marina, N, Tomarev, S. I, & Martin, K. R. (2010). Neuroprotective effects of intravitreal mesenchymal stem cell transplantation in experimental glaucoma. *Invest Ophthalmol Vis Sci.* , 51, 2051-2059.

[26] Kanatsu-shinohara, M, Lee, J, Inoue, K, Ogonuki, N, Miki, H, Toyokuni, S, Ikawa, M, Nakamura, T, Ogura, A, & Shinohara, T. (2008). Pluripotency of a single spermatogonial stem cell in mice. *Biol Reprod.* ;, 78, 681-687.

[27] Kawanabe, N, Murata, S, Murakami, K, Ishihara, Y, Hayano, S, Kurosaka, H, Kamioka, H, Takano-yamamoto, T, & Yamashiro, T. (2010). Isolation of multipotent stem cells in human periodontal ligament using stage-specific embryonic antigen-4. *Differentiation*, 79, 74-83.

[28] Kim, D, Kim, C. H, Moon, J. I, Chung, Y. G, Chang, M. Y, Han, B. S, Ko, S, Yang, E, Cha, K. Y, Lanza, R, & Kim, K. S. (2009). Generation of human induced pluripotent stem cells by direct delivery of reprogramming proteins. *Cell Stem Cell*, 4, 472-476.

[29] Kondo, M, Wagers, A. J, Manz, M. G, Prohaska, S. S, Scherer, D. C, Beilhack, G. F, Shizuru, J. A, & Weissman, I. L. (2003). Biology of hematopoietic stem cells and progenitors: implications for clinical application. *Annu Rev Immunol.* , 21, 759-806.

[30] Kuroda, Y, Kitada, M, Wakao, S, Nishikawa, K, Tanimura, Y, Makinoshima, H, Goda, M, Akashi, H, Inutsuka, A, Niwa, A, Shigemoto, T, Nabeshima, Y, Nakahata, T, Nabeshima, Y, Fujiyoshi, Y, & Dezawa, M. (2010). Unique multipotent cells in adult human mesenchymal cell populations. *Proc Natl Acad Sci U S A.* , 107, 8639-8643.

[31] Lister, R, Pelizzola, M, Kida, Y. S, Hawkins, R. D, Nery, J. R, Hon, G, Antosiewicz-bourget, J, Malley, R. O, Castanon, R, Klugman, S, Downes, M, Yu, R, Stewart, R, Ren, B, Thomson, J. A, Evans, R. M, & Ecker, J. R. (2011). Hotspots of aberrant epigenomic reprogramming in human induced pluripotent stem cells. *Nature*, 471, 68-73.

[32] Lysaght, T, & Campbell, A. V. (2011). Regulating autologous adult stem cells: the FDA steps up. *Cell Stem Cell*, 9, 393-396.

[33] Mayshar, Y, Ben-david, U, Lavon, N, Biancotti, J. C, Yakir, B, Clark, A. T, Plath, K, Lowry, W. E, & Benvenisty, N. (2010). Identification and classification of chromosomal aberrations in human induced pluripotent stem cells. *Cell Stem Cell*, 7, 521-531.

[34] Mödder, U. I, Roforth, M. M, Nicks, K. M, Peterson, J. M, Mccready, L. K, Monroe, D. G, & Khosla, S. (2012). Characterization of mesenchymal progenitor cells isolated from human bone marrow by negative selection. *Mödder, U.I., M.M. Roforth, K.M. Nicks, J.M. Peterson, L.K. McCready, D.G. Monroe, and S. Khosla. 2012. Characterization of mesenchymal progenitor cells isolated from human bone marrow by negative selection. Bone.* ., 50, 804-810.

[35] Mrozik, K, Gronthos, S, Shi, S, & Bartold, P. M. (2010). A method to isolate, purify, and characterize human periodontal ligament stem cells. *Methods Mol Biol.* , 666, 269-284.

[36] Notta, F, Doulatov, S, Laurenti, E, Poeppl, A, Jurisica, I, & Dick, J. E. (2011). Isolation of single human hematopoietic stem cells capable of long-term multilineage engraftment. *Science*, 333, 218-221.

[37] Panda, R. P, Barman, H. K, & Mohapatra, C. (2011). Isolation of enriched carp spermatogonial stem cells from Labeo rohita testis for in vitro propagation. *Theriogenology*, 76, 241-251.

[38] Perry, A. C. (2005). Progress in human somatic-cell nuclear transfer. *N Engl J Med.* , 353, 87-88.

[39] Pittenger, M. F, Mackay, A. M, Beck, S. C, Jaiswal, R. K, Douglas, R, Mosca, J. D, Moorman, M. A, Simonetti, D. W, Craig, S, & Marshak, D. R. (1999). Multilineage potential of adult human mesenchymal stem cells. *Science.* , 284, 143-147.

[40] Pralong, D, Trounson, A. O, & Verma, P. J. (2006). Cell fusion for reprogramming pluripotency: toward elimination of the pluripotent genome. *Stem Cell Rev.* , 2, 331-340.

[41] Riekstina, U, Cakstina, I, Parfejevs, V, Hoogduijn, M, Jankovskis, G, Muiznieks, I, Muceniece, R, & Ancans, J. (2009). Embryonic stem cell marker expression pattern in human mesenchymal stem cells derived from bone marrow, adipose tissue, heart and dermis. *Stem Cell Rev.* , 5, 378-386.

[42] Schwartz, S. D, Hubschman, J. P, Heilwell, G, Franco-cardenas, V, Pan, C. K, Ostrick, R. M, Mickunas, E, Gay, R, Klimanskaya, I, & Lanza, R. (2012). Embryonic stem cell trials for macular degeneration: a preliminary report. *Lancet.* , 379, 713-720.

[43] Seo, B. M, Miura, M, Gronthos, S, Bartold, P. M, Batouli, S, Brahim, J, Young, M, Robey, P. G, Wang, C. Y, & Shi, S. (2004). Investigation of multipotent postnatal stem cells from human periodontal ligament. *Lancet.* , 364, 149-155.

[44] Shi, Y, Desponts, C, Do, J. T, Hahm, H. S, Schöler, H. R, & Ding, S. (2008). Induction of pluripotent stem cells from mouse embryonic fibroblasts by Oct4 and Klf4 with small-molecule compounds. *Cell Stem Cell.* , 3, 568-574.

[45] Slack, J. M. (2008). Origin of stem cells in organogenesis. *Science.* , 322, 1498-1501.

[46] Strober, S, Spitzer, T. R, Lowsky, R, & Sykes, M. (2011). Translational studies in hematopoietic cell transplantation: treatment of hematologic malignancies as a stepping stone to tolerance induction. *Semin Immunol.* , 23, 273-281.

[47] Sung, H. J, Hong, S. C, Yoo, J. H, Oh, J. H, Shin, H. J, Choi, I. Y, Ahn, K. H, Kim, S. H, Park, Y, & Kim, B. S. (2010). Stemness evaluation of mesenchymal stem cells from placentas according to developmental stage: comparison to those from adult bone marrow. *J Korean Med Sci.* , 25, 1418-1426.

[48] Takahashi, K, Tanabe, K, Ohnuki, M, Narita, M, Ichisaka, T, Tomoda, K, & Yamanaka, S. (2007). Induction of pluripotent stem cells from adult human fibroblasts by defined factors. *Cell.* , 131, 861-872.

[49] Tang, C, Lee, A. S, Volkmer, J. P, Sahoo, D, Nag, D, Mosley, A. R, Inlay, M. A, Ardehali, R, Chavez, S. L, Pera, R. R, Behr, B, Wu, J. C, Weissman, I. L, & Drukker, M. (2011). An antibody against SSEA-5 glycan on human pluripotent stem cells enables removal of teratoma-forming cells. *Nat Biotechnol.*, 29, 829-834.

[50] Tárnok, A, Ulrich, H, & Bocsi, J. (2010). Phenotypes of stem cells from diverse origin. *Cytometry A.*, 77, 6-10.

[51] Trounson, A, Thakar, R. G, Lomax, G, & Gibbons, D. (2011). Clinical trials for stem cell therapies. *BMC Med.* 9:52.

[52] Wada, N, Menicanin, D, Shi, S, Bartold, P. M, & Gronthos, S. (2009). Immunomodulatory properties of human periodontal ligament stem cells. *J Cell Physiol.*, 219, 667-676.

[53] Wagers, A. J, & Weissman, I. L. (2004). Plasticity of adult stem cells. *Cell.*, 116, 639-648.

[54] Wang, J, Zhou, X, Cui, L, Yan, L, Liang, J, Cheng, X, Qiao, L, Shi, Y, Han, Z, Cao, Y, Han, Y, & Fan, D. (2010). The significance of CD14+ monocytes in peripheral blood stem cells for the treatment of rat liver cirrhosis. *Cytotherapy.*, 12, 1022-1034.

[55] Williams, A. R, & Hare, J. M. (2011). Mesenchymal stem cells: biology, pathophysiology, translational findings, and therapeutic implications for cardiac disease. *Circ Res.*, 109, 923-940.

[56] Wobus, A. M, Holzhausen, H, Jäkel, P, & Schöneich, J. (1984). Characterization of a pluripotent stem cell line derived from a mouse embryo. *Exp Cell Res.*, 152, 212-219.

[57] Yu, J, Vodyanik, M. A, Smuga-otto, K, Antosiewicz-bourget, J, Frane, J. L, Tian, S, Nie, J, Jonsdottir, G. A, Ruotti, V, Stewart, R, Slukvin, I. I, & Thomson, J. A. (2007). Induced pluripotent stem cell lines derived from human somatic cells. *Science.*, 318, 1917-1920.

[58] Zarzeczny, A, & Caulfield, T. (2009). Emerging ethical, legal and social issues associated with stem cell research & and the current role of the moral status of the embryo. *Stem Cell Rev.*, 5, 96-101.

[59] Zhao, Y, Wang, H, & Mazzone, T. (2006). Identification of stem cells from human umbilical cord blood with embryonic and hematopoietic characteristics. *Exp Cell Res.*, 312, 2454-2464.

[60] Zheng, K, Wu, X, Kaestner, K. H, & Wang, P. J. (2009). The pluripotency factor LIN28 marks undifferentiated spermatogonia in mouse. *BMC Dev Biol.* 9:38.

[61] Zipori, D. (2005). The stem state: plasticity is essential, whereas self-renewal and hierarchy are optional. *Stem Cells.*, 23, 719-726.

[62] Zulli, A, Buxton, B. F, Merrilees, M, & Hare, D. L. (2008). Human diseased arteries contain cells expressing leukocytic and embryonic stem cell markers. *Hum Pathol.*, 39, 657-665.

Chapter 3

De-Differentiation of Somatic Cells to a Pluripotent State

Cristiane V. Wenceslau, Irina Kerkis,
Nelson F. Lizier and Alexandre Kerkis

Additional information is available at the end of the chapter

http://dx.doi.org/10.5772/54372

1. Introduction

The gametes are highly specialized haploid cells that harbor genetic background of each individual. During fertilization, the fusion of female and male gametes occurs in order to produce zygote. These zygotes are diploid cells and have genetic material of both individuals. Zygotes start to divide and undergo further pre-implantation development through the formation of morula, blastocysts and finally fetus (Fig. 1). When intrauterine fetal development is finished, the organism is already formed and ready to birth. The zygotes can be considered primordial stem cells, which originate the whole organism through unequal divisions to produce blastomeres, the cells resulting by cleavage of a zygote. Sixteen blastomeres constitute a morula, the spherical embryonic mass surrounded by the zona pellucid, which further became a blastocyst. Blastocyst is a thin-walled hollow structure surrounded by trophoblasts layer that contains a cluster of cells called the inner cell mass (ICM) from which the embryo arises and the scientists isolate embryonic stem (ES) cells for *in vitro* cultivation and for study the process of differentiation. However, ES cells are pluripotent cells able to produce any cell type raise ethical concerns about the destruction of human embryo to produce stem cell lines. To get the better concept of pluripotent cells for stem cell based therapies the reprogramming of patient specific adult cells to embryonic stage was suggested (Takashi & Yamanaka,2006). Cell reprogramming is a process of de-differentiation of somatic cells into pluripotent state whereby they adopt features of ES cells. De-differentiation of adult cells can be achieved through i. somatic cell nuclear transfer; ii. cell fusion - somatic cell hybrids and; iii. production of induced pluripotent stem (iPS) cells through the activation of essential stemness genes (reprogramming factor), over-expression in fibroblasts and/or other adult cells. Small molecules and other technologies are also exploring to reprogram cells without the use of viral vectors primarily by Yamanaka. The methods used for cell de-differentiation induce the expression of genes that are not normally expressed in adult cell but are expressed in pluripotent stem cells, leading to the activation of pluripotent cell transcriptional networks. A cascade of transcriptional activity switch on the changes in gene expression profile in the adult cells, which begins to express a repertoire of genes that are commonly, identified in pluripotent ES cells. Following reprograming the adult cells undergo morphological changes and begin to grow as a tightly packed cluster of cells known as a colony, which mirrors how undifferentiated ES cells grow in culture. Both processes of reprograming and resulting pluripotency of reprogrammed cells vary significantly and elucidation of different approaches can clarify the reprograming process. In this chapter we will describe different methods of reprogramming of differentiated cells to pluripotent cells and the knowledge gain from each. Additionally, we try to provide a functional vision on reprogramming process and to analyze different types of stem cell niches produced by natural and reprogrammed cells. The better comprehension of stem cell niches will allow us to improve the reprogramming technology and to put more close in production of natural pluripotent stem cells using molecular biology approaches.

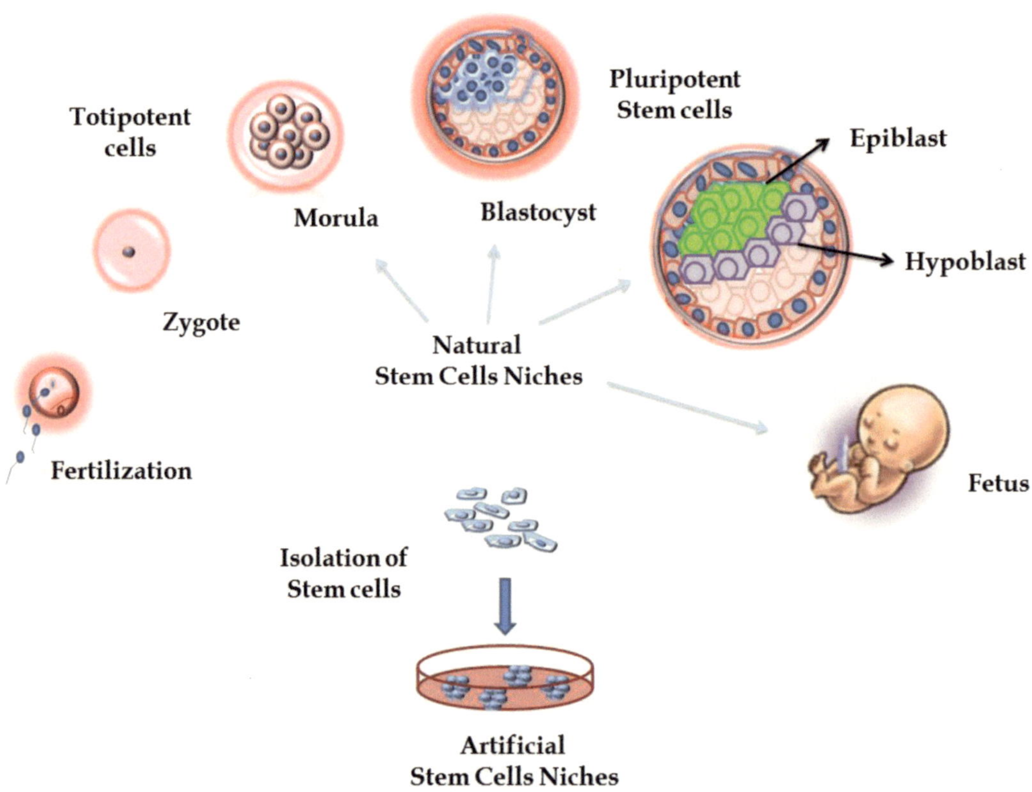

Figure 1. Early development of stem cells niches. According to current knowledge there are natural stem cells niches during development: morula, blastocyst, epiblast and fetus, and artificial stem cells niches: stem cell culture *in vitro*.

2. Natural *in vivo* pluripotent stem cell niches

2.1. Morula stem cell niche

Here we proposed that starting from morula, when zona pellucid arises, a first specific compartment called stem cell niche is formed. This niche can be defined as a microenvironment in which stem cells are found. Stem cell niche provides the both interaction between the cells and their interaction with local microenvironment, by which their fate regulates and occurs. In morula, stem cell niche consists of pluripotent stem cells that provide expression of specific transcription factor, such as POU domain transcription factor (Oct3/4) responsible for self-renewal capacity and pluripotency of these cells. In mammals morula first cell fate decisions is governed by key transcriptional factor: Oct3/4 (Palmieri et al., 1994). Oct3/4 is unique because it requires maintaining the pluripotency in both conditions *in vivo* and *in vitro* (Nichols et al., 1998) and it is essential for epigenetic reprogramming (Niwa et al., 2000). The depletion of transcription factors leads to increased expression of genes that are involved in the processes of development and cell differentiation (Niwa et al., 2000).

2.2. Blastocyst stem cell niche

The first lineage segregation is resulted in the formation of trophectoderm and ICM (Wobus, et al., 2005). Upon silencing of Oct3/4, a part of morula cells spontaneously inactivates the self-renewal process and start to differentiate into trophoblast cells, thus forming pluripotent stem cell niche in blastocyst. This blastocyst niche is a dynamic structure which follows developmental program of an organism in parallel with Oct3/4, expression of other transcriptional factor such as Nanog occurs in ICM. in early blastocyst (Nichols et al., 1998; Avilon et al., 2003).

2.3. Epiblast stem cell niche: Naive and primed pluripotent stem cells

In mice in late blastocyst transcription factor Sox2 starts to express in the cells of ICM in addition to GATA6 and Nanog, which lead to formation of two distinct populations: epiblast and hypoblast (Mitsui et al., 2003). These cell populations are considered the precursors of the primitive endoderm and the pluripotent epiblast (Morrissey et al., 1998). Recent studies suggest that stem cells in rodent epiblast have two distinct stable states of pluripotency: naïve and primed, thus establishing epiblast stem cell niche (Tesar et al., 2007; Nichols, 2009). According to these classification both of states exhibit features of bona fide pluripotent stem cells, such as have indefinite self-renewal, tri-germ layer potential and depend on expression of all three transcription factors, such as Oct3/4, Sox2 and Nanog (Tesar et al., 2009; Nichols, 2009; de Los Angeles et al., 2012). Naïve (more immature) pluripotent stem cells can be obtained from pre-implanted stage of embryo in rodents (Okamoto et al., 2003). These cells have both sex X chromosomes activated and are able to produce high-grade chimeras after their reintroduction into the host blastocyst. In contrast, in humans primed pluripotent ES cells are isolated from human pre-implantation blastocysts stage of development. In these cells one of female X chromosome is inactivated, albeit human ES cells are self-renewing and express key transcription factors and are able to form teratoma (Okamoto et al., 2003; Brons et al., 2007; Tesar et al., 2005). Studies of X chromosome inactivation in pre-implantation human embryos reported that *XIST* transcript accumulation on this chromosome occurs in the eight-cell stage embryo, however the identity of the cells, which show *XIST* accumulation is not clear. *In vitro* studies of *XIST* accumulation in human ES cells lines revealed three different patterns of X chromosome inactivation. The naïve state - with both active X chromosomes, intermediate state - with both *XIST* accumulation and last - state, when the cells never undergo X chromosome inactivation even under differentiation and *XIST* accumulation does not occur (Dvash and Fan, 2009). Therefore, using current technologies "true" pluripotent stem cells can be exclusively isolated from mouse and may be from some other rodents, which present similar pattern of early embryonic development with mice. It is not obvious if it is possible to obtain "true" human ES cells, once we cannot test their contribution into developing human embryo (due to ethic consideration). However, the lack of X chromosome reactivation indicates that probably these cells will never be able to reintegrate into early development events *in vivo* similar to rodent.

3. *In vitro* pluripotent stem cell niches

After isolation, pluripotent stem cells start to organize *in vitro* stem cell niche, which up to a certain degree simulates experience of these cells *in vivo*. *In vitro* these cells showed similar morphology with ICM, forming islands of juxtaposed cells and expressing pluripotent stem cell markers such as Oct3/4, Nanog, Sox2 (Tesar et al., 2005; Tesar et al., 2007). However, to distinguish *in vitro* naïve and primed pluripotent cells are difficult or even impossible task due to high heterogenety of pluripotent cell lines established *in vitro* (Brons et al., 2007; Tesar et al., 2007). Different factors may contribute to this heterogeneity, such as natural polymorphism of the cells, selection of colonies in vitro, which can be pluripotent at different degrees, cell culture conditions adopted in each work as well as pluripotent cells, itself, may produce an "imperfect" *in vitro* microenvironment again due to their natural heterogeneity.

4. Generation of artificial pluripotent stem cell – Reprogramming strategies

4.1. Reprogramming by means of differentiated cells nuclear transfer

Several strategies can be provided in order to reprogramming differentiated or committed somatic cell genome. One of these strategies is a nuclear transfer (NT) of differentiated cell nucleus to oocyte whose maternal DNA was removed (Campbell et al., 1996). This type of reprogramming uses the natural components without any previous genetic or molecular modification of nucleus–donor and oocyte-recipient. NT is relatively efficient and frequently depends on technical experience of researcher (Galli et al., 2012). There are two kinds of nuclear transfer trial: egg-NT involves the transfer of a single somatic nucleus to an unfertil-

http://dx.doi.org/10.5772/54372

ized enucleated oocyte and oocyte-NT involves the transplantation of multiple somatic cell nuclei into immature oocyte of amphibian. Nevertheless are important differences between the two types of nuclear transfer experiment. In oocyte-NT experiments extensive cell division take places and new functional cell types appear as soon as the nuclear transplant embryo start to develop. In this experiment somatic cell chromatin is directly reprogrammed to express pluripotency genes within a day. In contrast to oocyte-NT experiments, in egg-NT no new cell types are formed, and neither oocyte nor nuclei divide, however direct transition of reprogrammed nuclei that transcribe genes of pluripotency into differentiated cells occurs. Analysis of the mechanism of reprogramming in egg-NT experiments, which involves transcription pluripotency genes and others, is complicated owing to rapid DNA replication and numerous cell divisions (Halley–Scott et al., 2010; Julien et al., 2010).

The NT process leads to direct reprogramming of pluripotent stem cell and expression of such markers as Oct3/4, Nanog, and Sox2 that are silent in differentiated somatic cell nucleus. In general, the reactivation of silent pluripotency genes starts around 24 and 48 hours after NT (Halley–Scott et al., 2010; Julien et al., 2010; Byrne et al., 2003). Upon NT occurs the series of events when oocyte cytoplasm induces changes in the structure of donor chromatin toward pluripotent state, which became more appropriate for embryonic development. However, synchronization process which should happen between genomic DNA of donor cell and cytoplasm of recipient cell is complex and may affect significantly pluripotency of reprogrammed cells. Attempts to facilitate this reprogramming process have been made using chemicals that alter the methylation status of the chromatin, such as TSA (trichostatin A), azacytidine, scriptaid, either before or after NT. In the mouse, the use of TSA (a histone deacetylase inhibitor, HDACi) significantly increased the success rate of mouse cloning (Kishigami et al., 2007).

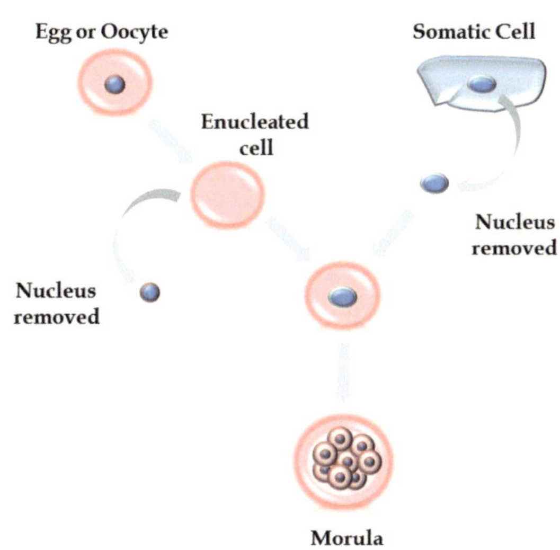

Figure 2. Stages of nuclear transfer. The nucleus is removed from an egg (or oocyte) and replaced by a nucleus from a donor cell (somatic cell).

In mammals, embryo obtained by NT and transferred into foster mother can result (or not) in full term development. The clones, obtained by NT method, are genetically identical to donor organism, which provide a nucleus. The sheep Dolly was the first successfully cloned farm animal. Dolly was obtained from NT of terminally differentiated mammary epithelial cell (Campbell et al., 1996). However the generation of animals by NT is not very efficient, once many clones are dying soon after implantation, and only few clones survive and born (Galli et al. 1999; Ritchie 2006). These clones frequently affected with severe abnormalities, they die prematurely and often obese. The survival rate of clones depends on species, on donor cell type, method of NT and varied significantly between different laboratories (Oback & Wells, 2002; Wilmut et al., 2002). However, pre-implantation development does not seem to be a problem (Ono et al., 2001; Ono et al., 2001a) the majority of the term losses occurs during the post implantation period and/or after birth. It has been reported in some experimental studies, that only 2-3% of the transferred embryos develop to term in mice (Ono et al., 2001a; Sakai et al., 2005). Over time the methods were improved and other species have been cloned with success from differentiated donor cells, such as cattle (Galli et al., 1999); mouse (Wakayama & Yanagimachi, 1999); pig (Polejaeva et al., 2000a); cat (Shin et al., 2002); goat (Keefer et al., 2002); mule (Woods et al., 2003); horse (Galli et al., 1999); rabbit (Challah-Jacques et al., 2003); rat (Zhou et al., 2003) and dog (Lee et al., 2005). In humans, the attempt to NT has been achieved using animal oocytes as recipients for human genetic material. The reprogramming of human somatic cell nuclei did not occur after NT into bovine and rabbit oocytes. These oocytes with human genome were not able to follow early embryonic development. The up-regulation of human pluripotency-associated genes did not occur. These data raised a question about the potential use of animal embryonic environment to generate patient-specific stem cells using NT technology. Ethical implications also should be taken in consideration (Chung et al., 2009).

4.2. Reprogramming by means of stem cells nuclear transfer

In 1998, Cibelli performed stem cells nuclear transfer (SCNT) using nucleus of bovine fibroblasts and enucleated bovine oocytes. They obtained 330 reconstructed oocytes, generated 37 cloned blastocysts, which served for isolation of 22 ES-like cell lines. These ES-like cells were injected into bovine oocytes, cultured cultured to produce embryos that further which were transferred into recipient females. In six out of seven calves at least one tissue originated from ES cell has been found. Other authors demonstrated the ability of karyoplast of ES cells induce Oct4 expression in the somatic genome (Tada, 2001).

In humans (Hall et al., 2007) and non-human primate (Mitalipov et al., 2002) the SCNT efficiency of blastocyst formation has typically been very low, thus suggesting a lack in or complete nuclear reprogramming. In order to overcome these difficulties modified SCNT approach was used to produce rhesus macaque blastocysts from adult skin fibroblasts and to isolate from this blastocyst two ES cell lines. This was achieved thought non-invasive approaches for meiotic spindle detection in oocytes and their removal using high-performance imaging. Spindle imaging system supports rapid and highly efficient real-time enucleation of primate oocytes. In this experiment spindle removal efficiency was 100%. The investigation of karyotype, microsatellite and single nucleotide polymorphisms (SNP) analyses confirmed that both ES cell lines were originated from SCNT embryos and were not from parthenotes. These ES cell lines demonstrated typical pluripotent cells morphology, self-renewal capacity and expression of stem cell markers. They were also transcriptionally similar to ES cells derived from fertilized blastocysts, and pluripotent, as demonstrated by the generation of several tissues from three germ layers after *in vivo* teratoma formation (Byrne 2007). Additionally, the experiments using mouse pluripotent primordial germ (PG) and ES cells as nuclei donors have also been performed using single-cell NT method. The results showed that embryos obtained from PG or ES cells NT method cannot develop and complete pre-implantation stage (Kato and Tsunoda, 1995). Possibly that long term *in vitro* culture can affect the karyotype of these cells accumulating chromosomal abnormalities, thus resulting in formation of abnormal embryos (Balbach et al., 2007).

The main goal of NT technology was to multiply the genotypes of high genetic value in farm animals and species, which are under the risk of extinction. Further, this technology was used as a tool for genome reprogramming of somatic differentiated cells into pluripotent state. The principles of cloning, which were developed by Willadsen (1986), are also important today. All the cloning studies provided the first experimental evidence for reprogramming (Kono et al., 1997; Gurdon, 2008). Currently, NT technologies can be applied in two different ways, to produce animal clones and to reprogram the nuclei of differentiated somatic cell, which can be used for basic research to analyze X chromosome inactivation or to study the dynamics of imprinting process during reprogramming and in some cases for preclinical evaluation of these cells in animal models (Hochedlinger and Jaenisch 2006). This technology yet holds medical interest to produce patient-specific stem cells, which can be used in cell therapy and regenerative medicine.

4.3. Reprogramming by means of early embryonic environment

The pluripotency, characteristic feature of ES cells, can be evaluated by their capacity to differentiate into cells of the three germ layers. More precisely, ES cells pluripotency can be evaluated by generation of chimaeras, organisms composed of cells from two or more individuals from the same or different species (Kaufman, 1981; Keller, 1995; Wobus, 2005). Production of human/animal chimaeras is a method currently in use to analyze developmental potency of mammalian ES in biomedical research (Behringer, 2007; Lensch et al., 2007). James et al (2006) showed for the first time that a nonhuman embryo surrogate environment could be used to study developmental potential of human ES cells as well as biological compatibility between human ES cells and the mouse ICM. Adult stem cells (ASC) are now seen as an alternative to ES cells, which can raise a number of ethical objections due requires destruction of human embryo. Populations of multipotent ASC that express ES cell markers, such as Oct3/4, Nanog and Sox2, presenting a differentiation capacity similar to that of ES cells *in vitro*, can be isolated from different fetal and adult animal and human tissues (Wenceslau et al., 2011). For example, we have reported the isolation of human immature dental pulp stem cells (hIDPSC) from deciduous (baby) teeth, which express the aforementioned pluripotent markers and can differentiate into several cell types *in vitro*, such as bone, cartilage, skeletal, smooth muscles and neurons (Kerkis et al., 2006; Lizier et al., 2012). We found that after their transplantation into adult mice, they engrafted within different mouse organs, such as the liver, heart, spleen, kidney and the brain. Although hIDPSC express pluripotent cell markers they present fibroblast-like morphology and were isolated from adult tissues (Kerkis et al., 2006; Lizier et al., 2012). We demonstrated that hIDPSC are truly multipotent cells, which were able to undergo further development similar to mouse ES cells in nonhuman embryo surrogate environment. These cells were able to contribute *in vitro* into ICM of mouse blastocyst, thus undergoing cell divisions, and *in vivo* into fetus development thus generating pretermed human/mouse chimaera, which a prerequisite to characterizing pluripotency similar for ES cells. In this study in order to analyze the ability of hIDPSC (46, XY) to contribute to ICM and trophectoderm of mouse early embryos, 6–8 cells stained with vital Vibrant fluorescent dye (Fig. 3A) were injected into the perivitelline space and/or the blastocell of 8 compacted morulae and 20 early blastocysts (Fig. 3B). After injection these cells have adopted similar size to those of the recipient mouse embryo. They proliferated in the recipient mouse embryonic environment and showed a contribution to the ICM and also to the trophoblast cell layer (Fig. 4A). To determine the developmental and pluripotent capacity of hIDPSC, six to eight stained cells were injected into the blastocele of 57 early blastocysts (Fig. 4B) and were immediately transferred to the uterus of five foster mothers. Three mice achieved pregnancy and, according to ethical recommendations, human/mouse chimaeras were collected before birth. The 18 d.p.c. mouse foetuses seemed to be well formed based on their morphological appearance (Fig. 4C).

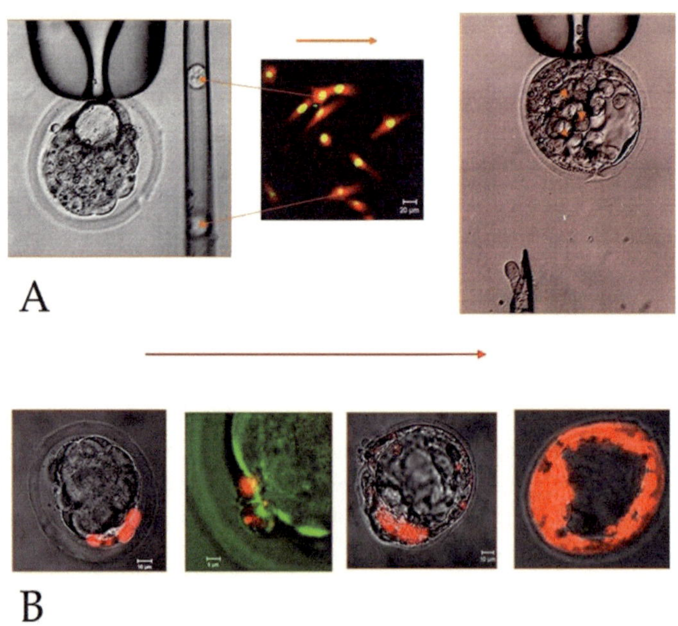

Figure 3. The hIDPSC injection in early embryonic environment. (A) hIDPSC stained with vital Vibrant fluorescent dye were injected into the perivitelline space and/or compacted morulae (B) hIDPSC showed a contribution to the ICM and also to the trophoblast cell layer.

Additionally to Vibrant fluorescent dye the anti-hIDPSC antibody, (this identifies exclusively hIDPSC), was used to detect the presence of these cells in 18 d.p.c. mouse foetuses). Strong fluorescent signals were observed in different organs of the chimaeras, such as the brain, liver, intestine and muscles (Fig. 4D and Fig. 5B). Using a variety of methods we demonstrated hIDPSC contribution to mouse embryos, which did not present any type of morphological deficiency (Fig. 5A). We were able to produce evidence, that these cells accomplished differentiation within local tissues, by the presence of human-specific tissue proteins, such as myosin and cytokeratin. Moreover, we used a specific antibody against human nuclei to confirm, again, that the cells were indeed of human origin (Siqueira da Fonseca et al., 2009). Little is known about the initial reprogramming events that occur after transference of ASC into mouse blastocysts (Yokoo et al., 2005). In our experiment, hIDPSC were capable of engrafting and proliferating inside mouse morulae and blastocysts and forming pretermed chimaeras. These cells contributed not only to ICM, as do human ES cells, but also to the trophoblast cell layer – without any embryo damage.

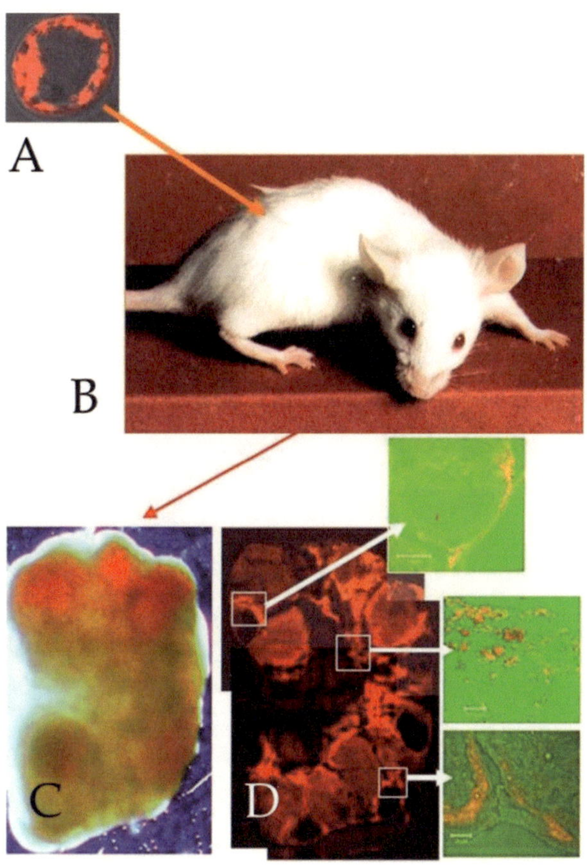

Figure 4. Developmental and pluripotent capacity of hIDPSC to generating pretermed human/mouse chimaera. Early chimera blastocyst (A) were transferred to the uterus of foster mother (B). Human/mouse chimaeras (C) were collected before birth and fluorescent signals were observed in different organs of the chimaeras (D).

Furthermore, hIDPSC integrated into host embryos and developed foetuses, undergoing the process of differentiation. Obviously that due to the difference in cell cycle dynamics between mouse and human cells, the number of human cells during mouse pre-natal development is decreased in comparison with hIDPSC contribution in ICM of blastocyst. However, it is not clear if hIDPSC can really undergo reprogramming into ES-like cells within nonhuman embryo surrogate environment.

Our finding suggests that expression of such pluripotent markers, as nanog and oct4 by hIDPSC is enough condition for these cells to contribute into different mouse tissues in early embryo-fetal development, to differentiate properly and to express human proteins within mouse fetal an immune privileged environment (Siqueira da Fonseca et al., 2009).

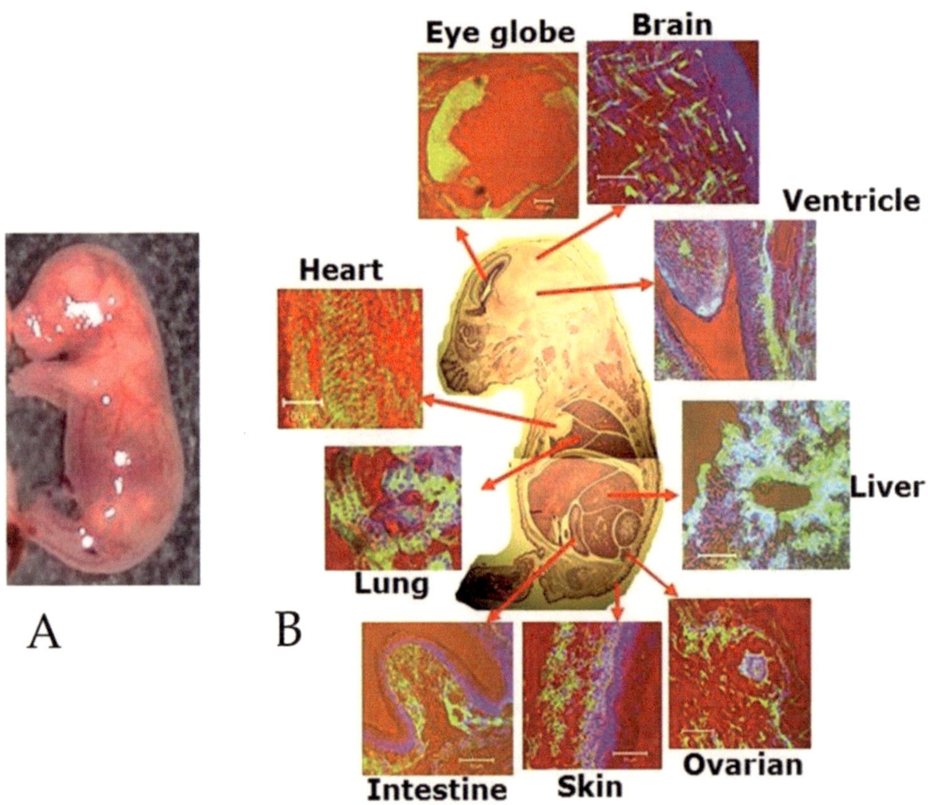

Figure 5. The hIDPSC contribution in pretermed human/mouse chimaera. (A) 18 d.p.c. mouse fetus. (B) Strong fluorescent signals were observed in different organs of the chimaeras, such as the brain, liver, intestine, muscles and others.

4.4. Reprogramming by means of cell fusion

First pluripotent hybrid cells have been isolated by fusion of pluripotent teratocarcinoma (TC) cells with differentiated somatic cells, which served as a tool for investigating the interaction between different genomes. These TC cells are similar to ES cells in morphology and gene expression pattern, thus maintaining variable levels of pluripotency, however not all TC cells able to generate chimaeras and to contribute to germ line (Papaioannou and Rossant 1983). These cells frequently have abnormal karyotype, such as loss of the Y chromosome, trisomy, deletions or translocations (Takagi et al. 1983, Rousset et al. 1983, Modlinski et al. 1990). The hybrid cells obtained from pluripotent TC cells and somatic cell partner, which express embryonic antigens, were able to produce teratomas containing derivatives of all three embryonic germ layers (Andrews and Goodfellow 1980, Atsumi et al., 1982; Rousset et al. 1983; Forejt et al., 1984; Takagi, 1983) and to form embryoid bodies (EBs) in suspension culture (Takagi, 1983). These hybrid cells showed also reactivation of particular genes after reprogramming (Miller and Ruddle, 1976, 1977; Andrews and Goodfellow, 1980; Rousset et al., 1983) and reactivation of inactive X chromosome originated from the somatic partner (McBurney and Adamson, 1976; McBurney and Strutt, 1980; Takagi et al., 1983, Takagi, 1988; Mise et al., 1996). However, pluripotent hybrids were obtained when lymphocytes or thymocytes, not fibroblasts, were used as the somatic parents in fusion (Rousset et al., 1979). These studies indicate that hybrid cells generated by ES cells and differentiated cells, which have less cytoplasm, seem to be more adequate systems to undergo reprogramming.

Matveeva et al. (1996) has obtained cultures of intraspecific embryonic hybrid cells by fusion of mouse ES cells, denominated HM-1 cells, which were derived from HPRT-deficient strain 129 mice (Magin et al., 1992) and characterized as highly pluripotent (Magin et al., 1992; Selfridge et al.1992) with splenocytes derived from an adult DD/c female. These hybrids were denominated as hybrid embryonic stem and somatic (HESS) cells and characterized as pluripotent and HPRT positive (Matveeva et al., 1996; 1998). Our group used three mouse hybrid clones HESS-1, HESS-2 and HESS-3 in order to study their karyotypes and investigate the influence of the karyotypes on the differentiation of these cells through the formation of embryonic bodies (Mittmann et al., 2002). The hybrid cells used in our study were near diploid (HESS-2 and HESS-3) and near tetraploid (HESS-1) and chromosome analysis showed different trisomies. The trisomies of chromosomes 1 and 11 were found in near diploid hybrids. These trisomies are probably typical of these pluripotent cells, and have previously been described in the mouse ES cells line (Crolla et al., 1990) and in TC cells (McBurney and Rogers, 1982). We found that the sex chromosome constitution in the HESS-2 line was predominantly XY, while in the HESS-3 line it was XO. Interesting that in HESS-2 and HESS-3 lines the segregated X chromosome was of embryonic origin. Indeed, it has been demonstrated by Ringertz and Savage (1976) that hybrids lose the chromosomes originating from differentiated, more slowly dividing cells. In our experiments, hybrids showed the capacity to form EBs *in vitro*, even at late passages (Fig. 6). The EBs formed by the hybrid cells could be considered as complex as those derived from the HM-1 line and the cystic-type EBs formed by pluripotent cells (Martin and Evans, 1975; Van der Kamp et al., 1984; Doetschman et al., 1985; Pease et al., 1990).

In the EBs derived from hybrids we observed haematopoietic-like cells, cells resembling skeletal and smooth muscle and others (Fig. 7). Cells of ectodermal origin (e.g. nerve cells) were not identified in EBs derived from hybrids. Our data shows that the 'embryonic' X chromosome may be lost in pluripotent hybrids, but reprogramming of the 'somatic' X chromosome may still occur, thus allowing restricted pluripotency. The normal karyotype may be a prerequisite for the efficient contribution of these cells to the germ line in transgenic and chimeric animals and for their ability to differentiate *in vitro* into a wide spectrum of cell types (Papaioannou et al., 1978; McBurney Rogers, 1982; Pease et al., 1990; Bronson et al., 1995; Liu et al., 1997; Suzuki et al., 1997). Therefore, we further tested the capacity of near diploid HESS-2 to differentiate *in vitro* in putative germ cells (GC) (Fig. 8). We demonstrated that two days after induction of differentiation by retinoic acid, the HESS-2 derived GC-like cells presented expression patterns of a gene set, involved in the progression of early stages of gametogenesis (Vasa, Stella, Dazl, Piwil 2, Tex14, Bmp8b, Tdrd1 and Rnf17). This finding is similar to previous descriptions of GC obtained *in vitro* from mouse ES cells (Hübner et al., 2003; Geijsen et al., 2004; Kerkis et al., 2007). HESS-2 generates GC *in vitro,* which were able to differentiation into sperm- and oocyte-like cells. These structures resembling the formation of presumptive oocytes appeared floating in the culture medium. FISH analyses indicate that several GC derived from HESS-2 hybrid cells were able to undergo sex chromosome reduction. The expression of ZP2 and ZP3, oocyte-specific markers, was also detected supporting our morphological observation. Hence our observations indicate that HESS-2 cells can progress into both female- and male- GC differentiation, however, the female developmental program could be achieved only in early stages (Lavaginolli et al., 2009).

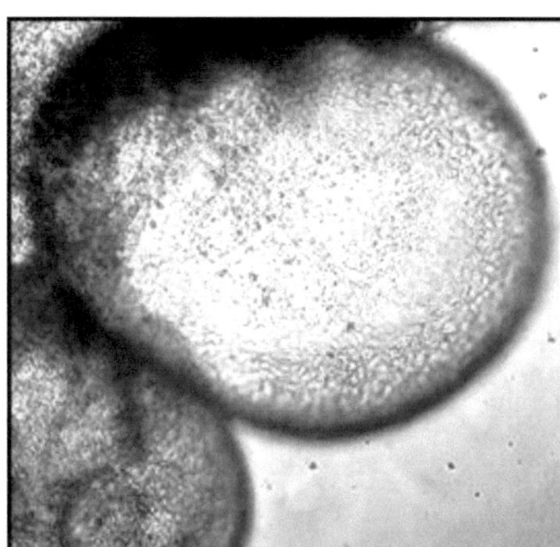

Figure 6. Cystic embryoid body - a globular cell cluster cultured from mouse ES cells.

Therefore, we demonstrated that near diploid somatic cell hybrids obtained by the fusion of ES cells with differentiated cell can be fully reprogrammed and able to produce *in vitro* even GCs. It is not likely that these cells will be able to generate live offspring after fertilization of normal oocyte due to abnormal karyotype. However, they represent an interesting model to study the influence of karyotype on the process of GC *in vitro* formation. More recently the reprograming of somatic cell nucleus after the fusion with induced pluripotent stem (iPS)

http://dx.doi.org/10.5772/54372

cells has been reported (Takahashi and Yamanaka, 2006). These iPS-somatic cell hybrids demonstrated the expression of markers of pluripotent cells, such as Oct4, SSEA-1, and alkaline phosphatase and were able to differentiate into multiple cell types similar to ES cells, thus confirming the reprogramming ability of iPS cells (Takahashi and Yamanaka, 2006).

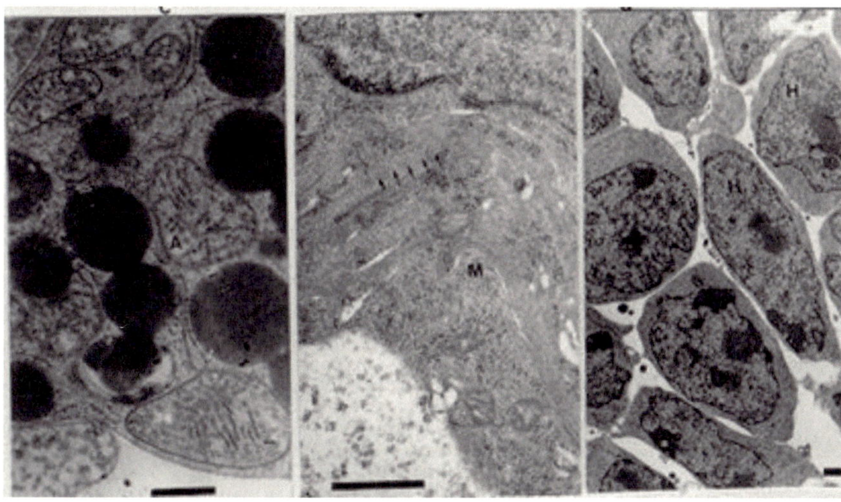

Figure 7. Electron microscopy demonstrates differentiation within cystic embryoid bodies derived from somatic cell hybrids (HESS-2).

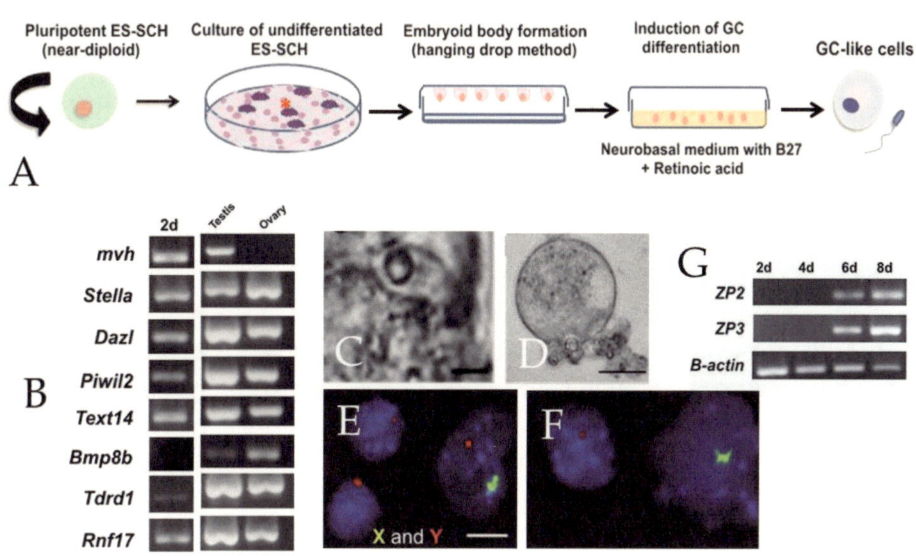

Figure 8. Germ cell derived in vitro from HESS-2. (A) Protocol of differentiation. (B) Expression of genes involved in the progression of early stages of gametogenesis during HESS-2 differentiation. (C) Sperm-like structure. (D) Oocyte-like structure. (E and F) Haploid cells with X or Y chromosomes. (G) Expression of oocyte-specific markers in oocyte-like structures obtained from HESS-2.

4.5. Reprogramming by means of Yamanaka's factors

The pluripotency manifests during short time of early mammalian development (Choen et al., 2011; Dejosez et al., 2012). Such powerful, pluripotent cells can be obtained *in vitro* from early embryo and they are very promising for the future of regenerative medicine and even for organ generation. However, in humans isolation of these cells implicates with ethical problem of embryo destruction. Thus the idea to obtain such pluripotent cells artificially took the minds of the researchers. The technologies to obtain such alternative pluripotent cells are growing continuously. Yamanaka 's group using the combination of different factors performed the first reprograming of mouse embryonic fibroblasts. The resulted iPS cells showed gene-protein expression of ES-cell markers, teratoma formation, differentiation into the tissues of three germ layers, beside chimaeras generation. This reprogramming strategy, using defined factors (i.e. Klf4, Oct4, Sox2, and c-Myc, termed "KOSM"), is conceptually and technically simple (Takahashi & Yamanaka, 2006). However, it is a low efficient and reproducibility process, which is influenced by several variables and also could affect the quality, such as completely or non-completely reprogrammed iPS cells. These variables are the age donor, cell type, different delivery systems and reprogramming cocktail choice, factors used for reprogramming (Daley et al., 2009). Currently, several strategies, based on genes, proteins, iRNA, as well as on different chemicals, are available for the reprogramming of somatic cells (Nakagawa et al., 2008; Yu et al., 2007). In the original method of iPS generation developed by Yamanaka's group used the moloney murine leukemia virus (MMLV) retrovirus for transgene expression (Takahashi and Yamanaka 2006). This vector has cloning capacity of around 8 kb allows delivery of genes into the genome of cells and expected to be silenced after reprogramming and induction of endogenous genes activation. The efficiency of iPS cells generated using MMLV retroviruses Expressing the KOSM set genes is around 0.1% in mouse embryonic fibroblasts and approximately 0.01% in human fibroblasts (Jahner et al., 1982; Stewart et al., 1982; Hotta et al., 2008). Lentiviral vector is also used in reprograming experiments thus exhibiting slightly higher (8–10 kb) cloning capacity and usually have higher infection efficiency than MMLV retroviruses (Blelloch et al., 2007). However, carcinogenesis may be caused by genomic integration of retro- or lentiviral fragments into host DNA (Varas et al., 2009) and use of c-Myc oncogene, which after reactivation might cause malignant tumor formation (Okita et al., 2007; Brambrink et al., 2008). Thus viral systems are still unsafe for therapeutic application. Therefore, a number of reports demonstrates that iPS cells can be generated by reducing the use of viral constructs and/or minimize viral integration through substitution of key reprogramming factors by chemical compounds or employing less differentiated cells, which already express endogenously one or more of the key pluripotency factors (Hota et al., 2008). Our group reprogrammed hIDPSC-fibroblast-like cells isolated from deciduous (baby) teeth, which express endogenous Oct3/4 and Nanog, using retroviral vector and four Yamanaka' s factors (Fig. 9).

Reprogrammed hIDPSC presented all key characteristics of pluripotent cells: formed juxtaposed colonies of ES-like morphology and produce teratoma with derivates of all three germ layers. These cells did not integrate retroviral vector in their genome and express lower levels of Oct4, Nanog and Sox 2. In contrast to iPS cells derived to fibroblast cell, the hIDPSC derived iPS cells were generated to in shorter time and presented higher efficiency of colonies formation And were able to form under iPS colonies feeder –free conditions conditions. For example, the time of fibroblasts reprogramming using retrovirus vectors takes 20–25 days (Aesen et al., 2008), while reprogramming of hIDPSC occurs only in eleven days after infection (Beltrão-Braga, 2011). These results suggest that age of donor and differentiation status of cell type used for reprograming may also affect reprogramming efficiency. Accordingly, Maherali and Hochedlinger et al., (2007) compared skin fibroblasts reprograming efficiency from two-month-old and two years-old mice. Older cells produced half as many iPS cell colonies as young skin fibroblasts. It has been shown that iPS cells have so-called epigenetic memory, which means that after reprograming their differentiation potential can reflect on their lineage commitment before reprograming. Therefore, hIDPSC showed strong neural commitment, which is due to their ectomesodermal origin. After reprograming strong neural commitment was evidenced within teratomas as well as spontaneous *in vitro* differentiation into neurons hIDPS-iPSC was also detected. It was expected that ordinary human adult cells reprogrammed as iPSC may revolutionize medicine by creating new therapies unique to individual patients. However, important questions have persisted about the safety of these cells, such as it is not clear the degree to which these cells are homologous to ES cells in respect of the genes expression pattern, differentiation capacities, epigenetics and in particular interest is the question whether iPSCs genetic material is altered during the reprogramming process. The researchers, which examined 22 different human iPSC lines obtained from seven research groups showed that these cells present 10 times more mutations than they expected to find. While some of the mutations appeared to be silent, the majority did change specific protein functions, including those in genes associated with causative effects in cancers. Anyway, the studies of iPSC provide an important new tool in the fight against human disease, but to use these cells directly in the clinic, we must ensure that they are safe.

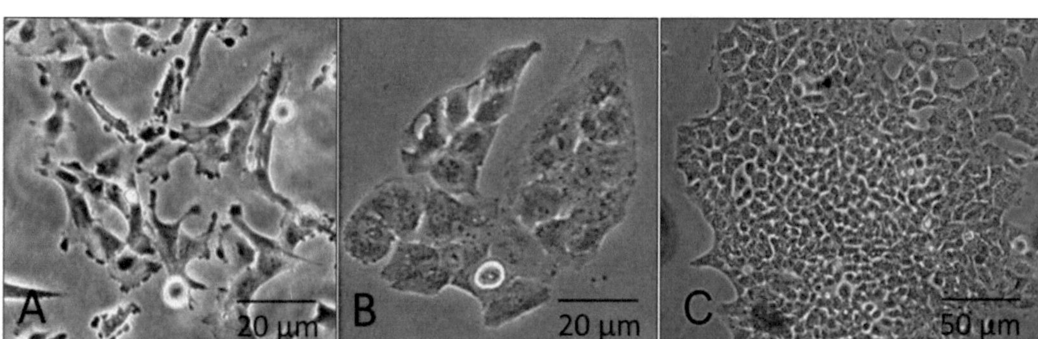

Figure 9. hIDPSC-derived iPS cell. (A) Representative figure of morphological characteristics of hIDPSC *in vitro* culturing. (B) iPS cell derivation were shown to be obtained under feeder-free condition on matrigel-coated dishes. (C) A typical hIDPSC-derived iPS cell colony. Light microscopy.

5. Final considerations

All reprogramming strategies are aimed at genomic reprogramming, which is a key biological process. It is still unknown, how many and what reprogramming factors, which initiate a cascade of reprogramming events, are involved in NT, SCNT, in cell fusion and even in iPS cell production. Yamanaka's study suggests that these factors may be mainly proteins of the nucleus; however the cytoplasm factors also should be taken in consideration. NT technique, which was used for Dolly the Sheep and many other species, has been abandon by many researches due to the low efficiency. Some researches try to use SCNT in stem cells research in order to obtain stem cells that are genetically matched to the donor organism. However, up to data no human ES cells were obtained using SCNT. Another limitation of this method is that resulting cells retain mitochondrial structures, which originally belonged to the egg. The great limitation of cell fusion technology is chromosome set composed by different genomes. Currently many scientists, which used all these methods moved to iPS cell production.

We started this chapter with simplified description of the concept of stem cell niches formation during early development. This conception lead to comprehension that such niches are very complex and composed by heterogeneous population of different somatic and stem cells. We know, that at least two different populations of pluripotent stem cells naïve and prime can be identified *in vivo* and isolated *in vitro* in rodents. In humans these two populations are difficult to identified and isolate. Additionally, the data on the pattern of X chromosome activation of *in vitro* cultured human ES cells suggest the existence of may be three such populations. In adult organism the number of stem cell niches increased dramatically, the examples are neuronal, hematopoietic, hair follicle, skeletal muscle, dental pulp and many other stem cells niches. In order to obtain stem cells of the most excellent quality the scientist try to re-create stem cell niche *in vitro*, which enables ad of control of culture conditions, including oxygen tension and hydrostatic pressure and various factors believed to be involved in self-renewing, division, migration, recruitment and lineage commitment of stem cells. Any strategies of reprogramming are closely related with the conception of stem cell niche, because in all strategies of reprogramming the nucleus or the cell with different developmental histories and from different cells niches are used. In order to translate the potential of reprogrammed cells into to the clinical reality our knowledge about reprogrammed stem cells microenvironment should be significantly improved.

Acknowledgements

The authors thank Dr. Thais M. C. Lavagnolli from Imperial College London and Dr. Simone A. S. da Fonseca from University of Sao Paulo for their contribution in our research.

Author details

Cristiane V. Wenceslau[1,2], Irina Kerkis[1,2], Nelson F. Lizier[1,2] and Alexandre Kerkis[1]

1 Laboratory of Genetics, Butantan Institute, Federal University of Sao Paulo, Brazil

2 Department of Morphology and Genetics, Federal University of Sao Paulo, Brazil

References

[1] Aasen, T. et al (2008) Efficient and rapid generation of induced pluripotent stem cells from human keratinocytes. Nature Biotech. 26, 1276–1284.

[2] Amps K et al. (2011) Screening ethnically diverse human embryonic stem cells identifies a chromosome 20 minimal amplicon conferring growth advantage. Nat Biotechnol. Nov 27;29(12):1132-44.

[3] Andrews PW, Goodfellow PN (1980) Antigen expression by somatic cell hybrids of a murine embryonal carcinoma cell with thymocytes and L cells. Somat Cell Genet 6: 271-284.

[4] Atsumi T, Shirayoshi Y, Takeichi M and Okada TS (1982) Nullipotent teratocarcinoma cells acquire the pluripotency for differentiation by fusion with somatic cells. Differentiation 23: 83-86.

[5] Avilion AA, Nicolis SK, Pevny LH, Perez L, Vivian N, Lovell-Badge R (2003) Multipotent cell lineages in early mouse development depend on SOX2 function Genes Dev; 17:126 – 140.

[6] Balbach ST, Jauch A, Böhm-Steuer B, Cavaleri FM, Han YM, Boiani M (2007) Chromosome stability differs in cloned mouse embryos and derivative ES cells. Dev Biol. 15; 308(2):309-21.

[7] Behringer RR (2007) Human-animal chimaeras in biomedical research. Cell Stem Cell 1, 259–262.

[8] Beltrão-Braga PCB, Pignatari GC, Maiorka PC, Lizier NF, Wenceslau CV, Miglino MA, Muotri AR, Kerkis I (2011) Feeder free derivation of induced pluripotent stem cells from human immature dental pulp stem cells. Cell Transplatation.

[9] Brambrink, T. et al (2008) Sequential expression of pluripotency markers during direct reprogramming of mouse somatic cells. Cell Stem Cell 2, 151–159.

[10] Brons, L.E. Smithers, M.W. Trotter, P. Rugg-Gunn, P. Sun, S.M. Chuva de Sousa Lps, S.K. Howlett, A. Clarkson, L. Ahrlund-Richter, R.A. Pedersen, L. Vallier (2007) Deri-

vation of pluripotent epiblast stem cells from mammalian embryos,Nature, 448 pp. 191–195.

[11] Bronson SK, Smithies O, Mascarello JT. High incidence of XXY and XYY males among the offspring of female chimeras from embryonic stem cells (1995) Proc Natl Acad Sci U S A. 11; 92(8):3120-3.

[12] Byrne A,.Pedersen DA, Clepper LL, Nelson M, Sanger WG, Gokhale S, D. P. Wolf1 DP, Mitalipov SM (2007) Producing primate embryonic stem cells by somatic cell nuclear transfer. Nature, 450.

[13] Byrne J A, Simonsson S, Western P S, Gurdon JB (2003) Nuclei of adult mammalian somatic cells are directly reprogrammed to oct-4 stem cell gene expression by amphibian oocytes. Curr. Biol. 13, 1206–1213.

[14] Campbell, K. H., McWhir, J., Ritchie, W. A. & Wilmut, I (1996) Sheep cloned by nuclear transfer from a cultured cell line. Nature 380, 64–66.

[15] Challah-Jacques M, Chesne P, Renard J P (2003) "Production of cloned rabbits by somatic nuclear transfer." Cloning Stem Cells; 5(4): 295-9.

[16] Chung Y, Bishop CE, Treff NR, Walker SJ, Sandler VM, Becker S, Klimanskaya I, Wun WS, Dunn R, Hall RM, Su J, Lu SJ, Maserati M,Choi YH, Scott R, Atala A, Dittman R, Lanza R (2009) Reprogramming of human somatic cells using human and animal. Cloning Stem Cells. Jun; 11(2):213-23.

[17] Cibelli JB, Stice SL, Golueke PJ, Kane JJ, Jerry J, Blackwell C, Ponce dLn, Robl JM Ž (1998) Cloned transgenic calves produced from non-quiescent fetal fibroblasts. Science 280:1256-1258.

[18] Cohen DE, Melton D (2011) Turning straw into gold: directing cell fate for regenerative medicine. Nat Rev Genet. Apr; 12(4):243-52.

[19] Crolla JA, Lakeman SK, Seller MJ (1990) The induction of tail malformations in trisomy 16 mouse fetuses heterozygous for the curly tail recessive gene. Genet Res. 55(1): 27-32.

[20] Daley G Q. et al (2009) Broader implications of defining standards for the pluripotency of iPSCs. Cell Stem Cell 4, 200–201.

[21] De Los Angeles A, Loh YH, Tesar PJ, Daley G Q (2012) Accessing naïve human pluripotency. Current Opinion in Genetics & Development. 22, 3, 272–282.

[22] Dejosez M, ZwakaTP (2012) Pluripotency and nuclear reprogramming. Annu Rev Biochem. 2012; 81:737-65.

[23] Doetschaman T, Eistetter H, Katz M, Schmidt W & Kemler R (1985) The in vitro development of blastocyst-derived embryonic stem cell lines: formation of visceral yolk sac, blood islands and myocardium. Journal of embryology and Experimental Morphology, 87:27-45.

[24] Dvash T, Fan G. (2009).Epigenetic regulation of X-inactivation in human embryonic stem cells. Epigenetics. 4(1):19-22.

[25] Evans, M& Kaufman, M (1981) Establishment in culture of pluripotent cells from mouse embryos. Natures, 292: 154-156.

[26] Forejt J, Gregorova S, Dohnal K , Nosek J (1984) Gene expression of differentiated parent in teratocarcinoma cell hybrids. Repression or reprogramming?Cell Diff 15: 229-234.

[27] Galli C, Duchi R, Moor RM, Lazzari G (1999) Mammalian leukocytes contain all the genetic information necessary for the development of a new individual. Cloning 1, 161–170.

[28] Galli C, Lagutina I, Perota A, Colleoni S, Duchi R, Lucchini F, Lazzari G (2012) Somatic Cell Nuclear Transfer and Transgenesis in Large Animals: Current and Future Insights. Reprod Dom Anim 47, (Suppl. 3), 2–11.

[29] Geijsen N, Horoschak M, Kim K., et al (2004) Derivation of embryonic germ cells and male gametes from embryonic stem cells. Nature 427, 148–154.

[30] Hall, V. J. et al (2007) Developmental competence of human in vitro aged oocytes as host cells for nuclear transfer. Hum. Reprod. 22, 52–62.

[31] Halley-Stott, R. P. et al (2010) Mammalian nuclear transplantation to germinal vesicle stage Xenopus oocytes—a method for quantitative transcriptional reprogramming. Methods 51, 56–65.

[32] Hotta A, Ellis J (2008) Retroviral vector silencing during iPS cell induction: an epigenetic beacon that signals distinct pluripotent states. J. Cell. Biochem. 105, 940–948.

[33] Hubner, K., Fuhrmann, G., Christenson, L.K., Kehler, J., De La Reinbold, R., Fuente,R., Wood, J., Strauss, J.F., Boiani, M., Scholer, H.R (2000) Derivation of oocytes from mouse embryonicstemcells.Science300,1251–1256.

[34] Jahner D. et al (1982) De novo methylation and expression of retroviral genomes during mouse embryogenesis. Nature 298, 623–628.

[35] James D, Noggle SA, Swigut T, Brivanlou AH (2006) Contribution of human embryonic stem cells to mouse blastocysts. Dev. Biol. 295, 90–102.

[36] Jullien J, Astrand C, Halley-Stott R P, Garrett N, Gurdon JB (2010) Characterization of somatic cell nuclear reprogramming by oocytes in which a linker histone is required for pluripotency gene reactivation. Proc. Natl Acad. Sci. USA 107, 5483–5488.

[37] Kato Y, Tsunoda Y (1995) Germ cell nuclei of male fetal mice can support development of chimeras to midgestation following serial transplantation. Development. 121(3):779-8.

[38] Keefer CL, Keyston R, Lazaris A, Bhatia B, Begin I, Bilodeau AS, Zhou FJ, Kafidi N, Wang B, Baldassarre H, Karatzas CN (2002) "Production of cloned goats after nuclear transfer using adult somatic cells". Biol Reprod., 66(1).

[39] Keller G.M (1995) In vitro differentiation of embryonic stem cells. Current Opnion in cell Biology, 7-862-829.

[40] Kerkis I, Kerkis A, Dozortsev D, Stukart-Parsons GC, Gomes Massironi SM, Pereira LV, Caplan AI, Cerruti HF (2006) Isolation and characterization of a population of immature dental pulp stem cells expressing OCT-4 and other embryonic stem cell markers. Cells Tissues Organs. 184(3-4):105-16.

[41] Kerkis A, Fonseca SAS, Serafim RC , Lavagnolli TMC, , Pereira VS, Abdelmassih S, Abdelmassih R, Kerkis I. (2007) In Vitro Differentiation of Male Mouse Embryonic Stem Cells into Both Presumptive Sperm Cells and Oocytes. Cloning and Stem Cells. 9, 4.

[42] Kishigami S, Bui HT, Wakayama S, Tokunaga K, Van Thuan N, Hikichi T, Mizutani E, Ohta H, Suetsugu R, Sata T, Wakayama T (2007) Successful mouse cloning of an outbred strain by trichostatin A treatment after somatic nuclear transfer. J Reprod Dev. 53(1):165-70.

[43] Lavagnolli TMC, Fonseca SAS, Serafim RC, Pereira VS, Santos EJC, Abdelmassih S, Kerkis A, and Kerkis I (2009) Presumptive germ cells derived from mouse pluripotent somatic cell hybrids. Differentiation. 78; 124–130.

[44] Lee BC, Kim MK, Jang G, Oh HJ, Yuda, F, Kim HJ, Shamim MH, Kim JJ, Kang SK, Schatten G, Hwang WS (2005) "Dogs cloned from adult somatic cells". Nature, 436: 641.

[45] Lensch MW, Schlaeger TM, Zon LI, Daley GQ (2007) Teratoma formation assays with human embryonic stem cells: a rationale for one type of human-animal chimaera. Cell Stem Cell, 253–258.

[46] Liu X, Wu H, Loring J, Hormuzdi S, Disteche CM, Bornstein P, Jaenisch R. (1997) Trisomy eight in ES cells is a common potential problem in gene targeting and interferes with germ line transmission. Dev Dyn. 209(1):85-91.

[47] Lizier NF, Kerkis A, Gomes CM, Hebling J, Oliveira CF, Caplan AI, Kerkis I (2012) Scaling-up of dental pulp stem cells isolated from multiple niches. PLoS One.; 7(6):e39885.

[48] Magin TM, McWhir J and Melton DW (1992) A new mouse embryonic stem cell line with good germ line contribution and gene targeting frequency. Nucleic Acid Res 20: 3795-3796.

[49] Maherali, N., Sridharan, R., Xie, W., Utikal, J., Eminli, S., Arnold, K., Stadtfeld, M., Yachechko, R., Tchieu, J., Jaenisch, R., et al. (2007) Directly reprogrammed fibroblasts

show global epigenetic remodeling and widespread tissue contribution. Cell Stem Cell 1, 55–70.

[50] Martin GR, Evans M.J (1975) Differentiration of clonallines of teratocarcinoma cells: formation of embryoids in vitro. Proceedings of national academy of Sciences. USA, 72: 1441-1445.

[51] Matveeva NM, Shilov XG, Bayborodin SI, Philimonenko VV, Rolinskaya IV and Serov OL (1996) Hybrid between mouse embryonic stem and somatic cells retain pluripotency. Proc Acad Sci Russia 349: 129-132.

[52] Matveeva NM, Shilov AG, Kaftanovskaya HM, Bayborodin SI, Fokina MM, Maximovsky LP, Zhelezova AI, Golubitsa AN and Serov OL (1998) In vitro and in vivo study of pluripotency in hybrid cells obtained by fusion of mouse embryonic stem cells with splenocytes. Mol Reprod Dev 50: 128-138.

[53] McBurney MW, Adamson ED (1976) Studies on the activity of the X chromosomes in female teratocarcinoma cells in culture. Cell. 9(1):57-70.

[54] McBurneyMW, Strutt BJ (1980) Genetic activity of X chromosomes in pluripotent female teratocarcinoma cells and their differentiated progeny. Cell 21: 357-364.

[55] McBurney MW, Rogers BJ (1982) Isolation of male embryonal carcinoma cells and their chromosome replication patterns. Dev Biol. 89(2):503-8.

[56] Miller RA, Ruddle FH (1976) Pluripotent teratocarcinoma-thymus somatic cell hybrids. Cell 9: 45-55.

[57] Mise N, Sado T, Tada S and Takagi N (1996) Activation of the inactive X chromosome induced by cell fusion between a murine EC and female somatic cell accompanies reproducible changes in the methylation pattern of the Xist gene. Exp Cell Res 223: 193-202.

[58] Mitalipov S M, Yeoman RR Nusser KD, Wolf DP (2002) Rhesus monkey embryos produced by nuclear transfer from embryonic blastomeres or somatic cells. Biol. Reprod. 66, 1367–1373.

[59] Mittmann J, Kerkis I, Kawashima C, Sukoyan M, Santos E, Kerkis A (2002) Differentiation of mouse embryonic stem cells and their hybrids during embryoid body formation. Gen. Mol. Biol. 25 (1), 103–11.

[60] Mitsui K, Tokuzawa Y, Itoh H, Segawa K, Murakami M, Takahashi K, Maruyama M,Maeda M, Yamanaka S (2003) The homeoprotein Nanog is required for maintenance of pluripotency in mouse epiblast and ES cells. Cell; 113:631–642.

[61] Modlinski JA, Gerhauser D, Lioli B,Winking H and Illmensee K (1990) Nuclear transfer from teratocarcinoma cells into mouse oocytes and eggs. Development 108: 337-348.

[62] Nakagawa M. et al (2008) Generation of induced pluripotent stem cells without Myc from mouse and human fibroblasts. Nature Biotech. 26, 101–106.

[63] Nichols J, Zevnik B, Anastassiadis K, Niwa H, Klewe-Nebenius D, Chambers I, Schöler H, Smith A (1998) Formation of pluripotent stem cells in the mammalian embryo depend on the POU transcription factor oct4. Cell 95:379:391

[64] Nichols J, A. Smith (2009) Naïve and primed pluripotent states. Cell Stem Cell, 4, pp. 487–492.

[65] Niwa H, Miyazaki J, and Smith AG (2000) Quantitative expression of oct-3/4 defines differentiation dedifferentiation or self-renewal of ES cell. Nat Genet. 24: 372-3726.

[66] Oback B, Wells D (2002) "Donor cells for nuclear cloning: many are called, but few are chosen". Cloning and Stem Cells, 4 (2): 147-168.

[67] Okamoto, A.P. Otte, C.D. Allis, D. Reinberg, E. Heard (2003) Epigenetic dynamics of imprinted X inactivation during early mouse development Science. 303; 644–649.

[68] Okita K, Ichisaka T, & Yamanaka S (2007) Generation of germline-competent induced pluripotent stem cells. Nature 448, 313–317.

[69] Ono, Y; Shimozawa, N.; Ito, M.; Kono, T. (2001a). "Cloned mice from fetal fibroblasts cells arrested at metaphase by serial nuclear transfer". Biol. Reprod., 64: 44-50.

[70] Ono, Y.; Shimozawa, N.; Muguruma, K.; Kimoto, S.; Hioki, K.; Tachibana, M.; Shinkai, Y.; Ito, M.; Kono, T. (2001b). "Production of cloned mice from embryonic stem cells arrested at metaphase". Reproduction, 122: 731-736.

[71] Palmieri SL, Peter W, Hess H, Scholer HR (1994) Oct-4 transcription factor is differentially expressed in the mouse embryo during establishment of the first two extraembryonic cell lineages involved in implantation. Dev Biol; 166:259 – 267.

[72] Papaioannou VE, Rossant J (1983) Effects of the embryonic environment on proliferation and differentiation of embryonal carcinoma cells. Cancer Surveys 2: 165-183.

[73] Papaioannou VE, Gardner RL, McBurney MW, Babinet C, Evans MJ (1978) Participation of cultured teratocarcinoma cells in mouse embryogenesis. J Embryol Exp Morphol. 44:93-104.

[74] Pease S, WilliamsRL (1990) Formation of germ line chimeras from embryonic stem cells maintained with recombinant leukemia inhibitory factor. Exp Cell Res. 190(2): 209-11.

[75] Polejaeva I A, Chen S-H, Vaught TD, Page R L, Mullins J, Ball S, Dai Y, Boone J, Walker S, Ayares DL, Colman A, Campbell' KHS (2000) "Cloned pigs produced by nuclear transfer from adult somatic cells". Nature, 407: 86-90.174

[76] Ritchie WA. (2006).Nuclear transfer in sheep. Methods Mol Biol. ;325:11-23

[77] Rousset J-P, Bucchini D, Jami J (1983) Hybrids between F9 nulli-potent teratocarcinoma and thymus cells produce multi-differentiated tumors in mice. Dev Biol 96: 331-336.

[78] Ringers N.R, Savage R.E (1976) Cell Hybrids. Acad. Press, New York, 417, ppl.

[79] Martin GR, Evans M.J (1975) Differentiration of clonallines of teratocarcinoma cells: formation of embryoids in vitro. Proceedings of national academy of Sciences. USA, 72: 1441-1445.

[80] Rousset J-P, Dubois P, Lasserre C, Aviles D, Fellous M, Jami J (1979) Phenotype and surface antigenes of mouse teratocarcinoma x fibroblast cell hybrids. Somat Cell Genet 5: 739-752.

[81] Sakai, R.R.; Tamashino, K.L.K.; Yamazaki, Y.; Yanagimachi, R (2005). "Cloning and assisted reproductive techniques: influence on early development and adult phenotype". Birth Defects Research, 75: 151-162.

[82] Selfridge J, Pow AM, McWhir J, Magin TM, Melton DW (1992) Gene targeting using a mouse HPRT mini/HPRT-deficient embryonic stem cell system: Inactivation of the mouse ERCC-1 gene. Somat Cell Mol Genet 18:325–336.

[83] Siqueira da Fonseca SA, Abdelmassih S, Serafim RC , Lavagnolli TMC Clemente Santos ES, Mendes CM, de Souza Pereira V, Ambrosio CE., Miglino MA., Visintin JR, Abdelmassih, Kerkis A, Kerkis I (2009) Human immature dental pulp stem cells' contribution to developing mouse embryos: production of human/mouse preterm chimaeras Cell Prolif. 42, 132–140.

[84] Shin T, Kraemer D, Pryor J, Liu L, Rugila J, Howe L, Buck S, Murphy K, Lyons L, Westhusin, M (2002) "A cat cloned by nuclear transplantation". Nature, 415(6874): 859.

[85] Stewart CL, Stuhlmann H, Jahner D & Jaenisch R (1982) De novo methylation, expression, and infectivity of retroviral genomes introduced into embryonal carcinoma cells. Proc. Natl Acad. Sci. USA 79, 4098–4102.

[86] Suzuki H, Kamada N, Ueda O, Jishage K, Kurihara H, Terauchi Y, Azuma S, Kadowaki T, Kodama T, Yazaki Y & Toyoda Y (1997) Germ-line contribution of embryonic stem cells in chimeric mice: influenceof karyotype and in vitro differentiation ability. Experimental Animals, 46: 17- 23.

[87] Tada M, Takahama Y, Abe§ K, Nakatsuji N ,Tada T (2001) Nuclear reprogramming of somatic cells by in vitro hybridization with ES cell, Current Biology. 11: 9.

[88] Takagi N (1988) Requirement of mitoses for the reversal of X-inactivation in cell hybrids between murine embryonal carcinoma and normal thymocytes. Expl Cell Res 175: 363-375.

[89] Takagi N, Yoshida MA, Sugawara O and Sasaki M (1983) Reversal of X inactivation in female mouse somatic cells hybridized with murine teratocarcinoma stem cells in vitro. Cell 34: 1053-1062.

[90] Takahashi, K. & Yamanaka, S (2006) Induction of pluripotent stem cells from mouse embryonic and adult fibroblast cultures by defined factors. Cell 126, 663–676.

[91] Tesar PJ, Chenoweth JG, Brook FA, Davies TJ, Evans EP, Mack DL, Gardner, RL, McKay RDG (2007) New cell lines from mouse epiblast share defining features with human embryonic stem cells . Nature. 488, 196–199.

[92] Tesar PJ (2005) Derivation of germ-line-competent embryonic stem cell lines from preblastocyst mouse embryos. Proc Natl Acad Sci U S A. 7; 102(23):8239-44.

[93] Van der Kamp AW, Roza-de Jongh EJ, Houwen RH, Magrane GG, van Dongen JM, EvansMJ (1984) Developmental characteristics of somatic cell hybrids between totipotent mouse teratocarcinoma and rat intestinal villus cells. Exp Cell Res. 154(1): 53-64, 1984

[94] Varas, F. et al (2009) Fibroblast-derived induced pluripotent stem cells show no common retroviral vector insertions. Stem Cells 27, 300–306.

[95] Wakayama T, Perry AC, Zuccotti M, Johnson KR, Yanagimachi R (1998) Full-term development of mice from enucleated oocytes injected with cumulus cell nuclei. Nature 394, 369–374.

[96] Wenceslau CV, Miglino MA, Martins DS, Ambrósio CE, Lizier NF, Pignatari GC,Kerkis I (2011) Mesenchymal progenitor cells from canine fetal tissues: yolk sac, liver and bone marrow. Tissue Eng Part A.

[97] Willadsen SM (1986) Nuclear transplantation in sheep embryos. Nature 320, 63–65. 441, 1061–1067.

[98] Wilmut I, Beaujeau N, de Sousa PA, Dinnyes A, King T J, Paterson LA, Wells DN, Young L E (2002) "Somatic cell nuclear transfer". Nature, 149: 583-586.

[99] Woods G L, White K L, Vanderwall DK, Li G-P, Aston KI, Bunch T D, Meerdo LM, Pate B J (2003) "A mule cloned from fetal cells by nuclear transfer". Science, 301(5636): 1063.

[100] Wobus, A.M; Boheler K (2005) Embryonic stem cells: prospects for developmental biology and cell therapy. Physiol rRev. 85: 635-678.

[101] Yu J. et al (2007) Induced pluripotent stem cell lines derived from human somatic cells. Science 318, 1917–1920.

[102] Yokoo T, Ohashi T, Shen JS, Sakurai K, Miyazaki Y, Utsunomiya Y, Takahashi M, Terada Y, Eto Y, Kawamura T, Osumi N, Hosoya T (2005) Human mesenchymal stem cells in rodent whole-embryo culture are reprogrammed to contribute to kidney tissues. Proc. Natl. Acad. Sci. USA 102, 3296–3300.

[103] Zhou Q, Renard JP, Le Friec G, Brochard V, Beaujean N, Cherifi Y, Fraichard A, Cozzi J (2003) "Generation of fertile cloned rats by regulating oocyte activation". Science, 302(5648): 1179.

Chapter 4

The Dark Side of Pluripotency – Cancer Stem Cell

Patricia Ng and Wang Cheng-I

Additional information is available at the end of the chapter

http://dx.doi.org/10.5772/54369

1. Introduction

Cancer stem cells are defined as cancer cells that show the two properties of stemness: unlimited self-renewal and, pluripotency or multipotency. These properties make cancer stem cell tumorigenic i.e. the ability to induce and sustain cancer.

The definition of cancer stem cell has been a topic of debate and has changed with time. Cancer stem cells were proposed in 1994 by John Dick and coworkers as the cells that initiated leukemia [1]. It was thought that this leukemic cell was derived from the mutation of a hematopoietic stem cell. Importantly, the term was used to distinguish a small subpopulation of leukemic cells that could initiate and maintain cancer from the rest of the leukemic cells that could not. Subsequently, it was also observed in other types of cancer that only a very small subpopulation of cancer cells had the ability to initiate cancer when transplanted into a new host [2–12]. This subpopulation of cancer cells was considered as cancer stem cells. The rest of the cancer cells, which ranged from progenitor to fully differentiated cancer cells, that formed the bulk of the cancer had limited proliferative capacity and hence could not initiate cancer when transplanted. Since cancer comprise a heterogeneous collection of cells, a unique set of cell surface markers that were expressed on cancer stem cells were used to define them.

The definition underwent revision when new experimental methods showed that turmorigenicity had been underassigned to a small group of cells due to limitations of the detection technique used. When different experimental approaches were undertaken, tumorigenicity was found to be widespread amongst phenotypically diverse cancer cells, resulting in a paradigm shift in the definition of cancer stem cells. Hence in 2006, the American Association of Cancer Research (AACR) defined a cancer stem cell as any cancer cell that possessed stem cell-like properties of unlimited self-renewal and multi/pluripotency. AACR specifically highlighted that the definition of a cancer stem cell does not imply that such cells are derived from the stem cells of the corresponding tissue. Also, a cancer stem cell does not have to be that initial cell in the body that caused cancer. For example, a differentiated cell that reacquires immortality through genetic mutations is considered a cancer stem cell. Thus any cancer cell that possesses or acquires stemness which results in unlimited tumorigenic potential is considered a cancer stem cell.

More recently, interesting data has emerged demonstrating that partially differentiated cancer cells, when exposed to a specific set of microenvironmental factors, can reacquire stemness [13]. Induction of stemness through this mechanism is reversible and could also result in epigenetic modifications, which then becomes heritable. This finding would again modify our understanding of the nature of cancer stem cell, suggesting that the cancer stem cell can be a dynamic and reversible entity.

In this section, experimental data shaping the identification and definition of cancer stem cells are presented in three parts. It begins with a description of early studies demonstrating that cancer stem cells were found to be a small and distinct subpopulation of cancer. This is followed by evidence suggesting that cancer stem cells can also be a highly common and heterogeneous population of cancer cells. Finally, evidence that cancer stem cell is a dynamic and reversible entity in cancer is discussed.

1.1. Cancer stem cells: A distinct subpopulation of cancer cells

The concept of cancer stem cell is not new. The first experimental evidence for the existence of cancer stem cell was in 1937 when Furth and Kahn injected a single leukemic cell from a mouse into an inbred mouse and transmitted leukemia [14]. At that time, it was unclear if every cancer cell or only a subpopulation of cancer cells possessed this ability to transmit. In 1994, a landmark experiment showed that only a subpopulation of cancer cells could transmit cancer [1]. John Dick and his group isolated cancer cells from patients with acute myeloid leukemia (AML) and separated these cells based on their expression of CD34 and CD38. In this study, transplanting half a million of CD34+CD38- cells into severe combined immunodeficiency (SCID) mice induced AML in mice within thirty days, while the same number of CD34+CD38+ cells did not induce any AML in mice. The subpopulation of cancer cells that could transmit cancer was termed SCID-leukemia initiating cells and was thought to be amongst CD34+CD38- cells.

To determine the amount of the SCID-leukemia initiating cells within the CD34+CD38- cell population, a quantitative transplantation approach was used [15]. The cancer cells were serially diluted and transplanted into NOD/SCID (non-obese diabetic/severe combined immunodeficiency) mice. The minimum dose required to cause leukemia was then determined. Based on this experiment, it was found that there was about one cancer stem cell per 5,000 CD34+CD38- cells. This would mean that within a population of a million cancer cells, there was about one cancer stem cell. The ability of cancer stem cell to self-renew was provided by experiments that used the same transplantation approach described above, i.e. using human leukemic cells and NOD/SCID mice [15]. In one of these experiments, the number of human cancer stem cells in mice was initially found to be about three in 16 million leukemia cells. However, after six weeks, human cancer stem cells had increased to about 100 cancer stem cells within a population of 20 million leukemia cells in the bone marrow of these mice. This indicated that SCID-leukemia initiating cells in the mice had multiplied from three to a 100 and therefore behaved like stem cells, possessing unlimited self-renewal ability.

Evidence that cancer stem cells could differentiate into the rest of the cancer cell population was provided by characterizing the CD34+CD38- cells after transplantation into NOD/SCID mice [15]. Flow cytometry analysis of human cells isolated from the bone marrow of mice showed that transplanting CD34+CD38- cells resulted in an increase in cancer cell population, of which 98% were positive for both CD34 and CD38. This differentiative capacity, together with self-renewal ability, led to the conclusions that cancer stem cells existed and formed a distinct subpopulation of cancer cells.

Following the identification of cancer stem cells in leukemia, a series of in vivo studies documenting the presence of cancer stem cells in other cancers came to light. These studies, summarized in Table 1, characterized human cancer-initiating cells by their surface markers and were based mostly on the NOD/SCID mouse xenotransplantation assay. The table highlights information regarding the frequency of expression of cancer-stem cell-associated markers in the cancer cells, and the estimated frequency of the cancer stem cells residing in the cell population that bears the cancer stem cell-associated markers. Based on these studies, it was estimated that cancer stem cell existed in, at most, one in ten thousand cancer cells. Importantly, it was shown that this small population of cancer stem cells had a distinct CD-phenotype, which when fully defined would serve as the address for accurate delivery of cytotoxic drugs.

In addition to the use of NOD/SCID mice, in vitro techniques that were previously used for the isolation of normal stem cells were also used to isolate cancer stem cells. These techniques included the formation of non-adherent spheroids in tissue culture method [10] and the exclusion of the fluorescent Hoechst dye method [16]. Both methods led to the identification of a subpopulation of cells that, when transplanted into mice, resulted in tumorigenicity. Hence both in vivo and in vitro studies suggested that tumorigenic cancer cells were stem-like in phenotype.

1.2. Cancer stem cells are not ALWAYS a distinct subpopulation of cancer cells

After a decade of using human cancer cells with NOD/SCID mouse as a model for cancer stem cell detection, there were concerns of incompatibility issues between the two species with regards to the cytokines and receptors involved in cancer stem cell research. Cytokines and receptors from different species could prevent critical interactions that were required for cancer cells to survive. Furthermore, the NOD/SCID mouse immune system, even though rendered compromised, could still mount some level of response to reject the human cells, thereby potentially resulting in erroneously lower count estimation of cancer stem cell population.

Markers for enrichment of cancer stem cells	Cancer type	Cancer stem cell detection assay	Ref.
$CD34^+ CD38^-$	Leukemia (0.02-2% of mononuclear cells)[a]	2 of 2 NOD/SCID mice (5,000 cells)[b]	[1,15]
$CD44^+$ Lin^-	Head and Neck cancer (0.1-42%)	5 of 7 NOD/SCID mice (5,000 cells)	[20]
$CD44^+ ESA^+ CD24^{-\,or\,low} Lin^-$	Breast cancer (2%)	4 of 4 NOD/SCID mice (200 cells)	[2,4]
$CD44^+ CD24^- ALDH1^+ Lin^-$	Breast cancer (0.1-1.2%)	NOD/SCID mice (20 cells)	[21]
$CD44^+ ESA^+ CD24^+$	Pancreatic cancer (0.2-0.8%)	6 of 12 NOD/SCID mice (100 cells)	[12]
$CD44^+ ESA^{high} CD166^+$	Colon cancer (1.2-16%)	1 of 2 NOD/SCID mice (150 cells)	[3]
$CD44^+ CD117^+$	Ovarian cancer (0.2%)	9 of 10 nude mice (100 cells)	[11]
$CD133^+$	Brain cancer (6-29%)	4 of 4 NOD/SCID mice (100 cells)	[10]
	Colon cancer (1.8-24.5%)	5 of 6 NOD/SCID mice (500 cells) 15 of 30 NOD/SCID mice (3,000 cells)	[6] [8]
	Lung cancer (0.32-22%)	4 of 4 SCID mice (10,000 cells)	[5]
	Pancreatic cancer (1.1-3.2%)	Nude mice (500 cells)	[22]
$ABCB5^+$	Melanoma (1.6-20.4%)	11 of 11 NOD/SCID mice (1,000,000 cells)	[9]

Table 1. Experiments using markers for the enrichment of human cancer stem cells and xenotransplation assay for the detection and quantification of human caner stem cells. An estimation of the population of cancer stem cell in a tumor is given based. Epithelial-specific antigen (ESA) and ATP-binding cassette B5 (ABCB5) are surface markers. Aldehyde dehydrogenase 1 (ALDH1) is an enzyme inside the cell. Lin refers to a collection of lineage markers CD2, CD3, CD10, CD16, CD18, CD31, CD64 and CD140b. Mice were condition by irradiation prior to receiving the transplantation. [a]Percentages of tumor cells expressing the selected markers. [b]Minimum number of surface-marker expressing cells required to induce cancer in at least 50% of the mice.

To address these concerns, alternative experimental models were used. The first model used mouse cancer cells instead of human cancer cells to circumvent the issue of cross species barrier. One of these experiments involved transplanting mouse leukemic cells from transgenic mice bearing the oncogene Myc with the immunoglobulin heavy chain enhancer. Just ten mouse leukemic cells were sufficient to induce cancer. Indeed, this experiment recapitulates the very first experiment in 1937 in which a single cancer cell from a chemically-induced cancer mouse was able to cause cancer in an inbred mouse [14,17]. This suggested that cancer stem cells were not necessarily a small population of cancer cells but rather could be more common than previously thought.

In the second model, human cancer cells continued to be used. However, they were transplanted into mice that were rendered even more severely immunocompromised than NOD/SCID mice [18]. In one such study where NOD/SCID ILR2γnull mice were used, 27% (and hence more than a quarter) of single cell transplantation of human melanoma cells into the mice resulted in cancer [18,19]. Importantly, this experiment showed that these cancer stem cells were not associated with any of the surface markers that were previously characterized (See Table 1). A total of 85 cell surface markers from these cancer stem cells were studied. Of these, 22 cell surface markers showed heterogeneous expression within the cancer cell population of which none had an association with the capacity for tumor initiation. For example, both CD133+ and CD133- cells were able to induce cancer [19]. In addition, the cancers that resulted from both CD133+ and CD133- cancer cells created a population of cancer cells that was heterogeneous in their expression of CD133. These findings implied that cancer stem cells were both common and heterogeneous. Hence, these 2 models showed that cancer stem cells were not a small, distinct subpopulation of cancer cells but rather a common and heterogeneous population of cancer cells in cancers such as melanoma.

Amidst the new findings that challenged the concept of cancer stem cells as a small, distinct subpopulation of cancer cells, there were experiments which still showed that the cancer stem cell subpopulation was indeed low, and not common and heterogeneous, even when syngenic mice were used [23]. Interestingly, human experiments (which could not have been conducted currently due to ethical reasons) provided evidence that when human cancer cells were transplanted back into the human subject from whom the cancer cells originated, the likelihood of cancer-initiation in the autologous human host is rare [24–26].

Taking all the evidence together, cancer stem cells are indeed a small and distinct subpopulation of cancer cells in some cancers, whereas in other cancers, cancer stem cells are common and heterogeneous. An alternative explanation that could account for these varied observations about cancer stem cells is that a cancer stem cell is not a static entity but rather a state that cancer cells can transform into.

1.3. Cancer stem cells are dynamic

Even more recently, it has been shown that a partially differentiated cancer cell, under the "right" microenvironmental influence, can reacquire stemness [13]. This finding is crucial in furthering our understanding of the cancer stem cell as a dynamic and reversible entity, rather than a static one.

In a study on colorectal cancer, differentiated colorectal cancer cells were able to dedifferentiate back into cancer cells with cancer stem cell phenotype after being exposed to hepatocyte growth factor (HGF) [27]. Upon exposure, these cells showed increased colony-forming ability (clonogenicity) and increased tumorigenicity. Biochemically, the cells exhibited an increase in the Wnt signaling pathway leading to the expression of β-catenin dependent genes. This finding is important as HGF is present within the natural microenvironment of colorectal cancer as it is normally produced by myofibroblasts which are prominent in the colorectal stroma. Hence, given the right microenvironment, non-tumorigenic cancer cells can become cancer stem cells.

Similarly, in a separate study using PDGF-induced glioma in mice, exposure to nitric oxide caused differentiated glioma cancer cells to transform into glioma cancer stem cells [25]. Again, nitric oxide is normally present in the natural microenvironment of gliomas as nitric oxide is produced by blood vessels. Hence glioma cancer cells in close proximity to brain blood vessels were able to re-acquire stem-like properties. These two recent studies presented reiterate the concept that cancer stem cells are dynamic - cancer cells are able to transform back into cancer stem cells given the right micorenvironmental conditions.

1.4. Conclusions

Cancer stem cells are cancer cells that have self-renewable and multi or pluripotent abilities. Our current understanding is that cancer stem cells can be a distinct subpopulation of cancer cells in certain cancers while in other cancers, they can be relatively common and heterogeneous. They are also dynamic in nature.

Understanding the defining characteristics of cancer stem cells is important as these have important therapeutic implications. In cancers in which the cancer stem cells form a distinct subpopulation, eliminating this subpopulatuion of cancer stem cells can potentially lead to a cure. In contrast, targeting one specific group of cancer stem cells in cancers in which the cancer stem cells are common and heterogeneous would be futile. In addition, learning more about the microenvironmental factors that promote the cancer stem cell state provides another interesting approach in finding a cure for cancer.

2. Cancer stem cell: The survivor

Chemotherapeutic agents against cancer are able to reduce tumor mass significantly but often a cure may not be achievable. In such cases, a cure is not possible due to a subpopulation of cells that are resistant to cancer drugs. The cancer stem cells amongst this resistant population then self-renew, proliferate and metastasize to cause relapse after treatment. In addition to understanding the defining characteristics of cancer stem cells for therapeutic purposes, a working knowledge of the molecular mechanisms of drug-resistance in cancer stem cells will empower researchers to better design new therapeutic agents that can overcome drug resistance. We will also explore the mechanism for metastasis in cancer stem cells, which serves as another potential therapeutic target.

http://dx.doi.org/10.5772/54369

2.1. Drug resistance

Normal stem cells have traits that confer high survival capacity under harsh environments. These include (1) cell quiescent, (2) active DNA-repair system, (3) expression of transporters that keeps toxic substances out, (4) high metabolism in detoxification, and (5) resistance to apoptosis. These mechanisms are thought to be employed by cancer stem cells, in addition to genetic mutations, to evade anti-cancer drugs.

2.1.1. Cell quiescence

Chemotherapeutic cancer drugs such as vincristine, vinblastine, paclitaxel and docitaxel works by arresting cancer cells in mitosis, thus leading to apoptosis [28]. One hypothesis to explain resistance of cancer stem cells to these drugs is that cancer stem cells are in a quiescent state. Indeed, quiescent cancer stem cells have been shown to exist in some cancers [29]. Moreover, drug resistance in slow-cycling cancer stem cell population has also been reported [30,31]. In addition to cell quiescence, cancer stem cells are likely to have other mechanisms for drug resistance as discussed below.

2.1.2. DNA-repair

Ionizing radiation and anti-cancer drugs that disrupt the genome kill cancer cells by targeting their DNA. Cancer stem cells have efficient DNA-repair systems that confer resistance to these anti-cancer agents. A study on glioblastoma demonstrated that cancer stem cells, identified by their expression of CD133, showed preferential activation of the DNA damage check point response resulting in an increase in their DNA repair capacity [32]. The study also shows that both in vitro as well as mouse brain samples of cancer cells have increased the proportion of CD133-positive cells to CD133-negative cells following radiotherapy. This suggested that the subpopulation of CD133-positive, i.e. cancer stem cells, had developed resistance to radiotherapy and were the cause of cancer relapse in the mouse.

2.1.3. Drug transporters

A third mechanism of drug resistance is the expression of transporters of the ATP-binding cassette (ABC) family. ABC transporters are efflux pumps that can actively expel a wide range of chemotherapeutic drugs from the cell. ABC transporters are expressed in both normal stem cells and cancer stem cells. Three members of this family of ABC transporters, ABCB1, ABCC1 and ABCG2 have been identified as the culprits of multidrug resistance in many cancers.

A study on neuroblastoma patients illustrates how cancer stem cells use the efflux transporter, ABCG2 to protect themselves from anti-cancer drugs. In this study, cells expressing ABCG2 were identified by the fluorescent Hoechst dye 33342, in flow cytometry, as a "side population" (SP) of cells that did not take up this dye. A previous study had shown that cancer stem cells reside in the SP fraction of neuroblastoma [10]. SP cells from neuroblastoma patients showed increased efflux of mitoxantrone when compared to non-SP cells. Also, treatment of neuroblastoma cell lines with mitoxantrone led to an increase in the proportion of SP cells to non-SP cells, suggesting that ABCG2 conferred a survival advantage to cancer stem cells [33]. Similarly, in acute myeloid leukemia, SP cells derived from mononuclear cells in bone marrow of patients showed an increase efflux of daunorubicin and mitoxantrone when compared to non-SP cells [34]. Taken together, these findings suggest that cancer stem cell uses efflux transporters to guard against anti-cancer drugs.

2.1.4. High detoxification activity

Aldehyde dehydrogenase I (ALDH1) is a detoxifying enzyme that oxidizes intracellular aldhydes and is a marker of normal stem cell. Cancer stem cells from acute myeloid leukemia and breast carcinoma are known to have high levels of ALDH1. [21]. In breast cancer patients undergoing chemotherapy with paclitaxel and epirubicin, it was found that the proportion of ALDH1-positive cancer cells increased significantly post treatment, resulting in treatment failure[35]. A high ALDH1 level is thus associated with poor clinical outcomes. This finding indicates that new therapeutic agents must be able to overcome the detoxification prowess of cancer stem cells in order to be effective.

2.1.5. Blockage of apoptosis

Blockage of apoptosis is a major mechanism for drug resistance as it offers protection against any therapy that results in cell destruction. This ability to prevent apoptosis from occurring in cancer stem cell is mediated by both inherent cellular factors and extrinsic microenvironmental factors.

Inherent cellular factors are important in blocking the apoptotic process. In a study on cancer stem cells (isolated via CD133) from glioblastoma, exposure of cancer cells to conventional chemotherapeutics such as temozolomide, carboplatin, paclitaxel and etoposide, showed that CD133-positive cells had higher viability compared to CD133-negative cells [36]. In contrast to CD133-negative cells, CD133-positive cells had higher mRNA levels of the anti-apoptotic proteins, such as B-cell lymphoma (Bcl) -2 and -XL proteins, inhibitors of apoptosis proteins (IAPs), FLICE-like inhibitory protein (FLIP) and Sirtuin 1 (SIRT1). In addition, CD133-positive cells had lower mRNA level of the pro-apoptotic protein, including Bcl-2 associated X protein (BAX). In a separate study on colon cancer, autocrine production of interleukin-4 (IL4) by CD133-positive colon cancer cells was found to prevent apoptosis of cancer stem cells from occurring when conventional chemotherapeutics and a recombinant protein called TNF-related apoptosis-inducing ligand (TRAIL) were used [37]. These findings show that cancer stem cells have an armament of proteins to protect themselves from undergoing apoptosis.

In some cancers, resistance to apoptosis is highly dependent on extrinsic microenvironmental factors. For example, culturing ovarian cancer cells under stem cell culture conditions led to formation of spheroid cultures of cells that were self-renewing and resistant to cisplatin and paclitaxel [38]. However, this resistance was lost once the cells were cultured under a different set of conditions. One extrinsic factor that has been recognized as the main cause for resistance to cancer therapeutics is hypoxia. The normal stem cell niche has been associated with a hypoxic microenvironment. Expression of hypoxia-inducible factors (HIF) are important as these factors regulate stem cell self-renewal and pluripotency [39]. Although not well studied in cancer stem cells per se, the role of HIF in regulating apoptosis has been shown in cancer cell cultures. HIF directly regulates the transcription of anti-apoptotic genes such as myeloid cell leukemia 1 (MCL-1) and B-cell lymphoma extra-large (BCL-XL) [40,41]. Hypoxia also results in a lower level of reactive oxygen species (ROS) in the cell. A lower level of ROS leads to a decrease in activation of caspase-8, a decrease in expression of pro-apoptotic receptor TRAIL-R2 and an increase expression of pro-survival proteins like cFLIP and BCL-2 [42,43].

Besides the extrinsic chemical factors described above, another type of extrinsic factor that blocks apoptosis is the stimulating ligands produced by neighboring cells. The hematopoietic niche has been found to confer resistance to leukemic cells via adhesion molecules such as integrins and soluble molecules of the Wnt pathway [44]. Wnt, Notch and Hedgehog are developmental regulatory molecules that are increasingly shown to be involved in cancer stem cell self-renewal, growth and differentiation [27,45–50]. Therapeutics targeting factors of these pathways are currently undergoing clinical trials and have shown promising results in eliminating cancer stem cells that are resistant to existing therapies.

2.2. Metastasis

Tumorigenicity is an essential characteristic that metastatic cancer cells must possess in order to initiate tumor formation after metastasizing to a distant site. Hence, it can be assumed that cancer stem cells are responsible for metastasis as they have tumorigenic properties by definition. Recent findings suggest that not all cancer stem cells have the capacity to metastasize. Rather, this capacity is confined to a subset of cancer stem cells. Cancer stem cells (identified by their CD 133 marker) isolated from pancreatic cancer patients were found to contain a subset of cells that expressed CXCR4, the receptor for stromal-cell derived factor 1 (SDF-1). These cancer stem cells that expressed CXCR4 were able to induce tumors in mice, spread via the blood circulation and cause liver metastasis. On the other hand, cancer stem cells that were CXCR4-negative were only able to induce tumors in mice, failing to spread and cause metastasis [22]. In light of this finding, therapies targeting this subset of cancer stem cells could prevent metastasis.

2.3. Conclusions

The multitude of research in cancer stem cell has deepened our understanding in this field. We present a schema (figure 1) that summarizes the literature in cancer stem cell research from a therapeutic perspective. This schema illustrates that targeting cancer cells with tumorigenic abilities, i.e. cancer stem cells, is not enough. It is the problematic subset of resistant cancer stem cells (outlined black in figure 1) that accounts for failure of current cancer therapies. Overcoming resistance in cancer stem cells is crucial and we have described several mechanisms that cancer stem cells use to stem our efforts for a cure. One innovative approach to eliminate resistant cancer stem cells is differentiation therapy, where cancer stem cells are made to differentiate, thereby losing their resistant capabilities [51].

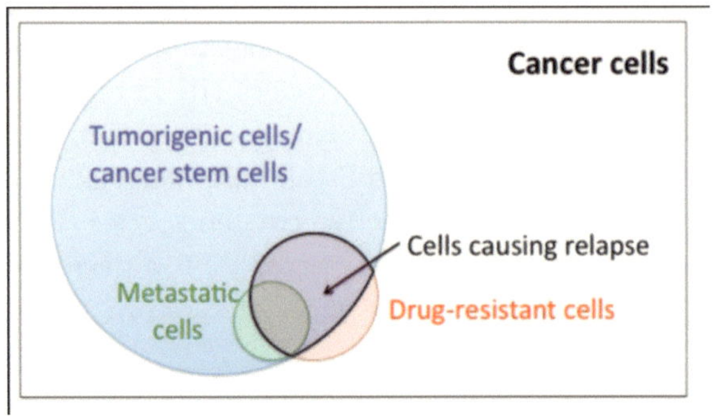

Figure 1. Properties of cancer cells. The subset of tumorigenic cells that are also drug-resistant should be targeted to ensure elimination of cancer.

3. Molecular targets for cancer therapy

3.1. Current clinical drug trials targeting cancer stem cells

Translating research from bench to bedside is perhaps the most challenging and rewarding part of science. The development of drugs against cancer stem cell is an exciting field with many different, innovative approaches. In this section, a review of the drugs that have already reached clinical trials is presented (Table 2).

One approach is the targeting of the cancer stem cell machinery. An example of this approach is the telomerase inhibitors. Telomeric inhibitors block replication and a clinical candidate Imetelstat have shown efficacy in cancer stem cells [52]. As a bonus, telomerase inhibitors are expected to also target the bulk of the tumor. Importantly, unlike normal stem cells, cancer stem cells express higher levels of telomerase [53]. Hence this could potentially be a drug that targets cancer stem cells without hurting normal stem cells.

In a second approach the targeting of the cancer stem cell phenotype, the immunogenic-response that ironically had been a problem to researchers in the detection of cancer stem cells using the mouse model, has become a solution against cancer stem cells. In a study in which cancer stem cells were injected into immunocompetent syngenic mice, cancer stem cells induced antitumor response more effectively than unselected cancer cells [54]. This finding is important and has led to the development of various clinical candidates by three different pharmaceutical companies. These candidates, all of which are currently in clinical trials, were developed based on cancer stem cell-associated proteins. These proteins serve as antigens to evoke an immune response against cancer stem cells. In essence, these proteins act as vaccines against cancer stem cells. Immunocellular Therapeutics has developed a dendritic-based vaccine comprising dendritic cells that were obtained from patients and primed in vitro by two CD133-peptides. This vaccine has just recently been approved for phase I clinical trials. Using the same approach, other clinical candidates were developed by two pharmaceutical companies. Instead of obtaining dendritic cells from patients, peptides of cancer stem cell antigens were injected directly into patients to prime the immune system against cancer stem cells. The peptides used in these vaccinations are found in both cancer stem cells and non-stem cancer cells. One of these antigens is Wilms' Tumor 1 (WT1). WT1 is a transcription factor that is expressed in leukemia. Although a direct association between WT1 and the leukemic stem cell has never been shown, WT1 however, has been associated with the CD34+CD38- cell population which is thought to harbor the hematopoietic stem cell [55] and also the leukemia stem cell [15].

Drug target	Drug name	Cancer	Stage	Company
Undisclosed	Cancer stem cell inhibitor BBI608	Colorectal cancer	Entering Phase III	Boston Biomedicals, Inc
Telomerase (inhibitor)	Imetelstat	Broad range	Phase II	Geron Corporation
CD-133	Dendritic cell-based vaccine ICT-121	Glioblastoma	Entering phase I	ImmunoCellular Therapeutics Ltd.
Focal adhesion kinase (inhibitor)	VS6063	Advance solid tumors	Phase I completed	Verastem and Pfeizer
Wilms Tumor 1	Peptides from Wilms Tumor 1 (FPI-01)	Leukemia and mesothelioma	Phase II	Formula Pharmaceuticals
EphA3	Human monoclonal antibody (KB004) binds EphA3	Leukemia	Phase I	KaloBios Pharmaceuticals, Inc.
Notch pathway	Anti-DLL4 (demcizumab)	Solid Tumors	Phase I	OncoMed
	Anti-Notch2/3 (OPM-59R5)	Solid tumors	Phase I	
Wnt pathway	Anti-Fzd7 (OMP-18R5, binds 5 Frizzled receptors)	Solid Tumors	Phase I	
	Truncated Frizzled 8-Fc fusion protein (OMP-54F28)	Advance solid tumor cancers	Phase I	
Undisclosed cancer stem cell antigen	Peptides vaccine (SL401 and SL701)	Advance leukemia and advance brain cancer	Phase I/II completed	Stemline Therapeutics

Table 2. Current clinical drug trials in cancer stem cell therapy

A third approach is the targeting of the cancer stem cell and its microenvironment. Anti-EphA3 antibody is a clinical candidate against cancer stem cell that has been developed by KaloBios. This antibody treatment is now in phase I trial. EphA3 expression is found in preB leukemia cell line and in a subset of samples from leukemia patients [56]. There is no direct evidence that links EphA3 expression to the leukemic stem cells, however, in an in-house study by KaloBios, incubating anti-EphA3 with cancer cells leads to the cancer cells losing their ability to form colonies in vitro, suggesting that the antibody was active against cancer stem cells. In addition, the antibody was also found to bind to EphA3 that was expressed on cancer vasculature cells as well as cancer stromal cells. The binding was reported to cause cell-cell repulsion, resulting in the destruction of new vessels and failure to establish a cancer stromal environment [57]. This strategy, which targets a protein that is found in cancer stem cell, cancer stromal cells and cancer vasculature cells, would be "killing-three-birds-with-one-stone". Other clinical candidates that are based on a similar approach have also been developed via targeting the Wnt pathway and the Notch Pathway instead [47–50].

3.2. Potential new targets – Insights from Pluripotent Stem cells

The discovery that transcription factors, namely, Oct4, Sox2, Klf4 and c-Myc, can induce pluripotency in a differentiated adult cell [58], accelerated the understanding of the molecular machinery driving pluripotent stem cells. Systems biology approaches based on these transcription factors generated genome-wide regulatory networks that are thought to be the supporting framework for an embryonic stem cell state. These data serves as a rich resource in furthering our understanding of the cancer stem cell.

In a recent study of regulatory networks in embryonic stem cell, analysis of the protein-protein interactions of key transcription factors and the downstream targeted genes revealed that the embryonic stem cell regulatory network can be divided into three independent modules (Figure 2). The three modules are the core module, the c-Myc module and the Polycomb Repressive Complex (PRC) module. The core module comprises genes that are regulated by the embryonic stem cell-specific transcription factor Oct4 and Oct4-associated proteins while the PRC module comprises genes that are repressed by the PRC. The Myc-module comprises genes that are regulated by c-Myc and its associated proteins. Proper functioning of all three modules are essential for having a normal pluripotent stem cell [59].

Using the 3-module model to study the genes expressed in bladder cancer and breast cancer samples, it was found that the Myc module was more active in cancers while that of the core module was more repressed, when compared to normal urothelium obtained from a distant site of the cancer [61,62]. This suggested that in cancer cells, the Myc module is re-activated but is not balanced by a core module. It should be noted that this comparison was done with the heterogeneous cancer cell population and not the cancer stem cell population. Repeating the same characterization analysis on cancer stem cell samples will likely highlight the key differences between the regulatory network of cancer stem cells and that of normal pluripotent stem cells. These differences could become potential targets for anti-cancer therapy.

Factors that are crucial for the maintenance of pluripotent stem cell have been found to be involved in cancer. Hypoxia-inducible factors (HIFs) have been found to be important in pluripotent stem cell [39,63]. Studies now show that HIFs could be the key factor in switching on the pluripotency machinery in cancer cells to form cancer stem cells [64]. In an experiment where glioma cells and cervical cells were exposed to HIFs, activation of the embryonic stem cell marker, Oct4, was observed [65]. Subjecting glioma cells to hypoxia resulted in an increase in the level of CD133 mRNA [65]. In samples from glioma patients, subjecting the CD133-positive fraction to hypoxia resulted in increased mRNA levels of OCT4, NANOG and cMYC. Interestingly, when CD133-positive and negative fractions were cultured under hypoxic conditions, embryonic stem cell-association gene expression and formation of neuroshpere were seen in both fractions [65]. Thus, low levels of oxygen promotes the transformation of cancer cells into cancer stem cells by activating the pluripotency machinery in cancer cells with expression and repression of modules that are similar in profile to embryonic cells. This should be taken into consideration when targeting cancer cells. Studies on HIF have shown that both HIF-1α and HIF-2α h are associated with cancer. HIF-1α has been shown to play a role in angiogenesis [66] and anti-angiogenesis therapies targeting HIF-1α have been undertaken [67,68]. In contrast, recent findings suggest that HIF-2α is involved in the triggering of stemness in cancer which in turn promotes cancer growth and aggressiveness [69]. Hence a potential pathway to target cancer stem cell will be the HIF-2α-mediated pathway.

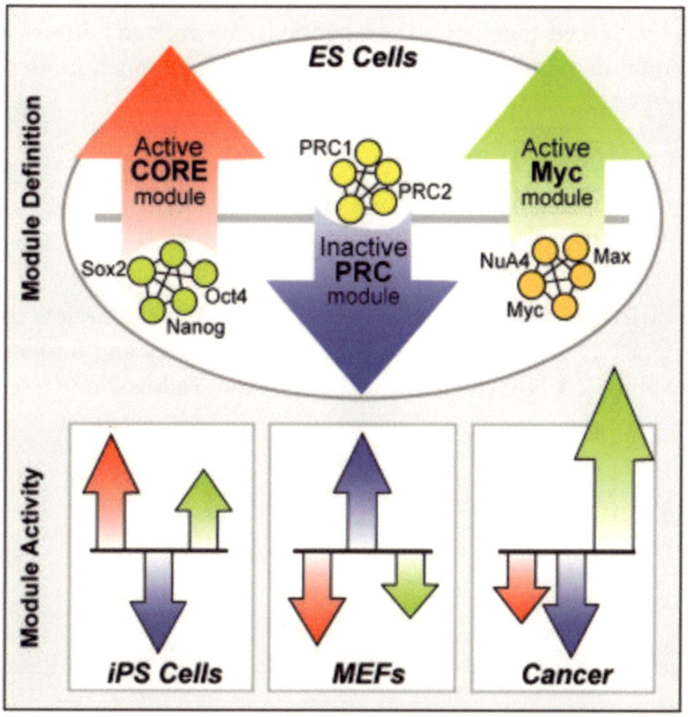

Figure 2. The three sets of genes that are activated or repressed by distinct sets of transcriptional regulators. Induce pluripotent stem (iPS) cells show similar gene expression profile as embryonic stem (ES) cells. Mouse embryonic fibroblast (MEF) represents a set of differentiated cells and shows a profile that is opposite to that of ES cells. Analysis of the profile in cancer highlights a pluripotency machinery in which the core module has been suppressed and the Myc module overactivated. Figure is a reprint of the graphical abstract provided by Kim et al [60] (Reprinted with permission from Cell Press)

Knowledge garnered from studies on pluripotent stem cell provides a rich resource for cancer stem cell research and paves the way in identifying novel key targets for cancer therapy. Targeting molecules or pathways specific to embryonic stem cells gives us the opportunity to kill cancer cells without harming innocent bystander cells.

4. Summary

Research in cancer is immense and complex as cancer is a diverse disease with a myriad of genetic mutations. A pressing practical concern in cancer therapeutics is the development of resistance of cancer cells to current treatment, resulting in failure of therapy and eventual death. In the last two decades, cancer stem cell hypothesis has emerged as the likely reason for this resistance in cancer. We now understand that cancer stem cells are present in different cancers. They can be a small, distinct population characterized by certain phenotypes in some cancers while heterogeneous and with no defining phenotypes in others. Cancer stem cells can also result from cancer cells under the influence of environmental factors such as hypoxia. They are also highly resistant to cancer drugs with several mechanisms employed for enhanced survival. Research into stem cells and pluripotency regulatory networks will provide further characterization and understanding of cancer stem cells. The information on cancer stem cells has pieced together a therapeutic framework to address cancer resistance with several potential therapies in clinical trials currently. So much more needs to be done in this field in our quest to conquer cancer totally.

Acknowledgements

The authors acknowledge Dr Jansen Koh for active discussion and review of the manuscript. Patricia Ng is supported by the Agency for Science, Technology and Research (A*STAR) Research Fellowship and the UNESCO-L'Oréal International Fellowship.

Author details

Patricia Ng and Wang Cheng-I

Singapore Immunology Network, Agency for Science, Technology and Research, Singapore

References

[1] Lapidot T, Sirard C, Vormoor J, Murdoch B, Hoang T, Caceres-Cortes J, et al. A cell initiating human acute myeloi leukaemia after transplation into SCID mice. Nature 1994;367:645–8.

[2] Al-Hajj M, Wicha MS, Benito-Hernandez A, Morrison SJ, Clarke MF. Prospective identification of tumorigenic breast cancer cells. Proceedings of the National Academy of Sciences of the United States of America 2003;100(7):3983–8.

[3] Dalerba P, Dylla SJ, Park I-K, Liu R, Wang X, Cho RW, et al. Phenotypic characterization of human colorectal cancer stem cells. Proceedings of the National Academy of Sciences of the United States of America 2007;104(24):10158–63.

[4] Dick JE. Breast cancer stem cells revealed. Proceedings of the National Academy of Sciences of the United States of America 2003;100(7):3547–9.

[5] Eramo a, Lotti F, Sette G, Pilozzi E, Biffoni M, Di Virgilio a, et al. Identification and expansion of the tumorigenic lung cancer stem cell population. Cell death and differentiation 2008;15(3):504–14.

[6] O'Brien C a, Pollett A, Gallinger S, Dick JE. A human colon cancer cell capable of initiating tumour growth in immunodeficient mice. Nature 2007;445(7123):106–10.

[7] Prince ME, Sivanandan R, Kaczorowski a, Wolf GT, Kaplan MJ, Dalerba P, et al. Identification of a subpopulation of cells with cancer stem cell properties in head and neck squamous cell carcinoma. Proceedings of the National Academy of Sciences of the United States of America 2007;104(3):973–8.

[8] Ricci-Vitiani L, Lombardi DG, Pilozzi E, Biffoni M, Todaro M, Peschle C, et al. Identification and expansion of human colon-cancer-initiating cells. Nature 2007;445(7123): 111–5.

[9] Schatton T, Murphy GF, Frank NY, Yamaura K, Waaga-Gasser AM, Gasser M, et al. Identification of cells initiating human melanomas. Nature 2008;451(7176):345–9.

[10] Singh SK, Hawkins C, Clarke ID, Squire JA, Bayani J, Hide T, et al. Identification of human brain tumour initiating cells. Nature 2004;432(November):396–401.

[11] Zhang S, Balch C, Chan MW, Lai H-C, Matei D, Schilder JM, et al. Identification and characterization of ovarian cancer-initiating cells from primary human tumors. Cancer research 2008;68(11):4311–20.

[12] Li C, Heidt DG, Dalerba P, Burant CF, Zhang L, Adsay V, et al. Identification of pancreatic cancer stem cells. Cancer research 2007;67(3):1030–7.

[13] Li Y, Laterra J. Cancer stem cells: distinct entities or dynamically regulated phenotypes? Cancer research 2012;72(3):576–80.

[14] Furth J, Kahn MC. The Transmission of Leukemia of Mice with a Single Cell. American journal of cancer 1937;31:276–82.

[15] Bonnet D, Dick JE. Human acute myeloid leukemia is organized as a hierarchy that originates from a primitive hematopoietic cell. Nature medicine 1997;3(7):730–7.

[16] Ho MM, Ng AV, Lam S, Hung JY. Side population in human lung cancer cell lines and tumors is enriched with stem-like cancer cells. Cancer research 2007;67(10):4827–33.

[17] Prince ME, Sivanandan R, Kaczorowski a, Wolf GT, Kaplan MJ, Dalerba P, et al. Identification of a subpopulation of cells with cancer stem cell properties in head and

neck squamous cell carcinoma. Proceedings of the National Academy of Sciences of the United States of America 2007;104(3):973–8.

[18] Ginestier C, Hur MH, Charafe-Jauffret E, Monville F, Dutcher J, Brown M, et al. ALDH1 is a marker of normal and malignant human mammary stem cells and a predictor of poor clinical outcome. Cell stem cell 2007;1(5):555–67.

[19] Hermann PC, Huber SL, Herrler T, Aicher A, Ellwart JW, Guba M, et al. Distinct populations of cancer stem cells determine tumor growth and metastatic activity in human pancreatic cancer. Cell stem cell 2007;1(3):313–23.

[20] Barnes WA, Furth J. A transmissible leukemia i n m i c e w i t h atypical cells.

[21] Quintana E, Shackleton M, Sabel MS, Fullen DR, Johnson TM, Morrison SJ. Efficient tumour formation by single human melanoma cells. Nature 2008;456(7222):593–8.

[22] Quintana E, Shackleton M, Foster HR, Fullen DR, Sabel MS, Johnson TM, et al. Phenotypic heterogeneity among tumorigenic melanoma cells from patients that is reversible and not hierarchically organized. Cancer cell 2010;18(5):510–23.

[23] Yilmaz OH, Valdez R, Theisen BK, Guo W, Ferguson DO, Wu H, et al. Pten dependence distinguishes haematopoietic stem cells from leukaemia-initiating cells. Nature 2006;441(7092):475–82.

[24] Southam CM, Brunschwig A. Quantitative studies of autotransplantation of human cancer. Cancer 1961;14:971–8.

[25] Koike a, Moore GE, Mendoza CB, Watne a L. Heterologous, Homologous, and Autologous Transplantation of Human Tumors. Cancer 1963;16:1065–71.

[26] Nadler SH, Moore GE. Autotransplantation of Human Cancer. JAMA®: the journal of the American Medical Association 1965;191:105–6.

[27] Vermeulen L, De Sousa E Melo F, van der Heijden M, Cameron K, de Jong JH, Borovski T, et al. Wnt activity defines colon cancer stem cells and is regulated by the microenvironment. Nature cell biology 2010;12(5):468–76.

[28] Ribatti D. Cancer stem cells and tumor angiogenesis. Cancer letters 2012;321(1):13–7.

[29] Moore N, Lyle S. Quiescent, slow-cycling stem cell populations in cancer: a review of the evidence and discussion of significance. Journal of oncology 2011;2011.

[30] Naumov GN, Townson JL, MacDonald IC, Wilson SM, Bramwell VH, Groom AC, et al. Ineffectiveness of Doxorubicin Treatment on Solitary Dormant Mammary Carcinoma Cells or Late-developing Metastases. Breast Cancer Research and Treatment 2003;82:199–206.

[31] Dembinski JL, Krauss S. Characterization and functional analysis of a slow cycling stem cell-like subpopulation in pancreas adenocarcinoma. Clinical & experimental metastasis 2009;26(7):611–23.

[32] Bao S, Wu Q, McLendon RE, Hao Y, Shi Q, Hjelmeland AB, et al. Glioma stem cells promote radioresistance by preferential activation of the DNA damage response. Nature 2006;444(7120):756–60.

[33] Hirschmann-Jax C, Foster a E, Wulf GG, Nuchtern JG, Jax TW, Gobel U, et al. A distinct "side population" of cells with high drug efflux capacity in human tumor cells. Proceedings of the National Academy of Sciences of the United States of America 2004;101(39):14228–33.

[34] Wulf GG, Wang R-Y, Kuehnle I, Weidner D, Marini F, Brenner MK, et al. A leukemic stem cell with intrinsic drug efflux capacity in acute myeloid leukemia. Blood 2001;98(4):1166–73.

[35] Tanei T, Morimoto K, Shimazu K, Kim SJ, Tanji Y, Taguchi T, et al. Association of breast cancer stem cells identified by aldehyde dehydrogenase 1 expression with resistance to sequential Paclitaxel and epirubicin-based chemotherapy for breast cancers. Clinical cancer research®: an official journal of the American Association for Cancer Research 2009;15(12):4234–41.

[36] Liu G, Yuan X, Zeng Z, Tunici P, Ng H, Abdulkadir IR, et al. Analysis of gene expression and chemoresistance of CD133+ cancer stem cells in glioblastoma. Molecular cancer 2006;5:67.

[37] Todaro M, Alea MP, Di Stefano AB, Cammareri P, Vermeulen L, Iovino F, et al. Colon cancer stem cells dictate tumor growth and resist cell death by production of interleukin-4. Cell stem cell 2007;1(4):389–402.

[38] Zhang S, Balch C, Chan MW, Lai H-C, Matei D, Schilder JM, et al. Identification and characterization of ovarian cancer-initiating cells from primary human tumors. Cancer research 2008;68(11):4311–20.

[39] Ji L, Liu Y, Yang C, Yue W, Shi S, Bai C, et al. Self-renewal and pluripotency is maintained in human embryonic stem cells by co-culture with human fetal liver stromal cells expressing hypoxia inducible factor 1alpha. Journal of cellular physiology 2009;221(1):54–66.

[40] Shroff EH, Snyder C, Chandel NS. Role of Bcl-2 Family Members in Anoxia Induced Cell Death. Cell Cycle 2007;6(April):807–9.

[41] Chen N, Chen X, Huang R, Zeng H, Gong J, Meng W, et al. BCL-xL is a target gene regulated by hypoxia-inducible factor-1{alpha}. The Journal of biological chemistry 2009;284(15):10004–12.

[42] Jie W, Yi J. Cancer cell killing via ROS. Cancer Biology & Therapy 2008;7(December): 1875–84.

[43] Reinehr R, Becker S, Eberle A, Grether-beck S, Häussinger D. Involvement of NADPH oxidase isoforms and SRC family kinases in CD95-dependent hepatocyte apoptosis. Journal of biological chemistry 2005;280:27179–94.

[44] De Toni F, Racaud-Sultan C, Chicanne G, Mas VM-D, Cariven C, Mesange F, et al. A crosstalk between the Wnt and the adhesion-dependent signaling pathways governs the chemosensitivity of acute myeloid leukemia. Oncogene 2006;25(22):3113–22.

[45] Xu Q, Yuan X, Liu G, Black KL, Yu JS. Hedgehog signaling regulates brain tumor-initiating cell proliferation and portends shorter survival for patients with PTEN-co-expressing glioblastomas. Stem cells (Dayton, Ohio) 2008;26(12):3018–26.

[46] Reya T, Morrison SJ, Clarke MF, Weissman IL. Stem cells, cancer, and cancer stem cells. Nature 2001;414(6859):105–11.

[47] Gurney A, Axelrod F, Bond CJ, Cain J, Chartier C, Donigan L, et al. Wnt pathway inhibition via the targeting of Frizzled receptors results in decreased growth and tumorigenicity of human tumors. Proceedings of the National Academy of Sciences of the United States of America 2012;109(29):11717–22.

[48] Gurney A, Hoey T. Anti-DLL4, a cancer therapeutic with multiple mechanisms of action. Vascular cell 2011;3(1):18.

[49] Fischer M, Yen W-C, Kapoun AM, Wang M, O'Young G, Lewicki J, et al. Anti-DLL4 inhibits growth and reduces tumor-initiating cell frequency in colorectal tumors with oncogenic KRAS mutations. Cancer research 2011;71(5):1520–5.

[50] Hoey T, Yen W-C, Axelrod F, Basi J, Donigian L, Dylla S, et al. DLL4 blockade inhibits tumor growth and reduces tumor-initiating cell frequency. Cell stem cell 2009;5(2):168–77.

[51] Leszczyniecka M, Roberts T, Dent P, Grant S, Fisher PB. Differentiation therapy of human cancer: basic science and clinical applications. Pharmacology & therapeutics 2001;90(2-3):105–56.

[52] Marian CO, Cho SK, McEllin BM, Maher E a, Hatanpaa KJ, Madden CJ, et al. The telomerase antagonist, imetelstat, efficiently targets glioblastoma tumor-initiating cells leading to decreased proliferation and tumor growth. Clinical cancer research◎: an official journal of the American Association for Cancer Research 2010;16(1):154–63.

[53] Shay JW, Wright WE. Telomeres and telomerase in normal and cancer stem cells. FEBS letters 2012;584(17):3819–25.

[54] Ning N, Pan Q, Zheng F, Teitz-Tennenbaum S, Egenti M, Yet J, et al. Cancer stem cell vaccination confers significant antitumor immunity. Cancer research 2012;72(7):1853–64.

[55] Baird PN, Simmons PJ. Expression of the Wilms' tumor gene (WT1) in normal hemopoiesis. Experimental hematology 1997;25(4):312–20.

[56] Dottori M, Down M, Hüttmann A, Fitzpatrick DR, Boyd AW, Dottori BM, et al. Methylation Regulates Expression in Hematopoietic Tumor Cells. 2012;3:2477–86.

[57] Garber K. Of Ephs and Ephrins®: Companies Target Guidance Molecules in Cancer. Journal of the national cancer institute 2010;102(22):1692–4.

[58] Takahashi K, Tanabe K, Ohnuki M, Narita M, Ichisaka T, Tomoda K, et al. Induction of pluripotent stem cells from adult human fibroblasts by defined factors. Cell 2007;131(5):861–72.

[59] Fazzio TG, Huff JT, Panning B. An RNAi screen of chromatin proteins identifies Tip60-p400 as a regulator of embryonic stem cell identity. Cell 2008;134(1):162–74.

[60] Kim J, Woo AJ, Chu J, Snow JW, Fujiwara Y, Kim CG, et al. A Myc Network Accounts for Similarities between Embryonic Stem and Cancer Cell Transcription Programs. 2010;:313–24.

[61] Sanchez-Carbayo M, Socci ND, Lozano J, Saint F, Cordon-Cardo C. Defining molecular profiles of poor outcome in patients with invasive bladder cancer using oligonucleotide microarrays. Journal of clinical oncology®: official journal of the American Society of Clinical Oncology 2006;24(5):778–89.

[62] Kim J, Orkin SH. Embryonic stem cell-specific signatures in cancer: insights into genomic regulatory networks and implications for medicine. Genome medicine 2011;3(11):75.

[63] Forristal CE, Wright KL, Hanley N a, Oreffo ROC, Houghton FD. Hypoxia inducible factors regulate pluripotency and proliferation in human embryonic stem cells cultured at reduced oxygen tensions. Reproduction Research 2010;139(1):85–97.

[64] Keith B, Simon MC. Hypoxia-inducible factors, stem cells, and cancer. Cell 2007;129(3):465–72.

[65] Mathieu J, Zhang Z, Zhou W, Wang AJ, Heddleston JM, Pinna CM a, et al. HIF induces human embryonic stem cell markers in cancer cells. Cancer research 2011;71(13): 4640–52.

[66] Semenza GL. HIF-1 and tumor progression: pathophysiology and therapeutics. Trends in molecular medicine 2002;8(4 Suppl):S62–7.

[67] Ferrara N, Hillan KJ, Gerber H-P, Novotny W. Discovery and development of bevacizumab, an anti-VEGF antibody for treating cancer. Nature reviews. Drug discovery 2004;3(5):391–400.

[68] Tan C, Noronha RGD, Roecker AJ, Pyrzynska B, Khwaja F, Zhang Z, et al. Identification of a Novel Small-Molecule Inhibitor of the Hypoxia-Inducible Factor 1 Pathway Hypoxia-Inducible Factor 1 Pathway. 2005;:605–12.

[69] Koh MY, Lemos R, Liu X, Powis G. The hypoxia-associated factor switches cells from HIF-1α- to HIF-2α-dependent signaling promoting stem cell characteristics, aggressive tumor growth and invasion. Cancer research 2011;71(11):4015–27.

Chapter 5

Conditions and Techniques for Mouse Embryonic Stem Cell Derivation and Culture

Kun-Hsiung Lee

Additional information is available at the end of the chapter

http://dx.doi.org/10.5772/55105

1. Introduction

Stem cells, characterized by their ability for self-renewal and differentiation, have been derived from the embryo and from various postnatal animal sources. They are usually classified according to their developmental potential. Totipotency is defined as the ability of a single cell to replicate and produce all differentiated cells in an entire organism, including extraembryonic tissues that will develop and differentiate into the fetal placenta and fetal membranes [1,2]. In plants, spores are totipotent cells. In some cases, cells can de-differentiate and regain their totipotency. For instance, a plant cutting or callus can be utilized to grow an entire functional plant [3]. In mammals, only the zygote and early blastomeres are totipotent cells [4-7]. In other words, an individual cell is capable to generate a functionally normal animal with fertile ability [8-10]. Mouse embryonic stem (ES) cells, typically derived from inner cell masses (ICMs) or corresponding earlier blastomeres or later epiblasts (develop to embryo proper), are an example of pluripotent cells that can self-renew and generate all types of body cells *in vivo* and *in vitro*, but cannot generate the extraembryonic trophoblast lineage [11-14]. Under some particular conditions, an ES cell-derived mouse with germline transmission can be generated routinely [15-21]. Multipotent cells, such as hematopoietic stem cells, can give rise all cell types within a particular lineage. Spermatogonial stem cells are unipotent stem cells, as they can only form sperm [22].

In recent years, major improvements in deriving mouse ES cell (ESC) lines have dramatically increased success rates. Therefore, this chapter reviews and discusses the conditions and techniques for derivation and cultivation of mouse ES cell (mESC) lines. Thereafter, a proposed novel and user-friendly protocol that is efficient, reproducible, easy to carry out and relatively cheap is presented.

2. Conditions for derivation of mouse embryonic stem cells (mESCs)

Since the first mouse ES cell lines were described [23,24], various empirical combinations of conditions and techniques for derivation and cultivation of mESCs from blastocysts and isolated ICMs have been developed [25-27]. Of which, selected batches of fetal bovine serum (FBS), inactivated STO (a SIM mouse embryonic fibroblast line resistant to 6-thioguanine and ouabain) or murine embryonic fibroblast (mEF) feeder cells, conditioned media, mouse strains, embryo status, and different small growth areas of wells to initiate cultivation are the principal concerns when deriving mESCs [27].

To support fetal growth and development, FBS contains mixed combinations of cell replication stimulators and cell differentiation inducers. Notably, FBS is a biological product, such that its biopotency to support mESCs varies from batch to batch. Therefore, to circumvent interference from differentiation factors and other disadvantages associated with FBS, chemically defined KnockOut™ serum replacement (KSR) [28] and N2B27 [29,30] were developed to replace FBS. That is, when culturing established mESCs, KSR and N2B27 are usually as effective as FBS. Unfortunately, embryos in the KSR ESC medium do not result in effective derivation of ESCs [28,31]. However, a chemically defined ESC medium containing differentiation inhibitors has much better efficiency than the FBS ESC medium when deriving mESCs [31].

Zygotes to hatched embryos and blastomeres, ICMs, or epiblasts of early-stage embryos can be used to establish mESCs [31,32]. These cells have extremely high capability for cell division and differentiation. Theoretically, inhibiting endogenous differentiation and maintaining or enhancing proliferation of pre-implantation embryos can be helpful for the establishment of ES cell lines. In 1988, researchers have found that leukemia inhibitory factor (LIF) assists in the derivation and maintenance of mESCs pluripotency [33,34]. However, when protocols and media containing LIF for mESCs derivation are applied to mouse strains other than 129s, efficiency declines from about 20% to less than 5% [35-38]. Furthermore, ESC media supplemented with LIF are not good for deriving ESCs other than mESCs [39,40]. Since then, regulatory mechanisms, pathways, and signal transduction of self-renewal, differentiation, proliferation, and apoptosis have been investigated [41,42]. Additionally, the corresponding synthesized inhibitors and/or stimulators/inducers/enhancers of stem cells [43-47] have been investigated intensively.

In the early 1990s, as the specificity of developed inhibitors was not sufficiently strong, their effectiveness in inhibiting differentiation and then helping to establish mESCs were not satisfactory and culturing results were inconsistent. More recent year, a breakthrough result was reported. Following the progresses, using the novel mitogen-activated protein kinase (MAPK)/extracellular signal-regulated kinase (ERK) kinase (MEK) inhibitor PD0325901 or SU5402 to eliminate differentiation-inducing signaling from MAPK and using the glycogen synthase kinase 3 (GSK3) inhibitor CHIR99021 to enhance ESC growth capacity and viability helped dramatically in establishing mESC lines [48]. Accordingly, ES cell lines of the second mammalian species (rat) with germline transmission have been reported [49,50]. The combination of MEK inhibitor PD0325901 and GSK3 inhibitor CHIR99021 (2i) also appears to improve the generation efficiency of induced pluripotent stem cells (iPSCs) [51].

To date, the success rate in establishing mESCs via whole embryos is usually more than 50%, regardless of the mouse strains used [31,52-58]. Further progress in the derivation of mESCs from whole early-stage embryos seems limited. Conversely, the success rate in establishing mESCs via isolated single blastomeres is relatively low and highly variable [31,59]. Although the success rates in some reports are approximately 30% [59,60], it cannot be the routine yet due to the variable results.

2.1. Fetal bovine serum (FBS) *vs.* serum replacement (SR)

To date, FBS, with its excellent nutrient mixture, remains the most important and universal component for propagating cells. Additionally, FBS contains growth factors that support ESCs. However, FBS also contains potential differentiation factors for ESCs [61-64]. Therefore, testing and then selecting batches of FBS to support the growth of undifferentiated ESCs is necessary. Otherwise, qualified ESC-grade FBS, which is more costly than conventional FBS, can be used.

To support mESCs, the biopotency of FBS (a biological product) varies from batch to batch. Additionally, its supply worldwide is sometimes limited and it is expensive. Furthermore, animal-originated materials risk introducing adventitious agents into a cell culture system. Therefore, to circumvent interference from differentiation factors and other disadvantages of FBS, chemically defined KSR was developed for use in place of FBS [28].

Although KSR is a commercial product, its formula remains unknown. With its well-defined chemical formula, N2B27, can replace KSR and achieve almost the same deriving and maintaining ESCs results [48,65]. Originally, N2B27 is an empirically mixed formulation that provides optimum cell viability and efficient neural differentiation [29,30].

Conventional ESC media usually contain 15–20% FBS or 10% FBS plus 10% newborn serum. For a chemically defined ESC medium, FBS can be replaced completely by KSR or N2B27; otherwise, 5–15% FBS can be replaced by SR. When changing FBS to a new batch, the ESCs sometimes have to adapt gradually to the new batch. For example, one can mix 50% old FBS with 50% new FBS and allow the ESCs to acclimatize to the new medium. Generally, ESCs can be changed easily from serum replacement (SR) to FBS ESC media by sequential adaptation with approximately three passages. Reversely, acclimatizing ES cells from the FBS to the SR ESC medium is sometimes difficult and can fail. When one is switching to a serum-free cell culture, sequential adaptation for approximately 4–5 passages is required.

Notably, FBS, a good buffer, is a complex solution that contains many chemicals and proteins with different molecular weights. Therefore, mESCs in FBS-free medium are more sensitive to extremes of pH, osmolarity, enzyme treatment, and mechanical forces. Furthermore, a 5- to 10-fold lower antibiotics concentration is used in an FBS-free medium because serum proteins typically bind a certain amount of the antibiotic; without these serum proteins, the antibiotic concentration may be toxic to mESCs.

After passing mESCs in a SR ESC medium, centrifugation is necessary to remove trypsin from the cell suspension to avoid further digestion of cells due to the lack of trypsin inhibitors in the medium. For convenience, enzymes that can stop digestion due to the decreased concentration will be a good alternative to trypsin. TrypLE™, a recombinant enzyme derived from microbial fermentation, can stay at room temperature (RT) before expiration. My laboratory routinely uses TrypLE™ Express to pass mESCs during derivation and maintenance with good outcomes [31].

2.2. Feeder cells

The ESCs are extremely sensitive to culture conditions, including properties of culture media and dishes, when maintaining pluripotency without differentiation. Unfortunately, commercially available plates and dishes are not adequate for ESCs.

In the 1970s, pluripotent teratocarcinoma (stem) cell lines were established after introduction of a cell feeder layer [66-68]. Feeder (helper) cells have since been used to help study teratocarcinoma stem cells and embryonal carcinoma (EC) cells. The STO feeder cells then demonstrated to help derive the first mESCs [23] and maintain mESCs at the undifferentiated state.

Feeder cells are usually inactivated via mitomycin C or γ-irradiation treatment. While γ-irradiation leads to breaks in DNA strands, mitomycin C has the extraordinary ability to crosslink DNA with high efficiency and is specific for the CpG sequence [69,70]. Although feeder cells are alive, they do not replicate but gene transcription and protein synthesis are not affected.

It has been speculated that feeder cells support embryos and ESCs attachment through the physical matrix [71]. Furthermore, feeder cells may release embryo trophic factors, reduce inhibitory or toxic factors in FBS, or may be beneficial by lowering concentrations of ions and/or glucose in medium, and thereby overcoming the developmental blockage of embryos mediated by the release of growth factors essential for activation of the embryonic genome and for normal embryonic development. Additionally, feeder cells may protect embryos from oxygen toxicity [63].

Various feeder cells, which differ in their ability to support ESCs, have been utilized to establish, propagate, and maintain the pluripotency of ESCs [35,72]. Conventionally, STO and mEF are the most popular feeder cells for deriving and maintaining ESCs. However, human foreskin fibroblast (hFF) feeder cells also support propagation and self-renewal of human [73-76] and mouse ESCs [31,77].

As feeder cells at earlier passages are used, their ability to support ESCs is increased [78]. Reports have demonstrated that mEF, STO, and hFF feeder cells secrete different growth factors to support ESCs [76,77,79-81].

To help establish mESCs, the STO, mEF, and hFF feeder cells might have roughly equal efficiency [31,82]. Recently, a study reported that hFF supported mESC self-renewal superiorly to mEFs due to the convenience. Using the hFF system, multiple lines of mESCs have been successfully derived without addition of exogenous LIF and any inhibitors. These mESCs have capacities to self-renew for a long period of time and to differentiate into various cell types of the three germ layers both *in vitro* and *in vivo* [76]. The STO is a cell line (ATCC No. CRL-1503) for unlimited propagation. The hFF may be a cell line (CCD-1112Sk, ATCC No. CRL-2429; Hs68, ATCC No. CRL-1635; HFF-1, ATCC No. SCRC-1041) or primary cells. The mEF is

http://dx.doi.org/10.5772/55105

primary cells harvested from fetuses approximately 12.5–13.5 days post coitum (dpc). However, mEF can be passed and propagated only for a short period [83-85]. Therefore, they must be prepared continuously. Additionally, the traits of mEFs differ from batch to batch, and quality control of mEFs, especially for mycoplasma contamination, may be the problems [86,87]. Moreover, hFFs are more durable than mEFs in that they remain in healthy condition more than 2 weeks after inactivation by radiation. In contrast, mEF deteriorates within 1 week after the inactivation [76].

To derive and maintain mESCs, feeder cells and ESC medium supplemented with LIF (typical concentration is 1,000 unit/mL) are usually chosen. The reason is due to the LIF produced by mEFs and STO is not enough to maintain ESC properties most of the time. However, recombinant murine LIF is expensive. Therefore, STO that expresses a high level of LIF (SNL76/7, ATCC No. SCRC-1049; SNLP 76/7-4, ATCC No. SCRC-1050) has been developed [88]. This cell line also can be used as a feeder layer supporting the derivation and growth of mESCs and iPSCs [89]. The disadvantage of the SNL76/7 as a feeder cell is the highly variable level of LIF in culturing medium, that might have different effects on mESCs [24,56,57,90-93].

2.3. Conditioned media

Both teratocarcinoma stem cells and EC cells were established in the 1970s [66,94-96]. Both pluripotent cells usually undergo extensive differentiation *in vivo* and *in vitro* to generate a wide variety of cell types [97]. Mouse teratocarcinoma stem cells can condition themselves. Therefore, medium conditioned with teratocarcinoma stem cells, which is equivalent to a 5-fold concentration of LIF [92], was used to help establish mESCs [24]. Thereafter, the propagation of mESCs in high densities reduces possible differentiation [26,98].

Feeder cells secrete many different factors, including growth factors, to support ESCs [76,77,79-81]. The recovered conditioned medium is a complex solution containing many unidentified chemicals. However, as conditioned media are exhausted media prepared in a batch-by-batch manner, their biopotency might vary. The question is whether conditioned medium is still needed to derive ES cells when chemically defined ESC media supplemented with differentiation inhibitors and growth factors are available.

A few different conditioned media have been used to establish mESCs. Of which, an FBS ESC medium conditioned by a rabbit fibroblast cell line transduced with genomic rabbit LIF (10 ng/mL) allows efficient derivation and maintenance of mESC lines from all 10 inbred mouse strains tested, including some that were presumed nonpermissive for mESC derivation [56]. This commercialized conditioned medium, RESGRO™ Culture Medium (Millipore), can establish and rescue established mESCs that have started drifting, and either generate low-percentage chimeras or lose their germline transmission capability [53,92].

2.4. Mouse strain and embryo status

Teratomas (benign) and teratocarcinomas (malignant) are tumors composed of an ectoderm, endoderm, and mesoderm mixtures of adult tissues [99]. Most are found in gonads and rare in mammals, including experimental animals. In the 1950s, mouse strain 129 had found an incidence of spontaneous testicular teratoma of about 1% [100]. After progress and refinement of cell culture techniques, most notably the introduction of the cell feeder layer, allowed the reliable cultures of pluripotent teratocarcinoma (stem) cell lines [66-68].

The mESCs were first derived from 129SvE [23] and then (ICR × SWR/J) F1 hybrid embryos [24]. However, due to the demand for animal models of human diseases, strains other than 129s have been used to establish new mESC lines. Unfortunately, the following mESCs were derived mainly from 129 strains due to the permissive nature of the genetic background [27,32,35,38]. The derivation of mESCs from blastocysts is a process that is often very inefficient, and even in the most favorable 129 strains, a success rate of 30% is regarded as high [26]. Derivation of mESCs is strongly mouse strain–dependent [32], and in practice the efficiency of derivation in strains other than 129 strains does not usually exceed 10% [36].

Strains other than 129s, such as FVB, CBA/Ca, and the non-obese diabetic (NOD) mouse [101], have the extremely low derivation rates of mESCs using conventional conditions. They are traditionally regarded as highly refractory (nonpermissive) for derivation of mESCs. Moreover, they also produce chimeras either incapable of germline transmission [102] or restricted in their germline competence [37].

Mouse strain C57BL/6 (B6) is not usually considered as a permissive strain for mESC derivation. Although the B6 mES cell lines have been available since the early 1990s [35,103], the efficiency in establishing B6 mESCs via FBS or KSR ESC medium containing LIF varies and is typically less than 10% [32,38,56,61,82,103-109]. When establishing C57BL/6J mESCs using FBS (15%) ESC medium containing LIF, the highest success rate was 40% (36/89) [56]. Notably, this study used mEF feeder cells coated on 96-well cell plates for initiation of blastocysts and thereafter digested ICM outgrowths cultured at 39°C under 5% CO_2 in an incubator. In establishing B6 mESCs using KSR (20%) ESC medium containing LIF, the derivation rate of (C57BL/6N × C57BL/6J) F1 mES-like colony can be as high as 40% (10 of 25 blastocysts) [109]. Other studies also demonstrated that the efficiency in establishing mESCs is significantly higher for the C57BL/6N strain (53%) than for the C57BL/6J strain (20%) [109]. Recently, my study showed that C57BL/6J blastocysts in KSR ESC medium did not lead to effective derivation of mESCs; however, the success rate in 2i medium was as high as 75%. Furthermore, the efficacy of 2i medium was also demonstrated when using morulae (60%) and 8-cell embryos (50%) for mESCs derivation [31].

Although previous results suggested that true ES cells can be derived from embryos explanted at any stage of preimplantation development in the mouse [110]. My laboratory was the first to report success in deriving mESCs from mouse zygotes. These zygote-derived mESCs are morphologically indistinguishable from mESCs derived from fertilized embryos and blastomeres. Moreover, the generation of germline transmitted chimeras confirmed that the established mES-like cells are pluripotent mESCs [31].

In recent years, the mESC lines of nonpermissive strains have been established routinely from 3.5 dpc blastocysts when SR ESC medium containing differentiation inhibitors and/or proliferation enhancers was used.

Typically, diploid male mESCs capable of a high percentage generation of chimeras with germline transmission are selected for further utilization. However, some unusual mESCs have been reported including the androgenetic [111], germline transmitted female [112], adult somatic cell nuclear transferred [113], XO [114], parthenogenetic [115], haploid [54], and androgenetic haploid [116] mESCs.

2.5. Single blastomere

To date, most available ES cell lines were derived from the outgrowth of ICMs of blastocysts. However, due to ethical concerns over the derivation of human ES (hES) cells for regenerative medicine, a single blastomere (usually from 8-cell embryos) has been utilized to derive ESCs. Unfortunately, conventional methods used to establish mESCs directly from an isolated single blastomere, which is extremely sensitive to culture conditions, are unsuccessful. Actually, most of the isolated blastomeres divide to form small sheets of cells with a trophoblastic-like morphology or small blastocysts with or without visible ICM-like cells [6,31,117]. In 1996, mES-like cells were first reported from single blastomeres [118]. Since then, no ESCs were established from blastomeres until 2006 [117].

The mESCs can condition themselves to inhibit or prevent differentiation. Therefore, blastomeres aggregated with established mESCs for initial co-culturing to help in the derivation of mESCs is a logically reasonable alternative. In this manner, the single blastomere-derived mESCs have been established but overall success rates are less than 5% when using the conventional FBS ESC medium [117-119]. In spite of that, a possibility may exist for deriving personalized hES cells without destruction of 8-cell embryos.

On 2004, the study showed that the KSR ESC medium do not support mES single cell culturing. Contrary, single mESCs were propagated without loss of pluripotency when the adrenocorticotropic hormone (ACTH) was added to KSR ESC medium [62]. The authors of that study hypothesized that ACTH may be integrated via a weak cross interaction with an unknown, non-physiological inhibitory G protein coupled receptor. A signaling system other than the cAMP-PKA pathway or PKA pathway may play an important role in propagation of mESCs [62]. Later, simple and efficient establishment of mES cell lines from a single blastomere of 2- to 8-cell embryos with KSR ESC medium containing ACTH fragments 1–24 (ACTH 1–24) on mEF feeders was reported [59].

Wakayama *et al.* (2007) developed a novel protocol and established mESCs via blastomeres and polar bodies. In their experiment, isolated B6D2F2 blastomeres were cultured on KSR (20%) ESC medium containing ACTH 1–24 in 96-well plates coated with mEFs. The mESC establishment rates were 33%, 8%, and 8% for blastomeres derived from 2-cell (1/2), late 4-cell (1/4), and 8-cell embryos (1/8), respectively. However, they did not aggregate blastomeres with other mESCs, suggesting that success was likely attributable to the KSR ESC medium containing ACTH 1–24 [59]. Other studies then demonstrated that mESCs can be established from 1/2, 1/4, 2/4, 1/8, 2/8, 3/8, and 4/8 blastomeres in medium containing ACTH 1–24 [31,120-122]. For blastomeres, as the developmental stage of embryos decreases and the number of isolated blastomeres used increases, the derivation efficiency of mESCs increases. Moreover, whole embryos always have better derivation efficiency than corresponding blastomeres. Furthermore, KSR ESC medium containing differentiation inhibitors and/or proliferation stimulators is better than FBS ESC medium in establishing mESCs.

For ethical concerns, such as maintaining the developmental potential of embryos and establishing corresponding hES cells, a 1/8 blastomere is one of the best candidates. Unfortunately, derivation efficiency of mESCs via a 1/8 blastomere is usually approximately 5–10%, such that it cannot be a standard protocol for the routine operation. Obviously, reliable and efficient protocols for ESCs derivation should be developed.

Previous studies have indicated poor derivation efficiency of mESCs from a 1/8 blastomere, partly due to the low division rate of single blastomeres when compared to that of their counterparts with a higher number of blastomeres (2/8, 3/8, and 4/8). Communication and adhesion between blastomeres, from which the derivation process begins, are likely important aspects to efficiently deriving mESC lines. Therefore, an approach consisting of chimeric E-cadherin (E-cad-Fc) adhesion to the blastomere surface has been devised to recreate the signaling produced by native E-cadherin between neighboring blastomeres inside an embryo. Via this approach, the 1/8 blastomere incubated with E-cad-Fc for only 24 h can significantly improve the mESC derivation efficiency from 2.2% to 33.6% [60]. To date, this novel method via 1/8 blastomere has the best derivation efficiency for mESCs. However, its reproducibility must be confirmed by other laboratories.

2.6. Pluripotent signaling pathways

In 1988, two studies demonstrated that LIF could assist in derivation and maintenance of the pluripotency of mESCs [33,34]. Unfortunately, following studies revealed that LIF is not as effective or good for mammals other than mice in establishing ESCs. These experimental results implied that different mammals might have different regulatory mechanisms for ESCs. Actually, distinct signaling pathways have been shown to regulate the pluripotency of mouse and human ESCs [41,42,123].

The regulatory mechanisms and signal transductions of self-renewal, differentiation, proliferation, and apoptosis [41,42,124], as well as the corresponding inhibitors of stem cells [43-47, 125] have been investigated (Figure 1).

To date, self-renewal, pluripotency, and the propagation signaling transduction pathways of ESCs includes the LIF/signal transducer and activation of transcription 3 (STAT3), Wnt/β-catenin, phosphatidylinositol 3-kinase (PI3K), bone morphogenic protein 4 (BMP4)/Smad1/5, and basic fibroblast growth factor (bFGF) [42,45]. Conversely, the Ras/Raf/MEK/ERK pathway is central to the signaling networks that govern proliferation, differentiation, and cell survival [126]. The active Ras/Raf/MEK/ERK pathways induce differentiation of mESCs [42,124,126]. The mESCs have high ERK activity when they undergo differentiation. Suppression of the ERK pathway promotes self-renewal of mESCs. Moreover, BMP4 activation inhibits differentiation of mESCs in medium containing LIF due to inhibitor of differentiation (Id) genes expression and ERK inactivation [127-129].

Of the many pathways, self-renewal of mESCs largely depends on LIF/interleukin 6 (IL-6) family members [130] and BMP4 [65,127]. LIF binds to a cell surface complex composed of the

http://dx.doi.org/10.5772/55105

LIF receptor and the transmembrane signaling molecule gp130, resulting in activation of transcription factor STAT3, which is essential and sufficient to promote self-renewal and inhibit mesoderm and endoderm differentiation of mESCs [125,130-132]. Additionally, Wnt signaling inhibits GSK3β and results in the protein stabilization of cytoplasmic β-catenin (β-Ctnn). The GSK3, a constitutively acting multi-functional serine threonine kinase, derives its name from its substrate glycogen synthase, a key enzyme involved in conversion of glucose to glycogen. Although GSK3 is kept inactive by phosphorylation, activated GSK3 enhances synthesis of glycogen and inhibits cell proliferation. The name GSK does not adequately describe the multitude of diverse substrates and functions attributed to GSK3. For instance, it is involved in various cellular processes, ranging from glycogen metabolism, insulin signaling, cell proliferation, neuronal function, and oncogenesis to embryonic development [133]. Additionally, GSK3 is one of the crucial molecules involved in regulation of the Wnt/β-catenin, Hedgehog, and Notch signaling pathways. The undifferentiated pluripotency of both mouse and human ESCs can be maintained by GSK3-specific inhibitor 6-bromoindirubin-3′-oxime (BIO), which prevents phosphorylation of β-Ctnn by GSK3β and activates Wnt [134,135]. The Wnt signaling is endogenously activated in mESCs and is down regulated upon differentiation [135]. The target genes of the Wnt signaling pathway, such as c-Myc and CyclinD1, promote cell proliferation and self-renewal. The LIF/STAT3 pathway combines with the Wnt/GSK3β/β-catenin pathway to enhance self-renewal by activation of pluripotency genes, including transcription factors Nanog, Oct3/4, and Klf4. Nevertheless, LIF also activates the MAPK/ERK pathway, which induces mESC differentiation [45].

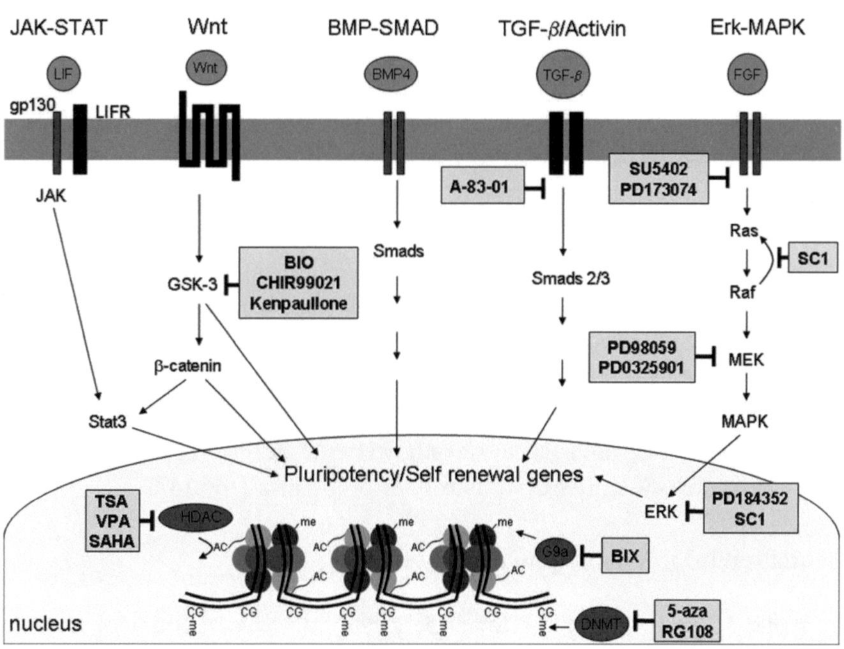

Figure 1. The use of signalling pathway inhibitors and chromatin modifiers for enhancing pluripotency. (Reproduced with permission from Sumer *et al.*, 2010. Theriogenology 74:525–533.)

As a protein, transforming growth factor beta (TGFβ) controls proliferation, cellular differentiation, and other functions in most cells. This protein is a secreted protein of cytokines. The TGFβ family is part of the large TGFβ superfamily, which has more than 40 members, including TGFβ, activin, nodal, BMPs, inhibins, anti-müllerian hormone, decapentaplegic, and Vg-1. All of these ligands are associated with ESCs. Although BMP4 is a member of the BMP family, it functions via a different signaling pathway with TGFβ. Notably, BMP4 induces expression of Id genes and inhibits MAPK signaling as well as neuroectoderm differentiation [65,127]. Although activation of STAT3 is sufficient for self-renewal of mESCs, a study showed that LIF-STAT3 do not maintain mESCs in serum-free ESC medium. The combination of LIF and BMP4 maintained the self-renewal of mESCs in the absence of both feeder cells and FBS [65,127]. Thus, BMP4 and LIF have synergistic effects on the self-renewal of mESCs [127]. In contrast to mESCs, BMP4 does not maintain the self-renewal of hES cells; rather, it induces trophoblast or primitive endoderm differentiation of hES cells [136].

The PI3K pathway is also important for proliferation, survival, and maintenance of pluripotency, as well as inhibiting apoptosis in ESCs. The ESC-expressed Ras (ERas) is specifically expressed in ESCs, stimulating PI3K. This PI3K activation promotes ESC proliferation [137] and self-renewal [138]. Inhibition of PI3K and Akt induces differentiation of mESCs in the presence of LIF and feeder cells [139], suggesting that PI3K/Akt signaling is necessary for maintenance of the pluripotency of ESCs.

The ESC can maintain its pluripotency with feeder cells. Exogenous supplementation of LIF is sufficient to sustain mESCs at undifferentiated state in a feeder cells free condition. However, LIF is insufficient for maintaining the pluripotency of hES cells. Human ESCs have been most commonly cultured in the presence of bFGF either on fibroblast feeder layers [140] or in fibroblast-conditioned medium. The bFGF signaling pathway appears to be important to the self-renewal of hES cells [141,142]. Reports have shown that bFGF (40 ng/mL) combined with noggin (inhibitor of BMP4) supports the undifferentiated proliferation of hES cells in the absence of feeder cells [143-145]. Furthermore, a high bFGF concentration (100 ng/ml) alone is sufficient to maintain human ESCs [145]. The post-implantation epiblast-derived stem cell (EpiSC) lines [14] express transcription factors that regulate pluripotency and robustly differentiate into the major somatic cell types as well as primordial germ cells [12]. Furthermore, the EpiSCs and hES cells share patterns of gene expression and signaling responses in the epiblasts [12]. In fact, the similarities (FGF and activin growth factors for self-renewal; *in vitro* pluripotency; chimera formation; and spontaneous trophoblast differentiation) between hES cells and mouse EpiSCs have led to the suggestion that hES cells are equivalent to early post-implantation epiblasts, rather than their ICM progenitor [146,147].

2.7. Differentiation inhibitors help to derive mouse ES cells

In the early 1990s, as the specificity of developed inhibitors was insufficient, the effectiveness of these inhibitors to inhibit differentiation and then help to establish mESCs was not satisfactory and outcomes are varied. Since the late 1990s, using MAPK/ERK kinases (MEKs) inhibitors PD098059 [128] and UO126 [41], or by dephosphorylating ERKs by mitogen-activated protein kinase phosphatase 3 (MKP-3) [41] enhanced the self-renewal of mESCs and inhibited their differentiation.

It was the first report showed that the combination of PD98059 (MEK 1 inhibitor) with LIF enhances the establishment of mESCs from the refractory CBA strain [148]. A combination of the MEK1/2 inhibitor, U0126, with LIF further promoted the efficiency of mESC derivation from CBA [129]. These small molecules, PD98059 and U0126, play positive roles in the self-renewal of mESCs, but they are incapable of maintaining the pluripotency of mESCs in long-term culture without LIF [48,128,135].

Notably, BMP4 inhibits both ERK and p38 MAPK pathways in mESCs. The inhibitors of the ERK and p38 MAPKs mimic the effect of BMP4 on mESCs. Inhibition of the p38 and MAPKs by SB203580 overcomes the roadblock in deriving mESCs from blastocysts lacking a functional Alk3, the BMP4 type-IA receptor [127].

The self-renewal of mESCs is generally dependent on multifactorial stimulation. To support the growth and development of fetuses, FBS contains cell replication stimulators and cell differentiation inducers. Serum and serum substitutes contain various inductive stimuli that may activate commitment and differentiation programs [49]. However, simple withdrawal of serum or other exogenous stimuli cannot prevent differentiation of ESCs due to endogenous autoinductive differentiation of fibroblast growth factor 4 (FGF4) via the MEK/ERK pathway [48,149]. Therefore, to suppress endogenous autoinductive differentiation and to maintain high viability and growth rates, one must inhibit the MEK/ERK pathway and/or provide LIF, or restrict the activity of GSK3, which acts mainly via the Wnt/β-catenin signaling pathway to suppress cellular biosynthetic capacity, and subsequent cell growth and viability [134,150]. Those studies have demonstrated that inhibition of GSK3 via BIO dramatically augment mESC derivation from isolated ICMs of both C57BL/6 (76%) [135] and BALB/c (31%) [134].

Through dual inhibition of the Ras GTPase-activating protein (RasGAP) and ERK1, a function-oriented and novel small molecular pluripotin, also called SC1, was developed. By using this novel pluripotin, one can propagate mESCs in an undifferentiated and pluripotent state under chemically defined conditions in the absence of feeder cells, FBS, and LIF. Moreover, long-term pluripotin-expanded mESCs can generate germline-transmitted chimeric mice [43]. By combining pluripotin and LIF for the derivation of mESCs, the successfully isolated mESCs from five strains of mice; efficiency was 57% for NOD-scid, 63% for SCID beige, 80% for CD-1, and 100% for two F1 strains from C57BL/6 x CD-1. Pluripotin combined with LIF improved the efficiency of mESC isolation by selectively maintaining Oct4-positive cells in outgrowths. This is the first work to efficiently derive mESCs from immunodeficient mice (NOD-scid on an NOD background and SCID beige on a BALB/c background) on refractory backgrounds [57].

Recently, a novel protocol involves an unusually long initial incubation of 12 days for blastocysts seeded in 12-well plates coated with mEFs and containing LIF and pluripotin-supplemented KSR ESC medium (15% KSR), which results in the formation of large spherical outgrowths. These outgrowths are morphologically distinct from classical ICM outgrowths and can be picked easily and trypsinized. Importantly, pluripotin needs to be omitted after the first trypsinization because it blocks the attachment of mES-like cells to the mEF feeder layer; its removal facilitates the formation of mESC colonies. In addition, pluripotin is harmful to the mEFs and it is not unusual that half of the mEFs die during the first week. However, this massive death of mEFs does not affect the mES cell derivation efficiency [55]. I also found that STO and Hs68 feeder cells were dying when culturing in KSR ESC medium supplemented with 3 μM pluripotin (unpublished observation). The modified protocol gives rise to mESCs (more then five passages) in a robust and reliable manner with an extremely high success rate of 94% (78/83) [55]. Surprisingly, 10 mESCs derived with 4 μM pluripotin showed the chromosomal instability. All of these cells generated weak chimeras. Thus, these lines are only suitable for *in vitro* analysis. In contrast, mES-cell lines derived with 2 μM pluripotin during the blastocyst outgrowth phase were generated with unusual high efficiency (100%) and these lines had a normal karyotype. In addition, strong chimeras could be derived from these mESCs [55]. Yang and colleagues (2009) derived mESCs with 3–5 μM pluripotin. Their mESCs showed some chromosomal abnormalities and were not proven to be germline competent [57].

Although the derivation of novel mESCs have been improved significantly via differentiation inhibitors and/or proliferation stimulators, no germline-transmitted ESCs other than the mESCs have been reported. In recent years, the MEK inhibitor PD0325901 or SU5402 has been used to eliminate differentiation-inducing signaling from MAPK and the GSK3 inhibitor CHIR99021 has been used to enhance mESC growth, enabling derivation and propagation of germline-competent mESCs from CBA and 129 strains in an N2B27 chemically defined ESC medium. These findings reveal that self-renewal is enabled by the elimination of differentiation-inducing signaling from MAPK. Additional inhibition of GSK3 consolidates biosynthetic capacity and suppresses residual differentiation. Complete bypass of cytokine signaling is confirmed by isolating *stat3*$^{-/-}$ mESCs [48]. Accordingly, ES cell lines of the second mammalian species (rat) with germline transmission have been established [49,50].

Recently, my report showed that two C57BL/6J mESCs were derived from two morulae in KSR ESC medium supplemented with 2i but without LIF [31]. These experimental results confirmed previous reports indicating that inhibitors that block the MEK/ERK differentiation pathway can support self-renewal of mESCs, even without LIF signaling [43,48].

The efficiency of establishing mESCs with 2i was not reported [48]. Recently, my experimental results showed that the efficiency in establishing C57BL/6J ES cells in KSR ESC medium supplemented with 2i and LIF could be high as 65% for morulae and 70% for blastocysts [31]. Thus, the efficacy of 2i in helping establish mESCs is confirmed.

2.8. Self-renewal and proliferation stimulator

Although a previous study showed that extrinsic stimuli are dispensable for derivation, propagation, and pluripotency of mESCs [48], the self-renewal and derivation of mESCs largely depends on LIF [130] and BMP4 [65,127]. When LIF is present at concentrations higher than those used in routine cultivation (500 unit/mL for the mEF feeder and 1,000 unit/mL for the STO feeder), the derivation efficiencies of mESCs from refractory strains are improved significantly [57,91,93]. Indeed, strain difference in response to soluble growth factors is evident from two original reports of mESC derivation. Evans and Kaufman (1981) established mESC lines from permissive 129SvE delayed blastocysts, solely by culturing them on STO feeder cells. Notably, Martin (1981) was unable to establish mESC lines from immunosurgically isolated ICMs of (ICR x SWR/J) F1 and (C3H x C57BL/6) F1 fully expanded blastocysts unless the teratocarcinoma stem cell-conditioned medium, equivalent to a 5-fold LIF concentration [92], was added. A previous report revealed that a mESC medium containing 10 ng/mL rabbit LIF or conditioned by a rabbit fibroblast cell line transduced with genomic rabbit LIF facilitates efficient derivation and maintenance of mESC lines (≥ 10 passages) from all 10 inbred mouse strains tested, including some that were presumed nonpermissive for ESC derivation [56]. A more recent study demonstrated that SR ESC medium containing high concentrations of LIF (2,500 and 5,000 unit/mL) could establish mES cell lines from C57BL/6, Balb/K, nonpermissive CBA/Ca, and NOD mice [92].

Additionally, another report shows that KSR ESC medium does not support single mESC culture, likely because this medium lacked some important growth factors or such undefined factors, such as "stem-cell autocrine factors," are secreted by mESCs themselves [62]. However, when KSR ESC medium is supplemented with ACTH, single mESCs are propagated and their pluripotency is maintained. Accordingly, mESC lines have been established efficiently using single blastomeres from 2- to 8-cell embryos in KSR ESC medium containing ACTH 1–24 [59].

The CHIR99021 pathway and proposed ACTH pathway likely differ. Therefore, synergetic effects of 2i (PD0325901 + CHIR99021) and ACTH 1–24 may exist in deriving mESCs. My experimental results reveal that neither STO nor Hs68 feeder cells coated on 10-μL droplets and cultured in KSR ESC medium supports effective derivation of mESCs from embryos or blastomeres of ICR or C57BL/6J. However, by supplementing KSR ESC medium with 2i or ACTH 1–24, efficiency in establishing mESCs increased dramatically. Experimental results also demonstrate that inhibiting cell differentiation and increasing cell growth/viability (in the presence of 2i) simultaneously is better than increasing only cell survival and/or proliferation (with ACTH 1–24) when deriving mESCs [31]. Additionally, experimental results also suggest that GSK3 inhibitor CHIR99021 and ACTH 1–24 likely have different pathways in synergistically enhancing the establishment of mESCs because 2i with ACTH 1–24 is much more effective than 2i or ACTH 1–24 alone [31].

2.9. Developing more powerful differentiation inhibitors

A previous study has demonstrated that blastomeres from 2- to 8-cell embryos developed into blastocysts within 3–5 days in KSR ESC medium containing ACTH 1–24 [59]. Another report indicates that 2- to 4-cell embryos and blastomeres, which developed into blastocysts in medium containing ACTH 1–24, PD98059 (MAP2K1 inhibitor) or SB203580 (MAPK14 inhibitor), yields developmental rates comparable to those of the control embryos [151]. My previous study showed that approximately 60% of 1.5–2.5 dpc denuded whole embryos developed into typical blastocysts or small blastocysts and approximately 80% of the corresponding blastomeres developed into small blastocysts within 2–4 days in KSR ESC medium containing 2i or ACTH 1–24. These experimental results imply that 2i and ACTH 1–24 might have little or no adverse effect on cell proliferation or the development of embryos and blastomeres. Furthermore, these observations indicate that 2i, MAP2K1, and MAPK14 inhibitors are not strong enough to inhibit differentiation of embryos and blastomeres entirely [31]. I also note that 0.3–3 μM pluripotin did not inhibit differentiation of ICR 2.5 dpc embryos and blastomeres completely (unpublished observation).

These experimental findings suggest that other more powerful inhibitors, via the Ras-MEK-ERK signaling pathway or other signaling pathways, may be worth developing to further enhance the success rate of deriving ESCs. Theoretically, chemical cocktails that completely inhibit endogenous differentiation, increase cell division, and decrease apoptosis of pluripotent cells should maximize the derivation efficiency of ESCs. Furthermore, is it possible to develop novel differentiation inhibitors that maintain the totipotency of very-early-stage blastomeres?

3. Novel efficient, reproducible, and user-friendly protocol for deriving mouse ES cells

An excellent protocol for deriving mESCs must be efficient, reproducible, easy to perform, and relatively cheap.

The following protocol, adopted and minor modified from my previous study [31], has been used in my laboratory for more than 3 years with reproducible high derivation efficiency (always more than 50%, occasionally reaching 100%) for 2.5 dpc whole embryos, regardless of which strains (ICR, B6, ICRB6F1, and B6CBF1) are tested. Moreover, all other pre-implantation embryos can be used. Although the same protocol can be used to derive mESCs via single blastomeres, derivation of mESC varies in efficiency, which depends on the origin of blastomeres (approximately 10–30%).

3.1. Growth area for initial cultivation

For initial cultivation of embryos, ICMs, and blastomeres to establish mESCs, different growth surface areas in multi-well cell plates are used. Of which, the 24-well and 96-well plates are used mainly. In practice, the amount of medium needed to half fill one well in a 24-well plate is approximately 0.5 mL and 150 μL for a 96-well plate. Both volumes are enormous to the mouse embryo (diameter, 85 μm; volume, 320 pL = 0.00032 μL) or 1/8 blastomere (diameter, 20 μm; volume, 4 pL = 0.000004 μL). Furthermore, the height of wells is not user-friendly when taking photographs or picking growing three-dimensional outgrowths. Therefore, to reduce the amount of expensive ESC media that is usually exchanged every other day, ease handling, and increase the disaggregation efficiency of growing three-dimensional outgrowths, different volumes of micro-droplets were tested. I hypothesize that homemade micro-droplets on cell culture dishes coated with feeder cells are as effective as cell plate wells for initial cultivation to establish mESCs. Finally, 10-μL and 20-μL droplets are chosen for initial and for disaggregated outgrowths cultivation, respectively. Up to 32 10-μL or 24 20-μL feeder droplets on a 60-mm cell culture dish covered with heavy weight paraffin oil can be prepared and used routinely. My culturing results show that 10-μL droplets support the growth of 2.5 dpc embryos for at least 7 days and 1/8 blastomeres for at least 10 days (Figure 2).

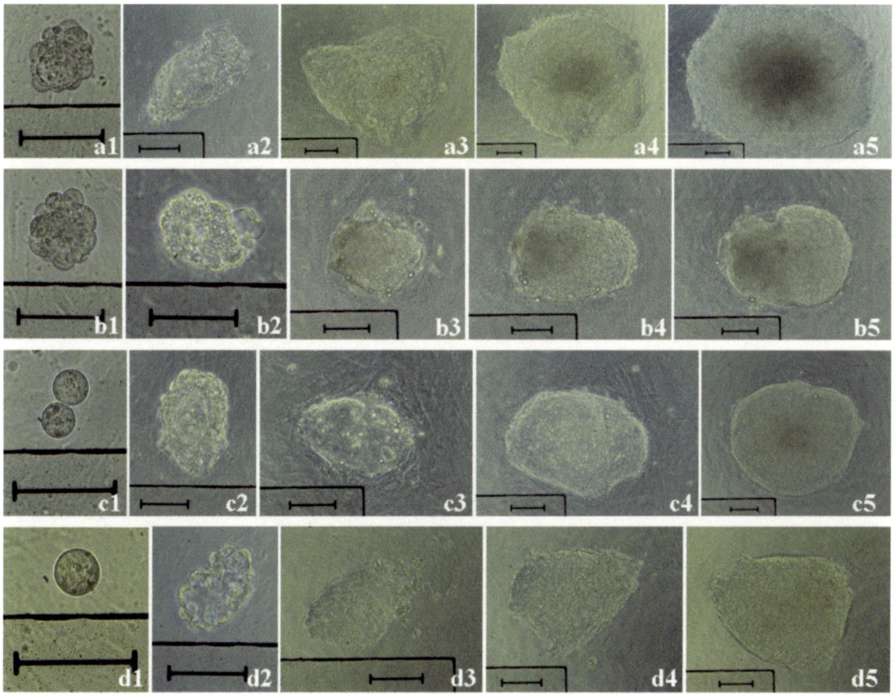

Figure 2. The 10-μL droplets are excellent for initial cultivation to derive mESCs. Embryos and blastomeres seeded in 10-μL droplets coated with Hs68 feeder cells in KnockOut™ serum replacement (KSR) ESC medium supplemented with 0.5 μM PD0325901 (MEK inhibitor), and 3 μM CHIR99021 (GSK3 inhibitor) (2i), 10 μM ACTH–24 (adenylyl cyclase inhibitor), as well as 1,000 unit/mL LIF support the growth of 2.5 dpc embryos for at least 7 days and 1/8 blastomeres for at least 10 days. a1, early blastocyst; b1, morula; c1, 2/16—isolated single blastomere that originated from late 8-cell embryos were divided; d1, 1/8—isolated single blastomere that originated from 8-cell embryos. a2–d2, cultured for 3 days after initiation of cultivation; a3–d3, cultured for 7 days; and a4–d4, cultured for 10 days; a5–d5, cultured for 12 days. The bar represents 100 μm.

The previous study reported that the culturing blastocysts allow to hatch and expand for approximately 6 days; however, the earlier trypsinization (at day 4 or 5) of outgrowths do not seem to affect efficiency but prolonged the time needed for mESCs derivation [58]. Very interestingly, blastocyst outgrowths can be cultured in 2 μM pluripotin-containing KSR ESC medium for 12 days (or up to 18 days) without losing their ability to form mESCs. The main advantage of a long initial cultivation is to greatly increase the number of cells, which including undifferentiated cells, before the first trypsinization. The derivation efficiency could be as high as 94% (78/83) [55].

Further, my experimental results demonstrated that mESCs are established efficiently from C57BL/6J denuded whole embryos culturing in a 10-μL droplet coated with Hs68 feeder cells of KSR ESC medium supplemented with 2i, ACTH-24, and LIF (Figure 3) [31].

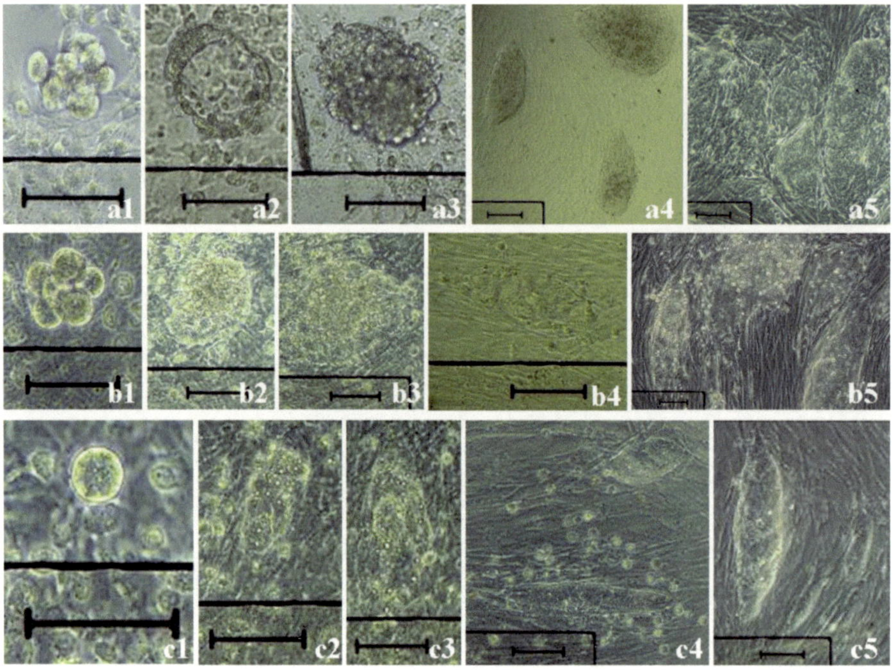

Figure 3. The C57BL/6J ES cells derived from 2.5-day post-coitum (dpc) denuded 8-cell embryos or single blastomeres cultured in 10-μL droplets. a1: An 8-cell embryo cultured in KSR ESC medium containing 0.5 μM PD0325901 (MEK inhibitor) and 3 μM CHIR99021 (GSK3 inhibitor) (2i) with STO feeders. Images show 2 (a2), 4 (a3), 11 (a4; passage #1, P1) and 14 (a5; ESC98B01, P2) days after initiation of cultivation. b1: An 8-cell embryo cultured in KSR ESC medium with Hs68 feeders. Images show 3 (b2), 4 (b3), 9 (b4; P1), and 18 (b5; ESC98B05, P4) days after initiation of cultivation. c1: A single blastomere isolated from an 8-cell embryo cultured in KSR ESC medium containing 2i with Hs68 feeders. Images show 3 (c2), 6 (c3), 9 (c4; P1), and 14 (c5; ESC98B04, P3) days after initiation of cultivation. The bar represents 100 μm. (Reproduced from Lee *et al.*, 2012. Stem Cells and Development 21:373–383.)

3.2. Feeder cells and medium

The cell lines of STO and Hs68 (Caucasian human newborn foreskin fibroblast; ATCC No. CRL-1635) as well as primary mEF cells can be used as feeder cells. These cells share the same growth medium, which comprises DMEM (glucose, 4.5 g/L) containing 10% FBS, penicillin (50 unit/mL), and streptomycin (50 μg/mL).

After mitotic inactivation by 10 μg/mL mitomycin C for 2–3 h, single trypsinized cells via 0.25% trypsin-EDTA are used to prepare feeder layers directly or are frozen until thawed for feeder preparation. Up to 32 10-μL feeder droplets on a 60-mm cell culture dish, covered with heavy weight paraffin oil, is prepared 1–2 days before denuded embryos or isolated blastomeres are seeded.

No difference existed between STO and Hs68 feeders in the derivation of mESCs [31]. However, as STO feeders occasionally detached and curled up during cultivation in 10 μL and 20 μL droplets, Hs68 is usually chosen as the feeder cells. However, propagation of Hs68 is slow

http://dx.doi.org/10.5772/55105

compared to that of STO. Therefore, STO feeder cells are adopted for the large growing areas in commercially available cell plates and dishes.

3.3. KSR ESC medium supplemented with 2i and ACTH 1–24

The basal KSR ESC medium contained KnockOut™ DMEM (glucose 4.5 g/L) supplemented with 20% KSR, 0.1 mM non-essential amino acids, 1.75 mM GlutaMAX™-I supplement, 0.1 mM β-mercaptoethanol, penicillin (15.62 unit/mL), streptomycin (15.62 μg/mL), and LIF (1,000 unit/mL). The KnockOut™ DMEM can be replaced by conventional DMEM or (KnockOut™) DMEM/F12.

Although media prepared from powder are far cheaper than using liquid directly, experiments always use embryo- or cell culture-tested grades liquid media or solutions, as they are more consistent and efficient for derivation of mESCs, especially when KSR ESC media are used in cultivations. The variable quality of homemade ultrapure ddH_2O might prove problematic. However, once the mESCs established, the media prepared from powder could be used for routine cultivation.

To enhance the successful derivation of mESCs, the basal KSR ESC medium is supplemented with 0.5 μM PD0325901 (MEK inhibitor), 3 μM CHIR99021 (GSK3 inhibitor) (2i) [48], and 10 μM ACTH 1–24 (adenylyl cyclase inhibitor) [59].

3.4. Derivation of mouse ES cells

Natural or superovulated 3.5–4.5 dpc morulae and blastocysts are used. Recovered embryos are washed and placed in the KSOM medium supplemented with 20.85 mM HEPES (HK) at RT until the next treatment. The zona pellucida of embryos is removed (denuded) in seconds using acidic Tyrode's solution. Blastomeres of 2- to 8-cell embryos are separated by incubating denuded embryos in 0.25% trypsin-EDTA for approximately 3–4 min in a 37°C incubator with humidified atmosphere of 5% CO_2 in air, followed by gentle pipetting using a mouth pipette. The blastomeres and denuded embryos or hatched (naked) blastocysts are washed and then plated into 10-μL feeder droplets (P0) in KSR ESC medium supplemented with 2i and ACTH 1–24, which is exchanged 1–2 h earlier. Embryos and blastomeres are cultured in a 37°C incubator with 5% CO_2 in air. Following attachment of embryos or blastomeres to the feeder cells, the media are exchanged at the second or third day. Thereafter, the media are exchanged every 1 to 2 days.

After culturing for 6 ± 2 d (for whole embryos) or 9 ± 2 d (for blastomeres), individual three-dimensional outgrowth is identified visually. The feeder cells and flat growing cells are removed mechanically by a mouth pipette. The outgrowth in the same droplet is washed with TrypLE™ Express once and then incubated with TrypLE™ Express in a 37°C incubator for approximately 13 ± 3 min. The outgrowth is washed with KSR ESC medium once and then disaggregated into clumps and single cells, which are reseeded onto fresh 20-μL droplets (P1). After 4 ± 1 d, only morphologically mES-like colonies are then passed (via TrypLE Express) to fresh 1 or 2 wells of 4-well cell plates (P2), and thereafter to either 4-well plates or 35-mm cell culture dishes (P3) depending on the number of cells. Once mES-like cells are growing in 4-

well plates and larger dishes, KSR ESC medium (2i + ACTH 1–24 is option) is used for subsequent propagation. At passages 5 ± 2, mES-like cells are frozen in FBS supplemented with 10% (v/v) dimethyl sulfoxide (DMSO). The typical duration of the above-described process of mESC derivation (from embryos to freezing of subconfluent 35 mm-dishes) is ranging from 14 to 20 days.

Practically, the first 2–3 passages are critical for successful derivation of mESCs. This protocol suggests that one use KSR ESC medium containing 2i + ACTH 1–24 only for the first 10-μL droplets for embryo outgrowth (P0) and second (for single cells and clumps of digested outgrowths, P1) and/or third 20-μL (for mES-like cells, P2) droplets. Once mES-like cells are growing on 4-well plates (P2–3), KSR ESC medium is used thereafter for all following cultivations.

The success rates in establishing B6 mESCs by this simple protocol are always greater than 50% for 2.5–4.5 dpc embryos. This efficiency is comparable to that achieved in two other studies [56,109] and much better than those in many other studies [32,38,61,82,103-108]. This proposed protocol has a simple layout, is easy to operate, is highly efficient, is reproducible, and can be an alternative method for establishing mES cell lines routinely.

4. Conclusions

Zygotes to hatched embryos and blastomeres, ICMs, and the epiblasts of early-stage pre-implantation embryos can be used to establish mESCs. Both embryos and blastomeres have an extremely high capability for cell division and differentiation. Theoretically, chemical cocktails that can completely inhibit endogenous differentiation, increase cell division, and decrease apoptosis of pre-implantation embryos can be helpful to maximize the derivation efficiency of ESCs.

Culturing pre-implantation embryos, no matter what strains, on a very small surface area coated with feeder cells in a chemically defined medium supplemented with differentiation inhibitors (*e.g.*, 2i) and/or proliferation enhancers/stimulators (*e.g.*, LIF and ACTH 1–24) can be used as a routine protocol to establish mESCs efficiently and reproducibly (always more than 50%, occasionally reaching 100%). Unfortunately, the same protocol when using 1/8 blastomeres to derive mESCs is merely acceptable (10–25%). Obviously, the possibility to increase the efficiency of deriving mESCs from whole embryos is limited and unexciting unless novel differentiation inhibitors or proliferation enhancers/stimulators reveal the effectiveness of using 1/8 blastomeres and embryos from species other than rodents.

Acknowledgements

I would like to thank the National Science Council of the Republic of China, Taiwan, for financially supporting my research under Contract Nos. NSC100-2321-B-059-001-MY3,

http://dx.doi.org/10.5772/55105

NSC99-2324-B-059-001, and NSC97-2317-B-059-004. Drs. C.H. Chen, C.k. Juang, C.F. Tu, and S.F. Guo as well as Ms. H.R. Chang and T.L. Hsu at the Animal Technology Institute Taiwan (ATIT) are thanked for their critical comments, technical assistance, and routine maintenance in the laboratory and mouse facility.

Author details

Kun-Hsiung Lee*

Address all correspondence to: khlee@mail.atit.org.tw

Division of Biotechnology, Animal Technology Institute Taiwan, Chunan, Miaoli, Taiwan

References

[1] Nagy A, Gertsenstein M, Vintersten K, Behringer R, Nagy A, Gertsenstein M, Vintersten K, Behringer R. Summary of mouse development, Manipulating the Mouse Embryo: A Laboratory Manual, 3 edition. New York: Cold Spring Harbor Lab. Press; 2003. p31-139.

[2] Rossant J, Tam PP. Blastocyst lineage formation, early embryonic asymmetries and axis patterning in the mouse. Development 2009;136: 701-713.

[3] Mitalipov S, Wolf D. Totipotency, pluripotency and nuclear reprogramming. Advances in Biochemical Engineering Biotechnology 2009;114: 185-199.

[4] Papaioannou VE, Ebert KM. Comparative aspects of embryo manipulation in mammals. In: Rossant J, Pedersen RA (eds), Experimental Approach to Mammalian Embryonic Development. Cambridge, UK: Cambridge Univ. Press; 1986. p67-96.

[5] Suwinska A, Czolowska R, Ozdzenski W, Tarkowski AK. Blastomeres of the mouse embryo lose totipotency after the fifth cleavage division: Expression of Cdx2 and Oct4 and developmental potential of inner and outer blastomeres of 16- and 32-cell embryos. Developmental Biology 2008;322: 133-144.

[6] Tarkowski AK, Wroblewska J. Development of blastomeres of mouse eggs isolated at the 4- and 8-cell stage. Journal of Embryology and Experimental Morphology 1967;18: 155-180.

[7] Ziomek CA, Johnson MH, Handyside AH. The developmental potential of mouse 16-cell blastomeres. Journal of Experimental Zoology 1982;221: 345-355.

[8] Willadsen SM, Polge C. Attempts to produce monozygotic quadruplets in cattle by blastomere separation. Veterinary Record 1981;108: 211-213.

[9] Fehilly CB, Willadsen SM. Embryo manipulation in farm animals. Oxford Reviews of Reproductive Biology 1986;8: 379-413.

[10] Willadsen SM. A method for culture of micromanipulated sheep embryos and its use to produce monozygotic twins. Nature 1979;277: 298-300.

[11] Beddington RS, Robertson EJ. An assessment of the developmental potential of embryonic stem cells in the midgestation mouse embryo. Development 1989;105: 733-737.

[12] Tesar PJ, Chenoweth JG, Brook FA, Davies TJ, Evans EP, Mack DL, Gardner RL, McKay RDG. New cell lines from mouse epiblast share defining features with human embryonic stem cells. Nature 2007;448: 196-199.

[13] Tanaka S, Kunath T, Hadjantonakis AK, Nagy A, Rossant J. Promotion of trophoblast stem cell proliferation by FGF4. Science 1998;282: 2072-2075.

[14] Brons, I. G, Smithers, L. E, Trotter M. W.B, Rugg-Gunn P, Sun B, Chuva de Sousa Lopes S, Howlett S. K, Clarkson A, hrlund-Richter L, Pedersen R. A, & Vallier, L. Derivation of pluripotent epiblast stem cells from mammalian embryos. Nature 2007;448: 191-195.

[15] Nagy A, Gocza E, Diaz EM, Prideaux VR, Ivanyi E, Markkula M, Rossant J. Embryonic stem cells alone are able to support fetal development in the mouse. Development 1990;110: 815-821.

[16] Nagy A, Rossant J, Nagy R, bramow-Newerly W, Roder JC. Derivation of completely cell culture-derived mice from early-passage embryonic stem cells. Proceedings of the National Academy of Sciences of the United States of America 1993;90: 8424-8428.

[17] Poueymirou WT, Auerbach W, Frendewey D, Hickey JF, Escaravage JM, Esau L, Dore AT, Stevens S, Adams NC, Dominguez MG, Gale NW, Yancopoulos GD, DeChiara TM, Valenzuela DM. F0 generation mice fully derived from gene-targeted embryonic stem cells allowing immediate phenotypic analyses. Nature Biotechnology 2007;25: 91-99.

[18] Lee, K-H. Methods to generate chimeric mice from embryonic stem cells. In: Kallos MS (ed), Embryonic Stem Cells: Basic Biology to Bioengineering. Rijeka: InTech; 2011. p193-212.

[19] Modlinski JA, Reed MA, Wagner TE, Karasiewicz J. Embryonic stem cells: Developmental capabilities and their possible use in mammalian embryo cloning. Animal Reproduction Science 1996;42: 437-446.

[20] Amano T, Kato Y, Tsunoda Y. Comparison of heat-treated and tetraploid blastocysts for the production of completely ES-cell-derived mice. Zygote 2001;9: 153-157.

[21] Iwasaki S, Kimura H. Further studies on thermal treatment of two-cell stage embryos to produce complete embryonic stem-cell-derived mice by cell-aggregation methods. Development, Growth & Differentiation 2003;45: 427-434.

[22] Jaenisch R, Young R. Stem cells, the molecular circuitry of pluripotency and nuclear reprogramming. Cell 2008;132: 567-582.

[23] Evans MJ, Kaufman MH. Establishment in culture of pluripotential cells from mouse embryos. Nature 1981;292: 154-156.

[24] Martin GR. Isolation of a pluripotent cell line from early mouse embryos cultured in medium conditioned by teratocarcinoma stem cells. Proceedings of the National Academy of Sciences of the United States of America 1981;78: 7634-7638.

[25] Nagy A, Gertsenstein M, Vintersten K, Behringer R, Nagy A, Gertsenstein M, Vintersten K, Behringer R. Isolation and culture of blastocyst-derived stem cell line, Manipulating the Mouse Embryo: A Laboratory Manual, 3 edition. New York: Cold Spring Harbor Lab. Press; 2003. p359-397.

[26] Robertson EJ. Embryo-derived stem cell lines. In: Robertson EJ (ed), Teratocarcinomas and Embryonic Stem Cells: A Practical Approach. Oxford UK: IRL Press; 1987. p71-112.

[27] Hogan B, Costantini F, Beddington R, Lacy E. Isolation, culture, and manipulation of embryonic stem cells, manipulating the Mouse Embryo: A Laboratory Manual, 2 edition. Plainview: Cold Spring Harbor Lab. Press; 1994. p253-290.

[28] Goldsborough MD, Tilkins ML, Price PJ, Lobo-Alfonso J, Morrison JR, Stevens ME, Meneses J, Pederson R, Koller R, Koller B, Latour A. Serum-free culture of murine embryonic stem (ES) cells. FOCUS (Gibco) 1998;20: 8-12.

[29] Ying QL, Smith AG. Defined conditions for neural commitment and differentiation. Methods in Enzymology 2003;365:327-41.: 327-341.

[30] Nichols J, Ying QL. Derivation and propagation of embryonic stem cells in serum- and feeder-free culture. Methods in Molecular Biology 2006;329: 91-98.

[31] Lee KH, Chuang CK, Guo SF, Tu CF. Simple and efficient derivation of mouse embryonic stem cell lines using differentiation inhibitors or proliferation stimulators. Stem Cells and Development 2012;21: 373-383.

[32] Brook FA, Gardner RL. The origin and efficient derivation of embryonic stem cells in the mouse. Proceedings of the National Academy of Sciences of the United States of America 1997;94: 5709-5712.

[33] Smith AG, Heath JK, Donaldson DD, Wong GG, Moreau J, Stahl M, Rogers D. Inhibition of pluripotential embryonic stem cell differentiation by purified polypeptides. Nature 1988;336: 688-690.

[34] Williams RL, Hilton DJ, Pease S, Willson TA, Stewart CL, Gearing DP, Wagner EF, Metcalf D, Nicola NA, Gough NM. Myeloid leukaemia inhibitory factor maintains the developmental potential of embryonic stem cells. Nature 1988;336: 684-687.

[35] Kawase E, Suemori H, Takahashi N, Okazaki K, Hashimoto K, Nakatsuji N. Strain difference in establishment of mouse embryonic stem (ES) cell lines. International Journal of Developmental Biology 1994;38: 385-390.

[36] McWhir J, Schnieke AE, Ansell R, Wallace H, Colman A, Scott AR, Kind AJ. Selective ablation of differentiated cells permits isolation of embryonic stem cell lines from murine embryos with a non-permissive genetic background. Nature Genetics 1996;14: 223-226.

[37] Nagafuchi S, Katsuta H, Kogawa K, Akashi T, Kondo S, Sakai Y, Tsukiyama T, Kitamura D, Niho Y, Watanabe T. Establishment of an embryonic stem (ES) cell line derived from a non-obese diabetic (NOD) mouse: In vivo differentiation into lymphocytes and potential for germ line transmission. FEBS Letters 1999;455: 101-104.

[38] Suzuki O, Matsuda J, Takano K, Yamamoto Y, Asano T, Naiki M, Kusanagi M. Effect of genetic background on establishment of mouse embryonic stem cells. Experimental Animals 1999;48: 213-216.

[39] Brevini, T. A. L, Antonini, S, Cillo, F, Crestan, M, & Gandolfi, F. Porcine embryonic stem cells: Facts, challenges and hopes. Theriogenology 2007;68: S1, 206-213.

[40] Keefer CL, Pant D, Blomberg L, Talbot NC. Challenges and prospects for the establishment of embryonic stem cell lines of domesticated ungulates. Animal Reproduction Science 2007;98: 147-168.

[41] Burdon T, Smith A, Savatier P. Signalling, cell cycle and pluripotency in embryonic stem cells. Trends in Cell Biology 2002;12: 432-438.

[42] Liu N, Lu M, Tian X, Han Z. Molecular mechanisms involved in self-renewal and pluripotency of embryonic stem cells. Journal of Cellular Physiology 2007;211: 279-286.

[43] Chen S, Do JT, Zhang Q, Yao S, Yan F, Peters EC, Scholer HR, Schultz PG, Ding S. Self-renewal of embryonic stem cells by a small molecule. Proceedings of the National Academy of Sciences of the United States of America 2006;103: 17266-17271.

[44] Emre N, Coleman R, Ding S. A chemical approach to stem cell biology. Current Opinion in Chemical Biology 2007;11: 252-258.

[45] Li W, Ding S. Small molecules that modulate embryonic stem cell fate and somatic cell reprogramming. Trends in Pharmacological Sciences 2010;31: 36-45.

[46] Sumer H, Liu J, Verma PJ. The use of signalling pathway inhibitors and chromatin modifiers for enhancing pluripotency. Theriogenology 2010;74: 525-533.

[47] Efe JA, Ding S. The evolving biology of small molecules: Controlling cell fate and identity. Philosophical Transactions of the Royal Society of London Series B: Biological Sciences 2011;366: 2208-2221.

[48] Ying QL, Wray J, Nichols J, Batlle-Morera L, Doble B, Woodgett J, Cohen P, Smith A. The ground state of embryonic stem cell self-renewal. Nature 2008;453: 519-523.

[49] Buehr M, Meek S, Blair K, Yang J, Ure J, Silva J, McLay R, Hall J, Ying QL, Smith A. Capture of authentic embryonic stem cells from rat blastocysts. Cell 2008;135: 1287-1298.

[50] Li P, Tong C, Mehrian-Shai R, Jia L, Wu N, Yan Y, Maxson RE, Schulze EN, Song H, Hsieh CL, Pera MF, Ying QL. Germline competent embryonic stem cells derived from rat blastocysts. Cell 2008;135: 1299-1310.

[51] Silva J, Barrandon O, Nichols J, Kawaguchi J, Theunissen TW, Smith A. Promotion of reprogramming to ground state pluripotency by signal inhibition. PLoS Biology 2008;6: 2237-2247.

[52] Hassani S-N, Totonchi M, Farrokhi A, Taei A, Larijani MR, Gourabi H, Baharvand H. Simultaneous suppression of TGF-β and ERK signaling contributes to the highly efficient and reproducible Generation of mouse embryonic stem cells from previously considered refractory and non-permissive strains. Stem Cell Reviews and Reports 2012;8: 472-481.

[53] Gertsenstein M, Nutter LM, Reid T, Pereira M, Stanford WL, Rossant J, Nagy A. Efficient generation of germ line transmitting chimeras from C57BL/6N ES cells by aggregation with outbred host embryos. PLoS ONE 2010;5: e11260.

[54] Leeb M, Wutz A. Derivation of haploid embryonic stem cells from mouse embryos. Nature 2011;479: 131-134.

[55] Pieters T, Haenebalcke L, Hochepied T, D'Hont, J, Haigh, J. J, Van Roy, F, & Van Hengel, J. Efficient and user-friendly pluripotin-based derivation of mouse embryonic stem cells. Stem Cell Reviews and Reports 2012;8: 768-778.

[56] Schoonjans L, Kreemers V, Danloy S, Moreadith RW, Laroche Y, Collen D. Improved generation of germline-competent embryonic stem cell lines from inbred mouse strains. Stem Cells 2003;21: 90-97.

[57] Yang W, Wei W, Shi C, Zhu J, Ying W, Shen Y, Ye X, Fang L, Duo S, Che J, Shen H, Ding S, Deng H. Pluripotin combined with leukemia inhibitory factor greatly promotes the derivation of embryonic stem cell lines from refractory strains. Stem Cells 2009;27: 383-389.

[58] Bryja V, Bonilla S, Cajanek L, Parish CL, Schwartz CM, Luo Y, Rao MS, Arenas E. An efficient method for the derivation of mouse embryonic stem cells. Stem Cells 2006;24: 844-849.

[59] Wakayama S, Hikichi T, Suetsugu R, Sakaide Y, Bui HT, Mizutani E, Wakayama T. Efficient establishment of mouse embryonic stem cell lines from single blastomeres and polar bodies. Stem Cells 2007;25: 986-993.

[60] Gonzalez S, Ibanez E, Santalo J. Influence of E-cadherin-mediated cell adhesion on mouse embryonic stem cells derivation from isolated blastomeres. Stem Cell Reviews and Reports 2011;7: 494-505.

[61] Cheng J, Dutra A, Takesono A, Garrett-Beal L, Schwartzberg PL. Improved generation of C57BL/6J mouse embryonic stem cells in a defined serum-free media. Genesis 2004;39: 100-104.

[62] Ogawa K, Matsui H, Ohtsuka S, Niwa H. A novel mechanism for regulating clonal propagation of mouse ES cells. Genes to Cells 2004;9: 471-477.

[63] Vanroose, G, Soom, A. V, Kruif, A. De. From co-culture to defined medium: State of the art and practical considerations. Reproduction in Domestic Animals 2001;36: 25-28.

[64] Mannello F, Tonti GA. No breakthroughs for human mesenchymal and embryonic stem cell culture: Conditioned medium, feeder layer, or feeder-free; medium with fetal calf serum, human serum, or enriched plasma; serum-free, serum replacement nonconditioned medium, or ad hoc formula? All that glitters is not gold! Stem Cells 2007;25: 1603-1609.

[65] Ying QL, Nichols J, Chambers I, Smith A. BMP induction of Id proteins suppresses differentiation and sustains embryonic stem cell self-renewal in collaboration with STAT3. Cell 2003;115: 281-292.

[66] Martin GR, Evans MJ. The morphology and growth of a pluripotent teratocarcinoma cell line and its derivatives in tissue culture. Cell 1974;2: 163-172.

[67] Martin GR, Evans MJ. Differentiation of clonal lines of teratocarcinoma cells: formation of embryoid bodies in vitro. Proceedings of the National Academy of Sciences of the United States of America 1975;72: 1441-1445.

[68] Martin GR, Evans MJ. Multiple differentiation of clonal teratocarcinoma stem sells following embryoid body formation in vitro. Cell 1975;6: 467–474.

[69] Park JH, Kim SJ, Oh EJ, Moon SY, Roh SI, Kim CG, Yoon HS. Establishment and maintenance of human embryonic stem cells on STO, a permanently growing cell line. Biology of Reproduction 2003;69: 2007-2014.

[70] Anderson RE, Pogue L, Troup GM, Standefer JC. A comparison of irradiation and mitomycin as blocking agents in the mixed lymphocyte reaction. Archives of Pathology & Laboratory Medicine 1984;108: 363-367.

[71] Li M, Ma W, Hou Y, Sun XF, Sun QY, Wang WH. Improved isolation and culture of embryonic stem cells from Chinese miniature pig. Journal of Reproduction and Development 2004;50: 237-244.

[72] Richards M, Tan S, Fong CY, Biswas A, Chan WK, Bongso A. Comparative evaluation of various human feeders for prolonged undifferentiated growth of human embryonic stem cells. Stem Cells 2003;21: 546-556.

[73] Amit M, Margulets V, Segev H, Shariki K, Laevsky I, Coleman R, Itskovitz-Eldor J. Human feeder layers for human embryonic stem cells. Biology of Reproduction 2003;68: 2150-2156.

[74] Ellerstrom C, Strehl R, Noaksson K, Hyllner J, Semb H. Facilitated expansion of human embryonic stem cells by single-cell enzymatic dissociation. Stem Cells 2007;25: 1690-1696.

[75] Hovatta O, Mikkola M, Gertow K, Stromberg AM, Inzunza J, Hreinsson J, Rozell B, Blennow E, Andang M, hrlund-Richter L. A culture system using human foreskin fibroblasts as feeder cells allows production of human embryonic stem cells. Human Reproduction 2003;18: 1404-1409.

[76] Ma Y, Gu J, Li C, Wei X, Tang F, Shi G, Jiang J, Kuang Y, Li J, Wang Z, Xie X, Jin Y. Human foreskin fibroblast produces interleukin-6 to support derivation and self-renewal of mouse embryonic stem cells. Stem Cell Research & Therapy 2012;3: 29.

[77] Meng, G. L, Zur Nieden, N. I, Liu, S. Y, Cormier, J. T, Kallos, M. S, & Rancourt, D. E. Properties of murine embryonic stem cells maintained on human foreskin fibroblasts without LIF. Molecular Reproduction and Development 2008;75: 614-622.

[78] Bryja V, Bonilla S, Arenas E. Derivation of mouse embryonic stem cells. Nature Protocols 2006;1: 2082-2087.

[79] Eiselleova L, Peterkova I, Neradil J, Slaninova I, Hampl A, Dvorak P. Comparative study of mouse and human feeder cells for human embryonic stem cells. International Journal of Developmental Biology 2008;52: 353-363.

[80] Villa-Diaz, LG, Pacut C, Slawny NA, Ding, J, O'Shea, K. S, & Smith, G. D. Analysis of the factors that limit the ability of feeder cells to maintain the undifferentiated state of human embryonic stem cells. Stem Cells and Development 2009;18: 641-651.

[81] Lim JW, Bodnar A. Proteome analysis of conditioned medium from mouse embryonic fibroblast feeder layers which support the growth of human embryonic stem cells. Proteomics 2002;2: 1187-1203.

[82] Suemori H, Nakatsuji N. Establishment of the embryo-derived stem (ES) cell lines from mouse blastocysts: Effects of the feeder cell layer. Development, Growth & Differentiation 1987;29: 133-139.

[83] Richards, M, Fong, C. Y, Chan, W. K, Wong, P. C, & Bongso A. Human feeders support prolonged undifferentiated growth of human inner cell masses and embryonic stem cells. Nature Biotechnology 2002;20: 933-936.

[84] Kim H, You S, Farris J, Kong BW, Christman SA, Foster LK, Foster DN. Expression profiles of p53-, $p16^{INK4a-}$, and telomere-regulating genes in replicative senescent primary human, mouse, and chicken fibroblast cells. Experimental Cell Research 2002;272: 199-208.

[85] Suda Y, Suzuki M, Ikawa Y, Aizawa S. Mouse embryonic stem cells exhibit indefinite proliferative potential. Journal of Cellular Physiology 1987;133: 197-201.

[86] Cobo F, Cortes JL, Cabrera C, Nieto A, Concha A. Microbiological contamination in stem cell cultures. Cell Biology International 2007;31: 991-995.

[87] Markoullis K, Bulian D, Holzlwimmer G, Quintanilla-Martinez L, Heiliger KJ, Zitzelsberger H, Scherb H, Mysliwietz J, Uphoff CC, Drexler HG, Adler T, Busch DH, Schmidt J, Mahabir E. Mycoplasma contamination of murine embryonic stem cells affects cell parameters, germline transmission and chimeric progeny. Transgenic Research 2009;18: 71-87.

[88] McMahon AP, Bradley A. The Wnt-1 (int-1) proto-oncogene is required for development of a large region of the mouse brain. Cell 1990;62: 1073-1085.

[89] Okita K, Ichisaka T, Yamanaka S. Generation of germline-competent induced pluripotent stem cells. Nature 2007;448: 313-317.

[90] Zandstra PW, Le HV, Daley GQ, Griffith LG, Lauffenburger DA. Leukemia inhibitory factor (LIF) concentration modulates embryonic stem cell self-renewal and differentiation independently of proliferation. Biotechnology and Bioengineering 2000;69: 607-617.

[91] Iijima S, Tanimoto Y, Mizuno S, Daitoku Y, Kunita S, Sugiyama F, Yagami K. Effect of different culture conditions on establishment of embryonic stem cells from BALB/cAJ and NZB/BINJ mice. Cellular Reprogramming 2010;12: 679-688.

[92] Davies TJ, Fairchild PJ. Optimization of protocols for derivation of mouse embryonic stem cell lines from refractory strains, including the non obese diabetic mouse. Stem Cells and Development 2012;21: 1688-1700.

[93] Baharvand H, Matthaei KI. Culture condition difference for establishment of new embryonic stem cell lines from the C57BL/6 and BALB/c mouse strains. In Vitro Cellular & Developmental Biology: Animal 2004;40: 76-81.

[94] Evans MJ. The isolation and properties of a clonal tissue culture strain of pluripotent mouse teratoma cells. Journal of Embryology and Experimental Morphology 1972;28: 163-176.

[95] Kahan BW, Ephrussi B. Developmental potentialities of clonal in vitro cultures of mouse testicular teratoma. Journal of the National Cancer Institute 1970;44: 1015-1036.

[96] Rosenthal MD, Wishnow RM, Sato GH. In vitro growth and differetiation of clonal populations of multipotential mouse cells derived from a transplantable testicular teratocarcinoma. Journal of the National Cancer Institute 1970;44: 1001-1014.

[97] Strickland S, Mahdavi V. The induction of differentiation in teratocarcinoma stem cells by retinoic acid. Cell 1978;15: 393-403.

[98] Hogan B, Costantini F, Beddington R, Lacy E, Hogan B, Costantini F, Beddington R, Lacy E. In vitro manipulation of preimplantation embryos, Manipulating the Mouse Embryo: A Laboratory Manual, 2 edition. Plainview: Cold Spring Harbor Lab. Press; 1994. p191-206.

[99] Solter D. From teratocarcinomas to embryonic stem cells and beyond: A history of embryonic stem cell research. Nature Reviews Genetics 2006;7: 319-327.

[100] Stevens LC, Little CC. Spontaneous testicular teratomas in an inbred strain of mice. Proceedings of the National Academy of Sciences of the United States of America 1954;40: 1080-1087.

[101] Ohta H, Ohinata Y, Ikawa M, Morioka Y, Sakaide Y, Saitou M, Kanagawa O, Wakayama T. Male germline and embryonic stem cell lines from NOD mice: Efficient derivation of GS cells from a nonpermissive strain for ES cell derivation. Biology of Reproduction 2009;81: 1147-1153.

[102] Roach ML, Stock JL, Byrum R, Koller BH, McNeish JD. A new embryonic stem cell line from DBA/1lacJ mice allows genetic modification in a murine model of human inflammation. Experimental Cell Research 1995;221: 520-525.

[103] Ledermann B, Burki K. Establishment of a germ-line competent C57BL/6 embryonic stem cell line. Experimental Cell Research 1991;197: 254-258.

[104] Auerbach W, Dunmore JH, Fairchild-Huntress V, Fang Q, Auerbach AB, Huszar D, Joyner AL. Establishment and chimera analysis of 129/SvEv- and C57BL/6-derived mouse embryonic stem cell lines. Biotechniques 2000;29: 1024-1032.

[105] Eistetter HR. Pluripotent embryonal stem cell lines can be established from disaggregated mouse morulae. Development, Growth & Differentiation 1989;31: 275-282.

[106] Keskintepe L, Norris K, Pacholczyk G, Dederscheck SM, Eroglu A. Derivation and comparison of C57BL/6 embryonic stem cells to a widely used 129 embryonic stem cell line. Transgenic Research 2007;16: 751-758.

[107] Lemckert FA, Sedgwick JD, Korner H. Gene targeting in C57BL/6 ES cells. Successful germ line transmission using recipient BALB/c blastocysts developmentally matured in vitro. Nucleic Acids Research 1997;25: 917-918.

[108] Shimizukawa R, Sakata A, Hirose M, Takahashi A, Iseki H, Liu Y, Kunita S, Sugiyama F, Yagami K. Establishment of a new embryonic stem cell line derived from C57BL/6 mouse expressing EGFP ubiquitously. Genesis 2005;42: 47-52.

[109] Tanimoto Y, Iijima S, Hasegawa Y, Suzuki Y, Daitoku Y, Mizuno S, Ishige T, Kudo T, Takahashi S, Kunita S, Sugiyama F, Yagami K. Embryonic stem cells derived from C57BL/6J and C57BL/6N mice. Comparative Medicine 2008;58: 347-352.

[110] Tesar PJ. Derivation of germ-line-competent embryonic stem cell lines from preblastocyst mouse embryos. Proceedings of the National Academy of Sciences of the United States of America 2005;102: 8239-8244.

[111] Mann JR, Gadi I, Harbison ML, Abbondanzo SJ, Stewart CL. Androgenetic mouse embryonic stem cells are pluripotent and cause skeletal defects in chimeras: Implications for genetic imprinting. Cell 1990;62: 251-260.

[112] Voss AK, Thomas T, Gruss P. Germ line chimeras from female ES cells. Experimental Cell Research 1997;230: 45-49.

[113] Wakayama, T, Tabar, V, Rodriguez, I, Perry, A. C. F, Studer, L, & Mombaerts, P. Differentiation of embryonic stem cell lines generated from adult somatic cells by nuclear transfer. Science 2001;292: 740-743.

[114] Eggan K, Rode A, Jentsch I, Samuel C, Hennek T, Tintrup H, Zevnik B, Erwin J, Loring J, Jackson-Grusby L, Speicher MR, Kuehn R, Jaenisch R. Male and female mice derived from the same embryonic stem cell clone by tetraploid embryo complementation. Nature Biotechnology 2002;20: 455-459.

[115] Chen Z, Liu Z, Huang J, Amano T, Li C, Cao S, Wu C, Liu B, Zhou L, Carter MG, Keefe DL, Yang X, Liu L. Birth of parthenote mice directly from parthenogenetic embryonic stem cells. Stem Cells 2009;27: 2136-2145.

[116] Li W, Shuai L, Wan H, Dong M, Wang M, Sang L, Feng C, Luo G-Z, Li T, Li X, Wang L, Zheng, Q.-Y, Sheng, C, Wu, H.-J, Liu, Z, Liu, L, Wang, L, Wang, X.-J, Zhao, X.-Y, Zhou Q. Androgenetic haploid embryonic stem cells produce live transgenic mice. Nature (2012)., 490, 407-411.

[117] Chung Y, Klimanskaya I, Becker S, Marh J, Lu SJ, Johnson J, Meisner L, Lanza R. Embryonic and extraembryonic stem cell lines derived from single mouse blastomeres. Nature 2006;439: 216-219.

[118] Delhaise F, Bralion V, Schuurbiers N, Dessy F. Establishment of an embryonic stem cell line from 8-cell stage mouse embryos. European Journal of morphology 1996;34: 237-243.

[119] Klimanskaya I, Chung Y, Becker S, Lu SJ, Lanza R. Human embryonic stem cell lines derived from single blastomeres. Nature 2006;444: 481-485.

[120] Lorthongpanich C, Yang SH, Piotrowska-Nitsche, K, Parnpai, R. & Chan, A. W. S. Development of single mouse blastomeres into blastocysts, outgrowths and the establishment of embryonic stem cells. Reproduction 2008;135: 805-813.

[121] Gonzalez S, Ibanez E, Santalo J. Establishment of mouse embryonic stem cells from isolated blastomeres and whole embryos using three derivation methods. Journal of Assisted Reproduction and Genetics 2010;27: 671-682.

[122] Gonzalez S, Ibanez E, Santalo J. Influence of early fate decisions at the two-cell stage on the derivation of mouse embryonic stem cell lines. Stem Cell Research 2011;7: 54-65.

[123] Zhang H, Wang ZZ. Mechanisms that mediate stem cell self-renewal and differentiation. Journal of Cellular Biochemistry 2008;103: 709-718.

[124] Boiani M, Scholer HR. Regulatory networks in embryo-derived pluripotent stem cells. Nature Reviews Molecular Cell Biology 2005;6: 872-881.

[125] Hirai H, Karian P, Kikyo N. Regulation of embryonic stem cell self-renewal and pluripotency by leukaemia inhibitory factor. Biochemical Journal 2011;438: 11-23.

[126] Kolch W. Meaningful relationships: The regulation of the Ras/Raf/MEK/ERK pathway by protein interactions. Biochemical Journal 2000;351: 289-305.

[127] Qi X, Li TG, Hao J, Hu J, Wang J, Simmons H, Miura S, Mishina Y, Zhao GQ. BMP4 supports self-renewal of embryonic stem cells by inhibiting mitogen-activated protein kinase pathways. Proceedings of the National Academy of Sciences of the United States of America 2004;101: 6027-6032.

[128] Burdon T, Stracey C, Chambers I, Nichols J, Smith A. Suppression of SHP-2 and ERK signalling promotes self-renewal of mouse embryonic stem cells. Developmental Biology 1999;210: 30-43.

[129] Lodge P, McWhir J, Gallagher E, Sang H. Increased gp130 signaling in combination with inhibition of the MEK/ERK pathway facilitates embryonic stem cell isolation from normally refractory murine CBA blastocysts. Cloning Stem Cells 2005;7: 2-7.

[130] Niwa H, Burdon T, Chambers I, Smith A. Self-renewal of pluripotent embryonic stem cells is mediated via activation of STAT3. Genes & Development 1998;12: 2048-2060.

[131] Smith AG, Heath JK, Donaldson DD, Wong GG, Moreau J, Stahl M, Rogers D. Inhibition of pluripotential embryonic stem cell differentiation by purified polypeptides. Nature 1988;336: 688-690.

[132] Matsuda T, Nakamura T, Nakao K, Arai T, Katsuki M, Heike T, Yokota T. STAT3 activation is sufficient to maintain an undifferentiated state of mouse embryonic stem cells. EMBO Journal 1999;18: 4261-4269.

[133] Rayasam GV, Tulasi VK, Sodhi R, Davis JA, Ray A. Glycogen synthase kinase 3: More than a namesake. British Journal of Pharmacology 2009;156: 885-898.

[134] Umehara H, Kimura T, Ohtsuka S, Nakamura T, Kitajima K, Ikawa M, Okabe M, Niwa H, Nakano T. Efficient derivation of embryonic stem cells by inhibition of glycogen synthase kinase-3. Stem Cells 2007;25: 2705-2711.

[135] Sato N, Meijer L, Skaltsounis L, Greengard P, Brivanlou AH. Maintenance of pluripotency in human and mouse embryonic stem cells through activation of Wnt signaling by a pharmacological GSK-3-specific inhibitor. Nature Medicine 2004;10: 55-63.

[136] Xu RH, Chen X, Li DS, Li R, Addicks GC, Glennon C, Zwaka TP, Thomson JA. BMP4 initiates human embryonic stem cell differentiation to trophoblast. Nature Biotechnology 2002;20: 1261-1264.

[137] Takahashi K, Mitsui K, Yamanaka S. Role of ERas in promoting tumour-like properties in mouse embryonic stem cells. Nature 2003;423: 541-545.

[138] Takahashi K, Murakami M, Yamanaka S. Role of the phosphoinositide 3-kinase pathway in mouse embryonic stem (ES) cells. Biochemical Society Transactions 2005;33: 1522-1525.

[139] Paling NR, Wheadon H, Bone HK, Welham MJ. Regulation of embryonic stem cell self-renewal by phosphoinositide 3-kinase-dependent signaling. Journal of Biological Chemistry 2004;279: 48063-48070.

[140] Thomson JA, Itskovitz-Eldor J, Shapiro SS, Waknitz MA, Swiergiel JJ, Marshall VS, Jones JM. Embryonic stem cell lines derived from human blastocysts. Science 1998;282: 1145-1147.

[141] Xu C. Feeder-free growth of undifferentiated human embryonic stem cells. Nature Biotechnology 2001;19: 971-974.

[142] Amit M. Clonally derived human embryonic stem cell lines maintain pluripotency and proliferative potential for prolonged periods of culture. Developmental Biology 2000;227: 271-278.

[143] Dvorak P, Dvorakova D, Koskova S, Vodinska M, Najvirtova M, Krekac D, Hampl A. Expression and potential role of fibroblast growth factor 2 and its receptors in human embryonic stem cells. Stem Cells 2005;23: 1200-1211.

[144] Wang G, Zhang H, Zhao Y, Li J, Cai J, Wang P, Meng S, Feng J, Miao C, Ding M, Li D, Deng H. Noggin and bFGF cooperate to maintain the pluripotency of human embryonic stem cells in the absence of feeder layers. Biochemical and Biophysical Research Communications 2005;330: 934-942.

[145] Xu RH, Peck RM, Li DS, Feng X, Ludwig T, Thomson JA. Basic FGF and suppression of BMP signaling sustain undifferentiated proliferation of human ES cells. Nature Methods 2005;2: 185-190.

[146] Stadtfeld M, Hochedlinger K. Induced pluripotency: History, mechanisms, and applications. Genes & Development 2010;24: 2239-2263.

[147] Rossant J. Stem cells and early lineage development. Cell 2008;132: 527-531.

[148] Buehr M, Smith A. Genesis of embryonic stem cells. Philosophical Transactions of the Royal Society of London Series B: Biological Sciences 2003;358: 1397-1402.

[149] Kunath T, Saba-El-Leil MK, Almousailleakh M, Wray J, Meloche S, Smith A. FGF stimulation of the Erk1/2 signalling cascade triggers transition of pluripotent embryonic stem cells from self-renewal to lineage commitment. Development 2007;134: 2895-2902.

[150] Frame S, Cohen P. GSK3 takes centre stage more than 20 years after its discovery. Biochemical Journal 2001;359: 1-16.

[151] Lorthongpanich C, Yang SH, Piotrowska-Nitsche K, Parnpai R, Chan AW. Chemical enhancement in embryo development and stem cell derivation from single blastomeres. Cloning Stem Cells 2008;10: 503-512.

Chapter 6

Human Testis–Derived Pluripotent Cells and Induced Pluripotent Stem Cells

Hideyuki Kobayashi, Koichi Nagao and
Koichi Nakajima

Additional information is available at the end of the chapter

http://dx.doi.org/10.5772/55570

1. Introduction

Pregnancy rates achieved by intercourse in normal human couples are 20-25% per month, 75% by six months, and 90% by one year [1]. However, 15% of couples of unknown fertility status are unable to conceive a baby after one year of intercourse without contraception. For 30% of these couples, their infertility can be attributed to a male factor alone; in an additional 20%, failure to conceive is explained by the presence of both male and female factors [2,3,4]. Among couples known to be infertile, a male factor is involved in 50% of the cases. The most common causes of male infertility include abnormal sperm production or function, impaired delivery of sperm, and overexposure to certain gonadotoxins in the environment. The pathogenesis of male infertility can be attributed to a disorder of germ-cell proliferation and differentiation or to somatic cell dysfunction [5].

The induction of spermatogenesis depends on the complementary actions of FSH and testosterone. FSH establishes the requisite Sertoli cell population. In the prepubertal primate, FSH alone can induce proliferation of Sertoli cells and spermatogonia, but this does not result in qualitatively and quantitatively normal spermatogenesis unless testosterone is simultaneously present [6] [7]. Testosterone affects the functional completion of meiosis and post-meiotic sperm differentiation and maturation. LH stimulates Leydig cells to produce testosterone. Although FSH appears to play a more dominant role in the maintenance of primate spermatogenesis than in its initiation, normal spermatogenesis is best maintained by the combined effects of FSH and LH [6].

The most severe form of male infertility is nonobstructive azoospermia, which is typically characterized by small-volume testes and elevated FSH. Patients with this disorder cannot produce biological children. Although microdissection testicular sperm extraction (micro-TESE) is used to treat patients with nonobstructive azoospermia [8], this technique does not have a good success rate. Therefore, new approaches are needed to develop treatments for male infertility.

Stem cells have the potential to differentiate into a variety of functional cell types in the body, and their discovery has given rise to the fields of regenerative medicine and cloning. Stem cells are regulated by the particular microenvironment in which they reside; these microenvironments are referred to as niches. Male germline stem cells can continuously produce sperm throughout adulthood, and investigators have sought to develop methods using stem cells to improve or restore fertility.

Embryonic stem cells (ESCs) have the potential to differentiate into nearly every cell type in the body. As the cells differentiate, they lose the ability to develop into different tissues. In contrast, specific tissues (gastrointestinal, integumentary, spermatogenic, and hematopoietic systems) maintain their regenerative capacity *in vivo*, and in fact, stem cells have been functionally identified in a wide range of adult tissues. These adult stem cells are believed to hold great promise for tissue generation in clinical settings. Here, we provide a summary of the therapeutic potential of stem cells for the rejuvenation of fertility in infertile males. Our hope is that future research will provide a range of options for the preservation of male fertility or the reversal of infertility.

2. Differentiation and characterization of human primordial germ cells

Human primordial germ cells (PGCs) can be isolated from tissues and their identity confirmed by observing their migratory activity in vitro [9]. Cultured human PGCs become human embryonic germ cells (hEGCs) in vitro, in the presence of feeder cells, leukemia inhibitory factor (LIF), and basic fibroblast growth factor (bFGF) [10]. hEGCs express alkaline phosphatase (AP), OCT4, SOX2, NANOG, stage specific embryonic antigen (SSEA)-3, SSEA-4, TRA-1-60, and TRA-1-81, which are pluripotent stem cell markers. In vivo, human PGCs do not express FGF4, SOX2 [11] [12], TRA-1-60, or TRA-1-81 [13] [14], which are expressed by hESCs or hEGCs in vitro. The molecular signature of human PGCs in vivo can be characterized as C-KIT$^+$, SOX2$^-$, TRA1-60$^-$, TRA1-81$^-$, and FGF4$^-$, in contrast with human pluripotent stem cell lines in vitro. (This information is summarized in Table 1.) However, the full complement of genes that are expressed specifically in human PGCs and their functions remain unclear.

3. Spermatogonial stem cells

Spermatogenesis is a complex and tightly regulated process in which a small pool of germline stem cells ultimately gives rise to spermatozoa [15]. These stem cells, called spermatogonial stem cells (SSCs) are found in the basal compartment of the seminiferous epithelium, where they adhere to the basement membrane. SSC self-renewal ensures the maintenance of

http://dx.doi.org/10.5772/55570

	hESC/hiPSC	hEGC	PGCs (early)	PGCs (late)
OCT4	+	+	+	+
NANOG	+	+	+	+/-
SOX2	+	+	-	?
SSEA1	-	+	+	+
SSEA3	+	+	?	?
SSEA4	+	+	+	+
TRA1-60	+	+	-	?
TRA1-81	+	+	-	?
VASA	-	?	-	+
C-KIT	-	?	+	+

Table 1. Markers of human pluripotent stem cells and germ cells.

the stem cell pool, while their differentiation generates a large number of germ cells. Therefore, a balance between SSC self-renewal and differentiation in the adult testis is essential to maintain normal spermatogenesis and fertility throughout life. SSCs need to reside in a unique environment, or niche, that provides the factors necessary for their survival and potency. In mice, Sertoli cells in the testis are a crucial component of the spermatogonial stem cell niche. They produce glial cell line-derived neurotrophic factor (GDNF), a distant member of the TGFβ family, which controls SSC self-renewal [16]. Several groups have reported that adding GDNF to freshly isolated germ cells in culture results in the proliferation of SSCs [17,18]. Other factors within the niche influence the fate of SSCs. One example is colony-stimulating factor 1 (CSF1), which is produced by Leydig cells and some peritubular myoid cells [19], and plays a role in SSC self-renewal (Figure 1).

The existence of SSCs was postulated almost 40 years ago on the basis of morphological studies [20] [21] [22] and observations of toxin-induced spermatogenic damage. The early studies of Clermont [23] [24] on human spermatogenesis revealed two types of spermatogonia, the A_{dark} and A_{pale} spermatogonia, which were differentiated by the staining pattern of their nucleus. Both cell types are generally considered stem cells [24,25]. A_{dark} spermatogonia function as reverse stem cells that rarely divide, but can be triggered to self-renew in the case of injury or disease, while A_{pale} spermatogonia are self-renewing stem cells [23,24,25,26]; they also divide into B spermatogonia, which further divide into spermatocytes [24].

In the last decade, molecular markers that can be used to identify and characterize human SSCs have been sought. A recent study reported that the expression of surface marker G protein coupled receptor 125 (GPR125) can be used in the isolation, characterization, and culture of putative human SSCs [27]. GPR125-positive spermatogonia are very rare, possibly limited to A_{dark} spermatogonia or a sub-population of A_{pale} spermatogonia. Human SSCs are also positive

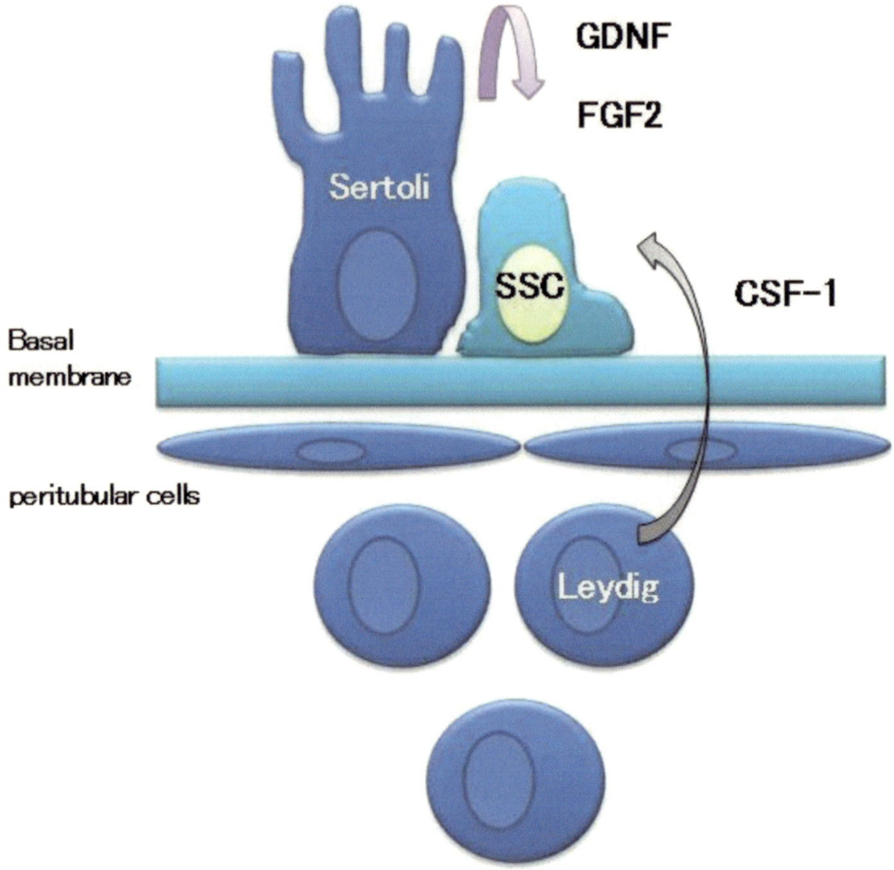

Figure 1. Diagram of the spermatogonial stem cell (SSC) niche showing that extrinsic factors drive SSC maintenance and self-renewal. SSCs and Sertoli cells are attached to the basement membrane. Sertoli cells produce glial cell line-derived neurotrophic factor (GDNF) and basic fibroblast growth factor (bFGF). Leydig cells and peritubular cells produce colony-stimulating factor-1 (CSF-1).

for some markers identified in mouse SSCs and other undifferentiated spermatogonia, including GFRA1, UCHL1 (PGP9.5), ZBTB16 (PLZF), and THY1 (CD90) [27,28]. We also have obtained evidence that THY1 is a potential surface marker for human SSCs [29].

Brinster and colleagues proved the existence of mouse SSCs by using unique approaches [30, 31]. These investigators transplanted cells obtained as testicular homogenates expressing the *LacZ* gene into the seminiferous tubules of otherwise sterile mice with a Sertoli-cell-only pathology. After three months, the transplanted spermatogonial stem cells had engrafted and colonized the seminiferous tubules. Spermatogenesis was restored.

The clinical implications of this work are enormous. The findings suggest that the isolation, enrichment, and cryopreservation of spermatogonial stem cells prior to chemotherapy or radiation therapy, with later autologous transplantation, may offer the potential for the subsequent restoration of fertility. The development of this technique will be especially important for survivors of childhood cancer. Adult patients can also bank sperm for cryopreservation. However, most couples would prefer a naturally conceived child. Work has progressed in many laboratories to partially enrich the spermatogonial stem cells of species ranging from mice to primates. Today, many urologists bank a testicular biopsy from patients about to undergo chemotherapy, with the expectation that technology will advance rapidly over the next 10 years and allow transplantation in the future.

4. Pluripotency of human testis–derived ESC–like cells

Previous studies have demonstrated that neonatal and adult germline stem cells (GSCs) can be self-reprogrammed into ESC-like cells, called germline-derived pluripotent stem cells [32,33,34,35]. In addition, Conrad et al. [36] reported that pluripotent cells can be derived from human testis, which those authors called human adult GSCs (haGSCs). Other research groups subsequently claimed that ESC-like cells could be obtained from cultures of human testicular cells [37,38,39]. Conrad and colleagues compared the global gene expression profile of hESCs and haGSCs, and concluded that the populations presented a similar gene expression profile, and thus, that the haGSCs were pluripotent. However, Ko et al. claimed that the gene expression profile of haGSCs differed substantially from the pluripotent profile of hESCs, determined by a number of laboratories [40]. For example, the haGSCs did not express NANOG, and had low OCT4 and SOX2 levels, but showed high levels of the fibroblast markers SNA12 and ACTA2 [40]. Ko and colleagues therefore suggested that the haGSCs originated from fibroblast cells, rather than from pluripotent tissue. They concluded that haGSCs were very similar to a human testicular fibroblast cell line (hTFCs) [40]. Conrad and colleagues argued that microarray data sets cannot be compared unless they are processed in parallel in the same experiment, suggesting that the similarity between haGSCs and hTFCs was inconclusive. However, studies on microarray results generated by different laboratories [41,42,43] have shown that findings from microarray analyses are comparable across multiple laboratories [44], particularly when a common platform and set of procedures are used. These findings justify the utility of microarray repositories, such as the GEO database [45], not only as data warehouses but also as resources for comparative and combinatory analyses of microarray data from different laboratories. In conclusion, the global gene expression analysis of haGSCs demonstrated that these cells resembled fibroblast hTFCs more than pluripotent hESCs.

5. Induced pluripotent stem (iPS) cells

The year 2006 saw the first description of mouse induced pluripotent stem cells (miPSCs), which were generated by the retrovirus-mediated transduction of four transcription factors (OCT3/4, SOX2, KLF4, and C-MYC) into mouse fibroblasts [46]. Human somatic cells can be reprogrammed to become human iPSCs via the introduction of a small set of genes, either those encoding OCT3/4, SOX2 and KLF4, with or without the addition of C-MYC, or an alternate combination of OCT3/4, SOX2, LIN28, and NANOG [47,48,49,50,51,52,53,54,55]. Human iPSCs (hiPSCs) have remarkable similarity to hESCs in terms of their morphology, in vitro characteristics, proliferation rate, gene expression, and ability to differentiate into mesoderm, endoderm, and ectoderm, both in vitro and in vivo, in teratoma assays [56,57].

In our laboratory, we induced iPS cells from adult human testicular tissue by introducing four transcription factors, OCT4, SOX2, KLF4, and C-MYC, using lentiviral vectors [58]. We also generated ES-like cells from 293FT cells by using OCT4, SOX2, NANOG, and LIN28 [59]. Finally, we generated iPS cells derived from the human testicular tissue of individuals with Klinefelter syndrome (KS, also called 47, XXY) [60].

6. Germline differentiation from ESCs and iPSCs in humans

Recent studies indicate that mouse [61,62,63,64,65] and human [66,67] [50,68,69,70,71] ESCs can differentiate in vitro into oocyte- or sperm-like cells. In particular, Clark et al. first reported the spontaneous differentiation of germ cells in embryoid bodies derived from human ESCs [66]. Male germline cells express specific RNA and protein markers, such as VASA. In 2009, Park et al. demonstrated that PGC-like cells can be differentiated from human iPSCs [50]. Subsequent reports on male germline differentiation from stem cells have used one of three approaches: (1) specific culture conditions, (2) manipulation of gene expression, and (3) purification of germ cells.

Culture conditions supporting differentiation into germline cells. Bucay et al. observed that as hESCs differentiate into putative germline cells, they also produce Sertoli-like support cells [69]. In addition, co-cultures of hESCs and hiPSCs with human fetal gonadal stromal cells [50], mouse Sertoli cells [72], or mouse embryonic fibroblasts [67] resulted in the increased efficiency of germ cell-like differentiation. Co-culture systems are used to mimic a suitable microenvironment for the growing germ cells. For the differentiation of germline-like cells from hESCs and hiPSCs, cytokines and other cell-signaling molecules are often added to the cultures. For example, BMP4 and other BMPs are added to promote PGC-like differentiation from hESCs and hiPSCs [73,74,75]. In addition, retinoic acid has been used to stimulate meiosis [75] [76]. Panula and colleagues reported the differentiation of fetal- and adult-derived iPSCs into germ cells, and showed that ~5% of human iPSCs differentiated into PGCs following induction with BMPs [77].

Manipulation of gene expression. By manipulating gene expression, researchers can regulate the cell lineage decisions of differentiating pluripotent stem cells. Overexpression of DAZL and VASA promotes PGC formation in differentiating human ESCs and iPSCs [78]. In addition, Kee and colleagues (2009) reported that hESCs differentiate into germline cells that initiate meiosis and progress to form haploid germ cells. These authors indicated that the overexpression of members of the DAZ gene family, DAZ, DAZL, and BOULE, promoted the progression of PGCs to meiosis and the production of haploid cells, a process that is unique to germ cell development [71].

Purification of germline cells. The isolation and purification of germline cells from stem cell cultures (ESCs and iPSCs) can be performed efficiently when specific antibodies for germ cell

http://dx.doi.org/10.5772/55570

surface markers are available. To purify PGC-like cells from differentiating human ESCs and iPSCs, cell sorting with specific antibodies for SSEA1 [68], SSEA1 and C-KIT [50] [79], or CXCR4 [69] has been effective. In particular, Eguizabal et al. (2011) published a straightforward protocol for germline cell purification that requires only three steps. First, human iPSCs and hESCs are allowed to differentiate for 3 weeks in a monolayer, in the absence of growth cytokines. Second, the cells are cultured for 3 weeks in the presence of retinoic acid. Finally, after these 6 weeks of differentiation, the cells are sorted for a specific combination of surface markers (CD49f++, CD9+, CD90-, and SSEA4-), and the isolated fraction is cultured in the presence of LIF, bFGF, Forskolin, and CYP26 inhibitor for 4 more weeks [76].

7. Germline differentiation from porcine iPSCs, non–human iPSCs

Despite their undoubted promise as sources of cells for tissue transplants, many roadblocks remain against using human ESCs clinically. Particularly troubling is the lack of tests for the efficacy of such therapies and the safety of transferring these cells in animals whose anatomy and physiology resemble those of humans better than mouse models do [80] [81] [82] [83] [84]. The pig is a potentially useful model in this regard, because of its similarities to humans in organ size, immunology, and whole animal physiology [85] [86] [87]. It was reported that porcine somatic cells can be reprogrammed to form piPSCs [88]. However, no reports on germline development from piPSCs have been published to date.

8. Conclusions

Research on stem cells has shown remarkable progress over the past 5 years. In particular, the development of human iPSCs has opened new avenues into the generation of an *in vitro* disease model of male infertility. However, improvements are still needed before stem cells can be used clinically. For the treatment and diagnosis of male infertility, future advances may enable spermatids to be differentiated from germline stem cells or iPS cells. In addition, by examining patient-specific iPSCs that are defective in their ability to generate germ cells and comparing their differentiation capacity with that of normal human ESCs and iPSCs, researchers can hope to uncover the nature of male infertility and to design new methods to reverse it.

Acknowledgements

This study was supported in part by a Grant-in-Aid for Young Scientists (B) of the Japan Society for the Promotion of Science (JSPS) and a Grant of the Strategic Research Foundation Grant-aided Project for Private schools at Heisei 23th from Ministry of Education, Culture, Sports, Science and Technology of Japan, 2011-2015.

Author details

Hideyuki Kobayashi*, Koichi Nagao and Koichi Nakajima

*Address all correspondence to: hideyukk@med.toho-u.ac.jp

Department of Urology, Toho University School of Medicine, Tokyo, Japan

References

[1] Spira, A. Epidemiology of human reproduction. Hum Reprod (1986). , 1, 111-5.

[2] MacLeod JHuman male infertility. Obstet Gynecol Surv (1971). , 26, 335-51.

[3] Mosher, W. D. Reproductive impairments in the United States, 1965-1982. Demography (1985). , 22, 415-30.

[4] Simmons, F. A. Human infertility. N Engl J Med (1956). contd., 255, 1140-6.

[5] Lue, Y, Erkkila, K, Liu, P. Y, Ma, K, Wang, C, Hikim, A. S, et al. Fate of bone marrow stem cells transplanted into the testis: potential implication for men with testicular failure. Am J Pathol (2007). , 170, 899-908.

[6] Nieschlag, E, Simoni, M, Gromoll, J, & Weinbauer, G. F. Role of FSH in the regulation of spermatogenesis: clinical aspects. Clin Endocrinol (Oxf) (1999). , 51, 139-46.

[7] Allan, C. M, Couse, J. F, Simanainen, U, Spaliviero, J, Jimenez, M, Rodriguez, K, et al. Estradiol induction of spermatogenesis is mediated via an estrogen receptor-{alpha} mechanism involving neuroendocrine activation of follicle-stimulating hormone secretion. Endocrinology (2010). , 151, 2800-10.

[8] Schlegel, P. N. Nonobstructive azoospermia: a revolutionary surgical approach and results. Semin Reprod Med (2009). , 27, 165-70.

[9] Kuwana, T, & Fujimoto, T. Active locomotion of human primordial germ cells in vitro. Anat Rec (1983). , 205, 21-6.

[10] Shamblott, M. J, Axelman, J, Wang, S, Bugg, E. M, Littlefield, J. W, Donovan, P. J, et al. Derivation of pluripotent stem cells from cultured human primordial germ cells. Proc Natl Acad Sci U S A (1998). , 95, 13726-31.

[11] Perrett, R. M, Turnpenny, L, Eckert, J. J, Shea, O, Sonne, M, & Cameron, S. B. IT et al. The early human germ cell lineage does not express SOX2 during in vivo development or upon in vitro culture. Biol Reprod (2008). , 78, 852-8.

[12] De Jong, J, Stoop, H, Gillis, A. J, & Van Gurp, R. J. van de Geijn GJ, Boer M et al. Differential expression of SOX17 and SOX2 in germ cells and stem cells has biological and clinical implications. J Pathol (2008). , 215, 21-30.

[13] Kerr, C. L, Hill, C. M, Blumenthal, P. D, & Gearhart, J. D. Expression of pluripotent stem cell markers in the human fetal testis. Stem Cells (2008). , 26, 412-21.

[14] Kerr, C. L, Hill, C. M, Blumenthal, P. D, & Gearhart, J. D. Expression of pluripotent stem cell markers in the human fetal ovary. Hum Reprod (2008). , 23, 589-99.

[15] De Rooij, D. G, & Russell, L. D. All you wanted to know about spermatogonia but were afraid to ask. J Androl (2000). , 21, 776-98.

[16] Meng, X, Lindahl, M, Hyvonen, M. E, Parvinen, M, De Rooij, D. G, Hess, M. W, et al. Regulation of cell fate decision of undifferentiated spermatogonia by GDNF. Science (2000). , 287, 1489-93.

[17] Kanatsu-shinohara, M, Ogonuki, N, Inoue, K, Miki, H, Ogura, A, Toyokuni, S, et al. Long-term proliferation in culture and germline transmission of mouse male germline stem cells. Biol Reprod (2003). , 69, 612-6.

[18] Kubota, H, Avarbock, M. R, & Brinster, R. L. Growth factors essential for self-renewal and expansion of mouse spermatogonial stem cells. Proc Natl Acad Sci U S A (2004). , 101, 16489-94.

[19] Oatley, J. M, Oatley, M. J, Avarbock, M. R, Tobias, J. W, & Brinster, R. L. Colony stimulating factor 1 is an extrinsic stimulator of mouse spermatogonial stem cell self-renewal. Development (2009). , 136, 1191-9.

[20] Huckins, C. The spermatogonial stem cell population in adult rats. 3. Evidence for a long-cycling population. Cell Tissue Kinet (1971). , 4, 335-49.

[21] Huckins, C. The spermatogonial stem cell population in adult rats. II. A radioautographic analysis of their cell cycle properties. Cell Tissue Kinet (1971). , 4, 313-34.

[22] Huckins, C. The spermatogonial stem cell population in adult rats. I. Their morphology, proliferation and maturation. Anat Rec (1971). , 169, 533-57.

[23] Clermont, Y. The cycle of the seminiferous epithelium in man. Am J Anat (1963). , 112, 35-51.

[24] Clermont, Y. Renewal of spermatogonia in man. Am J Anat (1966). , 118, 509-24.

[25] Clermont, Y. Spermatogenesis in man. A study of the spermatogonial population. Fertil Steril (1966). , 17, 705-21.

[26] Clermont, Y. Kinetics of spermatogenesis in mammals: seminiferous epithelium cycle and spermatogonial renewal. Physiol Rev (1972). , 52, 198-236.

[27] He, Z, Kokkinaki, M, Jiang, J, Dobrinski, I, & Dym, M. Isolation, characterization, and culture of human spermatogonia. Biol Reprod (2010). , 82, 363-72.

[28] Dym, M, Kokkinaki, M, & He, Z. Spermatogonial stem cells: mouse and human comparisons. Birth Defects Res C Embryo Today (2009). , 87, 27-34.

[29] Kobayashi, H, Nagao, K, Nakajima, K, Miura, K, & Ishii, N. Thy-1^+ cells isolated from adult human testicular tissues express human embryonic stem cell genes *OCT3/4* and *NANOG* and may include spermatogonial stem cells. Reprod Med Biol (2009).

[30] Brinster, R. L, & Avarbock, M. R. Germline transmission of donor haplotype following spermatogonial transplantation. Proc Natl Acad Sci U S A (1994). , 91, 11303-7.

[31] Brinster, R. L, & Zimmermann, J. W. Spermatogenesis following male germ-cell transplantation. Proc Natl Acad Sci U S A (1994). , 91, 11298-302.

[32] Ko, K, Tapia, N, Wu, G, Kim, J. B, Bravo, M. J, Sasse, P, et al. Induction of pluripotency in adult unipotent germline stem cells. Cell Stem Cell (2009). , 5, 87-96.

[33] Kanatsu-shinohara, M, Inoue, K, Lee, J, Yoshimoto, M, Ogonuki, N, Miki, H, et al. Generation of pluripotent stem cells from neonatal mouse testis. Cell (2004). , 119, 1001-12.

[34] Seandel, M, James, D, Shmelkov, S. V, Falciatori, I, Kim, J, Chavala, S, et al. Generation of functional multipotent adult stem cells from GPR125+ germline progenitors. Nature (2007). , 449, 346-50.

[35] Kanatsu-shinohara, M, Lee, J, Inoue, K, Ogonuki, N, Miki, H, Toyokuni, S, et al. Pluripotency of a single spermatogonial stem cell in mice. Biol Reprod (2008). , 78, 681-7.

[36] Conrad, S, Renninger, M, Hennenlotter, J, Wiesner, T, Just, L, Bonin, M, et al. Generation of pluripotent stem cells from adult human testis. Nature (2008). , 456, 344-9.

[37] Golestaneh, N, Kokkinaki, M, Pant, D, Jiang, J, Destefano, D, Fernandez-bueno, C, et al. Pluripotent stem cells derived from adult human testes. Stem Cells Dev (2009). , 18, 1115-26.

[38] Kossack, N, Meneses, J, Shefi, S, Nguyen, H. N, Chavez, S, Nicholas, C, et al. Isolation and characterization of pluripotent human spermatogonial stem cell-derived cells. Stem Cells (2009). , 27, 138-49.

[39] Mizrak, S. C, Chikhovskaya, J. V, Sadri-ardekani, H, Van Daalen, S, Korver, C. M, Hovingh, S. E, et al. Embryonic stem cell-like cells derived from adult human testis. Hum Reprod (2010). , 25, 158-67.

[40] Ko, K, Arauzo-bravo, M. J, Tapia, N, Kim, J, Lin, Q, Bernemann, C, et al. Human adult germline stem cells in question. Nature (2010). E1; discussion E3.

[41] Bammler, T, Beyer, R. P, Bhattacharya, S, Boorman, G. A, Boyles, A, Bradford, B. U, et al. Standardizing global gene expression analysis between laboratories and across platforms. Nat Methods (2005). , 2, 351-6.

[42] Larkin, J. E, Frank, B. C, Gavras, H, Sultana, R, & Quackenbush, J. Independence and reproducibility across microarray platforms. Nat Methods (2005). , 2, 337-44.

[43] Irizarry, R. A, Warren, D, Spencer, F, Kim, I. F, Biswal, S, Frank, B. C, et al. Multiple-laboratory comparison of microarray platforms. Nat Methods (2005). , 2, 345-50.

[44] Sherlock, G. Of fish and chips. Nat Methods (2005). , 2, 329-30.

[45] Edgar, R, Domrachev, M, & Lash, A. E. Gene Expression Omnibus: NCBI gene expression and hybridization array data repository. Nucleic Acids Res (2002). , 30, 207-10.

[46] Takahashi, K, & Yamanaka, S. Induction of pluripotent stem cells from mouse embryonic and adult fibroblast cultures by defined factors. Cell (2006). , 126, 663-76.

[47] Okita, K, Ichisaka, T, & Yamanaka, S. Generation of germline-competent induced pluripotent stem cells. Nature (2007). , 448, 313-7.

[48] Yu, J, Vodyanik, M. A, Smuga-otto, K, Antosiewicz-bourget, J, Frane, J. L, Tian, S, et al. Induced pluripotent stem cell lines derived from human somatic cells. Science (2007). , 318, 1917-20.

[49] Woltjen, K, Michael, I. P, Mohseni, P, Desai, R, Mileikovsky, M, Hamalainen, R, et al. piggyBac transposition reprograms fibroblasts to induced pluripotent stem cells. Nature (2009). , 458, 766-70.

[50] Park, T. S, Galic, Z, Conway, A. E, Lindgren, A, Van Handel, B. J, Magnusson, M, et al. Derivation of primordial germ cells from human embryonic and induced pluripotent stem cells is significantly improved by coculture with human fetal gonadal cells. Stem Cells (2009). , 27, 783-95.

[51] Zou, J, Maeder, M. L, Mali, P, Pruett-miller, S. M, Thibodeau-beganny, S, Chou, B. K, et al. Gene targeting of a disease-related gene in human induced pluripotent stem and embryonic stem cells. Cell Stem Cell (2009). , 5, 97-110.

[52] Nakagawa, M, Koyanagi, M, Tanabe, K, Takahashi, K, Ichisaka, T, Aoi, T, et al. Generation of induced pluripotent stem cells without Myc from mouse and human fibroblasts. Nat Biotechnol (2008). , 26, 101-6.

[53] Takahashi, K, Okita, K, Nakagawa, M, & Yamanaka, S. Induction of pluripotent stem cells from fibroblast cultures. Nat Protoc (2007). , 2, 3081-9.

[54] Liu, H, Zhu, F, Yong, J, Zhang, P, Hou, P, Li, H, et al. Generation of induced pluripotent stem cells from adult rhesus monkey fibroblasts. Cell Stem Cell (2008). , 3, 587-90.

[55] Takahashi, K, Tanabe, K, Ohnuki, M, Narita, M, Ichisaka, T, Tomoda, K, et al. Induction of pluripotent stem cells from adult human fibroblasts by defined factors. Cell (2007). , 131, 861-72.

[56] Yamanaka, S. Strategies and new developments in the generation of patient-specific pluripotent stem cells. Cell Stem Cell (2007). , 1, 39-49.

[57] Yamanaka, S. Pluripotency and nuclear reprogramming. Philos Trans R Soc Lond B Biol Sci (2008). , 363, 2079-87.

[58] Kobayashi, H, Nakajima, K, Oka, Y, Tai, T, Nagao, K, & Ishii, N. Reprogramming of adult human testicular cells by our transcription factors (OCT4, SOX2, KLF4, and C-MYC). Reprod Med Biol (2011). , 10, 105-112.

[59] Oka, Y, Nakajima, K, Nagao, K, Miura, K, Ishii, N, & Kobayashi, H. FT cells transduced with four transcription actors (OCT4, SOX2, NANOG, and LIN28) generate aberrant ES-like cells. J. Stem cell and Regenerative Medicine (2010). , 3, 149-156.

[60] Kobayashi, H. Pluripotent stem cells induced from testicular tissue with Klinefelter syndrome (47, XXY) by four ranscription factors (OCT4, SOX2, KLF4, and C-MYC). Methodological Advances in the Culture, Manipulation and Utilization of Embryonic Stem Cells for Basic and Practical Applications (2011). , 295-306.

[61] Hubner, K, Fuhrmann, G, Christenson, L. K, Kehler, J, Reinbold, R, De La Fuente, R, et al. Derivation of oocytes from mouse embryonic stem cells. Science (2003). , 300, 1251-6.

[62] Toyooka, Y, Tsunekawa, N, Akasu, R, & Noce, T. Embryonic stem cells can form germ cells in vitro. Proc Natl Acad Sci U S A (2003). , 100, 11457-62.

[63] Geijsen, N, Horoschak, M, Kim, K, Gribnau, J, Eggan, K, & Daley, G. Q. Derivation of embryonic germ cells and male gametes from embryonic stem cells. Nature (2004). , 427, 148-54.

[64] Qing, T, Shi, Y, Qin, H, Ye, X, Wei, W, Liu, H, et al. Induction of oocyte-like cells from mouse embryonic stem cells by co-culture with ovarian granulosa cells. Differentiation (2007). , 75, 902-11.

[65] Nagano, M. C. In vitro gamete derivation from pluripotent stem cells: progress and perspective. Biol Reprod (2007). , 76, 546-51.

[66] Clark, A. T, Bodnar, M. S, Fox, M, Rodriquez, R. T, Abeyta, M. J, Firpo, M. T, et al. Spontaneous differentiation of germ cells from human embryonic stem cells in vitro. Hum Mol Genet (2004). , 13, 727-39.

[67] West, F. D, Machacek, D. W, Boyd, N. L, Pandiyan, K, Robbins, K. R, & Stice, S. L. Enrichment and differentiation of human germ-like cells mediated by feeder cells and basic fibroblast growth factor signaling. Stem Cells (2008). , 26, 2768-76.

[68] Tilgner, K, Atkinson, S. P, Golebiewska, A, Stojkovic, M, Lako, M, & Armstrong, L. Isolation of primordial germ cells from differentiating human embryonic stem cells. Stem Cells (2008). , 26, 3075-85.

[69] Bucay, N, Yebra, M, Cirulli, V, Afrikanova, I, Kaido, T, Hayek, A, et al. A novel approach for the derivation of putative primordial germ cells and sertoli cells from human embryonic stem cells. Stem Cells (2009). , 27, 68-77.

[70] Tilgner, K, Atkinson, S. P, Yung, S, Golebiewska, A, Stojkovic, M, Moreno, R, et al. Expression of GFP under the control of the RNA helicase VASA permits fluorescence-activated cell sorting isolation of human primordial germ cells. Stem Cells (2010). , 28, 84-92.

[71] Kee, K, Angeles, V. T, Flores, M, & Nguyen, H. N. Reijo Pera RA. Human DAZL, DAZ and BOULE genes modulate primordial germ-cell and haploid gamete formation. Nature (2009). , 462, 222-5.

[72] Geens, M, & Sermon, K. D. Van de Velde H, Tournaye H. Sertoli cell-conditioned medium induces germ cell differentiation in human embryonic stem cells. J Assist Reprod Genet (2011). , 28, 471-80.

[73] West, F. D, Roche-rios, M. I, Abraham, S, Rao, R. R, Natrajan, M. S, Bacanamwo, M, et al. KIT ligand and bone morphogenetic protein signaling enhances human embryonic stem cell to germ-like cell differentiation. Hum Reprod (2010). , 25, 168-78.

[74] Kee, K, Gonsalves, J. M, Clark, A. T, & Pera, R. A. Bone morphogenetic proteins induce germ cell differentiation from human embryonic stem cells. Stem Cells Dev (2006). , 15, 831-7.

[75] Richards, M, Fong, C. Y, & Bongso, A. Comparative evaluation of different in vitro systems that stimulate germ cell differentiation in human embryonic stem cells. Fertil Steril (2010). , 93, 986-94.

[76] Eguizabal, C, Montserrat, N, Vassena, R, Barragan, M, Garreta, E, Garcia-quevedo, L, et al. Complete meiosis from human induced pluripotent stem cells. Stem Cells (2011). , 29, 1186-95.

[77] Panula, S, Medrano, J. V, Kee, K, Bergstrom, R, Nguyen, H. N, Byers, B, et al. Human germ cell differentiation from fetal- and adult-derived induced pluripotent stem cells. Hum Mol Genet (2011). , 20, 752-62.

[78] Medrano, J. V, Ramathal, C, Nguyen, H. N, & Simon, C. Reijo Pera RA. Divergent RNA-binding proteins, DAZL and VASA, induce meiotic progression in human germ cells derived in vitro. Stem Cells (2012). , 30, 441-51.

[79] Aflatoonian, B, Ruban, L, Jones, M, Aflatoonian, R, Fazeli, A, & Moore, H. D. In vitro post-meiotic germ cell development from human embryonic stem cells. Hum Reprod (2009). , 24, 3150-9.

[80] Vackova, I, Ungrova, A, & Lopes, F. Putative embryonic stem cell lines from pig embryos. J Reprod Dev (2007). , 53, 1137-49.

[81] Keefer, C. L, Pant, D, Blomberg, L, & Talbot, N. C. Challenges and prospects for the establishment of embryonic stem cell lines of domesticated ungulates. Anim Reprod Sci (2007). , 98, 147-68.

[82] Brevini, T. A, Antonini, S, Cillo, F, Crestan, M, & Gandolfi, F. Porcine embryonic stem cells: Facts, challenges and hopes. Theriogenology (2007). Suppl 1:S, 206-13.

[83] Talbot, N. C. Blomberg Le A. The pursuit of ES cell lines of domesticated ungulates. Stem Cell Rev (2008). , 4, 235-54.

[84] Hall, V. Porcine embryonic stem cells: a possible source for cell replacement therapy. Stem Cell Rev (2008). , 4, 275-82.

[85] Prather, R. S, Hawley, R. J, Carter, D. B, Lai, L, & Greenstein, J. L. Transgenic swine for biomedicine and agriculture. Theriogenology (2003). , 59, 115-23.

[86] Brandl, U, Michel, S, Erhardt, M, Brenner, P, Burdorf, L, Jockle, H, et al. Transgenic animals in experimental xenotransplantation models: orthotopic heart transplantation in the pig-to-baboon model. Transplant Proc (2007). , 39, 577-8.

[87] Piedrahita, J. A, & Mir, B. Cloning and transgenesis in mammals: implications for xenotransplantation. Am J Transplant (2004). Suppl , 6, 43-50.

[88] Ezashi, T, Telugu, B. P, Alexenko, A. P, Sachdev, S, Sinha, S, & Roberts, R. M. Derivation of induced pluripotent stem cells from pig somatic cells. Proc Natl Acad Sci U S A (2009). , 106, 10993-8.

Chapter 7

Generation of Induced Pluripotent Stem Cells from Dental Pulp Somatic Cells

Nelson F. Lizier, Irina Kerkis and
Cristiane V. Wenceslau

Additional information is available at the end of the chapter

http://dx.doi.org/10.5772/55856

1. Introduction

During early development, human dental pulp is originated from neural crest, which is a transient embryonic structure (Fig. 1). According to current knowledge, neural crest stem cells (NCSCs) have the capacity to self-renewal and display a developmental potential almost the same as embryonic stem (ES) cells (Kerkis and Caplan, 2012). These postmigratory NCSCs generate all craniofacial bones, the majority of the peripheral nervous system cells and tissues, as well as several non-neural cell types, such as smooth muscle cells of the cardiovascular system, pigment cells in the skin, cartilage, connective tissue, corneal epithelium and dental pulp among them. Although postmigratory, postnatal NCSCs are of restricted developmental potential they maintain functional characteristics resembling their embryonic counterparts and an ability to differentiate into a broad spectrum of cell types (Le Douarin et al., 2004, 2007, 2008; Dupin et al., 2007; Le Douarin & Dupin, 2003, 2012).

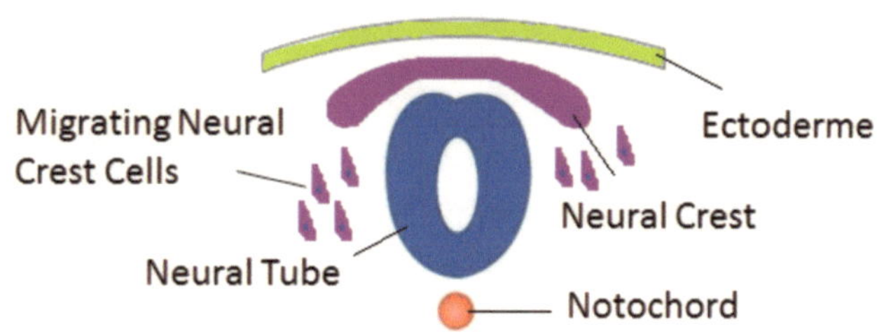

Figure 1. Early development of NCSCs. According to current knowledge, migrating neural crest cells are stem cells that display almost the same potential as ES cells.

The marathon of induced pluripotent stem cells (iPSC) started when Yamanaka in 2006, by forcing the expression of certain pluripotent genes in fibroblasts, reversed them into a pluripotent state similar to ES cells (Takahashi &Yamanaka, 2006). The main goal of iPSC generation is to create patient-specific cells, which would be advantageous for cell therapy due to immune compatibility (Ohnuki et al., 2009). Research involving the production of iPSC is being developed around the world. Production of iPSC opens new avenues for understanding human genetic diseases; embryogenesis and will likely have a great impact in drug screening and toxicological tests. However, fibroblasts, which were firstly used for iPSC production, present low efficiency and slow process of reprogramming. Moreover, these cells throughout all life are exposed to environmental factors, which can compromise their use as genetic models (Liu, 2008; McDevitt & Palecek, 2008; Nishikawa et al., 2008; Yu & Thomson; 2008; Zhao & Daley, 2008; Maherali & Hochedlinger, 2008; Ooi et al., 2012). Indeed, more immature somatic cells such as, postmigratory NCSCs, and adult stem cells isolated from young organism showed high efficiency of reprogramming (Zouboulis et al., 2008; Muchkaeva et al., 2012). Because of the possibility to isolate NCSCs from easily accessible tissue (e.g. baby teeth is discarded), the dental pulp derived somatic cells have become an ideal model system to study stem cell biology in diseases during different stages of the development (childhood, youth, middle-aged and old) with a special focus on non-invasive source of the cells for investigation of pediatric diseases (Kerkis & Caplan, 2012; Lizier et al., 2012).

Our group isolated and fully characterized human immature dental pulp stem cells (hIDPSC), which is a very attractive cell type, from deciduous teeth (baby teeth) (Kerkis et al., 2006; Lizier et al., 2012). The hIDPSC present fibroblast-like morphology, retain characteristics of adult multipotent stem cells and express at least one of three transcription factors: Oct4, Nanog and Sox2 (Kerkis et al., 2006; Lizier et al., 2012). We also used these cells as an alternative source for iPSC derivation (Beltrão-Braga et al., 2011) (Fig. 2).Different research groups derived iPSC from dental pulp fibroblasts and stem cells from young, middle aged and old patients. The difference was observed between the protocols and efficiency of iPSC generation in all these studies (Yan et al., 2010; Tamaoki et al., 2010; Oda et al., 2010; Beltrão-Braga et al., 2011). The present chapter is focused on comparative investigation of the methods and efficiency of iPSC generation from dental pulp stem cells and fibroblasts (control). Differentiation potential, assuredness and the future perspectives of the use of these iPSC derived from dental pulp stem cells in basic research and in biotechnology will also be broadly discussed.

2. Dental pulp somatic cells used for iPSC generation

Human dental tissues are rich in stem cells (Giordano et al., 2011; Kerkis & Caplan, 2012). Different research groups isolated and characterized several types of stem cells used for iPSC generation: (i) from apical papilla (Yan et al., 2010), (ii) from dental pulp of primary exfoliated deciduous teeth (Yan et al., 2010; Beltrão-Braga et al., 2011) and (iii) from wisdom teeth (Tamaoki et al., 2010; Oda et al., 2010). SCAP (stem cells from apical papilla) were obtained from tissue at the apex of a tooth root (Yan et al., 2010). SHED (stem cells from human exfoliated deciduous) and IDPSC (immature dental pulp stem cells) were derived from exfoliated

http://dx.doi.org/10.5772/55856

deciduous teeth (Miura et al., 2003; Kerkis et al., 2006). DPCs (dental pulp cells) and MStCs (mesenchymal/stromal cells) were isolated from human third molars by two independent groups (Takeda et al., 2008; Ikeda et al., 2008). All these cell types present fibroblast-like morphology and however differ in methods of isolation, show significant difference in expression pattern of stem cell markers and in purity of isolated population (Yan et al., 2010; Beltrão-Braga et al., 2011; Tamaoki et al., 2010; Oda et al., 2010).

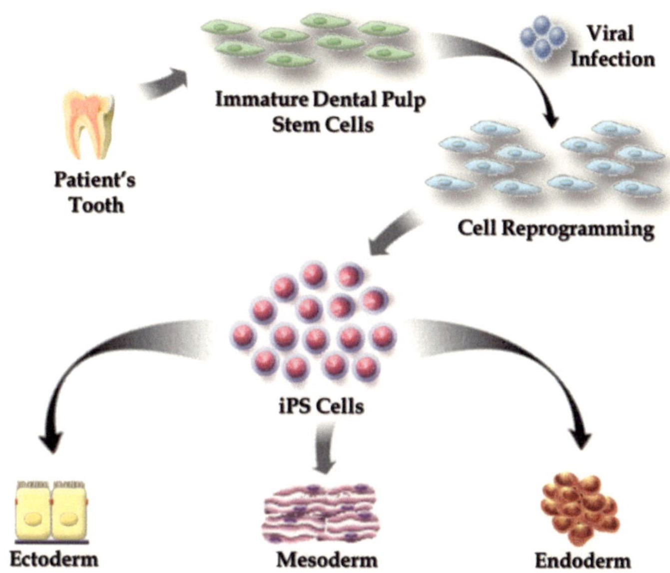

Figure 2. iPSC technology. The iPSC production can be induced by forcing the expression of certain pluripotent genes.

2.1. Transduction and reprogramming

For reprogramming of SHED/SCAP/DPSCs, heterogeneous primary human dental stem/ progenitor cell population at passages 2 and 3 were used (Yan et al., 2010). These populations were tested for their cell surface marker expression by flow cytometry and they were positive for STRO-1, CD146, CD73, CD90, CD105 and negative for CD14, CD34, and CD45, showing typical immunophenotype of mesenchymal stem cells (MSC) (Friedenstein et al., 1976; Caplan, 1991). The first study used four factor genes for reprogramming, such as c-Myc [Myc proto-oncogene protein], Klf4 [Krüppel-like factor], Oct4 [octamer-binding transcription factor 4], and Sox2 [(sex determining region Y)-box 2] into pLenti6.2/C-Lumio/V5-DEST vector system. Although the cells started to present morphological changes (fibroblastic to epithelial cell-like transition), the reprogramming process failed. Further, lentiviral vectors pSin-EF2-gene-Pur carrying 1 of the 4 factors Lin28 [Lin-28 homolog A], Nanog (Nanog homeobox), Oct4, and Sox2 were used and first ES-like colonies were obtained. To improve reprogramming efficiency, human genes c-Myc, Klf4, Oct4, and Sox2 were subcloned into the vector pMXs and produced retrovirus was used for second round of transduction. Human fibroblasts, used as a control in this study, were not able to undergo reprogramming under proposed conditions.

Retroviruses expressing four Oct3/4, Sox2, Klf4, and c-Myc or three (without c-Myc) factors were used for reprogramming DPCs from wisdom teeth and from human dermal fibroblasts (HDFs), which was performed according to the methods previously described (Takahashi et al., 2007). Another group, which used MStCs from wisdom teeth, also demonstrated successful reprogramming of these cells with pMXs retrovirus vectors containing three human Oct3/4, Sox2, and Klf4 factor genes (Oda et al., 2010).

IDPSC is a homogeneous population in respect of the expression of MSCs (Friedenstein et al., 1976; Caplan, 1991) markers, such as CD73, CD105, nestin and vimentin. Within IDPSC population, several cells also express Oct3/4 and Nanog (Kerkis et al., 2006; Lizier et al., 2012). To reprogram IDPSC, our group used four Yamanaka's factors (Klf4, Oct4, c-Myc and Sox2) and previously established protocol (Takahashi et al., 2007; Beltrão-Braga et al., 2011).

2.2. Formation of ES cell-like colonies and expansion

There are several important points that should be considered when iPSC are isolated and expanded: (i) the use of mouse embryonic fibroblasts (MEF) as a feeder layer, (ii) the efficiency of reprogramming and (iii) the efficiency of expansion (Takahashi & Yamanaka, 2006; Lewitzky & Yamanaka, 2007; Bilic & Belmonte, 2012). Isolation of iPSC on MEF limits the manipulation and further clinical application of these cells. Thus, isolation and expansion of iPSC without MEF is an important step, which avoids contamination of human cells with animal products. Efficiency of reprogramming depends on different factor, such as gene expression profile of cells, which were used in experiments. It has been shown that more immature cells undergo this process more efficiently, then committed or terminally differentiated cells (Zouboulis et al., 2008; Muchkaeva et al., 2012). And finally, during reprogramming, the cells receive different number of reprogramming factors and/or they did not respond equally to this process, therefore multiple ES-like did not complete reprogramming or non-ES cell-like colonies raised (Aasen et al., 2008; Marchetto et al., 2009). SHED/SCAP/DPSCs/DPCs/MStCs-derived iPSC were obtained using MEF as a feeder layer (Yan et al., 2010; Tamaoki et al., 2010; Oda et al., 2010). IDPSC-derived iPSC were shown to be obtained under both conditions: feeder-free on matrigel-coated dishes and on MEF (Fig. 3) (Beltrão-Braga et al., 2011). It seems that time-course of reprogramming of different cell types varied in accordance with age of cell donor, cell type and number of factors used. Thus, SHED/SCAP/DPSCs-derived iPSC, showed the formation of the first colonies ~2-3 weeks after gene transduction. DPCs-derived iPSC were reprogrammed in ~14 days, when 4 factors were used and in ~ 20-25 days, when reprogramming was performed with only 3 factors. MStCs-derived iPSC were reprogrammed in ~25 days, while IDPSC-derived iPSC demonstrated the formation of first colonies at ~ day 5-11. All studies demonstrated that efficiency of iPSC derivation from dental pulp tissues is higher than that from human dermal fibroblast and primary gingival fibroblasts (Yan et al., 2010; Tamaoki et al., 2010; Oda et al., 2010; Beltrão-Braga et al., 2011). Because the MStCs were a heterogeneous cell population, Oda and co-authors (2010) additionally used clonally expanded MStCs in reprogramming experiments. They observed that reprogramming

http://dx.doi.org/10.5772/55856

efficiency in clonally expanded MStCs was higher and it correlates with cell proliferative ability. The clones, which showed higher proliferative ability, demonstrated a rate of reprogramming ~ 30–100-fold higher than HDFs and ~ 7-fold higher than clones with lower proliferative ability. The IDPSC also present high reprogramming efficiency and no difference was observed between the cells from both donors. Therefore, difficulties in reprogramming SHED/SCAP/DPSCs can be related with cell heterogeneity of original populations (Yan et al., 2010).

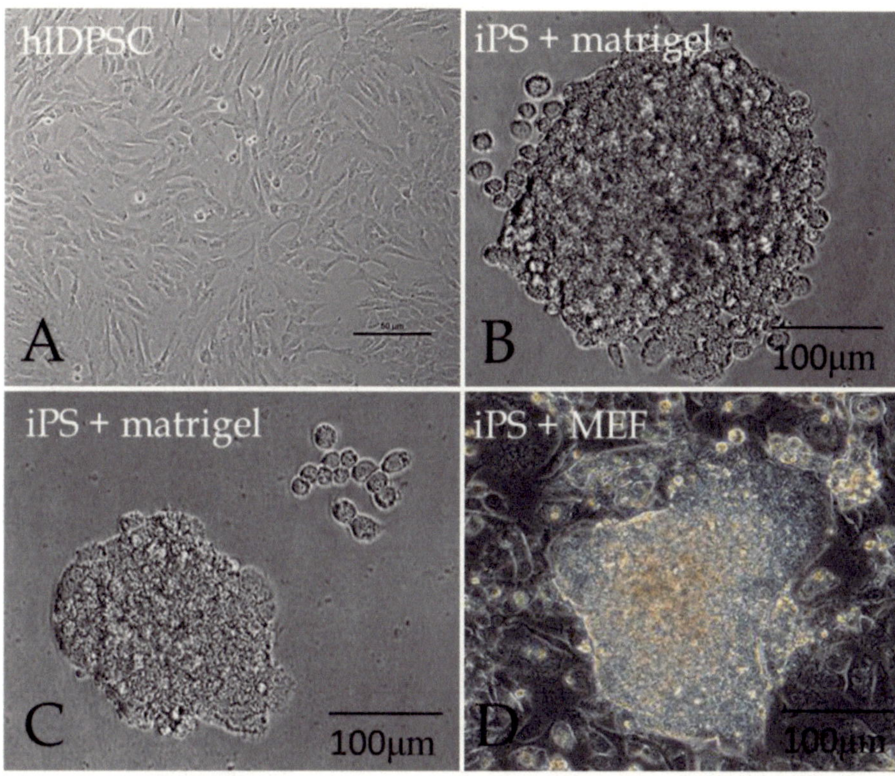

Figure 3. hIDPSC-derived iPSC. (A) Representative figure of morphological characteristics of hIDPSC *in vitro* culturing (light microscopy). (B-D) iPSC derivations were shown to be obtained under both conditions: feeder-free, on matrigel-coated dishes (B and C) and on MEF (D).

Furthermore, the aging process influences all organs, tissues and cells of organism. The studies showed that this factor is also important for cells reprogramming (Zouboulis et al., 2008; Banito et al., 2009). SHED/SCAP/DPSCs/DPCs/MStCs/IDPSC were isolated from young donors of variable ages 7, 10, 12, 13, 14 16, 19, 20 and 24 years old. The difference in efficiency of iPSC generation was observed between MStCs isolated from third molars of 10-, 13-, and 16-year-old donors. More efficient reprogramming was observed when MStCs from the 10-year-old donor were used. Similar observation was made by Tamaoki and co-wokers (2010). In our study we used IDPSC from 7 years old donors and we observed rapid and efficient reprogramming in both cell populations.

2.3. Provirus integration

Viral vectors are commonly used to deliver genetic material into cells, which can be performed *in vivo* (living organism) or *in vitro* (cell culture). Delivery of genes by a virus is efficient, however with respect to safety, it is ideal not to use lentivirus and retrovirus vectors, since they can integrate into the host DNA. As opposed to lentiviruses and retroviruses, adenoviral DNA does not integrate into the genome and are considered to be safer (Tamaoki et al., 2010). All studies performed with stem cells of dental tissue origin used lentivirus and/or retrovirus vectors for reprogramming. Yan and co-workers (2010) examined the presence of transgenes in the genome of iPSC clones isolating genomic DNA and generating primers specific for each transgene. They showed that the 4 factors were all integrated into the genome of the transduced SHED/SCAP/DPSC-iPSC. Other study, which used MStCs from wisdom teeth for reprogramming, did not provide any records about viral vectors integration (Tamaoki et al., 2010), while another group demonstrated retroviral silencing (Oda et al., 2010). Beltrão-Braga and co-authors (2011), which used retroviral vector for reprogramming, showed the lack of transgene expression by RT-PCR analyses in iPSC-derived from IDPSCs.

2.4. Characterization of SHED/SCAP/DPSCs/DPCs/MStCs/IDPSC-derived iPSC

2.4.1. Expression of pluripotent stem cell markers

As expected in all studies, the iPSC obtained from tissues of dental origin, which showed ES-like cells morphology, express key markers of pluripotent stem cells in an appropriate manner. Immunofluorescence study demonstrates uniform expression of these antigens in iPSC colonies derived from different types of dental stem cells. Transcription factor proteins as Oct3/4, Nanog, Sox2 demonstrate nuclear localization, while cell surface markers, such as stage specific embryonic antigen (SSEA) 3 and SSEA4, as well as cell surface antigens of human embryonic carcinoma cells (TRA-1-60 and TRA-1-81) show cell surface localization. Appropriate expression of transcription factors Klf4, c-Myc, Lin28, that were part of the transgene used for reprogramming, also was observed (Yan et al. 2010; Tamaoki et al., 2010; Oda et al., 2010; Beltrão-Braga et al., 2011) (Fig. 4).

2.4.2. Expression of molecular markers of pluripotent stem cells

Only one study performed quantitative PCR analysis before and after reprogramming for endogenous expression of Oct4, Nanog and Sox2 genes and compared the expression level of all these genes with those in pluripotent human ES cells. Albeit we revealed a tendency for increasing of expression of pluripotent factors Oct4, Nanog and Sox2, when compared to non-reprogrammed cells (18%, 1% and 2%, respectively), it was significantly lower 20% (Oct4), 10% (Nanog) and 40% (Sox2) in comparison with human ES cells (100% - Oct4, Nanog, Sox2) (Beltrão-Braga et al., 2011). Other studies did not provide any data about expression of these key markers in SHED/SCAP/DPSCs/DPCs/MStCs – derived iPSC in comparison with ES cells (Yan et al., 2010; Tamaoki et al., 2010; Oda et al., 2010). However, Oda and colleagues (2010) demonstrated that expression levels of Oct4, Nanog, Sox2, Klf4, c-Myc, Lin28 and P53 was

http://dx.doi.org/10.5772/55856

higher in iPSC derived from clonally isolated MStCs, when compared with parental cell lines, iPSC-derived from these lines and HDFs, used as a control.

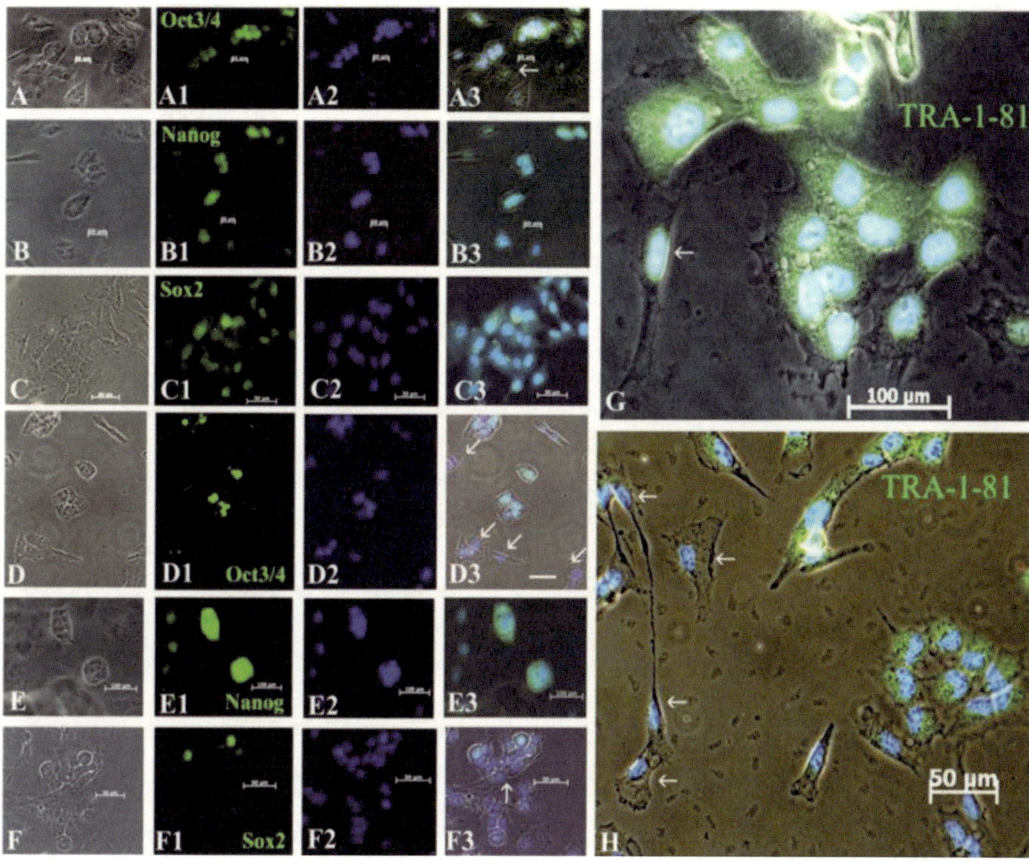

Figure 4. Expression of Oct3/4, Nanog, Sox2 and TRA-1-81 proteins in two lineages of hIDPS-IPSC five days after transduction with four factors. In A-C3,G) hIDPS-IPSC1 and D-F3,H) hIDPS-IPSC2 are presented, both showing multiple small colonies, which already express hallmarks of pluripotent cells, such as, A-A3) and D-D3) Oct3/4; B-B3) and E-E3) Nanog, C-C3) and F-F3) Sox2; G) and H) TRA-1-81, respectively. Nucleus stained with DAPI (blue). Note, that Oct3/4, Nanog, Sox2 present nuclear, while TRA-1-81 presents cytoplasm localization. Several cells, which did not present expression of these proteins and served as a control, are indicated by white arrows. Confocal Microscopy: A-F) Differential interference contrast (DIC); A1-F1 and A2-F2) Fluorescent microscopy (Fm); A3-F3, G, H) DIC+Fm. Scale Bars: A-D3, F-F3, H =50µm; E-E3,G=100µm.

Yan et al., (2010) quantified by real-time PCR the expression levels of endogenous Klf4 and c-Myc. Klf4 showed relative higher expression in DPC lines than in HDFs, however lower than in ES cells. Endogenous c-Myc expression in most DPC lines was also slightly higher than that in HDFs and in a few iPSC clones were close to ES cells. In contrast, Oda et al., (2010) observed low expression of Klf4 in high reprogramming cells, which was unexpected, once Klf4 is a reprogramming factor. Yan et al., (2010) showed that endogenous Klf4 expression level determined by real-time PCR did not completely correlate with the reprogramming efficiency of each DPCs (wisdom teeth) line. It is noteworthy that highly expression of KLF4 was previously reported in senescent cells and terminally differentiated cells (Shields et al., 1996; Conkright et al., 1999). Taken together, these data suggest that endogenous Klf4 expression may not be the single factor in charge for the reprogramming efficiency to MSCs derived from wisdom teeth.

2.4.3. Searching for new factors of reprogramming

Oda et al., (2010) tried to find the additional unknown factor(s) that could help in the cell reprogramming. They focused their study on practically two genes: PAXIP1 (or PTIP) and PARP. PAXIP1 acts as component of a histone H3 lysine four (H3K4) methyltransferase complex (Cho et al., 2007; Patel et al., 2007) and has a role in DNA double-strand break repair (van Attikum and Gasser, 2009). It was demonstrated that efficient reprogramming of pluripotent gene (Oct3/4, Sox2) expression is associated with H3K4 methylation in mouse somatic cell nuclei transplantation into amphibian oocytes (Murata et al., 2010). The expression of this gene was about 30% more in the high reprogramming cells than in low reprogramming as well as 3–4 times more in iPSC when compared with each parental cell line. PARP-1 belongs to PARP family being the most abundant member and is responsible for > 85% of nuclear PARP activity modifying histone structure through DNA-dependent "PARylation". Higher expression of PARP-1 was also seen after induction of reprogramming in cells derived from wisdom teeth. The authors supposed that due to possible conformational change of chromatin by direct/indirect actions of chromatin modification proteins such as PAXIP1 and possibly PARP-1, high iPSC generation clones may be accessible for reprogramming factors. However, further investigation is needed to illuminate the iPSC reprogramming mechanisms using these genes.

2.4.4. Methylation status of cytosine guanine dinucleotides (CpG)

The methylation status of CpG in the promoter regions of Nanog and Oct4 was examined using bisulfite DNA sequencing method in two studies (Yan et al., 2010; Oda et al., 2010) and of Nanog in one study (Tamaoki et al., 2010). They showed that parental MStCs from wisdom teeth were highly (Oct3/4) or partially (Nanog) methylated and the iPSC-derived from these cells were highly unmethylated, suggesting that these promoters were active after cells reprogramming. In contrast, the analysis of iPSC clones derived from DPSCs (wisdom teeth) and SHED (deciduous teeth) showed that Nanog promoter had similar or slightly higher number of methylated sites, than their non-transduced counterparts. The SHED-/DPSC-iPSC had less methylated sites of Oct3/4 promoter than the non-transduced cells (Yan et al., 2010).

2.4.5. Telomerase activity

Telomerase activity is known to be highly activated in ES cells in order to maintain the integrity of chromosome structure. After reprogramming, SHED- (deciduous teeth) SCAP-, and DPSC (wisdom teeth) - iPSC showed telomerase activity very close to ES cells and a lot more in comparison to their non-transduced counterparts (Yan et al., 2010). Parental DPSCs (wisdom teeth) showed low telomerase activity whereas in each iPSC telomerase activity was high (Oda et al., 2010).

http://dx.doi.org/10.5772/55856

2.4.6. Karyotype analysis of dental stem cell-derived iPSC

Karyotype study has been performed by all authors and demonstrated that karyotype of reprogrammed cells remained unchanged (Fig. 5). Overall, during reprogramming of stem cell from dental pulp, numerical and gross structural chromosomal abnormalities were not detected (Yan et al., 2010; Tamaoki et al., 2010; Oda et al., 2010; Beltrão-Braga et al., 2011).

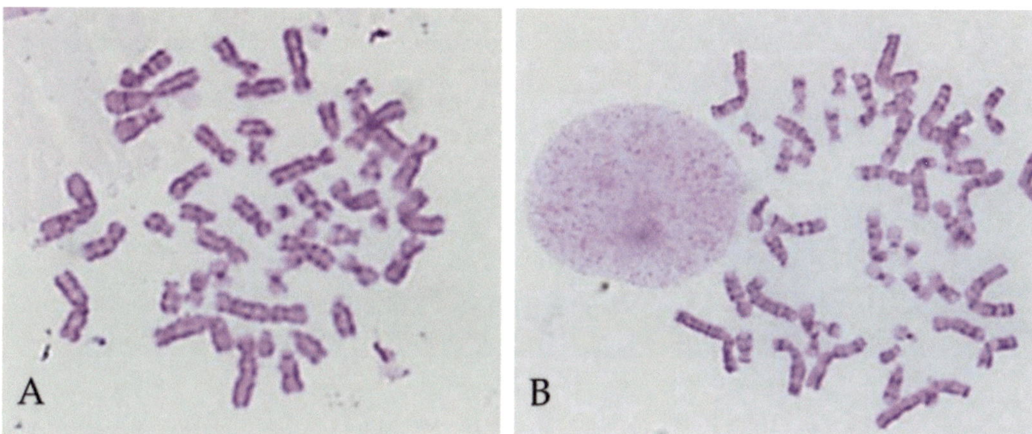

Figure 5. Representative figures of karyotype analysis of both hIDPSC and hIDPS-iPSC: Routine Giemsa staining did not reveal any numerical changes in chromosome number (A and B) and any chromosomal structural changes (B). Magnification 63X. Differential interference contrast (DIC)

2.5. Differentiation of SHED/SCAP/DPSCs/DPCs/MStCs/IDPSC-iPSC

2.5.1. Embryoid body formation and in vitro differentiation

Similar to human ES cells, iPSC require the formation of embryoid bodies (EB) in order to undergo *in vitro* differentiation into various cell types. All studies confirm successful differentiation of iPSC-derived from different types of dental stem cells into all three germ layers (Yan et al., 2010; Tamaoki et al., 2010; Oda et al., 2010; Beltrão-Braga et al., 2011). As expected, the majority of iPSC formed EBs with cystic cavities. The histological analysis demonstrates that EBs differentiated into different cell types of ectodermal, mesodermal, or endodermal origin, which was demonstrated using antibodies against specific proteins, which is expressed in each of three germ layers. It has been shown that after reprogramming, cells tend to maintain their original commitment. Dental pulp stem cells are multipotent stem cells derived from neural crest and they showed strong commitment into neural lineages. Therefore, neural differentiation is widely presented in all iPSC-derived from different types of dental stem cells. Under appropriate neurogenic culture medium, the EBs developed into the cells with neural-like morphology (Fig. 6), which express such markers as nestin, β-tubulin III (TUJ1), neuron-specific Enolase (NSE) and glial fibrillary acidic protein (GFAP) (Fig. 6).

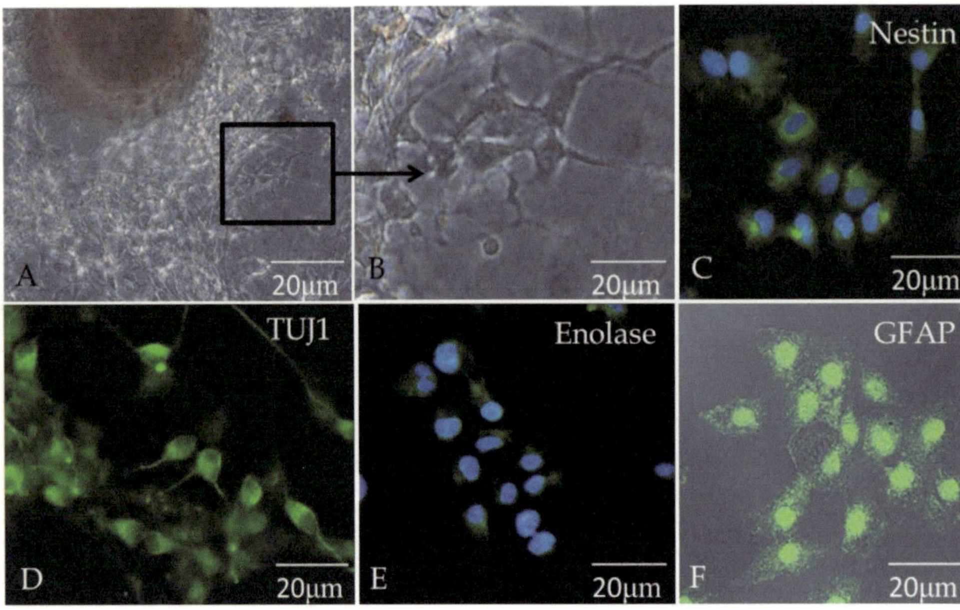

Figure 6. Representative figure of *in vitro* differentiation of hIDPS-iPSC. A) EBs adherent on Petri dish showing differentiation into neural-like cells after culturing in neurobasal+B27 medium. B) Same as in (A) showing neural-like cells in high magnification. C) Neural-like cells present positive immunostaining with anti-nestin antibody. D) positive immunostaining with anti-TUJ1 antibody. E) Neuron-specific enolase positive immunostaining (green) in hIDPS-iPSC derived neuronal cells. F) Morphological presentation of glial-like cells derived from hIDPS-iPSC. Positive immunostaining for (F) anti-GFAP antibody in glial-like cells. Nucleus stained with DAPI (blue). A, B= Differential interference contrast (DIC). C-E=Epi. F= DIC+Epi. Scale Bars: A-F=20µm.

2.5.2. Teratoma formation

To test the pluripotency, iPSC were injected into the testis or intramuscularly into the right and/or left hind leg of severe combined immunodeficient (SCID) mouse. Teratomas formation by SHED/SCAP/DPSCs/DPCs/MStCs–iPSC occurs of nine to eleven weeks after injection and histological examination of the tumor shows representative tissues of three embryonic germ layers, such as gut-like epithelium (endoderm), cartilage (mesoderm), and neuroepithelial rosettes (ectoderm) (Yan et al., 2010; Tamaoki et al., 2010; Oda et al., 2010). Teratomas obtained from IDPSC-iPSC were formed between 5 and 7 weeks after reprogrammed cells injection. The mice injected with parental IDPSC, as expected, did not form teratomas. We observed that teratomas were composed by tissues originated from three primary germ layers. Histological characterization of tumor masses showed that these teratomas includes ectodermal: primitive neural tissues, including neural tube and neural rosettes and retinal epithelium; mesodermal: muscle-like cells and gromerulus-like structures and endodermal tissues: respiratory or gastro-intestinal-like epithelium and glandular-like tissue formation (Fig. 7). Similar to *in vitro* differentiation, teratomas derived from IDPSC-iPS cells, display strong neuronal commitment forming rosette-, neurosphere- and neural tube-like structures. Neuron-specific enolase, which is a marker of

http://dx.doi.org/10.5772/55856

neurons and peripheral neuroendocrine tissue, and synaptophysin (a synaptic vesicle glycoprotein), which is found in neuroendocrine cells as well as virtually in all neurons that participate in synaptic transmission in the brain and spinal cord are expressed in neuronal cells produced by IDPSC-iPSC in teratomas (Beltrão-Braga et al., 2011).

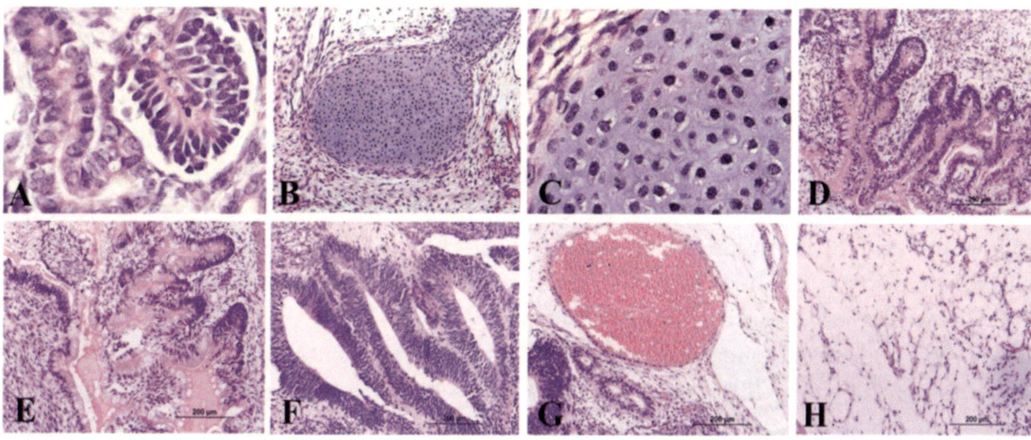

Figure 7. HE stained differentiated tissues from hIDPS-iPSC teratoma seven weeks after transplantation into nude mice right limb. A) Morphogenesis observed during differentiation of hIDPS-iPSC: glomerulus-like structure formation, with Bowman's capsule and convoluted tubule–like structures. B) Cartilage and in (C) Condrocytes-like cells (higher magnification). D) Respiratory-like epithelium. E) Gastrointestinal-like epithelium. F) Neural tubes-like structures. G) Blood vessel. H) Adipose-like tissue. Magnifications: A, C) 100x, B, D-G) 20x, Scale Bar (H) = 200μm.

2.6. HLA typing

Tamaoki and co-wokers (2010) determined the human leukocyte antigens (HLA) types of 107 dental pulp cells lines in the Japanese population and identified 2 cell lines with homozygous HLA types at all 3 loci (A, B, and DR) examined. They showed that in the Japanese population the frequencies of haplotypes of these 2 homozygous cell lines were estimated to be 8.7% and 1.5%, data provided by the Japanese Red Cross Society (http://www.bmdc.jrc.or.jp/stat.html). Using these frequencies, the coverage rate for a perfect match of iPSC lines, which were established from these 2 lines was calculated. The authors showed that iPSC lines established from these 2 homozygous cell lines would cover 16.6% and 3.0% of the Japanese population, respectively, which corresponds to approximately 20% of the Japanese population.

2.7. SHED/SCAP/DPSCs/DPCs/MStCs/IDPSC-iPSC bank and therapeutic use

One of the major challenge of pluripotent stem cells use in cell therapies is an immune-mediated rejection after transplantation. Today, this problem can be overcome by direct reprogramming of patients somatic cells and by creating an iPSC bank consisting of various HLA types thus providing therapeutic tool for the patients, which need cell transplantation free from immune-mediated rejection. Two works reported that the establishment of 50 unique stem cells lines, having homozygous alleles of the 3 HLA loci (A, B, and DR), would cover ~

90% of the Japanese population with a faultless match of these loci (Nakajima et al., 2007; Nakatsuji et al., 2008). Considering that iPSC derivation is a time consuming process and of elevated cost, it should be necessary for cell therapies and regenerative medicine to establish iPSC banks with a sufficient collection of HLA types, thus avoiding additional costs which are required for iPSC production for each individual patient.

2.7.1. Requirements of iPSC generation

In spite of optimistic prognosis in respect of how many iPSC should be produced in order to satisfy their immunological matching within definite human population, several requirements must be challenged before establishing iPSC bank. The principal requirement is a method of reprogramming, which should be safe. Therefore, three major concerns exist in the current reprogramming strategies for clinical applications: (i) the low reprogramming efficiency of human somatic cells makes it difficult to generate patient-specific iPSC, when a small amount of the cells of the patient is used; (ii) carcinogenesis may be caused by genomic integration of retro- or lentiviral fragments into host DNA; and (iii) Myc is an oncogene, which after reactivation might cause malignant tumor formation. Whereas iPSC can be generated by three transcription factors (Oct3/4, Sox2, and Klf4) without Myc, reprogramming efficiency are significantly reduced. Although, several methods of iPSC generation without viral integration have been reported; their efficiencies are extremely low in comparison with viral vectors used for induction of reprogramming (Okita et al., 2008; Stadtfeld et al., 2008; Fusaki et al., 2009; Kaji et al., 2009; Kim et al., 2009; Soldner et al., 2009; Woltjen et al., 2009; Yu et al., 2009; Yusa et al., 2009; Zhou et al., 2009). Next important issue is availability of donor cells, which can provide high efficiency in the generation of non-integrated human iPSC. Therefore, source of the cells also makes its own demand, such as, it should be easily accessible with minimum discomfort for the patient, the procedure of stem cell isolation should be non-invasive, the tissue should be easily processed, the cells should be rapidly proliferating and produced in sufficient quantities, these cells should be young and collected from healthy volunteers. Furthermore, the possibility of genetic abnormalities in donor cells due to ultraviolet (UV) irradiation should be minimized and finally, these cells would be able to be stored in liquid nitrogen for a long time without the loss of their prime characteristics.

Dental pulp stem cells from deciduous and wisdom teeth are an ideal source that meets the majority of aforementioned requirements. The loss of baby (deciduous) teeth occurs naturally and they can be removed with minimal discomfort to the patient during a routine visit to the dentist, in many clinics, as well as wisdom teeth. We also showed that not only cells, but also dental pulp can be cryopreserved and new cells can be obtained later, after thawing (Lizier et al., 2012). Therefore, frozen dental pulp does not require *in vitro* cultivation in order to produce the cells of donor until he needs these cells for iPSC production and/or clinical treatment. Additionally, several dental pulps from the same individual can be cryopreserved. This elevates a probability of successful MSCs isolation in high quantities. These cells are safer, once they can be used by first and second degree relatives and within all family (Kerkis and Caplan, 2012; Lizier et al., 2012).

2.7.2. Perspectives of iPSC therapeutic use

Currently, iPSC are used to understand human diseases, including Alzheimer's disease, Parkinson's disease, cardiovascular disease, diabetes, and amyotrophic lateral sclerosis (ALS), to develop and screen bioactive molecules - candidate to therapeutic drugs and to identify molecules or genes implicated in tissue regeneration. These *in vitro* studies enable researchers to understand fundamental principles of iPSC function and differentiation, which further will provide knowledge, necessary for therapeutic use of iPSC. Based on this knowledge multiple pre-clinical and clinical protocols will be produced optimizing iPSC transplantation in diverse animal and human diseases thus becoming a tool in cell-replacement therapy. Therefore, iPSC in the future may have tremendous clinical potential when highly efficient and safe protocols of generation of reprogrammed stem cells will be developed. Increasing our understanding of the molecular mechanisms that underlie reprogramming, we will be able to identify the cell types and methods of reprogramming, which will minimize DNA alterations, and conditions of iPSC cultivation that will allow widespread use of these cells in clinic. The scientists, however, should answer the question if iPSC are truly equivalent to human ES cells. Although iPSC potential for regenerative medicine is great, our current knowledge about iPSC variability, and utility must also increase greatly before iPSC became a standard tool for regenerative medicine.

3. Final considerations

Different dental tissues, which include apical papilla, primary exfoliated deciduous and permanent teeth, as well as wisdom teeth were used to derivate iPSC. The data obtained by different authors indicate that these tissues can be easily isolated and MSCs cells in sufficient quantities can be obtained. MSCs *in vivo* are reversibly arrested cells, which are localized in their niches maintaining their temporarily quiescent state. They differ from terminally differentiated cells by developmental path that involves a set of increasingly committed stages of specialization. The fact that these cells are undifferentiated cells suggests that their reprogramming will occur more easily than that of terminally differentiated cells. Indeed, all studies demonstrated that in SHED/SCAP/DPSCs/DPCs/MStCs/IDPSC the reprogramming process occurs more easily than in human fibroblasts used as a control, under the similar protocols and vectors used for transduction, which showed to be efficient. It was also possible to reprogram MSCs from wisdom teeth using only three factors (without c-Myc) avoiding future implications with potential risks of oncogene use (Yu et al., 2007). For the clinical applications of stem cells xenogeneic reagents pose the risk of a severe immune response, and the transmission of viral or bacterial infections, prions, and unidentified zoonosis. We demonstrated that isolation of IDPSC-iPSC can occur under feeder-free conditions on matrigel-coated dishes. Clinical stem cell therapy trials are ongoing, which request a strong focus on the safety and quality of *in vitro* expanded stem cell transplants. By replacing xenogeneic products with a defined xeno-free medium, the safety and quality of the cells with therapeutic potential may be enhanced significantly. Similar to human ES cells, dental MSCs derived iPSC form compact colony and retain immortal growth characteristics in culture. They express markers characteristic of pluripotency including Nanog, Oct-4, SSEA-3, SSEA-4, TRA-1-60, TRA-1-81, exhibit high telomerase activity and have stable karyotype. Controversies were observed in methylation status of CpG in the promoter regions of Nanog and Oct4 in dental tissue derived iPSC. Thus iPSC-derived from MStCs from wisdom teeth have their promoters of Oct4 and Nanog highly unmethylated after reprogramming (Yan et al., 2010; Oda et al., 2010). The SHED-/DPSC-iPSC had less methylated sites of Oct4 promoter than the non-transduced cells (Yan et al., 2010). What ensues at the molecular level during the reprogramming process, however, is not fully understood and is the current focus in iPSC research (Amabile and Meissner, 2009). Furthermore, dental tissue derived iPSC exhibits differentiation potential like human ES cells and can differentiate *in vitro* and *in vivo* into cells of all three primary germ layers (Takahashi et al., 2007; Yu et al., 2007).

In cell types with endogenously expression of one or more of the factors that induce pluripotency, such as neural cells that strongly express Sox2, pluripotency may be induced more easily or even with only a subset of factors (de Souza, 2010). In accordance, we observed that in hIDPSC, which express these factors, but at low level, the reprogramming was speedier, when compared with other dental tissue derived stem cells. During reprogramming, the integration the 4 factors into the genome of the transduced SHED/SCAP/DPSC-iPSC occurred (Yan et al., 2010). Currently, the nonintegrating reprogramming approaches, which include adenoviruses, plasmid- and episomal vector-based methods, and delivery of reprogramming factors directly as proteins have been developed. Additionally, other factors have been identified that can substitute the four Yamanaka's traditional transcription factors. Thus, Klf227 and Klf5 can replace Klf4, Sox1 and Sox3 can replace Sox2, and n-Myc and I-Myc can replace c-Myc (Nakagawa et al., 2008). Nr5a2 (Nuclear receptor subfamily 5, group A, member 2) can be used to substitute Oct-4 in the reprogramming of murine somatic cells (Heng et al., 2010). Some small molecules as the histone deacetylase inhibitor valproic acid can replace Klf4 and c-Myc for reprogramming human fibroblasts (Huangfu et al., 2008; Lin et al., 2009).

The creation of patient-specific stem cell lines is relevant for the study of basic biology, molecular mechanisms of various diseases, for drug discovery and for treating a number of human degenerative diseases without evoking immune rejection. HLA typing of DPC lines (Tamaoki et al., 2010) is of extreme importance because allows to limit the number of human iPSC, which should be obtained for each definite human population, thus avoiding unnecessary elevated costs of iPSC for cell therapies and regenerative medicine. So far, human iPSC have been used for the study of the reprogramming process itself and establishment of disease-specific cell lines and the differentiation of these cell lines into the different cell types affected by the disease, such as, spinal motor neurons, dopaminergic neurons and cardiomyocytes derived from patients suffering from amyotrophic lateral sclerosis (Dimos et al., 2008), spinal muscular atrophy (Ebert et al., 2009), sporadic Parkinson's disease (Soldner et al., 2009). Exploration of iPSC is still in its infancy, and understanding the true potential of these cells requires continued research, comprehension and profound comparisons with human ES cells.

http://dx.doi.org/10.5772/55856

Acknowledgements

The authors thank Dr. Alysson Muotri from University of California San Diego, Dr. Patricia Beltrao-Braga and Dr. Paulo Maiorka from University of Sao Paulo for their contribution in our research.

Author details

Nelson F. Lizier, Irina Kerkis and Cristiane V. Wenceslau

Laboratory of Genetics, Butantan Institute and Department of Morphology and Genetics, Federal University of Sao Paulo, Brazil

References

[1] Aasen T, Raya A, Barrero MJ, Garreta E, Consiglio A, Gonzalez F, Vassena R, Bilić J, Pekarik V, Tiscornia G, Edel M, Boué S, Izpisúa Belmonte JC. (2008) Efficient and rapid generation of induced pluripotent stem cells from human keratinocytes. Nat Biotechnol. Nov; 26(11):1276-84. Epub 2008 Oct 17.

[2] Amabile G, Meissner A. (2009) Induced pluripotent stem cells: current progress and potential for regenerative medicine. Trends Mol Med. Feb; 15(2):59-68. Epub 2009 Jan 21.

[3] Banito A, Rashid ST, Acosta JC, Li S, Pereira CF, Geti I, Pinho S, Silva JC, Azuara V, Walsh M, Vallier L, Gil J. (2009) Senescence impairs successful reprogramming to pluripotent stem cells. Genes Dev. Sep 15; 23(18):2134-9. Epub 2009 Aug 20.

[4] Beltrão-Braga PC, Pignatari GC, Maiorka PC, Oliveira NA, Lizier NF, Wenceslau CV, Miglino MA, Muotri AR, Kerkis I. (2011) Feeder-free derivation of induced pluripotent stem cells from human immature dental pulp stem cells. Cell Transplant.

[5] Bilic J, Izpisua Belmonte JC. (2012) Concise review: Induced pluripotent stem cells versus embryonic stem cells: close enough or yet too far apart? Stem Cells. Jan; 30(1): 33-41.

[6] Caplan AI. (1991) Mesenchymal stem cells. J Orthop Res. Sep; 9(5):641-50.

[7] Cho YW, Hong T, Hong S, Guo H, Yu H, Kim D, Guszczynski T, Dressler GR, Copeland TD, Kalkum M, Ge K. (2007) PTIP associates with MLL3- and MLL4-containing histone H3 lysine 4 methyltransferase complex. J Biol Chem. Jul 13; 282(28): 20395-406. Epub 2007 May 11.

[8] Conkright, M. D., Wani, M. A., Anderson, K. P., and Lingrel, J. B. (1999) A gene encoding an intestinal-enriched member of the Krüppel-like factor family expressed in intestinal epithelial cells. Nucleic Acids Res. 27, 1263–1270

[9] de Souza N. (2010) Primer: induced pluripotency. Nat Methods. Jan; 7(1):20-1.

[10] Dimos JT, Rodolfa KT, Niakan KK, Weisenthal LM, Mitsumoto H, Chung W, Croft GF, Saphier G, Leibel R, Goland R, Wichterle H, Henderson CE, Eggan K. (2008) Induced pluripotent stem cells generated from patients with ALS can be differentiated into motor neurons. Science. Aug 29; 321(5893):1218-21.

[11] Dupin E, Calloni G, Real C, Gonçalves-Trentin A, Le Douarin NM. (2007) Neural crest progenitors and stem cells. C R Biol. Jun-Jul; 330(6-7):521-9.

[12] Ebert AD, Yu J, Rose FF Jr, Mattis VB, Lorson CL, Thomson JA, Svendsen CN. (2009) Induced pluripotent stem cells from a spinal muscular atrophy patient. Nature. Jan 15; 457(7227):277-80. Epub 2008 Dec 21.

[13] Friedenstein AJ, Gorskaja JF, Kulagina NN. (1976) Fibroblast precursors in normal and irradiated mouse hematopoietic organs. Exp Hematol. Sep; 4(5):267-74.

[14] Fusaki N, Ban H, Nishiyama A, Saeki K, Hasegawa M. (2009) Efficient induction of transgene-free human pluripotent stem cells using a vector based on Sendai virus, an RNA virus that does not integrate into the host genome. Proc Jpn Acad 85: 348–362.

[15] Giordano G, La Monaca G, Annibali S, Cicconetti A, Ottolenghi L. (2011) Stem cells from oral niches: a review. Ann Stomatol (Roma). Jan; 2(1-2):3-8. Epub 2011 Jul 18.

[16] Heng JC, Feng B, Han J, Jiang J, Kraus P, Ng JH, Orlov YL, Huss M, Yang L, Lufkin T, Lim B, Ng HH. (2010) The nuclear receptor Nr5a2 can replace Oct4 in the reprogramming of murine somatic cells to pluripotent cells. Cell Stem Cell. Feb 5; 6(2): 167-74. Epub 2010 Jan 21.

[17] Huangfu D, Osafune K, Maehr R, Guo W, Eijkelenboom A, Chen S, Muhlestein W, Melton DA. (2008) Induction of pluripotent stem cells from primary human fibroblasts with only Oct4 and Sox2. Nat Biotechnol. Nov; 26(11):1269-75.

[18] Ikeda E, Yagi K, Kojima M, Yagyuu T, Ohshima A, Sobajima S, Tadokoro M, Katsube Y, Isoda K, Kondoh M, Kawase M, Go MJ, Adachi H, Yokota Y, Kirita T, Ohgushi H. (2008) Multipotent cells from the human third molar: feasibility of cell-based therapy for liver disease. Differentiation. May; 76(5):495-505.

[19] Kaji K, Norrby K, Paca A, Mileikovsky M, Mohseni P, Woltjen K. (2009) Virus-free induction of pluripotency and subsequent excision of reprogramming factors. Nature 458: 771–775.

[20] Kerkis I, Caplan AI. (2012) Stem cells in dental pulp of deciduous teeth. Tissue Eng Part B Rev. Apr;18(2):129-38

[21] Kerkis I, Kerkis A, Dozortsev D, Stukart-Parsons GC, Gomes Massironi SM, Pereira LV, Caplan AI, Cerruti HF (2006)Isolation and characterization of a population of im-

mature dental pulp stem cells expressing OCT-4 and other embryonic stem cell markers. Cells Tissues Organs; 184(3-4):105-16.

[22] Kim D, Kim CH, Moon JI, Chung YG, Chang MY, Han BS, Ko S, Yang E, Cha KY, Lanza R, et al. (2009) Generation of human induced pluripotent stem cells by direct delivery of reprogramming proteins. Cell Stem Cell 4: 472–476.

[23] Le Douarin NM, Brito JM, Creuzet S (2007) Role of the neural crest in face and brain development. Brain Res Rev. Oct; 55(2):237-47.

[24] Le Douarin NM, Calloni GW, Dupin E (2008) The stem cells of the neural crest. Cell Cycle 7: 1013–9.

[25] Le Douarin NM, Creuzet S, Dupin E (2004) Neural crest cell plasticity and its limits. Development 131: 4637-50

[26] Le Douarin NM, Dupin E (2003) Multipotentiality of the neural crest. Curr Opin Genet Dev 13: 529–36.

[27] Le Douarin NM, Dupin E. (2012)The neural crest in vertebrate evolution. Curr Opin Genet Dev. Jul 6. [Epub ahead of print]

[28] Lewitzky M, Yamanaka S. (2007) Reprogramming somatic cells towards pluripotency by defined factors. Curr Opin Biotechnol. Oct; 18(5):467-73.

[29] Lin T, Ambasudhan R, Yuan X, Li W, Hilcove S, Abujarour R, Lin X, Hahm HS, Hao E, Hayek A, et al. (2009) A chemical platform for improved induction of human iPSCs. Nat Methods 6: 805–808.

[30] Liu SV (2008) iPS cells: a more critical review. Stem Cells Dev. Jun;17(3):391-7.

[31] Lizier NF, Kerkis A, Gomes CM, Hebling J, Oliveira CF, Caplan AI, Kerkis I (2012) Scaling-up of dental pulp stem cells isolated from multiple niches. PLoS One; 7(6):e39885. Epub 2012 Jun 29.

[32] Maherali N, Hochedlinger K (2008) Induced pluripotency of mouse and human somatic cells. Cold Spring Harb Symp Quant Biol. 8; 73:157-62.

[33] Marchetto MC, Yeo GW, Kainohana O, Marsala M, Gage FH, Muotri AR. (2009) Transcriptional signature and memory retention of human-induced pluripotent stem cells. PLoS One. Sep 18; 4(9):e7076.

[34] McDevitt TC, Palecek SP (2008) Innovation in the culture and derivation of pluripotent human stem cells. Curr Opin Biotechnol. Oct; 19(5):527-33

[35] Miura, M., Gronthos, S., Zhao, M., Lu, B., Fisher, L.W., et al. (2003) SHED: Stem cells from human exfoliated deciduous teeth. Proc Natl Acad Sci U S A 100, 5807.

[36] Muchkaeva IA, Dashinimaev EB, Terskikh VV, Sukhanov YV, Vasiliev AV (2012) Molecular mechanisms of induced pluripotency. Acta Naturae. Jan; 4(1):12-22.

[37] Murata K, Kouzarides T, Bannister AJ, Gurdon JB (2010) Histone H3 lysine 4 methylation is associated with the transcriptional reprogramming efficiency of somatic nuclei by oocytes. Epigenetics Chromatin. Feb 4; 3(1):4.

[38] Nakagawa M, Koyanagi M, Tanabe K, Takahashi K, Ichisaka T, Aoi T, Okita K, Mochiduki Y, Takizawa N, Yamanaka S (2008) Generation of induced pluripotent stem cells without Myc from mouse and human fibroblasts. Nat Biotechnol. Jan; 26(1): 101-6. Epub 2007 Nov 30.

[39] Nakajima F, Tokunaga K, Nakatsuji N (2007) Human leukocyte antigen matching estimations in a hypothetical bank of human embryonic stem cell lines in the Japanese population for use in cell transplantation therapy. Stem Cells. Apr; 25(4):983-5. Epub 2006 Dec 21.

[40] Nakatsuji N, Nakajima F, Tokunaga K (2008) HLA-haplotype banking and iPS cells. Nat Biotechnol. Jul; 26(7):739-40.

[41] Nishikawa S, Goldstein RA, Nierras CR (2008) The promise of human induced pluripotent stem cells for research and therapy. Nat Rev Mol Cell Biol. Sep; 9(9):725-9.

[42] Oda Y, Yoshimura Y, Ohnishi H, Tadokoro M, Katsube Y, Sasao M, Kubo Y, Hattori K, Saito S, Horimoto K, Yuba S, Ohgushi H (2010) Induction of pluripotent stem cells from human third molar mesenchymal stromal cells. J Biol Chem. Sep 17; 285(38): 29270-8.

[43] Ohnuki M, Takahashi K, Yamanaka S (2009) Generation and characterization of human induced pluripotent stem cells. Curr Protoc Stem Cell Biol. Jun; Chapter 4: Unit 4A.2.

[44] Okita K, Nakagawa M, Hyenjong H, Ichisaka T, Yamanaka S (2008) Generation of mouse induced pluripotent stem cells without viral vectors. Science. Nov 7; 322(5903):949-53. Epub 2008 Oct 9.

[45] Ooi L, Sidhu K, Poljak A, Sutherland G, O'Connor MD, Sachdev P, Münch G (2012) Induced pluripotent stem cells as tools for disease modelling and drug discovery in Alzheimer's disease. J Neural Transm. Jun 13.

[46] Patel SR, Kim D, Levitan I, Dressler GR (2007) The BRCT-domain containing protein PTIP links PAX2 to a histone H3, lysine 4 methyltransferase complex. Dev Cell. Oct; 13(4):580-92.

[47] Shields, J. M., Christy, R. J., and Yang, V. W. (1996) Identification and characterization of a gene encoding a gut-enriched Krüppel-like factor expressed during growth arrest. J. Biol. Chem. 271, 20009–20017.

[48] Soldner F, Hockemeyer D, Beard C, Gao Q, Bell GW, Cook EG, Hargus G, Blak A, Cooper O, Mitalipova M, Isacson O, Jaenisch R (2009) Parkinson's disease patient-derived induced pluripotent stem cells free of viral reprogramming factors. Cell. Mar 6; 136(5):964-77.

[49] Stadtfeld M, Nagaya M, Utikal J, Weir G, Hochedlinger K (2008) Induced pluripotent stem cells generated without viral integration. Science 322: 945–949.

[50] Takahashi K, Tanabe K, Ohnuki M, Narita M, Ichisaka T, Tomoda K, Yamanaka S. (2007) Induction of pluripotent stem cells from adult human fibroblasts by defined factors. Cell 131: 861–872.

[51] Takahashi K, Yamanaka S (2006) Induction of pluripotent stem cells from mouse embryonic and adult fibroblast cultures by defined factors. Cell. 126(4): 663-76.

[52] Takeda T, Tezuka Y, Horiuchi M, Hosono K, Iida K, Hatakeyama D, Miyaki S, Kunisada T, Shibata T, Tezuka K. (2008) Characterization of dental pulp stem cells of human tooth germs. J Dent Res. Jul; 87(7):676-81.

[53] Tamaoki N, Takahashi K, Tanaka T, Ichisaka T, Aoki H, Takeda-Kawaguchi T, Iida K, Kunisada T, Shibata T, Yamanaka S, Tezuka K (2010) Dental pulp cells for induced pluripotent stem cell banking. J Dent Res. Aug; 89(8):773-8.

[54] van Attikum H, Gasser SM (2009) Crosstalk between histone modifications during the DNA damage response. Trends Cell Biol. May; 19(5):207-17. Review.

[55] Woltjen K, Michael IP, Mohseni P, Desai R, Mileikovsky M, Hamalainen R, Cowling R, Wang W, Liu P, Gertsenstein M, Stadtfeld and Hochedlinger et al. (2009) piggyBac transposition reprograms fibroblasts to induced pluripotent stem cells. Nature 458: 766–770

[56] Yan, X., Qin, H., Qu, C., Tuan, R.S., Shi, S., and Huang, G.T (2010) iPS cell reprogrammed from human mesenchymal-like stem/progenitor cells of dental tissue origin. Stem Cells Dev 19, 469.

[57] Yu J, Hu K, Smuga-Otto K, Tian S, Stewart R, Slukvin II, Thomson JA (2009) Human induced pluripotent stem cells free of vector and transgene sequences. Science 324: 797–801.

[58] Yu J, Thomson JA. (2008) Pluripotent stem cell lines. Genes Dev.; 22(15):1987-97.

[59] Yu J, Vodyanik MA, Smuga-Otto K, Antosiewicz-Bourget J, Frane JL, Tian S, Nie J, Jonsdottir GA, Ruotti V, Stewart R, et al. (2007) Induced pluripotent stem cell lines derived from human somatic cells. Science 318: 1917–1920.

[60] Yusa K, Rad R, Takeda J, Bradley A (2009) Generation of transgene-free induced pluripotent mouse stem cells by the piggyBac transposon. Nat Methods 6: 363–369.

[61] Zhao R, Daley GQ (2008) From fibroblasts to iPS cells: induced pluripotency by defined factors. J Cell Biochem.; 105(4): 949-55

[62] Zhou W, Freed CR (2009) Adenoviral gene delivery can reprogram human fibroblasts to induced pluripotent stem cells. Stem Cells 27: 2667–2674.

[63] Zouboulis CC, Adjaye J, Akamatsu H, Moe-Behrens G, Niemann C (2008) Human skin stem cells and the ageing process. Exp Gerontol. Nov; 43(11):986-97.

Section 2

Mechanistic Underpinning

Chapter 8

Molecular Mechanisms Underlying Pluripotency

Erhard Bieberich and Guanghu Wang

Additional information is available at the end of the chapter

http://dx.doi.org/10.5772/55596

1. Introduction

Pluripotency and self renewal are the two primary characteristics of pluripotent stem cells (PSCs) [1]-[4]. **Pluripotency** refers to the capacity of a single cell to give rise to any cell type of an embryo or an adult animal [5],[6]. A mammalian organism is developed from a single fertilized egg, the zygote, in an extremely ordered and error-proof fashion [7]. The zygote and the subsequent 2 to 4-cell stage blastomere are considered to be **totipotent** since they can give rise to the entire fetus, including the embryo and the extra embryonic tissue such as the placenta and the umbilical cord (Fig.1) [7]. As embryo development proceeds to 8-cell stage and beyond depending on the species, the cells in the blastomere gradually lose their totipotency. At about embryonic day 3.5 (E3.5) in mouse (about E5 in human) the blastomere compacts into a blastocyst in which two distinct cell populations reside. Cells in the outer layer of the blastocyst form the trophectoderm (TE) which eventually give rise to the extra embryonic tissue, trophoblast of the placenta, whereas cells in the inside of the blastocyst form the inner cell mass (ICM). The ICM then gives rise to additional two lineages of cells, the primitive endoderm (PrEn or hypoblast) and the primitive ectoderm (PrEc or epiblast) (Fig.1). The PrEn produces the secondary extra embryonic tissues, such as yolk sac, allantois and amnion, while the PrEc gives rise to all three germ layers of the embryo, namely the ectoderm, the mesoderm and the endoderm (Fig.1). Although the extra embryonic tissues are indispensible for mammalian embryonic development, it is the ICM derived PrEc (or epiblast) cells that form all the cells of an embryo and adult animal, thus these cells are defined as **pluripotent** [5],[8].

As the embryo implants into the uterus and development further commences to E5-E6.5 days in mouse, some of the post-implantation epiblast cells are found to maintain the capability of producing all derivatives of the three embryonic germ layers [1],[5]. The difference between cells derived from ICM of the pre-implantation blastocyst and those from post-plantation epiblast is that the ICM derived cells express stage-specific embryonic antigen 1 (SSEA1), give rise to all three embryonic germ layers, and most importantly, contribute to chimeric mouse and germ line transmission, while those from post-implantation epiblast do not express SSEA1 and do not contribute to chimeric mouse and germ line [1],[5],[9]-[11]. Thus the ICM derived cells are defined to be in a "naive (or ground, primordial)" state of pluripotency, and those from post-implantation epiblast are defined to be in a "primed (or refined)" state of pluripotency [1],[5],[12],[13]. Other *in vivo* sources of pluripotent cells include the germline cells extracted either from embryonic or adult male reproduction organs [5],[8] (Fig. 1).

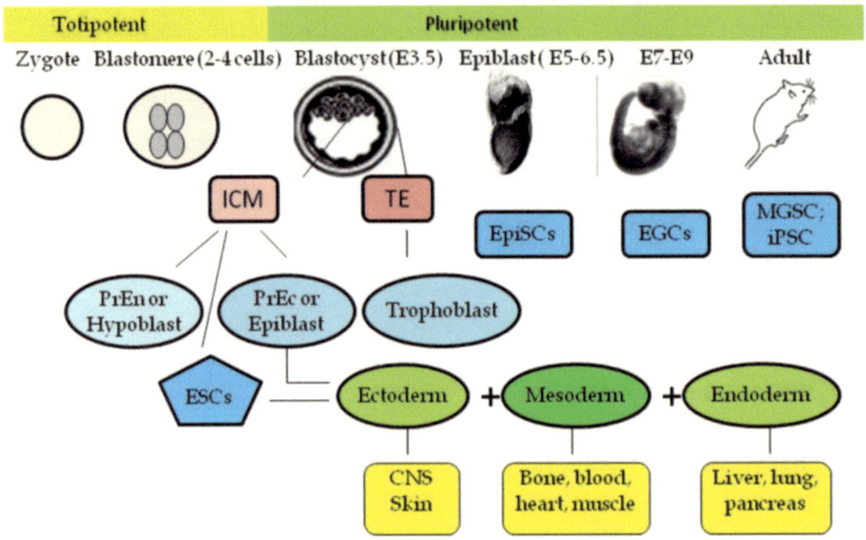

Figure 1. Early embryonic development and sources of PSCs. ICM, inner cell mass; TE, trophectoderm or Trophoblast; EpiSCs, epiblast stem cells; EGC, embryonic germ cells; iPSCs, induced pluripotent stem cells; MGSCs, male germ stem cells; PrEn, primitive endoderm; PrEc, primitive ectoderm; ESCs, embryonic stem cells.

Both the innate totipotency and pluripotency are transient developmental stages in the beginning of embryogenesis [5],[7]. Because of their finite number and transient nature, these cells are very challenging to study, although scientists have showed immense interest to understand them since they hold key answers to many aspects of biology and life.

Intriguingly, pluripotency can be captured or induced in cell cultures with defined growing conditions [14]-[16]. Mouse ESCs (mESCs) are one of the first and best-established ICM-derived cells (Fig.1 and Table 1). Well defined culture conditions allow mESCs to self renew infinitely while maintaining a pluripotent state *in vitro*, providing an invaluable source of cells for molecular studies and differentiation into a variety of desired cell types (Table 1) [17]-[21].

Despite the genetic similarities between mouse and human, it was until two decades later that the first human ESCs (hESCs) were established in cell culture [9]. These cells give rise to all lineages of the primary germ layers and form teratomas (Table 1). Although they were also derived from ICM of pre-implantation embryos, the hESCs demonstrate many striking differences from mESCs (discussed in detail below). Since germ line transmission studies cannot be performed with these cells, it is not known at what exact pluripotency state the hESCs are. However, hESCs resemble more closely mouse epiblast stem cells (mEpiSCs), which were derived from post-implantation epiblast [16],[22], indicating that the hESCs are probably in a primed state of pluripotency (Table 1).

Decoding the molecular basis of pluripotency and self renewal is fundamental to the understanding of stem cell biology, embryonic development, and clinical application of regenerative medicine. The *in vitro* culture of these PSCs, especially those from induced pluripotent stem cells (iPSCs), has provided an unprecedented tool to investigate deeper into the molecular mechanism governing pluripotency. Gradually we have uncovered that pluripotency is regulated by a complex network of factors, including transcription factors and epigenetic regulators, which trigger multiple signaling transduction pathways, such as the TGF-β pathway and Wnt pathway. Since the concise molecular mechanism controlling pluripotency varies among the different kinds of PSCs, we will first give a brief introduction of their properties.

	mECCs	mESCs	miPSCs	mEpiSCs	hESCs	hiPSCs
Origin	Teratoma	ICM of Blastocyst	Somatic cells	Late epiblast	ICM of Blastocyst	Somatic cells
Teratoma formation	Yes	Yes	Yes	Yes	Yes	Yes
Chimera and germ line contribution	Yes	Yes	Yes	No	Not determined	Not determined
Culture conditions	LIF, FBS	LIF, BMP4	LIF	Fgf2, Activin	Fgf2, Activin, MEF CM	Fgf2, Activin, MEF CM
Morphology	Domed shape	Domed shape	Domed shape	Flat	Flat	Flat
X chromosome	XaXa	XaXa	XaXa	XaXi	XaXi	XaXi
Pluripotency status	Naive state			Primed state	Not determined, possibly primed state	
Pluripotency factors	Oct4, Nanog, Sox2, Stat3, Klf2, Klf4,			Oct4, Nanog, Sox2		
Response to LIF	Self renewal and pluripotency			None		
Response to Fgf2	Differentiation			Self renewal and pluripotency		
Response to 2i	Self renewal and pluripotency			Differentiation and cell death		
References	7, 17-19	10-14, 22-28	32-35	3,9,16	4,15	51, 53, 54

Table 1. Properties of some PSCs

2. Properties of PSCs

Mouse ECCs: Mouse embryonic carcinoma cells (mECCs), the first PSCs established in cell culture, were derived in 1964 from teratomas from an inbred mouse line, which produces spontaneous testicular teratomas (Table 1) [14],[23]. These cells show many aspects of naive PSCs, such as the expression of antigen SSEA1, differentiation into all three germ layers when stimulated, and generation of chimeric mice when injected into blastocysts [14],[23]-[25]. However, since the ECCs carry many mutations, the chimeric mice derived from these cells develop spontaneous tumors [26].

Mouse ESCs: Based on the findings made from mECCs, derivation of mESCs directly from ICM of normal developing embryos became possible and faster (Fig.1). Two groups, Kaufman's and Martin's, isolated such cells in culture from the ICM of pre-implantation blastocysts using different protocols in 1981 [15],[27]. And it was Martin who coined the term ESCs [27].

mESCs satisfy all the characteristics of naive pluripotency with a normal karyotype, resembling their *in vivo* counterparts in terms of expressing the pluripotency factors Oct4, Sox2, and Nanog, and SSEA1 and alkaline phosphatase (AP) (Table 1). They can be differentiated into all derivatives of the three germ layers; grow in a dome-shaped morphology; display a high nuclei/cytoplasm ratio; and most importantly, form teratomas and give rise to germ line transmission when injected into blastocysts (Table 1) [17]-[21].

The cell cycle control in mESCs also seems to be unique. They have an unusually short G1 phase and no regulation at the G1–S transition, the presence of hyperphosphorylated retinoblastoma (RB) protein, and unresponsiveness to activity of cyclin-dependent kinase 4 (CDK4) [11],[28]. Epigenetically, mESCs possess a hypermethylated genome and both X-chromosomes are activated if isolated from female embryos [29]-[31].

To grow mESCs, a feeder cell layer of mouse embryonic fibroblasts is typically used, with medium containing ES qualified fetal bovine serum or knockout serum replacement, leukemia inhibitory factor (LIF), non-essential amino acid, and β-mercaptoethanol [20],[21],[32],[33]. LIF serves as the extrinsic factor for pluripotency and self renewal by activating the signal transducer and activator of transcription 3 (Stat3) pathway [21],[34],[35]. Later it was discovered that LIF and Bone morphogenic protein 4 (BMP4) can support mESCs pluripotency without serum (Table 1) [36],[37]. More recently, it has been demonstrated that mESCs can be derived and maintained using medium containing small molecule inhibitors of glycogen synthase kinase (GSK3) (which functions to activate the Wnt signaling pathway) and the mitogen-activated protein kinase (MAPK) signaling pathways, completely devoid of serum and extrinsic growth factors [5],[38]. This culture media with defined components is termed the 2i (2 inhibitors) system [5],[38].

Mouse EpiSCs: Mouse Epiblast Stem cells (mEpiSCs) have been derived from post-implantation blastocysts (E5–E6.5) (Fig.1) [16],[22]. These cells demonstrate the properties of self-renewal and pluripotency, but they cannot colonize the ICM of a blastocyst and produce germ line transmission [13],[16],[22]. Thus mEpiSCs are in the "primed" state of pluripotency. Furthermore, mEpiSCs express high levels of Oct4, Sox2, and Nanog but relatively low levels of other pluripotency factors that have been shown to be essential for mESCs, such as Klf4 and Stella [5],[13],[16]. This suggests fundamental differences in the mechanisms that maintain pluripotency in mESCs and mEpiSCs [8],[16].

Epigenetically, mEpiSCs display X chromosome inactivation (XCI) as well as stability of the genetic imprint [3],[5],[8]. This epigenetic status is shared with the late epiblast of the post-implantation embryo, which reinforces the similarity between cultured mEpiSCs and their *in vivo* counterpart. This similarity has been confirmed by gene expression profile experiments, which show that EpiSCs are closely related to the pluripotent cells located in the epiblast of a post-implantation embryo [16],[22]

Mouse EpiSCs also differ with mESCs regarding growing conditions, phenotypes and function. To keep them in a self renewing state, activin, fibroblast growth factor 2 (Fgf2), and transforming growth factor β (TGF-β) are needed, whereas LIF is dispensable and BMP4 leads to differentiation (Table 1) [16],[22]. Instead of growing in a dome-shaped morphology as mESCs, mEpiSCs exhibit a flattened shape and do not propagate well as single cells. Whereas cell cycle regulation in mEpiSCs remains to be investigated in detail, the doubling time of mEpiSCs is 18 hours, compared with only 10–14 hours doubling time of mESCs, suggesting that a normal G1–S transition occurs in mEpiSCs [16],[22].

Human ESCs: Like mESCs, human ESCs (hESCs) were isolated from the ICM of the pre-implantation blastocyst almost two decades after the isolation of mESCs [9]. hESCs possess the potential to differentiae into all three primary germ layers and to produce teratomas when injected into blastocysts [1],[9]. They express high levels of pluripotency factors Oct4, Nanog, and Sox2, and are positive for SSEA3/4 and AP. However, hESCs share multiple defining features with mouse EpiSCs rather than mESCs. These characteristics include flat morphology, dependence on FGF2/Activin signaling to self renew, inclination for XCI, and reduced tolerance to single-cell dissociation by trypsinization (Table 1). These molecular and biological similarities with mEpiSCs suggest that hESCs correspond, at least partially, to the primed pluripotency state rather than to the naive state.

iPSCs: In 2006, Shinya Yamanaka's research group at Kyoto University made a milestone achievement by converting adult mouse cells back to a ground pluripotent stem cell-like state through exogenous expression of only four transcription factors, Oct4, Sox2, Klf4, and c-Myc [39]. These miPSCs exhibit all characteristics of mESCs, including expression of pluripotency marker protein, activation of both X chromosomes, and most importantly, the ability to generate chimeric animals and contributing to germ line transmission (Table 1). Later on, similar cells were also induced from human somatic cells. hiPSCs resemble more hESCs and mEpiSCs than mESCs (Table 1)[40]-[42]. As their production efficiency rapidly improved, iPSCs have soon been able to compete with traditional embryonic and adult stem cells [39]-[50]. The primary advantages of iPSCs compared to other stem cells are: a) iPSCs can be created from the tissue of the same patient that will receive the transplantation, thus avoiding immune rejection, and b) the lack of ethical implications because cells are harvested from a consent individual. These patient-specific cells can be used to study diseases *in vitro*, to test drugs on a human model without ethical concerns, and to hopefully be used as a source of tissue replacement for diseased and damaged cells.

3. Embryonic Germ stem Cells (mEGC) and Male Germ Stem Cell (MGSCs)

PSCs have also been derived from reproductive system cells. When cultivated in adequate growth conditions, reproductive system cells generate ES-like stem cells (it is termed embryonic germ cells (EGCs) if isolated from mouse embryonic day 8.5 embryos [51], or male germ stem cells (MGSCs) if derived from postnatal male gonads [52]) (Fig.1). The EGCs and MGSCs are both naive stem cells, capable of generating all three embryonic germ layer cells, teratomas and chimeras. EGCs and MGSCs have also been derived from human sources, but their characteristics are not as well defined [53]-[55].

Next we will focus on the mESCs, which are in the naive state, and hESCs, which are probably in the primed state, to discuss the molecular mechanism of pluripotency maintenance.

4. Transcription factors regulatating pluripotency

An interplay of transcription factors and epigenetic factors participates in the maintenance of pluripotency of stem cells [34],[35],[56]-[62]. Among them Oct4 (or POU5F1), Nanog, and Sox2 are generally accepted as the core pluripotency factors, since they are vital to maintian the pluripotency of both the hESCs and mESCs, which are in a different pluripotency state [1],[8],[63]. These three factors also collectively bind to an array of genes that are essential for pluripotency and differentiation [1],[8],[63].

4.1. Core pluripotency factors and their transcription cotrol

The POU transcription factor Oct4 is a central player for stem cell pluripotency (Fig.2). Its expression is strictly confined to the totipotent, pluripotent, and germ cells during early development. *In vitro*, the cellular level of Oct4 must be tightly controlled to maintain the pluripotency status, up- or down-regulation by 50% leads to ESC differentiation [64],[65]. *In vivo*, Oct4 deletion in mice leads to ICM failure [57].

The homeoprotein Nanog is another central factor for pluripotency (Fig.2) [66]. The ICM in Nanog-deficient mice fails to generate epiblast and only produces endoderm-like cells [66]. Furthermore, ESCs derived from Nanog-deficient mice cannot maintain pluripotency and instead differentiate into extraembryonic endoderm lineages [66]. Mechanistically, Nanog functions by inhibiting NFκB and cooperating with Stat3 to inhibit cell differentiation in mESCs [66]-[68].

The third central factor is Sox2 (Fig.2) [3],[5],[63],[69]. Sox2 exhibits an expression pattern similar to that of Oct4 during development [70]. Genetic ablation studies indicate that silencing of *Sox2* affects a somewhat later stage of embryogenesis, possibly because of a stronger maternal contribution of Sox2 protein. Key feature of acute *Sox2* loss appears to be an inability to sustain appropriate Oct4 levels [70].

http://dx.doi.org/10.5772/55596

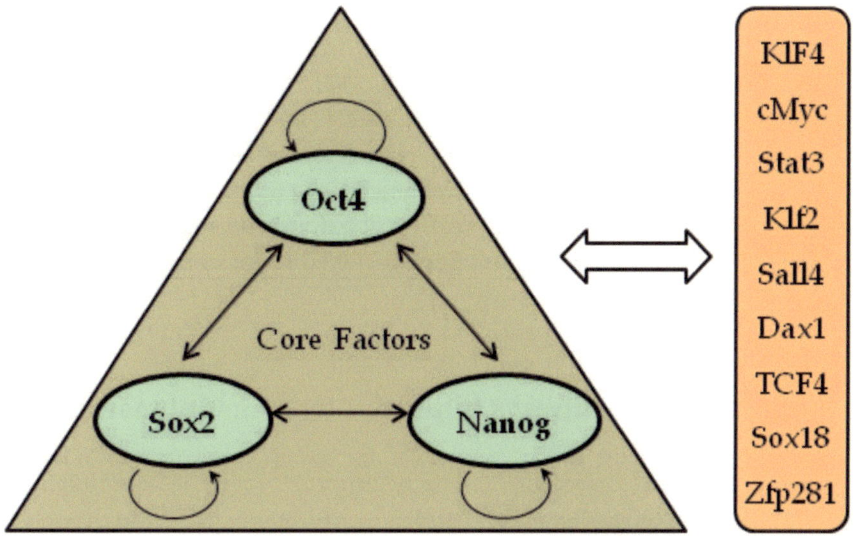

Figure 2. Schematic representation of the transcriptional regulation of core pluripotency factors and the extended factors

These three core factors do not function by themselves. Instead they are involved in a multiple-gene complex to regulate stem cell pluripotency (Fig.2). Their interacting partners have been extensively studied by coimmunoprecipitation (Co-IP) or chromatin immunoprecipitation (ChIP) assays in both mESCs and hESCs. Oct4 has been found to associate with Sox2, Nanog, Smad1, Stat3, TCF3, Rest, Hsp90, etc. [4],[21],[71]-[73]. Nanog seems to interact with Oct4, Smad1, Nac1, Zfp281, and Hsp90 [21],[72]-[74]. And Sox2 associates with Oct4, Nanog, Klf4, Rpa1, Sall4, and Npm1 [75],[76].

One unique aspect of the regulation of these core factors is that they act together to regulate their own promoters, forming an interconnected auto-regulatory feedback loop (Fig.2) [5], [51], [71]. Another unique aspect is that they co-occupy and active/enhance expression of other genes necessary to maintain ESC status, while contributing to repression of genes encoding differentiation signals (Fig.2) [5],[51],[71],[77]-[80]. For example, binding of Oct4 to a promoter region of a gene increases the likelihood of Nanog, Sox2, and other regulatory factors to bind to the same promoter [5],[51],[71],[77]-[80].

Apart from transcriptional control of these core factors, post-translational modifications also play an essential role. Oct4 has been reported to be ubiquitinated in differentiating mouse embryonic carcinoma cells but not in mESCs [81],[82]. Phosphorylation of Nanog promotes its interaction with the prolyl isomerase Pin1, leading to increased Nanog stability by suppressing its ubiquitination[83]. In addition, a recent report demonstrates that Hsp90, a molecular chaperone, associates with Oct4 and Nanog and maintains their cellular level, possibly through protecting them against degradation by the ubiquitin protesome pathway [21]. These studies demonstrate that post-translational modifications and protein stability of the pluripotency factors is also vital for stem cell pluripotency maintenance.

4.2. Extended network of regulatory factors

Besides the core pluripotency factors, many other transcription factors participate in the regulation of stem cell pluripotency, including cMyc, Klf2, Klf4, Stat3, Rex1, Sall4, Zfp281, and the proteins associated with the three core factors [1],[3],[5]. These transcription factors participate in the pluripotency regulation in a state or species-specific fashion. For example, Stat3 plays an important role in mESC pluripotency since target deletion of Stat3 resulted in early embryonic lethality, and ectopic expression of a dominant-negative Stat3 in ESCs leads to loss of pluripotency [34],[35]. However, Stat3 is not suficient to maintain the pluripotency of hESCs and mEpiSCs [21],[84].

5. Signal transduction pathways in pluripotency maintenance

Innate signal transduction pathways are crucially important for understanding the regulation of the stem cell pluripotency. Extensive efforts, including high throughput genetic and chemical screening, have been invested into identifying genes and pathways that affect the core pluripotency factors Oct4, Nanog, and Sox2, or their associated genes. We have now gradually obtained a glimpse of the intrinsic signalling pathways that are involved in the regulation of stem cell pluripotency and differentiation. These signallng pathways include the Wnt pathway, TGF-β pathway, LIF/Stat3 pathway, Fgf pathway, insulin geowth factor (IGF) pathway, Notch pathway, Hedgehog pathway, etc [3]-[5],[71],[74],[85],[86] (Fig.3). Among them, the Wnt and TGF-β pathways are most heavily studied and best understood in terms of their roles in stem cell pluripotency maintenance. Here, we will focus on discussing the Wnt pathway, TGF-β signaling, LIF/Stat3 pathway, and the Fgf pathway (Figs. 3 and 4).

5.1. WNT signaling pathway

The Wnt pathway plays an important role in tissue development by regulating a wide range of cellular processes such as proliferation, adhesion, morphology, and migration [87]-[91]. It consists of over 30 extracellular ligands that bind to Frizzled (FZD) and low-density lipoprotein receptor related protein (LRP) receptors at the cell surface (Fig.3) [91]. The Wnt ligands are able to activate both the canonical pathway and the non-canonical pathway [88],[92]. The activation of Wnt pathway in the canonical pathway results in the preservation of β-catenin and its subsequent nuclear translocation, which enables downstream gene activation by the TCF/LEF family transcription factors [87]-[95]. The non-canonical pathway is independent of β-catenin and involves the activation of several other signaling pathways, such as the JNK pathway [3]-[5],[88],[91],[92].

The Wnt signaling pathway is directly linked to the core transcriptional network of pluripotency and is demonstrated to be essential for self renewal and pluripotency of both naïve and primed PSCs, when LIF is absent. Evidences for this notion include: 1), Wnt signaling is activated in both mESCs and hESCs, and is down-regulated during differentiation [90]; 2), activation of the canonical Wnt pathway is required to maintain the expression level of core pluripotency factors Oct4 and Nanog, through which the self-renewal and pluripotency are sustained (Figs. 3 and 4); and 3), ectopic expressing of an constitutively active form of β-catenin maintains the expression levels of Oct4 and Nanog and thus self renewal and pluripotency in ESCs [96].

Moreover, Wnt signaling inhibits the differentiation of ESCs, especially to neural differentiation [90],[97],[98]. Mutation of Apc, an important mediator in the Wnt pathway, leads to impaired differentiation both *in vitro* and in teratomas [99]. Furthermore, ESCs with highly elevated β-catenin levels also have a compromised ability to differentiate [10].

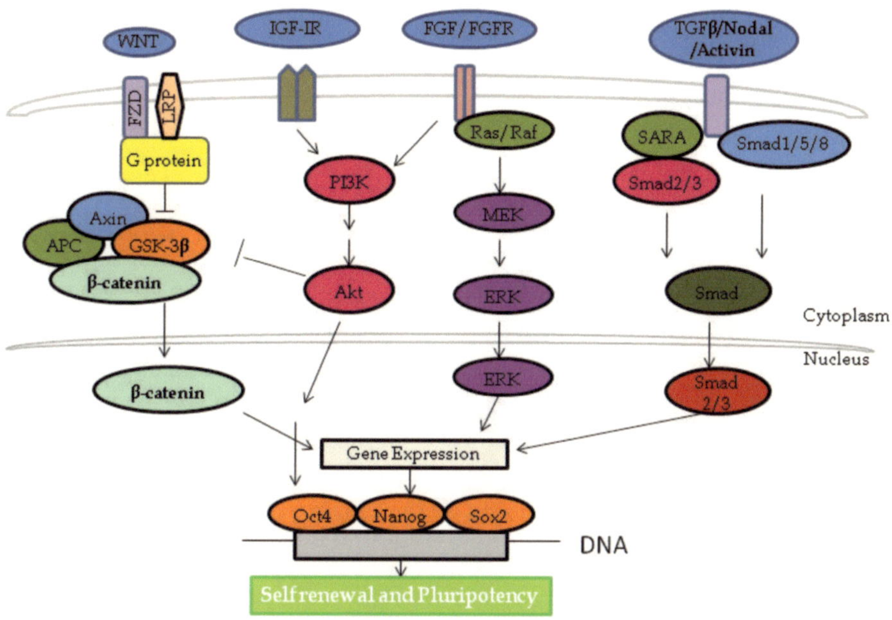

Figure 3. Signaling pathways regulating pluripotency of the primed stem cells, such as mEpiSCs and possibly hESCs

The role of the Wnt signaling in stem cell pluripotency is further confirmed by its down-stream effectors, the TCF/LEF family of transcription factors [72],[88],[92],[100],[101]. In ESCs, TCF3 is the most abundantly expressed member of this transcription factor family. TCF3-null ESCs have an increased resistance to differentiation and up-regulation of various Oct4 and Nanog-regulated genes [74],[102],[103]. Activation of Wnt converts TCF3 into an activator, elevating the expression of these same targets and suppressing differentiation [103]. TCF3 may also suppress the expression of Oct4 and Nanog, although its ability to activate these targets is unclear [104],[105].

However, Wnt signaling alone is not sufficient to support the ground state pluripotency [38]. It has been shown that inhibition of GSK3 in mESCs enhances growth capacity and suppresses neural differentiation, but it also promotes non-neural differentiation [37], [38]. To block differentiation of mESCs, the combination of a GSK3 inhibitor and an FGF-Erk inhibitor (the 2i system) [5],[38] is necessary.

5.2. TGF-β signaling pathway

The TGF-β signaling pathway plays a vital role in both the developmental and adult life of a mammalian organism by regulating many processes including apoptosis, proliferation, senescence, inflammation, cell fate, and tissue repair [94],[106],[107]. The TGF-β super family contains more than 30 growth factors including TGF-βs, BMPs, growth and differentiation factors (GDFs), Activin, and Nodal [2],[106],[108]. The canonical signaling cascade of TGF-β pathway involves the ligands of the TGF-β super family binding to cell surface receptors that activate the Smad proteins in the cytoplasm, which leads to their nuclear translocation and transcriptional activation of target genes [108]. The noncanonical TGF-β signaling includes intracellular signaling pathways activated by TGF-β family members that do not activate Smad proteins [108]. The TGF-β pathway can also be regulated by other key signaling pathways such as Wnt signaling pathways.

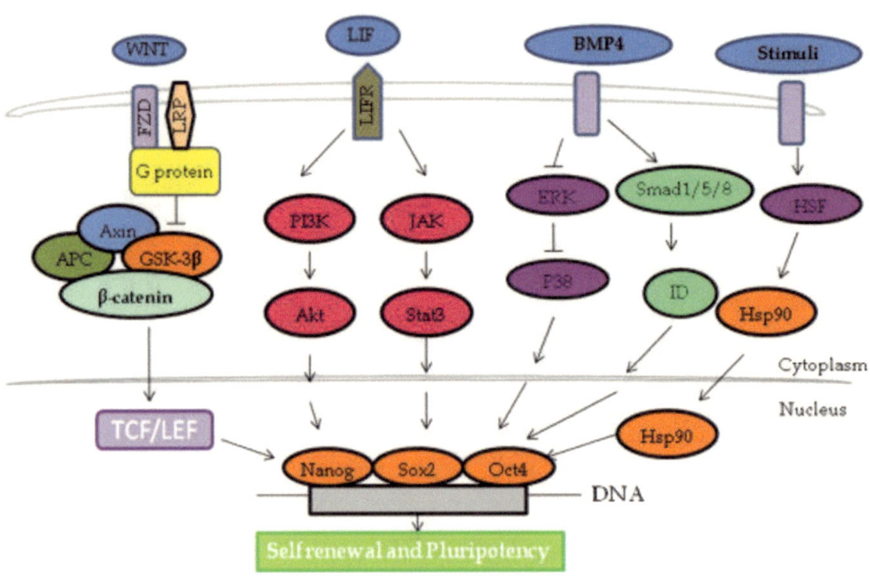

Figure 4. Signaling pathways regulating pluripotency of the naive stem cells, such as mESCs and miPSCs

All members of this family are important for stem cell pluripotency and self-renewal of both mESCs and hESCs, although the role of these signaling molecules appears to differ between the two types of cells [109]. In mESCs, BMP4 maintains self-renewal through inhibition of the MAPK/ERK pathway and the expression of Id protein [36],[37], and promotes mESC proliferation via an increase in Wnt expression (Fig.4) [86]. In contrast, BMP4 promotes hESC differentiation through down-regulation of Nanog and Oct4 [110]. Long-term maintenance of hESC pluripotency therefore requires down-regulation of BMP activity by Noggin and Fgf2 [85]. In hESCs, on the other hand, it is other members of the TGF-β super family that maintain their pluripotency. Phosphorylation and nuclear localization of Smad2 induced by TGF-β, Activin, or Nodal signaling was observed in undifferentiated hESCs and is decreased upon early differentiation (Fig.3) [111]. Activin A is demonstrated to be able to support long-term feeder-free culture and maintenance of pluripotency in hESCs by inducing the expression of Oct4 and Nanog, and suppressing BMP (Fig.3) [87]. Nodal expression also plays a role in the maintenance of human ES cell pluripotency through the inhibition of neuroectodermal differentiation, a default differentiation pathway of ESCs (Fig.3) [112]. Furthermore, inhibition of the TGF-β/Activin/Nodal pathways initiated differentiation and resulted in the decreased expression of stem cell marker proteins [111],[113].

In addition, Activin and Nodal signaling has been shown to promote mESC self-renewal in serum-free conditions [114]. It is therefore clear that TGF-β signaling plays an important role in the maintenance of self-renewal and pluripotency, although the exact mechanism of action for this family of growth factors appears to differ between family members, pluripotency state, and species (Figs. 3 and 4).

BMPs are also potent inhibitors of differentiation in mouse embryos. Knockdown of their down-stream mediator, Smad1 and Smad4, in mESCs leads to a change of the expression pattern of germ layer markers during differentiation [37],[115],[116]. TGF-β signaling also participates in the cell fate decision making of mESCs. Multiple cell lineages, including neural, hematopoietic, cardiomyogenic, and hepatic, have been found to be affected by the TGF-β family [115],[117]. For example, BMP4 regulates mesodermal cell commitment to the hematopoietic lineage and specifies blood lineages at the later stages of differentiation [118]-[120]. Another study found that BMP4 and Activin induce mesoderm differentiation into cardiac lineage [121]. In addition, BMP2-induced mesodermal and cardiac specification results in full cardiogenic differentiation, leading to an enrichment of cardiomyocytes within embryoid bodies [122]. This ability of the TGF-β family members to commit mESCs toward a mesodermal fate is thought to be due to Smad-mediated regulation of the Oct4 promoter, further implicating a role for Smad signaling in the regulation of the core self-renewal network in ESCs [123], [124]. Consistent with this notion, it is found that several Smad target genes overlap with genes bound by the key pluripotency factors, for example, Smad4-regulated genes have a substantial overlap with those of Sox2, NR0B1/Dax1, and Klf4 [116]. In addition, another study demonstrated that several Smad targets were mapped to Nanog, Oct4, and TCF3-bound genes [116].

Jak/Stat3 pathway The self renewal and pluripotency of mESCs are initially maintained by an extrinsic factor, leukemia inhibitory factor (LIF) [7]. The key downstream mediators of LIF are the Jak/Stat pathway [7]. Stat3 has multiple roles in the regulation of mESC pluripotency including gene activation, cell cycle regulation, and inhibition of differentiation pathways [35], [125]. The activation of the Stat3 pathway by LIF induces transcription of self-renewal and pluripotency genes such as Nanog [35],[125]. Furthermore, constitutively active Stat3 promotes mESC self-renewal in the absence of LIF [35],[125]. Stat3 has also been reported to function through the regulation of c-Myc and Klf family proteins [126],[127], although these target genes have not been shown to be completely sufficient to replace the effect of LIF. This ability of LIF-mediated activation of Stat3 to support the long-term self-renewal of mESCs *in vitro* has been supported *in vivo* by the requirement of this pathway in gp130-deficient blastocysts, an embryonic diapause case [128].

In addition to Stat3 homodimers, Stat1 is able to heterodimerize with Stat3 in mESCs. However, Stat1 is unlikely to be required for self-renewal and pluripotency as LIF still maintains undifferentiated growth of Stat1-deficient cells [129].

Although it is sufficient to maintain the pluripotency of mESCs, the Jak/Stat3 pathway does not appear to maintain pluripotency of hESCs, which are possibly in the primed state of pluripotency [90], indicating that the signaling pathways responsible for maintaining pluripotency is species (or pluripotency state)-specific.

5.3. Fgf and the MAPK pathway

Fgf2 (or basic Fgf) is the first growth factor identified as being crucial for hESC pluripotency maintenance and self renewal. It is widely accepted that a serum-free culture of hESCs on mouse feeder cells requires soluble Fgf2 [88],[107],[130],[131]. In hESCs, exogenous Fgf2 activates the ERK/MAPK pathway, which is thought to be necessary for the maintenance of pluripotency, although the mechanism of action is still unclear [130],[132],[133]. In contrast to hESCs, mESCs and miPSCs do not require the Fgf2 or the ERK/MAPK pathway for pluripotency and self-renewal (Table 1) [5],[11]. Actually, ERK signaling triggers mESCs to differentiate towards the primitive endoderm lineage (Table 1) [1],[134]. Interestingly, inhibition of ERK activity has been shown to enhance the efficiency of mESC derivation from mouse embryos [135]. The mechanism underlying this seems to be that ERK1/2 activation triggers mESCs to exit the self-renewal program and enter lineage differentiation [136]. As mentioned earlier, the direct consequence of this is that blocking the ERK/MAPK-mediated differentiation pathway can help the derivation and maintenance of naive state PSCs, such as mESCs.

6. Epigenetic factors regulating stem cell pluripotency

Apart from the aforementioned transcription factors, epigenetics factors have also been found to play a vital role in stem cell pluripotency. These mechanisms include covalent modification of histone, DNA methylation and acetylation, and non-coding RNAs [31],[62],[137],[138]. Here we will discuss the functions of noncoding RNAs and chromatin remodeling factors in stem cell pluripotency.

6.1. Noncoding RNAs

The best understood class of noncoding RNA is the family of microRNAs (miRNAs), short RNAs capable of destabilizing and repressing specific target RNAs. These miRNAs are generally generated by the enzymes Dicer and Dcgr8 [139]-[141]. As for their function in stem cell pluripotency, it has been shown that genetic ablation of these enzymes affects the cell cycle and differentiation of ESCs [139]-[141]. Furthermore, some specific miRNAs are involved in pluripotency regulation. For example, mir-302 and mir-290–295 bind directly to and modulate the core pluripotency factors Oct4, Sox2, and Nanog [6]. miR-145 represses the 3′ untranslated regions of Oct4, Sox2, and Klf4, thus increasing the amount of mir-145 leads to loss of pluripotency [142]. It is further demonstrated that Oct4 also binds to the promoter of mir-145 and suppresses its expression, forming a negative feedback loop involving mir-145, Oct4, Nanog, and Klf4 [142].

Another very important member of the miRNA family is let7 which has differentiation promoting activities itself, and also targets some of the pluripotency-associated genes [143], [144]. Let7 expression is negatively regulated by the RNA binding protein Lin28 [145]. Upon differentiation of pluripotent cells, Lin28 is down-regulated, resulting in stabilization and increase in the level of let7 [146]. This in turn provides the basis for establishment of negative feedback loops in which let7 expression is negatively regulated by the RNA binding protein Lin28 [145].

Recently, Oct4 has been shown to control and activate the expression of another type of noncoding RNA, the large intergenic noncoding RNAs [147]. Interestingly, knockdown of the expression of such RNAs caused growth defects and apoptosis, implying that these noncoding RNAs are involved in self-renewal and reprogramming of stem cells [147],[148].

6.2. Chromatin remodeling factors

Chromatin remodeling factors are recruited to the DNA to modify the density of the nucleosomes, thereby affecting gene expression [149],[150]. Some of these factors are essential for PSC viability, stability, and differentiation [31],[151],[152]. The ones known to have the most profound impact on ESC pluripotency are histone-modifying enzymes, such as Polycomb group (PcG) protein complexes, SetDB1, and Tip60-p400 [153],[154]. These enzymes repress genes that encode lineage-specific differentiation regulators by catalyzing methylation or ubiquitination of the histones in their promoters [153]-[155].

It is demonstrated that sumoylated SetDB1 binds to Oct4 and represses its expression [153], [156]. Loss of the Tip60-p400 complex affects ESC morphology and state [154]. The Tip60-p400 complex is shown to associate with active promoters in ESCs and appears to be recruited directly by the H3K4me3 mark and indirectly by Nanog [154]. Interestingly, the complex is also associated with nucleosomes with H3K4me3 at PcG-occupied genes encoding lineage specific regulators, where it apparently facilitates repression of these poised genes [154].

7. Conclusion

Our understanding of the nature of pluripotency has been formulated extensively by the recent development of different lines of PSCs, especially the iPSCs. Although differences exist between them, the naïve and primed PSCs share certain similarities. For example, they both express the core pluripotency factors, Oct4, Nanog, and Sox2. The core transcription factors frequently share enhancers and autoregulate themselves. They also collectively bind to the promoters of an expanded network of proteins, including pluripotency-associated factors and lineage-specific factors, to enhance or repress their gene expression, through which the fate of the cells is determined. The epigenetic studies have added another layer of complexity of the regulation of these core pluripotency factors and hence pluripotency. In addition, a recent study of our/my laboratory shows that Hsp90 maintains stem cell pluripotency by associating with and sustaining the cellular levels of Oct4 and Nanog, implying that the maturation or stability of these core pluripotency factors are crucially important for stem cell pluripotency [21].

Many of the methodologies to induce or convert somatic cells into PSCs involve using chemical inhibitors targeting specific pathways. This highlights the importance of understanding the roles of signaling pathways in stem cell pluripotency and self-renewal.

Furthermore, an in-depth understanding of pluripotency is highly applicable to regenerative medicine. Knowledge of their culture condition, state of pluripotency, and signal transduction pathways could greatly facilitate *in vitro* culture, manipulation, and differentiation, either from autologous or allogeneic sources. This knowledge will also guide a more effective generation of iPSCs, which will ultimately lead to individualized regenerative medicine.

Acknowledgements

This work is supported by a grant from American Heart Association and an intramural grant from Georgia Health Sciences University to GW, and in part by NIH (R01AG034389) and NSF (1121579) to EB. We are also thankful for the support by the Institute of Molecular Medicine and Genetics (under the directorship of Dr. Lin Mei) of Genetics of Georgia Health Sciences University (soon to be Georgia Regents University Augusta).

Author details

Erhard Bieberich and Guanghu Wang*

*Address all correspondence to: gwang@gru.edu

Institute of Molecular Medicine and Genetics, Medical College of Georgia, Georgia Regents University, Augusta, GA, USA

References

[1] Young, R. A. Control of the embryonic stem cell state. *Cell* 144, 940-954, doi:S0092-8674(11)00071-7 [pii]10.1016/j.cell.2011.01.032 (2011).

[2] Tiedemann, H., Asashima, M., Grunz, H. & Knochel, W. Pluripotent cells (stem cells) and their determination and differentiation in early vertebrate embryogenesis. *Dev Growth Differ* 43, 469-502, doi:dgd599 [pii] (2001).

[3] Xue, K., Ng, J. H. & Ng, H. H. Mapping the networks for pluripotency. *Philos Trans R Soc Lond B Biol Sci* 366, 2238-2246, doi:366/1575/2238 [pii]10.1098/rstb.2011.0005 (2011).

[4] Greenow, K. & Clarke, A. R. Controlling the stem cell compartment and regeneration in vivo: the role of pluripotency pathways. *Physiol Rev* 92, 75-99, doi:92/1/75 [pii]10.1152/physrev.00040.2010 (2012).

[5] Dejosez, M. & Zwaka, T. P. Pluripotency and nuclear reprogramming. *Annu Rev Biochem* 81, 737-765, doi:10.1146/annurev-biochem-052709-104948 (2012).

[6] Marson, A. *et al.* Connecting microRNA genes to the core transcriptional regulatory circuitry of embryonic stem cells. *Cell* 134, 521-533, doi:S0092-8674(08)00938-0 [pii]10.1016/j.cell.2008.07.020 (2008).

[7] Mitalipov, S. & Wolf, D. Totipotency, pluripotency and nuclear reprogramming. *Adv Biochem Eng Biotechnol* 114, 185-199, doi:10.1007/10_2008_45 (2009).

[8] Pauklin, S., Pedersen, R. A. & Vallier, L. Mouse pluripotent stem cells at a glance. *J Cell Sci* 124, 3727-3732, doi:124/22/3727 [pii]10.1242/jcs.074120 (2011).

[9] Thomson, J. A. *et al.* Embryonic stem cell lines derived from human blastocysts. *Science* 282, 1145-1147 (1998).

[10] Vescovi, A. L., Reynolds, B. A., Rietze, R. L. & Bjornson, C. Ground-breaking stem-cell work has been reproduced. *Nature* 447, 259 (2007).

[11] Burdon, T., Smith, A. & Savatier, P. Signalling, cell cycle and pluripotency in embryonic stem cells. *Trends Cell Biol* 12, 432-438, doi:S0962892402023528 [pii] (2002).

[12] Krishnamurthy, K., Wang, G., Silva, J., Condie, B. G. & Bieberich, E. Ceramide regulates atypical PKCzeta/lambda-mediated cell polarity in primitive ectoderm cells. A novel function of sphingolipids in morphogenesis. *J Biol Chem* 282, 3379-3390, doi:M607779200 [pii]10.1074/jbc.M607779200 (2007).

[13] Nichols, J. & Smith, A. Naive and primed pluripotent states. *Cell Stem Cell* 4, 487-492, doi:S1934-5909(09)00224-0 [pii]10.1016/j.stem.2009.05.015 (2009).

[14] Evans, M. Discovering pluripotency: 30 years of mouse embryonic stem cells. *Nat Rev Mol Cell Biol* 12, 680-686, doi:nrm3190 [pii]10.1038/nrm3190 (2011).

[15] Evans, M. J. & Kaufman, M. H. Establishment in culture of pluripotential cells from mouse embryos. *Nature* 292, 154-156 (1981).

[16] Tesar, P. J. *et al.* New cell lines from mouse epiblast share defining features with human embryonic stem cells. *Nature* 448, 196-199, doi:nature05972 [pii]10.1038/nature05972 (2007).

[17] Kuijk, E. W., Chuva de Sousa Lopes, S. M., Geijsen, N., Macklon, N. & Roelen, B. A. The different shades of mammalian pluripotent stem cells. *Hum Reprod Update* 17, 254-271, doi:dmq035 [pii]10.1093/humupd/dmq035 (2011).

[18] Ginis, I. *et al.* Differences between human and mouse embryonic stem cells. *Dev Biol* 269, 360-380, doi:10.1016/j.ydbio.2003.12.034S0012160604000193 [pii] (2004).

[19] Singh, S. K., Kagalwala, M. N., Parker-Thornburg, J., Adams, H. & Majumder, S. REST maintains self-renewal and pluripotency of embryonic stem cells. *Nature* 453, 223-227, doi:nature06863 [pii]10.1038/nature06863 (2008).

[20] Bieberich, E., Silva, J., Wang, G., Krishnamurthy, K. & Condie, B. G. Selective apoptosis of pluripotent mouse and human stem cells by novel ceramide analogues prevents teratoma formation and enriches for neural precursors in ES cell-derived neural transplants. *J Cell Biol* 167, 723-734 (2004).

[21] Bradley, E., Bieberich, E., Mivechi, N. F., Tangpisuthipongsa, D. & Wang, G. Regulation of embryonic stem cell pluripotency by heat shock protein 90. *Stem Cells* 30, 1624-1633, doi:10.1002/stem.1143 (2012).

[22] Brons, I. G. *et al.* Derivation of pluripotent epiblast stem cells from mammalian embryos. *Nature* 448, 191-195, doi:nature05950 [pii]10.1038/nature05950 (2007).

[23] Kleinsmith, L. J. & Pierce, G. B., Jr. Multipotentiality of Single Embryonal Carcinoma Cells. *Cancer Res* 24, 1544-1551 (1964).

[24] Gooi, H. C. *et al.* Stage-specific embryonic antigen involves alpha 1 goes to 3 fucosylated type 2 blood group chains. *Nature* 292, 156-158 (1981).

[25] Solter, D. & Knowles, B. B. Monoclonal antibody defining a stage-specific mouse embryonic antigen (SSEA-1). *Proc Natl Acad Sci U S A* 75, 5565-5569 (1978).

[26] Papaioannou, V. E., McBurney, M. W., Gardner, R. L. & Evans, M. J. Fate of teratocarcinoma cells injected into early mouse embryos. *Nature* 258, 70-73 (1975).

[27] Martin, G. R. Isolation of a pluripotent cell line from early mouse embryos cultured in medium conditioned by teratocarcinoma stem cells. *Proc Natl Acad Sci U S A* 78, 7634-7638 (1981).

[28] Savatier, P., Lapillonne, H., van Grunsven, L. A., Rudkin, B. B. & Samarut, J. Withdrawal of differentiation inhibitory activity/leukemia inhibitory factor up-regulates D-type cyclins and cyclin-dependent kinase inhibitors in mouse embryonic stem cells. *Oncogene* 12, 309-322 (1996).

[29] Minkovsky, A., Patel, S. & Plath, K. Concise review: Pluripotency and the transcriptional inactivation of the female Mammalian X chromosome. *Stem Cells* 30, 48-54, doi: 10.1002/stem.755 (2012).

[30] Bernstein, B. E., Meissner, A. & Lander, E. S. The mammalian epigenome. *Cell* 128, 669-681, doi:S0092-8674(07)00128-6 [pii]10.1016/j.cell.2007.01.033 (2007).

[31] Meissner, A. Epigenetic modifications in pluripotent and differentiated cells. *Nat Biotechnol* 28, 1079-1088, doi:nbt.1684 [pii]10.1038/nbt.1684 (2010).

[32] Wang, G. *et al.* Direct binding to ceramide activates protein kinase Czeta before the formation of a pro-apoptotic complex with PAR-4 in differentiating stem cells. *J Biol Chem* 280, 26415-26424 (2005).

[33] Bieberich, E., MacKinnon, S., Silva, J., Noggle, S. & Condie, B. G. Regulation of cell death in mitotic neural progenitor cells by asymmetric distribution of prostate apoptosis response 4 (PAR-4) and simultaneous elevation of endogenous ceramide. *J Cell Biol* 162, 469-479 (2003).

[34] Raz, R., Lee, C. K., Cannizzaro, L. A., d'Eustachio, P. & Levy, D. E. Essential role of STAT3 for embryonic stem cell pluripotency. *Proc Natl Acad Sci U S A* 96, 2846-2851 (1999).

[35] Niwa, H., Burdon, T., Chambers, I. & Smith, A. Self-renewal of pluripotent embryonic stem cells is mediated via activation of STAT3. *Genes Dev* 12, 2048-2060 (1998).

[36] Qi, X. *et al.* BMP4 supports self-renewal of embryonic stem cells by inhibiting mitogen-activated protein kinase pathways. *Proc Natl Acad Sci U S A* 101, 6027-6032, doi: 10.1073/pnas.04013671010401367101 [pii] (2004).

[37] Ying, Q. L., Nichols, J., Chambers, I. & Smith, A. BMP induction of Id proteins suppresses differentiation and sustains embryonic stem cell self-renewal in collaboration with STAT3. *Cell* 115, 281-292, doi:S009286740300847X [pii] (2003).

[38] Ying, Q. L. *et al.* The ground state of embryonic stem cell self-renewal. *Nature* 453, 519-523, doi:nature06968 [pii]10.1038/nature06968 (2008).

[39] Takahashi, K. & Yamanaka, S. Induction of pluripotent stem cells from mouse embryonic and adult fibroblast cultures by defined factors. *Cell* 126, 663-676, doi:S0092-8674(06)00976-7 [pii]10.1016/j.cell.2006.07.024 (2006).

[40] Takahashi, K. *et al.* Induction of pluripotent stem cells from adult human fibroblasts by defined factors. *Cell* 131, 861-872, doi:S0092-8674(07)01471-7 [pii]10.1016/j.cell.2007.11.019 (2007).

[41] Takahashi, K., Okita, K., Nakagawa, M. & Yamanaka, S. Induction of pluripotent stem cells from fibroblast cultures. *Nat Protoc* 2, 3081-3089, doi:nprot.2007.418 [pii]10.1038/nprot.2007.418 (2007).

[42] Yamanaka, S. Induction of pluripotent stem cells from mouse fibroblasts by four transcription factors. *Cell Prolif* 41 Suppl 1, 51-56, doi:CPR493 [pii]10.1111/j.1365-2184.2008.00493.x (2008).

[43] Cheng, D., Lei, L., Lu, Z., Li, Z. & Wang, H. [Induction and characterization of induced pluripotent stem (iPS) cells: a review]. *Sheng Wu Gong Cheng Xue Bao* 26, 421-430 (2010).

[44] Huangfu, D. *et al.* Induction of pluripotent stem cells by defined factors is greatly improved by small-molecule compounds. *Nat Biotechnol* 26, 795-797, doi:nbt1418 [pii]10.1038/nbt1418 (2008).

[45] Cho, H. J. *et al.* Induction of pluripotent stem cells from adult somatic cells by protein-based reprogramming without genetic manipulation. *Blood* 116, 386-395, doi:blood-2010-02-269589 [pii]10.1182/blood-2010-02-269589 (2010).

[46] Hjelm, B. E. *et al.* Induction of pluripotent stem cells from autopsy donor-derived somatic cells. *Neurosci Lett* 502, 219-224, doi:S0304-3940(11)01125-6 [pii]10.1016/j.neulet.2011.07.048 (2011).

[47] Okahara-Narita, J. *et al.* Induction of pluripotent stem cells from fetal and adult cynomolgus monkey fibroblasts using four human transcription factors. *Primates*, doi: 10.1007/s10329-011-0283-1 (2011).

[48] Oda, Y. *et al.* Induction of pluripotent stem cells from human third molar mesenchymal stromal cells. *J Biol Chem* 285, 29270-29278, doi:M109.055889 [pii]10.1074/jbc.M109.055889 (2010).

[49] Shi, Y. *et al.* Induction of pluripotent stem cells from mouse embryonic fibroblasts by Oct4 and Klf4 with small-molecule compounds. *Cell Stem Cell* 3, 568-574, doi:S1934-5909(08)00527-4 [pii] 10.1016/j.stem.2008.10.004 (2008).

[50] Huangfu, D. *et al.* Induction of pluripotent stem cells from primary human fibroblasts with only Oct4 and Sox2. *Nat Biotechnol* 26, 1269-1275, doi:nbt.1502 [pii]10.1038/nbt.1502 (2008).

[51] Surani, M. A. Reprogramming a somatic nucleus by trans-modification activity in germ cells. *Semin Cell Dev Biol* 10, 273-277, doi:10.1006/scdb.1998.0285S1084-9521(98)90285-3 [pii] (1999).

[52] Kanatsu-Shinohara, M. *et al.* Generation of pluripotent stem cells from neonatal mouse testis. *Cell* 119, 1001-1012, doi:S0092867404010578 [pii]10.1016/j.cell.2004.11.011 (2004).

[53] Shamblott, M. J. *et al.* Derivation of pluripotent stem cells from cultured human primordial germ cells. *Proc Natl Acad Sci U S A* 95, 13726-13731 (1998).

[54] Conrad, S. *et al.* Generation of pluripotent stem cells from adult human testis. *Nature* 456, 344-349, doi:nature07404 [pii]10.1038/nature07404 (2008).

[55] Ko, K. *et al.* Human adult germline stem cells in question. *Nature* 465, E1; discussion E3, doi:nature09089 [pii]10.1038/nature09089 (2010).

[56] Mountford, P., Nichols, J., Zevnik, B., O'Brien, C. & Smith, A. Maintenance of pluripotential embryonic stem cells by stem cell selection. *Reprod Fertil Dev* 10, 527-533 (1998).

[57] Nichols, J. *et al.* Formation of pluripotent stem cells in the mammalian embryo depends on the POU transcription factor Oct4. *Cell* 95, 379-391, doi:S0092-8674(00)81769-9 [pii] (1998).

[58] Kim, J. B. *et al.* Oct4-induced pluripotency in adult neural stem cells. *Cell* 136, 411-419, doi:S0092-8674(09)00071-3 [pii]10.1016/j.cell.2009.01.023 (2009).

[59] Torres, J. & Watt, F. M. Nanog maintains pluripotency of mouse embryonic stem cells by inhibiting NFkappaB and cooperating with Stat3. *Nat Cell Biol* 10, 194-201, doi:ncb1680 [pii]10.1038/ncb1680 (2008).

[60] Ebert, A. D. *et al.* Induced pluripotent stem cells from a spinal muscular atrophy patient. *Nature* 457, 277-280, doi:nature07677 [pii]10.1038/nature07677 (2009).

[61] Yu, J. *et al.* Human induced pluripotent stem cells free of vector and transgene sequences. *Science* 324, 797-801, doi:1172482 [pii]10.1126/science.1172482 (2009).

[62] Gan, Q., Yoshida, T., McDonald, O. G. & Owens, G. K. Concise review: epigenetic mechanisms contribute to pluripotency and cell lineage determination of embryonic stem cells. *Stem Cells* 25, 2-9, doi:2006-0383 [pii]10.1634/stemcells.2006-0383 (2007).

[63] Hanna, J. H., Saha, K. & Jaenisch, R. Pluripotency and cellular reprogramming: facts, hypotheses, unresolved issues. *Cell* 143, 508-525, doi:S0092-8674(10)01144-X [pii]10.1016/j.cell.2010.10.008 (2010).

[64] Niwa, H., Miyazaki, J. & Smith, A. G. Quantitative expression of Oct-3/4 defines differentiation, dedifferentiation or self-renewal of ES cells. *Nat Genet* 24, 372-376, doi: 10.1038/74199 (2000).

[65] Shimozaki, K., Nakashima, K., Niwa, H. & Taga, T. Involvement of Oct3/4 in the enhancement of neuronal differentiation of ES cells in neurogenesis-inducing cultures. *Development* 130, 2505-2512 (2003).

[66] Mitsui, K. *et al.* The homeoprotein Nanog is required for maintenance of pluripotency in mouse epiblast and ES cells. *Cell* 113, 631-642, doi:S0092867403003933 [pii] (2003).

[67] Chambers, I. *et al.* Functional expression cloning of Nanog, a pluripotency sustaining factor in embryonic stem cells. *Cell* 113, 643-655, doi:S0092867403003921 [pii] (2003).

[68] Suzuki, A. *et al.* Nanog binds to Smad1 and blocks bone morphogenetic protein-induced differentiation of embryonic stem cells. *Proc Natl Acad Sci U S A* 103, 10294-10299, doi:0506945103 [pii]10.1073/pnas.0506945103 (2006).

[69] Avilion, A. A. *et al.* Multipotent cell lineages in early mouse development depend on SOX2 function. *Genes Dev* 17, 126-140, doi:10.1101/gad.224503 (2003).

[70] Masui, S. *et al.* Pluripotency governed by Sox2 via regulation of Oct3/4 expression in mouse embryonic stem cells. *Nat Cell Biol* 9, 625-635, doi:ncb1589 [pii]10.1038/ncb1589 (2007).

[71] Chen, X. *et al.* Integration of external signaling pathways with the core transcriptional network in embryonic stem cells. *Cell* 133, 1106-1117, doi:S0092-8674(08)00617-X [pii]10.1016/j.cell.2008.04.043 (2008).

[72] van den Berg, D. L. *et al.* An Oct4-centered protein interaction network in embryonic stem cells. *Cell Stem Cell* 6, 369-381, doi:S1934-5909(10)00091-3 [pii]10.1016/j.stem.2010.02.014 (2010).

[73] Liang, J. *et al.* Nanog and Oct4 associate with unique transcriptional repression complexes in embryonic stem cells. *Nat Cell Biol* 10, 731-739, doi:ncb1736 [pii]10.1038/ncb1736 (2008).

[74] Boyer, L. A. *et al.* Core transcriptional regulatory circuitry in human embryonic stem cells. *Cell* 122, 947-956, doi:S0092-8674(05)00825-1 [pii]10.1016/j.cell.2005.08.020 (2005).

[75] Johansson, H. & Simonsson, S. Core transcription factors, Oct4, Sox2 and Nanog, individually form complexes with nucleophosmin (Npm1) to control embryonic stem (ES) cell fate determination. *Aging (Albany NY)* 2, 815-822, doi:100222 [pii] (2010).

[76] Gao, Z. *et al.* Determination of protein interactome of transcription factor sox2 in embryonic stem cells engineered for inducible expression of four reprogramming factors. *J Biol Chem* 287, 11384-11397, doi:M111.320143 [pii]10.1074/jbc.M111.320143 (2012).

[77] Ng, H. H. & Surani, M. A. The transcriptional and signalling networks of pluripotency. *Nat Cell Biol* 13, 490-496, doi:ncb0511-490 [pii]10.1038/ncb0511-490 (2011).

[78] Loh, Y. H. *et al.* The Oct4 and Nanog transcription network regulates pluripotency in mouse embryonic stem cells. *Nat Genet* 38, 431-440, doi:ng1760 [pii]10.1038/ng1760 (2006).

[79] Boyer, L. A. *et al.* Polycomb complexes repress developmental regulators in murine embryonic stem cells. *Nature* 441, 349-353, doi:nature04733 [pii]10.1038/nature04733 (2006).

[80] Lee, T. I. *et al.* Control of developmental regulators by Polycomb in human embryonic stem cells. *Cell* 125, 301-313, doi:S0092-8674(06)00384-9 [pii]10.1016/j.cell.2006.02.043 (2006).

[81] Liao, B. & Jin, Y. Wwp2 mediates Oct4 ubiquitination and its own auto-ubiquitination in a dosage-dependent manner. *Cell Res* 20, 332-344, doi:cr2009136 [pii]10.1038/cr.2009.136 (2010).

[82] Li, H. *et al.* Wwp2-mediated ubiquitination of the RNA polymerase II large subunit in mouse embryonic pluripotent stem cells. *Mol Cell Biol* 27, 5296-5305, doi:MCB.01667-06 [pii]10.1128/MCB.01667-06 (2007).

[83] Moretto-Zita, M. *et al.* Phosphorylation stabilizes Nanog by promoting its interaction with Pin1. *Proc Natl Acad Sci U S A* 107, 13312-13317, doi:1005847107 [pii]10.1073/pnas.1005847107 (2010).

[84] Daheron, L. *et al.* LIF/STAT3 signaling fails to maintain self-renewal of human embryonic stem cells. *Stem Cells* 22, 770-778, doi:10.1634/stemcells.22-5-770 22/5/770 [pii] (2004).

[85] Xu, R. H. *et al.* Basic FGF and suppression of BMP signaling sustain undifferentiated proliferation of human ES cells. *Nat Methods* 2, 185-190, doi:nmeth744 [pii] 10.1038/nmeth744 (2005).

[86] Lee, M. Y., Lim, H. W., Lee, S. H. & Han, H. J. Smad, PI3K/Akt, and Wnt-dependent signaling pathways are involved in BMP-4-induced ESC self-renewal. *Stem Cells* 27, 1858-1868, doi:10.1002/stem.124 (2009).

[87] Xiao, L., Yuan, X. & Sharkis, S. J. Activin A maintains self-renewal and regulates fibroblast growth factor, Wnt, and bone morphogenic protein pathways in human embryonic stem cells. *Stem Cells* 24, 1476-1486, doi:2005-0299 [pii] 10.1634/stemcells.2005-0299 (2006).

[88] Ding, V. M. *et al.* FGF-2 modulates Wnt signaling in undifferentiated hESC and iPS cells through activated PI3-K/GSK3beta signaling. *J Cell Physiol* 225, 417-428, doi:10.1002/jcp.22214 (2010).

[89] Kim, L. & Kimmel, A. R. GSK3 at the edge: regulation of developmental specification and cell polarization. *Curr Drug Targets* 7, 1411-1419 (2006).

[90] Sato, N., Meijer, L., Skaltsounis, L., Greengard, P. & Brivanlou, A. H. Maintenance of pluripotency in human and mouse embryonic stem cells through activation of Wnt signaling by a pharmacological GSK-3-specific inhibitor. *Nat Med* 10, 55-63, doi:10.1038/nm979 nm979 [pii] (2004).

[91] Bejsovec, A. Wnt pathway activation: new relations and locations. *Cell* 120, 11-14, doi:S0092867404012462 [pii] 10.1016/j.cell.2004.12.021 (2005).

[92] Widelitz, R. Wnt signaling through canonical and non-canonical pathways: recent progress. *Growth Factors* 23, 111-116, doi:T416183G77101V73 [pii] 10.1080/08977190500125746 (2005).

[93] Woodgett, J. R. Judging a protein by more than its name: GSK-3. *Sci STKE* 2001, RE12 (2001).

[94] Huang, J. *et al.* More synergetic cooperation of Yamanaka factors in induced pluripotent stem cells than in embryonic stem cells. *Cell Res* 19, 1127-1138, doi:cr2009106 [pii] 10.1038/cr.2009.106 (2009).

[95] Polakis, P. The oncogenic activation of beta-catenin. *Curr Opin Genet Dev* 9, 15-21 (1999).

[96] Takao, Y., Yokota, T. & Koide, H. Beta-catenin up-regulates Nanog expression through interaction with Oct-3/4 in embryonic stem cells. *Biochem Biophys Res Commun* 353, 699-705, doi:S0006-291X(06)02748-3 [pii] 10.1016/j.bbrc.2006.12.072 (2007).

[97] Haegele, L. *et al.* Wnt signalling inhibits neural differentiation of embryonic stem cells by controlling bone morphogenetic protein expression. *Mol Cell Neurosci* 24, 696-708, doi:S104474310300232X [pii] (2003).

[98] Aubert, J., Dunstan, H., Chambers, I. & Smith, A. Functional gene screening in embryonic stem cells implicates Wnt antagonism in neural differentiation. *Nat Biotechnol* 20, 1240-1245, doi:10.1038/nbt763 nbt763 [pii] (2002).

[99] Kielman, M. F. *et al.* Apc modulates embryonic stem-cell differentiation by controlling the dosage of beta-catenin signaling. *Nat Genet* 32, 594-605, doi:10.1038/ng1045 ng1045 [pii] (2002).

[100] Doble, B. W., Patel, S., Wood, G. A., Kockeritz, L. K. & Woodgett, J. R. Functional redundancy of GSK-3alpha and GSK-3beta in Wnt/beta-catenin signaling shown by using an allelic series of embryonic stem cell lines. *Dev Cell* 12, 957-971, doi:S1534-5807(07)00148-7 [pii] 10.1016/j.devcel.2007.04.001 (2007).

[101] Liu, Z., Scannell, D. R., Eisen, M. B. & Tjian, R. Control of embryonic stem cell lineage commitment by core promoter factor, TAF3. *Cell* 146, 720-731, doi:S0092-8674(11)00881-6 [pii] 10.1016/j.cell.2011.08.005 (2011).

[102] Sun, Y., Li, H., Yang, H., Rao, M. S. & Zhan, M. Mechanisms controlling embryonic stem cell self-renewal and differentiation. *Crit Rev Eukaryot Gene Expr* 16, 211-231, doi:5671614a1bbedf32,278d0bf81ca44a82 [pii] (2006).

[103] Cole, M. F., Johnstone, S. E., Newman, J. J., Kagey, M. H. & Young, R. A. Tcf3 is an integral component of the core regulatory circuitry of embryonic stem cells. *Genes Dev* 22, 746-755, doi:22/6/746 [pii] 10.1101/gad.1642408 (2008).

[104] Tam, W. L. *et al.* T-cell factor 3 regulates embryonic stem cell pluripotency and self-renewal by the transcriptional control of multiple lineage pathways. *Stem Cells* 26, 2019-2031, doi:2007-1115 [pii] 10.1634/stemcells.2007-1115 (2008).

[105] Pereira, L., Yi, F. & Merrill, B. J. Repression of Nanog gene transcription by Tcf3 limits embryonic stem cell self-renewal. *Mol Cell Biol* 26, 7479-7491, doi:MCB.00368-06 [pii] 10.1128/MCB.00368-06 (2006).

[106] Santibanez, J. F., Quintanilla, M. & Bernabeu, C. TGF-beta/TGF-beta receptor system and its role in physiological and pathological conditions. *Clin Sci (Lond)* 121, 233-251, doi:CS20110086 [pii] 10.1042/CS20110086 (2011).

[107] Xu, R. H. *et al.* NANOG is a direct target of TGFbeta/activin-mediated SMAD signaling in human ESCs. *Cell Stem Cell* 3, 196-206, doi:S1934-5909(08)00335-4 [pii] 10.1016/j.stem.2008.07.001 (2008).

[108] Wu, M. Y. & Hill, C. S. Tgf-beta superfamily signaling in embryonic development and homeostasis. *Dev Cell* 16, 329-343, doi:S1534-5807(09)00086-0 [pii] 10.1016/j.devcel.2009.02.012 (2009).

[109] Watabe, T. & Miyazono, K. Roles of TGF-beta family signaling in stem cell renewal and differentiation. *Cell Res* 19, 103-115, doi:cr2008323 [pii] 10.1038/cr.2008.323 (2009).

[110] Xu, R. H. *et al.* BMP4 initiates human embryonic stem cell differentiation to trophoblast. *Nat Biotechnol* 20, 1261-1264, doi:10.1038/nbt761 nbt761 [pii] (2002).

[111] James, D., Levine, A. J., Besser, D. & Hemmati-Brivanlou, A. TGFbeta/activin/nodal signaling is necessary for the maintenance of pluripotency in human embryonic stem cells. *Development* 132, 1273-1282, doi:dev.01706 [pii] 10.1242/dev.01706 (2005).

[112] Vallier, L., Reynolds, D. & Pedersen, R. A. Nodal inhibits differentiation of human embryonic stem cells along the neuroectodermal default pathway. *Dev Biol* 275, 403-421, doi:S0012-1606(04)00587-1 [pii] 10.1016/j.ydbio.2004.08.031 (2004).

[113] Vallier, L., Alexander, M. & Pedersen, R. A. Activin/Nodal and FGF pathways cooperate to maintain pluripotency of human embryonic stem cells. *J Cell Sci* 118, 4495-4509, doi:118/19/4495 [pii] 10.1242/jcs.02553 (2005).

[114] Ogawa, K. *et al.* Activin-Nodal signaling is involved in propagation of mouse embryonic stem cells. *J Cell Sci* 120, 55-65, doi:120/1/55 [pii] 10.1242/jcs.03296 (2007).

[115] Wilson, P. A. & Hemmati-Brivanlou, A. Induction of epidermis and inhibition of neural fate by Bmp-4. *Nature* 376, 331-333, doi:10.1038/376331a0 (1995).

[116] Fei, T. *et al.* Genome-wide mapping of SMAD target genes reveals the role of BMP signaling in embryonic stem cell fate determination. *Genome Res* 20, 36-44, doi:gr.092114.109 [pii] 10.1101/gr.092114.109 (2010).

[117] Yuasa, S. *et al.* Transient inhibition of BMP signaling by Noggin induces cardiomyocyte differentiation of mouse embryonic stem cells. *Nat Biotechnol* 23, 607-611, doi:nbt1093 [pii] 10.1038/nbt1093 (2005).

[118] Lengerke, C. *et al.* BMP and Wnt specify hematopoietic fate by activation of the Cdx-Hox pathway. *Cell Stem Cell* 2, 72-82, doi:S1934-5909(07)00239-1 [pii] 10.1016/j.stem.2007.10.022 (2008).

[119] Park, C. *et al.* A hierarchical order of factors in the generation of FLK1- and SCL-expressing hematopoietic and endothelial progenitors from embryonic stem cells. *Development* 131, 2749-2762, doi:10.1242/dev.01130 131/11/2749 [pii] (2004).

[120] Sadlon, T. J., Lewis, I. D. & D'Andrea, R. J. BMP4: its role in development of the hematopoietic system and potential as a hematopoietic growth factor. *Stem Cells* 22, 457-474, doi:22/4/457 [pii] 10.1634/stemcells.22-4-457 (2004).

[121] Laflamme, M. A. *et al.* Cardiomyocytes derived from human embryonic stem cells in pro-survival factors enhance function of infarcted rat hearts. *Nat Biotechnol* 25, 1015-1024, doi:nbt1327 [pii] 10.1038/nbt1327 (2007).

[122] Behfar, A. *et al.* Stem cell differentiation requires a paracrine pathway in the heart. *FASEB J* 16, 1558-1566, doi:10.1096/fj.02-0072com 16/12/1558 [pii] (2002).

[123] Zeineddine, D. *et al.* Oct-3/4 dose dependently regulates specification of embryonic stem cells toward a cardiac lineage and early heart development. *Dev Cell* 11, 535-546, doi:S1534-5807(06)00314-5 [pii] 10.1016/j.devcel.2006.07.013 (2006).

[124] Narazaki, G. *et al.* Directed and systematic differentiation of cardiovascular cells from mouse induced pluripotent stem cells. *Circulation* 118, 498-506, doi:CIRCULATIONAHA.108.769562 [pii] 10.1161/CIRCULATIONAHA.108.769562 (2008).

[125] Niwa, H., Ogawa, K., Shimosato, D. & Adachi, K. A parallel circuit of LIF signalling pathways maintains pluripotency of mouse ES cells. *Nature* 460, 118-122, doi:nature08113 [pii] 10.1038/nature08113 (2009).

[126] Li, Y. *et al.* Murine embryonic stem cell differentiation is promoted by SOCS-3 and inhibited by the zinc finger transcription factor Klf4. *Blood* 105, 635-637, doi:10.1182/blood-2004-07-2681 2004-07-2681 [pii] (2005).

[127] Cartwright, P. *et al.* LIF/STAT3 controls ES cell self-renewal and pluripotency by a Myc-dependent mechanism. *Development* 132, 885-896, doi:dev.01670 [pii] 10.1242/dev.01670 (2005).

[128] Nichols, J., Chambers, I., Taga, T. & Smith, A. Physiological rationale for responsiveness of mouse embryonic stem cells to gp130 cytokines. *Development* 128, 2333-2339 (2001).

[129] Durbin, J. E., Hackenmiller, R., Simon, M. C. & Levy, D. E. Targeted disruption of the mouse Stat1 gene results in compromised innate immunity to viral disease. *Cell* 84, 443-450, doi:S0092-8674(00)81289-1 [pii] (1996).

[130] Amit, M. *et al.* Clonally derived human embryonic stem cell lines maintain pluripotency and proliferative potential for prolonged periods of culture. *Dev Biol* 227, 271-278, doi:10.1006/dbio.2000.9912 S0012-1606(00)99912-3 [pii] (2000).

[131] Xu, C. *et al.* Feeder-free growth of undifferentiated human embryonic stem cells. *Nat Biotechnol* 19, 971-974, doi:10.1038/nbt1001-971 nbt1001-971 [pii] (2001).

[132] Dvorak, P. *et al.* Expression and potential role of fibroblast growth factor 2 and its receptors in human embryonic stem cells. *Stem Cells* 23, 1200-1211, doi:2004-0303 [pii] 10.1634/stemcells.2004-0303 (2005).

[133] Li, J. *et al.* MEK/ERK signaling contributes to the maintenance of human embryonic stem cell self-renewal. *Differentiation* 75, 299-307, doi:S0301-4681(09)60124-9 [pii] 10.1111/j.1432-0436.2006.00143.x (2007).

[134] Chazaud, C., Yamanaka, Y., Pawson, T. & Rossant, J. Early lineage segregation between epiblast and primitive endoderm in mouse blastocysts through the Grb2-MAPK pathway. *Dev Cell* 10, 615-624, doi:S1534-5807(06)00125-0 [pii] 10.1016/j.devcel.2006.02.020 (2006).

[135] Buehr, M. & Smith, A. Genesis of embryonic stem cells. *Philos Trans R Soc Lond B Biol Sci* 358, 1397-1402; discussion 1402, doi:10.1098/rstb.2003.1327 (2003).

[136] Kunath, T. *et al.* FGF stimulation of the Erk1/2 signalling cascade triggers transition of pluripotent embryonic stem cells from self-renewal to lineage commitment. *Development* 134, 2895-2902, doi:134/16/2895 [pii] 10.1242/dev.02880 (2007).

[137] Atkinson, S. & Armstrong, L. Epigenetics in embryonic stem cells: regulation of pluripotency and differentiation. *Cell Tissue Res* 331, 23-29, doi:10.1007/s00441-007-0536-x (2008).

[138] Marks, H. *et al.* The transcriptional and epigenomic foundations of ground state pluripotency. *Cell* 149, 590-604, doi:S0092-8674(12)00409-6 [pii] 10.1016/j.cell.2012.03.026 (2012).

[139] Murchison, E. P., Partridge, J. F., Tam, O. H., Cheloufi, S. & Hannon, G. J. Characterization of Dicer-deficient murine embryonic stem cells. *Proc Natl Acad Sci U S A* 102, 12135-12140, doi:0505479102 [pii] 10.1073/pnas.0505479102 (2005).

[140] Wang, Y., Medvid, R., Melton, C., Jaenisch, R. & Blelloch, R. DGCR8 is essential for microRNA biogenesis and silencing of embryonic stem cell self-renewal. *Nat Genet* 39, 380-385, doi:ng1969 [pii] 10.1038/ng1969 (2007).

[141] Melton, C., Judson, R. L. & Blelloch, R. Opposing microRNA families regulate self-renewal in mouse embryonic stem cells. *Nature* 463, 621-626, doi:nature08725 [pii] 10.1038/nature08725 (2010).

[142] Xu, N., Papagiannakopoulos, T., Pan, G., Thomson, J. A. & Kosik, K. S. MicroRNA-145 regulates OCT4, SOX2, and KLF4 and represses pluripotency in human embryonic stem cells. *Cell* 137, 647-658, doi:S0092-8674(09)00252-9 [pii] 10.1016/j.cell.2009.02.038 (2009).

[143] Helland, A. *et al.* Deregulation of MYCN, LIN28B and LET7 in a molecular subtype of aggressive high-grade serous ovarian cancers. *PLoS One* 6, e18064, doi:10.1371/journal.pone.0018064 (2011).

[144] Judson, R. L., Babiarz, J. E., Venere, M. & Blelloch, R. Embryonic stem cell-specific microRNAs promote induced pluripotency. *Nat Biotechnol* 27, 459-461, doi:nbt.1535 [pii] 10.1038/nbt.1535 (2009).

[145] Thornton, J. E. & Gregory, R. I. How does Lin28 let-7 control development and disease? *Trends Cell Biol*, doi:S0962-8924(12)00104-3 [pii] 10.1016/j.tcb.2012.06.001 (2012).

[146] Subramanyam, D. *et al.* Multiple targets of miR-302 and miR-372 promote reprogramming of human fibroblasts to induced pluripotent stem cells. *Nat Biotechnol* 29, 443-448, doi:nbt.1862 [pii] 10.1038/nbt.1862 (2011).

[147] Loewer, S. *et al.* Large intergenic non-coding RNA-RoR modulates reprogramming of human induced pluripotent stem cells. *Nat Genet* 42, 1113-1117, doi:ng.710 [pii] 10.1038/ng.710 (2010).

[148] Sheik Mohamed, J., Gaughwin, P. M., Lim, B., Robson, P. & Lipovich, L. Conserved long noncoding RNAs transcriptionally regulated by Oct4 and Nanog modulate pluripotency in mouse embryonic stem cells. *RNA* 16, 324-337, doi:rna.1441510 [pii] 10.1261/rna.1441510 (2010).

[149] Knezetic, J. A. & Luse, D. S. The presence of nucleosomes on a DNA template prevents initiation by RNA polymerase II in vitro. *Cell* 45, 95-104, doi: 0092-8674(86)90541-6 [pii] (1986).

[150] Li, B., Carey, M. & Workman, J. L. The role of chromatin during transcription. *Cell* 128, 707-719, doi:S0092-8674(07)00109-2 [pii] 10.1016/j.cell.2007.01.015 (2007).

[151] Kagey, M. H. *et al.* Mediator and cohesin connect gene expression and chromatin architecture. *Nature* 467, 430-435, doi:nature09380 [pii] 10.1038/nature09380 (2010).

[152] Leeb, M. *et al.* Polycomb complexes act redundantly to repress genomic repeats and genes. *Genes Dev* 24, 265-276, doi:24/3/265 [pii] 10.1101/gad.544410 (2010).

[153] Bilodeau, S., Kagey, M. H., Frampton, G. M., Rahl, P. B. & Young, R. A. SetDB1 contributes to repression of genes encoding developmental regulators and maintenance of ES cell state. *Genes Dev* 23, 2484-2489, doi:23/21/2484 [pii] 10.1101/gad.1837309 (2009).

[154] Fazzio, T. G., Huff, J. T. & Panning, B. An RNAi screen of chromatin proteins identifies Tip60-p400 as a regulator of embryonic stem cell identity. *Cell* 134, 162-174, doi:S0092-8674(08)00692-2 [pii] 10.1016/j.cell.2008.05.031 (2008).

[155] Stielow, B. *et al.* Identification of SUMO-dependent chromatin-associated transcriptional repression components by a genome-wide RNAi screen. *Mol Cell* 29, 742-754, doi:S1097-2765(08)00164-0 [pii] 10.1016/j.molcel.2007.12.032 (2008).

[156] Yuan, P. *et al.* Eset partners with Oct4 to restrict extraembryonic trophoblast lineage potential in embryonic stem cells. *Genes Dev* 23, 2507-2520, doi:23/21/2507 [pii] 10.1101/gad.1831909 (2009).

Chapter 9

Stem Cells and Epigenetic Reprogramming

Perla Cota, Mehdi Shafa and Derrick E. Rancourt

Additional information is available at the end of the chapter

http://dx.doi.org/10.5772/55983

1. Introduction

In general stem cells have to fulfill two characteristics: self-renewal and the ability to differentiate into different cell/tissue types. Depending on their limitations in differentiation (pluripotent vs. multipotent) stem cells can be divided in embryonic or adult stem cells, depending on their limitations in differentiation [1]. This chapter will focus only on embryonic stem cells (ESCs) and their cognate artificial derivatives known as induced pluripotent stem cells (iPSCs). Embryonic stem cells, have been the center of much attention because of their pluripotency or ability to differentiate into any cell type in the body [2,3]. Induced pluripotent stem cells (iPSCs) are reprogrammed into the pluripotent state by the introduction of exogenous factors. These factors change the potency state of terminally differentiated somatic cells to by interacting with cellular chromatin and protein/RNA networks with the somatic cell. Following reprogramming, the newly formed stem cell resembles the ESC [4]. The recent development of these artificial or "man-made" cells has delivered two key potential upsides: (a) the ability to avoid the ethical issues associated with embryo-derived cells, and (b) the ability to generate autologous (i.e. patient derived) cells for regenerative medicine, tissue engineering, and disease modeling purposes [4]. Compared to ESCs, which are derived from the limited resource of assisted fertility by-products, iPSCs can potentially provide an unlimited source of pluripotent cells.

One of the applications of iPSCs is the ability to model diseases for drug screening, toxicology testing, and cell therapy among others [4]. For basic biomedical research, cell culture has been a key element for every approach. However one drawback when studying human cells is that they have limited life span in culture. Many cell lines have not been faithfully adapted for growth *in vitro*. Hence the lack of accessible models of normal and pathologic tissue has left many important questions in human pathogenesis inaccessible [5]. In contrast due to their self-renewal and pluripotency patient derived iPSCs can be extremely useful for patient research and diagnostic purposes. Every iPSC that is compromised in disease can be restructured into tissue in culture giving researchers an unlimited source of cells/tissue for the study of the disease [6, 4].

When applied to cell therapy, autologous iPSCs are differentiated into a chosen cell type and then transplanted to the damaged tissue with the advantage that immune rejection can be avoided. Furthermore iPSCs be used as a conduit for somatic gene therapy. For example a disease-causing mutation can be repaired in patient iPSCs by homologous DNA recombination prior to transplantation [4]. A study on engineered mice that suffer from human sickle cell anemia showed that when applying hematopoietic progenitor cells produced from autologous iPSCs, animals were rescued from systemic hematological symptoms. In this case, for the production of the hematopoietic progenitor cells, a biopsy of adult fibroblast was taken from the afflicted mouse and reprogrammed into iPSCs. Derived iPSCs were repaired by homologous recombination. These cells were then differentiated to hematopoietic progenitor cells *in vitro* and transplanted back into the affected mouse [7].

Despite the success in animals, there are still some drawbacks with using iPSCs for human benefit. Since the derivation of the iPSCs commonly involves integrating viral vectors for introducing reprogramming factor, this represents a risk to the human health. Moreover certain epigenetic abnormalities in the iPSCs including the epigenetic memory of their donor cells could lead to mutation in prolonged culture [4]. Such epigenetic differences are one reason why research has begun to focus on the epigenetics of cellular reprograming. Although iPSCs are the functional equivalent of ESCs, epigenetic differences have been noted, including differences in gene expression, DNA methylation, histone marks and telomere/telomerase status [8, 9]. Moreover, researchers have also recognized a role for chromatin remodeling during reprogramming and have recently applied small molecules to circumvent epigenetic blocks and enhance reprogramming efficiency [10,11].

Given that there is a huge interest in using iPSCs, mainly in regenerative medicine; researchers want to understand the exact mechanism of reprogramming, as any error in this process could cause tumor formation once applied to patients. Understanding the fundamentals of this reprogramming process by comparing it to the pluripotent state of ESCs will give us many tools to be able to manipulate the reprogramming process within a controlled environment.

Since iPSC are being compared at all times to ESCs, a basic concept that must be kept in mind is that ESCs rely on a complex network of interacting pluripotency transcription factors, and different "epigenetic landscapes" in order to maintain their "open" chromatin to regulate either self-renewal or differentiation [1]. Moreover, when a somatic cell is subjected to reprogramming, it suffers large-scale epigenetic alterations, carried on as if they were different multiple layers of epigenetic events that control the expression and accommodation of important pluripotency transcription factors [1].

In this chapter, a deeper explanation about iPSCs together with the basic concepts of epigenetics and the different levels of regulation will be provided. Insight into some of the recently discovered epigenetic events of cellular reprogramming will be discussed.

http://dx.doi.org/10.5772/55983

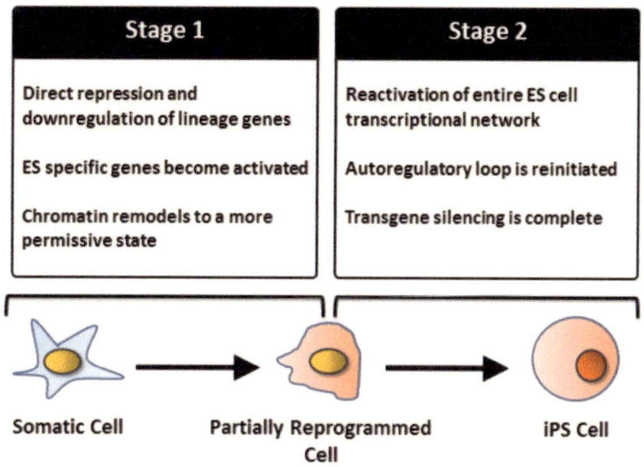

Figure 1. Two state process for reprogramming somatic cells (Adapted from 6)

2. Induced pluripotent stem cells

It was recently discovered that a terminally differentiated cell could be reprogrammed into an ESC-like cell using four transcription factors. Having pluripotent characteristics, these iPSCs are capable of becoming one of more than 200 cell types [12]. In order to be consider ESCs, they must fulfill certain criteria: (1) to express pluripotency factors such as Oct4, Sox2, Nanog, and SSEA1, however this criteria only apply for mouse ESCs, since in human ESCs SSEA3 and SSEA4 are expressed in stead however this criteria only apply for mouse ESCs, since in human ESCs SSEA3 and SSEA4 are expressed in stead (2) in female cells there must be the reactivation of the inactive X chromosome, (3) they should be able to differentiate into the three germ layers (ectoderm, endoderm and mesoderm) and in the case of mouse, be able to generate chimeras upon blastocyst implantation and pass through germline [13].

Takahashi and Yamanaka (2006) were the first that found a way to circumvent two of the most important drawbacks when using ESCs related to immune rejection and ethical. In their study they first hypothesized that the factors that play a role in maintaining ESC pluripotency could potentially turn somatic cells back into a pluripotent state. Starting with 24 candidate genes known to be involved in pluripotency and a herculean combinatorial effort they reduced this original 24 down to four factors: Oct3/4, Klf4, Sox2, and c-Myc able to reprogram mouse fibroblasts into an ESC-like state. Yamanaka and colleagues also reported the same results one year later for the human (2007) when using this same combination of factors [14].

After the four factor derivation of the iPSCs, much interest was focused on the process of somatic cell reprogramming. Although still not well understood, Scheper and Copray (2009) proposed one approach that divided reprogramming in two broad stages. First Oct4 and Sox2 repressed genes associated with the host cell lineage and reset the epigenome of the cell towards a permissive chromatin mode putting the cell in an embryonic-like state. The second stage allowed the reprogramming factors to reactivate the endogenous autoregulatory loop that triggers the pluripotency transcriptional network [6] (Figure 1).

After the first proposed cocktail to reprogram differentiated cells into iPSCs, many researchers started to ask the question of how these factors were interacting in order to modify existing epigenetic marks and return to a pluripotent state. To date it has been reported that differentiated cells have been successfully reprogrammed by substituting some of the factors such as Klf4 or c-Myc with other transcription factors such as Nanog or Lin28 or molecules (valproic acid or Wnt ligand). It tells us that there are different pathways involved in this process and that epigenetic enzymes are being activated in every case [9] all to one endpoint of pluripotency.

Waddington referred to epigenetics for the first time as genetic interactions that can affect the phenotype. Later, he proposed a model based on how cells followed a developmental differentiation path much like traveling down a series of canals that start from a fertilized totipotent embryo and ending up as a specific lineage committed cell [15]. In this model, cells committed to a specific lineage cannot be recommitted to another lineage or canal. However, with the recent milestone of iPSC generation, Yamanaka suggested that cells could be pushed back up the canal towards the pluripotent state. During reprogramming, cells can experience other events. They can be stopped by some epigenetic bump and remain incompletely reprogrammed. In this situation cells return to their specific lineage or transition to another lineage. Finally, instead of moving they can undergo apoptosis or cellular senescence. This model proposed by Yamanaka (2009) is known as the stochastic model of iPSC generation [16] (Figure 2).

Since the development of iPSCs, many researchers have focused their attention on the epigenetics changes that iPSCs acquire, together with the chromatin dynamics that occurs during cellular reprogramming. It has been already proposed that one way to ease cell destiny is by having less lineage epigenetic patterns [17]. The most used protocol for the production of iPSCs is the one that involves the application of the four transcription factors previously described by Yamanaka (2006) [15]. Thus initial studies have focused on how these four factors worked together to initiate the reprogramming cascade. In this regard it has been proposed that Oct4 and Sox2 are totally indispensable for reprogramming while Klf4 and c-Myc enhance the efficiency and alter the structure of the chromatin to enable Oct4 and Sox2 to target more genes that are important for pluripotency [15].

Finally, it is crucial to find the best method of reprogramming in order to approximate ESCs as much as possible. For this purpose, there are several variables that have to be taken into consideration in order to have reproducible and efficient reprogramming. First is the selection of reprogramming factors where the combination (and efficiency) can vary depending on the cell type. Second is the type of method used for factor delivery, be it viral vectors, RNA, protein, or small molecules, among others. Third is the selection of cell type, since the efficiency and kinetics reprogramming changes between cell types. Understanding how reprogramming factors coordinate the cascade to orchestrate reprogramming means it is important to know the right timing and stoichiometry for optimal reprogramming. Culture environment likewise is very important. Finally the selection of a method to identify and characterize iPSCs is very critical (Figure 3) [18].

http://dx.doi.org/10.5772/55983

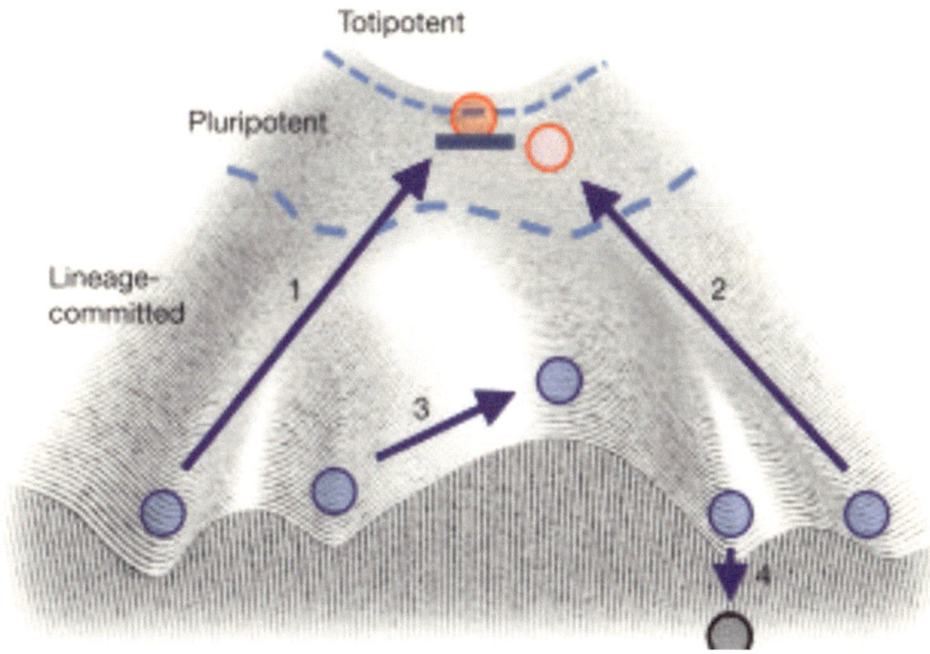

Figure 2. Stochastic model for iPSC generation. All of the cells initiate reprogramming, but only a few can achieve complete reprogramming. (Adapted from 16)

3. Epigenetics

Epigenetics is defined in general as heritable changes in gene expression that do not affect DNA-sequence [19, 20]. In the nucleus DNA is wrapped into a protein complex known as chromatin. This protein complex, known as the nucleosome, is formed by proteins called histones (H2A, H2B, H3, H4) (Figure 4) [21] into a structure resembling beads on a string. Histone H1 in turn play a role in assembling higher order chromatin structure by interacting with the "inter-bead" regions of chromation. Via changes in histone post-translational modifications (acetylation, methylation, ubiquitination, and phosphorylation), chromatin becomes very dynamic, controlling the expression or repression of specific genes in specific cells, as well as during the cell cycle or in response to environmental cues. These changes in histone via reversible post-translational marks (as well as reversible marks to primary DNA sequence) are considered to be epigenetic modifications. Additionally, changes in the nucleosome moving through DNA can be facilitated by chromatin remodeling enzymes [22, 21]. Histone modifications associated with active transcription, such as acetylation of histones 3 and 4 or di-/trimethylation of H3K4, are usually referred as euchromatin modifications. On the other hand there are the heterochromatin modifications which are characteristic to be on inactive regions such as H3K9me and H3k27me [23].

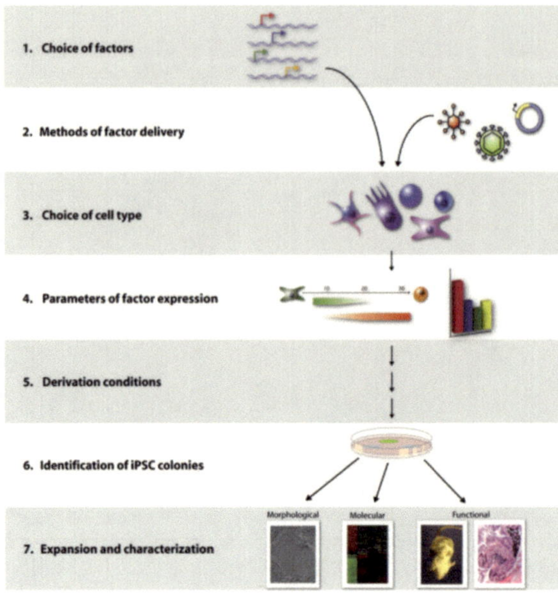

Figure 3. Overview of the iPSC Derivation Process (Adapted from 18)

4. Chromatin phases (epigenetic regulation)

- Chromatin Remodelling

Chromatin remodeling refers to the architectural change of chromatin by the movement of the nucleosomes along the DNA, giving rise to change the condensation of the chromatin. Protein complexes use ATP-hydrolysis to alter the histone DNA interaction, suggesting that there is a transient separation of the DNA from histone complexes, moving nucleosomes to a different position in the DNA or forming a DNA loop. These movements adjust the accessibility of DNA to transcription factors [23]. The many chromatin remodeling complexes are divided into families depending on their composition and biochemical activity. In this chapter, two of the most well studied ATP chromatin remodeling enzymes are discussed: SWI/SNF and CHD1 [24].

The basic assembling of the SWI/SNF chromatin remodeling enzyme in mammals (also known as BAF) is with the genes that code for the 9-12 subunits of the mammal SWI/SNF (mSWI/SNF) in combination with one catalytic ATPase subunit called brahma homolog (BRM or SMARCA2, BRM/SWI2-related gene 1[25]. mSWI/SNF uses the energy from ATPase hydrolysis to move along in the DNA. One way this works to move along is to bind the DNA into an internal site of the nucleosome, then pull it in order to weaken the nucleosome (Figure 5) [26].

One of the characteristics of mSWI/SNF chromatin remodelers is the subunit change during the transition from a pluripotent to a multipotent state and then from a multipotent state to differentiation. Ho and colleagues (2009) [27] did a whole genome study to observe the binding of mSWI/SNF in mouse ESCs and observed that the majority of the binding does not occur in transcriptional start sites, but at distal enhancers and silencer sites. In another study of Ho and collaborators, they showed that mSWI/SNF complex binds to promoters/enhancers of pluripotency transcription factors such as Oct4 and Sox2, in accordance with studies that have shown enhanced reprogramming when they add mSWI/SNF together with reprogramming factors [28].

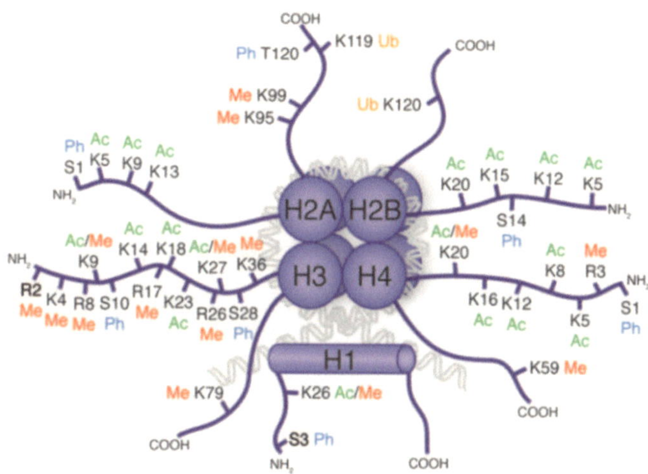

Figure 4. Nucleosome with histone posttranslational modifications (Adapted from 1)

The chromodomain helicase DNA binding protein 1 (CHD1) is well known for its remodeling activity in the maintenance of stemness. It also has main function in recognizing a substrate of transcription regulatory histone acetylation complex SAGA. CHD1 has been suggested to act as a molecular adaptor, which bring several epigenetic complexes together [29]. In ESCs, this adaptor has been suggested to be indispensable for the maintenance of pluripotent chromatin state where it is highly expressed when compared to differentiated cells. After knockdown of the CHD1 with RNAi, the pattern of diffuse ESCs heterochromatin disappears showing a higher amount of heterochromatin. In turn, CHD1 knockdown fibroblasts reprogrammed less efficiently [30]. The nature of CHD1 in pluripotent cells specifies that it can prevent the formation of heterochromatin foci [30]. CHD1 has also been reported to be one of the genes that activate Oct4, Sox2 and Nanog [31].

• Histone Modification

Histones are highly positively charged proteins that bind to DNA and have a major role in DNA packaging and gene expression. As mentioned earlier, they are subjected to a variety of post-translational modifications that alter the interaction of the histone protein with the bound DNA. These modifications include acetylation and methylation of the N-terminus tails as well as phosphorylation, poly ADP-ribosylation, ubiquitination and sumoylation. Differential modification of the core histones yields different chromatin structure. These patterns of modification form a kind of "histone code" that will ultimately govern gene expression [1].

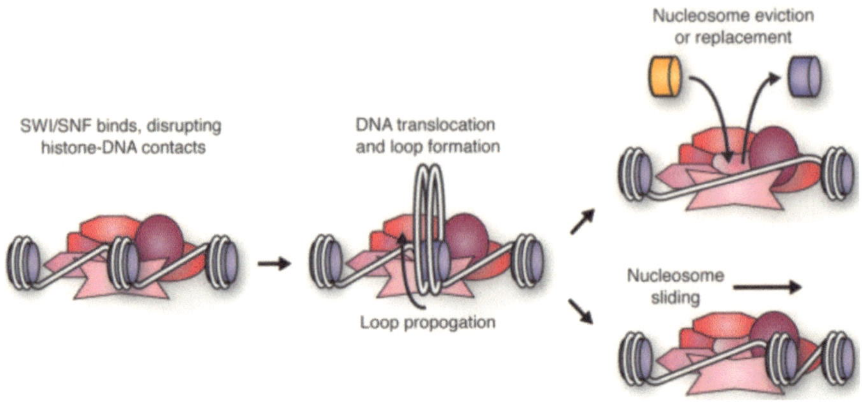

Figure 5. Chromatin Remodeling Enzyme SWI/SNFN (Adapted from 1)

Histone acetylation is the addition of acetyl moieties onto each of the histones of the nucleosome and is regulated by the activity of histone acetyltransferases (HAT) and histone deacetylases (HDAC). HATs and HDACs operate as coactivators and corepressors and together they dynamically change the activation and repression of genes in both a site specific as well as global manner. There are four families of HATs; Gcn5-related N-acetyltransferase (GNAT), MYST and p300/CBP. These HATs share highly similar motifs including an acetyl-CoA binding domain with the Arg/Gln-X-XGly-X-Gly/Ala sequence [32]. HAT activity and specificity are highly dependent on the complexes they form with other HATs and transcriptional co-activators. Lysine, found at the amino terminus of the histone, is the primary targeted site of acetylation. At physiological pH, lysine is positively charged and contributes to the overall positive charge of the histone. However, the amount of lysine acetylation is directly correlated with the accessibility of the amino terminus or "histone tail" [32]. More accessibility means greater degree of acetylation. Upon acetylation, the residue is neutralized, reducing the positive charge of the histone, decreasing the interaction with the negatively charged DNA and directly influencing chromatin structure. HDACs are broken down into two families; classical HDACs and NAD+ HDACs. Like their HAT counterparts, HDACs share a conserved active site [33] and also require the need to complex with co-repressors in order to function properly. Once bound, active HDACs serve to remove the acetyl moiety from the histone tail through a charge-relay system of residues found within the active site [33, 34]. Once removed, the histones bind tighter to the DNA as well as enabling tighter packing of adjacent histones leading to more transcriptional repression (Figure 6) [35].

Histone methylation is the addition of methyl groups onto lysine and arginine residues of histones in both transcriptionally active as well as silenced regions of the chromatin. The patterns of modification of lysine residues within histones are more defined as compared to arginine. Methylation is catalyzed by Histone methyltransferases (HMTs) and demethylation by histone demethylases (HDMs). Lysine residues can be methylated up to three times whereas arginine can only be methylated twice. Moreover, the symmetry of the methyl groups on each of the residues also plays an important role in the function of chromatin. HMT all share a common SET domain within their catalytic core [37]. Lysine HMTs are very well defined as

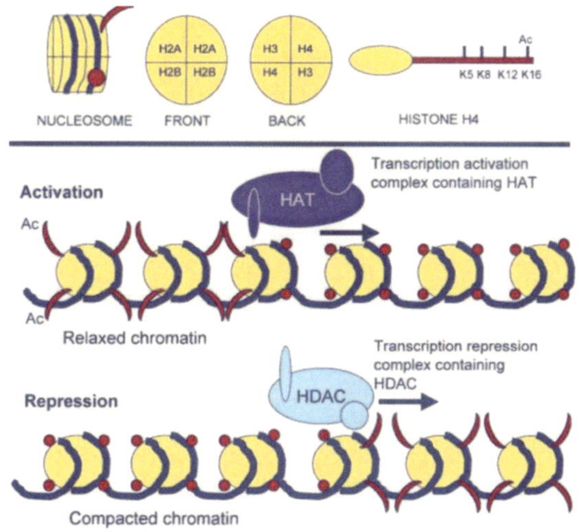

Figure 6. Histone acetylation and deacetylation mechanism (Adapted from 36)

compared to arginine HMT. Arginine N-methyltransferases (PRMT) are less defined in terms of the targeted sites of methylation, with multiple target residues in histone H3 and H4 [37].

The process of methylation does not alter the charge properties of the targeted residue, unlike those of acetylation. Instead, the addition of methyl groups serves as a recognition site for regulatory proteins to bind and elicit additional modification. This allows for a great deal of complexity depending on the target region of the MMT and the resultant recognition effector protein. In essence, it allows additional information to be encoded in the histones beyond just stearic and charge hindrance of acetylation and phosphorylation [38]. Thus, this places a great deal of importance on the proteins that interact with the methylated residue. There are a variety of motifs that are able to recognize both single and double methylated lysine residues and even one methylated lysine and methylated arginine. The basic conformation that recognizes single methylated lysine residues is a cage with polar and non-polar regions that envelops the methylated lysine residues [39]. A few of the common motifs include ankyrin repeats, chromodomain, MBT repeats, PHD finger, and double tudor [40]. An example of a motif that is able to recognize methylated lysine and arginine residues is RAG2-PHD. These varieties of recognition motifs underline the great deal of complexity behind methylated residues and even hints at potential cross talk between methylated lysine and arginine residues. There have been studies showing that the methylation of one residue, H3R2, precluded the recognition of a neighbouring methylated residue, H3K4me3 (Figure 7) [37].

Methylation of lysine and arginine residues has recently been discovered to also undergo demethylation via histone demehtylases (HDM). These enzymes are divided into two classes: amine oxidases, which are able to demethylate the first and second methyl lysine groups, and JmjC domain-containing proteins, which are able to demthylate all three methyl lysine groups. As well, it was also found recently that a JmjC domain-containing protein, JMJD6, was able to reverse arginine methylation [41]. HDM became a very key regulator of pluripotency after it was found that KDM3A and KMD4C are direct transcriptional targets of the pluripotency

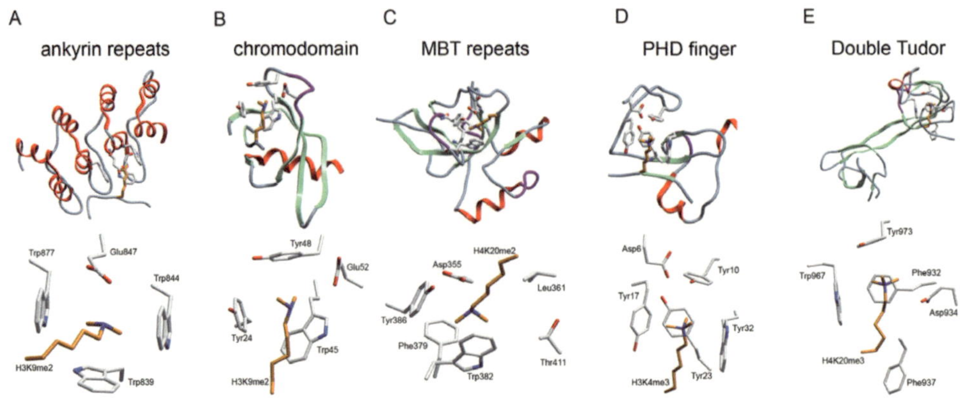

Figure 7. Methyl-lysine binding effector proteins (Adapted from 37)

promoting transcription factor Oct4 [42]. When these two enzymes were knocked down, the resulting cells lost the ability to self-renew as well as showed an altered morphology. Moreover, another high-ranking HDM JARID2 is highly expressed in ESCs but becomes rapidly downregulated upon differentiation [43]. The level of regulatory complexity of the genes, not only in ESCs or iPSCs, but also somatic cells, Hence, further examination is needed to elucidate how these enzymes contribute to the epigenetic regulation of genes.

- DNA methylation

DNA Methylation is the classical example of epigenetic regulation of gene expression. This process, catalyzed by DNA methyl transferase (DNMT) enzymes, involves the addition of a methyl group onto the carbon 5 position of cytosine residues within DNA, forming 5-methylcytosine. There are three main members of the DNMT family: Dmnt1, Dmnt3a and Dmnt3b. Dmnt1 is the best studied of the three and its primary role is to copy DNA methylation patterns during DNA synthesis as well as repair of DNA methylation patterns [44]. Dmnt3a and Dmnt3b are similar enzymes both in structure as well as function. These two DMNTs are capable of methylating native DNA, regardless of whether the DNA is in a replicative state or not [45]. Since they are able to write DNA methylation patterns onto "naked" DNA, they are termed *de novo* DMNTs.

Patterns of DNA methylation can be "read" through the recruitment of three different protein families: MBD, zinc-finger, and UHRF proteins. The most well-known are the MBD proteins, which interact with the DNA via a methyl-CpG-binding domain. Once the MBD proteins bind to the 5'site of the methylated cytosine, they repress transcription. Zinc finger proteins, like MBD proteins, also recognize and bind to methylcytosine, however they have a preference for consecutively methylated cytosine residues as well as non-methylated residues. Interestingly they are still able to repress transcription of DNA in a similar manner. Ubiquitin like containing PHD and RING finger Domain (UHRF) proteins use their intrinsic RING and SET DNA binding domains to interact with the methylated cytosine. However, the purpose of UHRF is not to repress transcription, but actually to aid DMNT, especially during DNA replication, in order to conserve and maintain the DNA methylation [46].

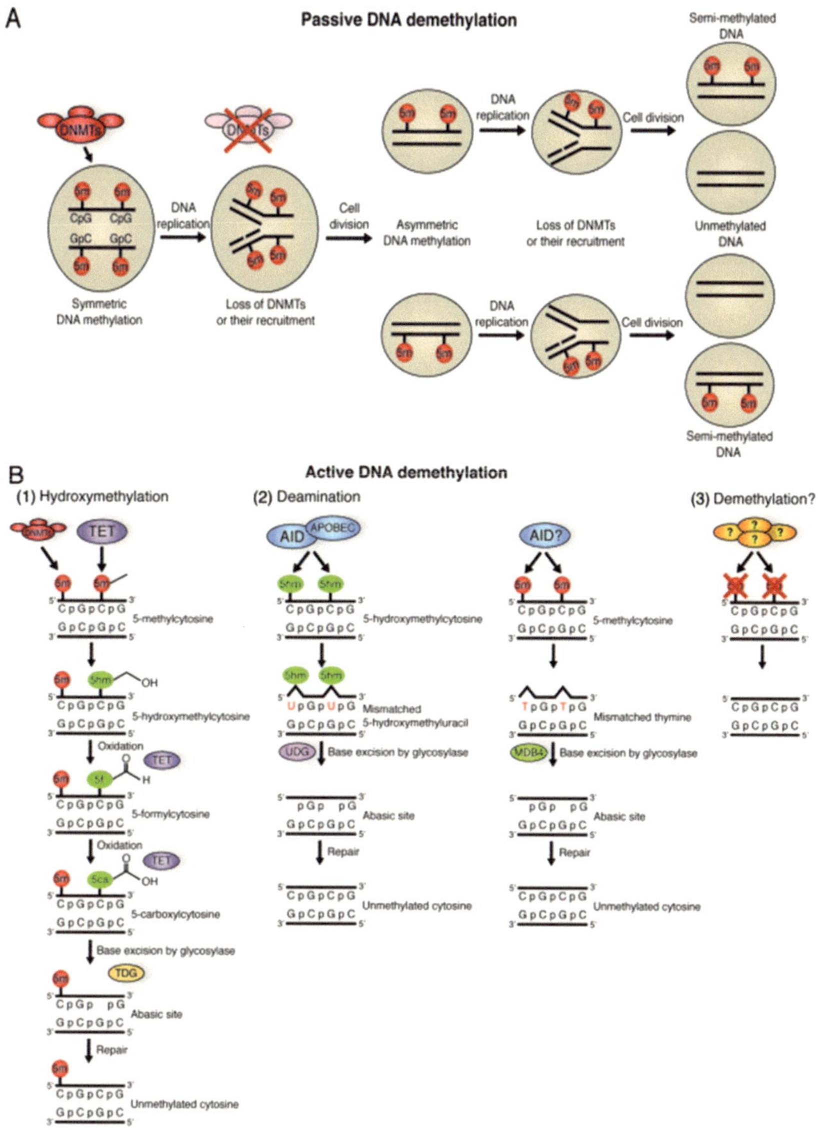

Figure 8. Active and passive DNA methylation mechanism (Adapted from 1)

DNA methylation can be achieved via two mechanisms, either actively or passively (Figure 8). Passive demethylation involves the inhibition of the DMNT protein during DNA replication and allows for newly synthesized cytosine to escape methyl imprinting from its parent DNA strand. This process usually occurs during cellular replication. Active demethylation can occur in both dividing and non-dividing cells [47]. Currently, there is no known enzyme that is able to remove the strong covalent bond of the methyl group from the cytosine residue. Instead, the methylated cytosine is thought to undergo a series of further modifications (AID/APOBEC) that ultimately change the 5mC into a thymine [48]. This elicits a base mismatch and activates the base excision repair pathway to replace the residue with a naked cytosine. Another proposed demethylation pathway involves the use of the ten-eleven translocation (Tet) enzymes. This family of proteins are able to add a hydroxyl group onto the methyl moiety of 5mC to form 5hmC. Once in this state, 5hmC can return to an unmodified cytosine residue through either further oxidation by Tet enzymes or deamination by AID/APOBEC. Unlike the deacetylation and demethylation of histones, DNA demethylation is much more complex and involves a number of enzymatic processes which has contributed to the great deal of debate about which pathway is more dominant [50].

5. Epigenetic reprogramming

In order to understand the interactions and mechanisms involved in reprogramming a differentiated cell into an iPSC, a great effort has been made to study the ESC pluripotent state, in particular the means by which pluripotency transcription factors interact with each other or with other proteins such as chromatin remodeling enzymes and histone modifying enzymes. Moreover researchers have focused on finding the networks in iPSCs once the endogenous pluripotency factors have been activated by the exogenous Yamanaka factors [13]. Understanding the interactions between the core pluripotency transcription factors and the previously mentioned epigenetic enzymes will provide some advantages to the iPSC field. One such advantage is the possible discovery of new cocktails that enhance reprogramming. In addition, it could explain the chronology of the epigenetic events for reprogramming on a molecular level. This section will cover some of the known molecular interactions among the pluripotent transcription factors and some of the epigenetic enzymes.

5.1. Pluripotency gene networks

The first event toward transition from a differentiated to an iPSC state is the establishment of a proper chromatin state. Once the cells have found the correct chromatin state, the second event is to maintain and inherit it as they divide and proliferate [14]. The natural state of an ESC chromatin is known as "open", where the heterochromatin is disperse and dynamic, which at the same time reflects a hyperactive transcriptional status [49]. The molecular structure for ESC to maintain pluripotency requires an interconnection of transcription factors with epigenetic proteins that are also interacting with the DNA. Due to fact that iPSCs are like ESC, they have to sustain the same molecular structure. In addition, they have to overcome an epigenetic barrier during the reprogramming process. The reprogramming process involves a chain reaction involving transcription factors, chromatin modifying enzymes and other histone related enzymes.

An approach of how reprogramming occurs, suggests that the first step maybe interaction of transcription factors with the naked DNA, via histone modifiers or together with chromatin remodeling factors [14]. There are not time points or an order to follow for each specific transcription factor. Certain transcription factors are able to interact with DNA or with a chromatin remodeling enzyme depending what gene is activating. This molecular mechanism is determined by the locus, the type of transcription factor and on the context [14]. Hence the function of the four Yamanaka factors, Oct4, Sox2, Klf4 and c-Myc, [13] is crucial.

It has been reported that Oct4 is indispensable in the reprogramming process. In some cases, such as what has been observed in neural stem cell reprogramming, the presence of Oct4 is sufficient for reprogramming in [51]. Moreover, as Oct4 can work alone, it also has a great effect in reprogramming when combined with Sox2. Oct4 and Sox2 form a heterodimer that interact with some promoters. In addition this heterodimer has been shown to interact with Nanog. Nanog is another transcription factor that participates in the ESCs regulatory circuitry together with Oct4 and Sox2 to maintain pluripotency [52]. In this context they activate transcription in a chromatin independent manner by interacting with transcriptional co-activators [53, 54]. Moreover, it has been shown in mouse ESCs that Oct4 and Nanog can repress gene expression through interaction with histone deacetylase such as Mta1 [55]. It has been elucidated that in the first stage of reprogramming, a cascade of differentiation genes are turned off, while pluripotency genes progressively become upregulated in order to push the differentiated cell toward an ESC-like state.

c-Myc is an important participant in recruiting multiple chromatin modifications, such as histone acetyltransferases (GCN5, p300) and histones deacetylases (HDACs). In this regard, c-Myc increases the methylation site H3K4me3 and the global acetylation [56]. In the reprogramming process c-Myc activates its target before other core pluripotency transcription factors are activated, facilitating the opening of the chromatin for other factors [57, 58]. An example of c-Myc's potential in opening chromatin is its association with Tip60-p400 complex, which acetylate and remodel nucleosomes respectively. p400 is a member of the Swi2/Snt2 family which is well known among the ATPase chromatin remodelling enzymes, exchanging histones H2AZ-H2B within nucleosomes [57]. It also functions to release paused RNA polymerase from about one-third of the genes that are being actively transcribed. This activity could enhance cellular reprogramming [59]. At the same time, the transcription factor Klf4, activates the transcription of Sox2 which participates in the pluripotency cascade [60].

5.2. Chromatin remodelling

Based on the Yamanaka's stochastic model (Figure 2), cells need to overcome the epigenetic barrier in order to become pluripotent [16, 17]. Nowadays, one of the major focuses in the iPSC field is to understand the epigenetic molecules that orchestrate chromatin remodelling in order to organize it into a pluripotent state similar to ESCs. While some somatic cell reprogramming mechanisms are being unveiled, many are still yet unknown [9].

- DNA methylation

Once the differentiated cells have been reprogrammed, the epigenetic marks in iPSCs resemble ESCs [61]. For a great amount of eukaryotic DNA methylation is a mark that serves to define different cellular functions such as X chromosome inactivation, aging, imprinting, genome stability, tissue specific gene regulation, and so on [63, 64]. DNA methylation is one of the epigenetic marks that is modified during reprogramming (Figure 9). In this case, the process of demethylation is most common taking the methyl group from the promoters of some genes that are responsible for pluripotency which in turn allows them to return to a pluripotent state.

Currently there is not too much evidence about the process of demethylation and the enzymes that catalyze this event. However, DNA demethylation events have been classified as passive or active. Passive DNA demethylation occurs during the process of DNA replication when maintenance methytransferases are inactive, and thus they are not able to methylate newly released strands [62]. On the other hand for active DNA demethylation, the main protagonists are enzymes that work regardless of DNA replication [62].

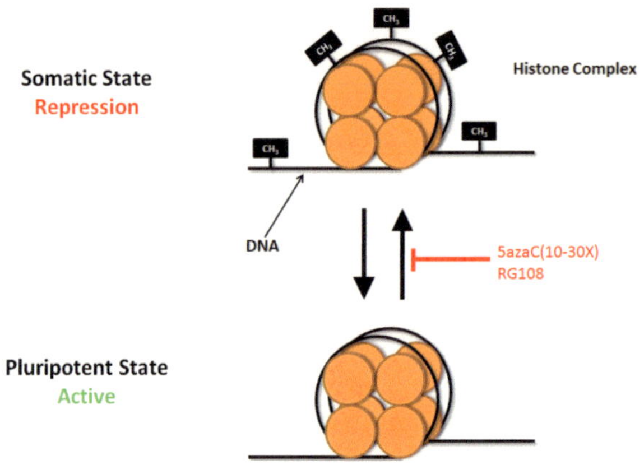

Figure 9. DNA demethylation as a reprogramming process when going to iPSC (Adapted from 13)

An example of an active DNA demethylation event is shown by Pereira and colleagues (2008), where they have studied the efficiency of reprogramming of human lymphocytes by fusing them with mouse ESCs. They found that one of the first events occurring was the demethylation of the Oct4 gene. They suggest that this event is a result of an active chromatin remodeling locus before its actual expression [65].

Two mechanisms have been proposed as candidates for the active DNA demethylation mechanism. The first one involves the deamination of 5-methylcytosine in DNA by an enzyme called activation-induce deaminase (AID) [62]. The other mechanism is based on the oxidization of the 5-methyl group (-CH_3) followed by conversion into 5-carboxylcytosine (-COOH). This conversion is catalyzed by the enzyme TET1 in a Fe(II) and α-ketoglutarate dependent reaction [66]. To date neither of these mechanisms has been proven *in vitro.* Hence there are still a number of unknown molecular mechanisms that govern the reprogramming process. Is there an active DNA demethylation? How are the DNS demethylating enzymes activated? Are they recruited by other processes of chromatin remodeling or do transcription factors initialize the process?

- Histone modifications

The most common marks in ESC and iPSCs are the active mark of H3k4me3 (histone 3, trimethylated at lysine 4) and the repressive mark H3k27me3 (histone 3, trimethylated at lysine

27). These marks in the histones occur by S-adenosylmethionine (SAM-dependent) protein methylation. Due to this bivalent mark in the histones, these cells have the capacity to activate or repress genes them in order to change their fate [67].

Since the histone-lysine N-methyltransferase (MLL) catalytic subunits are well known to introduce the H3k4me3 mark and activate transcription, this makes them potential regulators in reprogramming. An example of their activity is shown with Wdr5, a subunit in common with H3k4 methyltransferases. Wdr5 has been proposed to play an important role in maintaining pluripotency and has been proposed as one possible mechanism occurring during reprograming. Wdr5 is activated by exogenous Oct4 when mouse embryonic fibroblast are transfected with the four Yamanaka factors. Wdr5 directly binds to loci where self-renewal genes are encoded such as Oct4 and Nanog, in order to re-establish an H3k4 mark. Wdr5 is thus defined as an indispensable subunit which proportions H3k4 methylation [68]. Similarly, MLL interacts with some other chromatin remodeling enzymes, such as CHD1 and NURF, in order to achieve H3k4 methylation [29, 69].

It has been already mentioned that ESCs and iPSCs are known to have bivalent chromatin bearing both the active mark H3k4me3, and the repressive mark H3k27me3. One of the mechanism that controls the bivalency is through the activity of the polycomb proteins found in two major complexes PRC1 and PRC2 [70]. It is thought that PRC1 and PRC2 act as antagonists and are intrinsically involved in establishing the fate of ESC development. PRC2 is in charge of the H3k27me3 mark [71] and known to silence the HOX genes used and other regulators during ESC differentiation [72]. One of the basic mechanisms in ESC/iPSC differentiation is the demethylation of this H3k27me3 mark.

Utx demethylase has been reported to be a significant regulator of cellular reprogramming [73]. Utx is encoded by an X-chromosome gene and belongs to the small family of Jmjc proteins, mediating the demethylation of H3k27 tri- and di-methyl repressive chromatin marks. In this study, it was found that Utx was dispensable for the maintenance of pluripotency, since pluripotency marker expression was maintained in knockout ESC lines. However fibroblasts derived from Utx knockout mice failed to be reprogrammed. This result indicated that the absence of Utx prevented the demethylation of H3k27me3 marks needed to re-establish pluripotency *in vitro* [73].

H3K9me3 and H3k79me2 have also been reported to be important for the maintenance of pluripotency in ESCs. These marks are usually left by repressive methyltransferases via the identification of specific motifs in heterochromatin. Reprogramming has been facilitated by the inhibiting two methyltransferases (GLP and G9a) that methylate H3k9 [74].

Histone acetylation is another important histone mark that has usually been correlated with gene activation (Figure 10). This mark has been reported to transform chromatin during reprogramming. Little is known about histone acetyltransferases for pluripotency maintenance or reprogramming, however Tip60/p400 has been reported as a histone acetyltransferase important for maintaining the ESC state. Here Tip60/p400 also works as a chromatin remodeling enzyme, since it has a SWI2/SNF2 subunit [75].

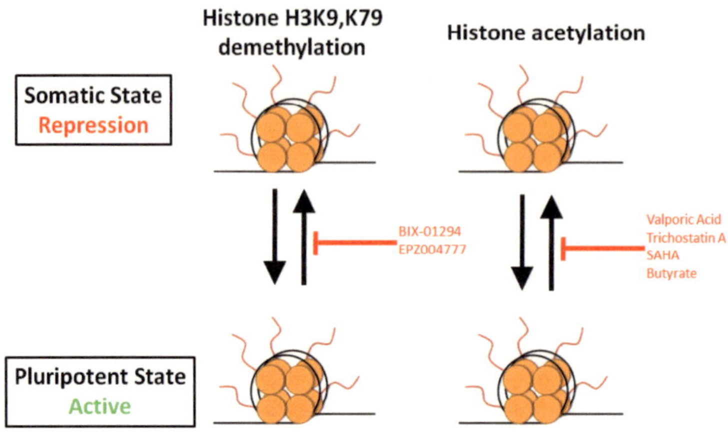

Figure 10. Histone acetylation as an activating mark for reprogramming. (Adapted from 13)

On the other hand the histone deacetylases (HDACs) are known to repress the expression of genes, therefore there is an increased interest on their inhibition. An example of their importance in reprogramming of somatic cells was shown by Hadas Hezroni and collaborators (2011) [76]. In this study they used hybrid cell lines by fusing mouse embryonic fibroblast with ESCs and found that low H3k9 acetylation correlated with low reprogramming capacity. When they tried to overcome this effect using histone deacetylase inhibitors, they found an increase in the reprogramming efficiency. They reported that genes involved in extracellular matrix (ECM) activity were enriched during reprograming and concluded that H3K9ac is a mark intrinsically related to pluripotency and that promoting its increase using HDACs inhibitors promote ECM activity, which co-relates positively affect pluripotency and self-renewal [76].

Most epigenetic reprogramming studies have focused on isolated chromatin marks, revealing the down regulation of somatic genes. However there are more than some marks that lead to an "open" dynamic chromatin. Anna Mattout and colleagues [77] presented a study where for the first time they showed chromatin dimensions as global changes occurring during reprogramming. They analyzed a battery of histone modifications (H3ac, H4ac, H4k5ac, H3k27ac, H3k4me3, H3k36me2, H3k9me3, and H3k27me3 also γH2AX, HP1α and lamin A, by immunofluorescence and biochemical fractionations comparing mouse ESCs to fully- and partially-reprogrammed mouse iPSCs. They first identified that H3k36me2, H4k5ac and H3k4me3 have the highest correlation with pluripotency. Later, they showed that most of the euchromatin/active marks (H3ac, H3k9ac, H3k27 ac, H4ac, H4k5ac, H3k4me3 and H3k36me2) are higher in the ESCs and fully reprogrammed iPSCs, whereas in partially reprogrammed cells these marks more closely resembled that of mouse embryonic fibroblasts. On the other hand they observed that the marks in heterochromatin, such as HP1α and H3k9me3 rearrange during reprogramming towards a more diffused pattern. This was seen in all of the cells lines including partially reprogrammed iPSCs. With these two phenomena they presented a time line suggesting that marks in heterochromatin start changing at a very early stage (by day 6 during reprogramming) compared to the histone changes occurring in the active euchromatin. They concluded that during reprogramming global histone heterochromatin defining marks start changing and spreading at an early stage of reprogramming as a form of physical rearrangement prior to the euchromatin epigenetic alterations which occur after day 7 (Figure 11) [77].

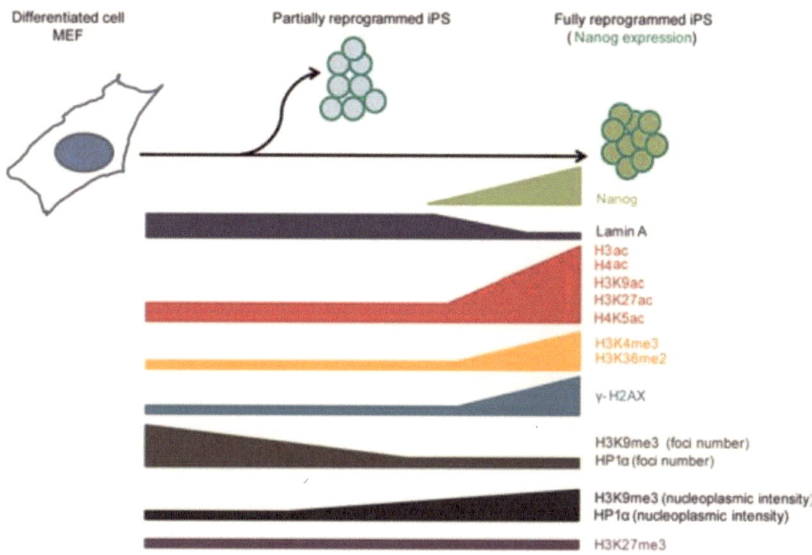

Figure 11. Global epigenetic changes in iPSCs (Adapted from 80).

- Chromatin remodeling

As previously mentioned, chromatin remodeling is caused by catalytic modification where ATPases use the energy from the ATP to move along in DNA. Thus they regulate gene expression by spacing nucleosome arrays, exchanging histone variants, disassembling or sliding the nucleosome [14]. One example of the importance of the chromatin remodeling enzymes is observed in a study of Brg, part of a family of DNA ATPases homologous to the catalytic subunit of yeast SWI2/SNT2 ATPase [78]. Brg is assembled to 11 other Brg/Brahma associated factors (BAFs). In ESCs, BAF complexes have an exclusive subunit which is called esBAF. The authors observed that esBAF facilitates STAT3 to access binding sites that will respond to LIF, which will further activate the pluripotency transcription factor Klf4 [78]. Therefore, it is thought that the LIF signaling pathway is dependent on prior chromatin remodeling [78]. A previous study showed that overexpressing of esBAF in addition to the four Yamanaka was able to acquire a euchromatic chromatin by increasing the kinetics of Oct4, Nanog and Rex1 promoter demethylation. This facilitated the accessibility of the reprogramming factors and hence the process was enhanced [28].

Moreover, Onder and colleagues (2012) [79] focused on the study of chromatin-modifying enzymes during the reprogramming process in iPSCs. They used a loss of function approach with shRNA where they selected 22 genes involved in DNA and histone methylation pathways. From their results, they found that inhibition of the histone methyltransferase DOT1L, reprogramming was enhanced resulting in more iPSCs colonies. DOT1L inhibition does not enhance the upregulation of the pluripotency gene network but can be used to substitute for

Klf4 and c-Myc during reprogramming. Inhibition of this molecule is associated with an increase of Nanog and Lin28, factors, which are necessary for reprogramming. Finally using ChIP-seq they found that the H3K79me2 mark was lost in genes that participate in epithelial mesenchymal transition. Among some of the mesenchymal regulators were SNAI1, SNAI2, ZEB1, ZEB2 and TGFB2, that at the same time where strongly repressed during reprogramming. Together with this, they also reported that epithelial genes such as CDH1 (E-cadherin) and OCLN were upregulated. The above was a clear example of how chromatin modifying enzymes are critical in the molecular process of reprogramming enhancing the cascade that begins with the four Yamanaka factors [79].

The above are just some of the studies of many that have been reported. They lead us to question whether remodeling enzymes are in charge of the major chromatin opening that occurs during reprogramming, or if histone marks lead the process.

5.3. MicroRNAs in reprogramming

MicroRNAs (miRs) are small RNAs involved in the inhibition of the gene expression by destabilizing target RNAs. They are usually formed by the proteins Dicer and Drosha with its cofactor Dgcr8. The importance of miRs arises from the observation that some miRs induce reprogramming of somatic cells into iPSCs [80]. Among some of the miRs found to positively regulate ESC pluripotency are: ESC cell-cycle regulating miR291a-3p, miR291b-3p, miR294, miR295 and miR302. Interestingly, miR302 has been reported regulate some of the epigenetic modifications that occur during reprogramming. miR302 is a family of four highly homologous microRNAs that are transcribed together and form a noncoding RNA cluster [81]. They are highly expressed in human ESCs and absent in differentiated cells. Lin and collaborators (2011) have focus in how the miR302 controls several enzymes that are involved in active demethylation [81]. MiR302 targets and represses AOF2/1 histone demethylases and MECP1/2 (methyl CpG binding proteins). At the same time it blocks cytosine methyltransferase 1 (DNMT1). During reprograming miR302 coordinates DNA demethylation, together with a histone methylation on the active mark H3k4 that will alter the chromatin structure and the gene activity. The example above highlights the critical role of miRs in founding and sustaining pluripotency in cells [82].

6. Conclusion

The biology of pluripotent stem cells is still in a very early stage; even understanding what is the best way to obtain a true embryonic stem cell remains unclear. During reprogramming a number of changes occur in the cell. These changes start usually by the stimulation of exogenous transcription factors that consequentially trigger a large number of other reactions: signalling, gene transcription, and most importantly epigenetic changes, including chromatin remodeling, histone modification, and DNA methylation [1, 12, 13].

During reprogramming, chromatin changes to an "open" dynamic configuration resembling the epigenetic landscape as in ESCs. In order to reach this configuration, the somatic cell has to interconnect transcription factors, chromatin, and histone modifier enzymes. What is the kinetics of this process? This is one of the first questions that puzzle most researchers. While there is firm understanding that the exogenous transcription factors are first to prompt these changes, reprogramming is not an efficient process. From the extensive interest in making reprogramming an efficient process, there have been a lot of remarkable results using different type of molecules that target chromatin enzymes. Thus understanding of the reprogramming process, including the timing of chromatin remodeling, interactions with transcription factors, increase or decrease of histone acetylation and most important, the precise interconnection of factors that break the epigenetic barrier, will give us a base line to design a better protocol for the develop of iPSCs.

There will come a point where researchers will manipulate chromatin kinetics in order to promote the reprogramming of somatic cells into iPSCs. This achievement will bring a cell that reprograms efficiently; in a short period of time, which will have an epigenetic signature identical to ESCs. In the future, with a better understanding of chromatin not only in reprogramming but also in differentiation, the iPSC field could become an area of synthetic biology. In any case, the iPSC field still has a long way to go before it is fruitful. A clearer understanding of the epigenetics of the reprogramming must come forward before iPSCs can be fully accepted for regenerative medicine.

Author details

Perla Cota, Mehdi Shafa and Derrick E. Rancourt*

*Address all correspondence to: rancourt@ucalgary.ca

Department of Biochemistry and Molecular Biology, University of Calgary, Calgary, AB, Canada

References

[1] Tollervey, J R, & Lunyak, V V. Epigenetics Judge, jury and executioner of stem cell fate. Landes Bioscience (2012). http://www-ncbi-nlm-nih-gov.ezproxy.lib.ucalgary.ca/pubmed?term=22805743

[2] Evans, M J, & Kaufman, M H. Establishment in culture of pluripotential cells from mouse embryos. Nature (1981). http://www-ncbi-nlm-nih-gov.ezproxy.lib.ucalgary.ca/pubmed/7242681, 5819, 154-6.

[3] Thomson, J. A, & Odorico, J. S. Human embryonic stem cell and embryonic germ cell lines. Trends Biotechnol (2000). http://www.ncbi.nlm.nih.gov/pubmed/10652509

[4] Verma, A. Verma N: Induced pluripotent stem cells and promises of neuroregenerative medicine. Neurol India (2011). http://www.ncbi.nlm.nih.gov/pubmed/21891933

[5] Park In-HArora N, Huo H, Maherali N, Ahfeldt T, Shimamura A, Lensch M W, Cowan C, Hochedlinger K and Daley G Q. Disease-Specific Induced Pluripotent Stem Cells. Cell (2008). http://www.ncbi.nlm.nih.gov/pubmed/18691744, 134, 1-10.

[6] Scheper, W, & Copray, S. The Molecular Mechanism of Induced Pluripotency: A two-Stage Switch Stem Cell. Rev and Rep (2009). http://www.ncbi.nlm.nih.gov/pubmed/19551525, 5, 204-223.

[7] Hanna, J, Wernig, M, Markoulaki, S, Sun, C-W, Meissner, A, Cassady, J P, Beard, C, Brambrink, T, Wu, L-C, Townes, T M, & Jaenisch, R. Treatment of Sickle Cell Anemia Mouse Model with iPS Cells Generated from Autologous Skin. Science (2007). http://www.ncbi.nlm.nih.gov/pubmed/18063756

[8] Maherali, N, Sridharan, R, Xie, W, Utikal, J, Eminli, S, Arnold, K, Stadfeld, M, Yachechko, R, Tchieu, J, Jaenisch, R, Plath, K, & Hochedlinger, K. Directly Reprogrammed Fibroblasts Show Global Epigenetic Remodeling and Widespread Tissue Contribution. Cell Stem Cell (2007). http://www.ncbi.nlm.nih.gov/pubmed/18371336, 1, 55-70.

[9] Hochedlinger, K, & Plath, K. Epigenetic reprogramming and induced pluripotency. Development (2009). http://www.ncbi.nlm.nih.gov/pubmed/19168672, 136(4), 509-23.

[10] Shafa, M, & Krawetz, R. and Rancourt D E: Returning to the stem state. Epigenetics of recapitulating pre-differentiation chromatin structure. BioEssays (2010). http://www.ncbi.nlm.nih.gov/pubmed/20652894, 32, 791-799.

[11] Shi, Y, Desponts, C, Do, J T, Hahm, H S, Schöler, H. R, & Ding, S. Induction of pluripotent stem cells from mouse embryonic fibroblasts by Oct4 and Klf4 with small-molecule compounds. Cell Stem Cell (2008). http://www.ncbi.nlm.nih.gov/pubmed/18983970

[12] Takahashi, K. Yamanaka S: Induction of Pluripotent Stem Cells from Mouse Embryonic and Adult Fibroblast Cultures by Defined Factors. Cell (2006). http://www.ncbi.nlm.nih.gov/pubmed/16904174, 126, 663-676.

[13] Ehrensberger, A H, & Svejstrup, J Q. Reprogramming chromatin. Critical Reviews in Biochemistry and Molecular Biology (2012). Early Online http://www.ncbi.nlm.nih.gov/pubmed/22757592, 1-19.

[14] Takahashi, K, Tanabe, K, Ohnuki, M, Narita, M, Ichisaka, T, Tomoda, K, & Yamanaka, S. Induction of pluripotent stem cells from adult human fibroblasts by defined factors. Cell (2007). http://www.ncbi.nlm.nih.gov/pubmed/18035408, 31(5), 861-72.

[15] Waddington, C. The strategy of the genes; a discussion of some aspects of theoretical biology. London, Allen and Unwin (1957).

[16] Yamanaka, S. Elite and stochastic models for induced pluripotent stem cells generation. Nature (2009). http://www.ncbi.nlm.nih.gov/pubmed/19571877, 460, 49-52.

[17] Okita, K, Ichisaka, T, & Yamanaka, S. Generation of germline-competent induced pluripotent stem cells. Nature (2007). http://www.nature.com/nature/journal/n7151/abs/nature05934.html

[18] Maherali, N, & Hochedlinger, K. Guidelines and Techniques for the Generation of Induced Pluripotent Stem Cells. Cell Stem Cell (2008). http://www.ncbi.nlm.nih.gov/pubmed/19041776, 3, 595-605.

[19] Wolffe, A P, & Matzke, M A. Epigenetics: regulation through repression. Science (1999). http://www.ncbi.nlm.nih.gov/pubmed/10521337, 288, 481-486.

[20] Strahl, B. D, & Allis, C. D. The language of covalent histone modifications. Nature (2000). http://www.ncbi.nlm.nih.gov/pubmed/10638745, 403, 41-45.

[21] Orkin, S H, & Hochedlinger, K. Chromatin connections to pluripotency and cellular reprogramming. Cell (2011). http://www.ncbi.nlm.nih.gov/pubmed/21663790

[22] Han, J W, & Yoon, Y-s. u. p. Epigenetic Landscape of pluripotent stem cells. Antioxidants & Redox Signaling (2012). http://www.ncbi.nlm.nih.gov/pubmed/22044221, 17, 205-223.

[23] Li, B, Carey, M, & Workman, J L. The Role of Chromatin during Transcription. Cell (2007). http://www.ncbi.nlm.nih.gov/pubmed/17320508, 128, 707-719.

[24] Hargreaves, D C, & Crabtree, G R. ATP-dependent chromatin remodeling genetics, genomics and mechanisms. Cell Res (2011). http://www.ncbi.nlm.nih.gov/pubmed/21358755, 21, 396-420.

[25] Phelan, M L, Sif, S, Narlikar, G J, & Kingston, R E. Reconstitution of a core chromatin remodeling complex from SWI/SNF subunits. Mol Cell (1999). http://www.ncbi.nlm.nih.gov/pubmed/10078207, 3, 247-253.

[26] Saha, A, Wittmeyer, J, & Cairns, B R. Chromatin remodeling through directional DNA translocation from an internal nucleosomal site. Nat Struct Mol Biol (2005). http://www.ncbi.nlm.nih.gov/pubmed/16086025, 12, 747-755.

[27] Ho, L, Ronan, J L, Wu, J, Staahl, B T, Chen, L, Kuo, a, Lessard, J, Nesvizhskii, A. I, Ranish, J, & Crabtree, G R. An embryonic stem cell chromatin remodeling complex, esBAF, is essential for embryonic stem cell self-renewal and pluripotency. Proc Natl Acad Sci USA (2009). http://www.pnas.org/content/early/2009/03/10/0812889106, 106, 5181-5186.

[28] Singhal, N, Graumann, J, Wu, G, Araúzo-bravo, M J, Han, D W, Greber, B, Gentile, L, Mann, M, & Schöler, H R. Chromatin-remodeling components of the BAF complex facilitates reprogramming. Cell (2010). http://www.cell.com/abstract/S0092-8674(10)00491-5, 141, 943-955.

[29] Sims, R J. rd, Chen C F, Santos-Rosa H, Kouzarides T, Patel S S, Reinberg D. Human but not yeast CHD1 binds directly and selectively to histone H3 methylated at lysine 4 via its tandem chromodomains. J Biol Chem (2005). http://www.ncbi.nlm.nih.gov/pubmed/16263726, 280, 41789-41792.

[30] Gaspar-maia, A. Alajem, Polesso F, Sridharan R, Mason M J, Heidersbach A, Ramalho-Santos J, McManus M T, Plath K, Meshorer E, Ramalho-Santos M. Chd1 regulates open chromatin and pluripotency of embryonic stem cells. Nature (2009). http://www.ncbi.nlm.nih.gov/pubmed/19587682, 460, 863-868.

[31] Chen, X, Xu, H, Yuan, P, Fang, F, Huss, M, Vega, V B, Wong, E, Orlove, Y L, Zhang, W, Jiang, J, Loh, Y H, Yeo, H C, Yeo, Z X, Narang, V, Govindarajan, K R, Leong, B, Shahab, A, Ruan, Y, Bourque, G, Sung, W K, Clarke, N D, Wei, C L, & Ng, H H. Integration of external signalling pathways with the core transcriptional network in embryonic stem cells. Cell (2008). http://www.ncbi.nlm.nih.gov/pubmed/18555785, 133, 1106-1117.

[32] Roth, S Y, Denu, J M, & Allis, C D. Histone Acetyltransferases. Annu. Rev. Biochem. (2001). http://www.ncbi.nlm.nih.gov/pubmed/11395403, 70, 81-120.

[33] Finnin, M S, Donigian, J R, Cohen, A, Richon, V M, Rifkind, R A, Marks, P A, Breslow, R, & Pavletich, N P. Structures of a histone deacetylase homologue bound to the TSA and SAHA inhibitors. Nature (1999). http://www.ncbi.nlm.nih.gov/pubmed/10490031, 401, 188-193.

[34] Buggy, J J, Sideris, M L, Mak, P, Lorimer, D D, Mcintosh, B, & Clark, J M. Cloning and characterization of a novel human histone deacetylase, HDAC8.Biochem J (2008). http://www.ncbi.nlm.nih.gov/pubmed/10926844, 350, 199-205.

[35] Wade, P A. Transcriptional control at regulatory checkpoints by histone deacetylases: molecular connections between cancer and chromatin. Hum. Mol.Genet (2001). http://hmg.oxfordjournals.org/content/10/7/693.short, 10, 693-698.

[36] De Ruijter, A, Van Gennip, M, Caron, A H, Kemp, H N, & Van Kuilenburg, S. A B P. Histone deacetylases (HDACs) : Characterization of the classical HDAC family. Biochem J (2003). Pt 3): 737-749. http://www.ncbi.nlm.nih.gov/pmc/articles/PMC1223209/

[37] Ng, S S, Yue, W W, Oppermann, U, & Klose, R J. Dynamic protein methylation in chromatin biology. Cell. Mol. Life Sci. (2009). http://www.ncbi.nlm.nih.gov/pmc/articles/PMC2794343/, 66, 407-422.

[38] Lan, F, Collins, R E, De Cegli, R, Alpatov, R, Horton, J R, Shi, X, Gozani, O, Cheng, X, & Shi, Y. Recognition of unmethylated histone H3 lysine 4 links BHC80 to LSD1-mediated gene repression. Nature (2007). http://www.ncbi.nlm.nih.gov/pubmed/17687328, 448, 718-722.

[39] Lee, J, Thompson, J R, Botuyan, M V, & Mer, G. Distinct binding modes specifies the recognition of methylated histones H3K4 and H4K20 by JMJD2A-tudor. Nat. Struct. Mol. Biol. (2008). http://www.ncbi.nlm.nih.gov/pubmed/18084306, 15, 109-111.

[40] Collins, R E, Northrop, J P, Horton, J R, Lee, D Y, Zhang, X, Stallcup, M R, & Cheng, X. The ankyrin repeats of G9a and GLP histone methyltransferases are mono- and dimethyllysine binding modules. Nat. Struct. Mol. Biol. (2008). http://www.ncbi.nlm.nih.gov/pubmed/18264113, 15, 245-250.

[41] Chang, B, Chen, Y, & Zhao, . . JMJD6 is a histone arginine demethylase. Science 2007; 318:444- 447. http://www.sciencemag.org/content/318/5849/444.short

[42] Loh, Y H, Zhang, W, Chen, X, George, J, Ng, H H, & Jmjd, a. and Jmjd2c histone H3 Lys 9 demethylases regulate self-renewal in embryonic stem cells. Genes Dev. (2007). http://www.ncbi.nlm.nih.gov/pubmed/17938240, 21, 2545-2557.

[43] Tsukada, Y, Fang, J, Erdjument-bromage, H, Warren, M E, Borchers, C H, Tempst, P, & Zhang, Y. Histone demethylation by a family of JmjC domain-containing proteins. Nature (2006). http://www.nature.com/nature/journal/n7078/abs/nature04433.html

[44] Leonhardt, H, Page, A W, Weier, H U, & Bestor, T H. A targeting sequence directs DNA methyltransferase to sites of DNA replication in mammalian nuclei. Cell (1992). http://www.ncbi.nlm.nih.gov/pubmed/1423634, 71, 865-873.

[45] Okano, M, Bell, D W, Haber, D A, & Li, E. DNA methyltransferases Dnmt3a and Dnmt3b are essential for de novo methylation and mammalian development. Cell (1999). http://www.ncbi.nlm.nih.gov/pubmed/10555141, 99, 247-257.

[46] Achour, M, Jacq, X, Rondé, P, Alhosin, M, Charlot, C, Chataigneau, T, Jeanblanc, M, Macaluso, M, Giordano, A, Hughes, A. D, Schini-kerth, V. B, & Bronner, C. The interaction of the SRA domain of ICBP90 with a novel domain of DNMT1 is involved in the regulation of VEGF gene expression. Oncogene (2008). http://www.ncbi.nlm.nih.gov/pubmed/17934516, 27, 2187-2197.

[47] Zhang, F, Pomerantz, J H, Sen, G, Palermo, A T, & Blau, H M. Active tissue-specific DNA demethylation conferred by somatic cell nuclei in stable heterokaryons. Proc Natl Acad Sci USA (2007). http://www.pnas.org/content/104/11/4395.short, 104, 4395-4400.

[48] Rai, K, Huggins, I J, James, S R, Karpf, A R, Jones, D A, & Cairns, B R. DNA demethylation in zebrafish involves the coupling of a deaminase, a glycosylase, and gadd45. Cell (2008). http://www.ncbi.nlm.nih.gov/pubmed/19109892, 135, 1201-1212.

[49] Efroni, S, Duttagupta, R, Cheng, J, Dehghani, H, Hoeppner, D J, Dash, C, & Bazett-jones, D P. Le Grice S, McKay R D, Buetow K, Gingeras T R, Misteli T, Mehorer E. Global transcription in pluripotent embryonic stem cells. Cell Stem Cell (2008). http://www.ncbi.nlm.nih.gov/pubmed/18462694, 2, 437-447.

[50] Moore, L D, Le, T, & Fan, G. DNA Methylation and Its Basic Function. Neuropsychopharmacology Reviews (2012). http://www.ncbi.nlm.nih.gov/pubmed/22781841, 1-16.

[51] Kim, J B, Greber, B, Araúzo-bravo, M J, Meyer, J, Park, K I, Zaehres, H, & Schöler, H R. Direct reprogramming of human neural stem cells by Oct4. Nature (2009). http://www.nature.com/nature/journal/n7264/abs/nature08436.html

[52] Boyer, L A, Lee, T I, Cole, M F, Johnstone, S E, Levine, S S, Zucker, J P, Guenther, M G, Kumar, R M, Murray, H L, Jenner, R G, Gifford, D K, Melton, D A, Jaenish, R, & Young, R A. Core Transcriptional regulatory circuitry in human embryonic stem cells. Cell (2005). http://www.ncbi.nlm.nih.gov/pubmed/16153702, 122, 947-956.

[53] Fong, Y W, Inouye, C, Yamaguchi, T, Cattoglio, C, Grubisic, I, & Tjian, R. A DNA repair complex functions as an Oct4/Sox2 coactivator in embryonic stem cells. Cell (2011). http://www.ncbi.nlm.nih.gov/pubmed/21962512, 147, 120-131.

[54] Levasseur, D N, Wang, J, Dorschner, M O, Stamatoyannopoulos, J A, & Orkin, S H. Oct4 dependence of chromatin structure within the extended Nanog locus in ES cells. Gene Dev (2008). http://www.ncbi.nlm.nih.gov/pmc/articles/PMC2259027/, 22, 575-580.

[55] Liang, J, Wan, M, Zhang, Y, Gu, P, Xin, H, Jung, S Y, Qin, J, Wong, J, Cooney, A J, Liu, D, & Songyang, Z. Nanog and Oct4 associate with unique transcriptional repression complexes in embryonic stem cells. Nat Cell Biol (2008). http://www.ncbi.nlm.nih.gov/pubmed/18454139, 10, 731-739.

[56] Lin, C H, Lin, C, Tanaka, H, Fero, M L, & Eisenman, R N. Gene regulation and epigenetic remodeling in murine embryonic stem cells by c-myc. PLoS ONE (2009). e7839. http://www.plosone.org/article/info%3Adoi%2F10.1371%2Fjournal.pone.0007839

[57] Kim, J, Woo, A J, Chu, J, Snow, J W, Fujiwara, Y, Kim, C G, Cantor, A B, & Orkin, S H. A Myc network accounts for similarities between embryonic stem and cancer cell transcription programs. Cell (2010). http://www.cell.com/retrieve/pii/S0092867410010585, 143, 313-324.

[58] Stadtfeld, M, & Hochedlinger, K. Induced pluripotency: history, mechanisms, and applications. Genes Dev. (2010). http://www.ncbi.nlm.nih.gov/pubmed/20952534, 24(20), 2239-2263.

[59] Rahl, P B, Lin, C Y, Seila, A C, Flynn, R A, Mccuine, S, Burge, C. B, & Sharp, P A. Young R A. c-Myc regulates transcriptional pause release. Cell (2010). http://www.ncbi.nlm.nih.gov/pmc/articles/PMC2864022/, 141, 432-445.

[60] Niwa, H, Ogawa, K, Shimosato, D, & Adachi, K. A parallel circuit of LIF signalling pathways maintains pluripotency of mouse ES cells. Nature (2009). http://www.ncbi.nlm.nih.gov/pubmed/19571885, 460, 118-122.

[61] Pawlak, M, & Jaenisch, R. De novo DNA methylation by Dnmt3a and Dnmt3b is dispensable for nuclear reprogramming of somatic cells to a pluripotent state. Genes &

Development (2011). http://genesdev.cshlp.org/content/25/10/1035.short, 25, 1035-1040.

[62] Zhu, J-K. Active DNA Demethylation Mediated by DNA Glycosylases. Annu Rev Genet (2009). http://www.ncbi.nlm.nih.gov/pubmed/19659441, 43, 143-166.

[63] Bender, J. DNA methylation and epigenetics. Annu Rev Plant Biol (2004). http://www.ncbi.nlm.nih.gov/pubmed/15725056, 55, 41-68.

[64] Bird, A. DNA methylation patterns and epigenetic memory. Genes Dev (2002). http://www.ncbi.nlm.nih.gov/pubmed/11782440, 16, 6-21.

[65] Pereira, C F, Terranova, R, Ryan, N. K, Santos, J, Morris, K J, Cui, W, Merkenschlager, M, & Fisher, A G. Heterokaryo-based reprogramming of human B lymphocytes for pluripotency requires Oct4 but not Sox2. PLoS Genet (2008). e1000170. http://www.ncbi.nlm.nih.gov/pubmed/18773085

[66] Ito, S, Shen, L, Dai, Q, Wu, S C, Collins, L B, Swenberg, J A, He, C, & Zhang, Y. Tet proteins can convert 5-methylcytosine to 5-formylcytosine and 5-carboxylcytosine. Science (2011). http://www.sciencemag.org/content/333/6047/1300.short, 333, 1300-1303.

[67] Bernstein, B E, Mikkelsen, T S, Xie, X, Kamal, M, Huebert, D J, Cuff, J, Fry, B, Meissner, A, Wernig, M, Plath, K, Jaenish, R, Wagschal, A, Feil, R, Schreiber, S L, & Lander, E S. A bivalent chromatin structure marks key developmental genes in embryonic stem cells. Cell (2006). http://www.ncbi.nlm.nih.gov/pubmed/16630819, 125, 315-326.

[68] Ang, Y S, Tsai, S Y, Lee, D F, Monk, J, Su, J, Ratnakumar, K, Ding, J, Ge, Y, Darr, H, Chang, B, Wang, J, Rendl, M, Berstein, E, Schaniel, C, & Lemischka, I R. Wdr5 mediates self-renewal and reprogramming via the embryonic stem cell core transcriptional network. Cell (2011). http://www.ncbi.nlm.nih.gov/pubmed/21477851, 145, 183-197.

[69] Wysocka, J, Swigut, T, Xiao, H, Milne, T A, Kwon, S Y, Landry, J, Kauer, M, Tackett, A J, Chair, B T, Badenhorst, P, Wu, C, & Allis, C D. A PHD finger of NURF couples histone H3 lysine 4 Trymethylation with chromatin remodeling. Nature (2006). http://www.ncbi.nlm.nih.gov/pubmed/16728976, 442, 86-90.

[70] Schuettengruber, B, Chourrout, D, Vervoort, M, Leblanc, B, & Cavalli, G. Genome regulation by polycomband trithorax proteins. Cell (2007). http://www.cell.com/abstract/S0092-8674(07)00190-0, 128, 735-745.

[71] Shen, X, Liu, Y, Hsu, Y J, Fujiwara, Y, Kim, J, Mao, X, Yuan, G. C, & Orkin, S H. EZH1 mediates methylation on histone H3 lysine 27 and complements EZH2 in maintaining stem cell identity and executing pluripotency. Mol Cell (2008). http://www.ncbi.nlm.nih.gov/pubmed/19026780, 32, 491-502.

[72] Boyer, L A, Plath, K, Seitlinger, J, Brambrink, T, Medeiros, L A, Lee, T I, Levine, S S, Wernig, M, Tajonar, A, Ray, M K, Bell, G W, Otte, A P, Vidal, M, Gifford, D K,

Young, R A, & Jaenish, R. Polycomb complexes repress developmental regulators in murine embryonic stem cells. Nature (2006). http://www.ncbi.nlm.nih.gov/pubmed/16625203, 441, 349-353.

[73] Mansour, A A, Gafni, O, Weinberger, L, Zviran, A, Ayyash, M, Rais, Y, Krupalnik, V, Zerbib, M, Amann-zalcenstein, D, Maza, I, Geula, S, Viukov, S, Holtzman, L, Pribluda, A, Canaani, E, Horn-saban, S, Amit, I, Novershtern, N, Hanna, J H, & The, H. K27 demethylase Utx regulates somatic and germ cell epigenetic reprogramming. Nature (2012). http://www.nature.com/nature/journal/vaop/ncurrent/full/nature11272.html

[74] Shi, Y, Do, J T, Desponts, C, Hahm, H S, Schöler, H R, & Ding, S. A combined chemical and genetic approach for the generation of induced pluripotent stem cells. Cell Stem Cell, (2008b). http://www.ncbi.nlm.nih.gov/pubmed/18522845, 2, 525-528.

[75] Doyon, Y, & Coté, J. The highly conserved and multifunctional NuA4 HAT complex. Curr Opin Genet Dev (2004). http://www.ncbi.nlm.nih.gov/pubmed/15196461, 14, 147-154.

[76] Hezroni, H, Tzchori, I, Davidi, A, Mattout, A, Biran, A, Nissim-rafinia, M, Westphal, H, & Meshorer, E. H. K9 histone acetylation predicts pluripotency and reprogramming capacity of ES cells. Nucelus (2011). http://www.ncbi.nlm.nih.gov/pubmed/21941115

[77] Mattout, A, Biran, A, & Meshorer, E. Global epigenetic changes during somatic cell reprogramming to iPS cells. J Mol Cell Biol (2011). http://www.ncbi.nlm.nih.gov/pubmed/22044880, 3, 341-350.

[78] Ho, L, Miller, E L, Ronan, J L, Ho, W Q, Jothi, R, & Crabtree, G R. esBAF facilitates pluripotency by conditioning the genome for LIF/STAT3 signalling and by regulating polycomb function. Nat Cell Biol (2011). http://www.ncbi.nlm.nih.gov/pubmed/21785422, 13, 903-913.

[79] Onder, T. T, Kara, N, Cherry, A, Sinha, A U, Zhu, N, Bernt, K M, Cahan, P, Marcarci, B O, Unternaehrer, J, Gupta, P B, Lander, E S, Armstrong, S A, & Daley, G Q. Chromatin-modifying enzymes as modulators of reprogramming. Nature (2012). http://www.ncbi.nlm.nih.gov/pubmed/22388813, 483, 598-602.

[80] Judson, R L, Babiarz, J E, Venere, M, & Blelloch, R. Embryonic stem cell-specific microRNAs promote induced pluripotency. Nat. Biotechnol (2009). http://www.ncbi.nlm.nih.gov/pubmed/19363475, 27, 459-461.

[81] Lin, S-L, Chang, D C, Lin, C H, Ying, S Y, Leu, D, & Wu, D T. S. Regulation of somatic cell reprogramming through inducible mir-302 expression. Nucleic Acids Research (2011). http://nar.oxfordjournals.org/content/39/3/1054.short, 39, 1054-1065.

[82] Suh, M R, Lee, Y, Kim, J Y, Kim, S K, Moon, S H, Lee, J Y, Cha, K Y, Chung, H M, Yoon, H S, & Moon, S Y. Human embryonic stem cells express a unique set of microRNAs. Dev. Biol. (2004). http://www.ncbi.nlm.nih.gov/pubmed/15183728, 270, 488-498.

Chapter 10

Epigenetic Reprogramming in Stem Cells

Richard L. Eckert, Yasin Kizilyer and Candace L. Kerr

Additional information is available at the end of the chapter

http://dx.doi.org/10.5772/56327

1. Introduction

This chapter will be focused on epigenetic mechanisms known to affect self-renewal and developmental potency of embryonic-like stem cells, and germ cells which mimic similar epigenetic signatures as pluripotent stem cells. Examples of epigenetic regulation have proven crucial for defining the stem cell state. In particular, a wealth of knowledge regarding stem cell-specific epigenetic modifications has occurred over the past decade with discoveries that include describing unique stem cell-specific chromosome structure, DNA and histone modifications and noncoding RNAs. The impact of these findings and the better understanding of epigenetic regulation in pluripotent stem cells provides a foundation for discovering mechanisms which regulate human development and differentiation in addition to those that can facilitate cellular reprogramming.

In eukaryotes, chromosomes consist of repeating chromatin units called nucleosomes, which encompass segments of DNA (~147 bp) wound around a central core of eight histone (**H**) proteins (two each of core histone proteins H2A, H2B, H3, and H4). These units are separated by a linker DNA associated with histone protein, H1. Post-translational modifications of these histone residues regulate gene expression [1, 2]. The types of modification of specific amino acids within these proteins include acetylation (*Ac*), mono- di- or tri-methylation (*Me1, 2, 3*), and ubiquitylation (**Ub**). Although these mechanism in general appear to work independently, recent evidence has demonstrated that crosstalk does exist between some of these modifications [1, 3]. Chromatin can be distinguished based on its anatomical structure as either heterochromatin or euchromatin. Heterochromatin is characterized by tightly packed nucleosomes that occur at centromeres, telomeres, and areas of repetitive DNA and is associated with low gene transcription while euchromatin constitute less compacted areas of chromatin and associated with active gene expression. In general, these areas of chromatin are non-randomly distributed within the nucleus and cell-type and cell-cycle dependent [4, 5].

In mammals, heterochromatin is associated with high levels of some methylation marks, including lysine (**K**)-9, K-27 and K-20 on histone (**H**) -3 (H3K9me3, H3K27me3, H3K20me3), low levels of acetylation, and its associated proteins including heterochromatin protein 1 (**HP1**) [6]. In contrast, euchromatin is associated with high levels of acetylation and methylation marks, such as K-4, K-36, and K79 on Histone 3 (H3K4, H3K36 and H3K7)(reviewed by [1]). These modifications occur through the activity of the following enzymes, histone acetyl transferases (**HATs**), histone deacetylases (**HDACs**), histone methyltransferases (**HMT**) and histone demethylases (**HDMT**). Further chromatin regulation is also possible through modifications of the histone tails which are suspected to produce a 'histone code' that initiate higher order chromatin folding [7].

2. Epigenetic control in pluripotent stem cells

2.1. Changes in chromatin ultrastructure in pluripotent stem cells

With the discovery of culturing embryonic stem cells (ESCs), several groups have been able to show the progression of global changes in the chromatin architecture of these cells. Through these studies, it has been shown that undifferentiated pluripotent stem cells contain less heterochromatic regions and express less chromatin structural proteins. Moreover, binding of these proteins (i.e. HP1α, lamin B) to heterochromatic regions is weaker compared to lineage-committed cells [8]. Additionally, pluripotent transcription factors and chromatin remodeling proteins are overexpressed in ESCs compared to more differentiated progenitor cells [9]. One study further showed that chromatin remodeler Chd1 knockdown results with heterochromatin accumulation and skewed differentiation in mouse ESCs, which suggests functional relevancy to the 'open' chromatin structure [10]. Together, these studies show that pluripotent ESCs has an open and hyperdynamic chromatin structure which transforms into a more compact, repressive-like, chromatin state during differentiation.

As ESC chromatin is more transcriptionally permissive, it is also more sensitive to nuclease activity. This may also be in part due to differences that are seen in the chromatin localization in the nucleus. For instance, one study using the DamID (DNA adenine methyltransferase identification) technique showed that pluripotency genes, including Oct4 and Nanog, move to the nuclear lamina and are silenced while lineage-specific genes disassociate from the lamina and are expressed. This was specifically shown during the differentiation of mouse ESCs into terminal astrocytes which demonstrated cell type-specific relocations of these areas during differentiation [11]. These areas near the nuclear periphery were called Lamina Associated Domains (**LADS**). Further study showed that these areas were enriched with repressive histone modifications, H3K9me2 and H3K27me3 which had tissue-specific distributions. Additionally, they consisted of few, minimally expressed genes, which were also marked by these repressive histone modifications [11-13].

Genome-wide ChIP analyses have also described other lamina associated domains with significant overlap with the LADS domains. These regions referred as **L**arge **O**rganized **C**hromatin domains of H3**K**9me2 or **LOCKS**, which are also hallmarked with increased

H3**K**9me2 marks, gene-poor and increased in size and abundance from ESCs to differentiated cells (from 4% genome coverage in mouse ESCs to at most 46%, in liver cells) [14]. Wen, Wu et al. described these LOCKs as large regions with K9 modifications up to 4.9 Mb that are conserved between human and mouse. Like LADS, LOCKS also show tissue-specific distributions and inversely correlate with gene expression. Specific knockouts of the H3K9 methyltransferase, G9a, abolished LOCK formation in mouse ESCs and caused gene derepression without any peripheral localization alteration [14, 15]. Although LADS and LOCKS are similar in their effects on gene expression and localization, the relationship between them and their function at the nuclear periphery is presently unclear.

2.2. Bivalency

Since the discovery of generating induced pluripotent stem (**iPS**) cells, the ability to reprogram a differentiated cell back towards a more embryonic-like state, it has been shown by a variety of laboratories that there is an extensive amount of epigenetic variability between different iPS lines and even among different clones. As a result, much research has been involved in understanding the global demethylation and methylation patterns in ESC to identify particular epigenetic marks and their effected genes to study their role in pluripotency and reprogramming. Methylation of DNA cytosine residues (**mCG**), particularly in CpG islands of promoters, is a well-established mechanism that represses gene transcription in adult cells [16]. Several studies have specifically shown in ESCs, that loss of DNA methyltransferases also compromises their ability to differentiate into mature cells without affecting their self-renewal [17].

In embryonic stem cells, lineage-specific gene expression program regulators are repressed, but poised for a rapid response to differentiate [18]. These areas of chromatin, have so called bivalent domains, consist of opposing chromatin marks; i.e. H3K4me for activation and H3K27me for silencing. In ESCs, these domains are believed to be responsible for preventing the transcription responsible for their early differentiation to a specific lineage while priming the area for activation when the appropriate cues are expressed. Consistent with this belief is the findings that the bulk of the protein-encoding genes of human ESCs, including transcriptionally inactive genes have H3K4me, H3K9ac and H3K14ac rich promoter regions in areas of the nucleosome adjacent and downstream of transcription start sites [19, 20]. Moreover, in ESCs, genes with bivalent gene promoters tend to have unmethylated CpG islands [21]. The initial step of active DNA demethylation in mammals occurs by the conversion of 5-methylcytosine of DNA (5mC) to 5-hydroxymethylcytosine (5hmC). A prime example of this in pluripotent stem cells has been shown in regulating the expression of the stem cell transcription factor, Nanog. Here, the demethylated state is critical for the upregulation of Nanog which is an essential regulator of ESC pluripotency and self-renewal, while its downregulation attributed to methylation of its promoter is required for ICM specification [22]. Recent studies demonstrate that demethylation of Nanog is in part contributed to the expression of the Tet methylcytosine dioxygenase 1 (TET1) enzyme which is a TET family member of enzymes that catalyze the conversion of 5mC to 5hmC. This enzyme has been found to demethylate Nanog promoter sites in mouse ES cells [23, 24]. Both TET1 and TET2 expression have also been shown to be regulated by Oct4 expression in mouse ES cells, downregulated following differentiation alongside other stem cell markers, and is induced concomitantly with 5-hmC during fibroblast reprogramming into iPS cells [25].

In addition to promoter regulation, methylC-Seq genome-wide analysis has also discovered novel types of DNA methylation regulation at non-CG sites (CHG and CHH sites where H = A, C, or T residues). These analyses showed that non-CG methylation accounted for 25% of the total ESC methylome and that these sites were more commonly found within gene bodies than within promoter sites [26]. Furthermore, the methylation of these sites was lost when differentiation was induced in ESCs, and restored during the generation of in induced pluripotent stem cells. This included many differentially methylated regions associated with genes involved in pluripotency and differentiation.

2.3. Polycomb and trithorax group proteins in pluripotent stem cells

Recent studies have established that developmental gene priming and bivalency are crucial for pluripotency whereby the chromatin of pluripotent stem cells are transcriptionally permissive, with normally silent DNA repeat regions, transcriptionally related histone modifications such as H3K9ac, H3K4me3, H3K36me3 and low stochastic transcription of lineage-restricted genes [8, 9]. The poised state is believed to inhibit the activity of RNA Polymerase II (**RNAP II**) and thereby deregulate elongation. In the poised state, RNAP II demonstrates high Ser5 phosphorylation and low levels of Ser2 phosphorylation which is in part controlled by a number of complexes involved in this process. For this, several groups or families of complexes involved in epigenetic regulation have been studied in pluripotent stem cells. These proteins include those which regulate histone modifications, DNA methylation and ATP-dependent chromatin remodelling and include the polycomb group (**PcG**) proteins, trithorax group (**TrxG**) proteins, and ATP-dependent enzymes of the BAF complex. In general, PcG proteins are usually associated with epigenetic gene silencing, while their antagonists TrxG and BAF complexes proteins are involved in epigenetic maintenance or activation of differentiation. The importance of these proteins in developmental regulation of gene expression is well-established [27], while, their functions in adult and pluripotent stem cells is only recently been understood [28]. This includes studies which show that PcG, TrxG, and other chromatin remodeling factors including ATP-dependent enzymes are interconnected in their roles to regulate pluripotency (see Figure 1) [29-33].

PcG, TrxG, and BAF complex associated genes are conserved from fly to man and are important in the regulation of organogenesis and development. PcG proteins were initially discovered as repressors of the Hox or homoeotic genes in *Drosophila,* while TrxG and BAP (BAF *Drosophilia* homolog) proteins maintained Hox gene expression in the appropriate spatial domains. Hox genes encode a family of evolutionarily conserved regulators, which are involved in establishing body segmentation patterns during the development of the fly. In mammalians, these proteins also regulate genes involved in development and differentiation.

PcG proteins produce two distinct protein complexes that act sequentially to regulate gene expression – the "Bmi-1 complex" also known as Polycomb Repressive Complexes (**PRC1**) and the "Eed complex" also known as **PRC2**. The PRC1 or Bmi-1 complex includes Bmi-1, Ring1A/B, Mph1/Rae28, Mel-18, M33, and Scmh1. The PRC2 or Eed complex includes Ezh1, Ezh2,

PHF1, MTF2, Eed, YY1, RBBP4, RBBP7, PCL1, PCL2, PCL3, JARID2, AEBP2, and PHF19 and Suz12 [34, 35]. Initial steps in stem cell reprogramming involves the recruitment of histone deacetylase by activity of the PRC2 complex, which causes local deacetylation of chromatin and subsequent methylation of K27 of histone H3. This H3K27 methylation then recruits the Bmi-1 complex to the site, which leads to the monoubiquitination of Lys119 histone H2A, and in turn suppresses gene expression [36]. The coordinative action of these two complexes plays an important role in the regulation and maintenance of gene expression during development and contributes to the epigenetic memory of stem cells [37, 38].

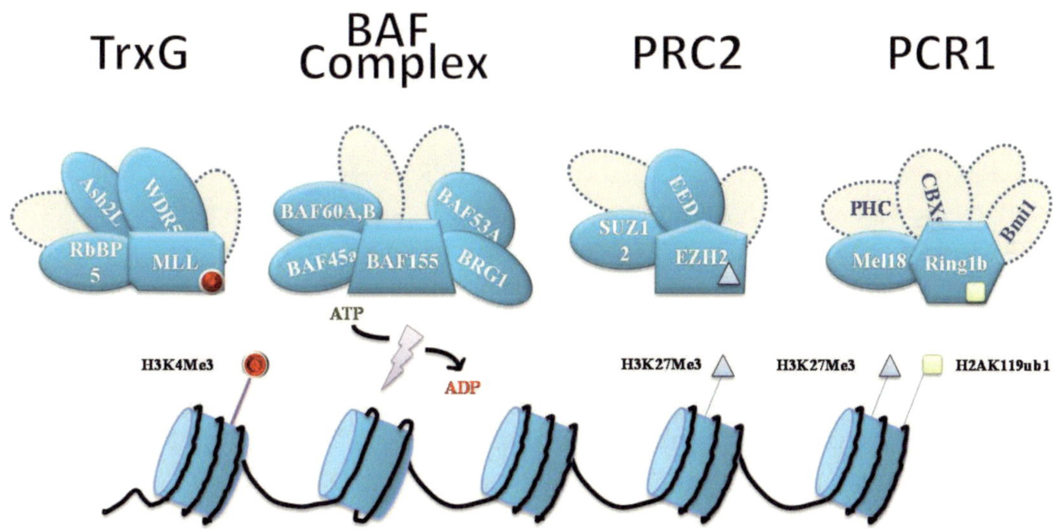

BAF complex > Nature Neuroscience 13, 1330–1337 (2010) doi:10.1038/nn.2671 Published online 26 October 2010
TrxG & PRCs : *Cell Stem Cell.* 2012 Jul 6;11(1):16-21. doi: 10.1016/j.stem.2012.06.005.

Figure 1. Chromatin remodeling factors of the TrxG, PRC1/2 and BAF complexes work together to regulate stem cell status. In pluripotent stem cells, genes necessary for lineage-specific regulation consist of 'bivalent' chromatin domains that contain repressed H3K27me3 marks, as well as active H3K4me3 marks. These genes are then 'primed' for rapid induction of expression upon receiving differentiation cues. Proteins of the TrxG family tri-methylate H3K4 leading to active chromatin marks. PRC2 activity leads to repressive tri-methylation of H3K27 and subsequent recruitment of PRC1 to the nucleosome region. Upon recruitment, PRC1 transfers a mono-ubiquitin residue to histone 2A (H2AK119). Together, the binding of PRC1 and the ubiquitylation of H2AK119ub silences gene expression. BAF complexes directly unwind nucleosomal DNA by using ATP and helicase-like subunits. Together, these complexes coexist and/or work hierarchically to regulate pluripotency and bivalency in stem cells.

Polycomb repressive complexes have been shown associated with many developmental regulator regions in ESCs, and many of the PcG repressed targets of ESCs are also 'bivalent' [30, 39]. For instance, PRC2 target genes have been shown to be preferentially turned on during ESC differentiation and that the pluripotent stem cell regulating genes Oct4, Sox2, and Nanog co-occupy a significant subset of these genes. Therefore, it has been suggested that the PRC2 complex represses a distinct group of developmental genes that have to be repressed to maintain pluripotency. This would promote a poised or primed state which could be readily activated during early differentiation [40]. For example, the histone methyltransferase Ezh2 is known to catalyze H3K27me3. In fact, bivalency domains at PRC2 regulated promoters are roughly five times more likely to become DNA methylated during differentiation than those with non-PRC2 regulated promoters [21] suggesting that the PRC2 complex plays a pivotal role in the switch for early lineage commitment [41]. Jarid2, a member of the Jumonji family of histone demethylases, has also been shown to play an important role in properly recruiting PRC1 and PRC2 and initiating the RNA Polymerase II activating form (Ser5P-RNAPII) [42] to bivalent loci to promote differentiation [43-45]. While Jarid2 is enzymatically inactive in ESCs, recent evidence has shown that Jarid2 is regulated by pluripotency factors in ESCs [43]. In null ESCs lacking Jarid2 expression were able to self-renew but unable to differentiate despite expressing appropriate PRC2 target genes demonstrating that transcriptional priming of bivalent genes in ESCs was dependent on Jarid2 expression.

In addition to the bivalent marks associated with PRC2 associated H3 modifications in pluripotent stem cells, bivalent genes in stem cells also exhibit repressive marks of H2AK119Ub1 at their promoter and throughout the coding region. For this reason, members of the PRC1 complex Ring1A and Ring1B proteins which catalyze H2AK119ub1 have also reveal a role in regulating pluripotency [30, 46]. For instance, in ESCs, double mutants of Ring1A/B demonstrate reduced levels of H2AK119Ub1, repression of known stemness genes (including Oct4 targets), increased expression of developmental regulator targets, and spontaneous differentiation. Furthermore, upon differentiation, Ring1A/B lost binding to their target loci suggesting that a Ring 1/B mediated complex functions downstream of the stem cell core transcriptional machinery to maintain the ESC state [47].

In addition to PcG complexes, core members of the TrxG and BAF chromatin remodeling complexes have also been shown to contribute to the bivalent mark in stem cells by acting in concert to establish and preserve H3K4me3 [48, 49]. Another core member, WD repeat domain 5 (**Wdr5**) of the TrxG complex, has also been associated with the undifferentiated state and shown to regulate self-renewal in mouse ESCs [50]. This study went further to show that Wdr5 expression can promote efficient derivation of pluripotent iPS cells. Along with PcG and TrxG complexes, other chromatin remodeling complexes have been shown to have crucial roles in maintaining pluripotency. For instance, it has been shown that ESCs contain a unique BAF complex, which distinguishes them from differentiated cell types. This ESC complex consists of BAF60a, BAF155 and BRG subunits in the absence of BAF60c, BAF170 or BRM [48, 51]. Components of this ESC BAF complex also appear in RNAi screens for pluripotent genes [52, 53]. In mouse ESCs, it was also shown that BRG removal causes decreased self-renewal and aberrant differentiation, whereas BRG withdrawal from differentiated cells had very little effect [51]. Additionally, the pluripotency regulator genes Oct4, Sox2 and Nanog, have been shown to be targeted by components of the ESC-specific BAF complex [48, 51, 54-56] as well as facilitate IPS cell reprogramming [33].

http://dx.doi.org/10.5772/56327

3. Epigenetic control in primordial germ cells and pluripotent embryonic germ cells

Primordial germ cells (**PGCs**) are the progenitors of sperm and egg. In humans, these cells first appear around the fifth week after conception and in mouse, during embryonic day (**E**) 5. At this time, PGCs emerge from differentiated epiblast cells which have already begun to undergo major epigenetic changes including DNA methylation and X inactivation [57]. Thus, epigenetic regulation or reprogramming must occur in PGCs in order to achieve an undifferentiated totipotent-like state [57, 58]. X activation and demethylation in PGCs is similar to that seen in the process of generating ESCs and reprogramming somatic cells into pluripotent iPS cells [59]. Together, these epigenetic changes in germ cell development is to prevent the transmission of aberrant epigenetic modifications to the next generation and to promote epigenetic equivalency in the germ line of male and female embryos, which is necessary for proper imprinting. This is the only time in which homologous chromosomes are epigenetically indistinguishable and in PGCs occurs primarily in the developing embryonic gonad [60, 61].

Key initiators of PGC induction in the epiblast include the Blimp1 (**B-lymphocyte-induced maturation protein 1**), also known as PR domain zinc finger protein 1 (**Prdm1**), Prdm14 and protein arginine methyltransferase 5 (**Prmt5**). These factors have been shown to initiate epigenetic reprogramming and induce repression of the somatic program during germ cell specification [62-64], and in a similar fashion to facilitate somatic cell reprogramming in iPS cells and in epiblast stem cell generation [65, 66]. Their effects occur through both direct and indirect targets. For instance, it is known that BLIMP1 associates with the arginine methyltransferase PRMT5 to reduce expression of Hox-family genes and other somatic genes in PGCs via H2A/H4 R3 methylation [67].

In addition to pluripotent associated genes, early germ-cell development and imprinted genes also undergo demethylation during this time. These include well-established postmigratory germ cell genes Mvh (also known as Ddx4), Sycp3 (synaptonemal complex protein 3) and Dazl (deleted in azoospermia-like). These genes are demethylated in germ cells and repressed in somatic cells. This demethylation occurs during the migration of PGCs into the gonad at CpG islands of their promoters as well as at differentially methylated regions (**DMRs**) of imprinted genes [68, 69]. Whether DNA methylation in PGCs is erased by an active or a passive mechanism is currently unclear. However, two DNA deaminases, activation-induced cytidine deaminase (**AID**) and apolipoprotein B mRNA editing enzyme, catalytic polypeptide 1 (**APOBEC1**) may demonstrate a key role in this process. While both enzymes have been shown *in vitro* to deaminate 5-methylcytosine in DNA in mouse PGCs, deficiencies in AID expression has also shown that it is essential for erasure of DNA methylation [70]. Both, Aid and Apobec1 are located in a cluster of genes which comprise Stella, Growth differentiation factor 3 (**GDF3**) and Nanog. Stella, Gdf3 and Nanog are all expressed in pluripotent tissues as well as in germ cells. [71]. While Stella is a known constituent of germ cell development, Gdf3 and Nanog have important roles in conferring stem-cell identity on ES cells. It has also been suggested that *in vivo* targeting in the zygote of AID to the methylated DMR of the imprinted gene H19 results in efficient and substantial demethylation of this region [72]. Together these findings suggest an important role of AID in facilitating demethylation and reprogramming the pluripotent state. Similar to AID, TET1 and TET2 may also play a facilitative role in PGC demethylation as both TET enzymes are expressed in mouse PGCs during imprint erasure, concurrent with 5hmC enrichment [73].

Another important epigenetic process required for germ cell development and cellular reprogramming to the pluripotent state involves the X chromosome. In female adult cells, one of the two X chromosomes is inactivated to compensate for the differences in gene expression between sexes. For this purpose, X chromosome inactivation is initiated in early embryos by noncoding X-inactive specific transcript (Xist) RNA followed by chromatin modifications on the inactive X chromosome which leads to stable gene repression in somatic cells. Likewise, reactivation of the X chromosome is required for the totipotency of the female blastocyst and germ cell development. Reactivation of the X chromosome also occurs to establish pluripotency in iPS cells. During development, epigenetic reprogramming or re-activation of the inactive X-chromosome commences in PGCs during their migration through the hindgut along their route to the developing gonads where imprint erasure is completed [74]. In mouse PGCs, decreased Xist expression, and the displacement of PcG repressor proteins EED and SUZ12, results in the loss of the inactive X associated histone modification, H3K27me3 [74]. In humans, PcG proteins YY1, EZH2, and EED have also been found in the ovarian follicles, oocytes and preimplantation embryos. YY1 and EZH2 transcripts were additionally detected in human metaphase II oocytes suggesting they may be play a similar role in human germ cell reprogramming [75].

In vitro, PGCs cultured under specific conditions can also demonstrate epigenetic reprogramming with pluripotent cell-like characteristics. In these cases, PGCs form pluripotent stem cell colonies called embryonic germ cells (EGCs) which have notable similarities in their epigenomes [58, 76-82], and like ESCs, EGCs have been shown to induce epigenetic reprogramming of somatic nucleus in hybrid cells [60].

4. MicroRNAs and stem cells

MicroRNAs (**miRNAs**) are a family of non-protein coding RNAs with transcripts of ~20–25 nucleotides that play essential roles in regulating gene expression (see [83-85]. A subset of miRNAs have been shown to be preferentially expressed in undifferentiated stem cells and for some, have been shown to play essential roles in pluripotency, proliferation, and modulation of expression patterns that are related to differentiation [86-89]. The promoter regions of these miRNAs are often occupied by the pluripotency transcription factors, including Oct4, Sox2, and Nanog [90]. In addition, during ESC differentiation, proteins modulated by miRNAs [91] have also been shown to be modulated by PcG proteins [92]. For instance, Marson *et al.* showed that approximately one quarter of the Oct4/Sox2/Nanog/Tcf3-occupied miRNAs belonged to a set of repressed miRNA genes bound by Suz12 in murine ES cells [90]. Here, the PRC2 complex protein SUZ12 was bound to a subset of inactive miRNAs controlling differentiation in mouse ESCs [90]. In this study, SUZ12 bound to the promoters of several miRNAs

http://dx.doi.org/10.5772/56327

associated with repressing differentiation in ESCs. In a similar fashion, another PRC2 associated factor, the transcription factor YY1 has also been shown to directly regulate miR-29 transcription through the recruitment of HDAC1 and EZH2 to the regulatory regions of the miR-29 promoter [93].

Studies in mice have shown that induction of neural differentiation in mouse ESCs with retinoic acid results in increased miR-134, miR-296, and miR-470 which in turn interact with the coding sequences of the pluripotency transcription factors Oct4, Sox2, and Nanog. These results suggest that through interaction of the miRNAs these pluripotent stem cells genes are downregulated thereby permitting differentiation to proceed [91]. Additionally, the ESC-specific miR-290 cluster has also been shown to regulate Oct4 methylation in differentiating ESCs [94]. Other studies have shown that mouse ESCs deficient in proteins of the miRNA processing apparatus such as Dicer, Drosha, DGCR8, and Ddx5 exhibit differentiation and developmental defects [95-97].

Interestingly, PcG proteins have been shown to be both regulators of miRNA expression as well as their targets. For instance, miRNA-101 has been shown to directly regulate the expression of the PRC protein EZH2 in highly aggressive cancers [98, 99].

5. Epigenetic regulation in progenitor and adult stem cells

Progenitor cells and adult stem cells are thought to be predecessors of pluripotent or multipotent stem cells that are generated during early differentiation. During their transition in development, bivalently marked stem cell genes can become either active, or inactive, or remain bivalent, dependent in part, on the activity of key enzymes which drive these chromatin modifications such as lysine demethylases (**KDMs**), histone deubiquitylases (**DUBs**), and DNA methyltransferases (**DNMTs**). Bivalent chromatin, although present in progenitor cells, is less frequent than in pluripotent stem cells. This has been shown in mesenchymal stem cells, hematopoietic stem cells and neural progenitors, in which cases the bivalency continues to resolve upon further differentiation [100]. During this process in pluripotent stem cells, active genes exhibit diminishing suppressive chromatin marks, an increase in H3K4me3, gain of H3K36me3 within coding sequences of DNA, and contain RNAP II that carries high Ser5 phosphorylation and low levels of Ser2 phosphorylation near promoter and within coding regions. Moreover, inactive genes show loss of active chromatin marks while retaining repressive ones, and in some cases gaining CpG methylation (mCpG).

Specific differences occur in the chromatin states between pluripotent stem cells, progenitor cells and more differentiated cell types which include active, repressed and poised states of chromatin. Several lines of evidence suggest that priming in the poised state enables genes to respond rapidly when differentiation cues are presented [30]. For example, during neural induction, several hundred genes including those required to maintain stem cell-ness become de novo mCG and therefore transcriptionally silenced. Furthermore, the observation was made that neural precursors that are derived from ESCs acquired more mCG than terminal neurons, suggesting that the transition from pluripotent to lineage-committed cells is associated with these changes [17, 21, 101, 102].

Polycomb group proteins also appear to play a unique role in defining the progenitor or adult stem cell state. It has been shown that the PRC1 complex protein Bmi-1 activates multiple pathways that are important for regulating the stem cell-like state. For example, it has been shown that Bmi-1 is potentially upregulated via the pluripotent stem cell marker SALL4 signaling and has been shown to regulate stem cell self-renewal by repressing Hox genes, as well as INK4a locus genes, p16INK4a and p19ARF. BMI1 has also been shown to facilitate stem cell-like features in adult stem cells such as increased telomerase activity, transcriptional factor GATA3, and NF-kB pathways. These pathways are associated with the prevention of senescence, differentiation and apoptosis, while promoting immortalization and proliferation (for review see [103]).

6. Epigenetic dysregulation in cancer stem cells

Cancer stem/initiating cells (CSC) have been defined as a subset of cancer cells that have clonal ability or self-renewal and are resilient against cancer therapies [104, 105]. As such CSCs are implicated in cancer initiation, metastasis, and recurrence of some cancers [106]. Although the most well established pluripotent stem cell genes OCT4, NANOG, cMYC and SOX2 are implicated in many poorly differentiated or metastatic cancers [107-109], they are not expressed in all and they are not all elevated concordantly. In addition, targets of NANOG, OCT4, SOX2, and c-MYC are often overexpressed in tumors that are poorly differentiated, more so than in those that are well differentiated [110]. These genes also play a significant role in the induction of pluripotency into iPS cells from differentiated cell types and are thus involved in regulating epigenetic reprogramming [111-113]. More specifically, it is found that c-MYC, which is also an oncogene is sufficient for the reactivation of ESC-like transcriptional program in both, normal and cancer cells [114]. Additionally, studies have shown that one of the inherent issues with generating iPS cells is their propensity to become cancer stem cell-like [115, 116]. Taken together, these results indicate that aberrant activation of an ESC or iPS-like transcriptional program might cause induction of pathological self-renewal in adult differentiated cells, characteristic of cancer stem cells.

Aberrant function of PcG proteins has also been established in the malignancy of various cancers [117]. This is not surprising as it is well known that polycomb complexes contribute to the epigenetic regulation of key networks associated with self-renewal [118], differentiation, and proliferation [92, 119-123]. These roles for polycombs have been demonstrated in cancer cells and normal stem cells [124] and more recently studied for their targeted function in CSCs [125]. For instance, there is much evidence that overexpression of the EZH2 polycomb gene occurs in multiple human malignancies (see [117, 126]). One study showed that this may in part be atributed to a genomic loss of miR-101 which has been shown to lead to increased EZH2 levels [99, 127]. Although how EZH2 contributes to carcinogenesis remains poorly defined, recent evidence suggests that overexpression of EZH2 can contribute to improper silencing of

http://dx.doi.org/10.5772/56327

tumor suppressor genes [121]. In this case, EZH2 was shown to target a pro-differentiation tumor suppressor gene, retinoic acid receptor β2 (**RARβ2**) [120], which is reduced or lost in many human malignancies.

7. Conclusion

The pluripotent stem cells have a chromatin that is hyperdynamic, with a preponderance of modified histones and chromatin remodelers that ensures low-level transcription and tight regulation. Losing pluripotency is accompanied with a more compact, repressive, chromatin structure, which leads to cellular differentiation. Chromatin architecture is regulated at multiple levels in conjunction with known pluripotent genes to constitute an interwoven pluripotency network. Although there are many gaps in our knowledge of how epigenetic modifications regulate the pluripotent state, it is known that PcG repressor proteins prevent the precocious expression of lineage-restricted gene expression in pluripotent stem cells and germ cells by contributing to a unique 'primed' bivalent state of the chromatin. Future studies will provide mechanistic insights into the signaling cues required to maintain this state and inhibit differentiation while iPS cells and adult stem cells provide a renewed opportunity to study the role of chromatin architecture for controlling the pluripotent state. This will include understanding the mechanisms that interplay between pluripotent transcription factors, epigenetic regulators, and miRNAs to balance self-renewal and differentiation, properties which regulate reprogramming and carcinogenesis.

Nomenclature

cdk, cyclin-dependent kinase; H2A-K119-Ub, ubiquitinylated histone H2A lysine 119; H3K27me3, tri-methylated histone H3 lysine K27; PcG, Polycomb group genes; ESC, embryonic stem cells; EGC, embryonic germ cells; PGC, primordial germ cells; iPS, induced pluripotent stem cells; CSC, cancer stem cells; RARβ2, retinoic acid receptor β2; Hh, Hedghog; KDMs, lysine demethylases; DUBs, histone deubiquitylases; DNMT, DNA methyltransferases; YY1, Ying Yang 1; EZH2 Enhancer of Zeste-2; EED, embryonic ectoderm development; GDF3, growth differentiation factor 3; DMR, differentially methylated regions; DAZL. deleted in azoospermia-like; Mvh, deadhead box 4; Sycp3, synaptonemal complex protein 3; E, embryonic day; RNAP II, RNA Polymerase II; trxG, trithorax group proteins; AID, activation-induced cytidine deaminase; **APOBEC1,** apolipoprotein B mRNA editing enzyme, catalytic polypeptide 1

Acknowledgements

This work was supported by NIH grants R01 AR053851 and R01 CA131074 to R Eckert and R21HD057487 awarded to C Kerr as well as by the State of Maryland Stem Cell Research Fund 2010-MSCRFI-0110-00 to R Eckert and 2007-MSCRFII-0159-00 awarded to C Kerr.

Author details

Richard L. Eckert[1,2,3,4], Yasin Kizilyer[1] and Candace L. Kerr[1*]

*Address all correspondence to: ckerr@som.umaryland.edu

1 Department of Biochemistry and Molecular Biology, University of Maryland School of Medicine, Baltimore, Maryland, USA

2 Department of Reproductive Sciences, University of Maryland School of Medicine, Baltimore, Maryland, USA

3 Department of Dermatology, University of Maryland School of Medicine, Baltimore, Maryland, USA

4 Department of Microbiology and Immunology, University of Maryland School of Medicine, Baltimore, Maryland, USA

References

[1] Kouzarides T. Chromatin modifications and their function. Cell. 2007 Feb 23;128(4): 693-705.

[2] Vermeulen M, Eberl HC, Matarese F, Marks H, Denissov S, Butter F, et al. Quantitative interaction proteomics and genome-wide profiling of epigenetic histone marks and their readers. Cell. 2010 Sep 17;142(6):967-80.

[3] Lee JS, Smith E, Shilatifard A. The language of histone crosstalk. Cell. 2010 Sep 3;142(5):682-5.

[4] Joffe B, Leonhardt H, Solovei I. Differentiation and large scale spatial organization of the genome. Curr Opin Genet Dev. 2010 Oct;20(5):562-9.

[5] Fisher CL, Fisher AG. Chromatin states in pluripotent, differentiated, and reprogrammed cells. Curr Opin Genet Dev. 2011 Apr;21(2):140-6.

[6] Kloc A, Martienssen R. RNAi, heterochromatin and the cell cycle. Trends Genet. 2008 Oct;24(10):511-7.

[7] Strahl BD, Allis CD. The language of covalent histone modifications. Nature. 2000 Jan 6;403(6765):41-5.

[8] Mattout A, Meshorer E. Chromatin plasticity and genome organization in pluripotent embryonic stem cells. Curr Opin Cell Biol. 2010 Jun;22(3):334-41.

[9] Efroni S, Duttagupta R, Cheng J, Dehghani H, Hoeppner DJ, Dash C, et al. Global transcription in pluripotent embryonic stem cells. Cell Stem Cell. 2008 May 8;2(5): 437-47.

[10] Gaspar-Maia A, Alajem A, Polesso F, Sridharan R, Mason MJ, Heidersbach A, et al. Chd1 regulates open chromatin and pluripotency of embryonic stem cells. Nature. 2009 Aug 13;460(7257):863-8.

[11] Peric-Hupkes D, Meuleman W, Pagie L, Bruggeman SW, Solovei I, Brugman W, et al. Molecular maps of the reorganization of genome-nuclear lamina interactions during differentiation. Mol Cell. 2010 May 28;38(4):603-13.

[12] Guelen L, Pagie L, Brasset E, Meuleman W, Faza MB, Talhout W, et al. Domain organization of human chromosomes revealed by mapping of nuclear lamina interactions. Nature. 2008 Jun 12;453(7197):948-51.

[13] Pickersgill H, Kalverda B, de Wit E, Talhout W, Fornerod M, van Steensel B. Characterization of the Drosophila melanogaster genome at the nuclear lamina. Nat Genet. 2006 Sep;38(9):1005-14.

[14] Wen B, Wu H, Shinkai Y, Irizarry RA, Feinberg AP. Large histone H3 lysine 9 dimethylated chromatin blocks distinguish differentiated from embryonic stem cells. Nat Genet. 2009 Feb;41(2):246-50.

[15] Yokochi T, Poduch K, Ryba T, Lu J, Hiratani I, Tachibana M, et al. G9a selectively represses a class of late-replicating genes at the nuclear periphery. Proc Natl Acad Sci U S A. 2009 Nov 17;106(46):19363-8.

[16] Cedar H, Bergman Y. Linking DNA methylation and histone modification: patterns and paradigms. Nat Rev Genet. 2009 May;10(5):295-304.

[17] Trowbridge JJ, Orkin SH. DNA methylation in adult stem cells: New insights into self-renewal. Epigenetics. 2010 Apr 31;5(3).

[18] Bernstein BE, Mikkelsen TS, Xie X, Kamal M, Huebert DJ, Cuff J, et al. A bivalent chromatin structure marks key developmental genes in embryonic stem cells. Cell. 2006 Apr 21;125(2):315-26.

[19] Guenther MG, Levine SS, Boyer LA, Jaenisch R, Young RA. A chromatin landmark and transcription initiation at most promoters in human cells. Cell. 2007 Jul 13;130(1): 77-88.

[20] Roelen BA, Lopes SM. Of stem cells and gametes: similarities and differences. Curr Med Chem. 2008;15(13):1249-56.

[21] Mohn F, Weber M, Rebhan M, Roloff TC, Richter J, Stadler MB, et al. Lineage-specific polycomb targets and de novo DNA methylation define restriction and potential of neuronal progenitors. Mol Cell. 2008 Jun 20;30(6):755-66.

[22] Ito S, D'Alessio AC, Taranova OV, Hong K, Sowers LC, Zhang Y. Role of Tet proteins in 5mC to 5hmC conversion, ES-cell self-renewal and inner cell mass specification. Nature. 2010 Aug 26;466(7310):1129-33.

[23] Ito S, Shen L, Dai Q, Wu SC, Collins LB, Swenberg JA, et al. Tet proteins can convert 5-methylcytosine to 5-formylcytosine and 5-carboxylcytosine. Science. 2010 Sep 2;333(6047):1300-3.

[24] Tahiliani M, Koh KP, Shen Y, Pastor WA, Bandukwala H, Brudno Y, et al. Conversion of 5-methylcytosine to 5-hydroxymethylcytosine in mammalian DNA by MLL partner TET1. Science. 2009 May 15;324(5929):930-5.

[25] Koh KP, Yabuuchi A, Rao S, Huang Y, Cunniff K, Nardone J, et al. Tet1 and Tet2 regulate 5-hydroxymethylcytosine production and cell lineage specification in mouse embryonic stem cells. Cell Stem Cell. 2011 Feb 4;8(2):200-13.

[26] Lister R, Pelizzola M, Dowen RH, Hawkins RD, Hon G, Tonti-Filippini J, et al. Human DNA methylomes at base resolution show widespread epigenomic differences. Nature. 2009 Nov 19;462(7271):315-22.

[27] Simon JA, Kingston RE. Mechanisms of polycomb gene silencing: knowns and unknowns. Nat Rev Mol Cell Biol. 2009 Oct;10(10):697-708.

[28] Sauvageau M, Sauvageau G. Polycomb group genes: keeping stem cell activity in balance. PLoS Biol. 2008 Apr 29;6(4):e113.

[29] Cole MF, Young RA. Mapping key features of transcriptional regulatory circuitry in embryonic stem cells. Cold Spring Harb Symp Quant Biol. 2008;73:183-93.

[30] Surface LE, Thornton SR, Boyer LA. Polycomb group proteins set the stage for early lineage commitment. Cell Stem Cell. 2010 Sep 3;7(3):288-98.

[31] Ho L, Crabtree GR. Chromatin remodelling during development. Nature. 2010 Jan 28;463(7280):474-84.

[32] Pietersen AM, van Lohuizen M. Stem cell regulation by polycomb repressors: postponing commitment. Curr Opin Cell Biol. 2008 Apr;20(2):201-7.

[33] Singhal N, Graumann J, Wu G, Arauzo-Bravo MJ, Han DW, Greber B, et al. Chromatin-Remodeling Components of the BAF Complex Facilitate Reprogramming. Cell. 2010 Jun 11;141(6):943-55.

[34] Sauvageau M, Sauvageau G. Polycomb group proteins: multi-faceted regulators of somatic stem cells and cancer. Cell Stem Cell. 2010 Sep 3;7(3):299-313.

[35] Wu H, D'Alessio AC, Ito S, Wang Z, Cui K, Zhao K, et al. Genome-wide analysis of 5-hydroxymethylcytosine distribution reveals its dual function in transcriptional regulation in mouse embryonic stem cells. Genes Dev. 2011 Apr 1;25(7):679-84.

[36] Fischle W, Wang Y, Jacobs SA, Kim Y, Allis CD, Khorasanizadeh S. Molecular basis for the discrimination of repressive methyl-lysine marks in histone H3 by Polycomb and HP1 chromodomains. Genes Dev. 2003 Aug 1;17(15):1870-81.

[37] Jacobs SA, Khorasanizadeh S. Structure of HP1 chromodomain bound to a lysine 9-methylated histone H3 tail. Science. 2002 Mar 15;295(5562):2080-3.

[38] Orlando V. Polycomb, epigenomes, and control of cell identity. Cell. 2003 Mar 7;112(5):599-606.

[39] Spivakov M, Fisher AG. Epigenetic signatures of stem-cell identity. Nat Rev Genet. 2007 Apr;8(4):263-71.

[40] Lee TI, Jenner RG, Boyer LA, Guenther MG, Levine SS, Kumar RM, et al. Control of developmental regulators by Polycomb in human embryonic stem cells. Cell. 2006 Apr 21;125(2):301-13.

[41] Landeira D, Fisher AG. Inactive yet indispensable: the tale of Jarid2. Trends Cell Biol. 2011 Feb;21(2):74-80.

[42] Brookes E, Pombo A. Modifications of RNA polymerase II are pivotal in regulating gene expression states. EMBO Rep. 2009 Nov;10(11):1213-9.

[43] Landeira D, Sauer S, Poot R, Dvorkina M, Mazzarella L, Jorgensen HF, et al. Jarid2 is a PRC2 component in embryonic stem cells required for multi-lineage differentiation and recruitment of PRC1 and RNA Polymerase II to developmental regulators. Nat Cell Biol. 2010 Jun;12(6):618-24.

[44] Pasini D, Cloos PA, Walfridsson J, Olsson L, Bukowski JP, Johansen JV, et al. JARID2 regulates binding of the Polycomb repressive complex 2 to target genes in ES cells. Nature. 2010 Mar 11;464(7286):306-10.

[45] Shen X, Kim W, Fujiwara Y, Simon MD, Liu Y, Mysliwiec MR, et al. Jumonji modulates polycomb activity and self-renewal versus differentiation of stem cells. Cell. 2009 Dec 24;139(7):1303-14.

[46] Eskeland R, Leeb M, Grimes GR, Kress C, Boyle S, Sproul D, et al. Ring1B compacts chromatin structure and represses gene expression independent of histone ubiquitination. Mol Cell. 2010 May 14;38(3):452-64.

[47] Endoh M, Endo TA, Endoh T, Fujimura Y, Ohara O, Toyoda T, et al. Polycomb group proteins Ring1A/B are functionally linked to the core transcriptional regulatory circuitry to maintain ES cell identity. Development. 2008 Apr;135(8):1513-24.

[48] Ho L, Jothi R, Ronan JL, Cui K, Zhao K, Crabtree GR. An embryonic stem cell chromatin remodeling complex, esBAF, is an essential component of the core pluripotency transcriptional network. Proc Natl Acad Sci U S A. 2009 Mar 31;106(13):5187-91.

[49] Lessard JA, Crabtree GR. Chromatin regulatory mechanisms in pluripotency. Annu Rev Cell Dev Biol. 2010 Nov 10;26:503-32.

[50] Ang YS, Tsai SY, Lee DF, Monk J, Su J, Ratnakumar K, et al. Wdr5 mediates self-renewal and reprogramming via the embryonic stem cell core transcriptional network. Cell. 2010 Apr 15;145(2):183-97.

[51] Ho L, Ronan JL, Wu J, Staahl BT, Chen L, Kuo A, et al. An embryonic stem cell chromatin remodeling complex, esBAF, is essential for embryonic stem cell self-renewal and pluripotency. Proc Natl Acad Sci U S A. 2009 Mar 31;106(13):5181-6.

[52] Fazzio TG, Huff JT, Panning B. An RNAi screen of chromatin proteins identifies Tip60-p400 as a regulator of embryonic stem cell identity. Cell. 2008 Jul 11;134(1): 162-74.

[53] Schaniel C, Ang YS, Ratnakumar K, Cormier C, James T, Bernstein E, et al. Smarcc1/Baf155 couples self-renewal gene repression with changes in chromatin structure in mouse embryonic stem cells. Stem Cells. 2009 Dec;27(12):2979-91.

[54] Liang J, Wan M, Zhang Y, Gu P, Xin H, Jung SY, et al. Nanog and Oct4 associate with unique transcriptional repression complexes in embryonic stem cells. Nat Cell Biol. 2008 Jun;10(6):731-9.

[55] Pardo M, Lang B, Yu L, Prosser H, Bradley A, Babu MM, et al. An expanded Oct4 interaction network: implications for stem cell biology, development, and disease. Cell Stem Cell. 2010 Apr 2;6(4):382-95.

[56] van den Berg DL, Snoek T, Mullin NP, Yates A, Bezstarosti K, Demmers J, et al. An Oct4-centered protein interaction network in embryonic stem cells. Cell Stem Cell. 2010 Apr 2;6(4):369-81.

[57] Hayashi K, Surani MA. Resetting the epigenome beyond pluripotency in the germline. Cell Stem Cell. 2009 Jun 5;4(6):493-8.

[58] Surani MA, Hayashi K, Hajkova P. Genetic and epigenetic regulators of pluripotency. Cell. 2007 Feb 23;128(4):747-62.

[59] Monk M, Boubelik M, Lehnert S. Temporal and regional changes in DNA methylation in the embryonic, extraembryonic and germ cell lineages during mouse embryo development. Development. 1987 Mar;99(3):371-82.

[60] Tada M, Tada T, Lefebvre L, Barton SC, Surani MA. Embryonic germ cells induce epigenetic reprogramming of somatic nucleus in hybrid cells. Embo J. 1997 Nov 3;16(21):6510-20.

[61] Hajkova P, Ancelin K, Waldmann T, Lacoste N, Lange UC, Cesari F, et al. Chromatin dynamics during epigenetic reprogramming in the mouse germ line. Nature. 2008 Apr 17;452(7189):877-81.

[62] Ohinata Y, Payer B, O'Carroll D, Ancelin K, Ono Y, Sano M, et al. Blimp1 is a critical determinant of the germ cell lineage in mice. Nature. 2005 Jul 14;436(7048):207-13.

[63] Vincent SD, Dunn NR, Sciammas R, Shapiro-Shalef M, Davis MM, Calame K, et al. The zinc finger transcriptional repressor Blimp1/Prdm1 is dispensable for early axis

formation but is required for specification of primordial germ cells in the mouse. Development. 2005 Mar;132(6):1315-25.

[64] Yamaji M, Seki Y, Kurimoto K, Yabuta Y, Yuasa M, Shigeta M, et al. Critical function of Prdm14 for the establishment of the germ cell lineage in mice. Nat Genet. 2008 Aug;40(8):1016-22.

[65] Gillich A, Bao S, Grabole N, Hayashi K, Trotter MW, Pasque V, et al. Epiblast stem cell-based system reveals reprogramming synergy of germline factors. Cell Stem Cell. 2012 Apr 6;10(4):425-39.

[66] Nagamatsu G, Kosaka T, Kawasumi M, Kinoshita T, Takubo K, Akiyama H, et al. A germ cell-specific gene, Prmt5, works in somatic cell reprogramming. J Biol Chem. 2011 Mar 25;286(12):10641-8.

[67] Ancelin K, Lange UC, Hajkova P, Schneider R, Bannister AJ, Kouzarides T, et al. Blimp1 associates with Prmt5 and directs histone arginine methylation in mouse germ cells. Nat Cell Biol. 2006;8(6):623-30.

[68] Maatouk DM, Kellam LD, Mann MR, Lei H, Li E, Bartolomei MS, et al. DNA methylation is a primary mechanism for silencing postmigratory primordial germ cell genes in both germ cell and somatic cell lineages. Development. 2006 Sep;133(17):3411-8.

[69] Reik W, Dean W. Back to the beginning. Nature. 2002 Nov 14;420(6912):127.

[70] Popp C, Dean W, Feng S, Cokus SJ, Andrews S, Pellegrini M, et al. Genome-wide erasure of DNA methylation in mouse primordial germ cells is affected by AID deficiency. Nature. 2010 Feb 25;463(7284):1101-5.

[71] Morgan HD, Dean W, Coker HA, Reik W, Petersen-Mahrt SK. Activation-induced cytidine deaminase deaminates 5-methylcytosine in DNA and is expressed in pluripotent tissues: implications for epigenetic reprogramming. J Biol Chem. 2004 Dec 10;279(50):52353-60.

[72] Reik W. Stability and flexibility of epigenetic gene regulation in mammalian development. Nature. 2007 May 24;447(7143):425-32.

[73] Hackett JA, Sengupta R, Zylicz JJ, Murakami K, Lee C, Down TA, et al. Germline DNA Demethylation Dynamics and Imprint Erasure Through 5-Hydroxymethylcytosine. Science. 2012 Dec 13.

[74] de Napoles M, Nesterova T, Brockdorff N. Early loss of Xist RNA expression and inactive X chromosome associated chromatin modification in developing primordial germ cells. PLoS One. 2007;2(9):e860.

[75] Hinkins M, Huntriss J, Miller D, Picton HM. Expression of Polycomb-group genes in human ovarian follicles, oocytes and preimplantation embryos. Reproduction. 2005 Dec;130(6):883-8.

[76] Labosky PA, Barlow DP, Hogan BL. Mouse embryonic germ (EG) cell lines: transmission through the germline and differences in the methylation imprint of insulin-like

growth factor 2 receptor (Igf2r) gene compared with embryonic stem (ES) cell lines. Development. 1994 Nov;120(11):3197-204.

[77] Tada T, Tada M, Hilton K, Barton SC, Sado T, Takagi N, et al. Epigenotype switching of imprintable loci in embryonic germ cells. Dev Genes Evol. 1998 Feb;207(8):551-61.

[78] Onyango P, Jiang S, Uejima H, Shamblott MJ, Gearhart JD, Cui H, et al. Monoallelic expression and methylation of imprinted genes in human and mouse embryonic germ cell lineages. Proc Natl Acad Sci U S A. 2002 Aug 6;99(16):10599-604.

[79] Matsui Y, Zsebo K, Hogan BL. Derivation of pluripotential embryonic stem cells from murine primordial germ cells in culture. Cell. 1992;70(5):841-7.

[80] Stewart CL, Gadi I, Bhatt H. Stem cells from primordial germ cells can reenter the germ line. Dev Biol. 1994 Feb;161(2):626-8.

[81] Shamblott MJ, Axelman J, Wang S, Bugg EM, Littlefield JW, Donovan PJ, et al. Derivation of pluripotent stem cells from cultured human primordial germ cells. Proceedings of the National Academy of Sciences of the United States of America. 1998;95(23):13726-31.

[82] Hiller M, Liu C, Blumenthal PD, Gearhart JD, Kerr CL. Bone morphogenetic protein 4 mediates human embryonic germ cell derivation. Stem Cells Dev. 2011 Feb;20(2): 351-61.

[83] Kashyap V, Rezende NC, Scotland KB, Shaffer SM, Persson JL, Gudas LJ, et al. Regulation of stem cell pluripotency and differentiation involves a mutual regulatory circuit of the NANOG, OCT4, and SOX2 pluripotency transcription factors with polycomb repressive complexes and stem cell microRNAs. Stem Cells Dev. 2009 Sep; 18(7):1093-108.

[84] Bartel DP. MicroRNAs: target recognition and regulatory functions. Cell. 2009 Jan 23;136(2):215-33.

[85] Gangaraju VK, Lin H. MicroRNAs: key regulators of stem cells. Nat Rev Mol Cell Biol. 2009 Feb;10(2):116-25.

[86] Houbaviy HB, Murray MF, Sharp PA. Embryonic stem cell-specific MicroRNAs. Dev Cell. 2003 Aug;5(2):351-8.

[87] Barroso-delJesus A, Romero-Lopez C, Lucena-Aguilar G, Melen GJ, Sanchez L, Ligero G, et al. Embryonic stem cell-specific miR302-367 cluster: human gene structure and functional characterization of its core promoter. Mol Cell Biol. 2008 Nov;28(21): 6609-19.

[88] Letzen B, Liu C, Thakor N, Gearhart J, All A, Kerr C. MicroRNA Expression Profiling of Oligodendrocyte Differentiation from Human Embryonic Stem Cells. PLoS one. 2010;5(5):e10480 http://www.ncbi.nlm.nih.gov/pmc/articles/PMC2864763/.

[89] Lakshmipathy U, Love B, Goff LA, Jornsten R, Graichen R, Hart RP, et al. MicroRNA Expression Pattern of Undifferentiated and Differentiated Human Embryonic Stem Cells. Stem Cells Dev. 2007 Dec;16(6):1003-16.

[90] Marson A, Levine SS, Cole MF, Frampton GM, Brambrink T, Johnstone S, et al. Connecting microRNA genes to the core transcriptional regulatory circuitry of embryonic stem cells. Cell. 2008 Aug 8;134(3):521-33.

[91] Tay Y, Zhang J, Thomson AM, Lim B, Rigoutsos I. MicroRNAs to Nanog, Oct4 and Sox2 coding regions modulate embryonic stem cell differentiation. Nature. 2008 Oct 23;455(7216):1124-8.

[92] Pasini D, Bracken AP, Helin K. Polycomb group proteins in cell cycle progression and cancer. Cell Cycle. 2004 Apr;3(4):396-400.

[93] Wang H, Garzon R, Sun H, Ladner KJ, Singh R, Dahlman J, et al. NF-kappaB-YY1-miR-29 regulatory circuitry in skeletal myogenesis and rhabdomyosarcoma. Cancer Cell. 2008 Nov 4;14(5):369-81.

[94] Sinkkonen L, Hugenschmidt T, Berninger P, Gaidatzis D, Mohn F, Artus-Revel CG, et al. MicroRNAs control de novo DNA methylation through regulation of transcriptional repressors in mouse embryonic stem cells. Nat Struct Mol Biol. 2008 Mar;15(3): 259-67.

[95] Kanellopoulou C, Muljo SA, Kung AL, Ganesan S, Drapkin R, Jenuwein T, et al. Dicer-deficient mouse embryonic stem cells are defective in differentiation and centromeric silencing. Genes Dev. 2005 Feb 15;19(4):489-501.

[96] Wang Y, Medvid R, Melton C, Jaenisch R, Blelloch R. DGCR8 is essential for microRNA biogenesis and silencing of embryonic stem cell self-renewal. Nat Genet. 2007 Mar;39(3):380-5.

[97] Fukuda T, Yamagata K, Fujiyama S, Matsumoto T, Koshida I, Yoshimura K, et al. DEAD-box RNA helicase subunits of the Drosha complex are required for processing of rRNA and a subset of microRNAs. Nat Cell Biol. 2007 May;9(5):604-11.

[98] Banerjee R, Mani RS, Russo N, Scanlon CS, Tsodikov A, Jing X, et al. The tumor suppressor gene rap1GAP is silenced by miR-101-mediated EZH2 overexpression in invasive squamous cell carcinoma. Oncogene. 2011 Oct 20;30(42):4339-49.

[99] Varambally S, Cao Q, Mani RS, Shankar S, Wang X, Ateeq B, et al. Genomic loss of microRNA-101 leads to overexpression of histone methyltransferase EZH2 in cancer. Science. 2008 Dec 12;322(5908):1695-9.

[100] Cui K, Zang C, Roh TY, Schones DE, Childs RW, Peng W, et al. Chromatin signatures in multipotent human hematopoietic stem cells indicate the fate of bivalent genes during differentiation. Cell Stem Cell. 2009 Jan 9;4(1):80-93.

[101] Collas P. Programming differentiation potential in mesenchymal stem cells. Epigenetics. 2010 Aug 16;5(6):476-82.

[102] Meissner A, Mikkelsen TS, Gu H, Wernig M, Hanna J, Sivachenko A, et al. Genome-scale DNA methylation maps of pluripotent and differentiated cells. Nature. 2008 Aug 7;454(7205):766-70.

[103] Jiang L, Li J, Song L. Bmi-1, stem cells and cancer. Acta Biochim Biophys Sin (Shanghai). 2009 Jul;41(7):527-34.

[104] Clarke MF, Fuller M. Stem cells and cancer: two faces of eve. Cell. 2006 Mar 24;124(6): 1111-5.

[105] Hill RP, Perris R. "Destemming" cancer stem cells. J Natl Cancer Inst. 2007 Oct 3;99(19):1435-40.

[106] Dalerba P, Cho RW, Clarke MF. Cancer stem cells: models and concepts. Annu Rev Med. 2007;58:267-84.

[107] Mongan NP, Martin KM, Gudas LJ. The putative human stem cell marker, Rex-1 (Zfp42): structural classification and expression in normal human epithelial and carcinoma cell cultures. Mol Carcinog. 2006 Dec;45(12):887-900.

[108] Chiou SH, Yu CC, Huang CY, Lin SC, Liu CJ, Tsai TH, et al. Positive correlations of Oct-4 and Nanog in oral cancer stem-like cells and high-grade oral squamous cell carcinoma. Clin Cancer Res. 2008 Jul 1;14(13):4085-95.

[109] Raman JD, Mongan NP, Liu L, Tickoo SK, Nanus DM, Scherr DS, et al. Decreased expression of the human stem cell marker, Rex-1 (zfp-42), in renal cell carcinoma. Carcinogenesis. 2006 Mar;27(3):499-507.

[110] Ben-Porath I, Thomson MW, Carey VJ, Ge R, Bell GW, Regev A, et al. An embryonic stem cell-like gene expression signature in poorly differentiated aggressive human tumors. Nat Genet. 2008 May;40(5):499-507.

[111] Takahashi K, Tanabe K, Ohnuki M, Narita M, Ichisaka T, Tomoda K, et al. Induction of pluripotent stem cells from adult human fibroblasts by defined factors. Cell. 2007 Nov 30;131(5):861-72.

[112] Yu J, Vodyanik M, Smuga-Otto K, Antosiewicz-Bourget J, Frane J, Tian S, et al. Induced pluripotent stem cell lines derived from human somatic cells. Science. 2007;318(5858):1917.

[113] Wernig M, Meissner A, Foreman R, Brambrink T, Ku M, Hochedlinger K, et al. In vitro reprogramming of fibroblasts into a pluripotent ES-cell-like state. Nature. 2007 Jul 19;448(7151):318-24.

[114] Wong DJ, Liu H, Ridky TW, Cassarino D, Segal E, Chang HY. Module map of stem cell genes guides creation of epithelial cancer stem cells. Cell Stem Cell. 2008 Apr 10;2(4):333-44.

[115] Ohm J, Mali P, Van Neste L, Berman D, Liang L, Pandiyan K, et al. Cancer-Related Epigenome Changes Associated with Reprogramming to Induced Pluripotent Stem Cells. Cancer Research. 2010;70(19):7662.

[116] Mali P, Chou BK, Yen J, Ye Z, Zou J, Dowey S, et al. Butyrate greatly enhances derivation of human induced pluripotent stem cells by promoting epigenetic remodeling and the expression of pluripotency-associated genes. Stem Cells. 2010 Apr;28(4): 713-20.

[117] Varambally S, Dhanasekaran SM, Zhou M, Barrette TR, Kumar-Sinha C, Sanda MG, et al. The polycomb group protein EZH2 is involved in progression of prostate cancer. Nature. 2002 Oct 10;419(6907):624-9.

[118] Kamminga LM, Bystrykh LV, de Boer A, Houwer S, Douma J, Weersing E, et al. The Polycomb group gene Ezh2 prevents hematopoietic stem cell exhaustion. Blood. 2006 Mar 1;107(5):2170-9.

[119] Pasini D, Bracken AP, Jensen MR, Lazzerini Denchi E, Helin K. Suz12 is essential for mouse development and for EZH2 histone methyltransferase activity. Embo J. 2004 Oct 13;23(20):4061-71.

[120] Bracken AP, Dietrich N, Pasini D, Hansen KH, Helin K. Genome-wide mapping of Polycomb target genes unravels their roles in cell fate transitions. Genes Dev. 2006 May 1;20(9):1123-36.

[121] Bracken AP, Kleine-Kohlbrecher D, Dietrich N, Pasini D, Gargiulo G, Beekman C, et al. The Polycomb group proteins bind throughout the INK4A-ARF locus and are dissociated in senescent cells. Genes Dev. 2007 Mar 1;21(5):525-30.

[122] Dietrich N, Bracken AP, Trinh E, Schjerling CK, Koseki H, Rappsilber J, et al. Bypass of senescence by the polycomb group protein CBX8 through direct binding to the INK4A-ARF locus. Embo J. 2007 Mar 21;26(6):1637-48.

[123] Bracken AP, Pasini D, Capra M, Prosperini E, Colli E, Helin K. EZH2 is downstream of the pRB-E2F pathway, essential for proliferation and amplified in cancer. Embo J. 2003 Oct 15;22(20):5323-35.

[124] Widschwendter M, Fiegl H, Egle D, Mueller-Holzner E, Spizzo G, Marth C, et al. Epigenetic stem cell signature in cancer. Nat Genet. 2007 Feb;39(2):157-8.

[125] Liu S, Dontu G, Mantle ID, Patel S, Ahn NS, Jackson KW, et al. Hedgehog signaling and Bmi-1 regulate self-renewal of normal and malignant human mammary stem cells. Cancer Res. 2006 Jun 15;66(12):6063-71.

[126] Raman JD, Mongan NP, Tickoo SK, Boorjian SA, Scherr DS, Gudas LJ. Increased expression of the polycomb group gene, EZH2, in transitional cell carcinoma of the bladder. Clin Cancer Res. 2005 Dec 15;11(24 Pt 1):8570-6.

[127] Friedman JM, Liang G, Liu CC, Wolff EM, Tsai YC, Ye W, et al. The putative tumor suppressor microRNA-101 modulates the cancer epigenome by repressing the polycomb group protein EZH2. Cancer Res. 2009 Mar 15;69(6):2623-9.

Chapter 11

Multiple Paths to Reprogramming

Antonio Lo Nigro and Catherine M. Verfaillie

Additional information is available at the end of the chapter

http://dx.doi.org/10.5772/55104

1. Introduction

For a long time, differentiation was considered a "one way process"; Conrad Waddington, in the 1950s, described cellular differentiation and development as a ball rolling towards different one-way ramified valleys, giving rise to specific cell fates, irreversibly [1]. However, in the last decades, a series of studies have shown that somatic cells and stem cells are more plastic than previously believed. Using different technical approaches, the epigenetic barriers imposed during development in differentiated cells can be erased, and cells can re-acquire pluripotency through a process, known as "reprogramming".

The first evidence came at the end of the 1950's from the pivotal experiments performed by J.B. Gordon in the zoology department at Oxford University [2]. At that time, embryologists, not aware of epigenetic regulation, i.e. the role of chromatin and its crucial modifications in cell fate determination, wondered whether development and cellular differentiation arise upon specific restriction of the genetic information contained in their nuclei. To answer this basic but intriguing question, Gordon used a technique, now known as somatic cellular nuclear transfer (SCNT) in *Xenopus laevis laevis*. For these experiments, nuclei from intestinal epithelial cells of albino tadpoles were transferred into unfertilized and enucleated wild-type frog oocytes. This resulted in the development of normal albino frogs, which in some cases were also fertile. They concluded from these studies that adult nuclei contain the genetic information necessary for the development of a frog. Moreover, cellular differentiation, during development, does not occur through loss of genetic information. These findings were exciting for the scientific community but at the same time controversial for two reasons: I) the efficiency with which a 'cloned' frog reached the adult stage was around 1% and II) the same technique did not work with mammalian cells [3].

For forty years, the scientific community was not able to use SCNT in other species. Finally, in 1997, by using the same technique, Ian Wilmut and colleagues, at the Roslin Institute in

Edinburgh, Scotland, succeeded in generating the sheep named "Dolly" by SCNT, further confirming that genetic modifications, leading to cellular differentiation, are not irreversible. Two key improvements in his technical strategy led to the first cloning of a mammal: unfertilized oocytes were used as recipients and donor cells were induced to exit from the normal cellular cycle, by serum withdrawal [4, 5].

One year later, Wakayama and colleagues [6] reported that SCNT also allowed the cloning of the most used animal model, the mouse. Again, another technical advance led to this progress: the use of an enucleation pipette, which allowed for the removal of the nucleus from the oocytes. This advance also allowed the conclusion that reprogramming factors are not oocyte-specific, meaning that SCNT can be done also using zygotes and fertilized eggs [7], and that the molecules responsible for reprogramming were present in the cell cytoplasm. In general, nuclear transfer (see Figure 1) involves two steps: a) de-differentiation of a somatic donor cell to an embryonic state and the *in vitro* maturation till the blastocyst stage (also known as therapeutic cloning); b) the further development of the cloned blastocyst, after the implantation in the maternal uterus (reproductive cloning).

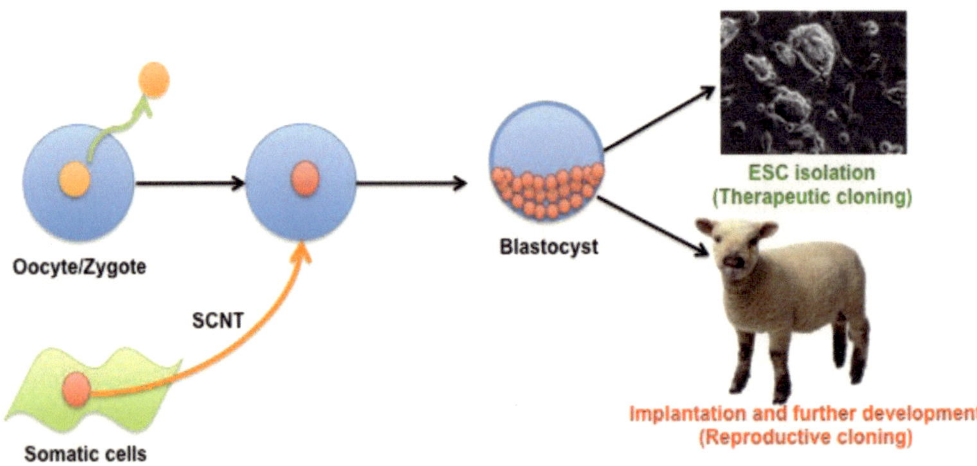

Figure 1. SCNT, Therapeutic and Reproductive cloning

Therapeutic cloning permits the derivation of nuclear transfer derived embryonic stem cells (ntESCs). Recently, the efficiency of isolation of ntESCs drastically increased, at least in mice, from 1% to 20% [8]. It has also been possible to derive similar cells in cats, dogs, wolves, goats and monkeys. Although the isolation of human ntESCs has been reported, this paper has been retracted later on [9]. Thus, the possibility of therapeutic cloning with human cells needs to still be demonstrated. However, the therapeutic cloning remains a promising technology for regenerative medicine, considering that ntESCs, from other species, were able to differentiate into all the cell types of an adult body.

Reproductive cloning is technically more difficult than therapeutic cloning, as it involves the further development *in vivo* in a pseudopregnant female. Embryos, derived after SCNT, develop till blastocyst stage with a good efficiency (20-50% depending on the species), but most of them die in the post-implantation stage, without reaching birth (1-5% of survival rate to birth, depending on the species).

Analysis of the cloned animals also showed several abnormalities: increased telomere shortening (which may have caused the premature death of Dolly), altered gene expression during development, prolonged gestation, fetal or placental edema, increased risk of obesity and cancer. The reasons for these pathologies remain not fully understood. The defects may be due to infidelity of the reprogramming: residual epigenetic memory of the donor cell may be present an/or imprinting of important developmental genes may be altered. Nevertheless, reproductive cloning remains attractive and may have potential implications in agriculture and industrial biotechnology. However, as it relates to humans, cloning (also therapeutic) remains controversial as theoretically, it may allow the cloning of a human being.

Evidence that differentiation is reversible also comes from another technique, known as cell fusion [10]. In cell fusion experiments, two or more cells can be fused together (by using polyethylene glycol (PEG) or electrofusion) to generate a single cell, called heterokaryon or hybrid. The larger or more dividing cell type is the "dominant" one, and the "recessive" cell will convert its gene expression profile to the one imposed by the dominant cell type. Obviously, alteration in the ratio of the two cell types during fusion will affect the final fate of the fused cells [11].

A heterokaryon, produced by inhibiting cell division, is a fused cell that becomes multinucleated and survives only short-term. If the cell cycle is not blocked, the fused cells will form a hybrid, because upon the first division the two different nuclei will become a single nucleus, having 4n chromosomes (see Figure 2). Its karyotype can be: 1) euploid, when fused cells are from same species (the two cell types have the same number of chromosomes, thus, their segregation will be balanced); 2) aneuploid, when cells fused are from different species (the two cell types have a different number of chromosomes, thus, they will be lost and/or rearranged).

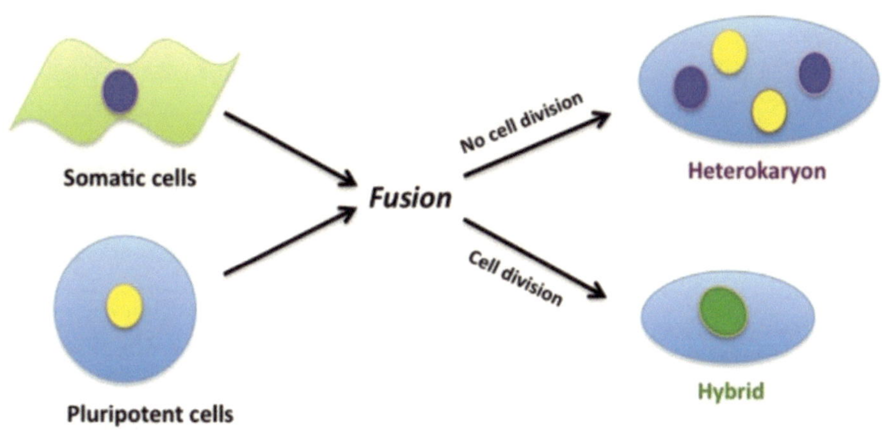

Figure 2. Cell fusion

Cell fusion experiments advanced medical knowledge on cell plasticity. In 1969, Harris et al. fused tumor cells with normal cells and demonstrated that there are trans-acting oncosuppressor genes. Upon fusion, malignancy was suppressed and this was not due to the loss of an oncogene, as after prolonged *in vitro* culture malignancy re-emerged [12]. Fibroblasts can be induced to produce albumin or melanin if fused with hepatocytes [13, 14] or melanocytes [15, 16], respectively.

In 1983, Blau et al. [17] produced for the first time heterokaryons from diploid human amniocytes fused with differentiated mouse muscle cells. She demonstrated that the heterokaryons express many human muscle-related genes and that this activation was mediated by factors present in the cytoplasm (as non-dividing heterokaryons have distinct nuclei). Similar heterokaryons with muscle cells can be produced not only by fusing them with amniocytes but also with cells of the three embryonic lineages (mesoderm, ectoderm, endoderm) [18].

In 1997, Surani, Tada and colleagues demonstrated, by producing proliferative hybrids that cell fusion not only "switches" the fate of different cell types but also "reprograms" them to a pluripotent state. Thymocytes from adult mice were fused to embryonic germ cells, pluripotent stem cells derived from primordial germ cells (PGCs). By using DNA sensitive restriction enzymes, they demonstrated that the genome of the somatic cell underwent a general demethylation, with reactivation of imprinted and non-imprinted genes, resembling the reprogramming events occurring in germ cell development [19].

They also fused female thymocytes, derived from Oct4-GFP mice, with mouse ESCs [17]. Two days after fusion, expression of GFP, from the thymocyte, was detected. The X chromosome, normally silenced in adult female cells, was reactivated. Moreover, hybrids had developmental potential, like ESCs, as they contributed to the three germ layers of chimeric animals, upon blastocyst aggregation [18]. Using the same approach, in 2005, Cowan succeeded in creating hybrids between human somatic cells and human ESCs [19].

This further elucidated that the differentiation state of cells is plastic and reversible; both SCNT and cell fusion experiments clearly demonstrated that it is possible to reset the epigenetic landscape of somatic cells. Despite all these studies were already present in the literature, the field of reprogramming only became jumpstarted in 2006 when Takahashi and Yamanaka [20] demonstrated that the overexpression of pluripotency-related transcription factors (TFs) can dedifferentiate adult fibroblasts to induced pluripotent stem cells (iPSCs), iPSCs strongly resemble ESCs. iPSC technology is an inefficient process, but differently from SCNT or cell fusion, may have in the near future therapeutic applications, including human disease modeling, drug screening and patient-specific cell therapy (see Figure 3).

After this publication [20], several studies demonstrated the potential of epigenetic reprogramming. Indeed, there is now evidence that use of different "cocktails" of TFs allows not only to redirect fibroblasts to an ESC-fate but also to a lineage-specific cell types/precursors, like cardiomyocytes, neuronal precursors, hepatocytes and blood cells, from a tissue different than the tissue from which the somatic cell was isolated [21-23].

http://dx.doi.org/10.5772/55104

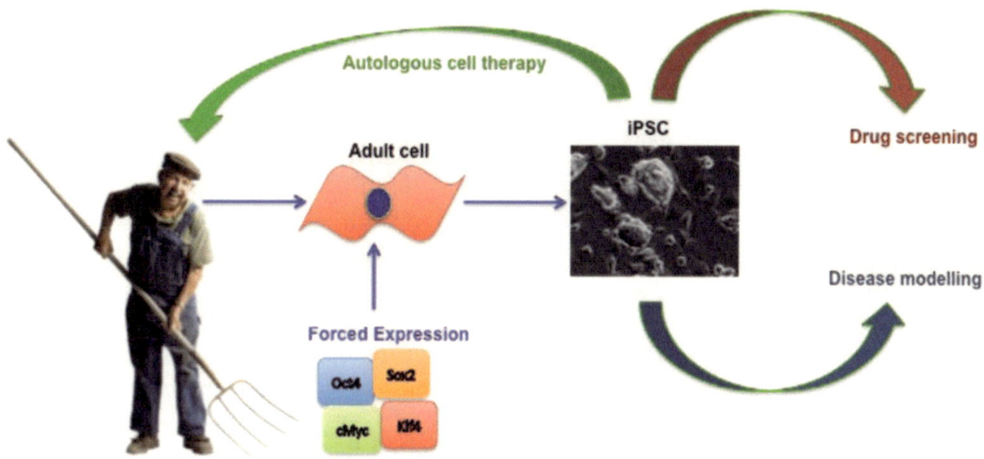

Figure 3. iPSC technology and applications

Finally, It has also become clear that cell plasticity and reprogramming may be partially achieved or enhanced by the culture microenvironment. An increasing number of studies is showing how small molecules, including epigenetic modifiers and signaling pathway modulators, play a crucial role in cell-fate determination [24]. All together, these studies highlight that culture media influences the epigenetic-state of the cells in which they are cultured and thus their features [25]. In this chapter, we will discuss:

1. *Reprogramming to the pluripotency-state*, describing transcription factors used in mouse and human, different methodologies and potentiality of iPSC technolology
2. *Lineage conversion*, illustrating the differentiated cell types/precursors obtained and the differences of this approach with the iPSC technology
3. *Culture mediated reprogramming*, providing the published data which highlight the influence of culture media and small molecules on stem cells fate and features

2. Reprogramming to the pluripotency-state

In 1987, two key discoveries highlighted how crucial the role is of some "master" TFs in cell fate determination. During *Drosophila Melanogaster* development, the gene *Antennopedia (Antp)* specifies the formation of the thoracic segment, which will then form the legs of the adult fly. Heat-induced overexpression of *Atnp,* at specific larval stage, led to the formation of additional legs instead of antennae [26]. Similarly, ectopic expression of *eyeless* (the homologous of *Pax6* in mice) caused the development of functional eyes on the wings, antennae and legs of drosophila [27].

Twenty years after these pivotal experiments, Shinya Yamanaka's group used the same TF-based technology to reprogram adult fibroblasts to pluripotent state [20]. Mouse embryonic and adult fibroblasts, transduced with retroviral vectors encoding for Oct4, Sox2, Klf4 and c-Myc and cultured in ESC-medium, erased their differentiated epigenetic state and reestablished the pluripotent state; these cells were named induced pluripotent stem cells (iPSCs). Murine iPSCs exhibited morphological and growth properties of ESCs, and expressed alkaline phosphatase and SSEA1. No differences, if compared with ESCs, could be detected in their methylation status, X activation status, embryoid body (EB) formation, *in vitro* differentiation capacity (ectodermal, mesodermal and endodermal), teratoma formation and *in vivo* developmental potential (contribution to the three germ layers of chimeric animals).

Noteworthy, pluripotent stem cells (PSCs) possess mechanisms that lead to the silencing of the integrated transgenes. Therefore, the expression of the four TFs is necessary for generating iPSCs but dispensable for maintenance of the iPSC fate; hence, pluripotency and selfrenewal capacity rely on the trans-activation of the endogenous genes, suggesting a true and complete reprogramming. Mouse iPSCs, like ESCs, were germline competent [28] and supported the development of a mice in tetraploid complementation assay [29]. In this assay, embryonic cells at the two cell-stage are fused together. This results in a tetraploid blastocyst in which just the extraembryonic tissues will further develop; by complementing the tetraploid embryo with normal diploid PSCs, it is possible to generate an individual, completely derived from the diploid PSCs. Interestingly, the same combination of TFs [30] or a somewhat different one (Oct4, Sox2, Nanog and Lin28) can be used for the reprogramming of human cells [31].

This discovery is groundbreaking because with iPSC technology, PSCs can now be induced/derived for autologous cell transplantation, avoiding immunological problems and ethical issues related to the use of human ESCs. In addition, iPSCs from patients carrying a disease can be derived and used to better understand the biological problem leading to the disease as well as for drug-screening.

2.1. Rationale behind iPSCs

The rationale for the selection of the genes for this "reprogramming" cocktail was obviously based on the studies, done in the preceding decade, aimed at understanding the network of TFs responsible for ESC pluripotency and selfrenewal.

The Oct4 (also known as Pou5f1) gene encodes for a TF that belongs to the POU homeodomain DNA binding domain family [32]. It plays a key role in the development and in maintenance of both ESCs self-renewal and pluripotency. Misregulation of its levels triggers loss of the ESC fate; a 50% loss of expression drives ESCs to trophectoderm while a 50% greater expression induces primitive endoderm or mesoderm [33]. Knockout (KO) experiments in mouse demonstrated lethality at the preimplantation stage *in vivo* and failure of ESC derivation *in vitro* [34].

Sox2 belongs to a family of TFs, having the high mobility group (HMG) DNA-binding domain, identified for the first time in the SRY (sex determining Y region) protein, which is the testis determining factor. In general, the genes from the Sox family are involved in different and crucial steps of mammalian development [35, 36]. Sox2 KO embryos die immediately after implantation [37]. Moreover, it is well known that Oct4 and Sox2 form a complex together, which regulates synergistically the trascription of among others, Fbx15, Fgf4 and Utf1 [38-41].

Nanog, (from Tír na nÓg, the Land of Ever-Young in the old Irish mythology) is another homeobox TF essential for pluripotency. Nanog expression can be detected in the late morula, in the ICM of the blastocyst and in the epiblast. Nanog knockout mice are lethal and the ICM fails to further progress to the epiblast-stage [42-44]. In contrast to Oct4, Nanog overexpression maintains ESC selferewal and pluripotency in a feeder-free and LIF-independent way.

Lin28 is a negative regulator of micro (mi)RNA processing. It blocks the posttranscriptional processing of several primary miRNA transcripts (pri-miRNAs). It is responsible for miRNA biogenesis in both cancer cells and ESCs; so it plays a key role in tumorigenesis and development. Lin28 KO mice have decreased weight at birth and increased postnatal lethality [45].

Different from the above genes, Klf4 and c-Myc are not ESC-specific but are required for their direct or indirect effect on cell proliferation. Of note, iPSCs can also be produced without c-Myc and this is clinically relevant, considering the oncogenic features of c-Myc [46].

Although the combination of Oct4, Sox2, Klf4 and c-Myc (OSKM) consistently allows the reprogramming, this it is not an efficient process (0.01-0.1%). For this reason, subsequent studies were focused on improving iPSC efficiency and since the first iPSC publication, many papers have reported several other genes which enhance the efficiency of iPSC generation.

Inclusion of Utf1, another TF involved in ESC pluripotency, together with the inhibition of p53, increases iPSC generation by 200-fold [47]. Similarly, other factors (like the SV40 large T antigen, SV40LT; the telomerase reverse transcriptase, TERT) and microRNAs (miRNAs) controlling cell proliferation, senescence and apoptosis also affects the efficiency and the speed of reprogramming [48-52]. Other studies have reported important roles for Sall4 [53], Esrrb (which can replace Klf4] [54] and Tbx3 (which improves the germline contribution) [55].

2.2. The donor cell type and epigenetic memory

The starting cell, used for reprogramming, is a key parameter that influences the kinetics, the efficiency and the quality of the iPSCs. Fibroblasts are the most commonly used somatic cells because they can be easily isolated. In mouse studies, embryonic fibroblasts (MEFs) are commonly used as iPSCs can be generated in 10-12 days; MEF-derived iPSC generation is therefore recommended for studies aimed at understanding the mechanisms underlying iPSC generation, as well as the TFs and the chemicals that may enhance this process.

To generate iPSCs from human foreskin fibroblasts (HFFs), three weeks are required and the efficiency is 100 fold less compared with human primary keratinocytes, in which reprogramming also occurs faster [56]. When using CD133$^+$ cord blood cells, iPSCs can be produced by overexpression of only Oct4 and Sox2. As cord blood banks exist, it is believed that this cell source may be useful to make an iPSC bank representing a wide panel of haplotypes for regenerative medicine [57].

Another crucial parameter is the differentiation status; Hematopoietic stem cells and progenitors have a higher efficiency of reprogramming than terminally differentiated B- and T-lymphocytes [58]. Similarly, $Sox2^{+}$ neural progenitor cells form iPSCs just by forced overexpression of Oct4 [59]. Many other cell types, such as adipose stem cells [60], dental pulp cells [61], oral mucosa cells [62] and peripheral blood cells [63] can also be used to generate iPSCs.

iPSCs from different origins have a similar, if not identical, gene expression profile in their pluripotent state. However, it has become clear that some genomic regions are differentially methylated [64]; they retain an epigenetic memory of the cell of origin and this is reflected in their differentiation capacity. For example, iPSCs generated from blood poorly differentiated into neuronal cells but had a higher capacity to differentiate into hematopoietic cells [65].

The cell of origin to be used for iPSC generation, also has impact on safety issues; iPSC lines, generated from tail tip fibroblasts, have shown a higher propensity to form teratomas than lines obtained from stomach, hepatocyte or MEF, due to the persistence of undifferentiated cells even after iPSC differentiation [66].

2.3. Methods for iPSC generation

The method of transgene delivery is a crucial factor in determining the efficiency but also the clinical relevance of iPSCs. Although initial reports were based on retroviral vectors, later publications described several other methods, which allow the generation of iPSCs. They can be divided into two main groups: integrative and non-integrative methods [67, 68]. Integrative methods are in general more efficient but they are less safe than the non-integrative methods, which are, however, inefficient.

When choosing the strategy of reprogramming, it is important to consider the aim of the study; integrative methods, the most efficient ones, should be used for elucidating mechanisms underlying iPSC generation, and TFs and chemicals that may enhance this process, while, non-integrative methods will be required to generate clinical-grade iPSCs.

2.3.1. Integrative method

2.3.1.1. Viral vector-based delivery

Mouse and human iPSCs were initially produced by transduction of Moloney murine leukemia virus (MMLV)-derived retroviral vectors. Vectors, based on this system, allow cargoes of up to 8Kb fragments, can efficiently infect (although only dividing cells) and are generally silenced in pluripotent cells [69].

Lentiviral vectors, derived from HIV, have also been used. Differently from the retroviral vectors, the latter have a higher cloning capacity (up to 10Kb of DNA) and can infect both dividing and non-dividing cells. However, transgenes introduced using lentiviral vectors are less-silenced and this can result in a more laborious identification of bona fide iPSC clones (having the transgenes silenced). The lentiviral vector system allowed the Tet-inducible expression of the transgenes in a tightly controlled way [70]. Subsequently, polycistronic lentiviral vectors, having the OSKM cDNA under the control of a unique promoter were used

[71]. This was possible by including in between the different cDNAs, the 2A self-deleting peptide. This permits the continuous translation of downstream cDNA after the release of the previous protein [72]. In general, viral vector-based methods are quite efficient and reproducible (>0.1% in mouse cells, <0.01 in human cells). However, clones generated by viral delivery are not clinically safe. The transgenes may be reactivated during iPSC differentiation; moreover long terminal repeats (LTR) may activate proto-oncogenes increasing their tumorigenicity [28, 73].

2.3.1.2. *Transposon delivery*

Another possible strategy for iPSC generation is the transient delivery of OSKM by Piggyback (PB) transposon [74]. This system consists of a donor vector, containing the cassette (OSKM), flanked by a 5′ and a 3′ inverted repeat, and a helper plasmid, expressing the PB transposase. When cotransfected, the cassette of the donor plasmid is pasted into the TTAA sequences present in the genome, but can be remobilized after the reprogramming [75, 76]. The PB transposon-mediated generation of iPSCs occurs with high efficiency and, among the integrative methods, this is the only one that allows a precise deletion of the cassette. However, alterations at the integration sites were found; therefore, sequences at the integrations sites must be verified.

2.3.2. *Non-integrative method*

2.3.2.1. *Viral vector-based delivery*

Adenoviral vectors do not integrate into the host genome and can, thus, be used for making iPSCs [77, 78]. Adenoviral vectors can carry up to 36kB and can infect both dividing and non-dividing cells. However, the efficiency of iPSC generation is extremely low (0.002 to 0.0001%), probably because the premature dilution of adenoviral vectors during cell replication.

A more efficient alternative has been reported by Fusaki and colleagues [79], using F-deficient Sendai viral vectors. Sendai viruses are minus strand RNA virus, which replicate their genome in the host cytoplasm. Because these viral vectors replicate ubiquitously, their efficiency of reprogramming is similar to retroviral vectors. However, to obtain viral vector-free iPSCs, elimination of the vector using temperature sensitive mutant or antiviral compounds is required.

2.3.2.2. *Episomal and minicircle vectors*

OriP/Epstain-Barr nuclear antigen-1-based (OriP/EBNA1) vectors can be transfected into host cells and maintained stably episomically (because they replicate during cell divisions through their oriP element) using a drug in the culture medium [80]. Yu and colleagues [81] used the combination of three OriP/EBNA1 vectors (having a combination of 10 reprogramming factors) to generate iPSCs from HFF. By removing the drug selection, episomal vectors are eliminated from proliferating iPSCs.

Another alternative are the minicircle vectors, that differently from the above vectors, are non-replicative [82]. These vectors have a better transfection efficiency than OriP/EBNA1, due to their reduced length (they lack the bacterial origin of replication and the antibiotic resistance gene). However, both strategies have a three-fold lower efficiency (<0.001%) than retroviral vector-based reprogramming.

2.3.2.3. *Protein/RNA based delivery*

Previous studies have shown that proteins can be directly delivered into cells by fusing them with peptides [83], which mediates their transduction, such as poly-arginine or the HIV transactivator of transcription (TAT). Zhou et al. [84] produced recombinant OSKM proteins fused with poly-arginine in *Escherichia coli* and generated iPSCs from Oct4-GFP MEF, including valproic acid (a histone deacetylase inhibitor) in the medium. Kim et al. [85], succeeded in reprogramming human neonatal fibroblasts by producing OSKM, fused to a Myc tag and nine arginines.

Similarly, *in vitro* transcribed ssRNA, modified by phosphatase treatment and by substituting cytidine and uridine for 5-methylcytidine and pseudouridine, can be delivered into different human cells [86]. This method also requires a recombinant B18R protein, which improves cell viability and protein stability. Differently from protein delivery, the latter strategy has a fast kinetics and a higher efficiency [0.01-0.1%, depending on the cell type). However, also in this case, a careful screening for mutations in different iPSC clones will be needed before an eventual clinical application.

Type of vector	**Method**	**Genomic Integration**	**Efficiency reported**
Viral	Retrovirus	+	4X
Viral	Lentivirus	+	3X
Viral	Adenovirus	-	1X
Viral	Sendai Virus	-	4X
DNA	Transposon	-	2X
DNA	Minicircle	-	X
DNA	Episomal plasmid	-	X
RNA	Recombinant RNA	-	3X
PROTEIN	Recombinant protein	-	X

4x= >0.1%; 3X= <0.1%; 2X= <0.01%;1X= <0.001%

Table 1. Comparison of different strategies for iPSC generation

2.4. iPSC technology, unsolved questions and emerging technologies

iPSCs were reproducibly derived from most, if not all, somatic tissues; however the efficiency reported is always less than 1%. It is the consensus of scientific community, that many more than 1% of the transfected/transduced cells start the reprogramming process. Using a live cell imaging approach, it was demonstrated [87] that almost all the transduced cells undergo symmetric divisions within 48 hours, retaining a fibroblast-morphology. At later stages, reprogramming cells undergo asymmetric divisions: one descendant becomes an iPSC while the other one undegoes cell death. Still unknown, stochastic and clonal events appear to control this process at later stages; in fact most of the cells do not complete the initiated process. Several studies have demonstrated key roles for demethylation [88], telomerase length [89] and mesenchymal to epithelial transition [90], during the reprogramming. A better understanding, especially of the later stochastic mechanisms, is still needed to fully understand and improve the efficiency of iPSC technology [91].

Another important question has been whether or not ESCs and iPSCs are similar and if not, whether differences are functionally important for their application. Conclusions from different studies are conflicting. Several papers have reported that there are remarkable differences in gene expression and DNA methylation [92, 93], while other studies, which included a large number of ESC and iPSC lines, concluded that it is quite difficult to distinguish between ESCs and iPSCs [94-96]. Considering that there are differences among different ESC lines [97], it is now believed that iPSCs clones have a higher variability than ESCs but that at least some iPSC clones are indistinguishable from ESCs.

Furthermore, the recent work of Young and colleagues [98] demonstrated that most of the genetic variability in between different iPSC clones is already present in the starting cell line and is thus not caused by the reprogramming process.

Interestingly, in the last five years, several studies have clearly demonstrated the potential of iPSC technology in regenerative medicine. Hanna et al [99] generated iPSCs from a humanized model of sickle cell anemia. After correcting the hemoglobin gene, by gene targeting, iPSCs were able to generate hematopoietic stem cells and to rescue the disease. Similarly, the potential of iPSCs for cell therapy was demonstrated for macular degeneration [100], Parkinson's disease [101], platelet deficiencies [102] and spinal cord injury [103, 104].

iPSCs derived from patients with specific diseases have been used for studying the mechanisms involved in these diseases and for drug screening [105, 106]. *In vitro* disease modeling is not only possible for monogenic disease but also for more complex polygenic diseases, having a late onset, like schizophrenia [107] and Alzheimer [108, 109].

As a result of the success with reprogramming of somatic cells to a pluripotency-state, lineage reprogramming (trans-differentiation) between different somatic cell types has also become a burgeoning field of research (see next section).

In conclusion, iPSC technology will, in the near future, have a drastic impact on science, regenerative medicine and business. Precise selection of "clean" clones, through the evaluation of their genomic and epigenetic integrity, as well as their gene expression profile, will be crucial for downstream applications. Despite these remaining hurdles, it is believed that clinical applications for iPSCs are not far off.

3. Lineage conversion

The discovery of iPSCs, together with previous experiments involving SCNT and cell fusion, clearly showed that differentiation is a reversible process and that cells are more 'plastic' than previously believed. Therefore, a new field, called lineage reprogramming, emerged rapidly in the last five years. Recent attempts have demonstrated that, the forced overexpression of TFs can also convert one cell type to another of the same or of other somatic germ layers. Lineage reprogramming depends on the capacity of certain TFs to overcome the existing epigenetic barriers and to rapidly initiate the new identity-specific gene network [110-113].

Examples of direct lineage conversion were described already in 1986; Davis, Lassar et al. [114, 115] converted different fibroblast lines into myogenic cells by overexpression of MyoD, a basic helix-loop-helix transcription factor, in combination with the demethylating agent, 5-azacytidine. Subsequent studies confirmed that myogenic conversion, as shown by presence of desmin and myosin heavy chain, could be achieved *in vitro* starting from a variety of cell types (adipose, melanoma, neuroblastoma and liver cell lines). However, the complete downregulation of the 'original' tissue-specific genes was only seen when starting with mesodermal cells and not with endodermal or ectodermal cell lines [116, 117].

A similar transdifferantiation was also seen in the blood system. The deletion of Pax5 in pro-B cells resulted in their switch into the T-cell lineage [118, 119]. Later on, the same group investigated this transdifferentiation more extensively; mature B cells were isolated from Pax5 knockout mice and transplanted back into a lymphocyte deficient recipient. Surprisingly, they could detect in the reconstituted mice donor pro B cells, which then gradually converted into T cells [120]. This demonstrated that lost of Pax5 led to a T cell phenotype through de-differentiation rather than direct transdifferentiation. Another example of direct conversion came from the work done by Graf and colleagues [121]; overexpression of C/EBPα *or* C/EBPβ, a basic leucine zipper TF binding CCAAT enhancers, induced a macrophage phenotype (as shown by Mac1 expression) in bone marrow or spleen-derived B cells. In the induced macrophages almost all the B cell genes analyzed were downregulated and cells acquired phagocytic function *in vitro*.

The above examples describe experimental conversions but there were also cases in which this conversion occurs naturally. Jarriault et al. [122], demonstrated that the epithelial rectal cell 'Y', migrates anterodorsally from the rectum to become a 'PDA' motor neuron. This conversion from Y to PDA is not direct but occurs through a de-differentiation state, in which the initial (Y-cell) and the final (PDA-cell) identity are not present [123]. In this section, we will describe the relevant cell types, recently, obtained by lineage reprogramming.

3.1. Conversion into mesodermal types

Seale and colleagues have recently found that Myf5[+] muscle precursors can convert into brown fat cells *in vivo* and *in vitro* [124], while studying the role of PRDM16 during development. Overexpression of PRDM16 differentiated primary mouse myoblasts with nearly 100% efficiency to brown fat. *Vice versa,* downregulation of PRDM16 in primary brown fat cells resulted in the expression of MyoD and Myogenin and in a myotube-like morphology. Interestingly, forced expression PRDM16 was not able to induce the same conversion in non-myogenic cell lines, like fibroblasts. Performing proteomic studies, they identified C/EBPα as a partner of PRDM16 in brown fat. Subsequently, they used combined overexpression of C/EBPα and PRDM16 [125] and demonstrated that mouse and human dermal fibroblasts could differentiate into brown-fat cells, which functional features (fat pad formation and glucose uptake after transplantation into mice).

Human dermal fibroblasts were converted into multipotent blood progenitors by just Oct4 overexpression [126] in combination with treatment with a hematopoietic permissive medium, containing growth factors and cytokines. Oct4 is a key TF for pluripotency but it is not expressed in the hematopoieic system [127]; probably, Oct4, in this case of lineage reprogramming, is mimicking the effect of Oct1 and Oct2, two other members of Pou family of TFs expressed in lymphoid development [128]. The induced progenitors express CD45 and express adult globin protein (unlike hematopoietic cells derived from human ESCs and iPSCs). Multipotent blood progenitors have myeloid, erythroid and megakaryocytic but not lymphoid potential, as shown by transplantation experiments.

The forced overexpression of TFs involved in cardiac development (Tbx5, Mef2c and Gata4) converts mouse cardiac and dermal fibroblasts into cardiomyocyte-like cells, termed induced cardiomyocyte (iCMs) [129]. Around 20% of the cells appear to be 'converted' in three days as measured by the expression of alpha-myosin heavy chain (αMHC), although one month is required for their complete maturation, which resulted in spontaneous beating capacity. Transplantation of iCMs, the day after the viral transduction, in injured hearts results in their engraftment and differentiation *in vivo*. Interestingly, the same strategy is able to convert cardiac fibroblasts *in vivo* [130]. When retroviral vectors, carrying the above factors, are injected after myocardial infarction, this results in the efficient conversion (>50%) of cardiac fibroblasts into functionally beating cardiomyocytes. Efficiency of cardiac-conversion is increased when Hand2 is added to the above cocktail of genes [131].

3.2. Conversion into endodermal types

The lineage reprogramming into β-cells is of particular interest, considering the potential for the treatment of diabetes. Zhou et al., [132] were able to *in vivo* convert exocrine acinar cells into functional β-like cells, combining three genes essential for pancreatic development (Ngn3, Pdx1 and MafA). Adenoviral vectors, carrying the pancreatic cocktail, were injected; again, conversion occurred in three day and efficiency of conversion was relatively high (20%). Analysis, one month later, showed that induced β-like cells produced insulin and rescued the hyperglycemic level after streptozotocin-treatment. However, the same combination of factors failed to reprogram myocytes *in vivo* and mouse embryonic fibroblast *in vitro*, indicating that additional factors will be needed to achieve this conversion from unrelated cell types.

More recently, mouse fibroblasts were converted into hepatocyte-like cells by overexpressing Hnf4α, FoxA1, FoxA2 and FoxA3 [133] or by Gata4, Hnf1α, Foxa3 together with $p19^{Arf}$ inactivation [134]. The reprogrammed cells were termed induced hepatocytes (iHeps) and had a gene expression profile and features (albumin production and cytochrome P450 activity), which closely resemble mature hepatocyte. iHeps *in vivo* were able to reconstitute hepatic tissues and to support hepatic function in the fumoaryl-acetate hydrolase ($FAH^{-/-}$) deficient mice.

3.3. Conversion into neuronal types

The conversion into neuronal types is, probably, the one that received more attention in the field of lineage reprogramming. The increasing attention is due to their possible application for the treatment of diseases involving the nervous system.

In 2010, Vierbuchen et al. [135] were the first to describe how overexpression of Ascl1, Mytl1 and Bm2 (also known as Pou3f2, again a member of Pou family) can convert embryonic and tail-tip fibroblasts into a mixed populations of induced neurons (iNs). iNs generate functional synapses with mouse cortical neurons and have action potentials; the detailed electrophysiological analysis showed that iNs contains mainly cells with features of glutamatergic neurons (with just a small percentage of GABAergic neurons). Remarkably, the addition of NeuroD1 to the above set of genes was necessary to achieve the same conversion in human cells [136]. The enriched cocktail of factors was able not only convert fibroblasts but also mouse hepatocyte into iNs [137].

Several groups, differently, converted fibroblasts into induced neural stem cells (iNSCs), that differently from the previous examples, can still self-renew and differentiate into different neuronal subtypes (multipotent). Different cocktail of factors and inductive media have been used to obtain multipotent neuronal stem cells from human and mouse fibroblasts: the group of Scholer [138] used Sox2, Klf4, c-Myc, together with Tcf3 and Brn4 (also known as Pou3f4); our group [139] by adding Zic3 to Oct4, Sox2 and Klf4; Ring et al. [140], by just overexpressing Sox2.

Different laboratories focused on a more direct conversion into specific neuronal subtypes, with a particular interest on neuronal cell types affected in neurodegenerative diseases. Two groups have been able to convert mouse and human fibroblasts into induced Dopaminergic Neurons (iDAs), the subtype affected in Parkinson's disease. The first laboratory [141] achieved this by adding FoxA2 and Lmx1a to Ascl1, Mytl1 and Bm2; the second [142] by overexpressing Ascl1, Lmx1a and Nurr1 (also known as Nr4a2). iDA cells, upon transplantation in mice, were capable to integrate into the host neuronal circuitry and express markers typical for mature dopaminergic neurons.

Lineage conversion was also achieved into spinal motor neurons, the subtypes involved in amyotrophic lateral sclerosis and spinal muscular atrophy. Conversion into induced Motor Neurons (iMNs) was achieved for both mouse and human fibroblasts; mouse embryonic fibroblasts were converted with Ascl1, Brn2, Mytl1, Lhx3, Ngn2, Isl1 and Hb9 whereas human cells also required NeuroD1 [143]. iMNs displayed markers, electrophysiological features and gene expression profile, which strongly resemble motor neurons. Moreover, iMNs engrafted into the developing chick spinal cord, forming axonal and dentritic projections toward the adjacent musculature.

Of note, Qiang et al [144] demonstrated that lineage reprogramming is also useful for drug screening and disease modeling. iNs, again with glutamatergic features, were induced by overexpressing Ascl1, Bm2, Mytl1 together with Zic1 and Olig2. iNs were produced from both healthy donors and Alzheimer's patients. iNs produced from patients displayed the typical accumulation of beta amyloid peptides (Aβ40 and Aβ42). Combining lineage reprogramming with gene-targeting technology, similar cells could also be used for autologous transplantation.

Starting cell type	Conversion into	Factors
(m) B-Cells	Macrophage-like cells	C/EBPα or β, PU.1 (121)
(m/h) dermal fibroblasts, myoblasts	Brown-fat cells	C/EBPα and PRDM16 (125)
(m) embryonic fibroblasts	Myoblasts	MyoD (114, 115)
(h) dermal fibroblasts	Multipotent blood progenitors	Oct4 (126)
(m) cardiac and tail tip fibroblasts	Cardiomyocytes	Tbx5, Mef2c, Gata4 ± Hand2 (129) (131)
(m) embryonic fibroblasts	Cardiomyocytes	Oct4, Sox2, Klf4 and cMyc (145)
(m) exocrine cells	β-like cells	Ngn3, Pdx1 and MafA (132)
(m) embryonic and dermal fibroblasts	Hepatocyte-like cells	Hnf4α, FoxA1, FoxA2 and FoxA3 (133)
(m) embryonic and tail tip fibroblasts	Hepatocyte-like cells	Gata4, Hnf1α, Foxa3 and $p19^{Arf}$ KD (134)
(m) embryonic fibroblasts	Neural progenitor cells	Oct4, Sox2, Klf4 and cMyc (146)
(m/h) fibroblasts	Neural progenitor cells	Sox2, Klf4, c-Myc,Tcf3 and Brn4 (138)
(h) fibroblasts	Neural progenitor cells	Oct4, Sox2, Klf4 and Zic3 (139)
(m) embryonic and (h) fetal fibroblasts	Neural progenitor cells	Sox2 (140)
(m) embryonic and (h) fetal, postnatal and dermal fibroblasts	Neurons	Ascl1, Mytl1, Bm2 and NeuroD1 (135) (136)
(m) tail and (h) embryonic fibroblasts	Dopaminergic neurons	Ascl1, Lmx1a and Nurr1 (142)
(h) embryonic and fetal lung fibroblasts	Dopaminergic neurons	Ascl1, Mytl1, Bm2, Lmx1a and FoxA2 (141)
(h) skin fibroblasts	Glutamatergic neurons	Ascl1, Bm2, Mytl1, Zic1 and Olig2 (144)
(m/h) embryonic fibroblasts	Motor neurons	Ascl1, Brn2, Mytl1, Lhx3, Ngn2, Isl1, Hb9 and NeuroD1 (143)

Table 2. Examples of lineage conversion. (m)= mouse (h) = human

3.4. Direct versus indirect strategy

Most of the examples, given in the previous section, describe the direct conversion from one cell type to another, in which the reprogramming is achieved without any intermediate state. However, other reports clearly demonstrated the possibility to achieve similar results, by using an alternative strategy, in which lineage conversion is indirect. Indirect conversion is achieved passing through a limited de-differentiation state by overexpressing Yamanaka factors for a shorter time. Like for the direct conversion, the indirect conversion is strongly dependent on the specific culture medium (growth factors and cytokines) given during the reprogramming phase.

The laboratory of Sheng Ding, at the Gladstone Institute of San Francisco, was the first to describe the possibility of lineage reprogramming through an indirect strategy. Short temporal overexpression of the Yamanaka factors induced a partial dedifferentiatied state, that allowed the subsequent conversion into cardiomyocytes-like cells by applying extracellular factors [145]. OSKM factors were overexpressed for six days in a medium free of signals necessary for pluripotency (i.e. leukemia inhibitory factor). After this short priming phase, cells were then cultured in media promoting cardiogenesis, i.e. cointaining BMP4. Three day after the cardiac induction, the expression of Nkx2.5, Gata4 and Flk1 (mid-stage markers of cardiac developments) could be detected. The further development into more mature cardiomyocytes, showing sarcomeric structures and cardiac features (expression of cardiac markers and cell-cell interaction) required at least two more days.

Interestingly, the authors also demonstrated that this indirect lineage conversion does not pass through a pluripotency-state, i.e. ESC/iPSC culture media in the induction phase drastically decrease the efficiency of conversion; *vice versa*, the addiction of a Jak inhibitor (which blocks the most important pluripotency-pathway) increased the efficiency of the process. The same group also demonstrated that a similar strategy induced expandable Neural Progenitor Cells (NPCs), having multipotent potential [146].

Both direct and indirect lineage conversions have pros and cons. The direct conversion, as in case of SCNT or cell fusions, occurs in hours-days. Induced cells are unipotent, are produced with a high efficiency, without the requirement of cell proliferation and with a lower risk for teratoma. The indirect strategy requires days-weeks and produces cells, which can be unipotent or multipotent. Cells induced by this strategy can be expanded but have a moderate risk for teratoma.

3.5. Mechanisms, differences with iPSC technology and unsolved questions

Reprogramming to the pluripotency-state occurs via a gradual and genome-wide de-differentiation, involving a first phase where epigenetic marks of differentiation are erased and a second phase in which the epigenetic marks of pluripotency are established to initiate the endogenous pluripotency-network. In lineage conversion, specific TFs are able to modulate cell fate in two different ways (direct or indirect), which does not involve a pluripotent-state and is associated with a lower tumor risk, still a major obstacle to achieve clinical applications with ESCs and iPSCs. In the direct conversion, ectopic TFs, involved in cell fate determination

http://dx.doi.org/10.5772/55104

or maintenance during embryonic development, overcome the pre-existing epigenetic marks and generate a new state. In the indirect conversion, the TFs, which allow the reprogramming to the pluripotency-state, are temporally overerexpressed together with fate-specific signals to convert original cell type into a new state. Differently than iPSC technology, the efficiencies are much higher (even 20 % in some cases) and the kinetics of conversion are rapid (a few day to a week maximum, and not two weeks to a month, like for iPSC, see Table 3).

Strategy	Efficiency	Kinetic	Potentiality	Expandibility	Tumorigenicity	Cell Proliferation
iPSC	Low	Weeks-Months	Pluripotent	Yes	High	Required
Direct conversion	High	Hours-Days	Unipotent	No	Low	Not required
Indirect conversion	High	Days-Weeks	Multi/Unipotent	Yes	Moderate	Required

Table 3. Comparison of different strategy of TF-based Reprogramming

As for iPSCs, many questions still remain unsolved in lineage conversion. It is not clear whether the new cell type, generated upon conversion, is a hybrid between the original and the new cell. It is intriguing that, in direct conversion, TFs erase partially or completely the previous epigenetic marks, without cell divisions (in which chromatin marks are lost) but it is totally unknown how this is possible. Remarkably, in both iPSCs and lineage conversion, efficiencies are lower with human cells, if compared with mouse. It is unknown whether this is due to the intrinsic karyotypic instability of mouse cells in culture or to molecular mechanisms.

4. Culture mediated reprogramming

Reprogramming to the pluripotency-state and lineage conversion are achieved through the forced expression of TFs. However, in the last decade, several reports have highlighted how culture medium per se can be responsible for (partial) reprogramming. Moreover, there is an increasing amout of evidences showing that small molecules, including epigenetic modifiers and signaling pathway inhibitors, enhance the efficiency and kinetics of reprogramming.

Epiblast stem cells (EpiSCs) are isolated from post-implantation embryos between E5.5-E7.5. EpiSCs are the post-implantation equivalent of ESCs; they still express Oct4, Nanog and Sox2 but express lower levels of Stella and Rex1 [147]. ESCs and EpiSCs have also different culture requirements and features. While ESC selfrenewal is LIF dependent, EpiSC proliferation requires bFGF and Activin signaling. EpiSC female lines, but not ESC lines, have one of the X chromosome inactive. Importantly, EpiSCs, differently from ESCs, do not have the ability to contribute to chimeras *in vivo*, when aggregated into recipient morula/blastocysts.

In 2009, Bao et al. [148] demonstrated that established EpiSC lines could de-differentiate/ revert into an ESC-like state by culturing EpiSCs in ESC medium (cointaining LIF) for 2-5 weeks. Once 'reverted' cells lost all the features of the original EpiSCs and acquired all ESC-characteristics (X was reactivated, growth was LIF-dependent and cells were capable to contribute to chimeras). This report showed that the simple manipulation of culture medium can dedifferentiate EpiSCs to a more primitive ESC-state but this is not the only case reported in literature.

In 2004, Kanatsu-Shinohara et al. [149] descibed that mouse germline stem cells (GSCs) isolated from neonatal testis reverted occasionally into cells with ESC-like colonies morphology within 4-7 weeks if cultured in LIF, epidermal growth factor (EGF), glial cell line-derived neurotrophic factor (GDNF) and fibroblast growth factor 2 (FGF2). The reverted cells were named multipotent germ stem cells (mGSs); they expressed not only Oct4 (already present in GSCs) but also Nanog and Sox2 at ESC-level. Analysis on mGSs showed the loss of spermatogonial properties (although the erasure of the androgenic imprinting was not complete) and the gain of ESC features (teratoma formation and contribution to chimeras with germline transmission). However, despite their siimilarity to ESCs, mGSs were not able to form offspring, after tetraploid complementation. Unipotent germline stem cells, but this time from adult testis, were converted into germline-derived pluripotent stem cells (gPSs) by Ko and colleagues [150]. Reprogrammed cells, like in the above case, were higly similar to ESCs but again, they could not form live animals in tetraploid complementation assay. The reason for this is most likely the residual persistence of androgenetic imprinting. The possibility to reprogram a germline stem cell into a cell with pluripotent features, even without the capacity of forming chimeric animal, is interesting because it might allow autologous cell therapy without embryo-manipulation. Similar conversions with mouse cells were also described by other laboratories [151, 152].

In 2008, Conrad and colleagues [153], showed that cells derived from human testis can be converted into cells with human ESC-like features. Cells isolated from human testis were cultured in GDNF-containing medium for 4 days and then selected based on the expression of CD49f and further selection on laminin matrix in medium cointaining LIF. 3-4 weeks later colonies with ESC-morphology appeared; human adult GSCs (haGSCs), like human ESCs, expressed SSEA4, TRA 1-60, TRA 1-81 and generated EBs and teratomas. However, a later report [154] questioned the previous finding of Conrad, arguing whether haGSCs really expressed Oct4, Nanog and Sox2; moreover, microarray data comparison further showed that haGSCs are similar to fibroblasts-derived from human testic biopsies but not to hESCs.

These studies strongly suggest that stem/progenitor cells derived from testis can to some extent be converted, by long-term culture, to cells with ESC-like properties, without any reprogramming factors; however, converted cells differ significantly from ESCs. The propensity of GSCs to be converted to ESC-like cells may depend on their Oct4 expression. Although gonads are the only place where Oct4 is functionally expressed in adult healthy-rodents [127], many reports described the isolation of $Oct4^{+}$ cells from rodents [155-169]. It remains to be determined whether culture mediated reprogramming is responsible for the Oct4 re-activation in such cell lines.

In 2002, our group [170] isolated multipotent adult progenitor cells (MAPCs) from rodent bone marrow (BM), upon prolonged culture at low density in a medium cointaining LIF, PDGF and EGF. Murine MAPCs differentiated *in vitro* into cells of the three germ layers and one murine line was also able to contribute to chimeric mice, although at low efficiency and without germline trasmission. Subsequently analysis on rodent MAPC lines [171, 172] showed a lineage marker profile (Oct4, Gata4, Gata6, Sox7 and SSEA1) found also in the nascent hypoblast of the blastocyst and in rat blastocyst-derived Extraembryonic Endoderm Precursor cells (XEN-Ps) [173]. Recently, we demonstrated that similar cells are not present in fresh BM but appear after prolonged *in vitro* culture. To ascertain whether the MAPC culture system reprograms BM cells to the equivalent of XEN-P, we, first, showed that rMAPC and XEN-P cells exhibit similar features under reciprocal culture conditions. Second, we reported, using the same MAPC medium, the quick and efficient isolation of new cell lines directly from blastocyst, which we termed Hypoblast Stem Cells (HypoSCs) and which strongly resemble XEN-P in features and developmental potential [174].

Moreover, specific culture media may also be responsible for the broader differentiation potential described for some adult stem cell types [175] and this should be more considered in stem cell research, especially when reaching clinical trials phases [176].

4.1. Small molecules in stem cells and reprogramming

Small molecules are acquiring, on a daily basis, more relevance in the stem cell field because they can control protein functions selectively, reversibly and in a tunable way. Strikingly many reports have also shown how pathway inhibitors and epigenetic modifiers play a crucial role in the reprogramming process [177]. In 2010, the group of Ding reported that human primary somatic cells can be reprogrammed into human iPSCs with only Oct4 and a cocktail of small molecules [178].

4.1.1. Signaling modulators

Mouse (m)ESCs were first isolated more then three decades ago [179, 180]. mESCs have been derived and cultured in LIF and bone morphogenetic protein (BMP, contained in the serum) to inhibit their differentiation [181]. However, the efficiency of mESCs derivation was low in general and almost not possible from some mouse strains (like C57BL/6). More recently, several reports have now demonstrated that mESC culture in MEF or feeder-free are heterogeneous and fluctuates between a pre-implantation ESC and a post-implantation EpiSC-state [182, 183].

Ying and colleagues [184] demonstrated that mESCs can be maintained in an homogenous ground-state without the requirements of external stimuli, provided by growth factors and/or feeders. This achievement was possible by using two signaling modulators that regulate pathways involved in mESC differentiation: *PD0325901*, which blocks the differentiation-inducing signalling from mitogen-activated protein kinase (MEK), inhibiting the phosphorylation of ERK1/2; and *CHIR99021*, which inhibits the glycogen synthase kinase 3 (GSK3) and decreases the phosphorylation of β-Catenin, supporting their growth and further suppressing residual differentiation. The isolation of mESCs, with the two inhibitors (2i), together with LIF, allows now the efficient derivation of ESCs regardless of the mouse strain as well as from rat for the first time [185]. ESC lines cultured in 2i and LIF can be clonally propagated without feeders and support superior chimerism and germline transmission. The two inhibitors have also been used to increase the efficiency of iPSCs generation [178].

Mesenchymal-to-epithelial transition (MET) is a reversible process which drives cells from a multipolar, spindle and motile mesenchymal shape to a planar and polarized epithelial shape. MET is an important process during embryo development but also in reprogramming; i.e. fibroblasts change shape towards an epithelial morphology at the early stage of iPSC generation. TGFβ pathway negatively regulates an epithelial phenotype. The block of TGFβ1-2-3 receptors, using *SB431542*, in combination with PD0325901, enhances both the kinetics and the efficiency of reprogramming, during iPSC generation [186].

Cellular senescence is a pathway that negatively interferes with reprogramming. Expression of OSKM increases oxidative stress and DNA damage, inducing senescence. *Vitamin C* (or ascorbic acid), is an important cofactor for metabolic processes but also has a strong antioxidant effect; i.e. Vitamin C reduces reactive oxidandant species (ROS). In iPSC reprogramming, Vitamin C enhances the conversion from a partial reprogrammed to a fully reprogrammed-state [187], capable of forming completely iPSC-derived mice in tetraploid complementation assay [188].

Stem cells have a different metabolism if compared to differentiated cells [189]. Stem cells have a strong energetic and metabolic demand to meet their self-renewal and to do this, they mainly rely on glycolysis followed by fermentation of lactic acid in the cytosol. Differently, differentiated cells mainly rely on a low rate of glycolysis followed by oxidation of pyruvate in the mitochondria, which results in the production of ROS. Consistent with this, *PS48*, an activator of 3-phosphoinositide dependent protein kinase-1 (PDK1) that activates the PI3/Akt pathway, results in the upregulation of glycolytic genes and strongly facilitates iPSC reprogramming [178].

4.1.2. Epigenetic modifiers

The structure of eukaryotic genome is higly organized; genomic DNA is wrapped around structural proteines, called histones. DNA and histones, together, form the chromatin. Protein complexes are responsible for chromatin modifications. Histones then determine the transcriptional status; i.e. in an open and closed form. In somatic cells, chromatin is mainly in a closed conformation, while in pluripotent cells, chromatin is in an open conformation and it is dynamically associated with chromatin proteins. Obviously, during iPSC generation, the chromatin must change from a somatic to a pluripotent state. Therefore, many small molecules, which modulates chromatin have been described to enhance the efficiency of reprogramming and even to substitute for some of the reprogramming factors.

Pluripotent stem cells have, in general, a more demethylated DNA, in comparison with somatic cells; in fact, *5-azacytidine* [5-aza) and *RG108*, two inhibitors of DNA methyltransferases (DNMTs), are responsible of DNA methylation and methylation maintanance, increases efficiency of reprogramming [190, 191].

G9a is an histone methyltransferase (HMTase), which induce silencing of Oct4, through methylation of H3K9. *BIX-01294*, an inhibitor of G9a, enhances reprogramming [192]. Recently, *parnate*, an inhibitor of LSD1, a H3K4 demethylase, was used to reprogram human somatic cells with only Oct4 and Klf4 [193].

http://dx.doi.org/10.5772/55104

Similarly, *Trichostatin A* [137], suberoylanilide hydroxamic acid [71] and valproic acid (VPA), inhibitors of histone deacetylases (HDACs) also increases efficiencies of reprogramming, even with only Oct4 and Sox2 overexpression. [190, 194].

Small molecules	Category	Effect on
PD0325901	Signaling modulators	MEK inhibitor
CHIR99021	Signaling modulators	WNT/β-Catenin
SB431542	Signaling modulators	MET
Vitamin C	Signaling modulators	Cellular Senescence
PS48	Signaling modulators	Glycolysis
5-aza, RG108	Epigenetic modifiers	DNMT inhibitor
BIX-01294	Epigenetic modifiers	HMTase inhibitor
Parnate	Epigenetic modifiers	LSD1 inhibitor
TSA, SAHA, VPA	Epigenetic modifiers	HDAC inhibitor

Table 4. Small molecules in reprogramming

5. Conclusions

The importance and the impact on society of reprogramming has been recently recognised by the Nobel Assembly at the Karolinka Institute of Stockholm, which co-awarded John Gurdon and Shiniya Yamanaka with the Nobel Price in Medicine 2012. Their outstanding reports demonstrated that cellular fate is plastic and that differentiation is a reversible process. Epigenetic markers imposed by development can be erased through the multiple pathways to reprogramming. This means the epigenetic landscape as described by Waddington should be revised, as balls are capable of rolling back up and over the hill. The SCNT and the forced expression of TFs show that somatic cells can re-acquire all the features, lost upon their differentiation. Adult somatic cells can be redirected to the pluripotency-state or can be converted into cells of another lineage.

Although the precise mechanism via which the phenotype of all these cells can be changed remains to be fully elucidated, the iPSC technology is drastically changing and boosting the stem cell field; it allows one to obtain pluripotent stem cells for autologous therapy, avoiding the problems of immune rejection as well as the ethical issues related to the use of human embryo for scientific purposes. The possibility to also obtain precursors, with restricted differentiation potential, may be another alternative to reach the bedside, as it is likely associated with lower tumorigenicity. It is also clear that culture conditions have such a significant effect on cell fate, not only during reprogramming but also in establishing the potential of different adult stem cells, that this should be kept in mind when comparing studies across laboratories, and definitely when contemplating clinical trials with cultured stem cells.

Acknowledgements

C.M.V. was supported by an Odysseus grant from FWO, Flanders, by a CoE and GOA KU Leuven award; A.L.N. was supported by an IWT grant, Flanders.

Author details

Antonio Lo Nigro* and Catherine M. Verfaillie

*Address all correspondence to: antonio.lonigro@med.kuleuven.be

*Address all correspondence to: catherine.verfaillie@med.kuleuven.be

Interdepartmental Stem Cell Institute Leuven (SCIL) & Department of Development and Regeneration, KU Leuven, Leuven, Belgium

References

[1] Waddington, C. H. The strategy of the genes; a discussion of some aspects of theoretical biology. London,: Allen & Unwin; (1957).

[2] Gurdon, J. B. The developmental capacity of nuclei taken from differentiating endoderm cells of Xenopus laevis. Journal of embryology and experimental morphology. (1960). Dec;, 8, 505-26.

[3] Gurdon, J. B. Adult frogs derived from the nuclei of single somatic cells. Developmental biology. (1962). Apr;, 4, 256-73.

[4] Campbell, K. H, et al. Sheep cloned by nuclear transfer from a cultured cell line. Nature. (1996). Mar 7;, 380(6569), 64-6.

[5] Wilmut, I, et al. Viable offspring derived from fetal and adult mammalian cells. Nature. (1997). Feb 27;, 385(6619), 810-3.

[6] Wakayama, T, et al. Full-term development of mice from enucleated oocytes injected with cumulus cell nuclei. Nature. (1998). Jul 23;, 394(6691), 369-74.

[7] Egli, D, et al. Developmental reprogramming after chromosome transfer into mitotic mouse zygotes. Nature. (2007). Jun 7;, 447(7145), 679-85.

[8] Yang, X, et al. Nuclear reprogramming of cloned embryos and its implications for therapeutic cloning. Nature genetics. (2007). Mar;, 39(3), 295-302.

[9] Kennedy, D. Editorial retraction. Science. (2006). Jan 20;311(5759):335.

[10] Yamanaka, S, et al. Nuclear reprogramming to a pluripotent state by three approaches. Nature. (2010). Jun 10;, 465(7299), 704-12.

[11] Pavlath, G. K, et al. Expression of muscle genes in heterokaryons depends on gene dosage. The Journal of cell biology. (1986). Jan;, 102(1), 124-30.

[12] Harris, H, et al. Suppression of malignancy by cell fusion. Nature. (1969). Jul 26;, 223(5204), 363-8.

[13] Peterson, J. A, et al. Expression of differentiated functions in hepatoma cell hybrids: induction of mouse albumin production in rat hepatoma-mouse fibroblast hybrids. Proceedings of the National Academy of Sciences of the United States of America. (1972). Mar;, 69(3), 571-5.

[14] Weiss, M. C, et al. Expression of differentiated functions in hepatoma cell hybrids: reappearance of tyrosine aminotransferase inducibility after the loss of chromosomes. Proceedings of the National Academy of Sciences of the United States of America. (1971). Dec;, 68(12), 3026-30.

[15] Davidson, R. L. Regulation of malanin synthesis in mammalian cells: effect of gene dosage on the expression of differentiation. Proceedings of the National Academy of Sciences of the United States of America. (1972). Apr;, 69(4), 951-5.

[16] Davidson, R. L, et al. Regulation of pigment synthesis in mammalian cells, as studied by somatic hybridization. Proceedings of the National Academy of Sciences of the United States of America. (1966). Nov;, 56(5), 1437-40.

[17] Tada, M, et al. Nuclear reprogramming of somatic cells by in vitro hybridization with ES cells. Current biology : CB. (2001). Oct 2;, 11(19), 1553-8.

[18] Ying, Q. L, et al. Changing potency by spontaneous fusion. Nature. (2002). Apr 4;, 416(6880), 545-8.

[19] Cowan, C. A, et al. Nuclear reprogramming of somatic cells after fusion with human embryonic stem cells. Science. (2005). Aug 26;, 309(5739), 1369-73.

[20] Takahashi, K, et al. Induction of pluripotent stem cells from mouse embryonic and adult fibroblast cultures by defined factors. Cell. (2006). Aug 25;, 126(4), 663-76.

[21] Vierbuchen, T, et al. Direct lineage conversions: unnatural but useful? Nature biotechnology. (2011). Oct;, 29(10), 892-907.

[22] Sancho-martinez, I, et al. Lineage conversion methodologies meet the reprogramming toolbox. Nature cell biology. (2012). Sep 3;, 14(9), 892-9.

[23] Asuelime, G. E, et al. A case of cellular alchemy: lineage reprogramming and its potential in regenerative medicine. Journal of molecular cell biology. (2012). Aug;, 4(4), 190-6.

[24] Yuan, X, et al. Small molecules in cellular reprogramming and differentiation. Progress in drug research Fortschritte der Arzneimittelforschung Progres des recherches pharmaceutiques. (2011). , 67, 253-66.

[25] Roobrouck, V. D, et al. Concise review: culture mediated changes in fate and/or potency of stem cells. Stem cells. (2011). Apr;, 29(4), 583-9.

[26] Schneuwly, S, et al. Redesigning the body plan of Drosophila by ectopic expression of the homoeotic gene Antennapedia. Nature. (1987). Feb 26-Mar 4;, 325(6107), 816-8.

[27] Gehring, W. J. The master control gene for morphogenesis and evolution of the eye. Genes to cells : devoted to molecular & cellular mechanisms. (1996). Jan;, 1(1), 11-5.

[28] Okita, K, et al. Generation of germline-competent induced pluripotent stem cells. Nature. (2007). Jul 19;, 448(7151), 313-7.

[29] Kang, L, et al. iPS cells can support full-term development of tetraploid blastocyst-complemented embryos. Cell stem cell. (2009). Aug 7;, 5(2), 135-8.

[30] Takahashi, K, et al. Induction of pluripotent stem cells from adult human fibroblasts by defined factors. Cell. (2007). Nov 30;, 131(5), 861-72.

[31] Yu, J, et al. Induced pluripotent stem cell lines derived from human somatic cells. Science. (2007). Dec 21;, 318(5858), 1917-20.

[32] Scholer, H. R, et al. Octamer binding proteins confer transcriptional activity in early mouse embryogenesis. The EMBO journal. (1989). Sep;, 8(9), 2551-7.

[33] Niwa, H, et al. Quantitative expression of Oct-3/4 defines differentiation, dedifferentiation or self-renewal of ES cells. Nature genetics. (2000). Apr;, 24(4), 372-6.

[34] Nichols, J, et al. Formation of pluripotent stem cells in the mammalian embryo depends on the POU transcription factor Oct4. Cell. (1998). Oct 30;, 95(3), 379-91.

[35] Gubbay, J, et al. A gene mapping to the sex-determining region of the mouse Y chromosome is a member of a novel family of embryonically expressed genes. Nature. (1990). Jul 19;, 346(6281), 245-50.

[36] Bowles, J, et al. Phylogeny of the SOX family of developmental transcription factors based on sequence and structural indicators. Developmental biology. (2000). Nov 15;, 227(2), 239-55.

[37] Avilion, A. A, et al. Multipotent cell lineages in early mouse development depend on SOX2 function. Genes & development. (2003). Jan 1;, 17(1), 126-40.

[38] Tokuzawa, Y, et al. Fbx15 is a novel target of Oct3/4 but is dispensable for embryonic stem cell self-renewal and mouse development. Molecular and cellular biology. (2003). Apr;, 23(8), 2699-708.

[39] Nishimoto, M, et al. The gene for the embryonic stem cell coactivator UTF1 carries a regulatory element which selectively interacts with a complex composed of Oct-3/4 and Sox-2. Molecular and cellular biology. (1999). Aug;, 19(8), 5453-65.

[40] Ambrosetti, D. C, et al. Modulation of the activity of multiple transcriptional activation domains by the DNA binding domains mediates the synergistic action of Sox2 and Oct-3 on the fibroblast growth factor-4 enhancer. The Journal of biological chemistry. (2000). Jul 28;, 275(30), 23387-97.

[41] Ambrosetti, D. C, et al. Synergistic activation of the fibroblast growth factor 4 enhancer by Sox2 and Oct-3 depends on protein-protein interactions facilitated by a specific spatial arrangement of factor binding sites. Molecular and cellular biology. (1997). Nov;, 17(11), 6321-9.

[42] Mitsui, K, et al. The homeoprotein Nanog is required for maintenance of pluripotency in mouse epiblast and ES cells. Cell. (2003). May 30;, 113(5), 631-42.

[43] Chambers, I, et al. Functional expression cloning of Nanog, a pluripotency sustaining factor in embryonic stem cells. Cell. (2003). May 30;, 113(5), 643-55.

[44] Silva, J, et al. Nanog is the gateway to the pluripotent ground state. Cell. (2009). Aug 21;, 138(4), 722-37.

[45] Viswanathan, S. R, et al. Selective blockade of microRNA processing by Lin28. Science. (2008). Apr 4;, 320(5872), 97-100.

[46] Nakagawa, M, et al. Generation of induced pluripotent stem cells without Myc from mouse and human fibroblasts. Nature biotechnology. (2008). Jan;, 26(1), 101-6.

[47] Zhao, Y, et al. Two supporting factors greatly improve the efficiency of human iPSC generation. Cell stem cell. (2008). Nov 6;, 3(5), 475-9.

[48] Li, H, et al. The Ink4/Arf locus is a barrier for iPS cell reprogramming. Nature. (2009). Aug 27;, 460(7259), 1136-9.

[49] Kawamura, T, et al. Linking the tumour suppressor pathway to somatic cell reprogramming. Nature. (2009). Aug 27;460(7259):1140-4., 53.

[50] Judson, R. L, et al. Embryonic stem cell-specific microRNAs promote induced pluripotency. Nature biotechnology. (2009). May;, 27(5), 459-61.

[51] Mallanna, S. K, et al. Emerging roles of microRNAs in the control of embryonic stem cells and the generation of induced pluripotent stem cells. Developmental biology. (2010). Aug 1;, 344(1), 16-25.

[52] Park, I. H, et al. Reprogramming of human somatic cells to pluripotency with defined factors. Nature. (2008). Jan 10;, 451(7175), 141-6.

[53] Tsubooka, N, et al. Roles of Sall4 in the generation of pluripotent stem cells from blastocysts and fibroblasts. Genes to cells : devoted to molecular & cellular mechanisms. (2009). Jun;, 14(6), 683-94.

[54] Feng, B, et al. Reprogramming of fibroblasts into induced pluripotent stem cells with orphan nuclear receptor Esrrb. Nature cell biology. (2009). Feb;, 11(2), 197-203.

[55] Han, J, et al. Tbx3 improves the germ-line competency of induced pluripotent stem cells. Nature. (2010). Feb 25;, 463(7284), 1096-100.

[56] Aasen, T, et al. Efficient and rapid generation of induced pluripotent stem cells from human keratinocytes. Nature biotechnology. (2008). Nov;, 26(11), 1276-84.

[57] Giorgetti, A, et al. Generation of induced pluripotent stem cells from human cord blood using OCT4 and SOX2. Cell stem cell. (2009). Oct 2;, 5(4), 353-7.

[58] Eminli, S, et al. Differentiation stage determines potential of hematopoietic cells for reprogramming into induced pluripotent stem cells. Nature genetics. (2009). Sep;, 41(9), 968-76.

[59] Kim, J. B, et al. Pluripotent stem cells induced from adult neural stem cells by reprogramming with two factors. Nature. (2008). Jul 31;, 454(7204), 646-50.

[60] Sun, N, et al. Feeder-free derivation of induced pluripotent stem cells from adult human adipose stem cells. Proceedings of the National Academy of Sciences of the United States of America. (2009). Sep 15;, 106(37), 15720-5.

[61] Tamaoki, N, et al. Dental pulp cells for induced pluripotent stem cell banking. Journal of dental research. (2010). Aug;, 89(8), 773-8.

[62] Miyoshi, K, et al. Generation of human induced pluripotent stem cells from oral mucosa. Journal of bioscience and bioengineering. (2010). Sep;, 110(3), 345-50.

[63] Loh, Y. H, et al. Generation of induced pluripotent stem cells from human blood. Blood. (2009). May 28;, 113(22), 5476-9.

[64] Doi, A, et al. Differential methylation of tissue- and cancer-specific CpG island shores distinguishes human induced pluripotent stem cells, embryonic stem cells and fibroblasts. Nature genetics. (2009). Dec;, 41(12), 1350-3.

[65] Kim, K, et al. Donor cell type can influence the epigenome and differentiation potential of human induced pluripotent stem cells. Nature biotechnology. (2011). Dec;, 29(12), 1117-9.

[66] Miura, K, et al. Variation in the safety of induced pluripotent stem cell lines. Nature biotechnology. (2009). Aug;, 27(8), 743-5.

[67] Gonzalez, F, et al. Methods for making induced pluripotent stem cells: reprogramming a la carte. Nature reviews Genetics. (2011). Apr;, 12(4), 231-42.

[68] Okita, K, et al. Induced pluripotent stem cells: opportunities and challenges. Philosophical transactions of the Royal Society of London Series B, Biological sciences. (2011). Aug 12;, 366(1575), 2198-207.

[69] Hotta, A, et al. Retroviral vector silencing during iPS cell induction: an epigenetic beacon that signals distinct pluripotent states. Journal of cellular biochemistry. (2008). Nov 1;, 105(4), 940-8.

[70] Wu, Z, et al. Generation of pig induced pluripotent stem cells with a drug-inducible system. Journal of molecular cell biology. (2009). Oct;, 1(1), 46-54.

[71] Carey, B. W, et al. Reprogramming of murine and human somatic cells using a single polycistronic vector. Proceedings of the National Academy of Sciences of the United States of America. (2009). Jan 6;, 106(1), 157-62.

[72] Ryan, M. D, et al. Virus-encoded proteinases of the picornavirus super-group. The Journal of general virology. (1997). Apr;78 (Pt 4):699-723.

[73] Hacein-bey-abina, S, et al. LMO2-associated clonal T cell proliferation in two patients after gene therapy for SCID-X1. Science. (2003). Oct 17;, 302(5644), 415-9.

[74] Lacoste, A, et al. An efficient and reversible transposable system for gene delivery and lineage-specific differentiation in human embryonic stem cells. Cell stem cell. (2009). Sep 4;, 5(3), 332-42.

[75] Woltjen, K, et al. piggyBac transposition reprograms fibroblasts to induced pluripotent stem cells. Nature. (2009). Apr 9;, 458(7239), 766-70.

[76] Yusa, K, et al. Generation of transgene-free induced pluripotent mouse stem cells by the piggyBac transposon. Nature methods. (2009). May;, 6(5), 363-9.

[77] Stadtfeld, M, et al. Induced pluripotent stem cells generated without viral integration. Science. (2008). Nov 7;, 322(5903), 945-9.

[78] Zhou, W, et al. Adenoviral gene delivery can reprogram human fibroblasts to induced pluripotent stem cells. Stem cells. (2009). Nov;, 27(11), 2667-74.

[79] Fusaki, N, et al. Efficient induction of transgene-free human pluripotent stem cells using a vector based on Sendai virus, an RNA virus that does not integrate into the host genome. Proceedings of the Japan Academy Series B, Physical and biological sciences. (2009). , 85(8), 348-62.

[80] Yates, J, et al. A cis-acting element from the Epstein-Barr viral genome that permits stable replication of recombinant plasmids in latently infected cells. Proceedings of the National Academy of Sciences of the United States of America. (1984). Jun;, 81(12), 3806-10.

[81] Yu, J, et al. Human induced pluripotent stem cells free of vector and transgene sequences. Science. (2009). May 8;, 324(5928), 797-801.

[82] Jia, F, et al. A nonviral minicircle vector for deriving human iPS cells. Nature methods. (2010). Mar;, 7(3), 197-9.

[83] Wadia, J. S, et al. Protein transduction technology. Current opinion in biotechnology. (2002). Feb;, 13(1), 52-6.

[84] Zhou, H, et al. Generation of induced pluripotent stem cells using recombinant proteins. Cell stem cell. (2009). May 8;, 4(5), 381-4.

[85] Kim, D, et al. Generation of human induced pluripotent stem cells by direct delivery of reprogramming proteins. Cell stem cell. (2009). Jun 5;, 4(6), 472-6.

[86] Warren, L, et al. Highly efficient reprogramming to pluripotency and directed differentiation of human cells with synthetic modified mRNA. Cell stem cell. (2010). Nov 5;, 7(5), 618-30.

[87] Araki, R, et al. Conversion of ancestral fibroblasts to induced pluripotent stem cells. Stem cells. (2010). Feb;, 28(2), 213-20.

[88] Rideout, W. M, et al. Nuclear cloning and epigenetic reprogramming of the genome. Science. (2001). Aug 10;, 293(5532), 1093-8.

[89] Marion, R. M, et al. Telomeres acquire embryonic stem cell characteristics in induced pluripotent stem cells. Cell stem cell. (2009). Feb 6;, 4(2), 141-54.

[90] Chen, J, et al. EMT and MET as paradigms for cell fate switching. Journal of molecular cell biology. (2012). Apr;, 4(2), 66-9.

[91] Yamanaka, S. Induced pluripotent stem cells: past, present, and future. Cell stem cell. (2012). Jun 14;, 10(6), 678-84.

[92] Chin, M. H, et al. Induced pluripotent stem cells and embryonic stem cells are distinguished by gene expression signatures. Cell stem cell. (2009). Jul 2;, 5(1), 111-23.

[93] Marchetto, M. C, et al. Transcriptional signature and memory retention of human-induced pluripotent stem cells. PloS one. (2009). e7076.

[94] Bock, C, et al. Reference Maps of human ES and iPS cell variation enable high-throughput characterization of pluripotent cell lines. Cell. (2011). Feb 4;, 144(3), 439-52.

[95] Guenther, M. G, et al. Chromatin structure and gene expression programs of human embryonic and induced pluripotent stem cells. Cell stem cell. (2010). Aug 6;, 7(2), 249-57.

[96] Newman, A. M, et al. Lab-specific gene expression signatures in pluripotent stem cells. Cell stem cell. (2010). Aug 6;, 7(2), 258-62.

[97] Osafune, K, et al. Marked differences in differentiation propensity among human embryonic stem cell lines. Nature biotechnology. (2008). Mar;, 26(3), 313-5.

[98] Young, M. A, et al. Background mutations in parental cells account for most of the genetic heterogeneity of induced pluripotent stem cells. Cell stem cell. (2012). May 4;, 10(5), 570-82.

[99] Hanna, J, et al. Treatment of sickle cell anemia mouse model with iPS cells generated from autologous skin. Science. (2007). Dec 21;, 318(5858), 1920-3.

[100] Okamoto, S, et al. Induction of retinal pigment epithelial cells from monkey iPS cells. Investigative ophthalmology & visual science. (2011). , 52(12), 8785-90.

[101] Kriks, S, et al. Dopamine neurons derived from human ES cells efficiently engraft in animal models of Parkinson's disease. Nature. (2011). Dec 22;, 480(7378), 547-51.

[102] Takayama, N, et al. Transient activation of c-MYC expression is critical for efficient platelet generation from human induced pluripotent stem cells. The Journal of experimental medicine. (2010). Dec 20;, 207(13), 2817-30.

[103] Nori, S, et al. Grafted human-induced pluripotent stem-cell-derived neurospheres promote motor functional recovery after spinal cord injury in mice. Proceedings of the National Academy of Sciences of the United States of America. (2011). Oct 4;, 108(40), 16825-30.

[104] Tsuji, O, et al. Therapeutic potential of appropriately evaluated safe-induced pluripotent stem cells for spinal cord injury. Proceedings of the National Academy of Sciences of the United States of America. (2010). Jul 13;, 107(28), 12704-9.

[105] Dimos, J. T, et al. Induced pluripotent stem cells generated from patients with ALS can be differentiated into motor neurons. Science. (2008). Aug 29;, 321(5893), 1218-21.

[106] Park, I. H, et al. Disease-specific induced pluripotent stem cells. Cell. (2008). Sep 5;, 134(5), 877-86.

[107] Brennand, K. J, et al. Modelling schizophrenia using human induced pluripotent stem cells. Nature. (2011). May 12;, 473(7346), 221-5.

[108] Israel, M. A, et al. Probing sporadic and familial Alzheimer's disease using induced pluripotent stem cells. Nature. (2012). Feb 9;, 482(7384), 216-20.

[109] Yahata, N, et al. Anti-Abeta drug screening platform using human iPS cell-derived neurons for the treatment of Alzheimer's disease. PloS one. (2011). e25788.

[110] Xu, J, et al. Pioneer factor interactions and unmethylated CpG dinucleotides mark silent tissue-specific enhancers in embryonic stem cells. Proceedings of the National Academy of Sciences of the United States of America. (2007). Jul 24;, 104(30), 12377-82.

[111] Xu, J, et al. Transcriptional competence and the active marking of tissue-specific enhancers by defined transcription factors in embryonic and induced pluripotent stem cells. Genes & development. (2009). Dec 15;, 23(24), 2824-38.

[112] Zaret, K. S, et al. Pioneer transcription factors: establishing competence for gene expression. Genes & development. (2011). Nov 1;, 25(21), 2227-41.

[113] Schafer, B. W, et al. Effect of cell history on response to helix-loop-helix family of myogenic regulators. Nature. (1990). Mar 29;, 344(6265), 454-8.

[114] Davis, R. L, et al. Expression of a single transfected cDNA converts fibroblasts to myoblasts. Cell. (1987). Dec 24;, 51(6), 987-1000.

[115] Lassar, A. B, et al. Transfection of a DNA locus that mediates the conversion of 10T1/2 fibroblasts to myoblasts. Cell. (1986). Dec 5;, 47(5), 649-56.

[116] Choi, J, et al. MyoD converts primary dermal fibroblasts, chondroblasts, smooth muscle, and retinal pigmented epithelial cells into striated mononucleated myoblasts and multinucleated myotubes. Proceedings of the National Academy of Sciences of the United States of America. (1990). Oct;, 87(20), 7988-92.

[117] Weintraub, H, et al. Activation of muscle-specific genes in pigment, nerve, fat, liver, and fibroblast cell lines by forced expression of MyoD. Proceedings of the National Academy of Sciences of the United States of America. (1989). Jul;, 86(14), 5434-8.

[118] Rolink, A. G, et al. Long-term in vivo reconstitution of T-cell development by Pax5-deficient B-cell progenitors. Nature. (1999). Oct 7;, 401(6753), 603-6.

[119] Mikkola, I, et al. Reversion of B cell commitment upon loss of Pax5 expression. Science. (2002). Jul 5;, 297(5578), 110-3.

[120] Cobaleda, C, et al. Conversion of mature B cells into T cells by dedifferentiation to uncommitted progenitors. Nature. (2007). Sep 27;, 449(7161), 473-7.

[121] Xie, H, et al. Stepwise reprogramming of B cells into macrophages. Cell. (2004). May 28;, 117(5), 663-76.

[122] Jarriault, S, et al. A Caenorhabditis elegans model for epithelial-neuronal transdifferentiation. Proceedings of the National Academy of Sciences of the United States of America. (2008). Mar 11;, 105(10), 3790-5.

[123] Richard, J. P, et al. Direct in vivo cellular reprogramming involves transition through discrete, non-pluripotent steps. Development. (2011). Apr;, 138(8), 1483-92.

[124] Seale, P, et al. PRDM16 controls a brown fat/skeletal muscle switch. Nature. (2008). Aug 21;, 454(7207), 961-7.

[125] Kajimura, S, et al. Initiation of myoblast to brown fat switch by a PRDM16-C/EBP-beta transcriptional complex. Nature. (2009). Aug 27;, 460(7259), 1154-8.

[126] Szabo, E, et al. Direct conversion of human fibroblasts to multilineage blood progenitors. Nature. (2010). Nov 25;, 468(7323), 521-6.

[127] Lengner, C. J, et al. Oct4 expression is not required for mouse somatic stem cell self-renewal. Cell stem cell. (2007). Oct 11;, 1(4), 403-15.

[128] Emslie, D, et al. Oct2 enhances antibody-secreting cell differentiation through regulation of IL-5 receptor alpha chain expression on activated B cells. The Journal of experimental medicine. (2008). Feb 18;, 205(2), 409-21.

[129] Ieda, M, et al. Direct reprogramming of fibroblasts into functional cardiomyocytes by defined factors. Cell. (2010). Aug 6;, 142(3), 375-86.

[130] Qian, L, et al. In vivo reprogramming of murine cardiac fibroblasts into induced cardiomyocytes. Nature. (2012). May 31;, 485(7400), 593-8.

[131] Song, K, et al. Heart repair by reprogramming non-myocytes with cardiac transcription factors. Nature. (2012). May 31;, 485(7400), 599-604.

[132] Zhou, Q, et al. In vivo reprogramming of adult pancreatic exocrine cells to beta-cells. Nature. (2008). Oct 2;, 455(7213), 627-32.

[133] Sekiya, S, et al. Direct conversion of mouse fibroblasts to hepatocyte-like cells by defined factors. Nature. (2011). Jul 21;, 475(7356), 390-3.

[134] Huang, P, et al. Induction of functional hepatocyte-like cells from mouse fibroblasts by defined factors. Nature. (2011). Jul 21;, 475(7356), 386-9.

[135] Vierbuchen, T, et al. Direct conversion of fibroblasts to functional neurons by defined factors. Nature. (2010). Feb 25;, 463(7284), 1035-41.

[136] Pang, Z. P, et al. Induction of human neuronal cells by defined transcription factors. Nature. (2011). Aug 11;, 476(7359), 220-3.

[137] Marro, S, et al. Direct lineage conversion of terminally differentiated hepatocytes to functional neurons. Cell stem cell. (2011). Oct 4;, 9(4), 374-82.

[138] Han, D. W, et al. Direct reprogramming of fibroblasts into neural stem cells by defined factors. Cell stem cell. (2012). Apr 6;, 10(4), 465-72.

[139] Kumar, A, et al. Zic3 induces conversion of human fibroblasts to stable neural progenitor-like cells. Journal of molecular cell biology. (2012). Aug;, 4(4), 252-5.

[140] Ring, K. L, et al. Direct reprogramming of mouse and human fibroblasts into multipotent neural stem cells with a single factor. Cell stem cell. (2012). Jul 6;, 11(1), 100-9.

[141] Pfisterer, U, et al. Direct conversion of human fibroblasts to dopaminergic neurons. Proceedings of the National Academy of Sciences of the United States of America. (2011). Jun 21;, 108(25), 10343-8.

[142] Caiazzo, M, et al. Direct generation of functional dopaminergic neurons from mouse and human fibroblasts. Nature. (2011). Aug 11;, 476(7359), 224-7.

[143] Son, E. Y, et al. Conversion of mouse and human fibroblasts into functional spinal motor neurons. Cell stem cell. (2011). Sep 2;, 9(3), 205-18.

[144] Qiang, L, et al. Directed conversion of Alzheimer's disease patient skin fibroblasts into functional neurons. Cell. (2011). Aug 5;, 146(3), 359-71.

[145] Efe, J. A, et al. Conversion of mouse fibroblasts into cardiomyocytes using a direct reprogramming strategy. Nature cell biology. (2011). Mar;, 13(3), 215-22.

[146] Kim, J, et al. Direct reprogramming of mouse fibroblasts to neural progenitors. Proceedings of the National Academy of Sciences of the United States of America. (2011). May 10;, 108(19), 7838-43.

[147] Brons, I. G, et al. Derivation of pluripotent epiblast stem cells from mammalian embryos. Nature. (2007). Jul 12;, 448(7150), 191-5.

[148] Bao, S, et al. Epigenetic reversion of post-implantation epiblast to pluripotent embryonic stem cells. Nature. (2009). Oct 29;, 461(7268), 1292-5.

[149] Kanatsu-shinohara, M, et al. Generation of pluripotent stem cells from neonatal mouse testis. Cell. (2004). Dec 29;, 119(7), 1001-12.

[150] Ko, K, et al. Induction of pluripotency in adult unipotent germline stem cells. Cell stem cell. (2009). Jul 2;, 5(1), 87-96.

[151] Guan, K, et al. Pluripotency of spermatogonial stem cells from adult mouse testis. Nature. (2006). Apr 27;, 440(7088), 1199-203.

[152] Seandel, M, et al. Generation of functional multipotent adult stem cells from GPR125+ germline progenitors. Nature. (2007). Sep 20;, 449(7160), 346-50.

[153] Conrad, S, et al. Generation of pluripotent stem cells from adult human testis. Nature. (2008). Nov 20;, 456(7220), 344-9.

[154] Ko, K, et al. Human adult germline stem cells in question. Nature. (2010). Jun 24;465(7301):E1; discussion E3.

[155] Anjos-afonso, F, et al. Nonhematopoietic/endothelial SSEA-1+ cells define the most primitive progenitors in the adult murine bone marrow mesenchymal compartment. Blood. (2007). Feb 1;, 109(3), 1298-306.

[156] Danner, S, et al. Derivation of oocyte-like cells from a clonal pancreatic stem cell line. Molecular human reproduction. (2007). Jan;, 13(1), 11-20.

[157] Goolsby, J, et al. Hematopoietic progenitors express neural genes. Proceedings of the National Academy of Sciences of the United States of America. (2003). Dec 9;, 100(25), 14926-31.

[158] Gupta, S, et al. Isolation and characterization of kidney-derived stem cells. Journal of the American Society of Nephrology : JASN. (2006). Nov;, 17(11), 3028-40.

[159] Johnson, J, et al. Oocyte generation in adult mammalian ovaries by putative germ cells in bone marrow and peripheral blood. Cell. (2005). Jul 29;, 122(2), 303-15.

[160] Kucia, M, et al. A population of very small embryonic-like (VSEL) CXCR4(+)SSEA-1(+)Oct-4+ stem cells identified in adult bone marrow. Leukemia : official journal of the Leukemia Society of America, Leukemia Research Fund, UK. (2006). May;, 20(5), 857-69.

[161] Lamoury, F. M, et al. Undifferentiated mouse mesenchymal stem cells spontaneously express neural and stem cell markers Oct-4 and Rex-1. Cytotherapy. (2006). , 8(3), 228-42.

[162] Ling, T. Y, et al. Identification of pulmonary Oct-4+ stem/progenitor cells and demonstration of their susceptibility to SARS coronavirus (SARS-CoV) infection in vitro. Proceedings of the National Academy of Sciences of the United States of America. (2006). Jun 20;, 103(25), 9530-5.

[163] Nayernia, K, et al. Derivation of male germ cells from bone marrow stem cells. Laboratory investigation; a journal of technical methods and pathology. (2006). Jul;, 86(7), 654-63.

[164] Pallante, B. A, et al. Bone marrow Oct3/4+ cells differentiate into cardiac myocytes via age-dependent paracrine mechanisms. Circulation research. (2007). Jan 5;100(1):e, 1-11.

[165] Redvers, R. P, et al. Side population in adult murine epidermis exhibits phenotypic and functional characteristics of keratinocyte stem cells. Proceedings of the National Academy of Sciences of the United States of America. (2006). Aug 29;, 103(35), 13168-73.

[166] Ren, H, et al. Proliferation and differentiation of bone marrow stromal cells under hypoxic conditions. Biochemical and biophysical research communications. (2006). Aug 18;, 347(1), 12-21.

[167] Romero-ramos, M, et al. Neuronal differentiation of stem cells isolated from adult muscle. Journal of neuroscience research. (2002). Sep 15;, 69(6), 894-907.

[168] Sagrinati, C, et al. Isolation and characterization of multipotent progenitor cells from the Bowman's capsule of adult human kidneys. Journal of the American Society of Nephrology : JASN. (2006). Sep;, 17(9), 2443-56.

[169] Wang, R, et al. Phenotypic analysis of c-Kit expression in epithelial monolayers derived from postnatal rat pancreatic islets. The Journal of endocrinology. (2004). Jul;, 182(1), 113-22.

[170] Jiang, Y, et al. Multipotent progenitor cells can be isolated from postnatal murine bone marrow, muscle, and brain. Experimental hematology. (2002). Aug;, 30(8), 896-904.

[171] Subramanian, K, et al. Isolation procedure and characterization of multipotent adult progenitor cells from rat bone marrow. Methods in molecular biology. (2010). , 636, 55-78.

[172] Ulloa-montoya, F, et al. Comparative transcriptome analysis of embryonic and adult stem cells with extended and limited differentiation capacity. Genome biology. (2007). R163.

[173] Debeb, B. G, et al. Isolation of Octexpressing extraembryonic endoderm precursor cell lines. PloS one. (2009). e7216., 4.

[174] Lo Nigro Aet al. MAPC culture conditions support the derivation of cells with nascent hypoblast features from bone marrow and blastocysts. Journal of molecular cell biology. (2012). Aug 9.

[175] Roobrouck, V. D, et al. Differentiation potential of human postnatal mesenchymal stem cells, mesoangioblasts, and multipotent adult progenitor cells reflected in their transcriptome and partially influenced by the culture conditions. Stem cells. (2011). May;, 29(5), 871-82.

[176] Fuchs, E. The impact of cell culture on stem cell research. Cell stem cell. (2012). Jun 14;, 10(6), 640-1.

[177] Nie, B, et al. Cellular reprogramming: a small molecule perspective. Current opinion in cell biology. (2012). Sep 6.

[178] Zhu, S, et al. Reprogramming of human primary somatic cells by OCT4 and chemical compounds. Cell stem cell. (2010). Dec 3;, 7(6), 651-5.

[179] Martin, G. R. Isolation of a pluripotent cell line from early mouse embryos cultured in medium conditioned by teratocarcinoma stem cells. Proceedings of the National Academy of Sciences of the United States of America. (1981). Dec;, 78(12), 7634-8.

[180] Evans, M. J, et al. Establishment in culture of pluripotential cells from mouse embryos. Nature. (1981). Jul 9;, 292(5819), 154-6.

[181] Ying, Q. L, et al. BMP induction of Id proteins suppresses differentiation and sustains embryonic stem cell self-renewal in collaboration with STAT3. Cell. (2003). Oct 31;, 115(3), 281-92.

[182] Hayashi, K, et al. Dynamic equilibrium and heterogeneity of mouse pluripotent stem cells with distinct functional and epigenetic states. Cell stem cell. (2008). Oct 9;, 3(4), 391-401.

[183] Toyooka, Y, et al. Identification and characterization of subpopulations in undifferentiated ES cell culture. Development. (2008). Mar;, 135(5), 909-18.

[184] Ying, Q. L, et al. The ground state of embryonic stem cell self-renewal. Nature. (2008). May 22;, 453(7194), 519-23.

[185] Buehr, M, et al. Capture of authentic embryonic stem cells from rat blastocysts. Cell. (2008). Dec 26;, 135(7), 1287-98.

[186] Lin, T, et al. A chemical platform for improved induction of human iPSCs. Nature methods. (2009). Nov;, 6(11), 805-8.

[187] Esteban, M. A, et al. Vitamin C enhances the generation of mouse and human induced pluripotent stem cells. Cell stem cell. (2010). Jan 8;, 6(1), 71-9.

[188] Stadtfeld, M, et al. Ascorbic acid prevents loss of Dlk1-Dio3 imprinting and facilitates generation of all-iPS cell mice from terminally differentiated B cells. Nature genetics. (2012). Apr;S1-2., 44(4), 398-405.

[189] Zhang, J, et al. UCP2 regulates energy metabolism and differentiation potential of human pluripotent stem cells. The EMBO journal. (2011). Dec 14;, 30(24), 4860-73.

[190] Huangfu, D, et al. Induction of pluripotent stem cells by defined factors is greatly improved by small-molecule compounds. Nature biotechnology. (2008). Jul;, 26(7), 795-7.

[191] Shi, Y, et al. Induction of pluripotent stem cells from mouse embryonic fibroblasts by Oct4 and Klf4 with small-molecule compounds. Cell stem cell. (2008). Nov 6;, 3(5), 568-74.

[192] Shi, Y, et al. A combined chemical and genetic approach for the generation of induced pluripotent stem cells. Cell stem cell. (2008). Jun 5;, 2(6), 525-8.

[193] Li, W, et al. Generation of human-induced pluripotent stem cells in the absence of exogenous Sox2. Stem cells. (2009). Dec;, 27(12), 2992-3000.

[194] Huangfu, D, et al. Induction of pluripotent stem cells from primary human fibroblasts with only Oct4 and Sox2. Nature biotechnology. (2008). Nov;, 26(11), 1269-75.

Chapter 12

Induced Pluripotent Stem Cells: Current and Emerging Technologies

Jacob Kimmel and Kiminobu Sugaya

Additional information is available at the end of the chapter

http://dx.doi.org/10.5772/55101

1. Introduction

Induced pluripotent stem (iPS) cells are somatic cells which have been imbued with pluripotent differentiation potential through some form of artificial treatment. On a general level, these treatments involve modifications in the expression of keystone genes associated with pluripotency in embryonic stem cells (ESCs) or their downstream expression products. Despite the conceptual simplicity of iPS cell technology, the 2006 development of the first iPS cell line by Yamanaka and Takahashi [1] has led to an exponential increase in the volume of pluripotency research and a new perspective from which to approach regenerative medicine.

iPS cells are a potential alternative to ESCs in therapeutic contexts, retaining the regenerative potential of ESCs inherent in pluripotent phenotypes, while bypassing some of the risks associated with ESC transplants. A number of studies have demonstrated that iPS cells and ESCs have effectively indistinguishable pluripotent capability, implying that iPS cells maintain the same therapeutic potential long associated with natural ESCs. However, unlike ESCs, iPS cells do not carry a risk of immunorejection due to their patient specific nature, and are not affected by the same ethical concerns as ESCs. As such, iPS cells may actually be preferable to ESCs in some therapeutic contexts due to reduced risk factors for the patient.

Since Yamanaka's hallmark 2006 paper and methodology, numerous iPS cell generation technologies have been developed. Most methods rely upon epigenetic expression of genes determined to be pluripotency regulators. Expression is most commonly induced through viral integration into the host genome, though other episomal methods do exist. Non-genetic induced pluripotency methods generally utilize the downstream expression products of the same keystone genes to generate the same effect as epigenetic expression, without requiring the host to transcribe and generate the products independently.

iPS cells promise a new paradigm in regenerative medicine. Developing iPS technologies have the potential to generate patient specific stem cells, for use in generating any target phenotype within the human body for transplant. In the research context as well, iPS cells have the potential to greatly advance existing disease models. Patient specific iPS cells could be used to create individualized disease models, potentially allowing for more specialized treatment of patients. Here, we discuss a number of the technologies in development seeking to fulfill these promises, as well as their potential applications in both therapeutic and research settings.

2. Canonical methodology

The seminal event in the development of iPS technology, Yamanaka and Takahashi's 2006 publication demonstrated for the first time that the pluripotent phenotype could be induced in somatic cells and was not exclusive to ESCs. In their initial approach, Yamanaka et al. screened 24 genes as potential candidates to induce pluripotency in mouse embryonic fibroblasts (MEFs). The candidate genes were chosen for their perceived roles in regulating pluripotencty in ESC cultures. From the initial candidates, four genes were eventually identified to be necessary for induction of pluripotency, each shown to play a role in ESC pluripotency regulation: Sox2 [2], Oct4 [3], Klf4, and Myc-c [4], often abbreviated as SOKM. MEF cultures were transduced using four pMXs-based retroviral vectors, each containing one of the target transcription factors. MEFs transduced with these factors formed colonies exhibiting ESC morphology and the pluripotent phenotype, as demonstrated by their differentiation ability and teratoma formation *in vivo.*

Groundbreaking as it was, this initial iPS technology had multiple issues preventing immediate use in downstream applications. Despite its effectiveness, the early SOKM method had a decidedly low efficacy [1, 5], inhibiting generation of large scale iPS cultures for use in potential clinical applications or in the laboratory. The viral integration method also presented challenges, as genome integration could lead to random gene reactivation within the iPS culture, potentially causing deleterious effects. Myc-c itself acts as a protooncogene, which led to tumorogenesis in 50% of mice chimeric mice derived using the SOKM method [6]. Compounded with the risk for random gene reactivation, the use of Myc-c could lead to tumorogenesis in potential patients.

3. Improved epigenetic methods

In the interim since the development of the initial induction methodology, many improvements and variations on the technology have been made. Most of these improved methods utilize a similar epigenetic pathway to that of the original study, relying upon the host culture to express downstream products which induce the pluripotent phenotype. We discuss in brief some of these improved epigenetic methods, their potential niche applications, and their delivery vectors.

http://dx.doi.org/10.5772/55101

3.1. Sox2, Oct4, Lin28, Nanog

Yu et. al. demonstrated in 2007 that iPS cells could be generated from pre-natal and post-natal fibroblasts without transduction of the protooncogene Myc-c, using a combination of Sox2, Oct4, Lin28, and Nanog (SOLN). Factors were selected based on their high expression in ESCs, in comparison to myeloid progenitors. Removal of Myc-c from the gene cocktail eliminated the risk of transcription factor induced tumorogenesis, overcoming one of the fundamental issues with Yamanaka and Takahashi's initial methodology. Additionally, Yu et. al. recognized the potential usefulness of Nanog in iPS technologies, noting that it could lead to an increased recovery rate for iPS cell clones generated using the SOLN method. This is potentially due to Nanog's action upstream of Oct4 and Sox2. Lin28 did not integrate in one iPS clone from each of the two cell lines tested, suggesting that while Lin28 may improve efficacy, it is not necessary for reprogramming [7].

3.2. iPS-S: Sox2, Oct4, Lin28, Nanog, Klf4, Myc-c

Combining the SOLN and SOKM transcription factor cocktails, Liao et al demonstrated in 2008 that the efficacy of transfection could be improved by using all 6 previously demonstrated transcription factors in a single transduction and deemed their method iPS-S. Combination of the 6 factors was attempted based on empirical speculation, and proved successful. Transduced colonies also developed more rapidly, within 17 days post-transduction, as opposed to 26 days using the standard SOLN factors. The iPS-S method also increased efficacy roughly 10 fold, which combined with the more rapid development of iPS colonies, partially addressed the inefficiency issues with the canonical iPS technology [8]. As with other transcription factor combinations utilizing Myc-c, the iPS-S method carries with it a risk of tumorogenesis due to random transgene reactivation, inhibiting the use of iPS-S in some applications.

3.3. Combined epigentic, small-compound, and endogenous expression approaches

One of the first approaches to reduce the number of factors required, it was demonstrated that treatment with the epigenetic small compound BIX-01294 could substitute transduction of Sox2 or Oct4, using the traditional SOKM combination, in neural progenitor cells (NPCs) [5]. The study was notable for multiple reasons, both the use of chemical conditions to remove transcription factors, and the reliance upon endogenous gene expression in the target somatic cell line. While reliance upon Sox2 expression in NPCs ultimately limits the applicability of the BIX-OKM combination, it set a precedent for use of endogenous gene expressions to reduce the number of necessary transcription factors in certain cell lines, potentially allowing for safer, more efficient iPS generation in specific contexts. This concept was further explored by Kim et. al., who demonstrated that Oct4 alone was capable of inducing pluripotency in neural stem cells (NSCs) due to their endogenous expression of Sox2, Myc-c, and Klf4 [9].

Shi et al. improved upon their original small compound approach, eliminating the need for Myc-c transfection and endogenous Sox2 expression. BIX-01294 and non-genetic calcium channel agonist BayK8644 were identified via a phenotypic compound screen of known drugs, and combined with the transduction of Oct4 and Klf4 (OK), were able to induce pluripotency in MEFs. The elimination of multiple transcription factors suggests that it may be possible to further replicate the effects of epigenetic transduction using chemical conditions, reducing the risk for random gene reactivation and potentially allowing for more controlled iPS generation temporality [10].

Year	Group	Vector	Transcription Factors
2006	Yamanaka et. al.	Retroviral	Sox2, Oct4, Klf4, Myc-c
2007	Yu et. al.	Lentiviral	Sox2, Oct4, LIN28, Nanog
2008	Shi et. al.	Retroviral, small-compound *	Oct4, Klf4, Myc-c, small-compound BIX-01294
2008	Shi et. al.	Retroviral, small-compound	Oct4, Klf4, small-compounds BIX-01294, BayK8644
2008	Liao et. al.	Lentiviral	Sox2, Oct4, Klf4, Myc-c, Lin28, Nanog
2008	Okita et. al.	Plasmid	Sox2, Oct4, Klf4, Myc-c
2009	Kim et. al.	Retroviral *	Oct4
2009	Fusaki et. al.	Sendai virus	Sox2, Oct4, Klf4, Myc-c
2010	Sugaya et. al.	Retroviral, plasmid	Nanog

Table 1. Epigenetic methods covered in this section: year of publication, vectors used, and required transcription factors. * Reliant upon endogenous expression of certain somatic cell phenotypes.

3.4. Nanog

While the majority of epigenetic approaches rely upon multiple transcription factors, chemical conditions, or endogenous expression, we patented technology capable of generating iPS cells through transfection of Nanog alone in 2006 [11]. Nanog is capable of inducing pluripotency without the aid of other factors due to its role upstream of Oct4 and Sox2. We demonstrated this interaction in bone marrow derived mesenchymal stem cells (MSCs), in which Nanog transfection successfully increased Sox2 and Oct4 levels [12]. Nanog has been demonstrated to induce pluripotency when delivered through lentiviral or plasmid vectors, providing both integrated and episomal gene expressions pathways.

Eliminating the need for multiple transcription factors has various benefits. As with other modified gene cocktails, the elimination of Myc-c greatly reduces the risk of tumorogenesis. Transfection of a single genetic factor may have higher efficacy than that of multiple factors and could lead to a lower overall cost per iPS cell generated. The improved efficiency and reduced cost of this method could allow for more rapid production of iPS cells for use in therapeutic treatments at a lower eventual cost to the patient.

3.5. Retroviral vectors

Beginning with the seminal paper by Yamanaka and Takahashi, the majority of improved epigenetic methods have utilized retroviral vectors to deliver their target transcription factors. In the context of induced pluripotency, retroviral vectors provide a number of distinct advantages, leading to their widespread use. Due to integration with the host genome, retroviral vectors are capable of generating stable iPS clones that maintain their phenotype

http://dx.doi.org/10.5772/55101

over time, unlike some episomal vectors. Retroviral technologies are very mature, allowing for rapid development of vectors and efficient production of vectors in the laboratory. While standard retroviral vectors are only capable of infecting dividing cells, the lentiviral subclass of retroviruses are indeed capable of infecting non-dividing cells, an important consideration when infecting cell types that divide rarely, such as neurons. A combination of these attributes makes retroviral vectors a highly functional candidate for iPS cell induction.

However, retroviral vectors and the lentiviral sub-class also have certain inherent risk factors. Most prominently, viral integration into the host genome can cause random gene reactivation, as discussed in section 2. Using the original SOKM transcription factors, this risk is exaggerated due to the protooncogenetic nature of Myc-c. Although various epigenetic methods have eliminated the necessity of Myc-c, random gene reactivation may still lead to tumorogenesis and deleterious effects in potential transplant patients. As a case study, an FDA clinical trial involving the retroviral transduction of non-protooncogenes led to the development of lymphoma in two patients [13]. Residual expression of transgenes may also lead to phenotypic expression differences between iPS cells and ESCs, leading to a less accurate model of human ESCs for research use or some clinical applications [7, 14]. To reduce the risk of random transgene reactivation and minimize remnant transgene expression, transgenes can be excised using a Cre/Lox system, as demonstrated by multiple groups [12, 14].

3.6. Plasmid vectors

Episomal factors, by definition, allow for the introduction of genetic factors without integration into the host genome. A lack of host genome integration inherently removes the risk of random transgene reactivation associated with viral vectors, but presents functional challenges in some contexts. The most common type of episomal vector in the context of iPS technology is the plasmid, a DNA library separate from the host's nucleic genome, first confirmed as a viable reprogramming vector using the original SOKM factors [15]. The plasmid method has several advantages, both in the laboratory and in downstream applications. Plasmids are a well-developed technology, are very easy to generate in great quantity in the laboratory, and have a relatively low cost-of-use compared to comparable viral vectors. For these reasons, plasmids are the favored vector in Yamanaka's laboratory [16].

The most prominent advantage of plasmids is the lack of integration inherent in episomal vectors. Although there is a potential for spontaneous integration of transgenes during the reprogramming process, iPS clones generated from plasmid vectors can be screened to select only integration-free clones [17]. As such, plasmid vectors are unaffected by issues related to transgene integration, such as residual transgene expression and random transgene reactivation. Although these advantages make plasmids a desirable vector for reprogramming, their efficacy remains well below that of viral integration, limiting the potential for large scale iPS cell generation using plasmids [18]. This reduced efficacy could potentially be due to the temporary nature of plasmids, and the speculated ongoing nature of the reprogramming process [18]; transcription factor expression may be reduced before the iPS reprogramming process is complete, altering the stoichiometric balance of factors and ending reprogramming

Vector	Advantages	Disadvantages
Retroviral	Genome integration allows single transduction iPS clone generation, well-developed technology, relatively efficient transduction rates	Genome integration may lead to random gene reactivation, cannot infect non-dividing cells, residual transgene expression concerns, can potentially induce immunogenicity
Lentiviral	Genome integration allows single transduction iPS clone generation, well-developed technology, relatively efficient transduction rates	Genome integration may lead to random gene reactivation, residual transgene expression concerns
Plasmid	Produces integration-free iPS clones, relatively low cost, volume production is easily scalable	Lower efficacy than viral integration methods, clones must be screened to check for integration
Sendai Virus	Produces integration-free iPS clones, relatively high efficacy	Pluripotent gene expression degrades over the course of 18-20 passages, clones must be screened for viral genome remnants

Table 2. Comparison of different epigenetic vector technologies

in cells that may have otherwise formed colonies. Depending upon the chosen transcription factor combination and somatic cell phenotype, plasmid transduction may also require multiple transfections to effectively reprogram cells, increasing the difficulty and labor-intensiveness of the technique.

3.7. Sendai virus

Sendai virus is widely known to replicate in the cytoplasm of host cells without integrating into the host genome. As such, it has been widely studied as an efficient expression vector and is known to effectively express transgene without integration [19-21]. Fusaki et. al. have demonstrated that a sendai virus vector carrying the four SOKM factors is sufficient to successfully generate iPS colonies [22]. Sendai virus reprogramming was shown to be as or more effective than traditional retroviral reprogramming, with a ~1% efficacy. Even though sendai viruses do not integrate into the genome, the persistence of a viral genome within iPS clones remains a concern for downstream applications. However, Fusaki et. al. were able to isolate clones that had no remnant presence of viral genomes. As such, the sendai viral vector is very attractive for use in downstream clinical applications. In the laboratory, sendai viral vectors leave something to be desired. Pluripotent gene expression of sendai induced iPS cells has been shown to degrade over the course of 18-20 passages, making long term iPS clone maintenance difficult.

4. Non-genetic reprograming methods

Complimenting research into genetic induction of pluripotency, a number of avenues into non-genetic iPS generation have been studied. Although non-integration epigenetic methods have been developed, many are inefficient, and cannot completely eliminate the possibility of genome alteration. Non-genetic induction removes the risk of genetic factor reactivation and consequent genetic modification inherent with these epigenetic methods. To circumvent genetic transfection, technologies have been developed which utilize downstream RNA and protein phases of the desired genetic factors to induce expression.

4.1. mRNA transduction

Warren et. al. have demonstrated that modified mRNAs transcribing for the four SOKM factor proteins are capable of reprogramming when passed into the cytosol of various human cell types with a catatonic delivery vehicle [23]. Initially, cytotoxicity of transfected mRNAs inhibited effective reprogramming, requiring modifications to the mRNA. In a novel approach, Warren et. al. modified the ribonucleotide bases of vector mRNAs by substituting 5-methyl-cytidine for cytidine and pseudouridine for uridine, reducing the immunogenicity of the mRNAs [24]. Combined with interferon inhibitor media supplements, the modifications allowed for generation of viable iPS clones.

mRNA induced pluripotent stem (RiPS) cell generation is highly efficient relative to other technologies, with an efficacy of 1.34% in Warren's initial study. However, the modified mRNAs are difficult to generate in the laboratory and the techniques are labor-intensive. Repeated mRNA administrations are also required, increasing the labor-intensive nature of the technique and complicating volume production of RiPS clones.

4.2. Protein transduction

Multiple groups have also demonstrated reprogramming utilizing the protein products of the SOKM factors [25, 26]. In order for the target proteins to pass through a lipid bilayer, both groups attached each target protein to a cell penetrating peptide (CPP). At this stage, cells are treated with CPP-conjugated proteins multiple times to ensure a continuous supply of reprogramming factors. The protein induced pluripotent stem (piPS) cell induction method is significantly less efficient than epigenetic methods, with an efficacy of ~0.001%. In addition to the transduction inefficiencies, the temporality of the process is relatively slow and the multiple treatment protocols are very labor intensive, making volume production of piPS clones difficult. Although inefficient, the piPS method does eliminate the risk of transgene reactivation and genome integration, just as the RiPS method.

5. Optimization of induction methods

5.1. Factors affecting efficacy

While each induction method has an inherent relative efficacy, it must be noted that a number of external factors affect reprogramming efficacy as well. Multiple groups have reported that O_2 concentrations play a role in reprogramming efficacy [27, 28], with hypoxia noted to increase efficiency. The presence of methylation inhibitors, such as 5'-azacytidine, in culture medium have also been noted increase efficiency [29, 30]. Hanna et. al. have also demonstrated that cell division rate plays a role in the kinetics of iPS induction [31]. Findings such as these suggest that the specific culture environment play a major role in pluripotency induction and may effect downstream development of iPS clones.

Of prominent concern, it has been reasoned that the stoichiometric abundances of reprogramming factors in relation to one another plays a role in reprogramming efficacy [18]. This rationale is based upon the differential effects of some pluripotency factors when expressed in different levels; for instance, expression of Oct4 and Sox2 at median levels can maintain pluripotency of ESCs, but overexpression of Oct4 can induce differentiation [32]. A similar action has been demonstrated in the context of iPS induction, in which a threefold increase of Oct4 increased efficacy, but further increases reduced the efficiency of reprogramming [33]. As such, the ability to monitor and manipulate the stoichiometric expression levels of transcription factors may play a role in selection of vectors and induction technologies in the future.

iPS cells have recently been shown to possess preferential differentiation based on their somatic cell origin, referred to as epigenetic memory [34-36]. It is believed that variations in DNA methylation status allow differentiation preferences to persist beyond the boundaries of reprogramming. It may be possible to exploit this epigenetic memory to increase the terminal differentiation efficiency of iPS cells based on the desired differentiated phenotype. By selecting cells of origin in the same lineage, or tissues known to have limited transdifferentiation ability into the target cell type, it may be possible to augment the efficacy of current induction protocols.

5.2. Application specific induced pluripotent stem cells

Until now, the majority of iPS cell research has sought to increase the efficacy at which stable pluripotent iPS clones could be developed. However, for the optimal production of a desired differentiated phenotype, solely optimizing the efficacy at which iPS clones can be developed may not be the best strategy. First touched on by Yamanaka in 2009 as the concept of "functional pluripotency" [6], it may be more effective to optimize for the generation of a target differentiated phenotype in the context of downstream applications. As such, a number of the efficacy factors mentioned above could be considered and optimized for each target phenotype and each downstream application.

Until such a time as reprogramming efficacy improves dramatically, the optimization of reprogramming in the context of specific downstream applications may be a way to increase efficiency. For each application, specific factors regulate the optimal induction method and environment, such as the acceptability of genome integration, the temporality in which desired phenotypes are needed, and the volume in which the target phenotype is required. Based on application specific factors such as these, it may be optimal to utilize various induction methods combined with an optimized set of efficacy conditions described above to generate iPS cell products on an application by application basis, rather than focusing solely on improving the generation of iPS clones.

6. Clinical applications

iPS cells can theoretically become any tissue in the body, which opens a number of possibilities for the use of iPS derived cells in graft and transplant based treatments. A key advantage of iPS cells is patient-specificity. iPS cells could be generated from a patient's own somatic cells and differentiated into the desired phenotype, allowing for an effectively autologous transplant which attenuates the risk of immunorejection. In the manner, iPS technology can be used as a pathway of sorts to generate desired tissues for transplant and tissue engineering applications (Figure 1).

6.1. Acute neurological damage

Induced pluripotent stem cell technologies have provided an exciting avenue for potential treatment of many neurological diseases, many of which have few treatment options at present. Among these disorders, acute neurological damage has an exceedingly direct treatment model through the iPS pathway. In many cases, such as stroke or spinal cord injury, direct transplantation of neuronal cells derived from patient-specific iPS cells to the damaged region could potentially aid in convalescence. Studies have already demonstrated functional recovery in spinal cord injury models of mice treated with iPS derived neuronal cells [37]. Groups have also confirmed functional recovery in peripheral nerve regions [38] and murine ischemia models [39, 40], with promising graft cell growth rates in ischemia models, and notable integration with existing neural networks. These findings show promise for the potential of iPS cell therapies in acute neurological damage conditions; however, further research is needed to ascertain the efficacy, safety, and long term effects of such transplantations.

6.2. Parkinson's disease

Some higher cognitive disorders could also be addressed using similar direct transplantation therapies. Parkinson's disease is perhaps the most direct of the higher cognitive disorders to address in this manner, as the primary cause of functional degradation can be traced to a single cell phenotype. The loss of dopamine secreting neurons in the *substantia nigra* region of the brain has been established as the leading cause of many Parkinson's symptoms, suggesting that direct replacement of lost dopamine secreting neurons through iPS cell derived neurons could aid in recovery. Through the use of various methods, multiple groups have efficiently differentiated iPS cells into dopaminergic neurons [41, 42], overcoming the first obstacle in the implementation of a transplantation therapy. In a rodent model, transplantation of dopamanergic neurons and other neuronal phenotypes into Parkinson's disease model were able to induce functional recovery [43].

These results demonstrate the potential of iPS cells to provide functional recovery in Parkinson's disease patients. However, further research is needed to establish the degree of recovery post-transplant, to improve the efficacy of transplantation, and to assess the long-term benefit of transplantation. It has been suggested that transplanted neuronal populations derived from iPS cells of hereditary Parkinson's patients may be inclined to exhibit similar degenerative phenotypes after implantation and this potential must be explored.

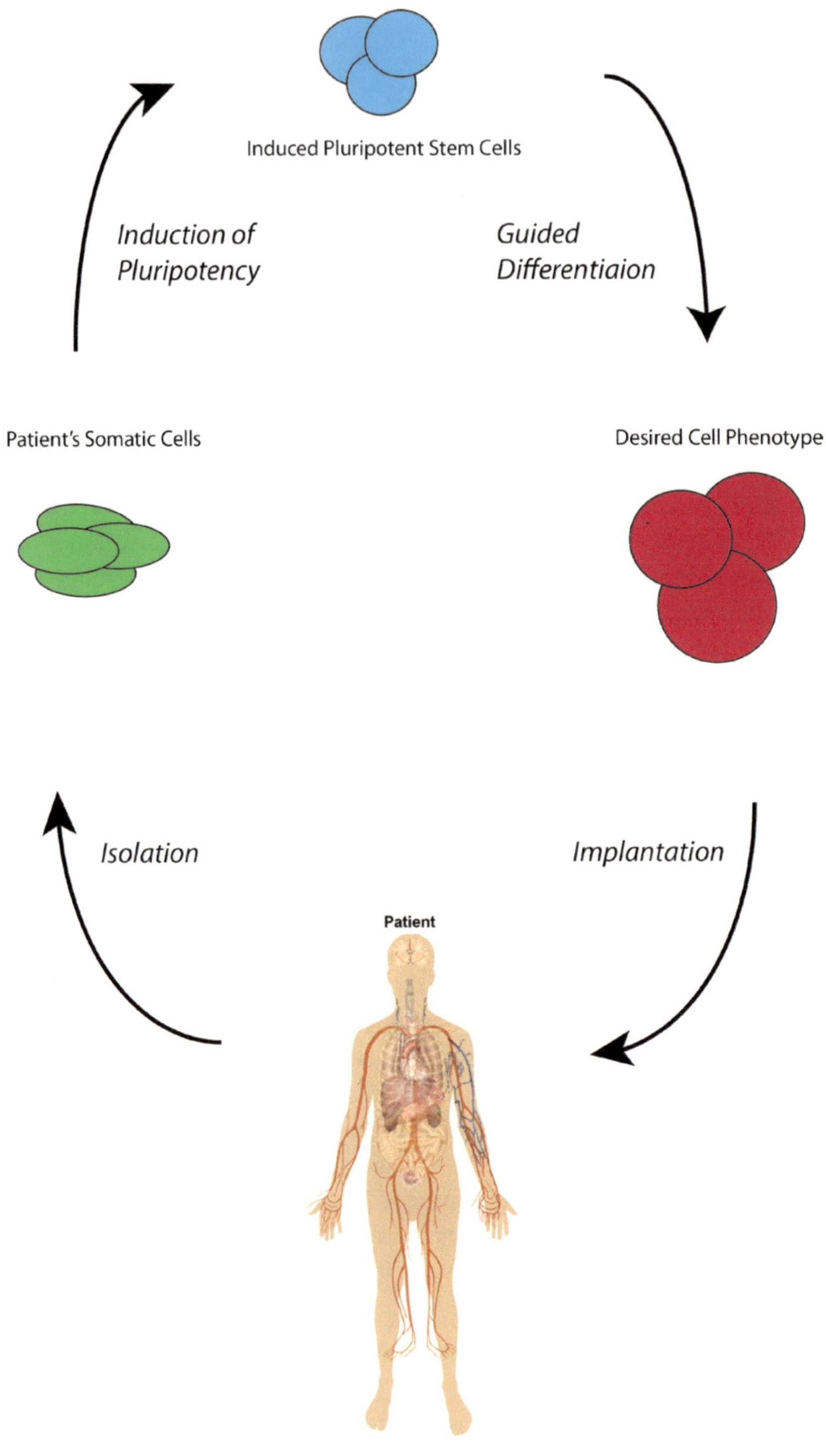

Figure 1. Induced pluripotent treatment pathway: Somatic cells are isolated from the patient, reprogrammed into iPS cells, and then differentiated into the target phenotype for treatment

6.3. Alzheimer's disease

Unlike Parkinson's disease, Alzheimer's disease cannot be traced to the loss of single cell phenotype in a distinct region. In Alzheimer's, damage is diffuse throughout the brain, forming neurofibrillary tangles characterized by high levels of amyloid precursor protein (APP) expression. It has been demonstrated that high levels of APP expression influence differentiation toward the glial phenotype [44], inhibiting direct replacement of neurons through non-terminally differentiated stem cells. As such, Alzheimer's does not lend itself to transplantation therapy as readily as acute neurological injury or Parkinson's. However, there is a potential that transplantation of terminally differentiated neuronal populations derived from iPS cells could have beneficial effects. Transplanted cells may not necessarily replace damaged neurons, but increased neurotrophic factor production from transplanted neuronal populations may have positive effects on patient phenotype.

6.4. Cardiovascular treatments

It has been demonstrated that pluripotent stem cells have the potential to differentiate into cardiomyocyte [45, 46]. Utilizing a number of various culture conditions, including co-culture with stromal cells and cytokine supplementation, differentiation into cardiomyocytes can be made relatively efficient [47]. There is a potential that transplantation of iPS derived cariomyocytes may be able to assist patients who have suffered a myocardial infraction, as has been demonstrated when transplanting other related phenotypes [48]. Studies in a murine model have shown that transplantation of ESC derived cariomyocytes mitigated the functional damage of myocardial infraction [49]. Due to the similarity of iPS cell and ESC phenotypes, there is a potential that similar results would be possible utilizing patient specific iPS cells as the source of cardiomyocytes.

6.5. Hemophilia

Hemophilia is caused by a genetic mutation that reduces the production of coagulant factor VIII or XI depending on the type. Therefore, it's possible that transplantation of iPS derived endothelial cells which express coagulant factors could correct the hemophilia phenotype in patients [50]. In a murine model, transplantation of iPS derived endothelial cells positive for factor VIII expression was able to mitigate the hemophilia A phenotype to a large degree. Endothelial cells were transplanted by injection directly into the liver of hemophiliac mice and functionality was assessed by a tail cutting assay. After treatment, mice with transplanted endothelial cells survived for 3+ months after tail cutting, while control mice died within hours. Factor VIII expression was increased to 8%-12% of normal, indicating that full restoration of factor VIII expression may not be necessary to effectively mitigate the hemophilia phenotype [51]. These findings show promise for the development of cell based therapies to treat hemophilia.

6.6. Blood supply

Blood supply shortages are an ever-present concern in many regions, leading to demand for additional sources of red blood cells (RBCs). iPS cells could theoretically be used to generate RBCs as a supplemental source and it has been demonstrated that iPS cells are capable of direct erythrocytic differentiation [52]. Although the technology exists, the use of iPS cells to generate RBCs may not always been practical due to the cost of iPS generation, culture, and subsequent differentiation. As such, until technologies are developed which allow for industrial scale iPS cell culture and differentiation, the use of iPS cells to augment the blood supply will be fairly limited. In certain circumstances, such as a patient in need of a rare blood type in advance of surgery, generation of iPS derived RBCs may be a viable option for treatment.

7. Research applications

7.1. Disease modeling

Accurate disease modeling is a biotechnological problem of fundamental importance. Most current disease models rely upon murine model organisms, which are capable of providing insight, but are less than ideal due to interspecies differences [53]. iPS technologies could allow for *in vitro* disease modeling, using cultures isolated from those suffering with a given condition. If widely applied, patient specific iPS cultures could potentially be created to analyze the nuances of a disease in a particular patient, determining which course of treatment would be best. Using skin fibroblasts isolated from a patient with spinal muscular atrophy, Ebert et. al. demonstrated that iPS derived motor neurons could be effectively grown in culture and maintained the disease phenotype of the patient [54]. These findings indicate that iPS cells derived from patients with genetic disorders may exhibit the disease phenotype, allowing for their use as a disease model.

Similar isolations have also occurred with Parkinson's patients, in which iPS clones were generated from patients and subsequently differentiated into dopamanergic neurons. However, in the context of Parkinson's disease, the disease phenotype was not as readily presented *in vitro* due to the relative age of the neurons. While cultured neurons have a lifespan in weeks, Parkinson's develops over a period of years due in conjunction with age related factors, possibly requiring a form of artificial stress treatment to accurately reproduce the phenotype *in vitro* [14]. However, early stage metabolic dysfunction has already been identified and corrected *in vitro* using neurons generated from familial Parkinson's patient derived iPS cells, indicating that some early stage phenotypes may be identifiable without full phenotypic replication [55]. Alzheimer's disease, like Parkinson's, is strongly influenced by a number of age related factor which complicate the creation of an accurate model. Recently, Shi et. al. demonstrated one potential approach to this problem by using iPS cells derived from Down syndrome patients. Down syndrome patients overexpress a gene known to encode for amyloid precursor protein (APP), a major component of the Alzheimer's phenotype. Cortical neurons generated from these iPS lines expressed amyloid aggregates and hyperphosphorylated tau protein, both hallmarks of the

Alzheimer's disease phenotype, after months in culture [56]. Utilizing a similar approach, it may be possible to emulate other age related disease phenotypes through variable gene expression, providing a second avenue from which to approach the issue. iPS line have also been derived from Huntington's patients, in which differentiated neurons maintained some portions of the Huntington's phenotype [57, 58]. CDKL5 mutant iPS lines have also been generated from Rett syndrome patients, and may allow for investigation of CDKL5's underlying mechanism within patient cells [59]. Amyotrophic lateral sclerosis (ALS) has also been effectively modeled using an iPS line derived from familial ALS patients [60].

Once generated, these disease models can provide insight into the underlying mechanisms of the disease. *In vitro* research of molecular level cellular mechanisms is much cheaper and more efficient than similar research in mammal models, potentially allowing for increased research throughput. Established *in vitro* models also remove confounding factors related to animal models, potentially making direct identification of mechanisms easier. In the context of phenotypic identification and the discovery of underlying mechanisms, it is important to consider the controls necessary for using these iPS derived disease models. Due to potential phenotypic differences in iPS clones, even from the same isolation, it would be necessary to generate models using multiple iPS lines from each patient in a diverse group. This spread would allow for adequate confirmation that the identified phenotype or mechanism is indeed consistent for all patients with the disease, rather than an artifact of reprogramming or a trait specific to a single individual [53].

While these results in summary are very promising, substantial challenges remain before iPS cell cultures can be used as disease models in every instance. Although diseases with limited temporal dependency, such as spinal muscular atrophy, and clear monogenic origin, such as Huntington's, are replicated relatively easily *in vitro*, there remain unsolved problems in replicating diseases influenced by multiple factors. As demonstrated in attempts to replicate the Parkinson's phenotype *in vitro*, time related factors can also play a large role in disease phenotype, complicating modeling. Other diseases, such as Alzheimer's, may be dependent upon cellular interactions between multiple cell phenotypes in addition to age related factors. The homogenous nature of iPS derived cell cultures complicates accurate replication of these interactions *in vitro*. In some cases, it may be possible to model some of these cell to cell interactions using coculture, as demonstrated in ALS models that incorporate both astrocytes and neurons [61]. Further research is needed to overcome these barriers before iPS cell based disease modeling can be exploited to its full potential.

7.2. Drug discovery

As a corollary to disease modeling, drug discovery is a promising research application for iPS cells. Developing new drugs is exceedingly expensive and many drug candidates are rejected in the final human trial stage due to toxicology concerns [46]. At present, 90% of all drugs candidates that enter clinical trials fail to be approved, leading to a low drug candidate to successful drug ratio [62]. If drugs could be screened for human toxicology earlier in the development cycle, a number of these candidates could be eliminated earlier, allowing for increased funding to more promising drugs. This redistribution of funding could eventually lead to more drug candidates developed in a more rapid fashion.

To assess for toxicology, iPS clones could be generated from a broad cross-section of potential patients, representing various patient backgrounds. Due to the immortalized nature of iPS cultures, these cells could be expanded and maintained indefinitely at relatively low expense to drug developers. As a consequence of effective cell storage technologies, an iPS clone bank would only expand overtime, allowing for the aggregation of clones generated during multiple studies. From these clones, tissues could be generated for toxicology testing early in the development cycle, potentially identifying toxic drug candidates before further testing takes place (Figure 2). In this manner, a diverse donor population could effectively provide each type of human tissue with a relatively small amount of tissue collection.

Using iPS disease models as described above, the effectiveness of new drug therapies could also be tested *in vitro.* The overall cost of testing using these *in vitro* models is less than that of animal modeling, and could allow large scale screening of potential drug candidates early in the development cycle. Due to the elimination of certain confounding factors present in animal models, drug testing in iPS derived disease models may also yield unique insights not demonstrated using traditional models. The iPS clone bank described above could be expanded to include similarly diverse clone populations from patients with a specific disease. Similar to its benefits in toxicology testing, an iPS clone bank could allow for testing on a broad cross-section of disease patients at a relatively low cost. Recently, studies have utilized iPS disease models to assess the efficacy of Alzheimer's disease drug candidates *in vitro* [56] and to successfully screen for new drugs to potentially treat ALS [60]. These studies demonstrate the potential for the use of iPS cells in the context of drug development, both to improve the efficiency of existing drug development pipelines and to screen for entirely new compounds in a relatively low cost model.

However, drug discovery and toxicology screening using iPS cells is limited by their ability to accurately replicate *in vivo* conditions. As discussed above, the homogenous nature of iPS cultures neglects many influential factors related to cellular interaction, and the temporally naive nature of iPS cultures neglects many age related factors. As such, further research is necessary before iPS derived tissues are suitable for use in toxicology testing. For drug therapy screening, the current state of disease models as discussed above is a limiting factor. Although not all diseases can be effectively modeled for screening today, some disorders that have well characterized iPS models may benefit from broad drug screening in the near future.

8. Conclusion: Challenges to the road ahead

A number of roadblocks remain before iPS cells are ready for the clinic. At present, there still remains a risk of teratoma formation in the event that a subpopulation of iPS cells is not terminally differentiated prior to transplantation. In the context of a patient-specific autologous treatment using iPS cells, methods must be developed by which iPS cells can be generated in sufficient quantity, reliably, and in a time frame appropriate for the targeted disease.

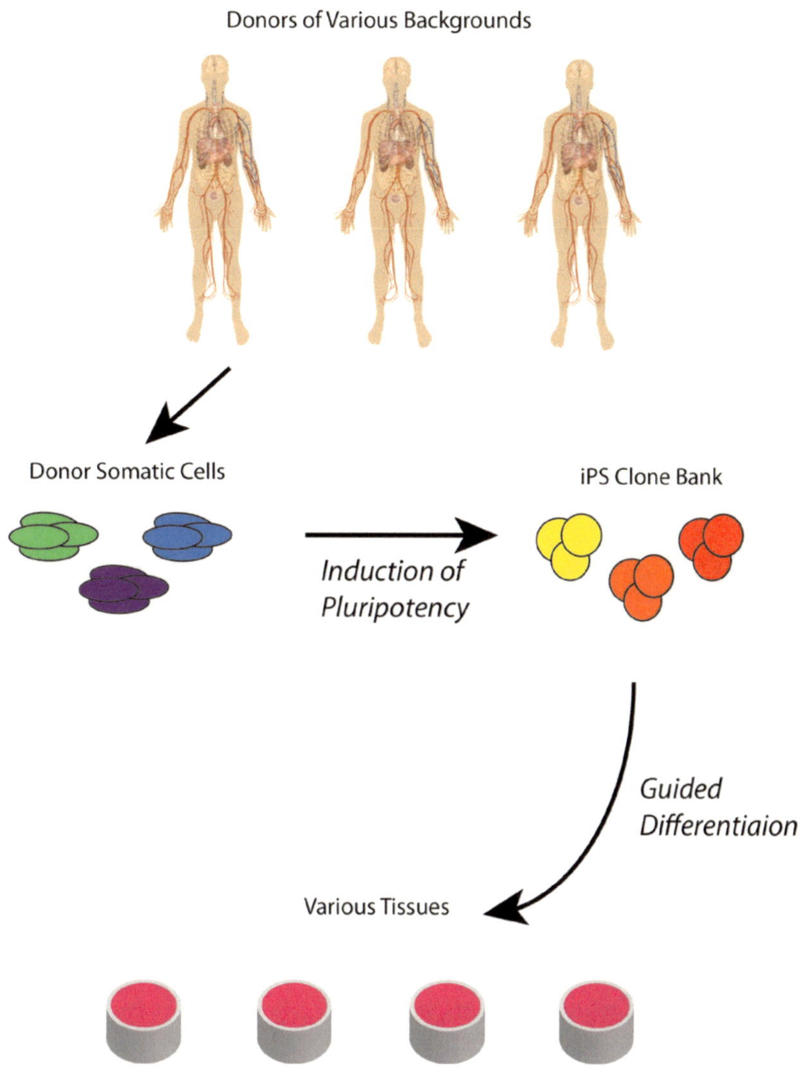

Figure 2. Drug discovery and toxicology workflow. Somatic cells are isolated from a broad cross section of donors, reprogrammed, and differentiated into relevant tissues for toxicology screening and drug testing

Efficiency remains an issue, especially with regards to technologies that do not integrate transgenes into the host genome. To address efficiency concerns, it is possible that application specific optimization of induction technologies could improve the efficacy of current induction technologies.

As disease models, iPS cells are limited by the neglect of several influential factors. Most prominently, the homogenous populations derived from iPS cells inherently neglect interactions between multiple cell phenotypes, and these interactions may be critical to understanding disease mechanisms [6]. iPS cells could potentially be differentiated into various cell types and cocultured to replicate interactions between cell types, but it may be difficult to generate an accurate interaction model, even with multiple cell types. iPS cell cultures also neglect various age related factors, which may be particularly problematic in modeling certain diseases. These same issues act as barriers to the use of iPS cells for drug discovery and toxicology screening, as both applications rely upon accurate iPS models of *in vivo* cellular activity.

Induced pluripotent stem cell technologies have progressed rapidly in recent years. Various induction methods have eliminated or reduced many of the fundamental issues with iPS cells, opening the door to a variety of possible applications. Though there remain a number of challenges facing the development of iPS cells in the clinic and the laboratory, the potential benefits to regenerative medicine are profound.

Author details

Jacob Kimmel and Kiminobu Sugaya

*Address all correspondence to: kiminobu.sugaya@ucf.edu

University of Central Florida, Orlando, FL, USA

References

[1] Yamanaka, S. and K. Takahashi, *Induction of Pluripotent Stem Cells from Mouse Embryonic and Adult Fibroblast Cultures by Defined Factors.* Cell, 2006. 126(4): p. 663-676.

[2] Avilion, A.A., et al., *Multipotent cell lineages in early mouse development depend on SOX2 function.* Genes & Development, 2003. 17(1): p. 126-140.

[3] Hong, Y., et al., *Activation of the mouse Oct4 promoter in medaka embryonic stem cells and its use for ablation of spontaneous differentiation.* Mechanisms of Development, 2004. 121(7–8): p. 933-943.

[4] Cartwright, P.P., et al., *LIF/STAT3 controls ES cell self-renewal and pluripotency by a Myc-dependent mechanism.* Development (Cambridge, England), 2005. 132(5): p. 885-896.

[5] Shi, Y., et al., *A Combined Chemical and Genetic Approach for the Generation of Induced Pluripotent Stem Cells.* Cell Stem Cell, 2008. 2(6): p. 525-528.

[6] Yamanaka, S., *A fresh look at iPS cells.* Cell, 2009. 137(1): p. 13-7.

[7] Junying Yu, et al., *Induced Pluripotent Stem Cell Lines Derived from Human Somatic Cells.* Science Express, 2007. 10: p. 1-4.

[8] Liao, J., et al., *Enhanced efficiency of generating induced pluripotent stem (iPS) cells from human somatic cells by a combination of six transcription factors.* Cell Research, 2008. 18: p. 600-603.

[9] Kim, J.B., et al., *Oct4-Induced Pluripotency in Adult Neural Stem Cells.* Cell, 2009. 136(3): p. 411-419.

[10] Shi, Y., et al., *Induction of pluripotent stem cells from mouse embryonic fibroblasts by Oct4 and Klf4 with small-molecule compounds.* Cell Stem Cell, 2008. 3(5): p. 568-74.

[11] Sugaya, K. and A. Alvarez, *Methods and materials for increasing potency of cells* U.P.a.T. Office, Editor 2006: USA.

[12] Alvarez, A., et al., *Nanog overexpression allows human mesenchymal stem cells to differentiate into neural cells——Nanog transdifferentiates mesenchymal stem cells.* Neuroscience and Medicine, 2010: p. 1-13.

[13] Hacein-Bey-Abina, S., et al., *LMO2-associated clonal T cell proliferation in two patients after gene therapy for SCID-X1.* Science, 2003. 302(5644): p. 415-9.

[14] Soldner, F., et al., *Parkinson's Disease Patient-Derived Induced Pluripotent Stem Cells Free of Viral Reprogramming Factors.* Cell, 2009. 136(5): p. 964-977.

[15] Okita, K., et al., *Generation of Mouse Induced Pluripotent Stem Cells Without Viral Vectors.* Science (Washington), 2008. 322(5903): p. 949-953.

[16] Yamanaka, S., *Induced Pluripotent Stem Cells: Past, Present, and Future.* Cell Stem Cell, 2012. 10(6): p. 678-684.

[17] Yu, J., et al., *Human induced pluripotent stem cells free of vector and transgene sequences.* Science, 2009. 324(5928): p. 797-801.

[18] Okita, K. and S. Yamanaka, *Induced pluripotent stem cells: opportunities and challenges.* Philos Trans R Soc Lond B Biol Sci, 2011. 366(1575): p. 2198-207.

[19] Armeanu, S., et al., *Cell cycle independent infection and gene transfer by recombinant Sendai viruses.* Journal of Virological Methods, 2003. 108(2): p. 229-233.

[20] Bernloehr, C., et al., *Efficient propagation of single gene deleted recombinant Sendai virus vectors.* Virus Research, 2004. 99(2): p. 193-197.

[21] Tokusumi, T., et al., *Recombinant Sendai viruses expressing different levels of a foreign reporter gene.* Virus Research, 2002. 86(1–2): p. 33-38.

[22] Fusaki, N., et al., *Efficient induction of transgene-free human pluripotent stem cells using a vector based on Sendai virus, an RNA virus that does not integrate into the host genome.* Proceedings of the Japan Academy, Series B. Physical and Biological Sciences, 2009. 85B(8): p. 348-362.

[23] Warren, L., et al., *Highly Efficient Reprogramming to Pluripotency and Directed Differentiation of Human Cells with Synthetic Modified mRNA.* Cell Stem Cell, 2010. 7(5): p. 618-630.

[24] Rosa, A. and A.H. Brivanlou, *Synthetic mRNAs: Powerful Tools for Reprogramming and Differentiation of Human Cells.* Cell Stem Cell, 2010. 7(5): p. 549-550.

[25] Zhao, T.T., et al., *Immunogenicity of induced pluripotent stem cells.* Nature, 2011. 474(7350): p. 212-215.

[26] Kim, D., et al., *Generation of Human Induced Pluripotent Stem Cells by Direct Delivery of Reprogramming Proteins.* Cell Stem Cell, 2009. 4(6): p. 472-476.

[27] Shimada, H., et al., *Accelerated generation of human induced pluripotent stem cells with retroviral transduction and chemical inhibitors under physiological hypoxia.* Biochemical and Biophysical Research Communications, 2012. 417(2): p. 659-664.

[28] Yoshida, Y., et al., *Hypoxia enhances the generation of induced pluripotent stem cells.* Cell Stem Cell, 2009. 5(3): p. 237-41.

[29] Lyssiotis, C.A., et al., *Reprogramming of murine fibroblasts to induced pluripotent stem cells with chemical complementation of Klf4.* Proc Natl Acad Sci U S A, 2009. 106(22): p. 8912-7.

[30] Huangfu, D., et al., *Induction of pluripotent stem cells by defined factors is greatly improved by small-molecule compounds.* Nat Biotechnol, 2008. 26(7): p. 795-797.

[31] Hanna, J., et al., *Direct cell reprogramming is a stochastic process amenable to acceleration.* Nature, 2009. 462(7273): p. 595-601.

[32] Loh, Y.H., et al., *The Oct4 and Nanog transcription network regulates pluripotency in mouse embryonic stem cells.* Nat Genet, 2006. 38(4): p. 431-40.

[33] Papapetrou, E.P., et al., *Stoichiometric and temporal requirements of Oct4, Sox2, Klf4, and c-Myc expression for efficient human iPSC induction and differentiation.* Proceedings of the National Academy of Sciences, USA, 2009. 106(31): p. 12759-12764.

[34] Bar-Nur, O., et al., *Epigenetic Memory and Preferential Lineage-Specific Differentiation in Induced Pluripotent Stem Cells Derived from Human Pancreatic Islet Beta Cells.* Cell Stem Cell, 2011. 9(1): p. 17-23.

[35] Hu, Q.Q., et al., *Memory in induced pluripotent stem cells: reprogrammed human retinal-pigmented epithelial cells show tendency for spontaneous redifferentiation.* Stem cells (Dayton, Ohio), 2010. 28(11): p. 1981-1991.

[36] Kim, K.K., et al., *Epigenetic memory in induced pluripotent stem cells.* Nature, 2010. 467(7313): p. 285-290.

[37] Nori, S., et al., *Transplantation of human iPS cell-derived neurospheres for the treatment of spinal cord injury in NOD-scid mice.* Neuroscience Research, 2010. 68, Supplement 1(0): p. e427.

[38] Uemura, T., et al., *Transplantation of induced pluripotent stem cell-derived neurospheres for peripheral nerve repair.* Biochemical and Biophysical Research Communications, 2012. 419(1): p. 130-135.

[39] Gomi, M., et al., *Functional recovery of the murine brain ischemia model using human induced pluripotent stem cell-derived telencephalic progenitors.* Brain Research, 2012. 1459(0): p. 52-60.

[40] Jensen, M.B., et al., *Survival and Differentiation of Transplanted Neural Stem Cells Derived from Human Induced Pluripotent Stem Cells in A Rat Stroke Model.* Journal of Stroke and Cerebrovascular Diseases, (0).

[41] Cooper, O., et al., *Differentiation of human ES and Parkinson's disease iPS cells into ventral midbrain dopaminergic neurons requires a high activity form of SHH, FGF8a and specific regionalization by retinoic acid.* Molecular and Cellular Neuroscience, 2010. 45(3): p. 258-266.

[42] Swistowski, A., et al., *Efficient generation of functional dopaminergic neurons from human induced pluripotent stem cells under defined conditions.* Stem Cells, 2010. 28(10): p. 1893-904.

[43] Wernig, M., et al., *Neurons derived from reprogrammed fibroblasts functionally integrate into the fetal brain and improve symptoms of rats with Parkinson's disease.* Proceedings of the National Academy of Sciences, USA, 2008. 105(15): p. 5856-5861.

[44] Kiminobu Sugaya, S.M., *How to approach alzheimer's disease with stem cell technologies.* Journal of Alzheimer's Disease, 2008. 15: p. 241–254.

[45] Taha, M.F. and V. Hedayati, *Isolation, identification and multipotential differentiation of mouse adipose tissue-derived stem cells.* Tissue Cell, 2010. 42(4): p. 211-6.

[46] Nsair, A. and W.R. MacLellan, *Induced pluripotent stem cells for regenerative cardiovascular therapies and biomedical discovery.* Advanced Drug Delivery Reviews, 2011. 63(4–5): p. 324-330.

[47] Yoshida, Y. and S. Yamanaka, *iPS cells: A source of cardiac regeneration.* Journal of Molecular and Cellular Cardiology, 2011. 50(2): p. 327-332.

[48] Galli, D., et al., *Mesoangioblasts, vessel-associated multipotent stem cells, repair the infarcted heart by multiple cellular mechanisms: a comparison with bone marrow progenitors, fibroblasts, and endothelial cells.* Arterioscler Thromb Vasc Biol, 2005. 25(4): p. 692-7.

[49] Christoforou, N.N., et al., *Implantation of mouse embryonic stem cell-derived cardiac progenitor cells preserves function of infarcted murine hearts.* PLoS ONE, 2010. 5(7): p. e11536-e11536.

[50] Liras, A., *Induced human pluripotent stem cells and advanced therapies: Future perspectives for the treatment of haemophilia?* Thrombosis Research, 2011. 128(1): p. 8-13.

[51] Xu, D.D., et al., *Phenotypic correction of murine hemophilia A using an iPS cell-based therapy.* Proceedings of the National Academy of Sciences of the United States of America, 2009. 106(3): p. 808-813.

[52] Peyrard, T., et al., *Banking of Pluripotent Adult Stem Cells as an Unlimited Source for Red Blood Cell Production: Potential Applications for Alloimmunized Patients and Rare Blood Challenges.* Transfusion Medicine Reviews, 2011. 25(3): p. 206-216.

[53] Saha, K. and R. Jaenisch, *Technical Challenges in Using Human Induced Pluripotent Stem Cells to Model Disease.* Cell Stem Cell, 2009. 5(6): p. 584-595.

[54] Ebert, A.D., et al., *Induced pluripotent stem cells from a spinal muscular atrophy patient.* Nature, 2009. 457(7227): p. 277-280.

[55] Cooper, O., et al., *Pharmacological Rescue of Mitochondrial Deficits in iPSC-Derived Neural Cells from Patients with Familial Parkinson's Disease.* Science Translational Medicine, 2012. 4(141): p. Article.

[56] Shi, Y., et al., *A Human Stem Cell Model of Early Alzheimer's Disease Pathology in Down Syndrome.* Science Translational Medicine, 2012. 4(124): p. 124ra29.

[57] Castiglioni, V., et al., *Induced pluripotent stem cell lines from Huntington's disease mice undergo neuronal differentiation while showing alterations in the lysosomal pathway.* Neurobiology of Disease, 2012. 46(1): p. 30-40.

[58] Camnasio, S., et al., *The first reported generation of several induced pluripotent stem cell lines from homozygous and heterozygous Huntington's disease patients demonstrates mutation related enhanced lysosomal activity.* Neurobiology of Disease, 2012. 46(1): p. 41-51.

[59] Amenduni, M., et al., *iPS cells to model CDKL5-related disorders.* Eur J Hum Genet, 2011. 19(12): p. 1246-55.

[60] Egawa, N., et al., *Drug Screening for ALS Using Patient-Specific Induced Pluripotent Stem Cells.* Science Translational Medicine, 2012. 4(145): p. 145ra104.

[61] Di Giorgio, F.P., et al., *Non-cell autonomous effect of glia on motor neurons in an embryonic stem cell-based ALS model.* Nat Neurosci, 2007. 10(5): p. 608-614.

[62] Rubin, L.L., *Stem Cells and Drug Discovery: The Beginning of a New Era?* Cell, 2008. 132(4): p. 549-552.

Chapter 13

Molecular Mechanisms of Embryonic Stem Cell Pluripotency

Guofeng Han, Hongtao Wang and Jijun Hao

Additional information is available at the end of the chapter

http://dx.doi.org/10.5772/54365

1. Introduction

Embryonic stem (ES) cells isolated from the inner cell mass (ICM) of blastocysts possess the defining pluroptency: unlimited self-renewal and giving rise to all cells of the organism[1, 2]. Thus, ES cells hold great promise for regenerative medicine to treat many diseases including heart failure, diabetes, Alzheimer's and Parkinson's disease by replacing the damaged cells with ES cell-derived healthy ones. The recent advent of induced pluripotent stem (iPS) cells reprogrammed from somatic cells has the potential to revolutionize the field of regenerative medicine since patient-derived iPS cells, in principle, circumvent the ethical problems and immune rejection associated with human ES cells[3]. Nevertheless, the future clinical translation of ES cells and iPS cells is facing numerous hurdles. Understanding the molecular mechanisms that impart ES cells with pluripotency may help address some of these challenges. The past few years have seen tremendous progress in understanding of mechanisms which govern ES cell pluripotency. In this chapter, we will review critical signaling and transcription factor networks that have been identified to maintain ES cell pluripotency.

2. Signaling pathways of ES cells

ES cells require extrinsic growth factors to maintain their pluripotency in culture. These extrinsic growth factors act on different signaling pathways to regulate intrinsic transcription factor networks to sustain ES cells in the undifferentiated state. The signaling pathways required to support pluripotency in mouse ES cell are distinct from those in human ES cells (Figure 1).

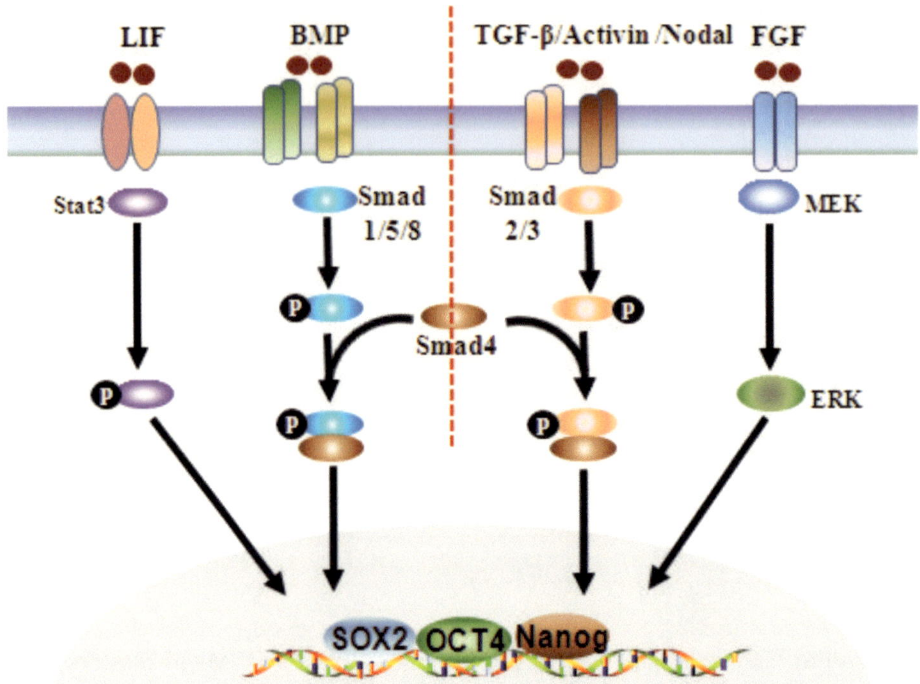

Figure 1. Exogenous growth factors signal through distinct signaling pathways to regulate transcription factors for ES cell pluripotency.

2.1. LIF/JAK/STAT3 pathway

Mouse ES cells were originally cultured on feeder layers derived from mouse embryonic fibroblasts (MEF). Later it was found that Leukaemia Inhibitory Factor (LIF), a member of the Interleukin-6 cytokines produced by MEFs, was the key factor to maintain pluripotency of mouse ES cells by inhibiting their differentiation[4]. Upon LIF binding, the LIF receptor recruits gp130 to form a heterodimer which subsequently activates Janus kinase (JAK) through transphosphorylation[5]. Activated JAK then phosphorylate gp130, creating a docking site to bind the SH2 domain of Signal Transducers and Activators of Transcription 3 (STAT3)[6-9]. Once STAT3 binds to the gp130 docking site, JAK then phosphorylates the recruited STAT3. Phosphorylated STAT3 forms a homodimer, which subsequently translocate into the nucleus, where it binds to gene enhancers to regulate target gene expression[10-12].

Although the LIF/JAK/STAT3 pathway has been well documented to maintain pluripotency of mouse ES cells in the presence of serum, the mechanisms by which activated STAT3 functions in this regard is poorly understood. Recently, studies in identification of STAT3 target genes have improved our understanding of activated STAT3 in maintaining pluripotency. Chen et al identified 718 STAT3-bound genomic sites that were co-occupied by pluripotency transcription markers (Oct4, Sox2 and Nanog) by using chromatin immunoprecipitation sequencing (ChIP-seq)[12]. In addition, Kidder and colleagues found that STAT3 target genes

http://dx.doi.org/10.5772/54365

enriched in ES cells were downregulated in differentiated cells by mapping STAT3 binding targets in mouse ES cells and differentiated embryoid bodies (EBs)[13]. Along with these results, it has been demonstrated that knocking down STAT3-target genes induces activation of endodermal and mesodermal genes, supporting the conclusion that STAT3 prevents mESC differentiation by suppressing lineage-specific genes[14].

Interestingly, the LIF receptor and gp130 are also expressed in human ES cells and human LIF can induce STAT3 phosphorylation and nuclear translocation in human ES cells. However, human LIF is unable to maintain the pluripotent state of human ESs, suggesting that mouse and human ES cells require distinct signaling mechanisms to govern their pluripotency[15].

2.2. TGF-β signaling

TGF-β superfamily consists of more than 40 members, including TGF-β, Activin, Nodal, and bone morphogenetic proteins (BMPs). The TGF-β members transduce signals by binding to heteromeric complexes of serine/threonine kinase receptors, type I and type II receptors, which subsequently activate intracellular Smad proteins. Smads 2 and 3 are specifically activated by activin, nodal and TGF-β ligands, whereas Smads 1, 5 and 8 are activated by BMP ligands[16, 17] (Figure 1). The TGF-β-related signaling pathways play complex roles in regulating the pluripotency and cell fate of ES cells.

2.2.1. BMP signaling pathway

Bone Morphogenetic Protein (BMP) is a subset of the TGF-β superfamily[18]. When BMP ligands bind to type II BMP receptors (BMPRII), BMPRII then recruits and phosphorylates type I BMP receptors (BMPRI). Activated type I receptors subsequently phosphorylate BMP-responsive SMAD1/5/8 which then forms a complex with SMAD4 and translocates into nucleus to regulate target gene expression (Figure 1). In mouse ES cells, LIF can substitute MEF feeder layers in maintaining pluripotency in the presence of animal serum by activating the transcription factor STAT3. However, in serum-free cultures, LIF is insufficient to block neural differentiation and maintain pluripotency. Recently, Ying et al reported that BMP was able to replace serum to maintain pluripotency of mouse ES cells in the presence of LIF. BMP has been shown to phosphorylate SMAD1/5 and activate inhibitors of differentiation (*Id*) genes, which block neural differentiation by antagonizing neurogenic transcription factors[19]. In the absence of MEF and serum, exogenous LIF, in combination with BMP4 proteins, can sufficiently maintain the pluripotency of mouse ES cells derived from "permissive" mouse strains.

In contrast to a maintenance role in mouse ES cell pluripotency, BMP has been shown to promote human ES cells differentiation to trophoblasts, and inhibiting BMP signaling with the BMP antagonist, Noggin, sustains the undifferentiated state of human ES cells[20, 21]. In consistence, dorsomorphin and DMH1, small molecule BMP inhibitors previously identified in our lab, were shown to promote long-term self-renewal an pluripotency of human ES cells, presumably by inhibiting BMP induced extraembryonic lineage differentiation[22-25].

2.2.2. *TGF-β/activin/nodal signaling pathway*

Although MEFs feeder layers were initially used to co-culture both mouse and human ES cells, signal factors secreted from MEFs to maintain pluripotency of the two types of ES cells are fundamentally different. Sato et al first discoveried that TGF-β and Nodal genes were highly expressed in undifferentiated human ES cells[26]. Beattie et al later reported that Activin A, a member of the TGF-β superfamily, was secreted by MEFs, and medium enriched with activin A can replace MEF feeder-layers or MEF-conditioned media to maintain human ES cells in an undifferentiated state[27]. In consistence, James et al demonstrated that the TGF-β/Activin/Nodal pathway was activated through the transcription factors Smad2/3 in undifferentiated human ES cells[28]. The notion that TGF-β/Activin/Nodal signaling supports human ES self-renewal and pluripotency is further supported by the fact that recombinant Activin or Nodal stimulation induces higher levels of pluripotent protein expression (Oct4 and Nanog), while inhibition of TGF-β/Activin/Nodal signaling with Lefty or Follistatin decreases expression of these pluripotent proteins in human ES cells[29, 30].

Recent studies have focused on understanding the molecular mechanisms of TGF-β/Activin/Nodal signaling in retaining human ES cells pluripotency. Xu and colleagues showed that TGF-β/Activin/Nodal signaling activated Smad2/3 which subsequently binds to the Nanog promoter in undifferentiated human ES cells to induce expression of Nanog, a pluripotent transcription factor[31]. Additionally, mutating the putative Smad-binding sites reduced the response of Nanog to modulation of TGF-β signaling[31]. Nanog was also shown to coordinate with Smad2 in a negative-feedback loop to inhibit human ES cell differentiation[32]. In contrast to its important role in maintaining human ES cell pluripotency, the TGF-β/Activin/Nodal signaling is not essential for pluripotency of mouse ES cells. Although this pathway was shown to be active in undifferentiated mouse ES cells as assessed by phosphorylation of smad 2/3, inhibition of smad 2/3 phosphorylation by SB431542 had no effect on the undifferentiated state of mouse ES cells[28]. However, the TGF-β/Activin/Nodal signaling may play a role in mouse ES proliferation. A recent study showed that Inhibition of TGF-β/Activin/Nodal signaling by Smad7 or SB-431542 dramatically decreased mouse ES cell proliferation without effect on their pluripotency[33].

2.2.3. *Growth and Differentiation factor 3 (GDF-3)*

GDF-3 is another TGF-beta superfamily member that plays opposite roles in mouse and human ES cells. GDF-3, which acts as a BMP antagonist by direct binding to BMP-4, is specifically expressed in the pluripotent state of both mouse and human ES cells[34]. Ectopic expression of GDF-3 leads to the maintenance of pluripotency in human ES cells, whereas a similar effect is observed in mouse ES cells when GDF-3 levels are decreased. In the absence of LIF, GDF-3-deficient mouse ES cells can still sustain pluripotent markers[34]. These results are consistent with previously discussed BMP signals which can promote pluripotency of mouse ES cells, but cause differentiation of human ES cells. Thus lower concentrations of BMP antagonists, such as GDF-3, may enhance pluripotency in mouse ES cells, whereas higher levels of GDF-3 may favor pluripotency of human ES cells by abrogating BMP signaling.

2.3. FGF/MEK signaling

The importance of Fibroblast growth factor (FGF) signaling for human ES cells pluripotency is highlighted by the facts that human ES cells are traditionally cultured in the presence of Fibroblast growth factors (FGFs) either on fibroblast feeder layers or in fibroblast-conditioned medium[35, 36]. Studies have demonstrated that all four FGF receptors (FGFR1, FGFR3 and FGFR4) and several components (SOS1, PTPN11 and RAF1) of their downstream activation cascade are significantly upregulated in undifferentiated human ES cells, in comparison to differentiated human ES cells[37-39]. In consistence, withdrawal of FGFs or inhibition of FGF signaling by a FGFR inhibitor, SU5402, rapidly induces human ES cell differentiation[40-42].

Although the pluirpotency maintenance role of exogenous FGFs in human ES cell has been known for a long time, the molecular mechanisms by which they function remain unclear. FGFs signal by binding to FGF receptors (FGFRs), and activate multiple signaling cascades, including Mitogen-Activated Protein Kinases (MAPKs), the Janus kinase/signal transducer and activator of transcription (Jak/Stat), phosphatidylinositol 3-kinase (PI3K) and phosphoinositide phospholipase C (PLCg) pathway[43]. Several studies have highlighted the FGF contribution to the maintenance of human ES cells mainly through the FGF/MEK pathway (Figure 1), [44, 45]. Studies have showed that FGF2 induces feeder layer cells to secret TGFβ1 and insulin-like growth factor 2 (IGF2), which can subsequently promote the undifferentiated state of human ES cells[46, 47]. Bendall et al further reported that the function of exogenous FGFs in promoting ES self-renewal could be replaced by addition of IGF2 alone, suggesting an indirect role of FGFs for human ES cell growth. However, this model was challenged in subsequent publications from Wang et al who reported that exogenous IGF2 alone was insufficient to maintain undifferentiated growth of human ES cells, and they proposed that FGFs may play a direct role in blocking caspase-activated apoptosis through anoikis in human ES cells[48]. Recently, Eiselleova and colleagues postulated a new model whereby endogenous FGF-2 signaling maintained the undifferentiated state and survival of human ESCs, while exogenous FGF-2 mainly suppress cell death and apoptosis genes, thus indirectly contributing to the maintenance of human ES cell pluripotency[49].

FGF signaling in mouse ES cells has also been extensively investigated. Mouse ES cells genetically deficient in Fgf4 and extracellular-signal regulated kinase 2 (Erk2) differentiate inefficiently. These results can be reproduced using inhibitors of FGF receptor and ERK, suggesting blockage of the FGF/MEK signaling pathway promotes mouse ES cell pluripotency[50-52]. Indeed, serum-free mouse ES cell medium supplemented with FGF/MEK inhibitors and LIF permits the derivation of mouse ES cells in the absence of feeders from strains normally considered non-permissive[53]. In addition, a recently identified compound, Pluripotin/SC1, has been shown to maintain mouse ES pluripotency by inhibiting ERK1 and activating the phophoinositide-3 kinase (PI3K) pathway through blocking RasGAP[54-56] [57, 58]. Although inhibition of FGF/MEK pathway can attenuate ES cell differentiation, it is insufficient to support mouse ES cell self-renewal. Combination of the MEK inhibitor PD0325901 with the Glycogen synthase kinase-3 (GSK-3) inhibitor CHIR99021 (known as 2i) can efficiently sustain the pluripotency of mouse ES cells in the absence of exogenous cytokines[59, 60]. Several groups demonstrated that improvement of mouse ES cell pluripotency by inhibition of GSK-3 occurred via Wnt/β-catenin signaling, whereas many others argued that GSK3 was likely to exert β-catenin independent effects in ES cells[59, 61-67].

As demonstrated above, human and mouse ES cells are both derived from blastocyst-stage embryos, but they require different biological signals for maintaining pluripotency. In general, mouse ES cells maintain their pluripotency by activating LIF/STAT3 and BMP signaling, while human ES cells require TGF-β/Nodal and FGF/MEK pathways. Interestingly, several pathways, such as BMP and FGF/MEK, have completely oppositing effects on maintaining the pluriotency of mouse and human ES cells. Activation of BMP signaling and inhibition of the FGF/MEK pathway promote mouse ES self-renewal, whereas inhibition of BMP signaling and activation of FGF/MEK pathway sustain human ES cell pluripotency. These distinct signaling effects on pluripotency may reflect intrinsic differences between mouse and human ES cells. Recent studies have demonstrated that conventional human ES cells do not represent the "ground or naïve state" of stemness, but rather a more developmentally mature "primed state" resembling mouse epiblast stem cells (mEpiSCs) found in the post-implantation, pre-gastrulation stage of embryos [68-74]. Conventional human ES cells exhibit numerous similarities to the mouse EpiSCs over mouse ES cells (Table 1). For instance, conventional human ES cells and mouse EpiSCs display flattened cell colonies and epigenetic X-chromosome inactivation (XiXa), and require Activin and FGF for pluripotency maintanince. In contrast, mouse ES cells exhibit dome-shaped colony morphology and epigenetic activation of both X-chromosome (XaXa), and require LIF/STAT3 signaling to promote self-renewal. Subsequent studies have demonstrated that the medium containing "2i" (MEK inhibitor and GSK-3 inhibitor), when supplemented with other factors (such as forskolin), can efficiently convert conventional human ES cells into a ground or "naïve" state with display of hallmark features of mouse ES cells. This medium can also maintain human ES cell pluriptoency at the naïve state [69, 70, 72, 75-78].

property	mESCs	mEpiSCs	hESCs	hiPSCs
morphology	domed	flattened	flattened	flattened
clonogenicity	high (single cells)	low (clumps)	low (clumps)	low (clumps)
response to LIF/Stat3	self-renewal	none	none	none
response to Activin/bFGF	differentiation	self-renewal	self-renewal	self-renewal
response to BMP	self-renewal	differentiation	differentiation	differentiation
XX status	XaXa	XaXi	XaXi	XaXi
teratoma	yes	yes	yes	yes
chimaera	yes	no	ND	ND

Table 1. Comparison of the properties of mouse ES cells (mESCs), mouse epiblast stem cells (mEpiSCs), human ES cells (hESCs) and human iPS cells (hiPSCs).

3. The regulatory network of pluripotency factors

ES cell pluripotency is conferred by a unique transcriptional network[79]. Early global transcriptional profiles and genetic studies have identified several critical transcription factors that are required for the pluripotency of ES cells, such as Oct4, Sox2, Nanog, Foxd3 and Id, etc [80-88]. Here we will mainly focus on Oct4, Sox2 and Nanog, three key transcription factors of the core pluripotency transcriptional network.

3.1. OCT4 and SOX2

OCT4 (also known as Oct3), a POU domain-containing transcription factor, was one of the first transcription factors identified as essential for both early embryo development and pluripotency maintenance in ES cells[84, 89]. The expression of Oct4 is activated at the 8-cell stage and is later restricted to the inner cell mass (ICM) and germ cells in early mouse embryogenesis *in vivo* [89-92]. Oct4 is highly expressed in both human and mouse ES cells, and its expression diminishes when these cells differentiate and lose pluripotency. Oct4 regulates a broad range of target genes including Fgf4, Utf1, Opn, Rex1/ Zfp42, Fbx15, Sox2 and Cdx2[93-95]. Repression of Oct4 activity in ES cells upregulates Cdx2 expression, leading to ES cell differentiation into trophectoderm[96]. Oct4 is also known to activate downstream genes by binding to enhancers carrying the octamer–sox motif (Oct–Sox enhancer), for synergistic activation with Sox2. In contrast with its target genes, little is known about Oct4 upstream regulators. The Oct4 promoter contains conserved distal and proximal enhancers that can either repress or activate its expression depending on the binding factors occupying these sites[97, 98]. The precise level of Oct4 is important for ES cell fate determination. Loss of Oct4 causes inappropriate differentiation of ES cells into trophectoderm, whereas overexpression of Oct4 results in differentiation into primitive endoderm and mesoderm[99, 100].

Sox2 is an HMG-box transcription factor that is detected in pluripotent cell lineages and the nervous system[101-103]. Inactivate Sox2 *in vivo* results in early embryonic lethality due to the failure of ICM maintenance[102]. Sox2 can form a complex with the Oct4 protein to occupy Oct–Sox enhancers to regulate target gene expression. Oct–Sox enhancers are found in the regulatory region of most of the genes that are specifically expressed in pluripotent stem cells, such as Oct4, Sox2, Nanog, Utf1, Lefty, Fgf4 and Fbx15[93, 94, 104-108].

3.2. Nanog

Nanog is another homeobox-containing transcription factor that is specifically expressed in pluripotent ES cells. The essential role of Nanog in maintaining the pluripotency of ES cells is highlighted by the facts that Nanog-deficient ES cells are prone to differentiation, whereas forced expression of Nanog partially renders ES cells self-renewal potential in the absence of LIF[85, 86, 109]. How Nanog regulates stem cell pluripotency remains entirely unknown. Studies have indicated that Nanog may maintain ES cell pluripotency by 1) downregulating downstream genes essential for cell differentiation such as Gata4 and Gata6 and 2) activating the expression of genes necessary for self-renewal such as Rex1 and Id[19, 85, 86]. Although it is widely accepted that Nanog, like Oct4 and Sox2, play a central role in pluripotency maintenance, this dogma has been challenged by a subsequent report that Nanog protein levels are undetectable in a fraction of ES cells that express Oct4, and the pure populations of Nanog−/− ES cells can be propagated without losing expression of other pluripotency markers[110].

Little is known about the mechanism by which Nanog is regulated in ES cells. Recently, Suzuki et al showed that Nanog expression was upregulated by BrachyuryT and STAT3 in mouse ES cells[111]. In human ES cells and in mouse EpiSCs, Vallier et al reported that Activin/Nodal signaling stimulated expression of Nanog, which in turn prevents FGF-induced neuroectoderm differentiation [112]. In addition, several studies indicated that the Oct4/Sox2 complex was directly bound to the Nanog promoter to regulate target gene expression [106, 107, 113]. Genomic studies have revealed that Oct4, Sox2, and Nanog frequently bind the same regulatory regions in undifferentiated mouse and human ESCs, and that these binding sites are often in close proximity to one another[113-116]. These results indicate that Oct4, Sox2, and Nanog may physically interact with each other and coordinately regulate target genes in some cases. Additionally, Goke and colleagues reported that combinatorial binding sites of the Oct4/Sox2/Nanog were more conserved between mouse and human ES cells than individual binding sites were [113, 114, 117-119].

4. Summary

Understanding the molecular mechanism of pluripotency can greatly expand our knowledge of ES cell biology and facilitate future stem cell clinical applications. In the past few years, we have seen tremendous advances in understanding ES cell pluripotency. Although mouse ES cells and conventional human ES cells require distinct signaling pathways to maintain pluripotency, they display similar gene expression profiles, activities of transcription factors (such as Oct4, Nanog and Sox2) and transcription factor networks. Our understanding of pluripotency has been further expanded by the advent of iPS cells and the very recent discovery that conventional human ES cells are more equivalent to mouse EpiSCs, but rather "naïve state" of mouse ES cells. Nevertheless, our knowledge of the molecular mechanisms of ES cell pluripotency is still very limited. For instance, it remains unknown how growth factors establish and control transcriptional networks to regulate pluripoency and how ES cells respond so precisely to exogenous cues. Given the rapid advance in ES cell biology, we anticipate the molecular mechanisms underlying pluripotency of ES cells will soon be uncovered and pluripotent stem cells, such as ES cells and iPS cells, will be widely used for clinical applications in the near future.

Acknowledgements

This work is funded by the seed fund of the Western University of Health Sciences.

http://dx.doi.org/10.5772/54365

Author details

Guofeng Han[1], Hongtao Wang[2] and Jijun Hao[3*]

*Address all correspondence to: jhao@westernu.edu

1 Department of Nephrology, Shanghai Jimin Hospital, Shanghai, China

2 Department of Burn and Cutaneous Surgery, XiJing Hospital, the Fourth Military Medical University, Xi'an, China

3 College of Veterinary Medicine, Western University of Health Sciences, Western University of Health Sciences, Pomona, CA, USA

References

[1] Evans MJ, Kaufman MH. Establishment in culture of pluripotential cells from mouse embryos. Nature. 1981;292(5819):154-6.

[2] Thomson JA, Itskovitz-Eldor J, Shapiro SS, Waknitz MA, Swiergiel JJ, Marshall VS, et al. Embryonic stem cell lines derived from human blastocysts. Science. 1998;282(5391):1145-7.

[3] Takahashi K, Yamanaka S. Induction of pluripotent stem cells from mouse embryonic and adult fibroblast cultures by defined factors. Cell. 2006;126(4):663-76.

[4] Smith AG, Heath JK, Donaldson DD, Wong GG, Moreau J, Stahl M, et al. Inhibition of pluripotential embryonic stem cell differentiation by purified polypeptides. Nature. 1988;336(6200):688-90.

[5] Niwa H, Burdon T, Chambers I, Smith A. Self-renewal of pluripotent embryonic stem cells is mediated via activation of STAT3. Genes Dev. 1998;12(13):2048-60.

[6] Stahl N, Farruggella TJ, Boulton TG, Zhong Z, Darnell JE, Jr., Yancopoulos GD. Choice of STATs and other substrates specified by modular tyrosine-based motifs in cytokine receptors. Science. 1995;267(5202):1349-53.

[7] Hemmann U, Gerhartz C, Heesel B, Sasse J, Kurapkat G, Grotzinger J, et al. Differential activation of acute phase response factor/STAT3 and Stat1 via the cytoplasmic domain of the interleukin 6 signal transducer gp130. II. Src homology SH2 domains define the specificity of stat factor activation. J Biol Chem. 1996;271(22):12999-3007.

[8] Gerhartz C, Heesel B, Sasse J, Hemmann U, Landgraf C, Schneider-Mergener J, et al. Differential activation of acute phase response factor/STAT3 and STAT1 via the cytoplasmic domain of the interleukin 6 signal transducer gp130. I. Definition of a novel phosphotyrosine motif mediating STAT1 activation. J Biol Chem. 1996;271(22):12991-8.

[9] Ihle JN, Kerr IM. Jaks and Stats in signaling by the cytokine receptor superfamily. Trends Genet. 1995;11(2):69-74.

[10] Auernhammer CJ, Melmed S. Leukemia-inhibitory factor-neuroimmune modulator of endocrine function. Endocr Rev. 2000;21(3):313-45.

[11] Reich NC, Liu L. Tracking STAT nuclear traffic. Nat Rev Immunol. 2006;6(8):602-12.

[12] Chen X, Xu H, Yuan P, Fang F, Huss M, Vega VB, et al. Integration of external signaling pathways with the core transcriptional network in embryonic stem cells. Cell. 2008;133(6):1106-17.

[13] Kidder BL, Yang J, Palmer S. STAT3 and c-Myc genome-wide promoter occupancy in embryonic stem cells. PLoS One. 2008;3(12):e3932.

[14] Bourillot PY, Aksoy I, Schreiber V, Wianny F, Schulz H, Hummel O, et al. Novel STAT3 target genes exert distinct roles in the inhibition of mesoderm and endoderm differentiation in cooperation with Nanog. Stem Cells. 2009;27(8):1760-71.

[15] Daheron L, Opitz SL, Zaehres H, Lensch MW, Andrews PW, Itskovitz-Eldor J, et al. LIF/STAT3 signaling fails to maintain self-renewal of human embryonic stem cells. Stem Cells. 2004;22(5):770-8.

[16] Wrana JL, Attisano L, Carcamo J, Zentella A, Doody J, Laiho M, et al. Tgf-Beta Signals through a Heteromeric Protein-Kinase Receptor Complex. Cell. 1992;71(6): 1003-14.

[17] Schmierer B, Hill CS. TGF beta-SMAD signal transduction: molecular specificity and functional flexibility. Nat Rev Mol Cell Bio. 2007;8(12):970-82.

[18] Sebald W, Nickel J, Zhang JL, Mueller TD. Molecular recognition in bone morphogenetic protein (BMP)/receptor interaction. Biol Chem. 2004;385(8):697-710.

[19] Ying QL, Nichols J, Chambers I, Smith A. BMP induction of Id proteins suppresses differentiation and sustains embryonic stem cell self-renewal in collaboration with STAT3. Cell. 2003;115(3):281-92.

[20] Xu RH, Chen X, Li DS, Li R, Addicks GC, Glennon C, et al. BMP4 initiates human embryonic stem cell differentiation to trophoblast. Nat Biotechnol. 2002;20(12): 1261-4.

[21] Xu RH, Peck RM, Li DS, Feng X, Ludwig T, Thomson JA. Basic FGF and suppression of BMP signaling sustain undifferentiated proliferation of human ES cells. Nat Methods. 2005;2(3):185-90.

[22] Yu PB, Hong CC, Sachidanandan C, Babitt JL, Deng DY, Hoyng SA, et al. Dorsomorphin inhibits BMP signals required for embryogenesis and iron metabolism. Nat Chem Biol. 2008;4(1):33-41.

[23] Hao J, Ho JN, Lewis JA, Karim KA, Daniels RN, Gentry PR, et al. In vivo structure-activity relationship study of dorsomorphin analogues identifies selective VEGF and BMP inhibitors. ACS Chem Biol. 2010;5(2):245-53.

[24] Gonzalez R, Lee JW, Snyder EY, Schultz PG. Dorsomorphin promotes human embryonic stem cell self-renewal. Angew Chem Int Ed Engl. 2011;50(15):3439-41.

[25] Hao J, Sawyer DB, Hatzopoulos AK, Hong CC. Recent Progress on Chemical Biology of Pluripotent Stem Cell Self-renewal, Reprogramming and Cardiomyogenesis. Rec Pat Regen Med. 2011;1(3):263-74.

[26] Sato N, Sanjuan IM, Heke M, Uchida M, Naef F, Brivanlou AH. Molecular signature of human embryonic stem cells and its comparison with the mouse. Dev Biol. 2003;260(2):404-13.

[27] Beattie GM, Lopez AD, Bucay N, Hinton A, Firpo MT, King CC, et al. Activin A maintains pluripotency of human embryonic stem cells in the absence of feeder layers. Stem Cells. 2005;23(4):489-95.

[28] James D, Levine AJ, Besser D, Hemmati-Brivanlou A. TGF beta/activin/nodal signaling is necessary for the maintenance of pluripotency in human embryonic stem. Development. 2005;132(6):1273-82.

[29] Xiao L, Yuan X, Sharkis SJ. Activin A maintains self-renewal and regulates fibroblast growth factor, Wnt, and bone morphogenic protein pathways in human embryonic stem cells. Stem Cells. 2006;24(6):1476-86.

[30] Vallier L, Alexander M, Pedersen RA. Activin/Nodal and FGF pathways cooperate to maintain pluripotency of human embryonic stem cells. J Cell Sci. 2005;118(Pt 19): 4495-509.

[31] Xu RH, Sampsell-Barron TL, Gu F, Root S, Peck RM, Pan GJ, et al. NANOG is a direct target of TGF beta/Activin-mediated SMAD signaling in human ESCs. Cell Stem Cell. 2008;3(2):196-206.

[32] Vallier L, Mendjan S, Brown S, Chng Z, Teo A, Smithers LE, et al. Activin/Nodal signalling maintains pluripotency by controlling Nanog expression. Development. 2009;136(8):1339-49.

[33] Ogawa K, Saito A, Matsui H, Suzuki H, Ohtsuka S, Shimosato D, et al. Activin-Nodal signaling is involved in propagation of mouse embryonic stem cells. J Cell Sci. 2007;120(Pt 1):55-65.

[34] Levine AJ, Brivanlou AH. GDF3, a BMP inhibitor, regulates cell fate in stem cells and early embryos. Development. 2006;133(2):209-16.

[35] Amit M, Carpenter MK, Inokuma MS, Chiu CP, Harris CP, Waknitz MA, et al. Clonally derived human embryonic stem cell lines maintain pluripotency and proliferative potential for prolonged periods of culture. Dev Biol. 2000;227(2):271-8.

[36] Xu C, Inokuma MS, Denham J, Golds K, Kundu P, Gold JD, et al. Feeder-free growth of undifferentiated human embryonic stem cells. Nat Biotechnol. 2001;19(10):971-4.

[37] Brandenberger R, Wei H, Zhang S, Lei S, Murage J, Fisk GJ, et al. Transcriptome characterization elucidates signaling networks that control human ES cell growth and differentiation. Nat Biotechnol. 2004;22(6):707-16.

[38] Dvash T, Mayshar Y, Darr H, McElhaney M, Barker D, Yanuka O, et al. Temporal gene expression during differentiation of human embryonic stem cells and embryoid bodies. Hum Reprod. 2004;19(12):2875-83.

[39] Ginis I, Luo YQ, Miura T, Thies S, Brandenberger R, Gerecht-Nir S, et al. Differences between human and mouse embryonic stem cells. Dev Biol. 2004;269(2):360-80.

[40] Kim SJ, Cheon SH, Yoo SJ, Kwon J, Park JH, Kim CG, et al. Contribution of the PI3K/Akt/PKB signal pathway to maintenance of self-renewal in human embryonic stem cells (Retracted Article. See vol 580, pg 1529, 2006). Febs Lett. 2005;579(2):534-40.

[41] Dvorak P. Basic fibroblast growth factor and its receptors in human embryonic stem cells. Folia Histochem Cyto. 2005;43(4):203-8.

[42] Dvorak P, Dvorakova D, Koskova S, Vodinska M, Najvirtova M, Krekac D, et al. Expression and potential role of fibroblast growth factor 2 and its receptors in human embryonic stem cells. Stem Cells. 2005;23(8):1200-11.

[43] Dailey L, Ambrosetti D, Mansukhani A, Basilico C. Mechanisms underlying differential responses to FGF signaling. Cytokine Growth Factor Rev. 2005;16(2):233-47.

[44] Kang HB, Kim JS, Kwon HJ, Nam KH, Youn HS, Sok DE, et al. Basic fibroblast growth factor activates ERK and induces c-fos in human embryonic stem cell line MizhES1. Stem Cells Dev. 2005;14(4):395-401.

[45] Li J, Wang GW, Wang CY, Zhao Y, Zhang H, Tan ZJ, et al. MEK/ERK signaling contributes to the maintenance of human embryonic stem cell self-renewal. Differentiation. 2007;75(4):299-307.

[46] Bendall SC, Stewart MH, Menendez P, George D, Vijayaragavan K, Werbowetski-Ogilvie T, et al. IGF and FGF cooperatively establish the regulatory stem cell niche of pluripotent human cells in vitro. Nature. 2007;448(7157):1015-21.

[47] Greber B, Lehrach H, Adjaye J. Fibroblast growth factor 2 modulates transforming growth factor beta signaling in mouse embryonic fibroblasts and human ESCs (hESCs) to support hESC self-renewal. Stem Cells. 2007;25(2):455-64.

[48] Wang XF, Lin G, Martins-Taylor K, Zeng H, Xu RH. Inhibition of Caspase-mediated Anoikis Is Critical for Basic Fibroblast Growth Factor-sustained Culture of Human Pluripotent Stem Cells. Journal of Biological Chemistry. 2009;284(49):34054-64.

[49] Eiselleova L, Matulka K, Kriz V, Kunova M, Schmidtova Z, Neradil J, et al. A Complex Role for FGF-2 in Self-Renewal, Survival, and Adhesion of Human Embryonic Stem Cells. Stem Cells. 2009;27(8):1847-57.

[50] Burdon T, Stracey C, Chambers I, Nichols J, Smith A. Suppression of SHP-2 and ERK signalling promotes self-renewal of mouse embryonic stem cells. Dev Biol. 1999;210(1):30-43.

[51] Kunath T, Saba-El-Leil MK, Almousailleakh M, Wray J, Meloche S, Smith A. FGF stimulation of the Erk1/2 signalling cascade triggers transition of pluripotent embryonic stem cells from self-renewal to lineage commitment. Development. 2007;134(16): 2895-902.

[52] Stavridis MP, Lunn JS, Collins BJ, Storey KG. A discrete period of FGF-induced Erk1/2 signalling is required for vertebrate neural specification. Development. 2007;134(16):2889-94.

[53] Batlle-Morera L, Smith A, Nichols J. Parameters Influencing Derivation of Embryonic Stem Cells From Murine Embryos. Genesis. 2008;46(12):758-67.

[54] Qi X, Li TG, Hao J, Hu J, Wang J, Simmons H, et al. BMP4 supports self-renewal of embryonic stem cells by inhibiting mitogen-activated protein kinase pathways. Proc Natl Acad Sci U S A. 2004;101(16):6027-32.

[55] Burdon T, Stracey C, Chambers I, Nichols J, Smith A. Suppression of SHP-2 and ERK signalling promotes self-renewal of mouse embryonic stem cells. Dev Biol. 1999;210(1):30-43.

[56] Paling NR, Wheadon H, Bone HK, Welham MJ. Regulation of embryonic stem cell self-renewal by phosphoinositide 3-kinase-dependent signaling. J Biol Chem. 2004;279(46):48063-70.

[57] Chen S, Do JT, Zhang Q, Yao S, Yan F, Peters EC, et al. Self-renewal of embryonic stem cells by a small molecule. Proc Natl Acad Sci U S A. 2006;103(46):17266-71.

[58] Chen S, Ding S, Yan F, Schultz P, inventors; Compounds that maintain pluripotency of embryonic stem cells patent US20100234400. 2010.

[59] Ying QL, Wray J, Nichols J, Batlle-Morera L, Doble B, Woodgett J, et al. The ground state of embryonic stem cell self-renewal. Nature. 2008;453(7194):519-23.

[60] Smith A, Ying Q, inventors; Culture medium containing kinase inhibitors, and uses thereof patent US20080014638. 2008.

[61] Aubert J, Dunstan H, Chambers I, Smith A. Functional gene screening in embryonic stem cells implicates Wnt antagonism in neural differentiation. Nat Biotechnol. 2002;20(12):1240-5.

[62] Sato N, Meijer L, Skaltsounis L, Greengard P, Brivanlou AH. Maintenance of pluripotency in human and mouse embryonic stem cells through activation of Wnt signaling by a pharmacological GSK-3-specific inhibitor. Nat Med. 2004;10(1):55-63.

[63] Ogawa K, Nishinakamura R, Iwamatsu Y, Shimosato D, Niwa H. Synergistic action of Wnt and LIF in maintaining pluripotency of mouse ES cells. Biochem Biophys Res Commun. 2006;343(1):159-66.

[64] Kielman MF, Rindapaa M, Gaspar C, van Poppel N, Breukel C, van Leeuwen S, et al. Apc modulates embryonic stem-cell differentiation by controlling the dosage of beta-catenin signaling. Nat Genet. 2002;32(4):594-605.

[65] Pereira L, Yi F, Merrill BJ. Repression of Nanog gene transcription by Tcf3 limits embryonic stem cell self-renewal. Mol Cell Biol. 2006;26(20):7479-91.

[66] Takao Y, Yokota T, Koide H. Beta-catenin up-regulates Nanog expression through interaction with Oct-3/4 in embryonic stem cells. Biochem Biophys Res Commun. 2007;353(3):699-705.

[67] Wray J, Kalkan T, Smith AG. The ground state of pluripotency. Biochem Soc Trans. 2010;38(4):1027-32.

[68] Bao SQ, Tang FC, Li XH, Hayashi K, Gillich A, Lao KQ, et al. Epigenetic reversion of post-implantation epiblast to pluripotent embryonic stem cells. Nature. 2009;461(7268):1292-5.

[69] Bao S, Tang F, Li X, Hayashi K, Gillich A, Lao K, et al. Epigenetic reversion of post-implantation epiblast to pluripotent embryonic stem cells. Nature. 2009;461(7268):1292-5.

[70] Hanna J, Cheng AW, Saha K, Kim J, Lengner CJ, Soldner F, et al. Human embryonic stem cells with biological and epigenetic characteristics similar to those of mouse ESCs. Proc Natl Acad Sci U S A. 2010;107(20):9222-7.

[71] Nichols J, Smith A. Naive and primed pluripotent states. Cell Stem Cell. 2009;4(6):487-92.

[72] Silva J, Barrandon O, Nichols J, Kawaguchi J, Theunissen TW, Smith A. Promotion of reprogramming to ground state pluripotency by signal inhibition. PLoS Biol. 2008;6(10):e253.

[73] Silva J, Nichols J, Theunissen TW, Guo G, van Oosten AL, Barrandon O, et al. Nanog is the gateway to the pluripotent ground state. Cell. 2009;138(4):722-37.

[74] Tesar PJ, Chenoweth JG, Brook FA, Davies TJ, Evans EP, Mack DL, et al. New cell lines from mouse epiblast share defining features with human embryonic stem cells. Nature. 2007;448(7150):196-9.

[75] Pera MF, Tam PP. Extrinsic regulation of pluripotent stem cells. Nature. 2010;465(7299):713-20.

[76] Tchieu J, Kuoy E, Chin MH, Trinh H, Patterson M, Sherman SP, et al. Female human iPSCs retain an inactive X chromosome. Cell Stem Cell. 2010;7(3):329-42.

[77] Hanna JH, Saha K, Jaenisch R. Pluripotency and cellular reprogramming: facts, hypotheses, unresolved issues. Cell. 2010;143(4):508-25.

[78] Lengner CJ, Gimelbrant AA, Erwin JA, Cheng AW, Guenther MG, Welstead GG, et al. Derivation of pre-X inactivation human embryonic stem cells under physiological oxygen concentrations. Cell. 2010;141(5):872-83.

[79] Jaenisch R, Young R. Stem cells, the molecular circuitry of pluripotency and nuclear reprogramming. Cell. 2008;132(4):567-82.

[80] Hollnagel A, Oehlmann V, Heymer J, Ruther U, Nordheim A. Id genes are direct targets of bone morphogenetic protein induction in embryonic stem cells. Journal of Biological Chemistry. 1999;274(28):19838-45.

[81] Ivanova NB, Dimos JT, Schaniel C, Hackney JA, Moore KA, Lemischka IR. A stem cell molecular signature. Science. 2002;298(5593):601-4.

[82] Ramalho-Santos M, Yoon S, Matsuzaki Y, Mulligan RC, Melton DA. "Stemness": transcriptional profiling of embryonic and adult stem cells. Science. 2002;298(5593): 597-600.

[83] Sato N, Sanjuan IM, Heke M, Uchida M, Naef F, Brivanlou AH. Molecular signature of human embryonic stem cells and its comparison with the mouse. Dev Biol. 2003;260(2):404-13.

[84] Nichols J, Zevnik B, Anastassiadis K, Niwa H, Klewe-Nebenius D, Chambers I, et al. Formation of pluripotent stem cells in the mammalian embryo depends on the POU transcription factor Oct4. Cell. 1998;95(3):379-91.

[85] Chambers I, Colby D, Robertson M, Nichols J, Lee S, Tweedie S, et al. Functional expression cloning of Nanog, a pluripotency sustaining factor in embryonic stem cells. Cell. 2003;113(5):643-55.

[86] Mitsui K, Tokuzawa Y, Itoh H, Segawa K, Murakami M, Takahashi K, et al. The homeoprotein Nanog is required for maintenance of pluripotency in mouse epiblast and ES cells. Cell. 2003;113(5):631-42.

[87] Liu Y, Labosky PA. Regulation of Embryonic Stem Cell Self-Renewal and Pluripotency by Foxd3. Stem Cells. 2008;26(10):2475-84.

[88] Hanna LA, Foreman RK, Tarasenko IA, Kessler DS, Labosky PA. Requirement for Foxd3 in maintaining pluripotent cells of the early mouse embryo. Gene Dev. 2002;16(20):2650-61.

[89] Rosner MH, Vigano MA, Ozato K, Timmons PM, Poirier F, Rigby PW, et al. A POU-domain transcription factor in early stem cells and germ cells of the mammalian embryo. Nature. 1990;345(6277):686-92.

[90] Scholer HR, Ruppert S, Suzuki N, Chowdhury K, Gruss P. New type of POU domain in germ line-specific protein Oct-4. Nature. 1990;344(6265):435-9.

[91] Okamoto K, Okazawa H, Okuda A, Sakai M, Muramatsu M, Hamada H. A Novel Octamer Binding Transcription Factor Is Differentially Expressed in Mouse Embryonic-Cells. Cell. 1990;60(3):461-72.

[92] Yeom YI, Ha HS, Balling R, Scholer HR, Artzt K. Structure, expression and chromosomal location of the Oct-4 gene. Mech Dev. 1991;35(3):171-9.

[93] Nishimoto M, Fukushima A, Okuda A, Muramatsu M. The gene for the embryonic stem cell coactivator UTF1 carries a regulatory element which selectively interacts with a complex composed of Oct-3/4 and Sox-2. Mol Cell Biol. 1999;19(8):5453-65.

[94] Tomioka M, Nishimoto M, Miyagi S, Katayanagi T, Fukui N, Niwa H, et al. Identification of Sox-2 regulatory region which is under the control of Oct-3/4-Sox-2 complex. Nucleic Acids Res. 2002;30(14):3202-13.

[95] Zeng XM, Miura T, Luo YQ, Bhattacharya B, Condie B, Chen J, et al. Properties of pluripotent human embryonic stem cells BG01 and BG02. Stem Cells. 2004;22(3): 292-312.

[96] Niwa H, Toyooka T, Shimosato D, Strumpf D, Takahashi K, Yagi R, et al. Interaction between Oct3/4 and Cdx2 determines trophectoderm differentiation. Cell. 2005;123(5):917-29.

[97] Ovitt CE, Scholer HR. The molecular biology of Oct-4 in the early mouse embryo. Mol Hum Reprod. 1998;4(11):1021-31.

[98] Pan GJ, Chang ZY, Scholer HR, Pei DQ. Stem cell pluripotency and transcription factor Oct4. Cell Res. 2002;12(5-6):321-9.

[99] Yeom YI, Fuhrmann G, Ovitt CE, Brehm A, Ohbo K, Gross M, et al. Germline regulatory element of Oct-4 specific for the totipotent cycle of embryonal cells. Development. 1996;122(3):881-94.

[100] Niwa H. Molecular mechanism to maintain stem cell renewal of ES cells. Cell Struct Funct. 2001;26(3):137-48.

[101] Li M, Pevny L, Lovell-Badge R, Smith A. Generation of purified neural precursors from embryonic stem cells by lineage selection. Curr Biol. 1998;8(17):971-4.

[102] Avilion AA, Nicolis SK, Pevny LH, Perez L, Vivian N, Lovell-Badge R. Multipotent cell lineages in early mouse development depend on SOX2 function. Gene Dev. 2003;17(1):126-40.

[103] Zappone MV, Galli R, Catena R, Meani N, De Biasi S, Mattei E, et al. Sox2 regulatory sequences direct expression of a (beta)-geo transgene to telencephalic neural stem cells and precursors of the mouse embryo, revealing regionalization of gene expression in CNS stem cells. Development. 2000;127(11):2367-82.

[104] Okumura-Nakanishi S, Saito M, Niwa H, Ishikawa F. Oct-3/4 and Sox2 regulate Oct-3/4 gene in embryonic stem cells. Journal of Biological Chemistry. 2005;280(7): 5307-17.

[105] Tokuzawa Y, Kaiho E, Maruyama M, Takahashi K, Mitsui K, Maeda M, et al. Fbx15 is a novel target of Oct3/4 but is dispensable for embryonic stem cell self-renewal and mouse development. Mol Cell Biol. 2003;23(8):2699-708.

[106] Rodda DJ, Chew JL, Lim LH, Loh YH, Wang B, Ng HH, et al. Transcriptional regulation of Nanog by Oct4 and Sox2. Journal of Biological Chemistry. 2005;280(26):24731-7.

[107] Kuroda T, Tada M, Kubota H, Kimura H, Hatano S, Suemori H, et al. Octamer and Sox elements are required for transcriptional cis regulation of Nanog gene expression. Mol Cell Biol. 2005;25(6):2475-85.

[108] Nakatake Y, Fukui N, Iwamatsu Y, Masui S, Takahashi K, Yagi R, et al. Klf4 cooperates with Oct3/4 and Sox2 to activate the Lefty1 core promoter in embryonic stem cells. Mol Cell Biol. 2006;26(20):7772-82.

[109] Ivanova N, Dobrin R, Lu R, Kotenko I, Levorse J, DeCoste C, et al. Dissecting self-renewal in stem cells with RNA interference. Nature. 2006;442(7102):533-8.

[110] Chambers I, Silva J, Colby D, Nichols J, Nijmeijer B, Robertson M, et al. Nanog safeguards pluripotency and mediates germline development. Nature. 2007;450(7173):1230-4.

[111] Suzuki A, Raya A, Kawakami Y, Morita M, Matsui T, Nakashima K, et al. Nanog binds to Smad1 and blocks bone morphogenetic protein-induced differentiation of embryonic stem cells. Proc Natl Acad Sci U S A. 2006;103(27):10294-9.

[112] Vallier L, Mendjan S, Brown S, Chng Z, Teo A, Smithers LE, et al. Activin/Nodal signalling maintains pluripotency by controlling Nanog expression. Development. 2009;136(8):1339-49.

[113] Boyer LA, Lee TI, Cole MF, Johnstone SE, Levine SS, Zucker JP, et al. Core transcriptional regulatory circuitry in human embryonic stem cells. Cell. 2005;122(6):947-56.

[114] Loh YH, Wu Q, Chew JL, Vega VB, Zhang W, Chen X, et al. The Oct4 and Nanog transcription network regulates pluripotency in mouse embryonic stem cells. Nat Genet. 2006;38(4):431-40.

[115] Mathur D, Danford TW, Boyer LA, Young RA, Gifford DK, Jaenisch R. Analysis of the mouse embryonic stem cell regulatory networks obtained by ChIP-chip and ChIP-PET. Genome Biol. 2008;9(8).

[116] Sharov AA, Masui S, Sharova LV, Piao Y, Aiba K, Matoba R, et al. Identification of Pou5f1, Sox2, and Nanog downstream target genes with statistical confidence by applying a novel algorithm to time course microarray and genome-wide chromatin immunoprecipitation data. BMC Genomics. 2008;9:269.

[117] Masui S, Nakatake Y, Toyooka Y, Shimosato D, Yagi R, Takahashi K, et al. Pluripotency governed by Sox2 via regulation of Oct3/4 expression in mouse embryonic stem cells. Nat Cell Biol. 2007;9(6):625-35.

[118] Wang J, Rao S, Chu J, Shen X, Levasseur DN, Theunissen TW, et al. A protein interaction network for pluripotency of embryonic stem cells. Nature. 2006;444(7117):364-8.

[119] Goke J, Jung M, Behrens S, Chavez L, O'Keeffe S, Timmermann B, et al. Combinatorial Binding in Human and Mouse Embryonic Stem Cells Identifies Conserved Enhancers Active in Early Embryonic Development. Plos Comput Biol. 2011;7(12).

Chapter 14

Epigenetic Instability in Embryonic Stem Cells

Takuro Horii and Izuho Hatada

Additional information is available at the end of the chapter

http://dx.doi.org/10.5772/54367

1. Introduction

Embryonic stem (ES) cells constitute a very important tool for regenerative medicine today. Human ES cells, in particular, are almost all derived from embryos obtained by *in vitro* fertilization (IVF) followed by *in vitro* culture (IVC); however, such *in vitro* manipulated embryos often show epigenetic abnormalities in imprinted genes that can lead to the development of various diseases. We recently reported that epigenetic differences occurred between ES cells derived from *in vivo* developed embryos (Vivo ES) and ES cells derived from *in vitro* manipulated embryos (Vitro ES) [1]. In addition, we found that the DNA methylation state of uniparental and somatic cell nuclear transfer (SCNT) ES cells exhibits epigenetic instability during *in vitro* culture [2]. In this chapter, we review studies that have examined the epigenetic instability of ES cells during generation and maintenance cultures, and discuss the candidate factors that may be responsible for this epigenetic instability.

2. Epigenetic regulation by DNA methylation

In vertebrate genomic DNA, the 5' cytosine residues in CpG sequences are often methylated [3]. DNA methylation plays an essential role in the normal development of mammalian embryos by regulating gene expression through genomic imprinting and X chromosome inactivation, and confers genomic stability [4-7]. In this chapter, we focus primarily on genomic imprinting, which is the preferential silencing of one of the parental alleles of a gene by epigenetic DNA methylation since epigenetic modifications to some imprinted genes cause diseases such as Beckwith-Wiedemann syndrome and Prader-Willie syndrome. For example, the expression level of the *H19* imprinted gene is regulated by an upstream differentially methylated region (DMR), and epigenetic alterations to the DMR result in Beckwith-Wiedemann syndrome [8-10]. The *H19* mRNA is transcribed from the unmethylated maternal allele but is not transcribed from the methylated paternal allele (Fig. 1). In contrast, DMRs of *Peg1* (Mest), *Snrpn* and *Igf2r* are methylated in the maternal allele and unmethylated in the paternal allele. Genomic imprinting is very stable except for the period when the reprogramming of genomic imprinting takes place in germline cells [11]. For the establishment and maintenance of DNA methylation, the cytosine-guanine (CpG) DNA methyltransferases (Dnmts), Dnmt1, Dnmt3a, and Dnmt3b, are the main factors that coordinately regulate CpG methylation in the genome [12-14]. Dnmt1 is involved in maintenance activity, while Dnmt3a and Dnmt3b are responsible primarily for the creation of new methylation patterns.

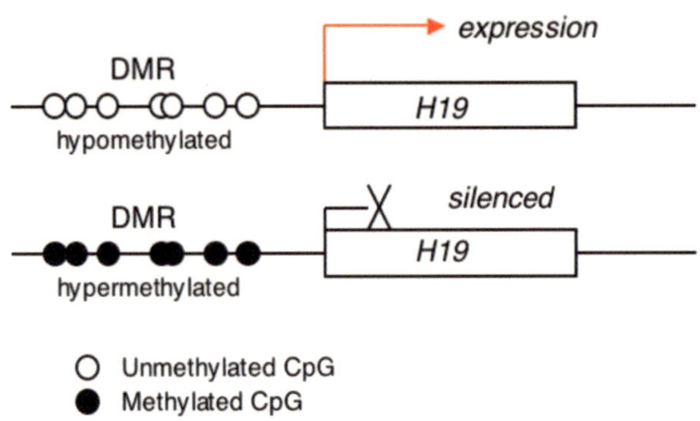

Figure 1. Regulation of gene expression in the *H19* imprinted gene.

3. Epigenetic instability in preimplantation embryos

In general, ES cells, especially human ES cells, are generated from blastocyst stage embryos that are produced by *in vitro* manipulations such as IVF and IVC. However, *in vitro* manipulated embryos may already possess epigenetic abnormalities because the culture conditions of fertilized embryos can influence the methylation state. For example, a sub-optimal culture medium (e.g., Whitten's medium) can cause aberrant genomic imprinting of the *H19* gene [15], and culture medium supplemented with fetal calf serum alters mRNA expression of imprinted genes [16]. Our recent study suggests that altered DNA methylation due to IVC conditions occurs not only in imprinted genes but also in genome-wide repetitive sequences, such as major and minor satellite sequences [17]. Thus, alteration of DNA methylation can occur in response to various factors, from the moment when embryos are collected from the oviducts or uterus.

4. Epigenetic instability in ES cells during prolonged culture

ES cells are established from the inner cell mass (ICM) of blastocyst stage embryos [18,19]. Once ES cell lines are established, they can be maintained for long periods of time and used for sever-

http://dx.doi.org/10.5772/54367

al applications. However, ES cells lose their pluripotency during prolonged *in vitro* culture [20]. Several studies indicate that the accumulation of epigenetic alterations over time is correlated with the loss of pluripotency in ES cells. Dean *et al.* reported that epigenetic alterations that occur in ES cells persist to later developmental stages and are associated with aberrant phenotypes in completely ES cell-derived mice [21]. Humpherys *et al.* show that variation in imprinted gene expression is observed in most cloned mice derived from ES cell donors, even those derived from ES cells of the same subclone [22]. Such epigenetic drift of imprinted genes was also observed in our experiments during prolonged culture of mouse ES cells (Fig. 2): DNA methylation of four imprinted genes, *Peg1, Snrpn, Igf2r* and *H19,* was unstable during cell culture (P3-30), even in the same cell line, over time. Minoguchi and Iba reported that retroviral DNA that is introduced into mouse ES cells is progressively silenced by DNA methylation; however, a substantial amount of retroviral DNA is reversibly reactivated by DNA demethylation [23]. Such epigenetic drift has also been observed in human ES cells, depending on the method of establishment and the culture conditions [24].

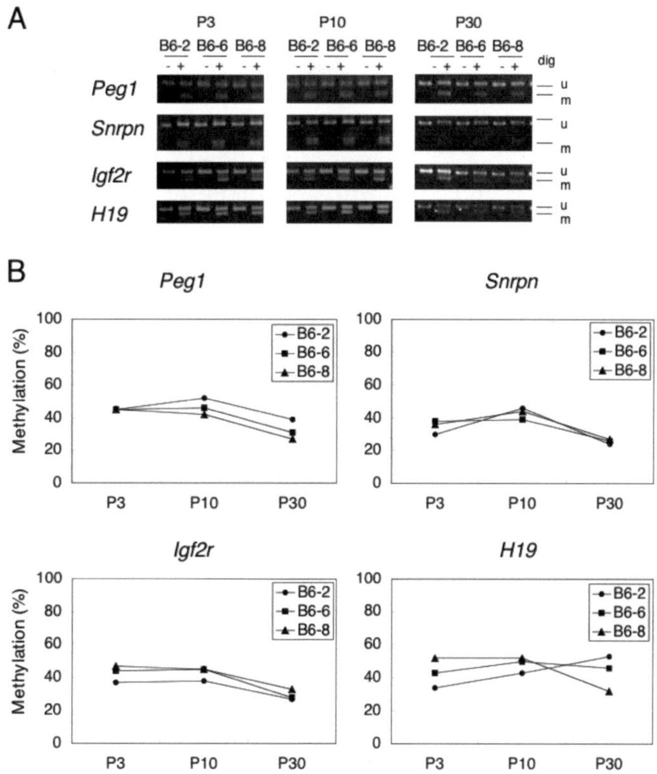

Figure 2. Epigenetic drift of imprinting methylations in fertilized embryo-derived ES cells. A. Combined bisulfite restriction analysis (COBRA) was conducted for three fertilized embryo-derived ES cell lines (B6-2, B6-6 and B6-8) during prolonged *in vitro* culture (P3, P10 and P30). The maternally methylated imprinted genes *Peg1*, *Snrpn* and *Igf2r*, and the paternally methylated imprinted gene, *H19*, were examined. B. Summary of imprinting methylations during prolonged culture of ES cells. dig, digestion by restriction enzymes; u, unmethylated PCR products; m, methylated PCR products.

5. Epigenetic differences between male and female ES cells

Large differences in epigenetic drift have been observed between male (XY) and female (XX) mouse ES cells. Global demethylation, including imprinted genes and satellite repeats, occurred more frequently in female ES cell lines compared to male ES cell lines [21, 25]. This global demethylation reflects the number and state of X chromosomes in ES cells. In general, both X chromosomes are active in female ES cells, whereas male ES cells have only one active X chromosome. The X chromosome state in female ES cells is thought to lead to downregulation of DNA methyltransferases (Dnmt3a and Dnmt3b) and, ultimately, to global hypomethylation [25]. Thus, DNA methylation of imprinted genes and repetitive sequences are gained or lost at high rates even in clonal populations of ES cells, and these alterations may have deleterious effects on phenotypes of ES cell-derived animals or tissues.

6. Epigenetic differences between *vivo* and *vitro* ES cells

6.1. Methylation state of *vivo* and *vitro* ES cells

In human ES cells, several studies have recently provided evidence for the efficient induction of endoderm, mesoderm, and ectoderm, and many of their downstream derivatives [26], and these reports offer broad possibilities for regenerative medicine. However, all human ES cell lines are established from *in vitro* manipulated embryos that often show abnormal genomic imprinting, which can lead to an increase in the frequency of diseases. Therefore, we have compared the methylation state of imprinted genes and the gene expression patterns of both Vivo and Vitro ES cell lines in mice [1].

Although the genomic imprinting is maintained during preimplantation development, normal imprinting can occasionally be disrupted in preimplantation embryos during IVC, resulting in biallelic expression of the *H19* gene [15,27]. To investigate whether Vitro ES cells take on abnormal imprinting from IVC blastocysts, we performed methylation analysis of the *H19* DMR for early passage (P2) cells (Fig. 3). COBRA analysis shows that the *H19* DMR is significantly demethylated in Vitro ES cells compared to Vivo ES cells. The *Igf2r* DMR2 also showed significant differences among Vitro vs. Vivo ES cells, but significant differences in the methylation of *Snrpn* and the major satellite repeats were not detected.

In additional experiments, both Vivo and Vitro ES cells were passaged several more times, and the methylation state of imprinted genes and satellite repeats was investigated at later passages (P5) (Fig. 3). Results from COBRA analysis at P5 showed no significant differences between Vivo and Vitro ES cells. Even Vivo ES cells exhibited highly demethylated alleles. In contrast, some Vitro ES cells had an almost normally methylated allele. This result indicates that the methylation state of ES cells at later passages depends more on the character of the individual cell lines than on the origin of the ES cells.

http://dx.doi.org/10.5772/54367

6.2. Gene expression of *vivo* and *vitro* ES cells

We assessed gene expression patterns in ES cells at early and late passages by quantitative real-time RT-PCR. The expression of *Oct3/4* mRNA, a pluripotent cell marker, was significantly higher in early passage Vivo ES cells than in Vitro ES cells, whereas other pluripotent marker genes, *Nanog* and *Stella*, showed no significant differences in expression levels between the two types of ES cells. Among the methylation-related genes, mRNA expression of the *de novo* DNA methyltransferase, Dnmt3b, was significantly higher in Vivo ES cells. Expression of growth arrest and DNA damage-inducible protein 45 beta (Gadd45*b*), which is a putative demethylation factor [28,29], is higher in Vitro ES cells. Thus, mRNA expression patterns of several methylation-related genes tended to shift, resulting in the promotion of demethylation and the inhibition of methylation in Vitro ES cells. In contrast, at later passages, no significant differences between Vivo and Vitro ES cells were found with respect to the pluripotent marker genes and methylation-related genes that were examined.

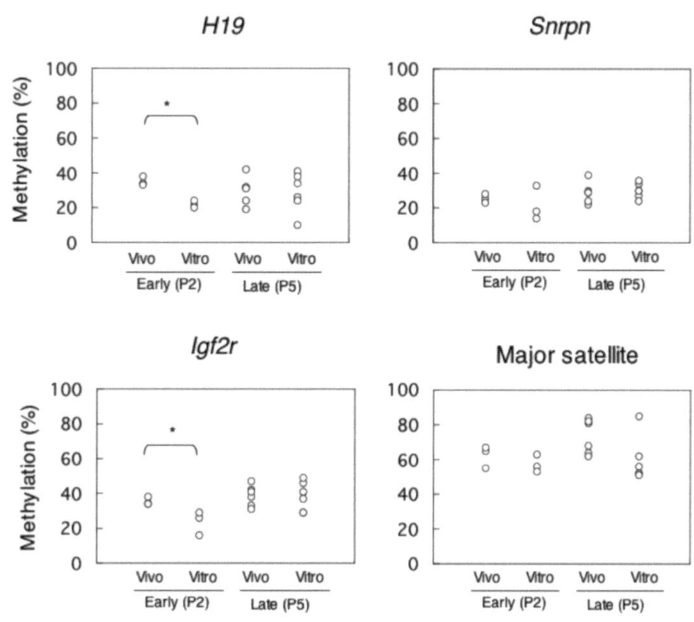

Figure 3. Epigenetic differences between Vivo and Vitro ES cells. DNA methylation status of imprinted genes, *H19*, *Snrpn* and *Igf2r*, and major satellite repeats were examined by COBRA in each ES cell line at an early passage (P2) and a later passage (P5). These graphs summarize previously reported data [1]. *, $P < 0.05$.

7. Epigenetic instability in SCNT and uniparental ES cells

7.1. SCNT ES cells

Maintenance of the normal epigenetic state in SCNT-ES cells is crucial for their use in therapeutic applications. We established two SCNT-ES cell lines from embryos that were produced by introducing mouse embryonic fibroblast (MEF) donor cells into enucleated oocytes. Only two ES cell lines were generated by SCNT, which give a small sample size to examine, but the DNA methylation state of imprinted genes seems to be more severely altered compared to normal ES cell lines at early passages (Fig. 2 and Fig. 4). The abnormal DNA methylation in SCNT-ES cells undergoes further changes during prolonged culture (P10 and P30). For example, the imprinting methylation of the *Snrpn* gene has been completely lost in both the Nt-1 and Nt-2 lines, and that of the *H19* gene has been completely lost in the Nt-1 line (Fig. 4). Chang *et al.* reported that the *H19* imprinted gene displays distinct abnormalities both in SCNT-ES and fertilized embryo-derived ES cell lines after long-term culture *in vitro*, and both exhibit indistinguishable DNA methylation patterns of the imprinted gene [30]. Nevertheless, methylation imprints vary widely in cultured donor cells and their derivative cloned mice, even across the same subclone of donor cells [22]. In fact, results from previous studies indicate that the methylation state of imprinted genes is frequently disrupted in SCNT embryos and their derivative cloned animals [31,32]. In addition, the process of nuclear transfer itself could alter the DNA methylation and gene expression [33]. Thus, the epigenetic marks in SCNT-ES cells may potentially be varied and altered compared to normal ES cells, at least in early passages.

7.2. Uniparental (parthenogenetic) ES cells

We and other groups have suggested that parthenogenetic ES (PgES) cells may be a pluripotent stem cell that could serve as a source of tissue for transplantation [34-36]. PgES cells do not require the destruction of viable biparental embryos as do normal ES cells. In addition, PgES cells do not need viruses or expression plasmids for the establishment of iPS cells. These are very powerful advantages for therapeutic applications. However, the biased epigenetic status and poor pluripotency of parthenogenetic cells are major issues to be overcome. PgES cells are established from parthenogenetic embryos that are produced by the artificial activation of the oocyte. Therefore, PgES cells that possess only maternal genomes could exhibit biallelic or silenced expression of imprinted genes, which causes poor pluripotency. Indeed, parthenogenetic embryos show poor growth and restricted tissue contribution in chimeras [37,38]. However, established PgES cells exhibit an improved contribution to chimeras, compared to chimeras derived from parthenogenetic embryos [39]. Recent reports have shown that loss of imprinting occurred in PgES cells and derivative somatic cells in chimeras and led to changes in the gene expression of imprinted genes and improved pluripotency [2,40]. For example, *Peg1* and *Snrpn* genes are originally silenced in parthenogenetic cells, whereas expression of these genes is elevated in PgES cells by demethylation of the DMR of each gene. PgES cell lines that were reprogrammed by loss of imprinting are closest to normal ES cell lines in

http://dx.doi.org/10.5772/54367

terms of gene expression pattern and pluripotency. Thus, reprogrammed PgES cells will provide a good tool for therapeutic applications. This is a case in which epigenetic instability in ES cells resulted in a desirable outcome. However, epigenetic instability in ES cells most often leads to undesirable results.

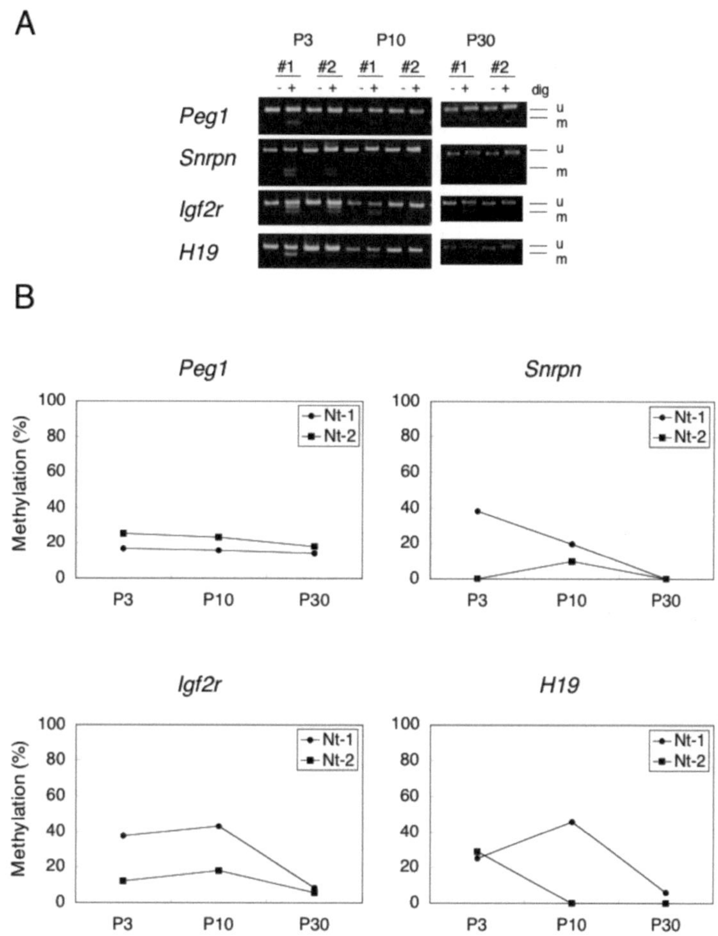

Figure 4. Epigenetic instability in SCNT-ES cells during prolonged culture. Methylation in two SCNT ES cell lines (Nt-1 and Nt-2) was examined by COBRA during prolonged *in vitro* culture (P3, P10 and P30).

8. Effect of altered DNA methylation on pluripotency and disease

In humans, a growing number of reports suggest that children born following ART have an increased risk of developing epigenetic diseases such as Beckwith-Wiedemann syndrome [41,42] and Angelman Syndrome [43], which are caused by epigenetic modifications of imprinted genes. In sheep, epigenetic changes in the *Igf2r* imprinted gene are associated with fetal overgrowth after IVC [44]. Genome-wide altered DNA methylation also causes epigenetic diseases. For example, genome-wide DNA hypomethylation is commonly observed in human cancers and schizophrenia, and occasionally induces tumors in mice [45-47]. Moreover, hypomethylation in the classical DNA satellites II and III, which are major components of constitutive heterochromatin, is found in ICF (immunodeficiency, centromeric instability, facial anomalies) syndrome in humans [48].

How do these abnormalities in ES cells affect chimeric mice or ES cell-derived tissues? Several studies have indicated that the accumulation of epigenetic alterations during prolonged culture causes a loss of pluripotency in ES cells [21,49]. In chimeras, prolonged culture of ES cells gives rise to abnormalities and frequently results in postnatal death of chimeras possessing a high ES cell contribution [20]. One reason for these problems could be that a loss of imprinting enhances tumorigenesis. In fact, mice derived from ES cells that had a global loss of DNA methylation display widespread cancer formation [50].

9. Candidate genes that cause altered DNA methylation

9.1. DNA methyltransferases

The most important factors for the maintenance of DNA methylation are the DNA methyltransferases. Three CpG DNA methyltransferases, Dnmt1, Dnmt3a and Dnmt3b, coordinately regulate CpG methylation in the genome [12-14]. Deletion of Dnmt1, Dnmt3a or Dnmt3b induces hypomethylation of genomic DNA [14,51], and forced expression of Dnmts causes genomic hypermethylation [52-54]. One of the Dnmt family members, Dnmt3L, is not expressed in differentiated somatic cells but is expressed in ES cells. Although Dnmt3L lacks the functional domains required for catalytic activity, overexpression or downregulation of Dnmt3L results in changes in DNA methylation in ES cells [55]. Thus, the upregulation or downregulation of Dnmts could cause epigenetic instability in ES cells. Indeed, hypomethylation in XX ES cells is associated with reduced levels of Dnmt3a and Dnmt3b, which is the result of both X chromosomes being in the active state [25]. Among Dnmts, a number of alternative splicing variants that lack the regulatory and/or catalytic regions have been reported. In particular, Dnmt3b has nearly 40 different isoforms generated by alternative splicing and/or alternative promoter usage. We recently reported that murine Dnmt3b lacking exon 6 (exon 5 in human) is highly expressed in *in vitro* manipulated embryos and their derivative ES cells that exhibit CpG hypomethylation [17]. Gopalakrishnan *et al.* reported that this isoform is expressed in tumor and iPS cells, and that ectopic overexpression resulted in repetitive element hypomethylation [56]. Similarly, forced expression of human specific DNMT3B4, which lacks a catalytic domain, induced DNA demethylation on satellite 2 in pericentromeric DNA [57]. These reports indicate that Dnmts have complex roles in the maintenance of the DNA methylation state. If this balance collapses, epigenetic instability will result.

http://dx.doi.org/10.5772/54367

9.2. Other methylation factors

Other new methylation factors are Stella (PGC7) and Zfp57. Stella (PGC7), a primordial germ cell and ES cell marker, protects against DNA demethylation in early embryogenesis [58]. Zfp57, a putative KRAB zinc finger protein, is also required for the post-fertilization maintenance of maternal and paternal methylation at multiple imprinted domains [59]. Reductions of the levels of these factors could induce hypomethylation of DNA in ES cells.

9.3. Active demethylation factors

Active DNA demethylation via the base excision repair pathway has recently been proposed in mammals. In zebrafish, the coupling of a deaminase (activation-induced cytidine deaminase, AID), a glycosylase (methyl-CpG binding domain protein 4, MBD4), and Gadd45 is involved in DNA demethylation [60]. In mammals, AID is indeed required for reprogramming of the somatic cell genome by demethylation of pluripotency genes in ES-somatic cell fusion [61]. Gadd45 also promotes epigenetic gene activation by repair-mediated demethylation in mammals [28,29]. A *Gadd45b* gene is activated in Vitro ES cells that possess hypomethylated imprinted genes and repetitive sequences [1]. Another recently proposed demethylation pathway is the conversion of 5-methylcytosine (5-mC) to 5-hydroxymethylcytosine (5-hmC) mediated by the Ten-eleven translocation (TET) proteins, which ultimately results in DNA demethylation [62-63]. In fact, the TET proteins (Tet1 and Tet2) that regulate 5-hmC production [64] are abundantly expressed in ES cells and may be a cause of epigenetic instability in ES cells.

9.4. Chromatin structure specific to ES cells

In ES cells, bivalent domains of chromatin, that regulate several key developmental genes, contain both repressive (histone H3 lysine 27 methylation) and activating (histone H3 lysine 4 methylation) histone modifications that are usually mutually exclusive [65]. Bivalent domains silence developmental genes in ES cells while preserving their potential to become activated upon initiation of specific differentiation programs. DNA methylation was thought to determine the chromatin structure; however, recent reports suggest that chromatin can affect DNA methylation and demethylation [66-67]. Therefore, bivalent chromatin modifications specific to ES cells could be associated with DNA methylation instability.

10. Conclusion

ES cells exhibit instabilities in DNA methylation that are correlated with the origin of the blastocysts from which they were derived (*in vivo, in vitro,* SCNT and uniparental), the culture conditions, sex, and prolonged culture. Epigenotyping of ES cells should be adopted as a prerequisite safety evaluation before their use in chimera production or therapeutic applications. Furthermore, genes associated with aberrant DNA methylation should be monitored in ES cell lines to ensure that the cells do not accumulate epigenetic instabilities.

Nomenclature

5-hmC, 5-hydroxymethylcytosine; 5-mC, 5-methylcytosine; AID, activation-induced cytidine deaminase; COBRA, Combined bisulfite restriction analysis; DMR, differentially methylated region; Dnmt, DNA methyltransferase; ES, embryonic stem; Gadd45, Growth arrest and DNA damage-inducible protein 45; ICM, inner cell mass; IVC, *in vitro* culture; IVF, *in vitro* fertilization; MBD4, methyl-CpG binding domain protein 4; PgES, parthenogenetic ES; SCNT, somatic cell nuclear transfer; TET, Ten-eleven translocation; Vitro ES, ES cells derived from in vitro manipulated embryos; Vivo ES, ES cells derived from embryos developed in vivo.

Acknowledgements

We thank Mr. Eikichi Yanagisawa, Dr. Sumiyo Morita, Ms. Mika Kimura, Mr. Daiki Tamura, Mr. Ryohei Kobayashi and Dr. Yasumitsu Nagao for technical assistance and helpful comments on the manuscript. This work was supported by grants from the Ministry of Education, Culture, Sports, Science and Technology of Japan; the Ministry of Health, Labour and Welfare of Japan; the National Institute of Biomedical Innovation; and the Takeda Science Foundation.

Author details

Takuro Horii and Izuho Hatada*

*Address all correspondence to: hatada@gunma-u.ac.jp

Biosignal Genome Resource Center, Gunma University, Maebashi, Japan

References

[1] Horii T, Yanagisawa E, Kimura M, Morita S, Hatada I. Epigenetic differences between embryonic stem cells generated from blastocysts developed in vitro and in vivo. Cellular Reprogramming 2010;12(5) 551-63.

[2] Horii T, Kimura M, Morita S, Nagao Y, Hatada I. Loss of genomic imprinting in mouse parthenogenetic embryonic stem cells. Stem Cells 2008;26(1) 79-88.

[3] Antequera F, Bird A. CpG islands in DNA methylation. In: Jost JP, Saluz HP (eds.) Molecular Biology and Biological Significance. Basel: Birkhauser Verlag; 1993. p169-85.

[4] Jaenisch R. DNA methylation and imprinting: why bother? Trends in Genetics 1997;13(8) 323-9.

[5] Jones P, Gonzalgo M. Altered DNA methylation and genome instability: a new pathway to cancer? Proceedings of National Academy Sciences of the United States of America 1997;94(6) 2103-5.

[6] Robertson KD, Wolffe AP. DNA methylation in health and disease. Nature Reviews Genetics 2000;1(1) 11–9.

[7] Surani MA. Imprinting and the initiation of gene silencing in the germline. Cell 1998;93(3) 309-12.

[8] Bartolomei MS, Zemel S, Tilghman SM. Parental imprinting of the mouse H19 gene. Nature 1991;351(6322) 153-5.

[9] Ferguson-Smith AC, Cattanach BM, Barton SC, Beechey CV, Surani MA. Embryological and molecular investigations of parental imprinting on mouse chromosome 7. Nature 1991;351(6328) 667-70.

[10] Pfeifer K. Mechanisms of genomic imprinting. American Journal of Human Genetics. 2000;67(4) 777-87.

[11] Szabo PE, and Mann JR. Biallelic expression of imprinted genes in the mouse germ line: implications for erasure, establishment, and mechanisms of genomic imprinting. Genes and Development 1995;9(15) 1857-68.

[12] Li E, Bestor TH, Jaenisch R. Targeted mutation of the DNA methyltransferase gene results in embryonic lethality. Cell 1992;69(6) 915-26.

[13] Okano M, Xie S, Li E. Cloning and characterization of a family of novel mammalian DNA (cytosine-5) methyltransferases. Nature Genetics 1998;19(3) 219-20.

[14] Okano, M., D. W. Bell, D. A. Haber, and E, Li. DNA methyltransferases Dnmt3a and Dnmt3b are essential for de novo methylation and mammalian development. Cell 1999;99(3) 247-57.

[15] Doherty AS, Mann MR, Tremblay KD, Bartolomei MS, Schultz RM. Differential effects of culture on imprinted H19 expression in the preimplantation mouse embryo. Biology of Reproduction 2000;62(6) 1526-35.

[16] Fernández-Gonzalez R, Moreira P, Bilbao A, Jiménez A, Pérez-Crespo M, Ramírez MA, Rodríguez De Fonseca F, Pintado B, Gutiérrez-Adán A. Long-term effect of in vitro culture of mouse embryos with serum on mRNA expression of imprinting genes, development, and behavior. Proceedings of National Academy Sciences of the United States of America 2004;101(16) 5880-5.

[17] Horii T, Suetake I, Yanagisawa E, Morita S, Kimura M, Nagao Y, Imai H, Tajima S, Hatada I. The Dnmt3b splice variant is specifically expressed in in vitro-manipulated

blastocysts and their derivative ES cells. Journal of Reproduction and Development. 2011;57(5) 579-85.

[18] Evans MJ, Kaufman MH. Establishment in culture of pluripotential cells from mouse embryos. Nature 1981;292(5819) 154-6.

[19] Martin GR. Isolation of a pluripotent cell line from early mouse embryos cultured in medium conditioned by teratocarcinoma stem cells. Proceedings of National Academy Sciences of the United States of America 1981;78(12) 7634-8.

[20] Nagy A, Gócza E, Diaz EM, Prideaux VR, Iványi E, Markkula M, Rossant J. Embryonic stem cells alone are able to support fetal development in the mouse. Development 1990;110(3) 815-21.

[21] Dean W, Bowden L, Aitchison A, Klose J, Moore T, Meneses JJ, Reik W, Feil R. Altered imprinted gene methylation and expression in completely ES cell-derived mouse fetuses: association with aberrant phenotypes. Development 1998;125(12) 2273-82.

[22] Humpherys D, Eggan K, Akutsu H, Hochedlinger K, Rideout WM 3rd, Biniszkiewicz D, Yanagimachi R, Jaenisch R. Epigenetic instability in ES cells and cloned mice. Science 2001;293(5527) 95-7.

[23] Minoguchi S, Iba H. Instability of retroviral DNA methylation in embryonic stem cells. Stem Cells 2008;26(5) 1166-73.

[24] Allegrucci C, Wu YZ, Thurston A, Denning CN, Priddle H, Mummery CL, Ward-van Oostwaard D, Andrews PW, Stojkovic M, Smith N, Parkin T, Jones ME, Warren G, Yu L, Brena RM, Plass C, Young LE. Restriction landmark genome scanning identifies culture-induced DNA methylation instability in the human embryonic stem cell epigenome. Human Molecular Genetics 2007;16(10) 1253-68.

[25] Zvetkova I, Apedaile A, Ramsahoye B, Mermoud JE, Crompton LA, John R, Feil R, Brockdorff N. Global hypomethylation of the genome in XX embryonic stem cells. Nature Genetics 2005;37(11), 1274-9.

[26] Murry CE, Keller G. Differentiation of embryonic stem cells to clinically relevant populations: lessons from embryonic development. Cell 2008;132(4) 661-80.

[27] Mann MR, Lee SS, Doherty AS, Verona RI, Nolen LD, Schultz RM, Bartolomei MS. Selective loss of imprinting in the placenta following preimplantation development in culture. Development 2004;131(15) 3727-35.

[28] Barreto G, Schafer A, Marhold J, Stach D, Swaminathan SK, Handa V, Doderlein G, Maltry N, Wu W, Lyko F, Niehrs C. Gadd45a promotes epigenetic gene activation by repair-mediated DNA demethylation. Nature 2007;445(7128) 671-5.

[29] Ma DK, Jang MH, Guo JU, Kitabatake Y, Chang ML, Pow-Anpongkul N, Flavell RA, Lu B, Ming GL, Song H. Neuronal activity-induced Gadd45b promotes epigenetic DNA demethylation and adult neurogenesis. Science 2009;323(5917) 1074-7.

[30] Chang G, Liu S, Wang F, Zhang Y, Kou Z, Chen D, Gao S. Differential methylation status of imprinted genes in nuclear transfer derived ES (NT-ES) cells. Genomics 2009;93(2) 112-9.

[31] Inoue K, Kohda T, Lee J, Ogonuki N, Mochida K, Noguchi Y, Tanemura K, Kaneko-Ishino T, Ishino F, Ogura A. Faithful expression of imprinted genes in cloned mice. Science 2002;295(5553) 297.

[32] Mann MR, Chung YG, Nolen LD, Verona RI, Latham KE, Bartolomei MS. Disruption of imprinted gene methylation and expression in cloned preimplantation stage mouse embryos. Biology of Reproduction 2003;69(3) 902-14.

[33] Hikichi T, Kohda T, Wakayama S, Ishino F, Wakayama T. Nuclear transfer alters the DNA methylation status of specific genes in fertilized and parthenogenetically activated mouse embryonic stem cells. Stem Cells 2008;26(3) 783-8.

[34] Cibelli JB, Grant KA, Chapman KB, Cunniff K, Worst T, Green HL, Walker SJ, Gutin PH, Vilner L, Tabar V, Dominko T, Kane J, Wettstein PJ, Lanza RP, Studer L, Vrana KE, West MD. Parthenogenetic stem cells in nonhuman primates. Science 2002;295(5556) 819.

[35] Kim K, Lerou P, Yabuuchi A, Lengerke C, Ng K, West J, Kirby A, Daly MJ, Daley GQ. Histocompatible embryonic stem cells by parthenogenesis. Science 2007;315(5811) 482-6.

[36] Horii T, Hatada I. Reprogrammed parthenogenetic ES cells - new choice for regenerative medicine. In: Atwood C.S. (ed.) Methodological Advances in the Culture, Manipulation and Utilization of Embryonic Stem Cells for Basic and Practical Applications. Rejeca: InTech; 2011. p221-36.

[37] Surani, M. A. H. and Barton, S. C. Development of gynogenetic eggs in the mouse: implications for parthenogenetic embryos. Science 1983;222(4627) 1034-1036.

[38] Fundele RH, Norris ML, Barton SC, Fehlau M, Howlett SK, Mills WE, Surani MA. Temporal and spatial selection against parthenogenetic cells during development of fetal chimeras. Development 1990;108(1) 203-11.

[39] Allen ND, Barton SC, Hilton K, Norris ML, Surani MA. A functional analysis of imprinting in parthenogenetic embryonic stem cells. Development 1994;120(6) 1473-82.

[40] Chen Z, Liu Z, Huang J, Amano T, Li C, Cao S, Wu C, Liu B, Zhou L, Carter MG, Keefe DL, Yang X, Liu L. Birth of parthenote mice directly from parthenogenetic embryonic stem cells. Stem Cells 2009;27(9) 2136-45.

[41] Maher ER, Brueton LA, Bowdin SC, Luharia A, Cooper W, Cole TR, Macdonald F, Sampson JR, Barratt CL, Reik W, Hawkins MM. Beckwith-Wiedemann syndrome

and assisted reproduction technology (ART). Journal of Medical Genetics 2003;40(1) 62-4.

[42] DeBaun MR, Niemitz EL, Feinberg AP. Association of in vitro fertilization with Beckwith-Wiedemann syndrome and epigenetic alterations of LIT1 and H19. American Journal of Human Genetics 2003;72(1) 156-60.

[43] Cox GF, Burger J, Lip V, Mau UA, Sperling K, Wu BL, Horsthemke B. Intracytoplasmic sperm injection may increase the risk of imprinting defects. American Journal of Human Genetics 2002;71(1) 162-164.

[44] Young LE, Fernandes K, McEvoy TG, Butterwith SC, Gutierrez CG, Carolan C, Broadbent PJ, Robinson JJ, Wilmut I, Sinclair KD. Epigenetic change in IGF2R is associated with fetal overgrowth after sheep embryo culture. Nature Genetics 2001;27(2) 153-4.

[45] Eden A, Gaudet F, Waghmare A, Jaenisch R. Chromosomal instability and tumors promoted by DNA hypomethylation. Science 2003;300(5618) 455.

[46] Howard G, Eiges R, Gaudet F, Jaenisch R, Eden A. Activation and transposition of endogenous retroviral elements in hypomethylation induced tumors in mice. Oncogene 2008;27(3) 404-8.

[47] Shimabukuro M, Sasaki T, Imamura A, Tsujita T, Fuke C, Umekage T, Tochigi M, Hiramatsu K, Miyazaki T, Oda T, Sugimoto J, Jinno Y, Okazaki Y. Global hypomethylation of peripheral leukocyte DNA in male patients with schizophrenia: a potential link between epigenetics and schizophrenia. Jounal of Psychiatric Research 2007;41(12) 1042-6.

[48] Jeanpierre M, Turleau C, Aurias A, Prieur M, Ledeist F, Fischer A, Viegas-Pequignot E. An embryonic-like methylation pattern of classical satellite DNA is observed in ICF syndrome. Human Molecular Genetics 1993;2(6) 731-5.

[49] Nagy A, Rossant J, Nagy R, Abramow-Newerly W, Roder JC. Derivation of completely cell culture-derived mice from early-passage embryonic stem cells. Proceedings of National Academy Sciences of the United States of America 1993;90(18) 8424-8.

[50] Holm TM, Jackson-Grusby L, Brambrink T, Yamada Y, Rideout WM 3rd, Jaenisch R. Global loss of imprinting leads to widespread tumorigenesis in adult mice. Cancer Cell 2005;8(4) 275-85.

[51] Li E, Beard C, Jaenisch R. Role for DNA methylation in genomic imprinting. Nature 1993;366(6453) 362-5.

[52] Vertino PM, Yen RW, Gao J, Baylin SB. De novo methylation of CpG island sequences in human fibroblasts overexpressing DNA (cytosine-5-)-methyltransferase. Molecular and Cellular Biology 1996;16(8) 4555-65.

[53] Hsieh CL. In vivo activity of murine de novo methyltransferases, Dnmt3a and Dnmt3b. Molecular and Cellular Biology 1999;19(12) 8211-8.

[54] Biniszkiewicz D, Gribnau J, Ramsahoye B, Gaudet F, Eggan K, Humpherys D, Mastrangelo MA, Jun Z, Walter J, Jaenisch R. Dnmt1 overexpression causes genomic hypermethylation, loss of imprinting, and embryonic lethality. Molecular and Cellular Biology 2002;22(7) 2124-35.

[55] Ooi SK, Wolf D, Hartung O, Agarwal S, Daley GQ, Goff SP, Bestor TH. Dynamic instability of genomic methylation patterns in pluripotent stem cells. Epigenetics and Chromatin 2010;3(1) 17.

[56] Gopalakrishnan S, Van Emburgh BO, Shan J, Su Z, Fields CR, Vieweg J, Hamazaki T, Schwartz PH, Terada N, Robertson KD. A novel DNMT3B splice variant expressed in tumor and pluripotent cells modulates genomic DNA methylation patterns and displays altered DNA binding. Molecular Cancer Research 2009;7(10) 1622-34.

[57] Saito Y, Kanai Y, Sakamoto M, Saito H, Ishii H, Hirohashi S. Overexpression of a splice variant of DNA methyltransferase 3b, DNMT3b4, associated with DNA hypomethylation on pericentromeric satellite regions during human hepatocarcinogenesis. Proceedings of National Academy Sciences of the United States of America 2002;99(15) 10060-5.

[58] Nakamura T, Arai Y, Umehara H, Masuhara M, Kimura T, Taniguchi H, Sekimoto T, Ikawa M, Yoneda Y, Okabe M, Tanaka S, Shiota K, Nakano T. PGC7/Stella protects against DNA demethylation in early embryogenesis. Nature Cell Biology 2007;9(1) 64-71.

[59] Li X, Ito M, Zhou F, Youngson N, Zuo X, Leder P, Ferguson-Smith AC. A maternal-zygotic effect gene, Zfp57, maintains both maternal and paternal imprints. Developmental Cell 2008;15(4) 547-57.

[60] Rai K, Huggins IJ, James SR, Karpf AR, Jones DA, Cairns BR. DNA demethylation in zebrafish involves the coupling of a deaminase, a glycosylase, and gadd45. Cell 2008;135(7) 1201-12.

[61] Bhutani N, Brady JJ, Damian M, Sacco A, Corbel SY, Blau HM. Reprogramming towards pluripotency requires AID-dependent DNA demethylation. Nature 2010;463(7284) 1042-7.

[62] Tahiliani M, Koh KP, Shen Y, Pastor WA, Bandukwala H, Brudno Y, Agarwal S, Iyer LM, Liu DR, Aravind L, Rao A. Conversion of 5-methylcytosine to 5-hydroxymethylcytosine in mammalian DNA by MLL partner TET1. Science 2009;324(5929) 930-5.

[63] Ito S, D'Alessio AC, Taranova OV, Hong K, Sowers LC, Zhang Y. Role of Tet proteins in 5mC to 5hmC conversion, ES-cell self-renewal and inner cell mass specification. Nature 2010;466(7310) 1129-33.

[64] Koh KP, Yabuuchi A, Rao S, Huang Y, Cunniff K, Nardone J, Laiho A, Tahiliani M, Sommer CA, Mostoslavsky G, Lahesmaa R, Orkin SH, Rodig SJ, Daley GQ, Rao A.

Tet1 and Tet2 regulate 5-hydroxymethylcytosine production and cell lineage specification in mouse embryonic stem cells. Cell Stem Cell 2011;8(2) 200-13.

[65] Bernstein BE, Mikkelsen TS, Xie X, Kamal M, Huebert DJ, Cuff J, Fry B, Meissner A, Wernig M, Plath K, Jaenisch R, Wagschal A, Feil R, Schreiber SL, Lander ES. A bivalent chromatin structure marks key developmental genes in embryonic stem cells. Cell 2006;125(2) 315-26.

[66] Tamaru H, Selker EU. A histone H3 methyltransferase controls DNA methylation in Neurospora crassa. Nature 2001;414(6861) 277-83.

[67] Jackson JP, Lindroth AM, Cao X, Jacobsen SE. Control of CpNpG DNA methylation by the KRYPTONITE histone H3 methyltransferase. Nature 2002;416(6880) 556-60.

[68] Bachman KE, Park BH, Rhee I, Rajagopalan H, Herman JG, Baylin SB, Kinzler KW, Vogelstein B. Histone modifications and silencing prior to DNA methylation of a tumor suppressor gene. Cancer Cell 2003;3(1) 89-95.

Chapter 15

Function of KLF4 in Stem Cell Biology

Ying Shi and Walden Ai

Additional information is available at the end of the chapter

http://dx.doi.org/10.5772/54370

1. Introduction

The Kruppel-like factor family is a group of zinc finger containing transcription factors, which are highly homologous with the Drosophila Kruppel protein. The feature that distinguishes the KLF family from other zinc finger containing transcriptional factors is the presence of three highly conserved C2H2 containing zinc finger motifs at the C-terminus [1-3]. These fingers enable KLFs to bind to the GC-box or CACCC-boxes on DNA with different affinities [4]. KLF4, as a member of KLF family, expresses in a wide range of tissues in mammals, and plays a critical role in regulating a diverse array of cellular processes including proliferation, differentiation, development, maintenance of normal tissue homeostasis and apoptosis. KLF4 can also acts either as a tumor suppressor or an oncogene depending on differing cellular context and cancer types.

The role that KLF4 plays in stem cell biology has attracted much more attention in recent years. For instance, in 2006, Takahashi K et al [5] reprogrammed somatic cells into pluripotent stem cells using KLF4 in combination with three other transcription factors: Oct4, Sox2 and c-Myc. Numerous recent literatures have further proved that KLF4 is essential for both embryonic stem (ES) cells self-renewal and maintenance, additionally our recent work revealed a critical role of KLF4 in maintenance of breast cancer stem cells [6]. Furthermore, we found that KLF4 is expressed in mouse skin hair follicle stem cells and such expression contributed to mouse cutaneous wound healing [7]. In this review, functions of KLF4 in stem cells, especially breast cancer stem cells and mouse hair follicle stem cells will be discussed, and the signaling pathways possibly involved will be addressed as well.

2. Identification and characterization of KLF4

Mouse KLF4 was first identified in 1996 independently by two groups and separately given two different names - GKLF (gut enriched Kruppel like factor):due to its high expression in the gastrointestinal tract [8], and EZF (epithelial zinc finger) since it was highly expressed in differentiated epithelial cells of the skin [9]. Human KLF4 cDNA was cloned from human umbilical vein endothelial cell cDNA library [10] and later renamed as KLF4 to avoid confusion.

The human KLF4 gene locus is mapped on chromosome 9q31 whereas mouse KLF4 is on chromosome 4B3. Mouse KLF4 has a single ORF of 1449 bp that encodes a polypeptide of 483 amino acids with a predicted molecular weight of 53 Kd; while human KLF4 has an ORF of 1444 bp coding for a 470 amino acid protein with an estimated molecular mass 50 Kd. At the amino acid level the human and mouse KLF4 are shown to have 91% sequence similarity. The three tandem zinc finger motifs are conserved completely in the human and mouse sequences. Except skin and colon [8, 9], KLF4 is also found in lung, testis, small intestine [8, 9], thymus [11], cornea [12], cardiac myocytes [13] and lymphocytes [14]. In testis, four KLF4 transcripts with alternative polyadenylation were found and they generated different RNA species in various testicular cells, strongly suggesting translational regulation of KLF4 in spermatogenesis [15, 16].

3. General functions of KLF4

3.1. Inhibition of cell proliferation

KLF4 is known to induce growth arrest, inhibiting cell proliferation by regulating the expression of key cell cycle genes. Elevated expression of KLF4 in NIH3T3 subjected to serum starvation [8] has been shown to inhibit DNA synthesis. Microarray analysis confirms that a number of genes were up- or down-regulated upon KLF4 induction, most of which are involved in cell cycle control [17]. For example, the expression of cell cycle inhibitor p21/Cip1 was elevated [18], while cell cycle promoter Cyclin D1 was depressed [19]. KLF4 has been shown to inhibit cell proliferation by blocking G1/S progression of the cell cycle and to mediate p53 dependent G1/S cell cycle arrest in response to DNA damage [20, 21]. Furthermore, KLF4 plays an important role in maintaining the integrity of the G2/M checkpoint following DNA damage. While wild type HCT 116 colon cancer cells were arrested at the G2/M phase checkpoint upon γ-irradiation, p53 -/- cells were able to enter M phase even after irradiation. It was observed that upon introduction of KLF4 into p53 -/- cells, the mitotic indices were considerably reduced and the Cyclin B1 levels were also risen [22]. These studies suggest that KLF4 is a critical factor in regulating entry of the cells into the mitotic phase. Finally, KLF4 was found both necessary and sufficient in preventing centrosome amplification following γ–irradiation-induced DNA damage by transcriptionally suppressing cyclin E expression [23].

3.2. Promotion of cell differentiation

Microarray analysis has shown that many keratin genes were upregulated on KLF4 induction, indicating its role in epithelial differentiation. Additionally, KLF4 has been reported to transactivate promoters of epithelial genes including CYP1A1 [24], laminin α 3A [25], laminin 1 [26], keratin 4 [27], keratin 19 [28]. Recent studies demonstrated that KLF4 plays a vital role in goblet cell differentiation in the intestine [29, 30], conjunctiva [31], and also in the formation of the epithelial barrier of the skin [32]. KLF4 null mice died one day after birth due to loss of barrier function of the skin. It appears that KLF4 influences the formation of the cornified envelope in the late-stage differentiation process that was supported by upregulation of Sprr2a, a cornified envelope gene, in KLF4 knockout mice. Two additional cornified envelope proteins: repetin (encoded by Rptn) and plasminogen activating inhibitor 2 (encoded by Planh2) were found later. KLF4 may regulate these genes resulting in an imbalance in cornified envelope assembly or composition, thereby altering the structural scaffold on which the lipid lamellae are organized. A differential role of KLF4 has also been reported in smooth muscle cells [33], monocytes [34], testes [15], T cells [11, 35] and murine tooth development [36].

3.3. Other functions

KLF4 is thought to be involved in chronic inflammatory disease since it has been shown to mediate proinflammatory signaling in human macrophages in vitro [37, 38] and regulate the expression of interleukin-10 in RAW264.7 macrophages [39]. KLF4 is also essential for differentiation of mouse inflammatory monocytes and involved in the differentiation of resident monocytes [34, 40]. The inflammation-selective effects of loss-of-KLF4 and gain-of-KLF4-induced monocytic differentiation in HL60 cells identify KLF4 as a key regulator of monocytic differentiation and a potential target for translational immune modulation [40]. KLF4 positively regulates human ghrelin expression [41], which is expressed in the gastrointestinal tract. In addition, it was found that KLF4 is an immediate early gene for Nerve Growth Factor [42]. A recent study showed that glutamatergic stimulation can trigger rapid elevation of KLF4 mRNA and protein levels, and that the over expression of KLF4 can regulate neuronal cell cycle proteins and sensitize neurons to NMDA-induced caspase-3 activity [43]. Another study demonstrated that KLF4 is involved in regulating the proliferation of CD8+ cells [44]. The transcription factor ELF4 directly activated the tumor suppressor KLF4 'downstream' of T cell antigen receptor signaling to induce cell cycle arrest in naive CD8+ T cells [44].

KLF4 has been implicated in the regulation of apoptosis [45, 46]. During DNA damage, cells can take two routes - either pass into the next phase overcoming the checkpoint or get arrested at the checkpoint and activates the repair machinery. As discussed previously, over expression of KLF4 in RKO colon cancer cells, when subjected to UV radiation, reduced the percentage of apoptotic cells [47]. In esophageal cancer cell lines, KLF4 has been shown to bind to the promoter and repress the activity of the surviving gene *in vivo* [48], which is necessary for caspase inactivation and therefore acts as a negative regulator of apoptosis.

4. KLF4 in stem cell biology

4.1. KLF4 function in embryonic stem cells

Embryonic stem (ES) cells are characterized by a self-renewal ability and pluripotency. Self-renewal is the capability of ES cells to be maintained in a proliferative state for prolonged periods of time, whereas pluripotency is the ability of ES cells to differentiate into a diverse array of specialized cell types. It has been shown that self renewal and maintenance of pluripotency in mouse ES cells requires leukemia inhibitory factor (LIF). LIF is a member of the IL6 cytokine family and is used to maintain ES cell cultures in an undifferentiated state through activation of the *Stat3* gene. Oct4, Sox2, and Nanog are all thought to be the master regulators of ES cell pluripotency. Although Oct4 and Sox2 are not direct targets of Stat3 [49], they have been identified as two essential transcription factors that form a heterodimer which binds to the Nanog promoter and regulates the expression of downstream genes that contribute to the maintenance of self-renewal [50]. KLF4 acts as a fast responding mediator to LIF-Stat3 signal changes, and directly binds to the promoter of Nanog to help Oct4 and Sox2 in regulating the expression of Nanog [51]. This observation confirms the critical role of KLF4 in ES cell self renewal as well as pluripotency.

4.2. KLF4 function in generation of induced pluripotent stem cells

ES cells are believed to hold great promise for regenerative medicine due to their unique ability to differentiate into any cell type. However, the application of human eggs or embryos encounters big ethical problems. This dilemma was broken in 2006 by Dr. Shinya Yamanaka's group. They picked four transcription factors, including Oct4, Sox2, c-Myc, and KLF4, to introduce into mouse embryonic fibroblasts via retroviral transfection [5]. The modified embryonic fibroblasts were found to be reprogrammed to a pluripotent state similar to that observed in ES cells. Later the finding was further confirmed by using either mouse or human adult fibroblasts [52-57]. The discovery of these "induced pluripotent stem cells" (iPS cells) was regarded as a great achievement in stem cell research and gave new insights into the feasibility of clinical application of stem cells.

A panel of assays has been performed to compare iPS cells with ES cells in morphology, surface marker expression, epigenetic status, formation of embryoid bodies *in vitro*, directed differentiation into neural cells and beating cardiomyocytes, teratoma formation *in vivo* and chimera contribution. The results indicated that iPS cells resemble ES cells by all measured criteria. Not only fibroblasts, but also other terminally differentiated cells can be reprogrammed to pluripotent cells [58]. After the introduction of pluripotency from terminally differentiated cells, the applications of the iPS cells have also been explored. By using a humanized sickle cell anemia mouse model, mice can be rescued after transplantation with hematopoietic progenitors obtained from autologous iPS cells *in vitro*. Mechanistically, the rescue was due to the correction of the human sickle hemoglobin allele by gene specific targeting. This report provides the first proof of principle for using iPS cells for disease treatment in mice [59] and demonstrates the therapeutic potential of iPS cells for human diseases.

Although iPS cells based on somatic cells avoid ethical issues, the use of oncogenes and retrovirus still raised safety concerns. For example, reactivation of the c-Myc retrovirus, increased tumorigenicity in the chimeras and progeny mice, hindering clinical applications [60]. Another problem is that iPS cells are refractory to differentiation and thereby increase the risk of immature teratoma formation after directed differentiation and transplantation into patients. Even if only a small portion of cells within each iPS cell clone shows impaired differentiation, then those cells might be sufficient to produce immature teratomas [61].

Nevertheless, the iPS cell technology potentially can overcome two important obstacles associated with human ES cells: immune rejection after transplantation and ethical concerns regarding the use of human embryos [61]. The advantage of iPS cell technology is that iPS cells can be generated using a few programming factors in any laboratory using standard techniques and equipment. Establishment of a stable and self-sustainable ES-specific transcriptional regulatory network is essential for reprogramming [62]. iPS cells still have the scope for clinical applications provided that proper ways are established to precisely evaluate each iPS cell clone and to select appropriate sub clones prior to clinical application.

4.3. KLF4 function in breast Cancer Stem Cells (CSC)

Cancer stem cells (CSCs) are a subpopulation of tumor cells that possess the stem cell properties of self renewal and differentiation, which allows them to generate the heterogeneous lineages of cancer cells that comprise the tumor. In 1997, a hierarchy in human acute myeloid leukemia cells was first reported, which improved the understanding of tumorigenesis and cast new light on cancer therapy [63]. CSCs in other types of hematological malignancies were identified later, and then CSC research was expanded to solid tumors shortly after. The identification of CSCs in solid tumors depends on specific biomarker. Recently, CSCs have been identified in numerous solid tumors, including pancreas [64], colon [65], prostate [66], bladder [67], lung [68] and breast cancer [69].

In breast cancer the first evidence of CSC was based on a combination of specific cell-surface antigen profile CD44+/CD24-/Lin- in 2003 [69]. More recently, aldehyde dehydrogenase (ALDH) was used as stem cell marker in a series of 577 breast carcinoma and 33 human breast cell lines [70]. ALDH is a detoxifying enzyme that oxidizes intracellular aldehydes and is thought to play a role in the differentiation of stem cells via the metabolism of retinal to retinoic acid [71]. Side population (SP) was also defined as a characteristic of breast CSC, which indicated an inherently high resistance to chemotherapeutic agents [72]. Since the CSCs have the capacity for self-renewal, differentiation into multiple cancer cell lineages, extensive proliferation as normal stem cells, and are responsible for tumor recurrence and chemotherapeutic resistance, it is necessary to figure out the key regulators and related signaling pathways that regulate the CSC in the process of carcinogenesis and tumor metastasis.

As discussed previously, KLF4 plays a critical role in ES self renewal and pluripotency, and is one of the four transcription factors creating iPS cells. Therefore, it's very worthy to explore the relationship between KLF4 and breast CSCs along with underlying mechanisms. Our recent work provides evidence for the first time that KLF4 is essential for the maintenance of breast CSCs and cell migration and invasion [7]. This evidence may offer important clues to understand how KLF4 promotes breast cancer development.

Earlier reports have shown that elevated KLF4 expression is detected in nearly 70% of breast carcinomas and that nuclear localization of KLF4 is associated with a more aggressive phenotype in early-stage breast cancer [73, 74]. However, the ability of KLF4 to initiate aggressive tumors in vivo has not been examined yet. Our study showed that KLF4 was highly expressed in CSC-enriched populations in mouse primary mammary tumor and human breast cancer cell lines (Figure 1). Knockdown of KLF4 in breast cancer cell MCF-7 and MDA-MB-231 inhibits cell migration, invasion and adhesion *in vitro*, and the self-renewal of breast CSCs (Figure 2). Tumor growth in mouse xenograft mode was suppressed as well (Figure 3), suggesting that KLF4 could act as an oncogenic protein in breast cancers.

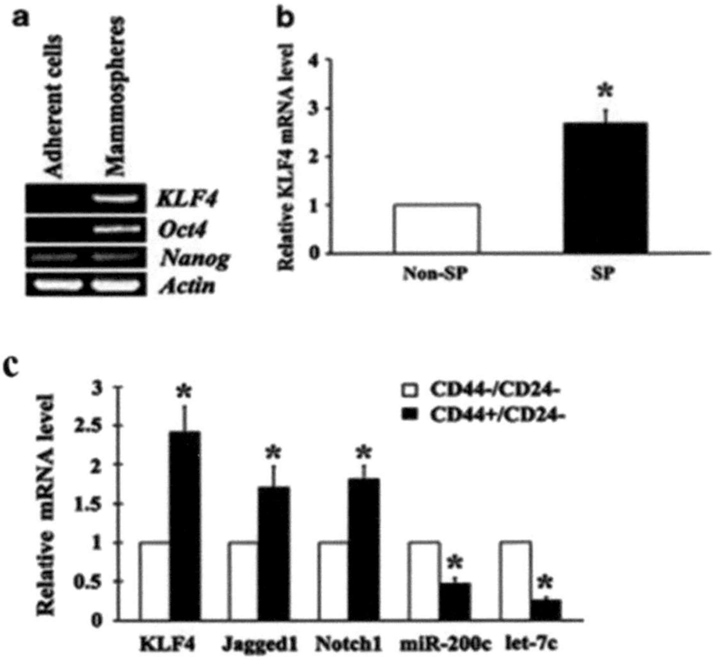

Figure 1. KLF4 was highly expressed in CSC-enriched population. (a) KLF4 expression was examined in adherent cells and mammospheres of primary tumors originated from MMTV-Neu transgenic mice. Oct4 and Nanog were used as positive and negative controls, respectively. (b) KLF4 expression was examined in SP and non-SP cells of MCF-7. The symbol * indicates $P<0.05$ vs non-SP cells group. (c) KLF4 expression was determined in CD44+/CD24- and CD44-/CD24- populations isolated by flow cytometry. The symbol * indicates $P<0.05$ vs CD44-/CD24- group.

The anti-proliferative function of KLF4 is associated with inhibition of cell cycle promoter cyclin-D1 [19] and activation of the cell-cycle inhibitor p21/Cip1 [18]. Since inactivation of either protein not only neutralizes the cytostatic effect of KLF4 but also collaborates with KLF4 in oncogenic transformation [75], thus further highlighting the importance of p21/Cip1. Although p21/Cip1 status might be a switch that determines the tumor suppressor or oncoprotein function of KLF4, the exact mechanism has not been elucidated yet. Moreover, a cellular mechanism by which KLF4 contributes to the aggressive characteristics of breast cancers remains unknown. Our current studies indicate that KLF4 is required for the maintenance of breast CSCs and the knockdown of KLF4 significantly decrease the self-renewal of breast CSCs by examining several different CSC markers. Notably KLF4 exerted an anti-apoptotic function in many cancer cell lines, so it is possible that the decreased CSC population upon KLF4 knockdown may be a result of the increased apoptosis mediated by KLF4 reduction. However, the fact that cell viability of KLF4 knockdown cells was comparable to that of the control cells would argue against this possibility. We have not performed limiting-dilution assays to determine the tumor-initiating capacities of CSC cells in non-obese diabetic/severe combined immunodeficiency mice yet, which is a traditional method in CSC studies. Nevertheless, our results not only provide additional experimental support for the important function of KLF4 in stem cell biology, but also are important for breast cancer studies. CSCs have been shown to foster blood vessel formation and promote cell motility. They are also resistant to chemotherapy and radiotherapy [76] and have been implicated in breast cancer metastasis that remains the number one cause of cancer-related mortality in women [77]. Our study suggested that overexpression of KLF4 was sufficient to drive cell migration and invasion. Additional studies on the mechanisms by which KLF4 maintains cancer stem cell phenotype will be very helpful to develop novel therapeutic strategies targeting KLF4 or the related signaling pathway to treat malignant breast cancer and metastasis.

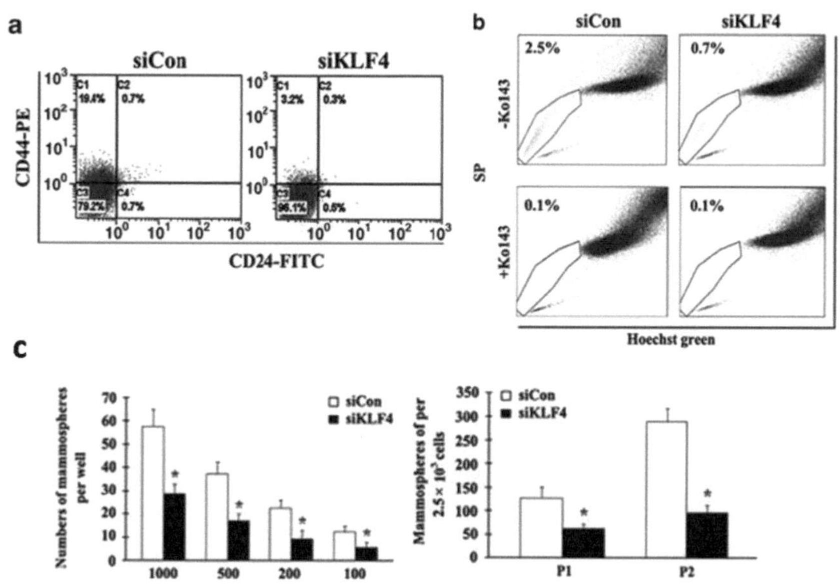

Figure 2. Knockdown of KLF4 resulted in a reduced stem cell population and decreased self-renewal of breast cancer stem cells. (a) Freshly isolated siCon and siKLF4 MCF-7 cells were labeled with CD24 (fluorescein isothiocyanate (FITC)) and CD44 (phycoerythrin (PE)) antibodies to identify CD44+/CD24-population using a FACSCalibur flow cytometer. (b) SP population in MCF-7 stable cells was determined by Hoechst 33342 efflux assays. (c) Left, MCF-7 cells (siCon and siKLF4) were grown in ultra-low attachment surface plates at a density of 1000, 500, 200, and 100 per well. Assays were conducted after 10 days (left). The symbol * indicates $P<0.05$ vs siCon group. Right, primary (P1) and secondary (P2) mammosphere formation under suspension culture conditions were evaluated in MCF-7 mammary tumor cell lines.

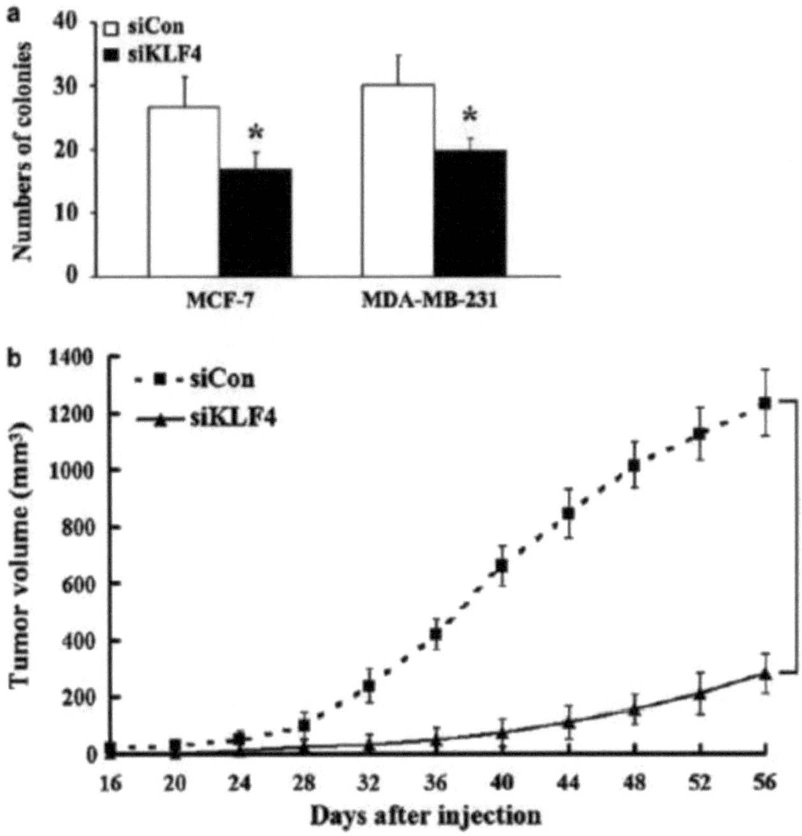

Figure 3. Knockdown of KLF4 reduced tumorigenesis in vitro and in vivo. (a) Colony-forming abilities of siCon and siKLF4 cells were assessed. The symbol * indicates $P<0.05$ vs siCon group. (b) Tumor growth curves were plotted for immunocompromised non-obese diabetic (NOD)/severe combined immunodeficiency (SCID) mice injected with KLF-knockdown (siKLF4, solid line) and control cells (siCon, dashed line). Data are shown as mean size ± s.e.m. of tumors in five mice per cell line.

The function of KLF4 in maintenance of CSCs has been further confirmed in our study by using Kenpaullone, a small molecule inhibitor of KLF4. Previous work has demonstrated that Kenpaullone is able to replace KLF4 in the reprogramming of primary and secondary fibroblasts, and that Kenpaullone-induced iPS cells display characteristics of pluripotent ES cells [78]. We tested KLF4 expression in Kenpaullone-treated breast cancer cell lines and found that it decreased at both of the mRNA and protein levels. Additional reporter assays showed that KLF4 promoter activity was significantly inhibited by Kenpaullone treatment, suggesting that Kenpaullone-mediated downregulation of KLF4 occurred at a transcriptional level. KLF4 downregulation was also accompanied by decreased expression of two previously reported down-stream targets [79, 80]: p53 and intestinal alkaline phosphatase. This further validates the regulation of KLF4 by Kenpaullone. Since a maximal downregulation of KLF4 was observed at a 4 h time point after Kenpaullone treatment, we postulate that KLF4 may be an early responsive gene after Kenpaullone treatment, and after this point, the expression of KLF4 gradually recovered. Kenpaullone-treated cells possessed phenotypes similar to KLF4 knockdown cells in our studies, which, from another point of view, confirmed the indispensable role of KLF4 in CSCs and extended a function of Kenpaullone from the induction of iPS cells to the maintenance of mammary CSCs.

Our research also indicates that KLF4 might promote epithelial-mesenchymal transition (EMT) in breast cancers. EMT is a unique process by which epithelial cells undergo remarkable morphological changes (leading to increased motility and invasion) and believed to be reminiscent of 'cancer stem-like cells', showing characteristics similar to many cancer systems [81, 82]. It has been reported that KLF4 interacts with transforming growth factor-β, a well established regulator of EMT [83], and β-catenin, one of the most important mesenchymal markers. Based on the pivotal role of KLF4 in CSCs, in combination with its links to the transforming growth factor-β signaling pathway, we highly suspected that KLF4 improved EMT in breast cancers. In our studies, KLF4 knockdown MCF-7 cells exhibited a well-spread morphology, with the majority of cells forming a rounded, epithelial-like form and aggregating together in groups, a typical characteristic of mesenchymal to epithelial transition [84] and a reversal of EMT. Fibronectin and vimentin, two critical mesenchymal-associated markers, were both decreased in KLF4 downregulated cells, which were consistent with reduced ability of migration and invasion of these cells. However, E-cadherin expression and localization, a hallmark of the EMT phenotype, showed no significant difference after KLF4 was knocked down. Contrary to our results though, KLF4 was reported to inhibit EMT in non-transformed MCF-10A cells by another group [85]. Our major argument is that MCF-10A cells are spontaneously transformed cells with no potential of tumorigenesis. Therefore, the results from MCF-10A cells may not be readily applicable to other mammary tumor cells. In their study, MDA-MB-231 tumor cells with KLF4 overexpression had also been used. However, results from our studies, using KLF4 knockdown and overexpression stable cells, supported a positive connection between KLF4 and EMT. Clearly, more studies are necessary to examine whether the difference of the two systems or the genetic background of specific MDA-MB-231 clones contributes to the discrepancies between the previously reported results and our current results.

4.4. KLF4 function in mouse hair follicle stem cells

Skin is renewed throughout life by proliferation of a multipotential stem cell population and terminal differentiation of stem cell progeny. Epidermal renewal is thought to be controlled by stem cells located either in the basal layer of the interfollicular epidermis (IFE) or in the deepest portion of permanent hair follicle called bulge [86]. Mouse hair follicle stem cells which reside in the hair follicle bulge are characterized by expression of CD34 and CD49 [87-89], retention of either DNA or histone labels over long periods [90, 91], and expression of Leucine-rich repeats and immunoglobin-like domain protein 1 (Lrig1) [92, 93]. Wound healing is an important response of skin in order that it might repair itself after an injury. Regeneration of epidermis after wounding involves activation, migration and proliferation of keratinocytes from both the surrounding epidermis and the adnexal structures such as hair follicles [94-96]. The discovery of properties of epidermal stem cells led to the hypothesis that these stem cells play a critical role in epidermal repair after wounding. Previous work has reported that bulge stem cells rapidly respond to wounding and migrate towards the IFE to help with the rapid hair-follicle regeneration, and that bulge-derived cells are transient amplifying cells committed to differentiation [93, 95, 97]. However, the role and contribution of keratinocytes derived from hair follicle bulge stem cells to cutaneous wound healing needs further elucidation.

It has been proven that KLF4 is essential for establishing the barrier function of skin. However, KLF4 expression and potential function in epidermal stem cells has not been studied before. In our current study, we have shown that KLF4 is likely expressed in mouse epidermal stem cells. A decreased number of hair bulge stem cells was observed in KLF4 knockout mice, which was accompanied by a decreased ability of colony formation from these cells when compared to those from control mice, suggesting that KLF4 may be required for the maintenance of skin hair follicle stem cells. Notably, KLF4 deficiency delayed the process of mouse cutaneous wound healing, during which KLF4-expressing multipotent cells migrated towards the wound area [6].

Using the wild type mice and KLF4/EGFP mouse model, we found that KLF4 was expressed in CD34+/CD49f+ bulge stem cell-enriched populations. However, KLF4 gene expression in CD34+/CD49f+/Lrig1+ cells was about 2.2 fold higher than in CD34+/CD49f-/Lrig1- cells sorted from wild-type mice. High levels of KLF4 expression in most differentiated, post mitotic skin epithelial cells [98] and low percentage of skin epidermal stem cells may be reasons why a difference has not been observed. Nevertheless, our studies collectively provide the first evidence that KLF4 was likely expressed in mouse hair follicle stem cells, especially in bulge stem cells.

The label retention cell (LRC) assay was used to confirm the quiescent nature of KLF4-expressing cells (Figure 4). Three-day-old KLF4/EGFP mice were injected with BrdU and left for an extended period. Twelve weeks later, the proportion of KLF4-positive cells in LRCs was 4.1%, suggesting that only a subset of these LRCs expressed KLF4. These results reveal a heterogeneous nature of LRCs. However, the difference between KLF4-expressing and KLF4-non-expressing LRCs and the related functional influence in wound healing still remain unknown. By lineage tracing to the KLF4/CreERTM/ Rosa26RLacZ mouse model, a multipotent and clonal nature of KLF4 expressing cells was identified as well (Figure 5). Our studies have also shown that KLF4 knockout decreased the population of CD34+/CD49f+ cells accompanied by reduced self-renewal ability of these cells. Together with the label retaining ability of KLF4 expressing cells, our results indicated KLF4 plays an important role in the homeostasis of skin bulge stem cells. In addition, expression of KLF4 in rare skin stem cells and in the bulk of differentiated keratinocytes may suggest that the functions of KLF4 in these populations are different. It has been reported that different KLF4 isoforms may exist and exhibit different functions in pancreatic cancer cell [99]. Characterization of different KLF4 isoforms and/or separation of distinct KLF4 expressing cells will be necessary for dissecting specific functions of KLF4 in skin homeostasis as well as pathogenesis including wound healing.

http://dx.doi.org/10.5772/54370

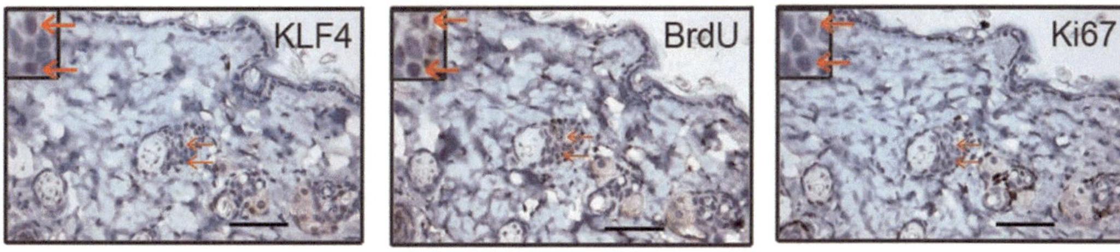

Figure 4. KLF4-expressing cells possessed label retaining property. 3-day-old KLF4/EGFP mice were injected with BrdU (75mg/kg) for 5 consecutive days. BrdU-positive cells were examined 3 months later by immunohistochemical staining. Anti-KLF4, anti-BrdU, and anti-Ki67 antibodies were used to stain consecutive slides. Insets show enlarged portion of the staining indicating co-localization of KLF4 and BrdU positive cells with no Ki67 signals (red arrows). Scale bars, 50 mm.

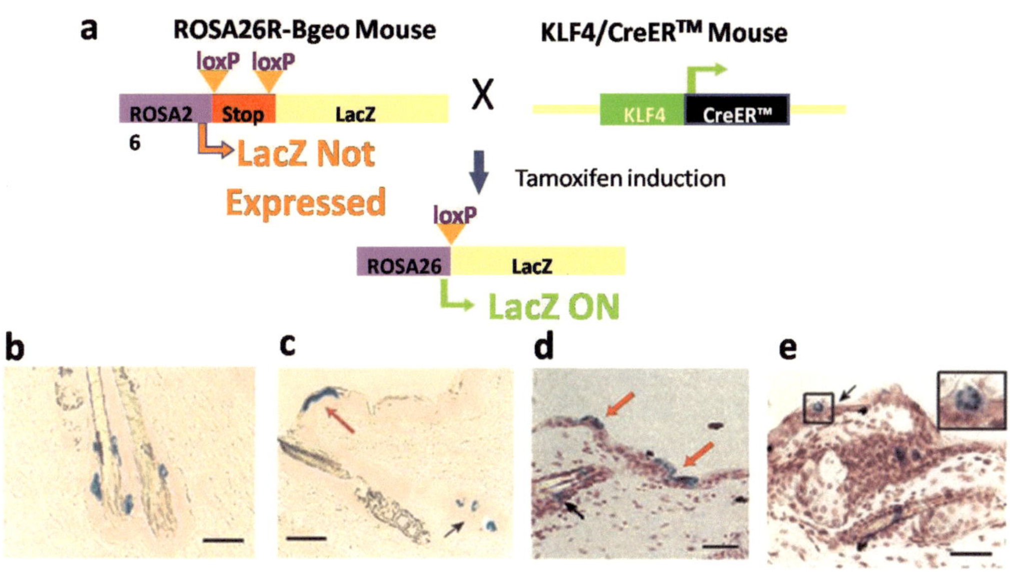

Figure 5. KLF4-expressing hair follicle stem cells were examined by lineage tracing. KLF4/CreERTM/Rosa26RLacZ mice were induced by tamoxifen (100mg/kg) for 5 consecutive days at 6-week-old (a).4 weeks later X-gal staining was performed. Potential KLF4 expression in interfollicular epidermis (shown by red arrows in c, d) and bulge area (b, and black arrows in c, d) was shown. A typical epithelial proliferation unit was shown in e (inset). Note that fixation was performed without xylene in a and b. Scale bars, 80 mm.

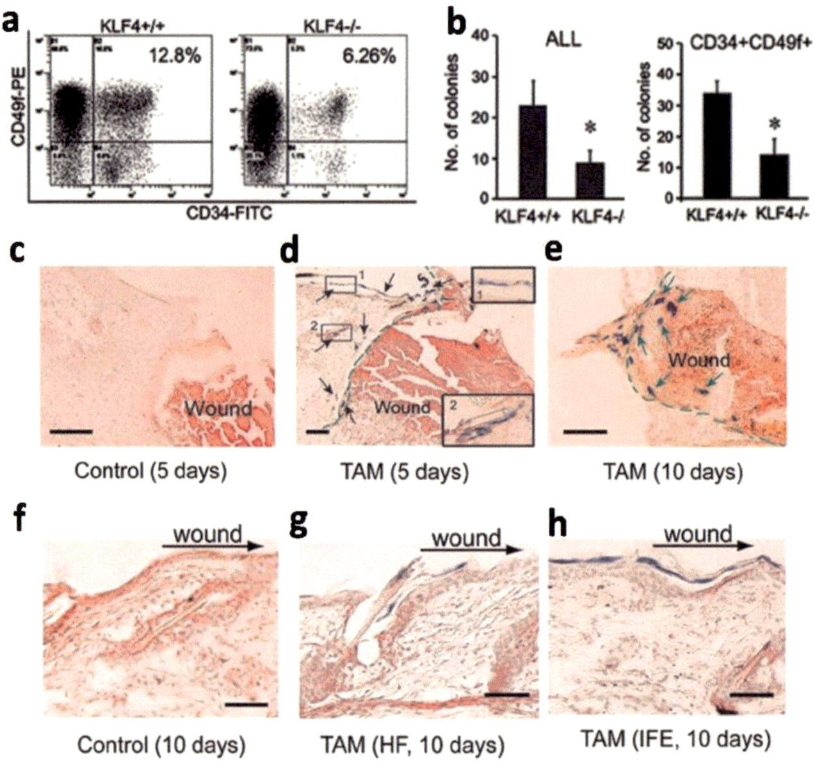

Figure 6. Knockout of KLF4 decreased hair follicle stem cell population and self-renewal potential *in vitro* and retarded would healing *in vivo*. (a) Dorsal skin keratinocytes isolatedfrom control (KLF4+/+) and KLF4 knockout (KLF4-/-) mice were analyzed by flow cytometry using mouse epidermal stem cell markers CD34 and CD49f. (b) Quantitation of the colony numbers from 2000 seeded keratinocytes. Data shown were the mean ± SM of three separate experiments. *$P < 0.05$ vs. control. 5mm wounds were introduced into the backs of KLF4/CreER^{TM}/Rosa26RLacZ mice 5 (c, d) or 10 days (e–h) after using control (c, f) or tamoxifen (d, e, g, h) induction and X-gal staining was performed. Blue strips on epidermis were shown in d (inset 1) and h. Blue cells was indicated by black arrows outside (d) and by green arrows inside (e) the conjunction of the wound (separated by dashed green lines). Inset 2 in d showed blue cells around hair follicles. Migration of KLF4 expressing multipotent cells from hair follicles (g) and interfollicular epidermis towards the wound area was detected similarly.. Scale bars, 80 mm.

Previous work has demonstrated that stem cells located in the bulge area [95] and isthmus [100] contribute to wound healing. Our work has shown that KLF4-expressing multipotent cells participate in re-epithelialization during cutaneous wound healing. It known that cutaneous wounds heal with an acute delay in re-epithelialization in the absence of hair follicles [101]. From our study we learned that KLF4 expression in possible hair follicle stem cells may contribute to the wound healing (Figure 6). We also observed that KLF4-expressing stem cells remained quiescent as evidenced by rarely detectable blue cells eight months after the cells were labeled. However, they were readily activated and detectable when the cutaneous wound occurred. This observation is consistent with a recent proposal for olfactory neural stem cells. In this pattern, stem cells within the LRC population serve as a reservoir of long-lived progenitors that remain largely quiescent during normal neuronal turnover or even after acute, selective loss of mature neurons; meanwhile previously identified progenitors are largely responsible for tissue maintenance. Surprisingly after extensive injuries that deplete resident neuronal precursors, these quiescent stem cells transiently proliferate and reconstitute the neuroepithelium to maintain homeostasis [102]. Moreover, KLF4 deficiency delayed the process of wound healing and cell migration. It has been proven that KLF4 is essential for establishing skin barrier function because KLF4 deficiency selectively perturbed the late-stage differentiation structures including the cornified envelope [32]. It is not clear though, whether the role of KLF4 in barrier function is also involved in wound healing in our setting. Finally, our wound healing model did not limit for contraction. Although this simple method allowed us to observe an obvious phenotype, more rigorous models should be used in the future in order to define the role of KLF4 in the complex wound healing process. Nonetheless, our results suggest a critical function of KLF4-expressing epidermal multipotent stem cells in cutaneous wound healing.

4.5. Signaling pathways regulating KLF4 and stem cell biology

Stem cells often reside in locations called stem cell niches. Specifically, stem cell niches are defined as particular locations or microenvironments that maintain the combined properties of stem cell self-renewal and multipotency [103]. A combination of genetic and molecular analyses has identified many factors that support stem cell niches that also control stem cell identity. These factors include components of Notch, Wnt, and Hedgehog signaling pathways, all of which KLF4 is thought to be involved in [104-106].

5. Notch signaling and KLF4

Notch signaling is involved in cell proliferation and apoptosis, which affects the development and function of many organs. The signal is initiated by interaction of a Notch receptor with a Notch ligand on an adjacent cell. Upon activation, Notch is cleaved, releasing intracellular domain of the Notch (ICN) through a cascade of proteolytic cleavages by the metalloprotease tumor necrosis factor-α-converting enzyme (TACE) and γ-secretase. ICN then translocates to the nucleus where it displaces corepressor complexes that are prebound with CSL. The following recruitment of coactivators, including Mastermind-like proteins and CBP/p300, then activates gene expression of downstream target genes [107].

It has been reported that altered Notch signaling affects the function of a variety of mammalian stem cells such as hematopoietic, intestinal, and skin stem cells, and intestinal stem cells in Drosophila and germ stem cells in C. elegans [103, 105, 108]. KLF4 is proposed as the downstream target of Notch signaling pathway and KLF4 promoter activity is inhibited by Notch, but the relationship between the Notch signaling pathway and KLF4 appears dependent on different cellular contexts. Our early work and that of others suggest that KLF4 is inhibited by Notch in the gastrointestinal tract [107, 109, 110]. Recently, downregulation of Notch1 gene expression in keratinocytes by KLF4 has also been reported [111].In our current study on breast CSCs, we found that the expression of Notch1, Notch2 and Jagged1 were significantly decreased in KLF4 knockdown cells, and upregulated by overexpression of KLF4. Unexpectedly, inhibition of the Notch pathway by CompE, a γ-secretase inhibitor, had no effect on stem cell numbers and self-renewal potential of breast cancer cells. This result suggested that the Notch signaling pathway is not required for KLF4-mediated maintenance of stem cells in breast cancer cells (Figure 7). On the other hand, inhibition of Notch signaling by CompE in KLF4-overexpressing cells led to decreased migration and invasion ability, which indicated that the Notch signaling pathway was responsible for KLF4-mediated mobility characteristics of breast cancer cells. These results are consistent with the role of Notch signaling as potent drivers during tumor progression and in converting polarized epithelial cells into motile, invasive cells [112]. However, in breast cancer cells, inhibitors of canonical Notch1 signaling suppressed the transformation induced by Notch1 whereas it had no effect on the transformation by KLF4, indicating KLF4-induced transformation requires Notch1, canonical Notch1 signaling is not required, and Notch1 may signal through a distinct pathway in cells with increased KLF4 activity. These results suggest that KLF4 could contribute to breast tumor progression by activating synthesis of Notch1 and by promoting signaling through a non-canonical Notch1 pathway [113].

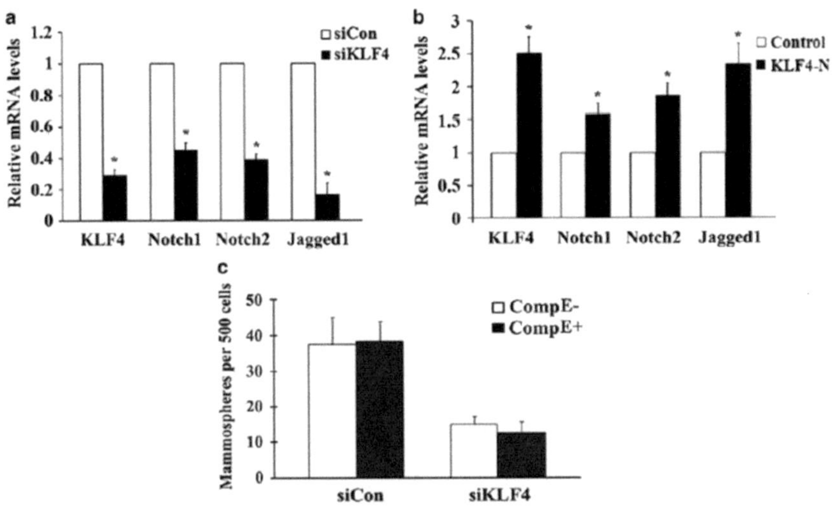

Figure 7. Notch signaling pathway is activated but not required for KLF4-mediated maintenance of stem cells in breast cancer cells (a) Levels of Notch1, Notch2 and Jagged1 expression in siCon and siKLF4 MCF-7 cells were detected by real-time PCR. The symbol * indicates $P<0.05$ vs siCon group. (b) Similar to (a) except that control and KLF4-N (KLF4 overexpression) MCF-7 cells were used. (c) MCF-7 cells (siCon and siKLF4) were seeded into ultra-low attachment surface plates and incubated with CompE at a concentration of 1 mM.

6. Wnt signaling and KLF4

Wnt signaling is an ancient and highly conserved system that is involved in embryogenesis, development, cell polarization, differentiation and proliferation [114-116]. Wnt signaling cascades have traditionally fallen into two categories: canonical and non-canonical, differentiated by their dependence on β-catenin. Canonical Wnt signaling is initiated when a Wnt ligand engages co-receptors of the Frizzled (Fzd) and low-density lipoprotein (LDL)-related protein (either Lrp5 or Lrp6), ultimately leading to β-catenin stabilization, nuclear translocation and activation of target genes. The canonical Wnt/β-catenin pathway plays a crucial role in stem and cancer stem cells' self-renewal and/or differentiation of skin, intestine and mammary gland [117].

In the absence of Wnt stimulus, β-catenin is held in an inactive state by a multimeric "destruction" complex comprised of adenomatous polyposis coli (APC), Axin, glycogen synthase kinase 3β (GSK3β) and casein kinase1α (CK1α) [118]. Nearly 90% of colon cancer harbors Wnt/β-catenin signaling mutations that result in β-catenin mutation. The most common type of mutation in colon cancer results in the inactivation of APC, thus driving constitutive activation of β-catenin [119-121]. KLF4 binds the transcriptional activation domain of β-catenin and inhibits β-catenin-mediated transcription in colorectal cancer cells, suggesting that the cross talk between KLF4 and β-catenin plays an important role in intestinal homeostasis and colorectal carcinogenesis [122]. A growing body of evidence illustrates a critical role of β-catenin in CSCs. For example, stem-like colon cells with a high level of β-catenin signaling have a much greater tumorigenic potential than counterpart cells with low β-catenin signaling [123]. The latest report shows that in stem cells and cancer cells, TERT, the enzymatic subunit of telomerase complex controlling telomere length, is directly regulated by β-catenin, and klf4 is required for β-catenin to localize to the *Tert* promoter [124].

In over 50% of clinical breast cancer cases a stabilization of β-catenin has been demonstrated. Inhibition of Wnt/β-catenin signaling in the mouse mammary gland blocks organ development and pregnancy-induced proliferation and heavily reduces the numbers of alveolar progenitor cells [125]. Wnt/β-catenin has also been implicated in mediating the radiation resistance of mouse mammary gland progenitor cells. Our recent study shows that KLF4 is required for maintenance of breast CSCs and for cell migration and invasion along with Notch signaling pathway [7]. However, the reaction of KLF4 and Wnt/β-catenin signaling in this setting still remains unknown and needs further investigation. Our other work showed that KLF4 contributes to cutaneous wound healing [6]. Additionally, the canonical Wnt signals are required in the normal skin to instruct bulge stem cells toward the hair cell fate [126], while in epidermal tumors, they control the maintenance of skin CSCs [84]. Therefore it is speculated that both of KLF4 and Wnt/β-catenin signaling are implicated in this process, and the relationship between them needs further investigation as well.

7. Hedgehog signaling and KLF4

Under normal conditions, HH signaling plays important roles in embryonic development and is also involved in tissue regeneration in adults [127, 128]. Activating events in the HH pathway are involved in numerous human cancers, including melanoma [129], glioma [130], and basal cell carcinoma (BCC) [131]. Mammalian HH signaling is initiated when one of three HH ligands (Sonic, Indian, and Desert HH) binds the dodecatransmembrane receptor Patched (Ptch1). Ligand/receptor interactions occur through an autocrine or paracrine manner, depending on the context. Receptor engagement results in activation of the heptatransmembrane Smoothened (Smo), which is held in an inactive state in the absence of a ligand. Smo activation in turn regulates the activity of transcription factors Gli1, Gli2 and Gli3. Gli1/2/3 function to regulate transcription of genes involved in HH signaling such as Gli1 and Ptch1, and importantly genes involved in epithelial-mesenchymaltransition (EMT), such as SNAIL1[127, 128].

HH-GLI signaling was found to modulate normal dorsal brain growth by controlling precursor proliferation [132]; it was also found to have an essential role in controlling the behavior of CD133+ glioma cancer stem cells [130]. However, HH pathway-driven tumorigenesis depends on canonical Wnt/β-catenin signaling in BCC [131]. Recently, CSC/tumor initiating cells (TIC) in human melanomas were found in a collection of human melanomas obtained from a broad spectrum of sites and stages by using non-adherent spheres and ALDH enzymatic activity. Both pharmacological inhibition of HH signaling by the SMO antagonist cyclopamine and GLI antagonist GANT61, and stable expression of shRNA targeting either SMO or GLI1 result in a significant decrease in melanoma stem cell self-renewal *in vitro* and a reduction in the number of ALDH high melanoma stem cells, indicating an essential role of the HH-GLI1 signaling in of melanoma CSC/TIC. Notably, melanomaspheres express not only high levels of Hedgehog pathway components, but also high levels of embryonic pluripotent stem cell factors Sox2, Nanog, Oct4 and KLF4 [129]. This is the first report that reveals a possible correlation of HH signaling and KLF4 in CSC, though the underlying mechanism appears entirely unknown.

8. Concluding remarks

Since the identification and characterization of KLF4 over 10 years ago, significant progression has been made to understand its biological function, including its role in cell proliferation, differentiation, apoptosis and maintenance of normal tissue homeostasis. However, a novel role of KLF4 in stem cell biology further opens a window to study KLF4 in a different area. KLF4 is believed to play a significant role in ES cell self-renewal and pluripotency. Notably, KLF4 collaborating with other transcription factors including Oct4, Sox2 and c-Myc, drives somatic cells into iPS cells. CSCs have been identified in various tumors, and KLF4 can be speculated to have similar functions in CSCs based on its function in ES cell [133]. Our work provides evidence for the first time that KLF4 is essential for the maintenance of breast CSC and cell migration and invasion, which may be helpful to develop new therapeutic strategies for breast cancer. Apart from just breast CSCs, our work also demonstrates that KLF4 is highly expressed in skin hair follicle stem cells and facilitates the process of cutaneous wound healing. Many papers have confirmed the underlying molecular mechanism that KLF4 exerts its action in stem cell biology by integration of different signaling pathways, including Notch, Wnt and HH. Notch signaling pathway is responsible for KLF4-mediated mobility characteristics of breast cancer cells, while Wnt/β-catenin signaling recruits KLF4 to regulate TERT expression in stem cells and cancer cells. As to HH signaling and KLF4, the research is still just beginning, but considering the crosstalk between Wnt/β-catenin and HH, it is very important to discern the communication between them. Nevertheless, understanding the signaling circuitries regulating stem cell fate decisions might provide important insights into novel therapeutic strategies for cancer and regeneration medicine.

Acknowledgements

The authors gratefully acknowledge Mr. Andrew Vaughan and Mr. Matthew Riester for critical reading and editing of the manuscript. This work was supported by NIH grants to W. A. (KO1DK069489 and RO3AR060987).

Author details

Ying Shi[1] and Walden Ai[2*]

*Address all correspondence to: Walden.Ai@uscmed.sc.edu

1 Department of Urology, Union Hospital, Tongji Medical College, Huazhong University of Science and Technology, Wuhan, Hubai, China

2 Department of Pathology, Microbiology, and Immunology, University of South Carolina School of Medicine, Columbia, SC, USA

References

[1] Bieker JJ.Kruppel-like factors: three fingers in many pies. Journal of Biological Chemistry 2001;276(37)34355-34358.

[2] Philipsen S, Suske G.A tale of three fingers: the family of mammalian Sp/XKLF transcription factors. Nucleic Acids Research 1999;27(15)2991-3000.

[3] Turner J, Crossley M.Mammalian Kruppel-like transcription factors: more than just a pretty finger. Trends In Biochemical Sciences 1999;24(6)236-240.

[4] Kaczynski J, Cook T, Urrutia R.Sp1- and Kruppel-like transcription factors. Genome Biology 2003;4(2)206.

[5] Takahashi K, Yamanaka S.Induction of pluripotent stem cells from mouse embryonic and adult fibroblast cultures by defined factors. Cell 2006;126(4)663-676.

[6] Li J, Zheng H, Wang J, Yu F, Morris RJ, et al.Expression of Kruppel-Like Factor KLF4 in Mouse Hair Follicle Stem Cells Contributes to Cutaneous Wound Healing. PLoS One 2012;7(6)e39663.

[7] Yu F, Li J, Chen H, Fu J, Ray S, et al.Kruppel-like factor 4 (KLF4) is required for maintenance of breast cancer stem cells and for cell migration and invasion. Oncogene 2011;30(18)2161-2172.

[8] Shields JM, Christy RJ, Yang VW.Identification and characterization of a gene encoding a gut-enriched Kruppel-like factor expressed during growth arrest. Journal of Biological Chemistry 1996;271(33)20009-20017.

[9] Garrett-Sinha LA, Eberspaecher H, Seldin MF, de Crombrugghe B.A gene for a novel zinc-finger protein expressed in differentiated epithelial cells and transiently in certain mesenchymal cells. Journal of Biological Chemistry 1996;271(49)31384-31390.

[10] Yet SF, McA'Nulty MM, Folta SC, Yen HW, Yoshizumi M, et al.Human EZF, a Kruppel-like zinc finger protein, is expressed in vascular endothelial cells and contains transcriptional activation and repression domains. Journal of Biological Chemistry 1998;273(2)1026-1031.

[11] Panigada M, Porcellini S, Sutti F, Doneda L, Pozzoli O, et al.GKLF in thymus epithelium as a developmentally regulated element of thymocyte-stroma cross-talk. Mechaniams of Development 1999;81(1-2)103-113.

[12] Chiambaretta F, De Graeve F, Turet G, Marceau G, Gain P, et al.Cell and tissue specific expression of human Kruppel-like transcription factors in human ocular surface. Molecular Vision 2004;10 901-909.

[13] Cullingford TE, Butler MJ, Marshall AK, Tham el L, Sugden PH, et al.Differential regulation of Kruppel-like factor family transcription factor expression in neonatal rat cardiac myocytes: effects of endothelin-1, oxidative stress and cytokines. Biochimica et Biophysica Acta 2008;1783(6)1229-1236.

[14] Fruman DA, Ferl GZ, An SS, Donahue AC, Satterthwaite AB, et al.Phosphoinositide 3-kinase and Bruton's tyrosine kinase regulate overlapping sets of genes in B lymphocytes. Proceedings of the National Academy of Sci ence of the United States of America 2002;99(1)359-364.

[15] Behr R, Kaestner KH.Developmental and cell type-specific expression of the zinc finger transcription factor Kruppel-like factor 4 (Klf4) in postnatal mouse testis. Mechaniams of Development 2002;115(1-2)167-169.

[16] Godmann M, Kromberg I, Mayer J, Behr R.The mouse Kruppel-like Factor 4 (Klf4) gene: four functional polyadenylation sites which are used in a cell-specific manner as revealed by testicular transcript analysis and multiple processed pseudogenes. Gene 2005;361(149-156.

[17] Chen X, Whitney EM, Gao SY, Yang VW.Transcriptional profiling of Kruppel-like factor 4 reveals a function in cell cycle regulation and epithelial differentiation. Journal of Molecular Biology 2003;326(3)665-677.

[18] Zhang W, Geiman DE, Shields JM, Dang DT, Mahatan CS, et al.The gut-enriched Kruppel-like factor (Kruppel-like factor 4) mediates the transactivating effect of p53 on the p21WAF1/Cip1 promoter. Journal of Biological Chemistry 2000;275(24)18391-18398.

[19] Shie JL, Chen ZY, Fu M, Pestell RG, Tseng CC.Gut-enriched Kruppel-like factor represses cyclin D1 promoter activity through Sp1 motif. Nucleic Acids Research 2000;28(15)2969-2976.

[20] Chen X, Johns DC, Geiman DE, Marban E, Dang DT, et al.Kruppel-like factor 4 (gut-enriched Kruppel-like factor) inhibits cell proliferation by blocking G1/S progression of the cell cycle. Journal of Biological Chemistry 2001;276(32)30423-30428.

[21] Yoon HS, Chen X, Yang VW.Kruppel-like factor 4 mediates p53-dependent G1/S cell cycle arrest in response to DNA damage. Journal of Biological Chemistry 2003;278(4)2101-2105.

[22] Yoon HS, Yang VW.Requirement of Kruppel-like factor 4 in preventing entry into mitosis following DNA damage. Journal of Biological Chemistry 2004;279(6)5035-5041.

[23] Yoon HS, Ghaleb AM, Nandan MO, Hisamuddin IM, Dalton WB, et al.Kruppel-like factor 4 prevents centrosome amplification following gamma-irradiation-induced DNA damage. Oncogene 2005;24(25)4017-4025.

[24] Zhang W, Shields JM, Sogawa K, Fujii-Kuriyama Y, Yang VW.The gut-enriched Kruppel-like factor suppresses the activity of the CYP1A1 promoter in an Sp1-dependent fashion. Journal of Biological Chemistry 1998;273(28)17917-17925.

[25] Miller KA, Eklund EA, Peddinghaus ML, Cao Z, Fernandes N, et al.Kruppel-like factor 4 regulates laminin alpha 3A expression in mammary epithelial cells. Journal of Biological Chemistry 2001;276(46)42863-42868.

[26] Higaki Y, Schullery D, Kawata Y, Shnyreva M, Abrass C, et al.Synergistic activation of the rat laminin gamma1 chain promoter by the gut-enriched Kruppel-like factor (GKLF/KLF4) and Sp1. Nucleic Acids Research 2002;30(11)2270-2279.

[27] Okano J, Opitz OG, Nakagawa H, Jenkins TD, Friedman SL, et al.The Kruppel-like transcriptional factors Zf9 and GKLF coactivate the human keratin 4 promoter and physically interact. Febs Letters 2000;473(1)95-100.

[28] Brembeck FH, Rustgi AK.The tissue-dependent keratin 19 gene transcription is regulated by GKLF/KLF4 and Sp1. Journal of Biological Chemistry 2000;275(36)28230-28239.

[29] Katz JP, Perreault N, Goldstein BG, Lee CS, Labosky PA, et al.The zinc-finger transcription factor Klf4 is required for terminal differentiation of goblet cells in the colon. Development 2002;129(11)2619-2628.

[30] Katz JP, Perreault N, Goldstein BG, Actman L, McNally SR, et al.Loss of Klf4 in mice causes altered proliferation and differentiation and precancerous changes in the adult stomach. Gastroenterology 2005;128(4)935-945.

[31] Swamynathan SK, Katz JP, Kaestner KH, Ashery-Padan R, Crawford MA, et al.Conditional deletion of the mouse Klf4 gene results in corneal epithelial fragility, stromal edema, and loss of conjunctival goblet cells. Molecular and Cellular Biology 2007;27(1)182-194.

[32] Segre JA, Bauer C, Fuchs E.Klf4 is a transcription factor required for establishing the barrier function of the skin. Nature Genetics 1999;22(4)356-360.

[33] Adam PJ, Regan CP, Hautmann MB, Owens GK.Positive- and negative-acting Kruppel-like transcription factors bind a transforming growth factor beta control element required for expression of the smooth muscle cell differentiation marker SM22alpha in vivo. Journal of Biological Chemistry 2000;275(48)37798-37806.

[34] Feinberg MW, Wara AK, Cao Z, Lebedeva MA, Rosenbauer F, et al.The Kruppel-like factor KLF4 is a critical regulator of monocyte differentiation. Embo Journal 2007;26(18)4138-4148.

[35] Dik WA, Pike-Overzet K, Weerkamp F, de Ridder D, de Haas EF, et al.New insights on human T cell development by quantitative T cell receptor gene rearrangement studies and gene expression profiling. Journal of Experimental Medicine 2005;201(11)1715-1723.

[36] Chen Z, Couble ML, Mouterfi N, Magloire H, Bleicher F.Spatial and temporal expression of KLF4 and KLF5 during murine tooth development. Archives of Oral Biology 2009;54(5)403-411.

[37] Feinberg MW, Cao Z, Wara AK, Lebedeva MA, Senbanerjee S, et al.Kruppel-like factor 4 is a mediator of proinflammatory signaling in macrophages. Journal of Biological Chemistry 2005;280(46)38247-38258.

[38] Liu J, Liu Y, Zhang H, Chen G, Wang K, et al.KLF4 promotes the expression, translocation, and releas eof HMGB1 in RAW264.7 macrophages in response to LPS. Shock 2008;30(3)260-266.

[39] Liu J, Zhang H, Liu Y, Wang K, Feng Y, et al.KLF4 regulates the expression of interleukin-10 in RAW264.7 macrophages. Biochemical and Biophysical Research Communications 2007;362(3)575-581.

[40] Alder JK, Georgantas RW, 3rd, Hildreth RL, Kaplan IM, Morisot S, et al.Kruppel-like factor 4 is essential for inflammatory monocyte differentiation in vivo. Journal of Immunology 2008;180(8)5645-5652.

[41] Lee HJ, Kang YM, Moon CS, Joe MK, Lim JH, et al.KLF4 positively regulates human ghrelin expression. Biochemical Journal 2009;420(3)403-411.

[42] Dijkmans TF, van Hooijdonk LW, Schouten TG, Kamphorst JT, Fitzsimons CP, et al.Identification of new Nerve Growth Factor-responsive immediate-early genes. Brain Research 2009;1249 19-33.

[43] Zhu S, Tai C, MacVicar BA, Jia W, Cynader MS.Glutamatergic stimulation triggers rapid Krupple-like factor 4 expression in neurons and the overexpression of KLF4 sensitizes neurons to NMDA-induced caspase-3 activity. Brain Research 2009;1250(49-62.

[44] Yamada T, Park CS, Mamonkin M, Lacorazza HD.Transcription factor ELF4 controls the proliferation and homing of CD8+ T cells via the Kruppel-like factors KLF4 and KLF2. Nature Immunology 2009;10(6)618-626.

[45] Wei D, Gong W, Kanai M, Schlunk C, Wang L, et al.Drastic down-regulation of Kruppel-like factor 4 expression is critical in human gastric cancer development and progression. Cancer Research 2005;65(7)2746-2754.

[46] Ghaleb AM, Katz JP, Kaestner KH, Du JX, Yang VW.Kruppel-like factor 4 exhibits antiapoptotic activity following gamma-radiation-induced DNA damage. Oncogene 2007;26(16)2365-2373.

[47] Dang DT, Chen X, Feng J, Torbenson M, Dang LH, et al.Overexpression of Kruppel-like factor 4 in the human colon cancer cell line RKO leads to reduced tumorigenecity. Oncogene 2003;22(22)3424-3430.

[48] Zhang G, Zhu H, Wang Y, Yang S, Liu M, et al.Kruppel-like factor 4 represses transcription of the survivin gene in esophageal cancer cell lines. Biological Chemistry 2009;390(5-6)463-469.

[49] Kidder BL, Yang J, Palmer S.Stat3 and c-Myc genome-wide promoter occupancy in embryonic stem cells. PLoS One 2008;3(12)e3932.

[50] Rodda DJ, Chew JL, Lim LH, Loh YH, Wang B, et al.Transcriptional regulation of nanog by OCT4 and SOX2. Journal of Biological Chemistry 2005;280(26)24731-24737.

[51] Zhang P, Andrianakos R, Yang Y, Liu C, Lu W.Kruppel-like factor 4 (Klf4) prevents embryonic stem (ES) cell differentiation by regulating Nanog gene expression. Journal of Biological Chemistry 2010;285(12)9180-9189.

[52] Wernig M, Meissner A, Foreman R, Brambrink T, Ku M, et al.In vitro reprogramming of fibroblasts into a pluripotent ES-cell-like state. Nature 2007;448(7151)318-324.

[53] Maherali N, Sridharan R, Xie W, Utikal J, Eminli S, et al.Directly reprogrammed fibroblasts show global epigenetic remodeling and widespread tissue contribution. Cell Stem Cell 2007;1(1)55-70.

[54] Yu J, Vodyanik MA, Smuga-Otto K, Antosiewicz-Bourget J, Frane JL, et al.Induced pluripotent stem cell lines derived from human somatic cells. Science 2007;318(5858)1917-1920.

[55] Takahashi K, Tanabe K, Ohnuki M, Narita M, Ichisaka T, et al.Induction of pluripotent stem cells from adult human fibroblasts by defined factors. Cell 2007;131(5)861-872.

[56] Park IH, Zhao R, West JA, Yabuuchi A, Huo H, et al.Reprogramming of human somatic cells to pluripotency with defined factors. Nature 2008;451(7175)141-146.

[57] Lowry WE, Richter L, Yachechko R, Pyle AD, Tchieu J, et al.Generation of human induced pluripotent stem cells from dermal fibroblasts. Proceedings of the National Academy of Sci ence of the United States of America 2008;105(8)2883-2888.

[58] Hanna J, Markoulaki S, Schorderet P, Carey BW, Beard C, et al.Direct reprogramming of terminally differentiated mature B lymphocytes to pluripotency. Cell 2008;133(2)250-264.

[59] Hanna J, Wernig M, Markoulaki S, Sun CW, Meissner A, et al.Treatment of sickle cell anemia mouse model with iPS cells generated from autologous skin. Science 2007;318(5858)1920-1923.

[60] Nakagawa M, Koyanagi M, Tanabe K, Takahashi K, Ichisaka T, et al.Generation of induced pluripotent stem cells without Myc from mouse and human fibroblasts. Nature Biotechnology 2008;26(1)101-106.

[61] Yamanaka S.A fresh look at iPS cells. Cell 2009;137(1)13-17.

[62] Zhao R, Daley GQ.From fibroblasts to iPS cells: induced pluripotency by defined factors. Journal of Cellular Biochemistry 2008;105(4)949-955.

[63] Bonnet D, Dick JE.Human acute myeloid leukemia is organized as a hierarchy that originates from a primitive hematopoietic cell. Nature Medicine 1997;3(7)730-737.

[64] Schatton T, Frank NY, Frank MH.Identification and targeting of cancer stem cells. Bioessays 2009;31(10)1038-1049.

[65] O'Brien CA, Pollett A, Gallinger S, Dick JE.A human colon cancer cell capable of initiating tumour growth in immunodeficient mice. Nature 2007;445(7123)106-110.

[66] Visvader JE, Lindeman GJ.Cancer stem cells in solid tumours: accumulating evidence and unresolved questions. Nature Reviews. Cancer 2008;8(10)755-768.

[67] Yang YM, Chang JW.Bladder cancer initiating cells (BCICs) are among EMA-CD44v6+ subset: novel methods for isolating undetermined cancer stem (initiating) cells. Cancer Investigation 2008;26(7)725-733.

[68] Eramo A, Haas TL, De Maria R.Lung cancer stem cells: tools and targets to fight lung cancer. Oncogene 2010;29(33)4625-4635.

[69] Al-Hajj M, Wicha MS, Benito-Hernandez A, Morrison SJ, Clarke MF.Prospective identification of tumorigenic breast cancer cells. Proceedings of the National Academy of Sci ence of the United States of America 2003;100(7)3983-3988.

[70] Charafe-Jauffret E, Ginestier C, Iovino F, Wicinski J, Cervera N, et al.Breast cancer cell lines contain functional cancer stem cells with metastatic capacity and a distinct molecular signature. Cancer Research 2009;69(4)1302-1313.

[71] Chute JP, Muramoto GG, Whitesides J, Colvin M, Safi R, et al.Inhibition of aldehyde dehydrogenase and retinoid signaling induces the expansion of human hematopoietic stem cells. Proceedings of the National Academy of Sci ence of the United States of America 2006;103(31)11707-11712.

[72] Hirschmann-Jax C, Foster AE, Wulf GG, Nuchtern JG, Jax TW, et al.A distinct "side population" of cells with high drug efflux capacity in human tumor cells. Proceedings of the National Academy of Sci ence of the United States of America 2004;101(39)14228-14233.

[73] Foster KW, Frost AR, McKie-Bell P, Lin CY, Engler JA, et al.Increase of GKLF messenger RNA and protein expression during progression of breast cancer. Cancer Research 2000;60(22)6488-6495.

[74] Pandya AY, Talley LI, Frost AR, Fitzgerald TJ, Trivedi V, et al.Nuclear localization of KLF4 is associated with an aggressive phenotype in early-stage breast cancer. Clinical Cancer Research 2004;10(8)2709-2719.

[75] Rowland BD, Bernards R, Peeper DS.The KLF4 tumour suppressor is a transcriptional repressor of p53 that acts as a context-dependent oncogene. Nature Cell Biology 2005;7(11)1074-1082.

[76] Charafe-Jauffret E, Monville F, Ginestier C, Dontu G, Birnbaum D, et al.Cancer stem cells in breast: current opinion and future challenges. Pathobiology 2008;75(2)75-84.

[77] Lawson JC, Blatch GL, Edkins AL.Cancer stem cells in breast cancer and metastasis. Breast Cancer Research and Treatment 2009;118(2)241-254.

[78] Lyssiotis CA, Foreman RK, Staerk J, Garcia M, Mathur D, et al.Reprogramming of murine fibroblasts to induced pluripotent stem cells with chemical complementation of Klf4. Proceedings of the National Academy of Sci ence of the United States of America 2009;106(22)8912-8917.

[79] Hinnebusch BF, Siddique A, Henderson JW, Malo MS, Zhang W, et al.Enterocyte differentiation marker intestinal alkaline phosphatase is a target gene of the gut-enriched Kruppel-like factor. American Journal of Physiology. Gastrointestinal and Liver Physiology 2004;286(1)G23-30.

[80] Wassmann S, Wassmann K, Jung A, Velten M, Knuefermann P, et al.Induction of p53 by GKLF is essential for inhibition of proliferation of vascular smooth muscle cells. Journal of Molecular and Cellular Cardiology 2007;43(3)301-307.

[81] Mani SA, Guo W, Liao MJ, Eaton EN, Ayyanan A, et al.The epithelial-mesenchymal transition generates cells with properties of stem cells. Cell 2008;133(4)704-715.

[82] Yang J, Weinberg RA.Epithelial-mesenchymal transition: at the crossroads of development and tumor metastasis. Developmental Cell 2008;14(6)818-829.

[83] Miettinen PJ, Ebner R, Lopez AR, Derynck R.TGF-beta induced transdifferentiation of mammary epithelial cells to mesenchymal cells: involvement of type I receptors. Journal of Cell Biology 1994;127(6 Pt 2)2021-2036.

[84] Malanchi I, Peinado H, Kassen D, Hussenet T, Metzger D, et al.Cutaneous cancer stem cell maintenance is dependent on beta-catenin signalling. Nature 2008;452(7187)650-653.

[85] Yori JL, Johnson E, Zhou G, Jain MK, Keri RA.Kruppel-like factor 4 inhibits epithelial-to-mesenchymal transition through regulation of E-cadherin gene expression. Journal of Biological Chemistry 2010;285(22)16854-16863.

[86] Cotsarelis G, Sun TT, Lavker RM.Label-retaining cells reside in the bulge area of pilosebaceous unit: implications for follicular stem cells, hair cycle, and skin carcinogenesis. Cell 1990;61(7)1329-1337.

[87] Trempus CS, Morris RJ, Bortner CD, Cotsarelis G, Faircloth RS, et al.Enrichment for living murine keratinocytes from the hair follicle bulge with the cell surface marker CD34. Journal of Investigative Dermatology 2003;120(4)501-511.

[88] Morris RJ, Liu Y, Marles L, Yang Z, Trempus C, et al.Capturing and profiling adult hair follicle stem cells. Nature Biotechnology 2004;22(4)411-417.

[89] Blanpain C, Lowry WE, Geoghegan A, Polak L, Fuchs E.Self-renewal, multipotency, and the existence of two cell populations within an epithelial stem cell niche. Cell 2004;118(5)635-648.

[90] Braun KM, Niemann C, Jensen UB, Sundberg JP, Silva-Vargas V, et al.Manipulation of stem cell proliferation and lineage commitment: visualisation of label-retaining cells in wholemounts of mouse epidermis. Development 2003;130(21)5241-5255.

[91] Tumbar T, Guasch G, Greco V, Blanpain C, Lowry WE, et al.Defining the epithelial stem cell niche in skin. Science 2004;303(5656)359-363.

[92] Jensen KB, Watt FM.Single-cell expression profiling of human epidermal stem and transit-amplifying cells: Lrig1 is a regulator of stem cell quiescence. Proceedings of the National Academy of Sci ence of the United States of America 2006;103(32)11958-11963.

[93] Jensen KB, Collins CA, Nascimento E, Tan DW, Frye M, et al.Lrig1 expression defines a distinct multipotent stem cell population in mammalian epidermis. Cell Stem Cell 2009;4(5)427-439.

[94] Taylor G, Lehrer MS, Jensen PJ, Sun TT, Lavker RM.Involvement of follicular stem cells in forming not only the follicle but also the epidermis. Cell 2000;102(4)451-461.

[95] Ito M, Liu Y, Yang Z, Nguyen J, Liang F, et al.Stem cells in the hair follicle bulge contribute to wound repair but not to homeostasis of the epidermis. Nature Medicine 2005;11(12)1351-1354.

[96] Levy V, Lindon C, Zheng Y, Harfe BD, Morgan BA.Epidermal stem cells arise from the hair follicle after wounding. Faseb Journal 2007;21(7)1358-1366.

[97] Nowak JA, Polak L, Pasolli HA, Fuchs E.Hair follicle stem cells are specified and function in early skin morphogenesis. Cell Stem Cell 2008;3(1)33-43.

[98] Dang DT, Pevsner J, Yang VW.The biology of the mammalian Kruppel-like family of transcription factors. The International Journal of Biochemistry & Cell Biology 2000;32(11-12)1103-1121.

[99] Wei D, Wang L, Kanai M, Jia Z, Le X, et al.KLF4alpha up-regulation promotes cell cycle progression and reduces survival time of patients with pancreatic cancer. Gastroenterology 2010;139(6)2135-2145.

[100] Snippert HJ, Haegebarth A, Kasper M, Jaks V, van Es JH, et al.Lgr6 marks stem cells in the hair follicle that generate all cell lineages of the skin. Science 2010;327(5971)1385-1389.

[101] Langton AK, Herrick SE, Headon DJ.An extended epidermal response heals cutaneous wounds in the absence of a hair follicle stem cell contribution. Journal of Investigative Dermatology 2008;128(5)1311-1318.

[102] Leung CT, Coulombe PA, Reed RR.Contribution of olfactory neural stem cells to tissue maintenance and regeneration. Nature Neuroscience 2007;10(6)720-726.

[103] Ohlstein B, Kai T, Decotto E, Spradling A.The stem cell niche: theme and variations. Current Opinion In Cell Biology 2004;16(6)693-699.

[104] Bray SJ.Notch signalling: a simple pathway becomes complex. Nature Reviews. Molecular Cell Biology 2006;7(9)678-689.

[105] Joseph NM, Morrison SJ.Toward an understanding of the physiological function of Mammalian stem cells. Developmental Cell 2005;9(2)173-183.

[106] Keith B, Simon MC.Hypoxia-inducible factors, stem cells, and cancer. Cell 2007;129(3)465-472.

[107] Zheng H, Pritchard DM, Yang X, Bennett E, Liu G, et al.KLF4 gene expression is inhibited by the notch signaling pathway that controls goblet cell differentiation in mouse gastrointestinal tract. American Journal of Physiology. Gastrointestinal and Liver Physiology 2009;296(3)G490-498.

[108] Ohlstein B, Spradling A.The adult Drosophila posterior midgut is maintained by pluripotent stem cells. Nature 2006;439(7075)470-474.

[109] Ghaleb AM, Aggarwal G, Bialkowska AB, Nandan MO, Yang VW.Notch inhibits expression of the Kruppel-like factor 4 tumor suppressor in the intestinal epithelium. Moecularl Cancer Research: MCR 2008;6(12)1920-1927.

[110] Real PJ, Tosello V, Palomero T, Castillo M, Hernando E, et al.Gamma-secretase inhibitors reverse glucocorticoid resistance in T cell acute lymphoblastic leukemia. Nature Medicine 2009;15(1)50-58.

[111] Lambertini C, Pantano S, Dotto GP.Differential control of Notch1 gene transcription by Klf4 and Sp3 transcription factors in normal versus cancer-derived keratinocytes. PLoS One 2010;5(4)e10369.

[112] Sahlgren C, Gustafsson MV, Jin S, Poellinger L, Lendahl U.Notch signaling mediates hypoxia-induced tumor cell migration and invasion. Proceedings of the National Academy of Sci ence of the United States of America 2008;105(17)6392-6397.

[113] Liu Z, Teng L, Bailey SK, Frost AR, Bland KI, et al.Epithelial transformation by KLF4 requires Notch1 but not canonical Notch1 signaling. Cancer Biology & Therapy 2009;8(19)1840-1851.

[114] Cadigan KM, Nusse R.Wnt signaling: a common theme in animal development. Genes & Development 1997;11(24)3286-3305.

[115] Clevers H.Wnt/beta-catenin signaling in development and disease. Cell 2006;127(3)469-480.

[116] van de Wetering M, Sancho E, Verweij C, de Lau W, Oving I, et al.The beta-catenin/TCF-4 complex imposes a crypt progenitor phenotype on colorectal cancer cells. Cell 2002;111(2)241-250.

[117] Curtin JC, Lorenzi MV.Drug discovery approaches to target Wnt signaling in cancer stem cells. Oncotarget 2010;1(7)563-577.

[118] MacDonald BT, Tamai K, He X.Wnt/beta-catenin signaling: components, mechanisms, and diseases. Developmental Cell 2009;17(1)9-26.

[119] Groden J, Thliveris A, Samowitz W, Carlson M, Gelbert L, et al.Identification and characterization of the familial adenomatous polyposis coli gene. Cell 1991;66(3)589-600.

[120] Kinzler KW, Nilbert MC, Su LK, Vogelstein B, Bryan TM, et al.Identification of FAP locus genes from chromosome 5q21. Science 1991;253(5020)661-665.

[121] Morin PJ, Sparks AB, Korinek V, Barker N, Clevers H, et al.Activation of beta-catenin-Tcf signaling in colon cancer by mutations in beta-catenin or APC. Science 1997;275(5307)1787-1790.

[122] Zhang W, Chen X, Kato Y, Evans PM, Yuan S, et al.Novel cross talk of Kruppel-like factor 4 and beta-catenin regulates normal intestinal homeostasis and tumor repression. Molecular and Cellular Biology 2006;26(6)2055-2064.

[123] Vermeulen L, De Sousa EMF, van der Heijden M, Cameron K, de Jong JH, et al.Wnt activity defines colon cancer stem cells and is regulated by the microenvironment. Nature Cell Biology 2010;12(5)468-476.

[124] Hoffmeyer K, Raggioli A, Rudloff S, Anton R, Hierholzer A, et al.Wnt/beta-catenin signaling regulates telomerase in stem cells and cancer cells. Science 2012;336(6088)1549-1554.

[125] Brennan KR, Brown AM.Wnt proteins in mammary development and cancer. Journalof Mammary Gland Biology and Neoplasia 2004;9(2)119-131.

[126] Grigoryan T, Wend P, Klaus A, Birchmeier W.Deciphering the function of canonical Wnt signals in development and disease: conditional loss- and gain-of-function mutations of beta-catenin in mice. Genes & Development 2008;22(17)2308-2341.

[127] Ingham PW, McMahon AP.Hedgehog signaling in animal development: paradigms and principles. Genes & Development 2001;15(23)3059-3087.

[128] Varjosalo M, Taipale J.Hedgehog: functions and mechanisms. Genes & Development 2008;22(18)2454-2472.

[129] Santini R, Vinci MC, Pandolfi S, Penachioni JY, Montagnani V, et al.HEDGEHOG-GLI Signaling Drives Self-Renewal and Tumorigenicity of Human Melanoma-Initiating Cells. Stem Cells 2012;

[130] Clement V, Sanchez P, de Tribolet N, Radovanovic I, Ruiz i Altaba A.HEDGEHOG-GLI1 signaling regulates human glioma growth, cancer stem cell self-renewal, and tumorigenicity. Current Biology 2007;17(2)165-172.

[131] Yang SH, Andl T, Grachtchouk V, Wang A, Liu J, et al.Pathological responses to oncogenic Hedgehog signaling in skin are dependent on canonical Wnt/beta3-catenin signaling. Nature Genetics 2008;40(9)1130-1135.

[132] Dahmane N, Sanchez P, Gitton Y, Palma V, Sun T, et al.The Sonic Hedgehog-Gli pathway regulates dorsal brain growth and tumorigenesis. Development 2001;128(24)5201-5212.

[133] Schoenhals M, Kassambara A, De Vos J, Hose D, Moreaux J, et al.Embryonic stem cell markers expression in cancers. Biochemical and Biophysical Research Communications 2009;383(2)157-162.

Chapter 16

β1,4-Galactosyltransferases, Potential Modifiers of Stem Cell Pluripotency and Differentiation

Michael Wassler

Additional information is available at the end of the chapter

http://dx.doi.org/10.5772/54376

1. Introduction

The ability of embryonic stem cells to self renew and, at a given signal, give rise to the multifaceted cell types normally observed in the body, is highly dependent on the complex interplay between both intrinsic (inside the cell) and extrinsic (outside of the cell) factors. Despite progress in analyzing the genome, proteome, and the transcriptome, challenges still exists to find the most efficient and specific conditions in which human embryonic stem cells (hESC) can maintain pluripotency and or/can be efficiently directed to differentiate towards a homogenous cell type. In a stem cell niche, the integrity of the cell matrix and the manifold of different cell-cell interactions and the ability of the cells to respond to a variety of cytokine cues from both interstitial fluids and from extracellular matrices, are crucial factors in giving the right signal signals to the cells internal machinery, in a space (spatio) and time (temporal) manner during different developmental stages. One of these molecules is the glycan. A glycan is a polysaccharide or oligosaccharide, that is attached to a glucoconjugate such as glycoprotein, glycolipid, and proteoglycan. Cell surface glycoproteins are abundant and constitute approximately 50% of all proteins in nature. For many years, the biological function of glycosylation in stem cell behavior/homeostasis was overlooked and thought of as a more or less redundant process with applications only limited to the identification and sorting of cells at different stages of pluripotency and during formation of induced pluripotent stem cells (iPSC). Markers such as stage specific embryonic antigen (SSEA1 and -3/4) and the tumor rejection antigens (TRA-1-60 and TRA-1-81) have been used to analyze the pluripotency and differentiation stages of embryonic stem cells and induced pluripotent stem cell (iPSC).The research of how glycosylation can impact stem cells has long been hampered by the structural complexities of glycosylation and the difficulties to identify and purify the enzymes, glycosyltransferases, responsible for these processes. This problem is partly due to the fact that glycans are not encoded directly from the genome but rather depends on the collaboration of a limited number of both glycosyltransferases and glycosidases, whose expression are reliant upon both intracellular as well as extracellular changes. Furthermore, glycosyltransferases are expressed differentially between many cell types and disease states in a spatio- temporal manner during development. In this review, I will summarize research on what is known for glycosyltransferases in stem cell pluripotency and differentiation. I will specifically focus on one glycosyltransferase, N-acetylglucosamin β1,4- Galactosyltransferase 1 (β4Gal-T1), a unique galactosyltrasferase implicated in a variety of cellular processes such as cell-cell and cell-matrix adhesion, apoptosis, proliferation and differentiation, to mention a few. I will discuss its regulation and potential mechanism(s) in cell-cell, cell-matrix and cytokine signaling pathways. Finally, in the last section, I will talk about some diseases related to galactosyltransferase deficiency. All in all, this chapter is intended to evoke more interest in the field of stem cell glycobiology, both for the layman as well as for the bench scientist. Ultimately, the goal of this review is to encourage future research to find alternative therapeutic modalities for glycoprotein related diseases, such as cancer, congenital disease and even Alzheimer's.

2. What is glycosyltransferases?

Glycosyltransferases (GTs; EC 2.4.x.y) constitute a large protein family of about 200-300 enzymes that are involved in the biosynthesis of glycans. GTs are type II transmembrane proteins with large carboxy-terminal globular catalytic domains, that face the luminal side of the Golgi complex, and a short cytoplasmic domain. The sequential action of GTs results in the formation of both linear as well as highly branched glycan structures that are present in both prokaryotes and eukaryotes. Mammalian GTs utilize a variety of uridine diphosphate activated (UDP) sugars as donors: UDP-glucose, UDP-galactose, UDP-GlcNAc, UDP-GalNAc, UDP-xylose, UDP-glucuronic acid, GDP-mannose, GDP-fucose, and CMP-sialic acid. Glycosyl transfer can occur on protein residues, usually to asparagine, to give N-linked glycoproteins and on tyrosine, serine, or threonine to give O-linked glycoproteins [1]. The first step in N-linked glycosylation occurs in the endoplamic reticulum (ER) in which a "high mannose" oligosaccharide branch is added to an Asparagine (Asn) residue in the protein backbone (N-linkage). Another type of glycan linkage is the *O*-linked glycosylation, which occurs through serine/threonin residues in the protein back bone during transport within the Golgie complex [2]. Other GTs are responsible for extensive branching of glycan structures such as the galactosyltransferase family (GalTs) [3] which together with glycosidases give rise to more "complex" type sugar chains (Figure 1). These processes creates oligosaccharide structures of enormous diversity and whose functions spans from cell adhesion, inflammation, cancer metastasis, stem cell proliferation and development [4]. This exciting area of biology has resulted in an intensive research to unveil the function of individual GTs in during stem cell pluripotency and differentiation. Several studies have implicated a variety of GTs in stem cell biology, some of which are presented below:

http://dx.doi.org/10.5772/54376

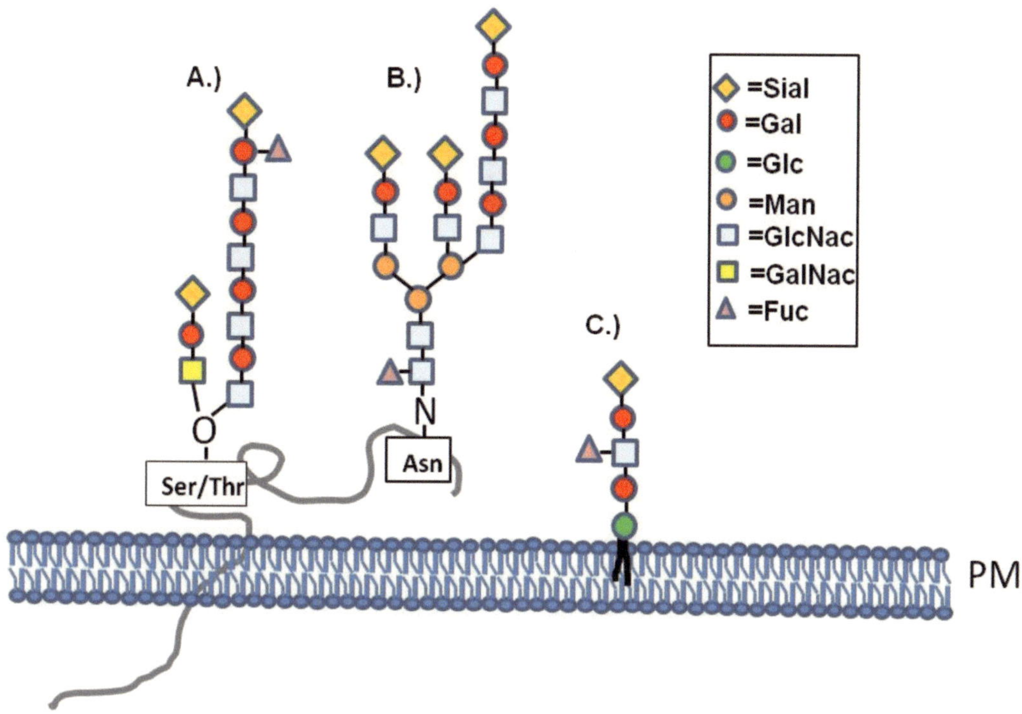

Figure 1. General view of an O-linked (A), and a (B) "complex" N-linked cell surface glycoprotein. A lactosylceramide-glycolipid (LacCer) (C) is also shown, located at the upper leaflet of the plasma membrane (PM). Ser; Serine, Thr;Threonine, Asn;Asparagine Sial;Sialic acid, Gal;Galactose, Glc;Glucose,Man;Mannose, GlcNac; N-Acetylglucoseamine, GalTNAc; N-Acetylglucoseamine, Fuc;Fucose

1. N-acetylglucosaminyl-1 phosphate transferase (GPT): The first steps in N-linked glycan synthesis begins on both the cytosolic and luminal side of the endoplasmic reticulum where nine mannosyl residues are sequentially added to a poly-isoprenoid lipid, dolychylmonophosphate by the activity of N-acetylglucosaminyl-1 phosphate transferase (GPT) and a number of mannosyltransferases. One inhibitor to GPT, tunicamycin (TM), inhibits N-linked glycosylation and has been reported affect cell proliferation, neu-vascularization and cancer progression, due to induced cell death from ER stress [4].

2. βGalNAc-T3: The cell surface glycan epitope LacdiNac (GalNac-β4GlcNAc) has been shown to be an important glycosylation modification of leukemia inhibitor factor receptor (LIFR) and its co-receptor, gp130.The addition of LacdiNac epitopes to LIFR was dependent on a specific transferase, β-3-N-acetyl-Galactosyl transferase 3 (βGalNac-T3). This modification is crucial for the localization of LIF to lipid rafts/ calveolar components, such as caveolin-1, in order to enhance its activity. Mouse and human stem cells (mESC, hESC) differ from each other in some aspects on how they respond to cytokines necessary for pluripotency. hESCs seem to be at a later developmental stage than mESCs, because of their independency of the LIF pathway for self renewal. Interestingly, the level of βGalNac-T3 was much lower in human versus mouse embryonic stem

indicating that LacdNac play an important role for adopting stem cells from a primed state (already programmed for germ line specification) to a more naïve state, e g fully pluripotent cells[5]

3. Ext1 and Ext2: Heparan sulphate is a large sulphated oligosaccharide chain located on proteoglycans impacting both the stability of pluripotency and differentiation into neural stem cell lineage. Ext1 and Ext2 encodes two bifunctional endoplasmic reticulum-resident type II transmembrane glycosyltransferase that are involved in the chain elongation and modification of HS biosynthesis. HS on embryonic stem cells has been shown to exhibit a lower amount of sulfated glycosaminoglycans relative to differentiated cells indicating that the ratio between nonsulphated versus sulphated HS is important in stem cell pluripotency [6-8]

4. O-GlcNac Transferase (OGT): O-GlcNAcylation is a O-β-glycosidic attachment of a single N-acetyl glucosamine to a serine or threonin residue in nucleoplasmic proteins. Some of these proteins are represented by the transcription factors Oct4, Klf4, Sox and Nanog, which are involved in the pluripotency network in stem cell self renewal and in the core proteins responsible for the production of induced pluripotent stem cells (iPSCs). Recently it was discovered that this specific O-linked modification of Oct4 and Sox was crucial for their transcriptional activities. Two enzymes are responsible for O-GlcNAcylation: O-GlcNac Transferase (OGT) adds the modification and O-glucNAcase removes it [9].

3. β-1,4-Galactosyltransferases

β-1,4-Galactosyltransferases (β4GalTs) are type II membrane proteins of the glycosyltransferase family that have the exclusive specificity to transfer an active UDP-galactose in a β1,4 linkage to acceptor sugars such as N-acetylglucosamine (GlcNAc), Glucose (Glc), Galactose (Gal) and even Xylose (Xyl). Each β4-GalTs have a distinct function in the biosynthesis of different glycoconjugates and disaccharide structures.The most common structure, the Galβ1-β4GlcNAc, or N-Acetyllactosamine, exists as disaccharide repeats within linear or branched poly-N-acetyllactoseamine chains, but also at the terminal ends of oligosaccharide chains where they become sialyllated. These structures are formed by a combined action of UDP-GlcNac:Mannosyl N-acetylglucosaminyltransferases and β-1,4-galactosyltransferases (β4GalTs) [10]. The first galactosyltransferase, β4GalT-1, was cloned in 1986 due to its function of transfer galactosyl residues to β-1,4-linked GlcNac found in glycoconjugates [11]. Targeted inactivation of mouse β4Gal-T1 gene, however, revealed that both tissue and serum glycoproteins still contained residual β4GalT-1 activity towards glycoprotein acceptors [10]. To date there are currently seven members of the β4GalT gene family designated β4Gal-T1-T7. Even though, β4Gal-T1 to -T6 shares various homologies (30-50%) to β4GalT-1 at the amino acid level, their substrate affinities and end products appear to be slightly different, depending on nature of the branched oligosaccharide structure tissue expression and the cellular milieu for the enzymes, e.g. lipid -rich environment [12, 13]. Both β4Gal-T1 and β4Gal-T2 preferentially transfer galactose to the GlcNacβ1-2Manα and the GlcNAcβ1-4Man1-3 branch. β4Gal-T4 and β4Gal-T5 catalyzes the addition of galactose to GlcNAcβ1-6Man and the GlcNAcβ1-4 Man, respectively (Figure 2). The β4Gal-T1, β4Gal-T2, and β4Gal-T3 can also transfer galactose residues to tetra-antenna oligosaccharides. In addition being involved in glycoconjugate synthesis, β4Gal-T2, -3, -4 and -6, are also important catalysts for glycolipid biosynthesis. β4Gal-T2 and -3 prefers a glycolipid intermediate, Lc3Cer, as a substrate and β4Gal-T4 uses GlcNac-6-sulphate, a common constituent of keratin sulphate, as a substrate [14]. β4Gal-T6 has been shown to have Lactosyl Ceramide synthase activity. Finally, β4Gal-T7, transfers a Galactose to an O-linked Xylose on a serine residue to start the synthesis of the linker region between glycosaminoglycans (GAG) and proteoglycans [15]. A general summary or the chromosomal location, tissue expression, glycosidic linkage and potential biological function of currently known β, 4-GalTs is summarized in Table 1.

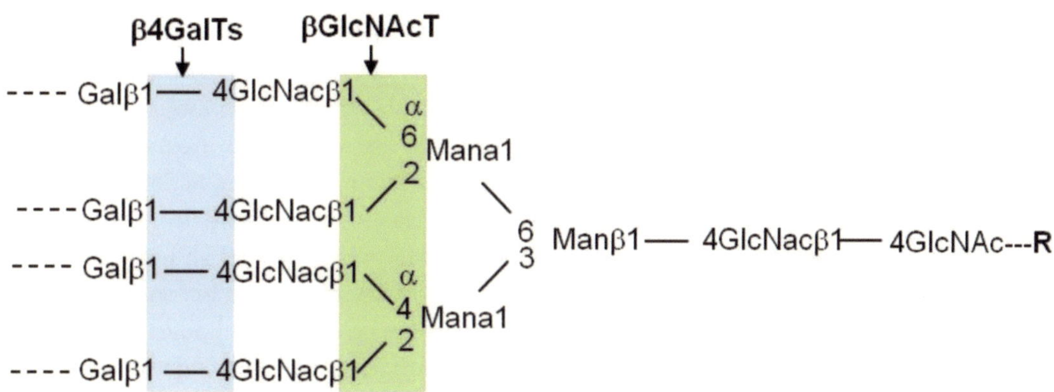

Figure 2. An example of a tetra antenna structure in a complex-type N-glycan. The numbers indicate the glycosidic linkages. The arrows and the boxed areas represent the bonds catalyzed by β1,4-galactosyltransferase (β4GalT, blue area) and βGlcNAc Transferases (βGlcNAcT, green area), respectively. Gal; Galactose, Man: Mannose, GlcNAc; N-Acetyl-glucosamin, R; glycoprotein back bone.

4. β-1,4-Galactosyltransferase 1 (β4Gal-T1)

One member of the β4galactosyltransferase family, that has got increased attention in stem cell biology, is the β4Gal-T1. β4Gal-T1 catalyze the transfer of galactose (Gal) from uridine diphosphate-galactose (UDP-Gal) to terminal N-Acetylglucosamine (GlcNac) residues of oligosaccharide chains in a β1,4 linkage, to form N-acetyllactosamine. β4Gal-T1 and βal-T2 are unique among the β4galactosyltransferases (β4GalTs) genes that they form a heterodimer with alpha-lactalbumin and changes substrate specificity from GlcNac towards Glucose (Glc) as a substrate, forming lactose, a very common protein in the mammary glands. Interestingly, β4Gal-T1 is constitutively expressed. However, apart from β4Gal-T1, β4Gal-T2 is only expressed in fetal brain. β4GalT-2 is a key regulator of glycosylation of the proteins involved in neuronal development [16] and is responsible for the synthesis of complex-type N-linked oligosaccharides in many glycoproteins, as well as the carbohydrate moieties of glycolipids. Like the β4Gal-T1 enzyme, its substrate specificity is affected by alpha-lactalbumin but is not expressed in lactating mammary tissue Apart from the other βGalTs, βGal-T1 encodes two protein isoforms produced by differential translation initiation at the 5′ end of the mRNA transcript: a long isoform, containing a 24 amino acid cytoplasmic domain, and a short isoform with only an 11 amino acid domain [24]. Both isoforms are localized to trans-Golgi network and are able to function as glycoprotein processing enzymes (Fig.3). However, a small fraction of the long isoform of βGaT-1, preferentially targets the cell surface of various cells [25]. The specific signal sequence in β4GalTs that regulate the differential localization between cell surface and the Golgi complex, has been shown to consist of a short N-terminal hydrophobic sequence in the cytoplasmic domain, adjacent to the plasma membrane. This observation was further extended by the findings that the 13 amino acid sequence in the cytoplasmic domain of long Gal-T1, could be phophorylated by p58 (CDK11), a GalT1 associated and cell cycle related Serine/Threonin kinase and, hence, could act as a retention signal for β4Gal-T1 in the Golgi complex [26, 27, 28, 38, 55] (Fig.3). Apart from being involved in a variety of physiological activities, such as, for example mouse gamete interaction, neurite extension, epithelial mesenchymal transition and neural crest cell migration [29], cell surface GalT1 is also responsible for late morula compaction during development [30]. For more than a decade ago, β4Gal-T1 was found to facilitate cell migration on laminin 1, an important constituent of the extra cellular matrix (ECM) and during development [31, 32]. Furthermore, addition of β4Gal-T1 perturbants to F9 embryonic carcinoma led to an arrest in cell growth and morphological changes of embryoid bodies (EB) during differentiation [33]. Eckstein et. al., showed that cell surface β4Gal-T1 needed to associate with intact actin cytoskeleton in order for its cell surface activity [34] Interestingly, the intracellular domain of long form of β4Gal-T1 has been shown to bind to an array of signal transduction molecules such as a trimeric G-proteins (Gi) [35], Src Suppressed C-kinase Substrate (SSECKs) [36, 37], CDK11 (p58) [26, 38] and a novel ubiquitin conjugating enzyme [39]. The β4Gal-T1 interaction with SSeCKS was detected using the two hybrid system with the amino terminal 13 amino acid long cytoplasmic domain of β4GalT-1 [37]. The β4Gal-T1 association with SSeCKS is interesting since both proteins show similar subcellular distributions and share important cellular functions, such as cell proliferation, actin dynamics, and cell migration during development [36, 40]. For example, ectopic expression of both cell surface β4Gal-T1 and SSeCKS has been reported to induce a transient tyrosine phosphorylation of focal adhesion kinase (FAK) and rearrangement of filamentous actin [41]. Furthermore, SSeCKs also control the G1 to S phase progression through regulation of cyclin D1 expression and localization. Since SSeCKS is a scaffolding molecule that can binds to several signaling proteins, such as PKC, Rho family members, and FAK, to mention a few, it is possible that most effects attributed to cell surface GalT1 in stem cell growh and differentiation may be mediated through SSeCKS. However it is unclear if cell surface β4GalT-1 performs in a similar manner as a lectin for its biological function [42, 105] or whether it utilizes its enzymatic activity to modify and release its galactosylated product [31].

http://dx.doi.org/10.5772/54376

(β4GalTs) (Chrom. #)	Expression	Glycosidic linkage and Acceptor substrates	Function in stem cell, cancer and/or development	References
β4Gal-T1 (9p13)	Heart, liver, lung, testis, ovary, placenta, fetal brain	Galβ1-4GlcNac-R	Morula compaction, cell growth, laminin dependent migration	[17], [18]
β4Gal-T2 (1p32-33)	Restricted in brain, testis	Galβ1-4GlcNac-R Glycolipid	Neuronal development, spermatogenic differentiation	[16], [18]
β4Gal-T3 (1q23)	Constitutively expressed, high in fetal brain.	Galβ1-4GlcNac-R Glycolipid	N/A	[14]
β4Gal-T4 (3q13.3)	Testis, ovary, placenta, pancreas	Galβ1-4GlcNac-R GlcNac-6-sulphate LacCer	Testicular development, tumor metastasis, keratin sulfate synthesis.	[19], [14], [20, 21]
β4Gal-T5 (11)	heart, lung, liver, kidney, testis, Restricted in brain	Galβ1-4GlcNac-R LacCer	Self renewal of glioma cells, astrocytoma, extraembryonic development	[18], [22]
β4Gal-T6 (18q)	Restricted to adult brain	Galβ1-4GlcNac-R LacCer	Extra embryonic development	[18, 19]
β4Gal-T7 (5q35.1-35.3)	Heart, Brain, Placenta, Liver, kidney, pancreas	GlcAβ1-3Galβ1-3Galβ1-4 Xylβ1-R	Glycosaminoglycan (GAG) biosynthesis	[15, 23]

Table 1. Table depicting the chromosomal region, the glycosidic linkage, substrate, and the function for the β4Galactosyltransferase family, related to stem cells and development.

5. β4GalTs in cancer

Glycosylation of cell surface glycoproteins and glycolipids changes dramatically upon the malignant transformation of cells [43]. β4GalTs have been reported to be increased in a fair amount of cancer. However, is not currently known if the elevated expression of β4GalTs contributes to the induction of cancer/malignancy, by affecting the cell surface landscape of glycans, or is an indirect effect of cancer progression or metasisis. β4Gal-T1 has been detected in highly metastatic lung cancer by transcription factor E1AF activation of the β4Gal-T1 promoter [17, 51]. Furthermore, siRNA interference of surface β4Gal-T1 function, inhibited cell adhesion on laminin, the invasive potential in vitro,and tyrosine phosphorylation of focal adhesion kinase [17]. The relative level of β4Gal-T1 has been reported to be important in melanoma invasiveness. For example, increasing cell surface β4Gal-T1 expression in cells of low metastatic potential promoted their invasive potential [44]. Other β4GalTs such as β4Gal-T5, function as an important growth regulator in glioma cells using both the E1AF and Sp1 transcription factors for its metastatic potential [17, 45]. Furthermore, clinically over expressed β4Gal-T4 and β4Gal-T6 have been shown to increase E2F1 and cyclin D3 transcription in colorectal cancer, respectively [18, 19]. Moreover, β4Gal-T1, -T2 and -T5 levels are higher in astrocytoma [18]. The expression of the β4Gal-T5 gene has also been shown to be regulated by transcription factors Sp1 and Ets-1 in cancer cells. Both these transcription factors regulate the gene expression levels of not only glycosyltransferases, but also key molecules involved in tumor growth, invasion and metastisis. Finally, small molecules that increase expression of GalTs could have beneficial effects during treatment of various cancer forms [45].

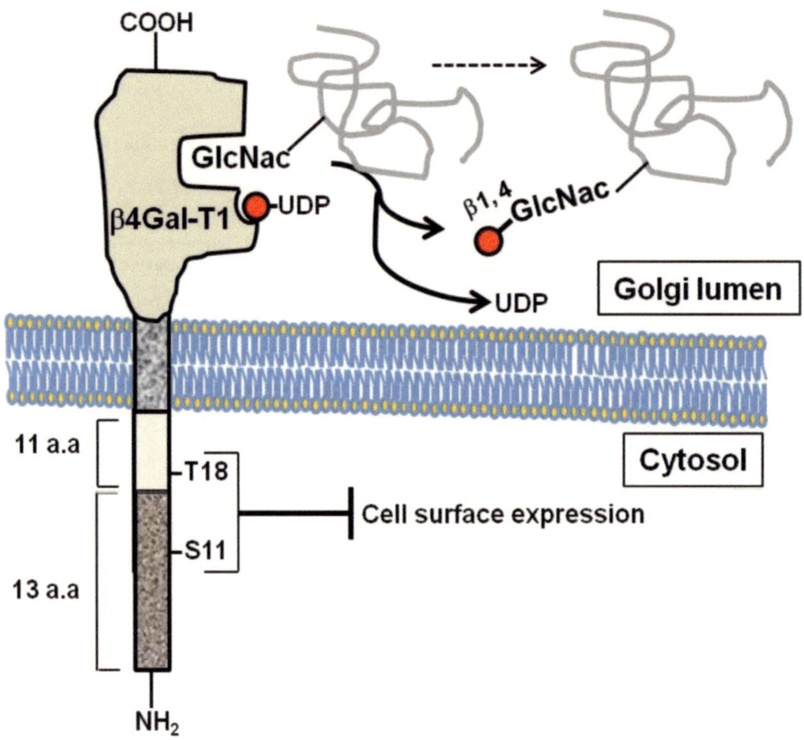

Figure 3. The long isomer of β1,4Galactosyltransferase 1 (β4GalT-1). β4GalT-1 catalyzes the transfer of UDP-galactose (red circle) to a terminal N-Acetylglycosamine (GlcNAc) residue in a newly synthesized glycoprotein in the golgie lumen. The cytosolic domain of the long β4Gal-T1 consists of 11 amino acids (a.a) together with a 13 a.a extension (24 a.a in total). Phosphorylation of Serine 11 (S11) and/or Theonine 18 (T18) in the cytoplasmic domain negatively regulate the localization and function GalT-1 as a cell surface receptor. The figure is not in scale.

6. β4Gal-T1 in cell cycle

The observation that some, or maybe all, of the β4GalTs have relevancy in cancer progression and/or metastasis, has highlighted the idea that stem cell pluripotency and differentiation may also depend on N-glycan structures [46]. One decisive factor in pluripotency and stem cell differentiation is the speed by which cells goes through the G1 phase in the cell cycle [47]. The cell cycle in pluripotent stem cells is remarkable for the shortness of the G1 phase, permitting rapid proliferation and reducing the duration of differentiation signal sensitivity associated with G1 phase. Changes in the length of G1 phase are understood to accompany the differentiation of human embryonic stem cells (hESCs), but the timing and extent of such changes are poorly defined. Terminally differentiated cells usually have a longer G1 phase than those of stem cell and progenitor cells. Understanding the early steps governing the differentiation of hESCs will facilitate better control over differentiation for regenerative medicine and drug discovery applications. To avoid that cells with genetic aberrations are expanded in the population, stem cells have adapted to their harsh environment by shutting off specific checkpoints normally activated in somatic cells. This will result in cell death as a default pathway for stem cells exhibiting chromosomal deveations, without slowing down proliferation of otherwise healthy cells. Since the upstream promoter region of the 4.1 kb β-GalT1 transcript is mainly occupied by the Sp1 transcriptional factor, GalT1 was long believed to be another "house keeping" gene. However, several laboratories have shown that β4GalT-1 is regulated during cell cycle [28, 48, 49]. Interestingly, experiments in F9 embryonic carcinoma cells and in 3T3 cells have indicated that the cell surface bound and the Golgi related forms of β4GalT-1 are regulated differently, in which the long form is induced much earlier than the short and Golgie bound form. β4Gal-T1 showed the highest activity during G1-S phase and during interphase of the cell cycle [50]. There are many transcription factors important during the G1-S transition. The E2F family members of transcription factors serve as key regulators of the cell cycle progression by inducing activators of S-phase related genes. Normally, during the onset of G1/S transition in cell cycle, the cyclic dependent kinases (CDKs) phosphorylate the retina blastoma (Rb) protein, resulting in a conformational change in Rb and subsequent release of active E2F from the Rb-E2F complex. This event results in transcription of G1-phase activating proteins such as e.g Cyclin D3. Interestingly, E2F1, one of the best characterized members of this family, also binds to a promoter element in β4Gal-T1 transcript and positively regulates its activity. Moreover, cells subjected to a short hairpin RNA (shRNA) to β4Gal-T1 became less responsive for E2F1 activation [51].The effect of E2F1 on the expression of the other family members of β4GalTs, however, (β4Gal-T1, -T7) has not been exclusively determined. Another cell cycle related protein that has been found to regulate β4GalT1 expression is the p16 protein. This protein is a product of a tumor suppressor gene called CDKN2A that inhibits the cyclin-dependent kinases (CDK)-4 and 6 which are responsible for the G1 checkpoint in cell cycle. Transfection of A549 human lung cancer with p16 led to down regulation of βGalT1 activity [53]. Thus, inactivation of either p16 or pRb function allows the cells to enter the S-phase only after a brief pause at the G1 checkpoint, leading to accelerated cell proliferation. Similar results for GalT1 expression was obtained in hepatocarcinoma SMMC-7721 cells after blocking endogenous activity of TGFβ, a known regulator of the G1 to S-phase transition of cell cycle, by arresting cells in G1 phase [54]. Over expressing β4Gal-T1 has also been shown to exasperate cyclohexamid induced apoptosis of [45]. This process is partly dependent on the activity of the CDK11(p58), a CDK11 family Ser/Thr kinase, a G2/M specific protein that contributes to regulation of cell cycle [55]. Recently GalT1 has been shown to interact with CDK11(p58) [26, 38] where it has an important function during cell cycle in stem cells progression [28, 56]. Furhtermore, β4Gal-T1 contributes to HBx-induced cell cycle progression In hepatoma cells [57]. All these findings have led to the conclusion that β4Gal-T1 may be directly or indirectly connected to cell cycle progression and could be a potential reason for the growth impeded phenotype observed earlier in knock out β4Gal-T1 mice [52]

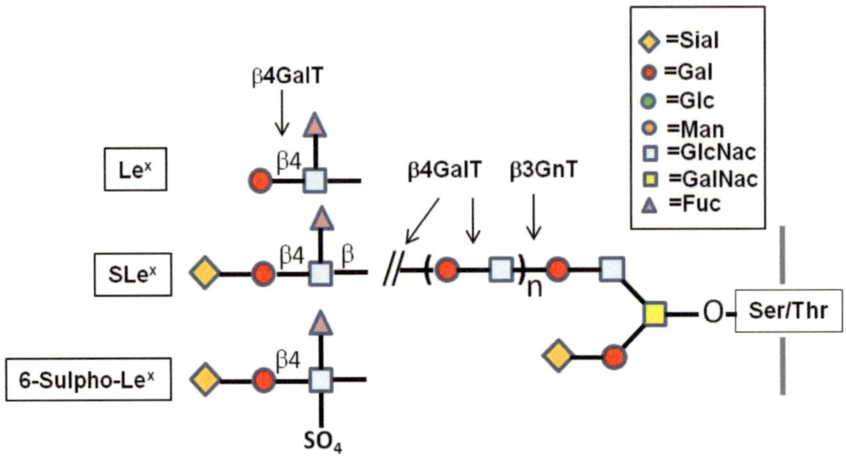

Figure 4. Biosynthesis of a core 2 O-glycan with Lewis X, (LeX,SSEA-1), Sialyl-Lewis (SLex) or 6-sulpho Sialyl (6-Sulpho-Lex synthesized at the terminus of poly-N-Acetyllactosamine chains. The action of β4Ga|Ts and β3GnT are indicated with arrows. Sial;Sia|ic acid, Gal;Ga|act0se, G|c;G|uc0se, Man;Mann0se, G|cNac; N-Acetylglucoseamine, Ga|TNAc; N-Acetylgalactoseamine, β4Ga|act0syltransferases; β4GalT, β1,3-N- Acetylglucosaminyltransferase; β3GnT

7. β4GalTs involvement in Lewis X, glycosphingolipids and embryoglycans

Lewis X: As mentioned in the beginning of this chapter, β4GalTs are Important for the synthesis of linear or branched poly-N-acetyllactoseamines chains. They are attached to N-glycan, O-Glycans or glycolipids and are synthesized by the repeating and alternate action of N-acetylglucosaminyltransferases (β3GnT or β4GnT) and β4Gal-T1 [58]. These structures often carry various functional epitopes important in stem cell homeostasis and inflammation [59]. One of these antigen is called the Lewis X antigen (Le^x) and constitutes the core structure from which other antigens are synthesized. Le^x epitope consists of a trisaccharide, Galβ1-4(Fucα1-3) GlcNAcβ1 which is produced by the action of β4Gal-T1 and α-1,3-Fucosyltransferase (FUT). Other examples of epitopes formed from this core, are the Sialyl-Lewis (SLe^x) and 6-sulpho Sialyl (6-Sulpho-Lex) epitopes (Figure 3), in which the latter involve the activity of β4Gal-T4 (Table 1). These epitopes are implicated in biospecific interactions with selectins and other glycan-binding proteins during inflammatory processes [59] as well as in important regulatory functions during development [60]. Also, Le^x structures has been implicated in specific differentiation, such as myocardial differentiation from embryonic stem cells [60, 61].

Glycosphingolipids: Glycosphingolipids, or sometimes called glycolipids (GLS) have been found in the upper leaflet of the plasma membrane in both lower and higher eukaryotic

http://dx.doi.org/10.5772/54376

sources. Several members of β4GalT family seem to be important enzymes in the synthesis of GSL [62]. The basic structure for GLS is a monosaccharide, usually glucose, attached directly to a ceramide molecule, mediated through the action of ceramide glucosyltransferase (Ugcg), resulting in a glycosylceramide (glucocerebroside;GlcCer) (Figure 5). βGalT-2 then transfer a UDP-Galactose to the GlcCer moiety, forming Lactosylceramide (LacCer) [62] (Figure 5). A variety of structural subclasses of GLS may then be synthesized from LacCer by the addition of other mono and disaccharides, resulting in the synthesis of structural subclasses of GLS such as ganglio-, lacto/neolacto-, globo,- -isoglo, and ganglioseries-series [63]. Many of these structures are important for various biological functions, such as for example cell growth, myocardial differentiation cell migration and during development of the nervous system[60, 61, 64]. When the Le^x epitope is attached to a lactosylceramid it is identical to stage specific antigen (SSEA-1). This antigen is highly regulated during embryogenesis, expressed at the morula stage in embryos and is considered to function as a cell-cell interaction ligand in the compaction process [65].

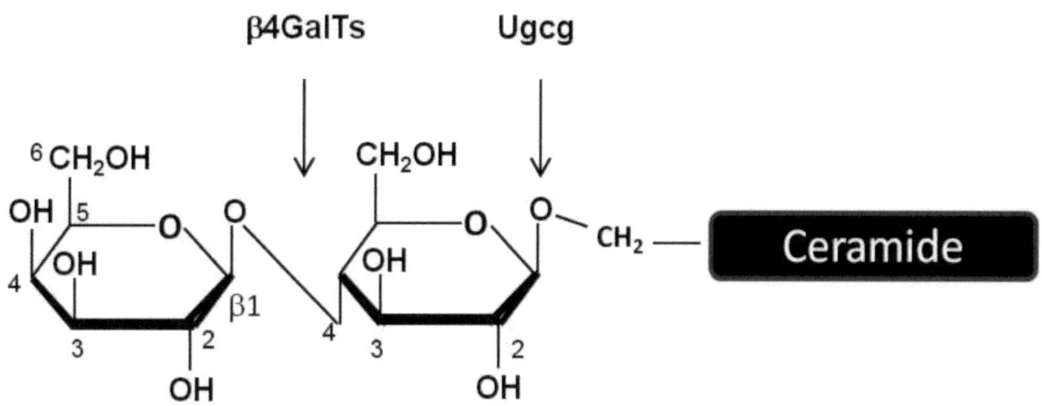

Figure 5. Core 2 structure of the glycoplipid, Lactosylceramide (LacCer) synthesized by UDP-glucose ceramide glucosyl transferase (Ugcg) and by β1,4Ga|act0sy|transferases (β4Ga|Ts) forming the β1,4-glycosidic linkage to ceramide.

Embryoglycans: Most developmentally regulated epitopes identified on embryonal carcinoma cells and murine preimplantation embryos are associated with a glycoprotein-bound and large glycans, called embryoglycans. Embryoglycans consists of linear of branched poly-N-acetyllactoseamines with high molecular weight that carries a number of different developmentally regulated carbohydrate epitopes, such as e. g. Le^x, described above (Figure 6). Apart from the mouse, where SSEA-1 is abundant from the 8-cell morula stage, SSEA-1 in human is not expressed until the germ cell line and in neural stem cells. Interestingly, β4Gal-T1 is expressed during the morula stage and has been shown to affect the compaction process [30]. Furthermore, human ES cells express SSEA-3 and -4 SSEA-1. SSEA-1 is also expressed in undifferentiated F9 teratocarcinoma cells. After induction of differentiation the expression of SSEA-1 decreases. This is caused by the upregulation of alpha-1,3-galactosyltransferase that is responsible for masking of the Le^x structure [66, 67]. The stage specific embryonic antigens 3 and 4, (SSEA-3,-4) are from the globo-series of glycosphingolipids (GL-5 and GL-7) and have not been found on linear poly-N-lactosamines [68].

Glycoseaminoglycans (GAG): GAGs are long unbranched polysaccharides containing a repeating disaccharide unit. The disaccharide units contain either one of two modified sugars, *N*-acetylgalactosamine (GalNAc) or *N*-acetylglucosamine (GlcNAc), and a uronic acid such as glucuronate or iduronate, forming heparin sulphate and hondroitin sulphate, respectively [69]. GAGs are highly negatively charged molecules, and are located primarily on the surface of cells or in the extracellular matrix (ECM). GAGs are normally attached to soluble or membranes bound core proteins to form proteoglycans which carries various carbohydrate markers expressed on early embryonic cells [60]. In the few past years it has become clear that many growth factor such as EGF and FGF has been shown to bind specific pentasaccharides within GAGs efficiently affect signaling during development [70]. The integrity of proteoglycans is important. One of the β4galactosyltransferase, β4Gal-T4, is one has recently been shown to be involved in the biosynthesis of keratin sulphate (KS), in which TRA-1-60 and TRA-1-80 epitopes are found, [14]. Furthermore, β4GalT-7 is involved in the synthesis of the GAG linkage region to proteoglycans, by catalyzing the transfer UDP-Gal to an *O*-linked Xylose/Ser residue in the sequence, GlcAcβ1-3Galβ1-3Galβ1-4Xylβ1-O-ser [23].

8. βGalTs and ESC signaling pathways

A number of reports have suggested β4GalTs to be direct or indirect mediators and regulators of cytokine signaling during stem cell and/or cancer development. As discusses below, many signal transduction pathways, such as EGF, FGF, Wnt and the Notch pathway, that utilize Le^x-containing carbohydrates are potential targets for aberrations in β4GalTs activities:

8.1. Epidermal Growth Factor (EGF)

EGF is involved in the regulation of cell proliferation and exerts its effects in the target cells by binding to the plasma membrane located EGF receptor. The EGF receptor is a transmembrane protein tyrosine kinase. Binding of EGF to the receptor causes activation of receptor autophosphorylation, which is essential for the interaction of the receptor with its cytosolic substrates. In mouse embryonic stem cells (mESC), EGF has been shown to stimulate proliferation of mouse ES cells via PLC/PKC, Ca2+influx and p44/42 MAPK signal pathway through EGF tyrosine kinase phosphorylation [71]. Altering the core components of *N*-linked glycans will change the EGF binding, the transport and the receptor endocytosis meanwhile substitution of the outer chain or terminal glucosides have been shown to affect the phosphorylation state and the dimerization of the receptor [72, 73]. Cell surface βGalT1 has been suggested to associate with and disrupt autophophorylation of EGF receptor Hinton et. al, showed that when a dominant negative form of long β4GalT-1 was over expressed in F9 embryonic carcinoma cells, the endogenous and active cell surface GalT-1 is displaced from its association to actin cytoskeleton. This inhibition of cell surface β4GalT-1 resulted in increased tyrosine phophorylation of the EGF receptor and attenuated cell proliferation, while the shorter form of βGal-T1 did not have any effect [48]. These results implies that cell surface β4Gal-T1 has an inhibitory effect on EGF activity. Later, several groups substantiated this observation by showing that knock-down of β4GalT1 activity in SMMC7721 hepatocarcinoma cells, elevated the autophosporylation of EGFR. Reversibly, the level of tyrosine phosphorylation was attenuated if cell surface βGal-T1 was over expressed [74]. Interestingly, EGF treatment of HeLa cells has been shown to increase the β4Gal-T1 mRNA level, suggesting that β4GalT1 also act in a negative feedback loop on EGF activity [17]. In another elegant experiment, using mutant Chinese hamster ovary cells (CHO), where the levels of six beta β4Galactosyltransferases (βGalT1-6) were reduced, the protein level of active and surface-located EGFR was greatly attenuated without affecting the transcriptional level and activity of EGF receptor [75]. β4Gal-T1 has also been shown to positively affect EGFR activity. Isoprenaline, a β-adrenergic receptor has a dramatic growth stimulating activation on the salivary glands of rat and mice, eventually leading to hyperplastic and hypertrophic gland enlargement. This effect has been suggested to be mediated in part by cell surface β4Gal-T1 by mimicking EGF receptor mediated receptor ligand binding and activation [76]. In any case, the specific β4GalT1 binding site on the EGF receptor has not, as yet, been investigated but it is possible that the recently discovered extracellular location of O-linked GlcNac moieties on the EGF receptor, could act as a recognition signal, as has been observed for other membrane anchored extracellular proteins, such as Notch and Dumpy receptor [77, 78]. In this scenario, β4GalT1 could act as a lectin like molecule, using its substrate, GlcNAc [79, 80].There are also possibilities for other, more indirect and β4Gal-T1 dependent effects on EGF receptor function, such as the ganglioside GM3. The synthesis of this glycolipid is dependent on β4Gal-T2 activity, and has been shown to inhibit ligand-induced tyrosine phophorylation of EGF receptor through its sialyllactose carbohydrate moiety by interacting with the GlcNAc termini [72, 81].

8.2. FGF-2

Fibroblast growth factor (FGF) functions as a natural inducer of mesoderm, regulator of cell differentiation and autocrine modulator of cell growth and transformation of various cell types. FGF is activated by ligand-receptor interaction that results in tyrosine phosphorylation of the intracellular domain of the FGF receptor [82]. FGF-2 is often used as a key player in regulating self renewal and proliferation of human embryonic stem cells. Recently FGF-2 has been shown to regulate the transition from one pluripotent state to another. It has been speculated that human embryonic stem cells, due to their precautious ability to differentiate in culture, are identical to a later or "primed" developmental stage of mouse embryonic stem cells, EpiESC. LIF signaling is dispensable for this state, but instead relies on FGF signaling. Inhibition of FGF signaling with inhibitors in the presence of human LIF can "rescue" human embryonic stem cells from a primed state to a more naïve state, e g full pluripotency [83].This difference is still unclear but there are indication that extracellular proteoglycans, such as heparin sulphate (HSPG) acts as key co-activators of FGF receptors. Furthermore, during development, oligosaccharides from embryoglycans are often shed into the extra cellular environment where they can influence cytokine and mitogen signaling. Lewis x epitopes on embryoglycans acts as a recognition molecule for FGF2 and plays an active role in the formation of FGF ligand receptor complexes. Free and soluble sulphated Lewis X was most prominent to activate the FGF-2 mitogenic acitivity [84, 85]. Also exogenous and free glycolipids in the form of gangliosides, can interact with the FGF-2. Gangliosides are derivatives of LacCer with a neuraminic acid (NeuAc) attached to the core, and seem to have dual roles in affecting both EGF and FGF proliferative action; soluble gangliosides and sulphated heparin act in a negative manner meanwhile membrane bound gangliosides increase the receptor activity. It seems clear that the close interplay between Le^x epitopes, adhesion molecules and cytokines has an important impact on the efficiency by which ligands are presented, and ultimately results in receptor oligomerization of the receptors and signalling [70]. It is therefore possible that β4GalTs could mediate some aspect of FGF receptor signalling, as described below.

8.3. Wnt pathway

The Wnt family of growth/differentiation factors has important developmental roles in embryonic stem cells. They act through the complex of Frizzled receptor and LPR co-receptor with effect on β-catenin transcriptional activity [86]. Similarly to EGF, the activity of Wnt also depends on association with HSPG for activity. HSPG is a rich source for developmentally regulated Le^x epitopes. Furthermore, Wnt-1 has been shown to interact directly with Le^x epitopes [87]. These observations suggests that surface bound and secreted Le^x have a regulatory function in stabilizing the stem cell niche, where they binds to and present appropriate factors, important for cell proliferation and self renewal.

8.4. Notch pathway

In a stem cell niche, stem cells and a variety of progenitor cells have to receive both temporal and spatial signals in order to differentiate or stay pluripotent. Also, during development and differentiation, cells have to decipher their precise localization in the dorso-ventral plane in order to form distinct and proper boundaries with other cell types in the tissue. These processes are governed by the Notch/ Delta system [88]. Notch is an essential developmental glycoprotein that plays key roles in both growth control and cell fate decisions. It is a transmembrane glycoprotein with a large extracellular domain made up of 29-36 EGF repeats, which can contain both N-linked and O-linked EGF repeats [90]. When Notch receptor is activated by a ligand on adjacent cells it is proteolytically cleaved, disposing the extracellular domain, followed by a second cleavage resulting in the released of the intracellular domain into the cytosol where it translocates to the nucleus and activates the transcription of numerous developmental genes. There are two ligands to Notch receptor, Delta and Jagged. Even though Notch receptor is ubiquitously expressed, Delta and Jagged are not usually located in the same cells but rather in different parts of the tissue during development where they exert their effect dependent on cell type and/or the environment. To avoid ubiquitous activation, Notch undergoes a post translational modification in which Fucose is first attached to certain EGF repeats on the extracellular domain of the receptor by *O*-Fuco-

http://dx.doi.org/10.5772/54376

syltransferase (O-FucT1). An N-acetyl glucosamine (GlcNac) and a Galactose (Gal) residue are then sequentially added to the fucosyl residue by the action of Fringe, a O-fucose β1,3-N-Acetyl glucosaminyl transferase and β4Gal-T1, respectively. The addition of Gal is necessary for the enhancement of Delta dependent signaling but not sufficient for the inhibition of Jagged induced Notch activation [89]. Recently, another layer of regulation of Delta induced Notch signaling was discovered in which the two Fringe genes, Lunatic Fringe (LFNG) and Manic Fringe (MFNG), seem to exhibit differential activity toward Delta dependent Notch activation. Gal was required for enhancement of Notch activation through LFNG and inhibited the enhancement of Delta induced signaling [90, 91]. Apart from O-linked Fucosylation, an O-linked GlcNAc modification of Notch EGF repeats was recently discovered [77]. Although the O-GlcNac modification is known to regulate a wide range of cellular processes, the list of known modified proteins has previously been limited to intracellular proteins in animals. Thus, this novel finding predicts a distinct glycosylation process associated with a novel regulatory mechanism for Notch receptor activity that may include a variety of βGalTs [77]. Furthermore, continuous hypoxic culturing conditions have been shown to activate Notch signaling to allow long-term propagation of human embryonic stem cells without spontaneous differentiation. Stem cells isolated and cultured under low oxygen tension (hypoxia) condition have been shown to maintain a stable pluripotency potential because of Notch activation [92]. Recently, it was also shown that β4GalT1 derived Lewis X epitopes on N-linked glycans was necessary for Notch activity and in the propagation of neural stem cells (NSC) [93].

9. β4GalTs deficiency in fish

It has been a challenge to get a consensus of the mechanisms by which complex carbohydrates control aspects of mammalian development and early differentiation. Some of the information has been available from knock-down experiment of individual galactosyltransferases. However, since many carbohydrate functions during early development in mammals are confined to " *in utero*", further analysis of the physiological effects of galactosyltransferases has not been possible. An attractive model using a more efficient "high-throughput " a assay system, is the zebrafish system. **β4Gal-T1:** The zebrafish β4Gal-T1 has the highest sequence homology to β4Gal-T1 among the human β4GalT family. β4Gal-T1 morpholino treated embryos had a truncated anterior-posterior axis phenotype that was a result of a defect in convergent extension [94]. Convergent extension is a developmental process that relies on coordinated cell migration to elongate and narrow a field of cells. Laminin is an extracellular substrate for cell surface β4Gal-T1 and constitutes one of the major components of the basement membrane upon which cell adhesion and migration occur during development [29]. Interestingly, in the mopholino treated embryos, laminin was hypo-galactosylated and hence could explain the decreased in ectodermal cell migration of [94]. **β4Gal-T2:** Tonoyama, et al. showed that β4Gal-T2 was indispensable for mediolateral cell intercalation and thus extension movement during gastrulation [95]. The specific substrates for β4Gal-T2 activity in glycoproteins responsible for these effects are currently not known but has been speculated to be related with N-glycosylated FGF receptor signaling. FGF signaling pathway is dependent on its N-glycans in the interaction with heparin co-receptor, regulating the efficiency of signaling [96]. **β4Gal-T5:** Transforming Growth factor (TGFβ) and bone morphogenic protein (BMP) are polypeptide members of the transforming growth factor beta (TGF) super family of cytokines. They are both secreted protein that performs many cellular functions, including the control of cell growth, cell proliferation, cell differentiation and apoptosis. In this context, knock-down of β4Gal-T5 using morpholino-injected zebrafish resulted in embryos with an elongated dorso-ventral axis and a defective tail bud [97]. This effect was suggested to be mediated through a decreased BMP-2 (a TGFβ family member) binding to proteoglycan due to defective glycosylation, and subsequent attenuation of SMAD signaling.

10. β4GalTs deficiency in mouse and human

Many diseases such as disorders of blood clotting, congenital disorder of glycosylation, diseases of blood vessels, cancer, angiogenesis essential for breast and other solid tumor progression and metastasis, are all associated with a dysfunctional *N*-glycan expression. The expression of many galactosyltransferases is under control of cytokines and could therefore become altered in various disease states. In order to find physiological functions for each galactosyl transferases, researchers have used both mouse and rat knock-out models. **β4Ga-T1:** β4Ga-T1 was the first galactosyltransferase that indicated potential relevance in physiology. About 50% of β4Gal-T1, knock-out mice died prematurely because of pituitary deficiency [10].The surviving animals showed growth retardation, elevated proliferation of skin epidermis, and delayed wound healing due to attenuated leukocyte recruitment and infiltration [59]. Recently, some diseases in humans due to aberrations in β4Gal-T1 have emerged. For example, congenital disorders of glycosylation (CDGs) comprise a group of inherited disorders associated with psychomotor and mental disorders. One of these groups, CDGII, comprises all defects in trimming and elongation of N-linked oligosaccharides. CDGIId fall into a group in which β4Gal-T1 is mutated in its catalytic domain. This resulted in an aberrant translation product that was 15 kDa shorter than normal. Since β4GalT-1 has been shown to be is important during the early development of the brain, the phenotype from this mutation is mental retardation [98]. **β4Gal-T5:** Furthermore, knock-out β4Gal-T5 in mouse resulted in growth retardation and early lethality of embryos due to hematopoietic and/or placental defects [99]. Also the expression of β4Gal-T5 strongly increased during embryonic stem (ES) cell differentiation [22]. Both β4Gal-T5 and β4Gal-T6 are lactosylceramid synthases. However, β4Gal-T5 is more restricted to the early embryogenisis than β4Gal-T6, which is more limited to adult brain. β4GalT-5 deficient animals showed abnormal extra embryonic structures that led to embryonic lethal phenotype at day E10.5. **β4Gal-T7:** A rare genetic mutation of β4Gal-T7, believed to be the consequence of two missense mutations in the active domain resulted defective GAG chain formation [15] gives rise to Ehlers-Danlos disease. This is a disorder in which patients exhibit phenotypes such as aged appear-

http://dx.doi.org/10.5772/54376

ance, developmental delay, dwarfism, craniofacial disproportion, delayed wound healing, loose skin, and general ostopenia [15, 100].

11. Potential treatments

The involvement of β4GalTs in cancer, inflammation and during development / stem cell homeostasis has encouraged research to come up with new modalities that can either boost or inhibit the expression/activity of endogenous glycosyltransferases. I will briefly discuss potential therapeutic models for treatment that will inhibit or activate specific galactosyltransferases.

11.1. Protein ubiquitination

A potential regulator of a galactosyltransferase, GTAP, was discovered 2008 in a two hybrid screen of a mouse embryonic library, using the cytoplasmic domain of cell surface Gal-T1 as bait. Ectopically expressed GTAP down regulated the expression of cell surface bound GalT-1 and negatively affected both laminin dependent stem cell migration and embryonic body formation during differentiation. GTAP is an ubiqutin conjugating enzyme that is expressed during early development of the inner cell mass and in embryonic stem cells but also in highly proliferative tissues, such as, such as kidney, lung and testis. This effect was not due to a proteasome dependent degradation of βGal-T1 but an increase of ubiquitin dependent lysosomal activity. So far this is the only report on ubiquitin related regulation of a cell surface galactosyltranferase and may be important for the development of more effective and specific inhibitors of various glycosyltransferases in glycan related diseases. The only known ubiquitin/proteaseome regulated system of glycans so far, is the endoplasmic reticulum assisted degradation (ERAD).This system helps cells to avoid stress and cell death by degradation of missfolded proteins in the ER [101]

11.2. Analogues to GalT donor and acceptor

A limited number of GalT-1 inhibitors have been described. Most of them have been analogues of either the donor substrate (e.g Gal) or the acceptor (GlcNac) molecules to galactosyltransferases. E. g. a modified GlcNac acceptor, called compound 612, was recently discovered showing differential affinities for β4Gal-T1 and β4Gal-T5, two galactosyl transferases with similar acceptor specificities [102]. Also, in contrast to other β4galactosyltransferases, β4Gal-T7 has the ability to bind, but not actively transfer Mannose or GalNAc to an acceptor substrate, implying that these donors can be used as potential inhibitors to GAG synthesis [103]

11.3. Lectins

During recent years, several laboratories, using specific cell lines that either over express or lack different glycosyltransferases in combination with high density lectin microarrays.

In order to entangle the mechanism by which the cellular glycome can influence stem cell pluripotency and differentiation. Lectins are proteins that bind to particular carbohydrate epitopes in a similar manner as an antibody. Glycans are located at the cell surface where many signal transduction pathways, cell-cell interaction and cell-to cell recognition are constantly active. Interactions between glycans and endogenous lectins may influence self renewal, maintenance of pluripotency and differentiaon of iPS/ESC. Such an approach has already been tested in which synthetic substrates, mimicking endogenous lectins, can facilitate the formation of induced pluripotent cell(iPSC) and help sustain long term culture of human ESCs [104]

12. Conclusion and perspectives

It is clear that both N-linked and O-linked glycans are implicated in many intricate and complex processes during development, differentiation and in many diseases. For many years glycosyltransferases were thought of as just redundant enzymes acting solely in the ER and Golgie, creating oligosaccharide structure mostly important for transport and solubility of secreted proteins. However, in the last decades, the functions of glycosyl transferases have been expanded to involve receptor oligomerization, antigen presentation, endocytosis, ligand-receptor binding, and even signal transduction. These observations have attracted attention in the stem cell biology field. Several markers for pluripotency, such as Lewis X antigen, e.g. SSEA-1, -3 and -4, and the keratin sulphate related markers, TRA-1-60 and TRA-1-80, are all dependent on functional galactosylation for their synthesis and functionality. The levels and modifications of these embryonic derived antigens are changing upon differentiation. These markers have mainly been used, and are still used, as markers for isolation and propagation of different stem cell populations. With recent technological advances and the development of more efficient lectin microarrays and HPLC systems, more and more details of the functional and structural requirements of early epitopes during stem cell self renewal and differentiation, are emerging. These techniques, combined with specific knock- down models and ectopical expression of individual galactosyltransferases, would eventually reveal the molecular mechanisms by which glycans influence stem cell and cancer progression. The complex interplay between members of the galactosyltransferase family, does not only affect the core structures of glycans but are also extensively involved in the synthesis of other bioactive compounds, such as glycolipids and the Lexis X antigens that affect a variety of biological systems spanning from cell migration to signal transduction. The presence of the long form of β4Gal-T1 at the cell surface raises many interesting questions on how this receptor, or maybe other glycosyltranferases as well, can influence so many different signal transduction pathways in the regulation of cell cycle, cell death, proliferation and differentiation. Apart from being located to the Golgi complex, where it is responsible for creating complex oligosaccharide structures on proteoglycans and glycolipid, the cell surface β4Gal-Ts also affect intracellular signal transduction pathways. As seen in Figure 6, cell surface β4GalTs can indirectly affect many cell specific functions because of its involvement in the synthesis of glycolipids, embryoglycans and many embryonal epitopes,

http://dx.doi.org/10.5772/54376

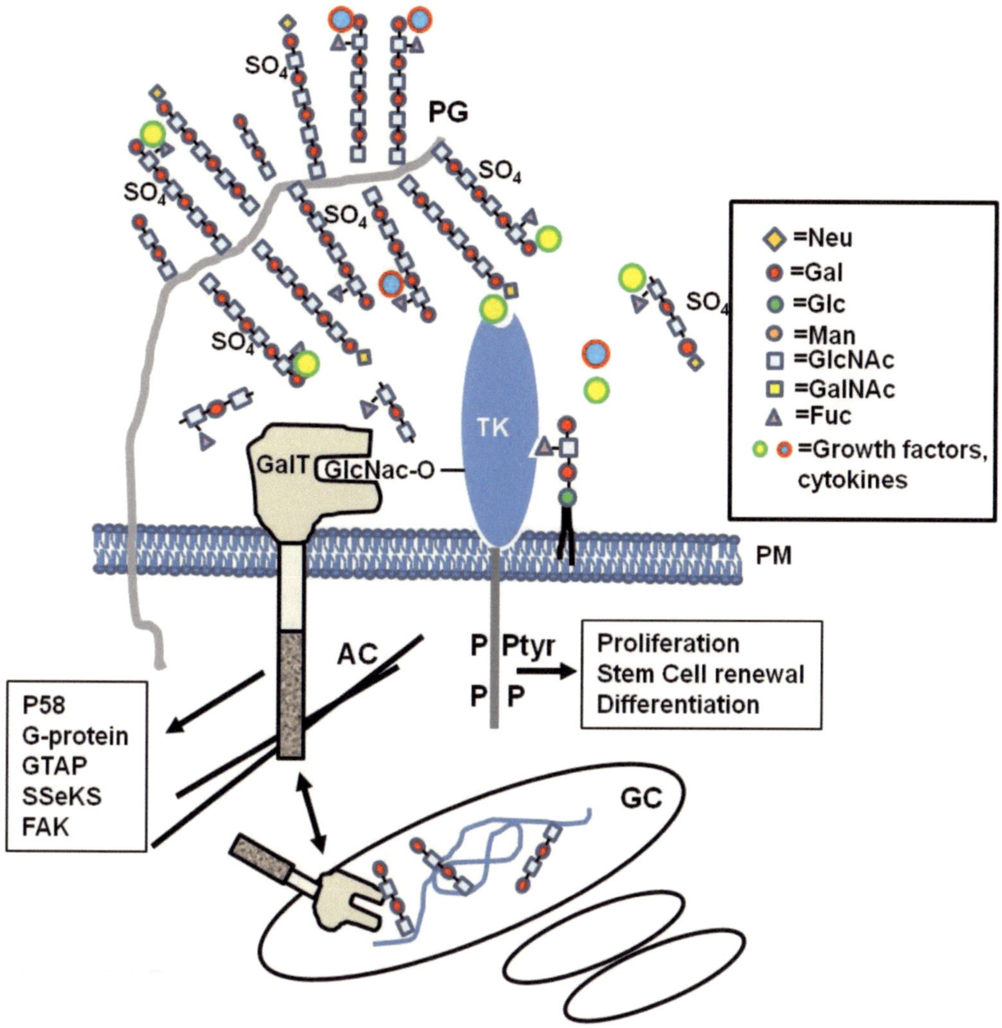

Figure 6. Schematic view of cell surface β4GalTs potential functions. Cell surface as well as Golgi bound long β34Galactosyltransferase (GalTs) can influence stem cell homeostatis. TK; Tyrosin kinase. AC; actin, GC; Golgi complex, GL;Glycolipid, PM; Plasma membrane, Ptyr; Tyrosine phosphorylation, PG; Proteoglycan, S04; sulphate, Neu;Neuramic acid, Gal;Galactose, Glc;Glucose, Man;Mannose, GalNAc; N-Acetylgalactosamine, GalNAc; N-Acetylgalactoseamine, Fuc;Fucose.

such Lewis X antigens. These complexes will either stabilize growth factor or cytokine-receptor complexes or, after shedded into the extracellular matrix during differentiation, inhibit receptor function. A change in galactosyltransferase activity could therefore indirectly affect the stem cell nitch by hinder effective glycolipid, proteoglycan/GAG synthesis and signal transduction through tyrosin kinase (TK) receptors. Secondly, apart from binding to the extracellular matrix, such as laminin, the cell surface β4Gal-T1 could also act directly as a lectin-like molecule that bind to tyrosine receptors (EGF, FGF or Notch), either on the same cells, or on adjacent cells, as long as a terminal GlcNAc are presented. This could either create a block or enhancement of the TK receptor- ligand complexes, or even hinder dimerization and activation of the receptors. Furthermore, the β4GalT-receptor binding could lead to aggregation of cell surface β4Gal-T1, increasing its association to actin, and subsequently lead to increase in intracellular signal transduction through FAK, SSeCKS and other signalling molecules. In this scenario, it is plausible that β4GalTs, control a myriad regulatory feedback loops. It is clear that so much more of the biological function of GalTs has to be understood in order to unravel attractive and potential therapies for cancer and in regenerative medicine.

Author details

Michael Wassler*

Address all correspondence to: Michael.Wassler@uth.tmc.edu

The center for cardiovascular Biology and Atherosclerosis Research, Department of Internal Medicine, The University of Texas Medical School at Houston, Houston, TX, USA

References

[1] Taniguchi, N. and H. Korekane, Branched N-glycans and their implications for cell adhesion, signaling and clinical applications for cancer biomarkers and in therapeutics. BMB Rep., 2011. 44 (12) p. 772-81.

[2] Weerapana, E. and B. Imperiali, Asparagine-linked protein glycosylation: from eukaryotic to prokaryotic systems. Glycobiology, 2006. 16(6): p. 91R-101R.

[3] Amado, M., et al., Identification and characterization of large galactosyltransferase gene families: galactosyltransferases for all functions. Biochim Biophys Acta, 1999. 1473(1): p. 35-53.

[4] Banerjee, D.K., N-glycans in cell survival and death: Cross-talk between glycosyltransferases. Biochim Biophys Acta, 2012. 1820(9): p. 1338-46.

[5] Sasaki, N., et al., LacdiNAc (GalNAcbeta1-4GlcNAc) contributes to self-renewal of mouse embryonic stem cells by regulating leukemia inhibitory factor/STAT3 signaling. Stem Cells, 2011. 29(4): p. 641-50.

[6] Kraushaar, D.C., Y. Yamaguchi, and L. Wang, Heparan sulfate is required for embryonic stem cells to exit from self-renewal. J Biol Chem, 2010. 285(8): p. 5907-16.

[7] Sasaki, N., et al., Heparan sulfate regulates self-renewal and pluripotency of embryonic stem cells. J Biol Chem, 2008. 283(6): p. 3594-606.

[8] Smith, R.A., et al., Glycosaminoglycans as regulators of stem cell differentiation. Biochem Soc Trans, 2011. 39(1): p. 383-7.

[9] Jang, H., et al., O-GlcNAc Regulates Pluripotency and Reprogramming by Directly Acting on Core Components of the Pluripotency Network. Cell Stem Cell, 2012. 11(1): p. 62-74.

[10] Furukawa, K. and T. Sato, Beta-1,4-galactosylation of N-glycans is a complex process. Biochim Biophys Acta, 1999. 1473(1): p. 54-66.

[11] Shaper, N.L., et al., Bovine galactosyltransferase: identification of a clone by direct immunological screening of a cDNA expression library. Proc Natl Acad Sci U S A, 1986. 83(6): p. 1573-7.

[12] Guo, S., et al., Galactosylation of N-linked oligosaccharides by human beta-1,4-galactosyltransferases I, II, III, IV, V, and VI expressed in Sf-9 cells. Glycobiology, 2001. 11(10): p. 813-20.

[13] Lee, J., et al., Chinese hamster ovary (CHO) cells may express six beta 4-galactosyltransferases (beta 4GalTs). Consequences of the loss of functional beta 4GalT-1, beta 4GalT-6, or both in CHO glycosylation mutants. J Biol Chem, 2001. 276(17): p. 13924-34.

[14] Seko, A., et al., Beta 1,4-galactosyltransferase (beta 4GalT)-IV is specific for GlcNAc 6-O-sulfate. Beta 4GalT-IV acts on keratan sulfate-related glycans and a precursor glycan of 6-sulfosialyl-Lewis X. J Biol Chem, 2003. 278(11): p. 9150-8.

[15] Bui, C., et al., Molecular characterization of beta1,4-galactosyltransferase 7 genetic mutations linked to the progeroid form of Ehlers-Danlos syndrome (EDS). FEBS Lett, 2010. 584(18): p. 3962-8.

[16] Sasaki, N., et al., beta4GalT-II is a key regulator of glycosylation of the proteins involved in neuronal development. Biochem Biophys Res Commun, 2005. 333(1): p. 131-7.

[17] Zhu, X., et al., Elevated beta1,4-galactosyltransferase I in highly metastatic human lung cancer cells. Identification of E1AF as important transcription activator. J Biol Chem, 2005. 280(13): p. 12503-16.

[18] Xu, S., et al., Over-expression of beta-1,4-galactosyltransferase I, II, and V in human astrocytoma. J Cancer Res Clin Oncol, 2001. 127(8): p. 502-6.

[19] Chen, W.S., et al., Tumor beta-1,4-galactosyltransferase IV overexpression is closely associated with colorectal cancer metastasis and poor prognosis. Clin Cancer Res, 2005. 11(24 Pt 1): p. 8615-22.

[20] Chatterjee, S., A. Kolmakova, and M. Rajesh, Regulation of lactosylceramide synthase (glucosylceramide beta1-->4 galactosyltransferase); implication as a drug target. Curr Drug Targets, 2008. 9(4): p. 272-81.

[21] Schwientek, T., et al., Cloning of a novel member of the UDP-galactose:beta-N-acetylglucosamine beta1,4-galactosyltransferase family, beta4Gal-T4, involved in glycosphingolipid biosynthesis. J Biol Chem, 1998. 273(45): p. 29331-40.

[22] Nishie, T., et al., Beta4-galactosyltransferase-5 is a lactosylceramide synthase essential for mouse extra-embryonic development. Glycobiology, 2010. 20(10): p. 1311-22.

[23] Almeida, R., et al., Cloning and expression of a proteoglycan UDP-galactose:beta-xylose beta1,4-galactosyltransferase I. A seventh member of the human beta4-galactosyltransferase gene family. J Biol Chem, 1999. 274(37): p. 26165-71.

[24] Shur, B.D., S. Evans, and Q. Lu, Cell surface galactosyltransferase: current issues. Glycoconj J, 1998. 15(6): p. 537-48.

[25] Shur, B.D., Cell surface beta 1,4 galactosyltransferase: twenty years later. Glycobiology, 1991. 1(6): p. 563-75.

[26] Bunnell, B.A., D.E. Adams, and V.J. Kidd, Transient expression of a p58 protein kinase cDNA enhances mammalian glycosyltransferase activity. Biochem Biophys Res Commun, 1990. 171(1): p. 196-203.

[27] Hathaway, H.J., et al., Mutational analysis of the cytoplasmic domain of beta1,4-galactosyltransferase I: influence of phosphorylation on cell surface expression. J Cell Sci, 2003. 116(Pt 21): p. 4319-30.

[28] Zhang, S.W., et al., Effect of p58GTA on beta-1,4-galactosyltransferase 1 activity and cell-cycle in human hepatocarcinoma cells. Mol Cell Biochem, 2001. 221(1-2): p. 161-8.

[29] Shur, B.D., Glycosyltransferases as cell adhesion molecules. Curr Opin Cell Biol, 1993. 5(5): p. 854-63.

[30] Bayna, E.M., J.H. Shaper, and B.D. Shur, Temporally specific involvement of cell surface beta-1,4 galactosyltransferase during mouse embryo morula compaction. Cell, 1988. 53(1): p. 145-57.

[31] Begovac, P.C., et al., Evidence that cell surface beta 1,4-galactosyltransferase spontaneously galactosylates an underlying laminin substrate during fibroblast migration. J Biol Chem, 1994. 269(50): p. 31793-9.

[32] Maillet, C.M. and B.D. Shur, Uvomorulin, LAMP-1, and laminin are substrates for cell surface beta-1,4-galactosyltransferase on F9 embryonal carcinoma cells: comparisons between wild-type and mutant 5.51 att- cells. Exp Cell Res, 1993. 208(1): p. 282-95.

[33] Maillet, C.M. and B.D. Shur, Perturbing cell surface beta-(1,4)-galactosyltransferase on F9 embryonal carcinoma cells arrests cell growth and induces laminin synthesis. J Cell Sci, 1994. 107 (Pt 6): p. 1713-24.

[34] Eckstein, D.J. and B.D. Shur, Cell surface beta-1,4-galactosyltransferase is associated with the detergent-insoluble cytoskeleton on migrating mesenchymal cells. Exp Cell Res, 1992. 201(1): p. 83-90.

[35] Gong, X., et al., Activation of a G protein complex by aggregation of beta-1,4-galactosyltransferase on the surface of sperm. Science, 1995. 269(5231): p. 1718-21.

[36] Lin, X., P. Nelson, and I.H. Gelman, SSeCKS, a major protein kinase C substrate with tumor suppressor activity, regulates G(1)-->S progression by controlling the expression and cellular compartmentalization of cyclin D. Mol Cell Biol, 2000. 20(19): p. 7259-72.

[37] Wassler, M.J., et al., Functional interaction between the SSeCKS scaffolding protein and the cytoplasmic domain of beta1,4-galactosyltransferase. J Cell Sci, 2001. 114(Pt 12): p. 2291-300.

[38] Bunnell, B.A., et al., Increased expression of a 58-kDa protein kinase leads to changes in the CHO cell cycle. Proc Natl Acad Sci U S A, 1990. 87(19): p. 7467-71.

[39] Wassler, M.J., et al., Characterization of a novel ubiquitin-conjugating enzyme that regulates beta1,4-galactosyltransferase-1 in embryonic stem cells. Stem Cells, 2008. 26(8): p. 2006-18.

[40] Gelman, I.H., E. Tombler, and J. Vargas, Jr., A role for SSeCKS, a major protein kinase C substrate with tumour suppressor activity, in cytoskeletal architecture, formation of migratory processes, and cell migration during embryogenesis. Histochem J, 2000. 32(1): p. 13-26.

[41] Wassler, M.J. and B.D. Shur, Clustering of cell surface (beta)1,4-galactosyltransferase I induces transient tyrosine phosphorylation of focal adhesion kinase and loss of stress fibers. J Cell Sci, 2000. 113 Pt 2: p. 237-45.

[42] Zeng, F.Y., et al., Differential response of the epidermal growth factor receptor tyrosine kinase activity to several plant and mammalian lectins. Mol Cell Biochem, 1995. 142(2): p. 117-24.

[43] Dennis, J.W., M. Granovsky, and C.E. Warren, Glycoprotein glycosylation and cancer progression. Biochim Biophys Acta, 1999. 1473(1): p. 21-34.

[44] Johnson, F.M. and B.D. Shur, The level of cell surface beta1,4-galactosyltransferase I influences the invasive potential of murine melanoma cells. J Cell Sci, 1999. 112 (Pt 16): p. 2785-95.

[45] Shen, J., et al., Two specific inhibitors of the phosphatidylinositol 3-kinase LY294002 and wortmannin up-regulate beta1,4-galactosyltransferase I and thus sensitize SMMC-7721 human hepatocarcinoma cells to cycloheximide-induced apoptosis. Mol Cell Biochem, 2007. 304(1-2): p. 361-7.

[46] Satomaa, T., et al., The N-glycome of human embryonic stem cells. BMC Cell Biol, 2009. 10: p. 42.

[47] Calder, A., et al., Lengthened G1 Phase Indicates Differentiation Status In Human Embryonic Stem Cells. Stem Cells Dev, 2012.

[48] Hinton, D.A., S.C. Evans, and B.D. Shur, Altering the expression of cell surface beta 1,4-galactosyltransferase modulates cell growth. Exp Cell Res, 1995. 219(2): p. 640-9.

[49] Pouncey, L., et al., Beta 1-4-galactosyltransferase gene expression is regulated during entry into the cell cycle and during the cell cycle. Somat Cell Mol Genet, 1991. 17(5): p. 435-43.

[50] Wu, G.Q., S.M. Jiang, and J.X. Gu, Studies on the beta(l-4)Galactosyltransferase of Cell Surface in Induced HL60 Cells. Sheng Wu Hua Xue Yu Sheng Wu Wu Li Xue Bao (Shanghai), 1997. 29(4): p. 402-408.

[51] Wei, Y., et al., Regulation of the beta1,4-Galactosyltransferase I promoter by E2F1. J Biochem, 2010. 148(3): p. 263-71.

[52] Asano, M., et al., Growth retardation and early death of beta-1,4-galactosyltransferase knockout mice with augmented proliferation and abnormal differentiation of epithelial cells. EMBO J, 1997. 16(8): p. 1850-7.

[53] Zhang, S.W., et al., Down-regulation of beta1,4-galactosyltransferase gene expression by cell-cycle suppressor gene p16. Biochim Biophys Acta, 1999. 1444(1): p. 49-54.

[54] Zhang, S.W., et al., Effect of suppression of TGF-beta1 expression on cell-cycle and gene expression of beta-1,4-galactosyltransferase 1 in human hepatocarcinoma cells. Biochem Biophys Res Commun, 2000. 273(3): p. 833-8.

[55] Li, Z., et al., Downregulation of beta1,4-galactosyltransferase 1 inhibits CDK11(p58)-mediated apoptosis induced by cycloheximide. Biochem Biophys Res Commun, 2005. 327(2): p. 628-36.

[56] Ohno, S., et al., Polypyrimidine tract-binding protein regulates the cell cycle through IRES-dependent translation of CDK11(p58) in mouse embryonic stem cells. Cell Cycle, 2011. 10(21): p. 3706-13.

[57] Wei, Y., et al., Identification of beta-1,4-galactosyltransferase I as a target gene of HBx-induced cell cycle progression of hepatoma cell. J Hepatol, 2008. 49(6): p. 1029-37.

[58] Ujita, M., et al., Poly-N-acetyllactosamine synthesis in branched N-glycans is controlled by complemental branch specificity of I-extension enzyme and beta1,4-galactosyltransferase I. J Biol Chem, 1999. 274(24): p. 16717-26.

[59] Asano, M., et al., Impaired selectin-ligand biosynthesis and reduced inflammatory responses in beta-1,4-galactosyltransferase-I-deficient mice. Blood, 2003. 102(5): p. 1678-85.

[60] Muramatsu, T. and H. Muramatsu, Carbohydrate antigens expressed on stem cells and early embryonic cells. Glycoconj J, 2004. 21(1-2): p. 41-5.

[61] Sudou, A., et al., Le(X) structure enhances myocardial differentiation from embryonic stem cells. Cell Struct Funct, 1997. 22(2): p. 247-51.

[62] Ma, R., et al., Post-translational and transcriptional regulation of glycolipid glycosyltransferase genes in apoptotic breast carcinoma cells: VII. Studied by DNA-microarray after treatment with L-PPMP. Glycoconj J, 2009. 26(6): p. 647-61.

[63] Biellmann, F., et al., The Lc3-synthase gene B3gnt5 is essential to pre-implantation development of the murine embryo. BMC Dev Biol, 2008. 8: p. 109.

[64] Yamashita, T., et al., A vital role for glycosphingolipid synthesis during development and differentiation. Proc Natl Acad Sci U S A, 1999. 96(16): p. 9142-7.

[65] Rastan, S., et al., Cell interactions in preimplantation embryos: evidence for involvement of saccharides of the poly-N-acetyllactosamine series. J Embryol Exp Morphol, 1985. 87: p. 115-28.

[66] Cho, S.K., et al., Transcriptional regulation of alpha1,3-galactosyltransferase in embryonal carcinoma cells by retinoic acid. Masking of Lewis X antigens by alpha-galactosylation. J Biol Chem, 1996. 271(6): p. 3238-46.

[67] Patil, S.A., et al., Scaling down the size and increasing the throughput of glycosyltransferase assays: activity changes on stem cell differentiation. Anal Biochem, 2012. 425(2): p. 135-44.

[68] Liang, Y.J., et al., Changes in glycosphingolipid composition during differentiation of human embryonic stem cells to ectodermal or endodermal lineages. Stem Cells, 2011. 29(12): p. 1995-2004.

[69] Hacker, U., K. Nybakken, and N. Perrimon, Heparan sulphate proteoglycans: the sweet side of development. Nat Rev Mol Cell Biol, 2005. 6(7): p. 530-41.

[70] Raman, R., V. Sasisekharan, and R. Sasisekharan, Structural insights into biological roles of protein-glycosaminoglycan interactions. Chem Biol, 2005. 12(3): p. 267-77.

[71] Heo, J.S., Y.J. Lee, and H.J. Han, EGF stimulates proliferation of mouse embryonic stem cells: involvement of Ca2+ influx and p44/42 MAPKs. Am J Physiol Cell Physiol, 2006. 290(1): p. C123-33.

[72] Kawashima, N., et al., Tyrosine kinase activity of epidermal growth factor receptor is regulated by GM3 binding through carbohydrate to carbohydrate interactions. J Biol Chem, 2009. 284(10): p. 6147-55.

[73] Takahashi, M., et al., Role of N-glycans in growth factor signaling. Glycoconj J, 2004. 20(3): p. 207-12.

[74] Li, Z., et al., Cell surface beta 1, 4-galactosyltransferase 1 promotes apoptosis by inhibiting epidermal growth factor receptor pathway. Mol Cell Biochem, 2006. 291(1-2): p. 69-76.

[75] Gabius, H.J., et al., Down-regulation of the epidermal growth factor receptor by altering N-glycosylation: emerging role of beta1,4-galactosyltransferases. Anticancer Res, 2012. 32(5): p. 1565-72.

[76] Purushotham, K.R., et al., A novel mechanism for isoprenaline-stimulated proliferation of rat parotid acinar cells involving the epidermal growth factor receptor and cell surface galactosyltransferase. Biochem J, 1992. 284 (Pt 3): p. 767-76.

[77] Matsuura, A., et al., O-linked N-acetylglucosamine is present on the extracellular domain of notch receptors. J Biol Chem, 2008. 283(51): p. 35486-95.

[78] Sakaidani, Y., et al., O-linked-N-acetylglucosamine on extracellular protein domains mediates epithelial cell-matrix interactions. Nat Commun, 2011. 2: p. 583.

[79] Hazan, R., L. Krushel, and K.L. Crossin, EGF receptor-mediated signals are differentially modulated by concanavalin A. J Cell Physiol, 1995. 162(1): p. 74-85.

[80] Hebert, E., Endogenous lectins as cell surface transducers. Biosci Rep, 2000. 20(4): p. 213-37.

[81] Rusnati, M., et al., Interaction of fibroblast growth factor-2 (FGF-2) with free gangliosides: biochemical characterization and biological consequences in endothelial cell cultures. Mol Biol Cell, 1999. 10(2): p. 313-27.

[82] Basilico, C. and D. Moscatelli, The FGF family of growth factors and oncogenes. Adv Cancer Res, 1992. 59: p. 115-65.

[83] Lanner, F. and J. Rossant, The role of FGF/Erk signaling in pluripotent cells. Development. 2010.137 (20) p. 3351-60.

[84] Dvorak, P., et al., Embryoglycan ectodomains regulate biological activity of FGF-2 to embryonic stem cells. J Cell Sci, 1998. 111 (Pt 19): p. 2945-52.

[85] Jirmanova, L., et al., O-linked carbohydrates are required for FGF-2-mediated proliferation of mouse embryonic cells. Int J Dev Biol, 1999. 43(6): p. 555-62.

[86] Katoh, M., WNT signaling pathway and stem cell signaling network. Clin Cancer Res, 2007. 13(14): p. 4042-5.

[87] Capela, A. and S. Temple, LeX is expressed by principle progenitor cells in the embryonic nervous system, is secreted into their environment and binds Wnt-1. Dev Biol, 2006. 291(2): p. 300-13.

[88] Haltiwanger, R.S., Regulation of signal transduction pathways in development by glycosylation. Curr Opin Struct Biol, 2002. 12(5): p. 593-8.

[89] Chen, J., D.J. Moloney, and P. Stanley, Fringe modulation of Jagged1-induced Notch signaling requires the action of beta 4galactosyltransferase-1. Proc Natl Acad Sci U S A, 2001. 98(24): p. 13716-21.

[90] Haltiwanger, R.S. and P. Stanley, Modulation of receptor signaling by glycosylation: fringe is an O-fucose-beta1,3-N-acetylglucosaminyltransferase. Biochim Biophys Acta, 2002. 1573(3): p. 328-35.

[91] Hou, X., Y. Tashima, and P. Stanley, Galactose differentially modulates lunatic and manic fringe effects on Delta1-induced NOTCH signaling. J Biol Chem, 2012. 287(1): p. 474-83.

[92] Prasad, S.M., et al., Continuous hypoxic culturing maintains activation of Notch and allows long-term propagation of human embryonic stem cells without spontaneous differentiation. Cell Prolif, 2009. 42(1): p. 63-74.

[93] Yagi, H., et al., Lewis X-carrying N-Glycans Regulate the Proliferation of Mouse Embryonic Neural Stem Cells via the Notch Signaling Pathway. J Biol Chem, 2012. 287(29): p. 24356-64.

[94] Machingo, Q.J., A. Fritz, and B.D. Shur, A beta1,4-galactosyltransferase is required for convergent extension movements in zebrafish. Dev Biol, 2006. 297(2): p. 471-82.

[95] Tonoyama, Y., et al., Essential role of beta-1,4-galactosyltransferase 2 during medaka (Oryzias latipes) gastrulation. Mech Dev, 2009. 126(7): p. 580-94.

[96] Duchesne, L., et al., N-glycosylation of fibroblast growth factor receptor 1 regulates ligand and heparan sulfate co-receptor binding. J Biol Chem, 2006. 281(37): p. 27178-89.

[97] Machingo, Q.J., A. Fritz, and B.D. Shur, A beta1,4-galactosyltransferase is required for Bmp2-dependent patterning of the dorsoventral axis during zebrafish embryogenesis. Development, 2006. 133(11): p. 2233-41.

[98] Hansske, B., et al., Deficiency of UDP-galactose:N-acetylglucosamine beta-1,4-galactosyltransferase I causes the congenital disorder of glycosylation type IId. J Clin Invest, 2002. 109(6): p. 725-33.

[99] Kumagai, T., et al., Early lethality of beta-1,4-galactosyltransferase V-mutant mice by growth retardation. Biochem Biophys Res Commun, 2009. 379(2): p. 456-9.

[100] Talhaoui, I., et al., Identification of key functional residues in the active site of human {beta}1,4-galactosyltransferase 7: a major enzyme in the glycosaminoglycan synthesis pathway. J Biol Chem, 2010. 285(48): p. 37342-58.

[101] Mallinger, A., et al., Using a ubiquitin ligase as an unfolded protein sensor. Biochem Biophys Res Commun, 2012. 418(1): p. 44-8.

[102] Gao, Y., et al., Specificity of beta1,4-galactosyltransferase inhibition by 2-naphthyl 2-butanamido-2-deoxy-1-thio-beta-D-glucopyranoside. Glycoconj J, 2010. 27(7-9): p. 673-84.

[103] Daligault, F., et al., Thermodynamic insights into the structural basis governing the donor substrate recognition by human beta1,4-galactosyltransferase 7. Biochem J, 2009. 418(3): p. 605-14.

[104] Tateno, H., et al., Glycome diagnosis of human induced pluripotent stem cells using lectin microarray. J Biol Chem, 2011. 286(23): p. 20345-53.

[105] Roth, S., E.J. McGuire, and S. Roseman, Evidence for cell-surface glycosyltransferases. Their potential role in cellular recognition. J Cell Biol, 1971. 51(21): p. 536-47.

Section 3

Therapeutic Implications and Ethical Concerns

Chapter 17

Advances in Stem Cell Therapies

Joel Sng and Thomas Lufkin

Additional information is available at the end of the chapter

http://dx.doi.org/10.5772/55061

1. Introduction

Four key milestones have to be realized for the ideal customized stem cell therapy to be successful. First, stem cells utilized in these therapies have to be genetically stable and epigenetically regulated to ensure the safety of stem cells employed in any future therapies. This is essential to ensure that patients undergoing stem cell therapy are not exposed to increased risks of tumorigenesis and other mutagenic diseases. Second, stem cells should be able to evade the innate immune response of patients, possibly via the secretion of immuno-suppressive molecules that inhibit immune responses or by displaying host cellular recognition markers. The survival of transplanted stem cells is crucial for the design of an effective therapy. Additionally the ability of transplanted stem cells to evade immune detection and inflammatory responses will prevent undesired symptoms such as graft-versus-host-disease in patients. Third, stem cells employed in these therapies should be location specific. These stem cells should possess specific homing cell surface markers that will allow them to locate and migrate to specific localities. This will ensure that stem cells used in therapies will only accumulate in diseased tissues for targeted therapeutic effect, and not in other healthy regions where detrimental non-specific interactions might occur. Finally, the stem cells used in these therapies should be functionally specific and disease relevant. Transplanted stem cells should be designed to restore a healthy phenotype in patients. These cells should be able to restore organ and tissue function in regenerative therapies, either directly by replicating to replace damaged portions of these organs and tissues and/or indirectly by secreting therapeutic molecules to mediate their functional restoration. These stem cells should also be epigenetically primed for specific functions to ensure that they are able to reverse the effects of treated diseases while minimizing unwanted side effects.

Stem cells are commonly classified into three broad categories based on how they were derived. Embryonic stem cells (ESCs) are stem cells that are isolated from the inner cell mass of the early developing embryo. Adult stem cells assist in the natural regeneration and repair in developed organisms and can be purified from their tissues. Induced pluripotent stem cells (iPSCs) are artificially derived stem cells that are formed via various genetic and epigenetic reprogramming procedures. Of the three broad categories of stem cells, adult stem cells are most widely utilized in clinical trials and experimental therapies worldwide. Most adult stem cells are multipotent and differentiate to form only a limited subset of cell types. Hence these stem cells are commonly classified according to their developmental commitment or tissue source. Examples of adult stem cells include mesenchymal stem cells (MSCs), neural stem cells (NSCs), hematopoietic stem cells, inner ear stem cells, mammary stem cells, endothelial stem cells, intestinal stem cells, and testicular stem cells.

Adult stem cells present the first success of human experimental stem cell therapy. There are several reasons why adult stem cells therapies are currently more successful than ESC and iPSC therapies. Firstly, stem cell therapies involving adult stem cells are often autotransplants with minimal potential for immune rejection. These adult stem cells can be harvested directly from individual patients before being utilized as transplants. Hence these adult stem cells will exhibit host cell recognition molecules unlike ESCs and iPSCs that may provoke an immune response when used in therapies. Secondly since most adult stem cells therapies involve minimally processed cells, there a reduced possibility of genetic mutation or chromosome aberration occurring compared to ESCs and iPSCs that have to be cultured extensively in vitro before their use in therapies. Thirdly, adult stem cells do not readily form tumors when introduced into patients and are considered to be safer than ESCs and iPSCs that display greater carcinogenic potential. Finally, the use of adult stem cells in therapies is not considered to be controversial as they can be readily extracted from patient tissues and do not require the destruction of embryos to derive stable cell lines unlike ESCs. These key advantages of adult stem cells have led to their wider utilization in research and various clinical trials compared to ESCs and iPSCs.

2. Mesenchymal stem cell therapy

Mesenchymal stem cells (MSCs) are one of the first multipotent adult stem cells to be utilized in stem cell therapies. These stem cells have the ability to differentiate and form bone, cartilage, and adipose tissues. While the bone marrow is the most common source of MSCs for therapeutic purposes, they can also be found in adipose and synovial tissue, skeletal muscles, peripheral blood, breast milk, and the umbilical cord [1, 2].

While these stem cells are commonly referred to as MSCs, they actually form a heterogeneous population of cells as evidenced by differences in proliferative capacity, differentiation potential, cellular markers, and morphology. For example, MSCs derived from the bone marrow (M-MSC) have lower proliferative capacity, followed by adipose tissue MSCs (A-MSC) and umbilical cord blood MSCs (U-MSC) which have the highest proliferative capacity [3]. MSCs also have differing differential potentials. For example, bone marrows MSCs have a higher chondrogenic potential while adipose MSCs have a lower chondrogenic potential [4, 5]. In addition both bone marrow and adipose MSCs readily form cells with adipogenic phenotypes unlike umbilical cord blood MSC that display a lower capacity to form adipocytes [3, 6]. MSCs also express different cellular markers. For example A-MSCs express CD34, CD49d, and CD54 at higher levels than M-MSCs while M-MSCs and U-MSCs express higher levels of CD 106 than A-MSCs [7]. M-MSCs and A-MSCs also have higher levels of CD90 and CD105 expression when compared to U-MSCs [3]. The morphology of MSCs can also differ significantly and even MSCs from the same source display heterogeneous morphologies. Various descriptions of MSCs in the literature include spindle shaped, round, fibroblastoid cells, flattened cells, and blanket cells [8, 9]. Further studies to understand these inherent differences in various subpopulations of MSCs could lead to an improved understanding of how epigenetic differences regulate stem cell differentiation fates, homing to specific recognition sites, proliferation rates, and senescence.

M-MSCs are currently the most widely used stem cells in clinical trials and therapies. Both autograft and allograft M-MSCs have been extensively tested for their therapeutic safety and effectiveness in alleviating the symptoms of several diseases. One of the key reasons for the success of M-MSCs therapy is because these cells possess intrinsic immunomodulatory properties that enable M-MSCs to inhibit and evade potential immune rejection when transplanted [10]. M-MSCs are able to inhibit the maturation and function of various immune cells including dendritic cells, natural killer cells, B cells, and T lymphocytes [11]. Additionally M-MSCs have reduced immunogenicity due to their minimal expression of surface MHC II proteins and the lack of T cell stimulatory proteins like CD80 and CD86 [12]. Another important reason for the early success of M-MSCs based therapies is that MSCs have low tumorigenic potential and are safer than therapies based on ESCs or iPSCs which display robust tumorigenicity [13].

Due to these intrinsic advantages of MSCs, clinical trials can be conducted to evaluate their safety and effectiveness in treating various diseases. For example, the safety and effectiveness of M-MSC transplantation for joint cartilage repair has been evaluated in several studies. In a clinical trial involving 41 patients studied over a period of between 5 to 137 months, M-MSC transplantation did not contribute to increased risk of tumors or infection [14]. Another study has reported the potential for M-MSC regenerative knee therapy to induce cartilage and meniscus growth and increase range of motion [15]. These results are supported in a larger scale M-MSC transplantation study involving 339 patients which reported no increased risk of tumor formation and a significant improvement of knee function in transplant patients [16]. While further clinical trials have to be conducted to verify these preliminary results, the successes of these initial clinical trials indicate that M-MSC therapy is likely to be safe and can catalyze cartilage repair.

The effectiveness of M-MSC and other MSC therapies for various autoimmune diseases has also been studied in several small clinical trials. These autoimmune diseases include multiple sclerosis, Crohn's disease, scleroderma, and systemic lupus erythematosus [17]. The causes of many of these autoimmune diseases are not well understood and it is likely that while the patients suffer from similar symptoms, contributing disease factors may vary significantly between patients. However the application of a generic MSC transplantation therapy was successful in alleviating the symptoms of these patients in several clinical trials.

Multiple sclerosis is a debilitating autoimmune disease caused by immune mediated damage of neural myelin sheath. Progressive neural damage results in many disabling symptoms including the loss of balance, vision and memory. M-MSC clinical trials for multiple sclerosis therapy have provided limited preliminary data indicating that M-MSC transplantation is safe, inhibits the progress of multiple sclerosis through immune regulated neuroprotection, and can repair limited damage to the CNS [18, 19]. For example, in a preliminary phase 2 clinical trial involving 10 patients diagnosed with progressive multiple sclerosis, autologous infusion of externally expanded M-MSCs was shown to improve visual acuity and increase optic nerve area without any major side-effects [18].

Crohn's disease is a chronic autoimmune bowel disease characterized by inflammation of the gastrointestinal tract. In severe cases, this uncontrolled immune response may result in infection, hemorrhage, and intestinal fistulas. M-MSC clinical trials involving patients suffering from Crohn's disease have sought to harness the innate immunomodulatory capacity of MSCs to mitigate abnormal immune response in these patients and determine the safety of any potential therapies. In two phase one clinical trials a total of 22 adult Crohn's disease patients were enrolled to investigate the effects of M-MSC therapy. In the first trial it was determined that while autologous M-MSC infusion therapy did not result in adverse side effects, it only had a modest impact in alleviating the autoimmune response in these patients [20]. In the second trial in vitro expanded M-MSCs were directly injected into the intestinal wall and lumen [21]. When M-MSCs were directly injected, they were able to inhibit inflammation locally and mediate healing of intestinal tissue in these regions.

Scleroderma is an autoimmune connective tissue disorder characterized by accumulation of collagen in the skin, heart, kidneys or lungs. This buildup of collagen may lead to skin ulcers, pulmonary fibrosis, heart and kidney failure. Exploratory M-MSC clinical trials involving patients suffering from Scleroderma have sought to harness the regenerative and immunomodulatory capacity of MSCs to initiate ulcer healing and prevent organ failure while evaluating the safety of these therapies. In two separate phase one clinical trials a total of 7 adult scleroderma patients were enrolled to determine the effects of M-MSC therapy. In the first trial allogeneic transplantation of donor M-MSC was performed via intravenous infusion and was associated with possible pericardial calcification and increased risk of cardiac impairment. While patients in this trial displayed a slight improvement in MRSS score and healing of skin ulcers, the effects were on occasion only temporary and the disease regressed in some patients [22]. The second trial involved autologous transplantation of either M-MSC or peripheral stem cells in patients via intramuscular injection [23]. This local stem cell therapy was able to induce healing of skin ulcers in these patients and improved endothelial function of blood vessels.

Systemic lupus erythematosus (SLE) is a chronic autoimmune disorder that can affect the kidney, lung, brain, and other organs. Severe SLE may result in kidney failure, stroke, and inflammation of blood vessels. M-MSC clinical trials in SLE patients have attempted to treat progression of SLE symptoms by harnessing the immunomodulatory properties of MSCs. In two clinical trials a total of 19 patients suffering from SLE were treated with M-MSC transplants to determine if MSC therapy is safe and effective in reversing the symptoms of SLE patients.

In the first clinical trial SLE patients were treated with allogeneic M-MSC infusion [24]. Treatment with donor M-MSCs was shown to restore kidney function and reverse the progression of SLE. The second clinical trial involved a larger group of patients and provided additional evidence that M-MSC therapy could mitigate the symptoms of SLE in patients [25].

Other studies have also attempted to verify the effectiveness of M-MSC therapies for various diseases. M-MSC therapy has been shown to improve liver function in patients suffering from liver cirrhosis by encouraging hepatocyte proliferation [26]. The co-transplantation of M-MSCs and kidney transplants for patients with kidney failure can reduce the risk of acute transplant rejection and improve transplant function in treated patients [27]. M-MSC therapy can also catalyze functional recovery and improve survival rates in ischemic stroke patients [28, 29]. These clinical studies provide preliminary evidence that M-MSC therapy is safe and the regenerative properties of these stem cells can be harnessed to treat a wide variety of diseases.

To develop the ideal next generation stem cell therapy, it is necessary to evaluate currently available therapies to identify their current limitations and suggest areas for improvement. Next generation stem cell therapies will have to fulfill the four key milestones (safety, immune evasion, location specificity, and disease relevancy) of customized stem cell therapy. Human M-MSC therapy has been extensively studied in multiple experiments and clinical trials and is an ideal candidate for evaluation against these key milestones.

Firstly the safety of M-MSC therapies must be considered. Multiple clinical trials mentioned previously involving the infusion and injection of both autologous and allogeneic M-MSCs for therapeutic purposes stated that patients were generally not exposed to increased risks of cancer or other serious side-effects. However, a study stated that the infusion of M-MSC may lead to pericardial calcification and increased risk of cardiac impairment in some patients [22]. As currently completed clinical trials often only involve a relatively small patient population or are only conducted over a brief period of time, the risks of M-MSC therapies may not be fully understood and more studies have to be conducted to ensure that the benefits of these therapies outweigh their potential risks. Another source of concern is the use of in vitro cultured M-MSCs in therapies. M-MSCs exist naturally in low concentrations in the human bone marrow, and often have to be concentrated and expanded in vitro media to provide sufficient numbers of stem cells for therapeutic purposes. This process may expose M-MSCs to xenogeneic antigens such as in fetal calf serum in the media. Culture of M-MSCs in vitro also exposes cells to an atmospheric oxygen concentration of 21% that is radically different from physiological conditions of 1-7% [30]. These in vitro culture conditions may affect the genetic and epigenetic stability of these stem cells resulting in an increased chance of mutagenesis. In an effort to resolve these potential issues, several studies have attempted to identify the ideal M-MSC culture media. From these studies, it has been proposed that human platelet lysate can be used as a viable substitute to fetal calf serum to reduce unnecessary exposure to xenogeneic antigens [31]. M-MSCs should also be cultured in low oxygen concentrations of approximately 3% to reduce oxidative stress and telomere shortening and increase the proliferative lifespan and genetic stability of in vitro M-MSCs [32]. The implementation of these protocols will provide M-MSCs with culture conditions that are more similar to the M-MSC native environment and minimize the impact of in vitro expansion on the genetic and epigenetic stability of M-MSCs. In conclusion, while some doubts about the safety of M-MSC based therapies remain, various clinical trials and experiments have indicated that the use of minimally expanded M-MSCs is relatively safe for patients, especially when coupled with the latest M-MSC expansion protocols.

Secondly the ability of M-MSCs to evade immune detection must be accessed. No incidents of acute immune rejection were reported in the various clinical trials involving autologous and allogeneic transplants of M-MSCs. This could possibly be attributed to the fact that autologous M-MSCs are extracted from the treated patients and present host cellular recognition markers. Additionally M-MSCs have reduced immunogenicity due to the naturally low expression levels of surface MHC II proteins and the lack of other T cell stimulatory proteins like CD80 and CD 86 in M-MSCs. The multifaceted immunomodulatory capacity of M-MSCs must also be considered. Various studies have indicated that M-MSCs are able to inhibit the proliferation of T lymphocytes possibly via the activation of regulatory T cells and secretion of immuno-suppressive factors like transforming growth factor beta1 and hepatocyte growth factor [33-35]. The inhibition of T lymphocytes that are essential for the recognition and destruction of foreign transplants contributes to the ability of M-MSCs to evade immune detection. M-MSCs can also interfere with the development and function of antigen-presenting dendritic cells. Soluble factors secreted by M-MSCs can inhibit differentiation of monocytes to dendritic cells and suppress the production of cytokines [36]. M-MSCs can also affect the function of mature dendritic cells by suppressing the expression of various presentation and co-stimulatory molecules like CD1a, CD80, CD83, and CD86 [37]. This impedes dendritic cells from inducing T cells and B cells and prevents resistance of foreign transplants from developing. Finally M-MSCs can also inhibit the proliferation of B cells stimulated with anti-CD40 monoclonal antibody and IL-4 by halting the G_0/G_1 cell cycle phase [38, 39]. The data from these studies indicate that M-MSCs are able to efficiently evade the innate immune response of patients via various mechanisms of cellular recognition and immunosuppression.

Thirdly the location specificity of M-MSCs employed in various therapies should be considered. While M-MSC clinical trials discussed previously indicate that M-MSC therapy is able to alleviate the conditions of various autoimmune diseases and induce cartilage repair, infusion of M-MSCs resulted in non-specific distribution of these cells within the patient. Non-specific infusion of M-MSCs resulted in a distribution of these cells in various organs including the heart muscle, liver, kidney, skin, and lung. This may result in undesirable side-effects such as pericardial calcification and increased risk of cardiac impairment in patients as described in a clinical study [22]. Hence further research is required to design a stem cell therapy that is more specific to the injury location. In the ideal therapy, stem cells could be engineered with receptors for mobilization to the location of injury. Alternatively, stem cells could be integrated within a scaffold that would then be implanted into patients to improve the specificity of these therapies. The direct injection of M-MSCs near sites of injury may also provide increased specificity to these therapies.

The fourth consideration is stem cells utilized in these therapies should be disease relevant. Disease relevant stem cells should be epigenetically primed to treat specific underlying causes of disease in each patient. M-MSCs utilized in these clinical trials are not disease relevant and

http://dx.doi.org/10.5772/55061

cure or alleviate various disease symptoms through their general immunomodulatory and regenerative properties. While non-specific M-MSCs may still be a viable therapy for a wide range of diseases, the lack of specificity in these therapies may result in potentially lethal consequences. For example, the general immunosuppressive properties of M-MSCs can increase the severity of breast cancer by increasing the concentration of regulatory T cells and inhibiting the innate immune response against cancer cells [35, 40]. M-MSCs could also secrete soluble factors that accelerate tumor growth, such as through the activation of the phosphatidylinositol-3-kinase/Akt signaling pathway which can prevent apoptosis and induce proliferation of cancer cells [41]. Hence additional studies have to be performed to understand how stem cells can be epigenetically reprogrammed to enhance their specificity for disease treatment and reduce undesirable side effects.

In conclusion, it can be seen that while current clinical data demonstrates that M-MSC based therapies are relatively safe and M-MSC transplants can evade immune detection and survive in patients, these therapies rely on the general immunosuppressive and regenerative properties of M-MSCs and are neither specific nor disease relevant. Hence although the utilization of M-MSC based therapies may potentially result in cures for various diseases, more research is necessary for developing the ideal stem cell therapy.

3. Other MSC and adult stem cell therapies

While the bone marrow is the most commonly mentioned source of MSCs, MSCs can also be extracted from other sources including adipose and synovial tissue, skeletal muscles, peripheral blood, breast milk, and the umbilical cord [1, 2]. In particular, adipose MSCs (A-MSCs) have been increasingly studied because these cells can be readily purified from adipose tissue via liposuction and is a relatively non-invasive procedure compared to bone marrow extraction of M-MSCs [42, 43]. A-MSCs have similar immunomodulatory effects compared to M-MSCs and can be utilized for treatment of similar diseases such as scleroderma [44, 45]. A-MSCs also possess a similar capacity to regenerate cartilage and bone tissues and mediate some symptoms in patients with osteonecrosis and osteoarthritis [46]. Hence the discovery of A-MSCs provides patients with an alternative source of MSCs in the event that they are unable to undergo M-MSC extraction.

MSCs are also present in human and animal synovial fluid. These synovial MSCs (S-MSCs) have a greater ability to proliferate and differentiate compared to other MSCs and can form osteoblasts, adipocytes, chondrocytes, and neurons [47, 48]. S-MSCs also possess greater cartilage regenerative potential than other MSCs with 60% of S-MSCs placed on cartilage defects attaching to the defect within 10 minutes [49]. S-MSCs posses similar regenerative potential as M-MSCs and can also initiate regeneration of the nucleus pulposus in the damaged rabbit intervertebral disc by suppressing inflammation and inducing the synthesis of type II collagen which acts as a supportive framework for nucleus pulposus repair [50]. In addition S-MSCs can be readily harvested via punch biopsy [9]. The greater innate proliferative ability of S-MSCs and the relative ease of obtaining S-MSCs indicate that it may be an excellent source of MSCs for future regenerative therapies

Neural Stem Cells (NSCs) have also been studied in an attempt to harness their regenerative potential for therapeutic purposes. NSCs can be found in various tissues including the bone marrow and striatum [51, 52] and their regenerative properties have been assessed by both NSC transplantation and endogenous NSC functional studies. NSCs can initiate axon remyelination, neuroprotection, proliferation of oligodendrocyte progenitors, and functional recovery when transplanted into mice experimental autoimmune encephalomyelitis (EAE) models of multiple sclerosis [53]. NSCs also possess similar immunomodulatory properties as MSCs. For example, NSCs can inhibit dendritic cell and antigen-specific T cell maturation through the release of morphogens such as bone morphogenetic protein 4 [54]. Additionally, NSCs can suppress T-cell proliferation through the release of prostaglandin E2 and nitric oxide [55]. This innate immunomodulatory property of NSCs has been harnessed to induce stable pancreatic islet graft function in mice, without the need for long-term immunosuppression [56]. The immunosuppressive potential of NSC can also be enhanced by engineering NSCs to produce anti-inflammatory cytokines such as IL-10 [57]. Engineered NSC transplants have greater therapeutic potential than ordinary NSCs and give rise to enhanced functional recovery of EAE mice.

Functional studies of endogenous NSCs have revealed the complex regulatory pathways governing in vivo neuronal regeneration. While neural stem cell niches exist in the subventricular zone and the subgranular zone of the hippocampal dentate gyrus [58], NSCs in these niches are unable to initiate spontaneous neural regeneration in many diseases. Hence recent research has been concerned with elucidating the regulators of neurogenesis and repair. For example neurogenesis can be initiated by suppressing Olig2 resulting in increased neurogenesis for brain injury repair [59]. Other molecular regulators of neurogenesis include morphogens like Shh and Wnt, transcription factors like Sox2, growth factors like Fibroblast Growth Factor family, and cell surface molecules like Notch1 [60]. An improved understanding of the molecular pathways that regulate the differentiation, mobilization, and proliferation of endogenous NSCs and the development of molecular tools to manipulate these pathways may lead to the development of novel minimally invasive regenerative therapies.

Other adult stem cells that have been evaluated for therapeutic use include hematopoietic stem cells, inner ear stem cells, mammary stem cells, intestinal stem cells, and adult germline stem cells. Hematopoietic stem cells are multipotent and can form various blood cells such as those from the lymphoid and myeloid lineages. Allogeneic hematopoietic stem cell transplantation (HSCT) therapy in leukemia patients can lead to remission by inducing an immune antitumor response [61]. HSCT has also been utilized to cure other diseases including sickle cell anemia, acquired aplastic anemia and thalassemia [62, 63]. HSCT can also halt neurological deterioration in X-linked adrenoleukodystrophy patients [64]. HSCT is also useful for alleviating symptoms of Hurler Syndrome and other lysosomal storage diseases and these grafts can replace metabolic enzymes that are lacking in host cells.

Inner ear stem cells are important progenitors of auditory hair cells and exist endogenously in both the utricular sensory epithelium and the dorsal epithelium of the cochlear canal [65, 66]. An improved understanding of molecular regulatory pathways in these stem cells could lead to the development of regenerative therapies for treating hearing impairment. Ongoing studies have revealed that the over-expression of *SKP2* can induce proliferation of non-sensory cells that can differentiate to form hair cells through the co-expression of *Atoh1* [67]. Developmental studies have also provided insight into the *Notch* signaling pathway, and its influence on the lateral-inhibition mediation differentiation of hair cells [68, 69]. Further studies could lead to the development of a viable stem cell therapy for regenerating auditory hair cells and a cure for hearing impairment.

Mammary stem cells are indispensible in the formation of mammary glands and can possess the capacity to form myoepithelial cells, alveolar epithelial cells, and ductal epithelial cells [70, 71]. The deregulation of various signaling pathways including the Notch, Wnt, and Hedgehog pathways in mammary stem cells has been implicated in breast cancer development [71, 72]. These studies could lead to the development of anti-cancer drugs that target specific signaling pathways.

Intestinal stem cells are multipotent progenitors of the intestinal epithelial cell lineages. Studies of intestinal stem cells have revealed the role of the Notch and Wnt signaling pathways in intestinal stem cell maintenance, differentiation, and proliferation and how deregulation of these pathways can promote intestinal carcinogenesis [73, 74]. An impaired differential capacity of intestinal stem cells has also been linked to inflammatory bowel diseases like Crohn's disease and ulcerative colitis [75]. Future studies based on these discoveries could lead to more effective cures for these diseases.

Adult germline stem cells are essential for gamete generation and can be derived from testis spermatogonial cells. These stem cells are pluripotent and share characteristics similar to ESCs [76]. The pluripotent nature of these stem cells may allow the development of regenerative therapies not possible with other multipotent adult stem cells. Adult germline stem cell transplantation can also be utilized for fertility restoration in animals [77-79]. This regenerative ability could be utilized for maintaining the fertility of patients undergoing radiotherapy, chemotherapy and other therapies that may cause infertility.

A comparison to determine how epigenetic differences inherent to these different classes of adult stem cells lead to a wide variation in differentiation, homing, proliferation, and immunomodulation capacities will enable the design of novel stem cell therapies for specific diseases. The differentiation potentials of adult stem cells can vary widely, for instance hematopoietic stem cells tend to form cells from the lymphoid and myeloid lineages, while neural stem cells tend to form neural cells like neurons, oligodendrocytes and astrocytes. Differences in differential predisposition also exist within a similar class of stem cells. A-MSCs tend to form adipocytes and cardiomyocytes, while M-MSCs form chondrocytes more readily [4, 5]. Further studies to map epigenetic differences between these stem cell populations will reveal how differentiation is regulated. This will lead to an improved ability to prime and select optimal stem cell transplants for disease therapy. For example, a better understanding of underlying pro-chondrogenic factors will enable the engineering of stem cells specialized in cartilage regeneration. Detailed studies of other intra-population epigenetic variations will also lead to better understanding of how these differences lead to differences in other properties of stem cells and augment the safety, effectiveness, and specificity of stem cell therapies. In conclusion, it can be seen that while the safety of adult stem cell therapies remains a key concern especially in the less studied stem cells, the innate immunomodulatory and regenerative capacity of adult stem cells can be exploited for curing a wide range of diseases.

4. Induced pluripotent stem cells

The discovery of iPSCs has led to a revolution in stem cell research. The ability to reprogram adult somatic cells to iPSCs using an increasing array of novel vectors and strategies has opened up a myriad of possibilities for therapeutic stem cell development. iPSC based therapies possess several advantages over adult stem cell and ESC based therapies. First, since iPSCs can be derived from patients like adult stem cells, they will exhibit host cellular recognition markers and can evade immune rejection more readily than ESCs. Additionally since iPSC lines can be derived from patients, they do not face the ethical concerns associated with ESC derivation. The use of iPSCs is also advantageous because iPSCs can be modified to produce desired cell phenotypes that may not be naturally available in adult stem cell and ESC populations. Hence iPSCs can be customized for treating specific diseases unlike other stem cells whose curative properties tend to be more general.

However currently available iPSCs face several limitations that prevent their use in patient therapies. First the iPSC derivation process commonly involves the use of viral vectors such as lentiviral and retroviral vectors which results in the integration of viral DNA in iPSCs [80, 81]. Second, many iPSC derivation processes involve the over-expression or integration of proto-oncogenes such as *Oct4*, *c-Myc*, and *Sox2* [82-86]. Third, iPSC cultures are genetically unstable and contain numerous genetic abnormalities including protein coding mutations, copy number variations, and chromosomal aberrations [87-89]. Fourth, the iPSC reprogramming process may be incomplete and iPSCs can retain epigenetic memory from parental somatic cells [90, 91]. Finally, the transformation efficiency of adult somatic cells to iPSCs is inefficient (0.001% - 4.4%) and only a small fraction of adult somatic cells can be transformed to iPSCs via existing methods.

These current shortcomings hinder the development of iPSCs suitable for patient therapy. The use of viral reprogramming vectors, over-expression of proto-oncogenes, and sub-optimal culture conditions contribute to widespread genetic mutation and increased tumorigenic potential in iPSCs. Consequently, iPSCs can readily form tumors in immune deficient mice and mice derived from iPSCs have a high chance of developing tumors [82, 92]. To overcome these limitations, new methods for iPSC reprogramming were developed to enhance the genetic integrity of iPSCs. Advances in reprogramming enabled the generation of iPSCs without *c-Myc* and mice without tumors could be derived from these iPSCs [93]. Additionally, non-integrating viral vectors like adenoviruses were used to prevent the introduction of foreign viral DNA into iPSCs [94]. More recently iPSCs have been generated via transfection of modified mRNA, this DNA free method results in a higher efficiency of iPSC generation compared to previous methods and does not introduce any exogenous DNA into reprogrammed cells [95]. Optimized iPSC culture and reprogramming conditions will also be essential for maintaining genetic stability and increasing transformation

http://dx.doi.org/10.5772/55061

efficiency. High atmospheric oxygen concentration (~21%) exposes stem cells to increased oxidative stress and DNA damage [32]. Lowering the oxygen concentration to 5% can improve iPSC generation efficiency and genetic stability of stem cells [96]. The addition of vitamin C and other antioxidants can also improve the efficiency of iPSC generation by preventing the accumulation of reactive oxygen species and promoting epigenetic modifications required for reprogramming to occur [97]. Reprogramming and iPSC culture maintenance also requires precise manipulation of other medium conditions. iPSC progenitors have to be cultivated in conditions that facilitate their survival but these original conditions may have to be modified to enhance reprogramming efficiency and maintain iPSC populations [98, 99]. The search for improving the efficiency of iPSC generation has also led to the use of miRNA sequences in reprogramming. Viral aided *miR302/367* cell reprogramming can reprogram fibroblasts to iPSCs with up to 10% efficiency [100].

The reprogramming of adult somatic cells to ideal iPSCs will involve a complex epigenomic transformation of the cellular epigenome to resemble ESC epigenetics. However iPSCs derived with current procedures retain unique epigenetic signatures that differ from the ESC epigenome. Some common epigenetic differences include variations in DNA methylation at CpG islands and histone modifications [101, 102]. The epigenetic memory of iPSCs is an artifact from the reprogramming process and parental cell epigenetics and can affect the differentiation predisposition of iPSCs [90]. This iPSC epigenetic signature can also be transmitted to successive generations of iPSCs and their differentiated progeny [102]. A failure to reset the epigenetic memory of iPSCs to more closely resemble the epigenetic ground state of ESCs could affect the function and safety of differentiated cells derived from these iPSC lines.

The issue of residual iPSC epigenetic memory can be partially addressed. For example, somatic cell nuclear transfer has been proposed as a viable method for resetting epigenetic memory [91]. Sodium butyrate, a short-chain fatty acid, could also assist in programming iPSCs closer to the epigenetic ground state by inhibiting histone deacetylase, directing the acetylation of specific genes, and encouraging stem cell renewal [103, 104]. DNA methyltransferase inhibitors can also be used to direct DNA methylation at specific sites for more complete epigenetic reprogramming [105]. The iPSC culture environment can also be manipulated to achieve a desired epigenetic state. For example, reducing culture oxygen concentration to 2% can induce epigenetic modifications that increase the expression of the retinal genes *Six3* and *Lhx2* in iPSCs while an oxygen concentration of 5% can increase the efficiency of iPSC regeneration [96, 106]. However despite the increased availability of tools for epigenetic modification, more studies are required to determine the ideal epigenomic ground state for therapeutic stem cells. Comprehensive epigenetic mapping of ESCs and adult stem cells could provide important clues and enable the development of improved experimental procedures for reprogramming adult somatic cells to mirror this ideal epigenomic ground state.

When evaluated against the four key milestones, iPSCs are clearly inferior to M-MSCs and other adult stem cells currently being evaluated in experimental therapies. First, iPSCs are neither genetically nor epigenetically stable, this inherent property of iPSCs, along with the integration of proto-oncogenes as a byproduct of some iPSC reprogramming procedures results in increased propensity for tumorigenesis in vivo. Hence the safety of iPSC based therapies remains a key concern and must be resolved before they can be tested in clinical therapies. Second, while iPSCs can be derived from patients and should be able to evade patient immune response, abnormal expression of genes in iPSCs and their differentiated progeny has been shown to induce immune responses in recipients [107]. A possible consequence of the genetic and epigenetic instability, the inability of iPSCs to evade the innate immune response of patients could lead to the rejection of iPSC grafts. Similarly, genomic and epigenetic instability of iPSCs will frustrate efforts for developing iPSCs with specific function and homing abilities. Hence more research is required before iPSCs suitable for use in patient therapies will be available.

While current limitations of iPSC technology forestalls their direct use in patient therapy, the versatility of iPSCs and their ease of derivation from patients has enabled their use in disease modeling and in vitro drug screening. iPSCs can be derived from patients affected by various diseases including LEOPARD syndrome, Schizophrenia, and X-linked adrenoleukodystrophy and used in drug and functional tests [108, 109]. This has enabled the molecular pathways and genetic mutations that cause these diseases to be studied in greater detail and led to the development of new therapies for patients. Hence the discovery of iPSCs continues to contribute to an improved knowledge of the underlying molecular mechanisms of various diseases and catalyze the development of novel drugs for their treatment.

5. Embryonic stem cells

The first breakthrough technique for isolating and growing human ESCs *in vitro* was developed at the University of Wisconsin-Madison in 1998 [110]. Since then, interest in developing more efficient methods for deriving ESCs and research into potential therapies involving ESCs has increased exponentially. ESC based therapies possess several natural advantages over other stem cell therapies. Since ESCs are directly derived from the developing embryo, they possess greater innate pluripotent capacity compared to most adult stem cells and could potentially be used in a wider range of therapies. Additionally, while the pluripotent potential and number of autogenic adult stem cells available may decline as patients age, ESC based therapies do not share the same limitation and a potentially limitless source of stem cells can be derived and cultured from blastocysts. Finally ESCs occur naturally in the inner cell mass of blastocysts and can be easily derived with minimal genetic or epigenetic manipulation unlike iPSCs.

The effectiveness and safety of ESCs therapies for treating various medical conditions including spinal cord injury, Stargardt's disease, and macular degeneration have been tested in animal and human clinical trials [111-115]. These studies demonstrate that it is relatively easy to obtain high quality and pathogen free human ESC cells, stimulate hESCs to form pure populations of differentiated cells for transplantation, and obtain sufficient quantities of cells for transplantation. Some studies of human ESC based transplants have also demonstrated that ESC is potentially safe and can be conducted with minimal risk of teratoma formation and graft rejection. The results of these animal and human ESC based clinical trials also indicate that human ESC transplantation can rescue animal models of retinal degeneration, Stargardt, and spinal cord injury, and catalyze limited visual improvement in human macular degeneration patients.

Despite these apparent advantages of human ESCs, its use in research and medical therapy has been fairly controversial historically as the derivation of human ESC lines requires the destruction of human embryos. Pro-life advocates have strongly opposed the destruction of embryos for research on the basis that human life begins when a human egg cell is fertilized, and the belief that human life is inviolable. More recently, these objections have been partially overcome through the development of human ESC derivation procedures that do not require embryo destruction [116] and the use of surplus frozen embryos from in vitro fertilization clinics.

When evaluated against the four key milestones, it can be determined that while ESCs are potentially safer than iPSCs, several key concerns continue to forestall their wider use in human clinical trials. First while some studies of human ESC based transplants in animals and humans have suggested that there is minimal risk of teratoma formation or uncontrolled proliferation of transplanted cells, other studies contend that *in vitro* culture conditions can result in potentially hazardous epigenetic modifications [117]. Second since ESC transplants are allogeneic, there is a higher likelihood of immune rejection compared to autographs of adult stem cells. Third ESC transplants potentially share similar location specificity limitations as adult stem cell transplants. This limitation has to be addressed for the development of a viable next generation stem cell therapy. Fourth since ESCs have a higher innate pluripotent capacity than adult stem cells and iPSCs, it may be easier to obtain pure populations of functionally specific and disease relevant transplant cells from ESC lines. In conclusion, while more clinical trials will be required to assess the viability of ESC therapy, studies have indicated that ESC therapy seems to offer a promising alternative for treating currently incurable diseases.

6. The promise of transdifferentiation therapy

Transdifferentiation is the direct conversion of one cell type to another without the involvement of an intermediate pluripotent state. Transdifferentiation could be a viable alternative therapy to stem cell therapies and relatively abundant adult somatic cells like fibroblasts and adipocytes could be harvested from patients and directly converted by transdifferentiation to neurons or cardiomyocytes before being used as autologous grafts in regenerative therapies. Transdifferentiation is advantageous compared to adult stem cell therapy because cell grafts could be designed specifically for each disease therapy resulting in improved functional and positional specificity. Transdifferentiation is also advantageous compared to iPSC based therapy because conversion to a desired cell type is a one step process requiring lesser epigenetic modification, and is a more rapid and direct process than dedifferentiation to form iPSCs followed by controlled differentiation into the desired cell type.

Transdifferentiation can be a relatively spontaneous process such as the in vitro transdifferentiation of chick retinal cells to lens cells [118]. It can also be induced via the guided expression of various molecular factors and genes. For example overexpression of *Atoh1* can induce the transdifferentiation of non-sensory supporting cells in the cochlea to auditory hair cells [67]. The expression of the microRNAs *miR-9/9** and *miR-124* can also induce the transdifferentiation of human fibroblasts to functional neurons [119]. However despite initial successes, currently available methods for inducing transdifferentiation remain too inefficient in vivo therapeutic purposes and further research is required to improve the process before it can be considered as a viable therapeutic alternative.

7. Summary

The use of stem cells for therapeutic purposes will be increasingly widespread as improved knowledge leads to the development of safer and more effective therapies. Stem cells derived from patients have also been successfully used in disease modeling and therapy evaluation. Further studies will enable the innate regenerative and immunomodulatory properties of stem cells to be harnessed more effectively for treating a larger variety of diseases and injury. The study and use of adult stem cells will continue to play a pivotal role in the ongoing search for novel therapies due to their availability and safety, while further developments in iPSC and ESC derivation and cultivation processes will be required before they can be used in therapies. An improved understanding of genetic and epigenetic control continues to be a prerequisite for the development of an ideal stem cell therapy. Finally further improvements in inducing direct transdifferentiation of adult non-stem cells to the desired cell types may be an alternative regenerative therapy that may circumvent the use of stem cells entirely.

Author details

Joel Sng and Thomas Lufkin*

*Address all correspondence to: lufkin@gis.a-star.edu.sg

Stem Cell and Developmental Biology Genome Institute of Singapore, Singapore

References

[1] Zhang, Y, et al. Mechanisms underlying the osteo- and adipo-differentiation of human mesenchymal stem cells. ScientificWorldJournal, (2012). , 793823.

[2] Patki, S, et al. Human breast milk is a rich source of multipotent mesenchymal stem cells. Hum Cell, (2010). , 35-40.

[3] Kern, S, et al. Comparative analysis of mesenchymal stem cells from bone marrow, umbilical cord blood, or adipose tissue. Stem Cells, (2006). , 1294-1301.

[4] Yang, J, et al. Differentiation potential of human mesenchymal stem cells derived from adipose tissue and bone marrow to sinus node-like cells. Mol Med Report, (2012). , 108-113.

[5] Huang, J. I, et al. Chondrogenic potential of progenitor cells derived from human bone marrow and adipose tissue: a patient-matched comparison. J Orthop Res, (2005). , 1383-1389.

[6] Chang, Y. J, et al. Disparate mesenchyme-lineage tendencies in mesenchymal stem cells from human bone marrow and umbilical cord blood. Stem Cells, (2006). , 679-685.

[7] De Ugarte, D. A, et al. Differential expression of stem cell mobilization-associated molecules on multi-lineage cells from adipose tissue and bone marrow. Immunol Lett, (2003). , 267-270.

[8] Pevsner-fischer, M, Levin, S, & Zipori, D. The origins of mesenchymal stromal cell heterogeneity. Stem Cell Rev, (2011). , 560-568.

[9] Sakaguchi, Y, et al. Comparison of human stem cells derived from various mesenchymal tissues: superiority of synovium as a cell source. Arthritis Rheum, (2005). , 2521-2529.

[10] Nauta, A. J, & Fibbe, W. E. Immunomodulatory properties of mesenchymal stromal cells. Blood, (2007). , 3499-3506.

[11] Zhao, S, et al. Immunomodulatory properties of mesenchymal stromal cells and their therapeutic consequences for immune-mediated disorders. Stem Cells Dev, (2010). , 607-614.

[12] Barry, F. P, et al. Immunogenicity of adult mesenchymal stem cells: lessons from the fetal allograft. Stem Cells Dev, (2005). , 252-265.

[13] Knoepfler, P. S. Deconstructing stem cell tumorigenicity: a roadmap to safe regenerative medicine. Stem Cells, (2009). , 1050-1056.

[14] Wakitani, S, et al. Safety of autologous bone marrow-derived mesenchymal stem cell transplantation for cartilage repair in 41 patients with 45 joints followed for up to 11 years and 5 months. J Tissue Eng Regen Med, (2011). , 146-150.

[15] Centeno, C. J, et al. Increased knee cartilage volume in degenerative joint disease using percutaneously implanted, autologous mesenchymal stem cells. Pain Physician, (2008). , 343-353.

[16] Centeno, C. J, et al. Safety and complications reporting update on the re-implantation of culture-expanded mesenchymal stem cells using autologous platelet lysate technique. Curr Stem Cell Res Ther, (2011). , 368-378.

[17] Tyndall, A. Successes and failures of stem cell transplantation in autoimmune diseases. Hematology Am Soc Hematol Educ Program, (2011). , 280-284.

[18] Connick, P, et al. Autologous mesenchymal stem cells for the treatment of secondary progressive multiple sclerosis: an open-label phase 2a proof-of-concept study. Lancet Neurol, (2012). , 150-156.

[19] Freedman, M. S, et al. The therapeutic potential of mesenchymal stem cell transplantation as a treatment for multiple sclerosis: consensus report of the International MSCT Study Group. Mult Scler, (2010). , 503-510.

[20] Duijvestein, M, et al. Autologous bone marrow-derived mesenchymal stromal cell treatment for refractory luminal Crohn's disease: results of a phase I study. Gut, (2010). , 1662-1669.

[21] Ciccocioppo, R, et al. Autologous bone marrow-derived mesenchymal stromal cells in the treatment of fistulising Crohn's disease. Gut, (2011). , 788-798.

[22] Keyszer, G, et al. Treatment of severe progressive systemic sclerosis with transplantation of mesenchymal stromal cells from allogeneic related donors: report of five cases. Arthritis Rheum, (2011). , 2540-2542.

[23] Nevskaya, T, et al. Autologous progenitor cell implantation as a novel therapeutic intervention for ischaemic digits in systemic sclerosis. Rheumatology (Oxford), (2009). , 61-64.

[24] Sun, L, et al. Mesenchymal stem cell transplantation reverses multiorgan dysfunction in systemic lupus erythematosus mice and humans. Stem Cells, (2009). , 1421-1432.

[25] Liang, J, et al. Allogenic mesenchymal stem cells transplantation in refractory systemic lupus erythematosus: a pilot clinical study. Ann Rheum Dis, (2010). , 1423-1429.

[26] Terai, S, et al. Timeline for development of autologous bone marrow infusion (ABMi) therapy and perspective for future stem cell therapy. J Gastroenterol, (2012).

[27] Tan, J, et al. Induction therapy with autologous mesenchymal stem cells in living-related kidney transplants: a randomized controlled trial. JAMA, (2012). , 1169-1177.

[28] Lee, J. S, et al. A long-term follow-up study of intravenous autologous mesenchymal stem cell transplantation in patients with ischemic stroke. Stem Cells, (2010). , 1099-1106.

[29] Bang, O. Y, et al. Autologous mesenchymal stem cell transplantation in stroke patients. Ann Neurol, (2005). , 874-882.

[30] Fehrer, C, et al. Reduced oxygen tension attenuates differentiation capacity of human mesenchymal stem cells and prolongs their lifespan. Aging Cell, (2007). , 745-757.

[31] Perez-ilzarbe, M, et al. Comparison of ex vivo expansion culture conditions of mesenchymal stem cells for human cell therapy. Transfusion, (2009). , 1901-1910.

[32] Estrada, J. C, et al. Culture of human mesenchymal stem cells at low oxygen tension improves growth and genetic stability by activating glycolysis. Cell Death Differ, (2012). , 743-755.

[33] Rasmusson, I, et al. Mesenchymal stem cells inhibit the formation of cytotoxic T lymphocytes, but not activated cytotoxic T lymphocytes or natural killer cells. Transplantation, (2003). , 1208-1213.

[34] Di NicolaM., et al., Human bone marrow stromal cells suppress T-lymphocyte proliferation induced by cellular or nonspecific mitogenic stimuli. Blood, (2002). , 3838-3843.

[35] Djouad, F, et al. Immunosuppressive effect of mesenchymal stem cells favors tumor growth in allogeneic animals. Blood, (2003). , 3837-3844.

[36] Nauta, A. J, et al. Mesenchymal stem cells inhibit generation and function of both CD34+-derived and monocyte-derived dendritic cells. J Immunol, (2006). , 2080-2087.

[37] Jiang, X. X, et al. Human mesenchymal stem cells inhibit differentiation and function of monocyte-derived dendritic cells. Blood, (2005). , 4120-4126.

[38] Corcione, A, et al. Human mesenchymal stem cells modulate B-cell functions. Blood, (2006). , 367-372.

[39] Glennie, S, et al. Bone marrow mesenchymal stem cells induce division arrest anergy of activated T cells. Blood, (2005). , 2821-2827.

[40] Patel, S. A, et al. Mesenchymal stem cells protect breast cancer cells through regulatory T cells: role of mesenchymal stem cell-derived TGF-beta. J Immunol, (2010). , 5885-5894.

[41] Torsvik, A, & Bjerkvig, R. Mesenchymal stem cell signaling in cancer progression. Cancer Treat Rev, (2012).

[42] Schreml, S, et al. Harvesting human adipose tissue-derived adult stem cells: resection versus liposuction. Cytotherapy, (2009). , 947-957.

[43] Dubois, S. G, et al. Isolation of human adipose-derived stem cells from biopsies and liposuction specimens. Methods Mol Biol, (2008). , 69-79.

[44] Lee, J. M, et al. Comparison of immunomodulatory effects of placenta mesenchymal stem cells with bone marrow and adipose mesenchymal stem cells. Int Immunopharmacol, (2012). , 219-224.

[45] Scuderi, N, et al. Human Adipose-Derived Stem Cells for Cell-Based Therapies in the Treatment of Systemic Sclerosis. Cell Transplant, (2012).

[46] Pak, J. Regeneration of human bones in hip osteonecrosis and human cartilage in knee osteoarthritis with autologous adipose-tissue-derived stem cells: a case series. J Med Case Reports, (2011). , 296.

[47] Koyama, N, et al. Pluripotency of mesenchymal cells derived from synovial fluid in patients with temporomandibular joint disorder. Life Sci, (2011). , 741-747.

[48] Ju, Y. J, et al. Synovial mesenchymal stem cells accelerate early remodeling of tendon-bone healing. Cell Tissue Res, (2008). , 469-478.

[49] Sekiya, I, et al. Articular cartilage regeneration with synovial mesenchymal stem cells]. Clin Calcium, (2011). , 879-889.

[50] Miyamoto, T, et al. Intradiscal transplantation of synovial mesenchymal stem cells prevents intervertebral disc degeneration through suppression of matrix metalloproteinase-related genes in nucleus pulposus cells in rabbits. Arthritis Res Ther, (2010). , R206.

[51] Yang, J, et al. Evaluation of bone marrow- and brain-derived neural stem cells in therapy of central nervous system autoimmunity. Am J Pathol, (2010). , 1989-2001.

[52] Reynolds, B. A, & Weiss, S. Generation of neurons and astrocytes from isolated cells of the adult mammalian central nervous system. Science, (1992). , 1707-1710.

[53] Pluchino, S, et al. Injection of adult neurospheres induces recovery in a chronic model of multiple sclerosis. Nature, (2003). , 688-694.

[54] Pluchino, S, et al. Immune regulatory neural stem/precursor cells protect from central nervous system autoimmunity by restraining dendritic cell function. PLoS One, (2009). , e5959.

[55] Wang, L, et al. Neural stem/progenitor cells modulate immune responses by suppressing T lymphocytes with nitric oxide and prostaglandin E2. Exp Neurol, (2009). , 177-183.

[56] Melzi, R, et al. Co-graft of allogeneic immune regulatory neural stem cells (NPC) and pancreatic islets mediates tolerance, while inducing NPC-derived tumors in mice. PLoS One, (2010). , e10357.

[57] Yang, J, et al. Adult neural stem cells expressing IL-10 confer potent immunomodulation and remyelination in experimental autoimmune encephalitis. J Clin Invest, (2009). , 3678-3691.

[58] Conover, J. C, & Notti, R. Q. The neural stem cell niche. Cell Tissue Res, (2008). , 211-224.

[59] Buffo, A, et al. Expression pattern of the transcription factor Olig2 in response to brain injuries: implications for neuronal repair. Proc Natl Acad Sci U S A, (2005). , 18183-18188.

[60] Saha, B, Jaber, M, & Gaillard, A. Potentials of endogenous neural stem cells in cortical repair. Front Cell Neurosci, (2012). , 14.

[61] Nishida, T, et al. Development of tumor-reactive T cells after nonmyeloablative allogeneic hematopoietic stem cell transplant for chronic lymphocytic leukemia. Clin Cancer Res, (2009). , 4759-4768.

[62] Lucarelli, G, et al. Hematopoietic stem cell transplantation in thalassemia and sickle cell anemia. Cold Spring Harb Perspect Med, (2012). , a011825.

[63] Kim, H, et al. Allogeneic Hematopoietic Stem Cell Transplant for Adults over 40 Years Old with Acquired Aplastic Anemia. Biol Blood Marrow Transplant, (2012).

[64] Shapiro, E, et al. Long-term effect of bone-marrow transplantation for childhood-onset cerebral X-linked adrenoleukodystrophy. Lancet, (2000). , 713-718.

[65] Breuskin, I, et al. Strategies to regenerate hair cells: identification of progenitors and critical genes. Hear Res, (2008). , 1-10.

[66] Li, H, Liu, H, & Heller, S. Pluripotent stem cells from the adult mouse inner ear. Nat Med, (2003). , 1293-1299.

[67] Minoda, R, et al. Manipulating cell cycle regulation in the mature cochlea. Hear Res, (2007). , 44-51.

[68] Dabdoub, A, et al. Sox2 signaling in prosensory domain specification and subsequent hair cell differentiation in the developing cochlea. Proc Natl Acad Sci U S A, (2008). , 18396-18401.

[69] Kiernan, A. E, et al. The Notch ligands DLL1 and JAG2 act synergistically to regulate hair cell development in the mammalian inner ear. Development, (2005). , 4353-4362.

[70] Shackleton, M, et al. Generation of a functional mammary gland from a single stem cell. Nature, (2006). , 84-88.

[71] Liu, S, Dontu, G, & Wicha, M. S. Mammary stem cells, self-renewal pathways, and carcinogenesis. Breast Cancer Res, (2005). , 86-95.

[72] Ercan, C, Van Diest, P. J, & Vooijs, M. Mammary development and breast cancer: the role of stem cells. Curr Mol Med, (2011). , 270-285.

[73] Katoh, M. Notch signaling in gastrointestinal tract (review). Int J Oncol, (2007). , 247-251.

[74] Crosnier, C, Stamataki, D, & Lewis, J. Organizing cell renewal in the intestine: stem cells, signals and combinatorial control. Nat Rev Genet, (2006). , 349-359.

[75] Gersemann, M, Stange, E. F, & Wehkamp, J. From intestinal stem cells to inflammatory bowel diseases. World J Gastroenterol, (2011). , 3198-3203.

[76] Conrad, S, et al. Generation of pluripotent stem cells from adult human testis. Nature, (2008). , 344-349.

[77] Dobrinski, I. Germ cell transplantation and testis tissue xenografting in domestic animals. Anim Reprod Sci, (2005). , 137-145.

[78] Honaramooz, A, et al. Fertility and germline transmission of donor haplotype following germ cell transplantation in immunocompetent goats. Biol Reprod, (2003). , 1260-1264.

[79] Brinster, R. L, & Avarbock, M. R. Germline transmission of donor haplotype following spermatogonial transplantation. Proc Natl Acad Sci U S A, (1994). , 11303-11307.

[80] Stadtfeld, M, Brennand, K, & Hochedlinger, K. Reprogramming of pancreatic beta cells into induced pluripotent stem cells. Curr Biol, (2008). , 890-894.

[81] Takahashi, K, & Yamanaka, S. Induction of pluripotent stem cells from mouse embryonic and adult fibroblast cultures by defined factors. Cell, (2006). , 663-676.

[82] Takahashi, K, et al. Induction of pluripotent stem cells from adult human fibroblasts by defined factors. Cell, (2007). , 861-872.

[83] Yu, J, et al. Induced pluripotent stem cell lines derived from human somatic cells. Science, (2007). , 1917-1920.

[84] Neumann, J, et al. SOX2 expression correlates with lymph-node metastases and distant spread in right-sided colon cancer. BMC Cancer, (2011). , 518.

[85] De Nigris, F, et al. c-Myc oncoprotein: cell cycle-related events and new therapeutic challenges in cancer and cardiovascular diseases. Cell Cycle, (2003). , 325-328.

[86] Wang, P, et al. The POU homeodomain protein OCT3 as a potential transcriptional activator for fibroblast growth factor-4 (FGF-4) in human breast cancer cells. Biochem J, (2003). Pt 1): , 199-205.

[87] Gore, A, et al. Somatic coding mutations in human induced pluripotent stem cells. Nature, (2011). , 63-67.

[88] Hussein, S. M, et al. Copy number variation and selection during reprogramming to pluripotency. Nature, (2011). , 58-62.

[89] Mayshar, Y, et al. Identification and classification of chromosomal aberrations in human induced pluripotent stem cells. Cell Stem Cell, (2010). , 521-531.

[90] Bar-nur, O, et al. Epigenetic memory and preferential lineage-specific differentiation in induced pluripotent stem cells derived from human pancreatic islet beta cells. Cell Stem Cell, (2011). , 17-23.

[91] Kim, K, et al. Epigenetic memory in induced pluripotent stem cells. Nature, (2010). , 285-290.

[92] Okita, K, Ichisaka, T, & Yamanaka, S. Generation of germline-competent induced pluripotent stem cells. Nature, (2007). , 313-317.

[93] Nakagawa, M, et al. Generation of induced pluripotent stem cells without Myc from mouse and human fibroblasts. Nat Biotechnol, (2008). , 101-106.

[94] Stadtfeld, M, et al. Induced pluripotent stem cells generated without viral integration. Science, (2008). , 945-949.

[95] Warren, L, et al. Highly efficient reprogramming to pluripotency and directed differentiation of human cells with synthetic modified mRNA. Cell Stem Cell, (2010). , 618-630.

[96] Yoshida, Y, et al. Hypoxia enhances the generation of induced pluripotent stem cells. Cell Stem Cell, (2009). , 237-241.

[97] Esteban, M. A, et al. Vitamin C enhances the generation of mouse and human induced pluripotent stem cells. Cell Stem Cell, (2010). , 71-79.

[98] Maherali, N, & Hochedlinger, K. Guidelines and techniques for the generation of induced pluripotent stem cells. Cell Stem Cell, (2008). , 595-605.

[99] Wernig, M, et al. A drug-inducible transgenic system for direct reprogramming of multiple somatic cell types. Nat Biotechnol, (2008). , 916-924.

[100] Anokye-danso, F, et al. Highly efficient miRNA-mediated reprogramming of mouse and human somatic cells to pluripotency. Cell Stem Cell, (2011). , 376-388.

[101] Doi, A, et al. Differential methylation of tissue- and cancer-specific CpG island shores distinguishes human induced pluripotent stem cells, embryonic stem cells and fibroblasts. Nat Genet, (2009). , 1350-1353.

[102] Lister, R, et al. Hotspots of aberrant epigenomic reprogramming in human induced pluripotent stem cells. Nature, (2011). , 68-73.

[103] Govindarajan, N, et al. Sodium butyrate improves memory function in an Alzheimer's disease mouse model when administered at an advanced stage of disease progression. J Alzheimers Dis, (2011). , 187-197.

[104] Ware, C. B, et al. Histone deacetylase inhibition elicits an evolutionarily conserved self-renewal program in embryonic stem cells. Cell Stem Cell, (2009). , 359-369.

[105] Hobley, G, et al. Development of rationally designed DNA N6 adenine methyltransferase inhibitors. Bioorg Med Chem Lett, (2012). , 3079-3082.

[106] Bae, D, et al. Hypoxia enhances the generation of retinal progenitor cells from human induced pluripotent and embryonic stem cells. Stem Cells Dev, (2012). , 1344-1355.

[107] Zhao, T, et al. Immunogenicity of induced pluripotent stem cells. Nature, (2011). , 212-215.

[108] Robinton, D. A, & Daley, G. Q. The promise of induced pluripotent stem cells in research and therapy. Nature, (2012). , 295-305.

[109] Carvajal-vergara, X, et al. Patient-specific induced pluripotent stem-cell-derived models of LEOPARD syndrome. Nature, (2010). , 808-812.

[110] Thomson, J. A, et al. Embryonic stem cell lines derived from human blastocysts. Science, (1998). , 1145-1147.

[111] Schwartz, S. D, et al. Embryonic stem cell trials for macular degeneration: a preliminary report. Lancet, (2012). , 713-720.

[112] Lu, B, et al. Long-term safety and function of RPE from human embryonic stem cells in preclinical models of macular degeneration. Stem Cells, (2009). , 2126-2135.

[113] Lukovic, D, et al. Concise review: human pluripotent stem cells in the treatment of spinal cord injury. Stem Cells, (2012). , 1787-1792.

[114] Wichterle, H, et al. Directed differentiation of embryonic stem cells into motor neurons. Cell, (2002). , 385-397.

[115] All, A. H, et al. Human Embryonic Stem Cell-Derived Oligodendrocyte Progenitors Aid in Functional Recovery of Sensory Pathways following Contusive Spinal Cord Injury. PLoS One, (2012). , e47645.

[116] Klimanskaya, I, et al. Human embryonic stem cell lines derived from single blastomeres. Nature, (2006). , 481-485.

[117] Tompkins, J. D, et al. Epigenetic stability, adaptability, and reversibility in human embryonic stem cells. Proc Natl Acad Sci U S A, (2012). , 12544-12549.

[118] Eguchi, G, & Okada, T. S. Differentiation of lens tissue from the progeny of chick retinal pigment cells cultured in vitro: a demonstration of a switch of cell types in clonal cell culture. Proc Natl Acad Sci U S A, (1973). , 1495-1499.

[119] Yoo, A. S, et al. MicroRNA-mediated conversion of human fibroblasts to neurons. Nature, (2011). , 228-231.

Chapter 18

Embryonic Stem Cell Therapy – From Bench to Bed

Laura E. Sperling

Additional information is available at the end of the chapter

http://dx.doi.org/10.5772/54368

1. Introduction

The term stem cell includes a large class of cells defined by their ability to give rise to various mature progeny while maintaining the capacity to self-renew. Embryonic stem cells (ESCs) were first isolated from the inner mass of late blastocysts in mice by Sir Martin J. Evans and Matthew Kaufman (Evans & Kaufman, 1981) and independently by Gail R. Martin (Martin, 1981). Later, it became possible to obtain ESCs from non-human primates and humans. In 1998, James Thomson and his team reported the first successful derivation of human ESC lines (Thomson *et al.*, 1998), thus extending the great potential of ESCs by providing the opportunity to develop stem cell-based therapies for human disease.

Embryonic stem cells are pluripotent, a term that defines the ability of a cell to differentiate into cells of all three germ layers. There are different types of mammalian pluripotent stem cells: embryonic stem cells derived from pre-implantation embryos (Evans & Kaufman, 1981), embryonic carcinoma (EC) cells, the stem cells of testicular tumors (Stevens, 1966; Evans, 1972), epiblast stem cells (EpiSCs) derived from the late epiblast layer of post-implantation embryos (Brons *et al.*, 2007), and embryonic germ (EG) cells derived from primordial germ cells (PGCs) of the post-implantation embryo (Matsui *et al.*, 1992; Stewart *et al.*, 1994).

Besides isolating pluripotent cells from different embryonic tissues, various experimental methods are available nowadays for inducing pluripotency *in vitro*. These methods include cloning by somatic cell nuclear transfer (SCNT), cellular fusion with embryonic stem cells, the induction of parthenogenesis, and direct reprogramming by addition of reprogramming transcription factors. SCNT is done by replacing the oocyte genome at metaphase II of meiosis with a somatic cell nucleus. Although somatic cell reprogramming has been achieved in several mammalian species (Wilmut *et al.*, 1997), this seems to be very difficult to achieve in humans. Only in 2011 Noggle et al. (Noggle *et al.*, 2011) succeeded to generate human pluripotent cells by using SCNT. However, their study revealed that the classical SCNT consistently leads to developmental arrest. The activated human oocytes develop to the blastocyst stage only when the somatic cell genome is merely added and the oocyte genome is not removed. Human stem cells derived from these blastocysts contain both a haploid genome derived from the oocyte and a diploid somatic cell genome reprogrammed to a pluripotent state (Noggle *et al.*, 2011). However, the SCNT raises some ethical concerns regarding the use of human eggs. It has also been reported that somatic cells could be reprogrammed by fusion with ES cells (Do *et al.*, 2006). These cells offer a good alternative to SCNT, especially for studying the mechanisms of reprogramming, but are thought to be less interesting for therapies due to the presence of the nuclei of stem cells in the hybrids and their instability. Human ESC lines derived from parthenogenetic blastocysts obtained by artificial activation of an oocyte have been obtained (Turovets *et al.*, 2011). Their immune-matching advantage, combined with the advantage of derivation from nonviable human embryos makes these cells a good source for cell-based transplantation therapy. However, one of the most exciting reports in reprogramming was the generation of iPSCs from terminally differentiated somatic cells by transduction of four transcription factors (*OCT4*, *SOX2*, *KLF4* and *c-MYC*) into fibroblasts (Takahashi & Yamanaka, 2006).

By using various biological reagents (e.g. growth factors) (Schuldiner *et al.*, 2000), ESCs can be differentiated in the laboratory into a range of different cell types, including neurons, glia, cardiomyocytes, islet beta cells, hepatocytes, hematopoietic progenitors and retinal pigment epithelium. The ESC ability to give rise to many different cell types is the reason that makes them very good candidates for cellular therapies. Many of the diseases that place the greatest burden on society are, at their root, diseases of cellular deficiency. Diabetes, stroke, heart diseases, hematological and neurodegenerative disorders, blindness, spinal cord injury, osteoarthritis, and kidney failure all result from the absence of one or more populations of cells that the body is unable to replace. Three basic methods have been developed to promote differentiation of ESCs: (1) the formation of three-dimensional aggregates known as embryoid bodies (EBs) (Itskovitz-Eldor *et al.*, 2000), (2) the culture of ESCs as monolayers on extracellular matrix proteins, and (3) the culture of ESCs directly on supportive stromal layers (Kawasaki *et al.*, 2000; Murry & Keller, 2008). However, the controlled differentiation of ESCs is rather difficult to optimize due to the use of serum in the culture media and difficulty to select differentiated cells. In this chapter I will focus on the differentiation of ESCs into the ectodermal lineage and on the two in 2012 ongoing clinical trials involving transplantation of ESCs derivates into eye and spinal cord.

2. Treatment of eye diseases

Retinal degenerative diseases that target photoreceptors or the adjacent retinal pigment epithelium (RPE) affect millions of people worldwide. Age-related macular degeneration (AMD) is a late-onset, complex disorder of the eye with a multi-factorial etiology in elderly (Katta *et al.*, 2009). Being the third leading cause of blindness worldwide, it accounts for 8.7% of blind persons globally. AMD results in progressive and irreversible loss of central vision affecting the macula of the eye and involves the RPE, Bruch's membrane (BM) and choriocapillaries (Katta *et al.*, 2009). Other retinal diseases with limited conventional treatments include Stargardt's macular dystrophy (SMD) and retinitis pigmentosa (RP). SMD is the most common early-onset macular degeneration disease, usually manifesting in people between ages 10 to 20. Initially there is an abnormal deposit of lipofuscin (yellow–brown granules of pigment that manifest with age) in the RPE. The RPE eventually degrades, which leads to photoreceptor loss, causing a decrease in central vision (Rowland *et al.*, 2012). In attempts to develop cell-based therapies for blinding diseases, two different approaches have to be distinguished. The first is a more direct approach of implanting appropriate retinal or RPE precursor cells, with the hope that they may integrate autonomously into the remaining (and diseased) target tissue. The second strategy counts on a lesser degree of cell autonomy within the diseased environment. Therefore, in this case, the bioengineer will first reconstruct a piece of retina or RPE tissue *in vitro*, which then can be implanted into the lesioned or diseased location (Layer *et al.*, 2010). This approach is called tissue engineering.

Restoration of vision has focused up to now on transplantation of neural progenitor cells (NPCs) and retinal pigmented epithelium (RPE) to the retina. The retinal pigment epithelium (RPE) is a monolayer of pigmented cells forming a part of the blood/retina barrier and plays crucial roles in the maintenance and function of the retina and its photoreceptors (Strauss, 2005). The apical membrane of the RPE is associated with the rod and cone photoreceptors of the retina. The basal side of the RPE faces Bruch's membrane, thereby separating the NR from the blood. The RPE absorbs light energy to increase visual sensitivity and protect against photooxidation, transports nutrients and ions between the photoreceptors at its apical surface and the choriocapillaries at its basal surface, phagocytoses photoreceptor outer segments, according to a daily circadian cycle, to relieve the photoreceptors of light-induced free radicals. The RPE secretes a variety of growth factors, such as the neuroprotective-antiangiogenic pigment epithelium-derived factor (PEDF) which is released to the neural retina, and the vasoprotective-angiogenic vascular endothelial growth factor (VEGF) that is secreted to the choroid (Layer *et al.*, 2010). With these diverse functions of the RPE it is not surprising that dysfunction and loss of RPE leads to degeneration of photoreceptors several diseases such as age-related macular degeneration (AMD), retinitis pigmentosa and Stargardt´s disease.

2.1. Preclinical work

Cell transplantation is a novel therapeutic strategy to restore visual responses. Human embryonic stem cells (hESCs) may serve as an unlimited source of RPE cells and photoreceptors for transplantation in different blinding conditions.

hESC studies have focused on the derivation of subsets of retinal cell populations (Meyer *et al.*, 2009), with emphasis on the production of either retinal progenitors (Banin *et al.*, 2006; Lamba *et al.*, 2006), or more mature cells such as retinal pigment epithelium (RPE) (Klimanskaya *et al.*, 2004) or photoreceptors (Osakada *et al.*, 2008).

Several groups have demonstrated that differentiating hESCs mimic the stepwise development of retinal cells *in vivo* (Meyer *et al.*, 2009). Furthermore, hESCs appear to respond to secreted morphogens in a manner predicted by studies of vertebrate neural induction and retinogenesis. In particular, blockade of bone morphogenetic protein and canonical Wnt signaling is known to be important for neural and retinal patterning, and many retinal differentiation protocols call for antagonists of one or both of these pathways to be included in the culture medium (Gamm & Meyer, 2010). Furthermore, the differentiation toward neural and further toward RPE fate is augmented by nicotinamide and Activin A (Idelson *et al.*, 2009). Several hESC lines actually generate neuroectodermal progenitors by spontaneous differentiation, without the addition of specific factors. RPE cells for example, were being isolated from several spontaneously differentiating human ES cell lines (Klimanskaya *et al.*, 2004). In their hands (Klimanskaya *et al.*, 2004), RPE-like differentiation occurred independently of the presence of serum. RPE cells reliably appeared in cultures grown in the presence or absence of FBS without significant variations in RPE number or time of appearance. The independence of this differentiation pathway on either coculture or extracellular matrix suggests the involvement of other differentiation cues, such as potential autocrine factors produced by differentiating hES cells. The hES-derived RPE-like cells expressed the same makers as RPE cells, e.g. RPE65 protein and CRALBP (Alge *et al.*, 2003; Klimanskaya *et al.*, 2004).

So far, it has been shown that transplanted postmitotic photoreceptor precursors are able to functionally integrate into the adult mouse neural retina. However, photoreceptors are neurons and they need to form synaptic connections in order to be functional. This makes the cell therapy with photoreceptors more challenging when compared to RPE cells. Interestingly, a group from Japan (Eiraku *et al.*, 2011) could obtain formation of a fully stratified neural retina from by using a three dimensional ESCs culture system. The 3D organoids would open up new avenues for the transplantation of artificial retinal tissue sheets, rather than simple cell grafting.

2.2. Clinical trial

Until shortly, the most relevant clinical studies currently being conducted in patients with retinal degeneration were fetal retinal sheet transplants (Radtke *et al.*, 2008). This strategy has its basis on the fact that immature retinal sheet extends cell processes and forms synaptic connections with the degenerate host retina. The underlying principle is that the inner retinal neurons of the host remain intact and therefore only require synaptic connections with photoreceptors for visual function to be restored. One big problem for the application of photoreceptor cell transplantation is that an appropriate source of the precursor cells is required.

Advanced Cell Technology and Jules Stein Eye Institute at UCLA started two prospective clinical studies to establish the safety and tolerability of subretinal transplantation of human ESC-derived retinal pigment epithelium (RPE) in patients with Stargardt's macular dystrophy (clinical trial identifier-NCT01469832) and dry age-related macular degeneration (clinical trial identifier-NCT01344993) — the leading cause of blindness in the developed world (Schwartz *et al.*, 2012). The studies are in phase I/II, where only the safety and tolerability of human ESC-derived RPE cells is assessed. The team of researchers from ACT and UCLA reported their preliminary work in two patients, one with AMD, the other with Stargardt's macular dystrophy, being the first to publish data on the use of human ESC-derived cells in the clinic (Schwartz *et al.*, 2012).

One of the rationales behind using the eye for cell therapy is that the eye represents an immuno-privileged site. The failure of the immune system to elicit an immune response in this and other such sites constitutes the hallmark of the immune privilege status (Hori *et al.*, 2010). The remarkably successful field of corneal transplantation in clinical practice is undoubtedly associated with corneal immune privilege. The subretinal space is protected by a blood–ocular barrier and the ocular fluids contain a potpourri of immunosuppressive and immunoregulatory factors that suppress T-cell proliferation and secretion of proinflammatory cytokines and inhibiting of both the cellular and humoral immune responses (Niederkorn, 2002).

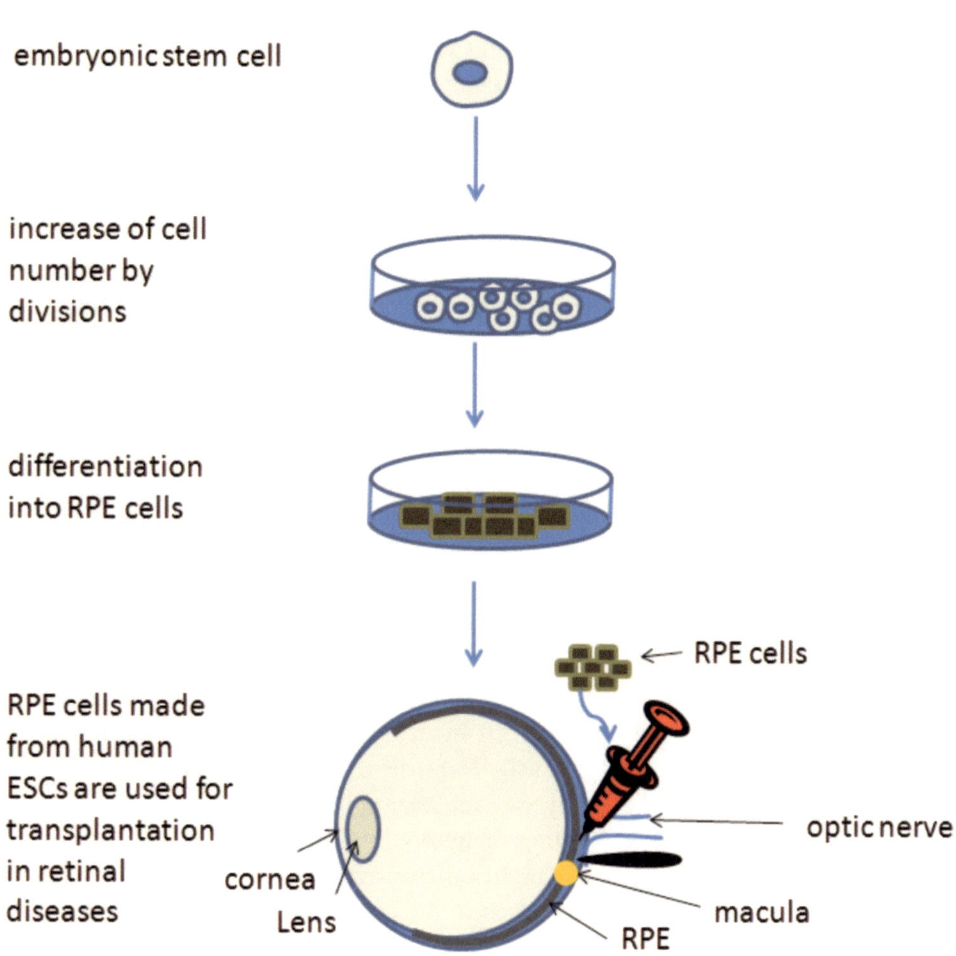

Figure 1. Scheme of procedure for replacing damaged retinal pigment epithelium cells.

Two patients enrolled in the clinical trial in order to test the safety of such cell transplantations. 50 000 viable RPE cells differentiated from the hESC line MA09 (Klimanskaya *et al.*, 2006) by embryoid body formation were injected into the subretinal space of each patient's eye (see Fig. 1 for schematic overview). The cells were resuspended in phosphate buffered saline (PBS) and delivered in a region of pericentral macula that was not completely lost to the disease. The authors thought that engraftment of the cells into a completely atrophic macula was unlikely due to the loss of Bruch's membrane. The primary outcome was positive: none of the concerns related to stem cell transplantations as teratomas, rejection, or inflammation were observed. The transplanted cells attached to Bruch's membrane and persisted for the duration of the observation period. This was however possible only in one of the two patients. Moreover, clear functional visual improvement was noted in the patient with Stargardt's macular dystrophy.

This is the first peer reviewed study that uses human ESCs for cell therapy. Although their report is preliminary, in only two patients, and with a short-term follow-up, the results are impressive - especially considering the progressive nature of both diseases (Atala, 2012).

3. Treatment of spinal cord injury

More than a decade ago, spinal-cord injury meant confinement to a wheelchair and a lifetime of medical care. Published incidence rates for traumatic spinal-cord injury in the USA range between 28 and 55 per million people, with about 10 000 new cases reported every year. Causes include motor vehicle accidents (36–48%), violence (5–29%), falls (17–21%), and recreational activities (7–16%) (McDonald & Sadowsky, 2002). The primary injury (the initial insult) is usually due to the mechanical trauma and includes traction and compression forces. Neural elements are compressed by fractured and displaced bone fragments, disc material, and ligaments and leads to injuries on both the central and peripheral nervous systems. Blood vessels are damaged, axons disrupted and cell membranes broken. Micro-haemorrhages occur within minutes in the central grey matter and spread out over the next few hours. Within minutes, the spinal cord swells to occupy the entire diameter of the spinal canal at the injury level. Secondary ischaemia results when cord swelling exceeds venous blood pressure. The more destructive phase of secondary injury is, however, more responsible for cell death and functional deficits. Hemorrhage, edema, ischaemia, release of toxic chemicals from disrupted neural membranes, and electrolyte shifts trigger a secondary injury cascade that substantially compounds initial mechanical damage by harming or killing neighbouring cells (McDonald & Sadowsky, 2002). Glutamate plays a key part in a highly disruptive process known as excitotoxicity. It was demonstrated that glutamate, released during injury, damages oligodendocytes (Domercq *et al.*, 2005). Oligodendrocytes express glutamate receptors as NMDA (Karadottir *et al.*, 2005) and AMPA/kainate receptors (Domercq *et al.*, 1999). Up to now, the primary approach in treatment is limitation of secondary injury by removal of damaging bone, disc, and ligament fragments to decompress the swollen cord, followed by the administration of the steroid methyl-prednisolone (Bracken *et al.*, 1990).

http://dx.doi.org/10.5772/54368

There are many repair strategies in spinal cord injury, as prevention of cell death by anti-glutamatergic drugs, promotion of axonal regeneration, compensation of the lost myelination or cell replacement therapy (McDonald *et al.*, 2002; McDonald & Sadowsky, 2002). Different sources and types of cells, including stem/progenitor cells (embryonic stem cells, neural progenitor cells, bone marrow mesenchymal cells) and non-stem cells (olfactory ensheathing cells [OECs] and Schwann cells) have been, and/or are being tested in clinical trials for spinal cord injury (Fehlings & Vawda, 2011).

3.1. Differentiation to oligodendrocytes

As mentioned before in the case of spinal cord injury, diseases of the nervous system involve proliferation of astrocytes and loss of oligodendrocytes (OLN) and the protective myelin sheath they produce. Transplantation of oligodendrocyte precursors in different animals systems show that these precursors can myelinate axons (Groves *et al.*, 1993). Thus, derivation of oligodendrocytes from ESCs has been an important goal for cell replacement therapy. The most common protocols involve an initial differentiation step to neural progenitors (Reubinoff *et al.*, 2001), followed by expansion, further differentiation, and selection. These protocols follow the differentiation steps that take place *in vivo*. During development, oligodendrocytes differentiate from precursors, which migrate and proliferate, through immature oligodendrocytes, which send out processes seeking axons to myelinate, to mature myelinating oligodendrocytes that form myelin sheaths. The precursor cells are morphologically bipolar (when migrating) or stellate (after migration). These initially differentiate into immature cells that put out processes seeking axons to myelinate, and eventually form mature cells with parallel processes myelinating up to 30 different axons (Karadottir & Attwell, 2007).

Oligodendrocytes were first efficiently derived from mouse ESCs (Brustle *et al.*, 1999), where ESCs were aggregated to embryoid bodies and plated in a defined medium that favors the survival of ES cell–derived neural precursors, followed by the expansion of progenitors in culture medium containing FGF2 and EGF, and a switch to FGF2 and PDGF to yield bipotential glial progenitors (Brustle *et al.*, 1999). These glial progenitors were transplanted into the spinal cords of rats with a genetic deficiency in myelin production, yielding myelinated fibers in the majority of animals (Learish *et al.*, 1999). Human ESCs were first shown to differentiate into oligodendrocytes by Zhang et al., 2001, who used a similar strategy involving FGF treatment followed by growth as neurospheres (Zhang *et al.*, 2001). They reported the differentiation of neural precursors into neurons, astrocytes and oligodendrocytes. However, no human oligodendrocytes were detected after transplantation of neural precursors into the brains of newborn mice, although human neurons and some astrocytes were found to have engrafted (Zhang *et al.*, 2001).

The first detailed protocol for directed differentiation of oligodendrocytes from human ESCs was published in 2005 and involved the induction of neural lineage by retinoic acid treatment, followed by expansion and selection in various media containing the differentiation factors triiodothyroidin hormone, FGF2, EGF, and insulin (Nistor *et al.*, 2005). After 42 days of culture, the desired cells were found in yellow spheroids, which upon differentiation as low-density monolayers formed 85%–95% oligodendrocytes expressing typical markers as GalC, RIP, and O4. Human embryonic stem cell (hESC)-derived oligodendrocytes were able to integrate, differentiate and display a functional myelinating phenotype following transplantation into the shiverer mutant mouse (Nistor *et al.*, 2005). Recently, other protocols were developed for generation of oligodendrocytes from ESCs. The Neman and de Vellis (Neman & de Vellis, 2012) laboratory has reported usage of defined serum-free media together with morphogens, as retinoic acid and sonic hedgehog, to devise a new method to derive a pure population of OLN from ESCs. These experiments show that human oligodendrocytes can be generated in large numbers and used to restore myelination under some circumstances in mice.

3.2. Clinical trial

In October 2010 the world's first clinical trial using human embryonic stem cells began, using ESCs converted into OLN precursor cells. The feasibility of the treatment was proofed by a wide range of pre-clinical studies that have shown that human oligodendrocyte progenitor cells implanted after spinal cord injury in rodent models show functional improvement (Keirstead, 2005; Keirstead *et al.*, 2005; Sharp *et al.*, 2010). Geron of Menlo Park, California, is the biotech company that received FDA approval to proceed with clinical trials that transplant cells derived from embryonic stem cells into the spinal cord (Alper, 2009). This company has pioneered translational research into human ESC therapies. The Geron trial (trial identification number NCT01217008), which was originally approved by the FDA, but then halted due to concerns of abnormal cyst formation, was reinitiated and approved for phase I clinical trials in the U.S. in October 2010. The trial was suspended following news that animals in a dose-escalation study developed microscopic cysts in regenerating tissue sites. In november 2011 Geron announced that it is dropping its entire program owing to financial concerns and started looking for partners for stem cell treatments and decided to not further invest in the clinical trials involving treatments with ESCs.

The trial was planned to involve treating ten patients who have suffered a complete thoracic-level spinal cord injury in a phase 1 multicenter trial. The pioneering therapy is Geron's 'GRNOPC1 product', which contains hES cell–derived oligodendrocyte progenitor cells that have demonstrated remyelinating and nerve growth–stimulating properties. In the human SCI lesion site, it is hoped that OLN precursors will work as a "combination therapy" - phenotypically replacing lost oligodendrocytes and hence remyelinating axons that have become demyelinated during injury, as well as secreting neurotrophic factors to establish a repair environment in the lesion (Hatch *et al.*, 2009). The ESCs were differentiated into OLN precursors (Hatch *et al.*, 2009) and one injection of 2 million GRNOPC1 cells was administered within 2 weeks in patients with thoracic spinal cord injury (Fig. 2). No serious adverse effects were observed in the 2 patients enrolled, only one of the patients experienced some side effects due to the immunosupression (Watson & Yeung, 2011). However, the data generated by Geron for the FDA are not published and no preliminary report on the safety of their product is available up to now.

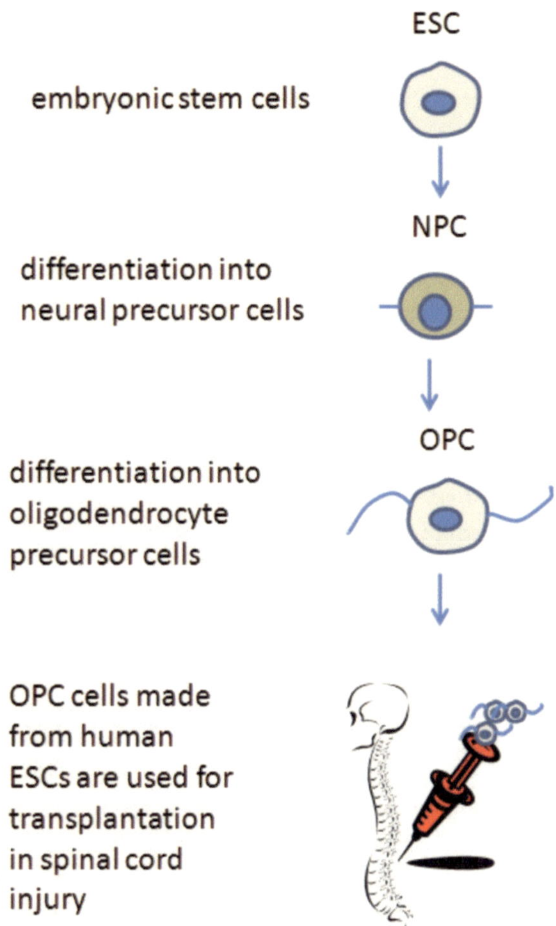

Figure 2. Scheme of procedure for treating spinal cord injury with human ESCs derived oligodendrocyte precursor cells.

4. Embryonic stem cells and tumorigenesis

The major safety concerns for the use of hESCs are related to the achievement of xenobiotic-free culture conditions, avoidance of genetic abnormalities, development of good differentiation and selection protocols, and the avoidance of the immune rejection. Moreover, the unlimited proliferative capacity of ESCs is a disadvantage in clinical applications because this could cause tumor formation upon transplantation. When implanted in an undifferentiated state, ESCs cause teratoma, a tumor type that consists of different kinds of differentiated cells. Teratomas are encapsulated, usually benign tumors that can occur naturally, but there is the fear, based on some animal studies, that some proportion of the cells derived from ESCs injected into the body could drift from their intended developmental pathway. Teratoma formation was reported in various cases when mouse

ESCs-derived cells like insulin producing islets (Fujikawa *et al.*, 2005), ESC-derived cardiomyocytes (Cao *et al.*, 2006), and ESC-derived neurons (Schuldiner *et al.*, 2001) were transplanted into immunosuppressed mice even though there was successful engraftment and functional improvement. When undifferentiated human ESCs were injected into the hind limb muscles or under the kidney capsule of SCID mice, teratomas were readily formed after 8–12 weeks (Richards *et al.*, 2002). Evidence of tumor formation has also been observed in differentiated hESC derivatives transplanted *in vivo* (Roy *et al.*, 2006). In another study, successful hESC-derived neuronal engraftment in a Parkinsonian rat model did not yield teratomas after 12 weeks (Ben-Hur *et al.*, 2004). When hESC-derived osteocytes or cardiomyocytes were transplanted into the bone or heart of severe combined immunodeficient mice (SCID), there was also no teratoma production within 1 month after injection (Bielby *et al.*, 2004; Laflamme *et al.*, 2007). It seems that the longer hESCs are differentiated *in vitro*, the risk of teratoma formation appears to be reduced. Certain sites appear to favor the growth of teratomas, while others do not, confirming a phenomenon already described that tumorigenesis of ESCs is site dependent. For example the rate of teratoma formation with hESCs in immunodeficient mice was subcutaneously 25–100%, intratesticularly 60%, intramuscularly 12.5% and under the kidney capsule 100% (Prokhorova *et al.*, 2009). Furthermore, tumor formation in the lung and thymus had the highest probability of teratoma formation while the pancreas was partially site-privileged (Shih *et al.*, 2007). Shih et al. observed an aggressive growth of tumors when human ESCs were injected into engrafted human fetal tissues in SCID mice (Shih *et al.*, 2007).

The simplest way to slow or even eliminate the tumorigenicity of normal stem cells prior to transplantation may be to take advantage of pluripotency by partially differentiating them into progenitors. Therefore, a promising proposed method for making stem cell-based regenerative medicine therapies safer may seem paradoxical: to not transplant stem cells at all into patients. The idea is to use the stem cells to produce progenitor or precursor cells of the desired lineage and then transplant progenitors purified by sorting (Knoepfler, 2009). This approach was presented in this chapter and is actually used in the clinical trial with oligodendrocyte progenitor cells. However, not only the embryonic stem cells, but also the implanted precursor cells seem to form teratoma in some cases. A group of Israeli researchers reported that a boy with ataxia telangiectasia who had received several fetal neural stem cell transplants developed teratomas in his brain and spinal cord four years after treatment (Amariglio *et al.*, 2009). For this reason is very important to achieve a 100% pure population of differentiated cells when using ESCs for cell therapy.

Currently, the only way to ensure that teratomas do not form is to differentiate the ESCs in advance, enrich for the desired cell type, and screen for the presence of undifferentiated cells. The elimination of undifferentiated hESCs may best be achieved by (1) destroying the remaining undifferentiated hESCs in the differentiated tissue population with specific agents or antibodies, (2) separating or removing the undifferentiated hESCs from the differentiated cell population, (3) eliminating pluripotent cells during the differentiation process, and (4) inducing further differentiation of left-over rogue undifferentiated hESCs (Bongso *et al.*, 2008). It is also very important to develop very good and reliable methods to detect residual ESCs contamination in ESCs derived cells prior to clinical application. In their review, Fong et al. (Fong *et al.*, 2010) presented some available methods for the elimination of undifferentiated ESCs. These included single cell propagation with encapsulation, usage of density gradients, MACS and FACS, usage of tumor privileged sites, usage of antibodies against undifferentiated ESCs, prolonged differentiation in vitro before transplantation or destruction of teratoma after engraftment. However, because differentiation is not an on/off process, it is probably the best to use a combination of these methods in order to do safe cell therapy.

5. Embryonic stem cells versus induced pluripotent stem cells in clinics

Induced pluripotent cells (iPS) are generated by re-engineering mature, fully differentiated cells (e.g. human skin fibroblasts) by modifying the cells with a set of transgenes (Takahashi & Yamanaka, 2006; Takahashi *et al.*, 2007). Induced pluripotent stem cells, created by turning back the developmental clock on adult tissues, display similar gene-expression patterns to ESCs, and can produce various tissues in the human body. However, iPS cells have a major advantage over ESCs; they can be obtained directly from the individual that has to be treated. Thus, as a source of cells for therapy, they are able to avoid the immunocompatibility issues. Furthermore, the utilization of these stem cells in both clinical and basic research studies does not face ethical and political issues that otherwise surround the use of embryonic stem cells.

During the last years various studies reported the differentiation of iPS cells to various types of cells in vitro and these cells were used for cellular therapy in various mouse models (Wernig *et al.*, 2008; Saha & Jaenisch, 2009).

However, before bringing these cells into the clinics, their safety should be tested. For example, the initial enthusiasm related to bringing iPS cells into clinics dampened when it was shown that these cells develop teratoma more efficiently than ESCs (Gutierrez-Aranda *et al.*, 2010). It was also shown that iPS retain the epigenetic memory of the cells from which they are derived; this fact makes them to preferentially differentiate into the cell lineage from which they came from. Future clinical applications will demand new techniques for generating factor-free iPS cells such as virus-free or DNA-free approaches at acceptable efficiencies. There are also other disadvantages in using iPS cells in the clinics. Usually, they are made by integrating retroviruses into the cells as shuttle for the reprogramming factors. This problem may be solved by transient gene transfer or by delivering the pluripotency factors in protein form (Murry & Keller, 2008). The second is that iPS cells are not an ''off-the-shelf'' product and would likely only be produced after the patient becomes ill, precluding their use in the acute phase of the disease (Murry & Keller, 2008). Quality control is will also be difficult and expensive, because a separate batch of iPS cells would have to be made for each patient.

6. Conclusion

There is no doubt that after the hurdles are overcome, hESC-derived cells have a promising future for transplantation therapy given the versatility of these cells. It is very encouraging to see that clinical trials involving the use of hESCs have begun, and that extensive efforts are underway to efficiently, and safely differentiate hESCs into specific cell types.

Acknowledgements

I would like to thank Paul G. Layer for carefully reading the manuscript.

Author details

Laura E. Sperling

Technische Universität Darmstadt, Fachbereich Biologie, Entwicklungsbiologie & Neurogenetik, Darmstadt, Germany

References

[1] Alge, C.S., Suppmann, S., Priglinger, S.G., Neubauer, A.S., May, C.A., Hauck, S., Welge-Lussen, U., Ueffing, M. & Kampik, A. (2003) Comparative proteome analysis of native differentiated and cultured dedifferentiated human RPE cells. Investigative ophthalmology & visual science, 44, 3629-3641.

[2] Alper, J. (2009) Geron gets green light for human trial of ES cell-derived product. Nature biotechnology, 27, 213-214.

[3] Amariglio, N., Hirshberg, A., Scheithauer, B.W., Cohen, Y., Loewenthal, R., Trakhtenbrot, L., Paz, N., Koren-Michowitz, M., Waldman, D., Leider-Trejo, L., Toren, A., Constantini, S. & Rechavi, G. (2009) Donor-derived brain tumor following neural stem cell transplantation in an ataxia telangiectasia patient. PLoS medicine, 6, e1000029.

[4] Atala, A. (2012) Human embryonic stem cells: early hints on safety and efficacy. Lancet, 379, 689-690.

[5] Banin, E., Obolensky, A., Idelson, M., Hemo, I., Reinhardtz, E., Pikarsky, E., Ben-Hur, T. & Reubinoff, B. (2006) Retinal incorporation and differentiation of neural precursors derived from human embryonic stem cells. Stem Cells, 24, 246-257.

[6] Ben-Hur, T., Idelson, M., Khaner, H., Pera, M., Reinhartz, E., Itzik, A. & Reubinoff, B.E. (2004) Transplantation of human embryonic stem cell-derived neural progenitors improves behavioral deficit in Parkinsonian rats. Stem Cells, 22, 1246-1255.

[7] Bielby, R.C., Boccaccini, A.R., Polak, J.M. & Buttery, L.D. (2004) In vitro differentiation and in vivo mineralization of osteogenic cells derived from human embryonic stem cells. Tissue engineering, 10, 1518-1525.

[8] Bongso, A., Fong, C.Y. & Gauthaman, K. (2008) Taking stem cells to the clinic: Major challenges. Journal of cellular biochemistry, 105, 1352-1360.

[9] Bracken, M.B., Shepard, M.J., Collins, W.F., Holford, T.R., Young, W., Baskin, D.S., Eisenberg, H.M., Flamm, E., Leo-Summers, L., Maroon, J. & et al. (1990) A randomized, controlled trial of methylprednisolone or naloxone in the treatment of acute spinal-cord injury. Results of the Second National Acute Spinal Cord Injury Study. The New England journal of medicine, 322, 1405-1411.

[10] Brons, I.G., Smithers, L.E., Trotter, M.W., Rugg-Gunn, P., Sun, B., Chuva de Sousa Lopes, S.M., Howlett, S.K., Clarkson, A., Ahrlund-Richter, L., Pedersen, R.A. & Vallier, L. (2007) Derivation of pluripotent epiblast stem cells from mammalian embryos. Nature, 448, 191-195.

[11] Brustle, O., Jones, K.N., Learish, R.D., Karram, K., Choudhary, K., Wiestler, O.D., Duncan, I.D. & McKay, R.D. (1999) Embryonic stem cell-derived glial precursors: a source of myelinating transplants. Science, 285, 754-756.

[12] Cao, F., Lin, S., Xie, X., Ray, P., Patel, M., Zhang, X., Drukker, M., Dylla, S.J., Connolly, A.J., Chen, X., Weissman, I.L., Gambhir, S.S. & Wu, J.C. (2006) In vivo visualization of embryonic stem cell survival, proliferation, and migration after cardiac delivery. Circulation, 113, 1005-1014.

[13] Do, J.T., Han, D.W. & Scholer, H.R. (2006) Reprogramming somatic gene activity by fusion with pluripotent cells. Stem cell reviews, 2, 257-264.

[14] Domercq, M., Etxebarria, E., Perez-Samartin, A. & Matute, C. (2005) Excitotoxic oligodendrocyte death and axonal damage induced by glutamate transporter inhibition. Glia, 52, 36-46.

[15] Domercq, M., Sanchez-Gomez, M.V., Areso, P. & Matute, C. (1999) Expression of glutamate transporters in rat optic nerve oligodendrocytes. The European journal of neuroscience, 11, 2226-2236.

[16] Eiraku, M., Takata, N., Ishibashi, H., Kawada, M., Sakakura, E., Okuda, S., Sekiguchi, K., Adachi, T. & Sasai, Y. (2011) Self-organizing optic-cup morphogenesis in three-dimensional culture. Nature, 472, 51-56.

[17] Evans, M.J. (1972) The isolation and properties of a clonal tissue culture strain of pluripotent mouse teratoma cells. Journal of embryology and experimental morphology, 28, 163-176.

[18] Evans, M.J. & Kaufman, M.H. (1981) Establishment in culture of pluripotential cells from mouse embryos. Nature, 292, 154-156.

[19] Fehlings, M.G. & Vawda, R. (2011) Cellular treatments for spinal cord injury: the time is right for clinical trials. Neurotherapeutics : the journal of the American Society for Experimental NeuroTherapeutics, 8, 704-720.

[20] Fong, C.Y., Gauthaman, K. & Bongso, A. (2010) Teratomas from pluripotent stem cells: A clinical hurdle. Journal of cellular biochemistry, 111, 769-781.

[21] Fujikawa, T., Oh, S.H., Pi, L., Hatch, H.M., Shupe, T. & Petersen, B.E. (2005) Teratoma formation leads to failure of treatment for type I diabetes using embryonic stem cell-derived insulin-producing cells. The American journal of pathology, 166, 1781-1791.

[22] Gamm, D.M. & Meyer, J.S. (2010) Directed differentiation of human induced pluripotent stem cells: a retina perspective. Regenerative medicine, 5, 315-317.

[23] Groves, A.K., Barnett, S.C., Franklin, R.J., Crang, A.J., Mayer, M., Blakemore, W.F. & Noble, M. (1993) Repair of demyelinated lesions by transplantation of purified O-2A progenitor cells. Nature, 362, 453-455.

[24] Gutierrez-Aranda, I., Ramos-Mejia, V., Bueno, C., Munoz-Lopez, M., Real, P.J., Macia, A., Sanchez, L., Ligero, G., Garcia-Parez, J.L. & Menendez, P. (2010) Human induced pluripotent stem cells develop teratoma more efficiently and faster than human embryonic stem cells regardless the site of injection. Stem Cells, 28, 1568-1570.

[25] Hatch, M.N., Nistor, G. & Keirstead, H.S. (2009) Derivation of high-purity oligodendroglial progenitors. Methods Mol Biol, 549, 59-75.

[26] Hori, J., Vega, J.L. & Masli, S. (2010) Review of ocular immune privilege in the year 2010: modifying the immune privilege of the eye. Ocular immunology and inflammation, 18, 325-333.

[27] Idelson, M., Alper, R., Obolensky, A., Ben-Shushan, E., Hemo, I., Yachimovich-Cohen, N., Khaner, H., Smith, Y., Wiser, O., Gropp, M., Cohen, M.A., Even-Ram, S., Berman-Zaken, Y., Matzrafi, L., Rechavi, G., Banin, E. & Reubinoff, B. (2009) Directed differentiation of human embryonic stem cells into functional retinal pigment epithelium cells. Cell stem cell, 5, 396-408.

[28] Itskovitz-Eldor, J., Schuldiner, M., Karsenti, D., Eden, A., Yanuka, O., Amit, M., Soreq, H. & Benvenisty, N. (2000) Differentiation of human embryonic stem cells into embryoid bodies compromising the three embryonic germ layers. Mol Med, 6, 88-95.

[29] Karadottir, R. & Attwell, D. (2007) Neurotransmitter receptors in the life and death of oligodendrocytes. Neuroscience, 145, 1426-1438.

[30] Karadottir, R., Cavelier, P., Bergersen, L.H. & Attwell, D. (2005) NMDA receptors are expressed in oligodendrocytes and activated in ischaemia. Nature, 438, 1162-1166.

[31] Katta, S., Kaur, I. & Chakrabarti, S. (2009) The molecular genetic basis of age-related macular degeneration: an overview. J Genet, 88, 425-449.

[32] Kawasaki, H., Mizuseki, K., Nishikawa, S., Kaneko, S., Kuwana, Y., Nakanishi, S., Nishikawa, S.I. & Sasai, Y. (2000) Induction of midbrain dopaminergic neurons from ES cells by stromal cell-derived inducing activity. Neuron, 28, 31-40.

[33] Keirstead, H.S. (2005) Stem cells for the treatment of myelin loss. Trends in neurosciences, 28, 677-683.

[34] Keirstead, H.S., Nistor, G., Bernal, G., Totoiu, M., Cloutier, F., Sharp, K. & Steward, O. (2005) Human embryonic stem cell-derived oligodendrocyte progenitor cell transplants remyelinate and restore locomotion after spinal cord injury. The Journal of neuroscience : the official journal of the Society for Neuroscience, 25, 4694-4705.

[35] Klimanskaya, I., Chung, Y., Becker, S., Lu, S.J. & Lanza, R. (2006) Human embryonic stem cell lines derived from single blastomeres. Nature, 444, 481-485.

[36] Klimanskaya, I., Hipp, J., Rezai, K.A., West, M., Atala, A. & Lanza, R. (2004) Derivation and comparative assessment of retinal pigment epithelium from human embryonic stem cells using transcriptomics. Cloning and stem cells, 6, 217-245.

[37] Knoepfler, P.S. (2009) Deconstructing stem cell tumorigenicity: a roadmap to safe regenerative medicine. Stem Cells, 27, 1050-1056.

[38] Laflamme, M.A., Zbinden, S., Epstein, S.E. & Murry, C.E. (2007) Cell-based therapy for myocardial ischemia and infarction: pathophysiological mechanisms. Annual review of pathology, 2, 307-339.

[39] Layer P.G., Araki, M. & Vogel-Höpker, A. (2010) New concepts for reconstruction of retinal and pigment epithelial tissues. *Expert Review of Ophthalmology*, 5, No. 4, 523-543.

[40] Lamba, D.A., Karl, M.O., Ware, C.B. & Reh, T.A. (2006) Efficient generation of retinal progenitor cells from human embryonic stem cells. Proceedings of the National Academy of Sciences of the United States of America, 103, 12769-12774.

[41] Learish, R.D., Brustle, O., Zhang, S.C. & Duncan, I.D. (1999) Intraventricular transplantation of oligodendrocyte progenitors into a fetal myelin mutant results in widespread formation of myelin. Annals of neurology, 46, 716-722.

[42] Martin, G.R. (1981) Isolation of a pluripotent cell line from early mouse embryos cultured in medium conditioned by teratocarcinoma stem cells. Proceedings of the National Academy of Sciences of the United States of America, 78, 7634-7638.

[43] Matsui, Y., Zsebo, K. & Hogan, B.L. (1992) Derivation of pluripotential embryonic stem cells from murine primordial germ cells in culture. Cell, 70, 841-847.

[44] McDonald, J.W., Becker, D., Sadowsky, C.L., Jane, J.A., Sr., Conturo, T.E. & Schultz, L.M. (2002) Late recovery following spinal cord injury. Case report and review of the literature. Journal of neurosurgery, 97, 252-265.

[45] McDonald, J.W. & Sadowsky, C. (2002) Spinal-cord injury. Lancet, 359, 417-425.

[46] Meyer, J.S., Shearer, R.L., Capowski, E.E., Wright, L.S., Wallace, K.A., McMillan, E.L., Zhang, S.C. & Gamm, D.M. (2009) Modeling early retinal development with human embryonic and induced pluripotent stem cells. Proceedings of the National Academy of Sciences of the United States of America, 106, 16698-16703.

[47] Murry, C.E. & Keller, G. (2008) Differentiation of embryonic stem cells to clinically relevant populations: lessons from embryonic development. Cell, 132, 661-680.

[48] Neman, J. & de Vellis, J. (2012) A method for deriving homogenous population of oligodendrocytes from mouse embryonic stem cells. Developmental neurobiology, 72, 777-788.

[49] Niederkorn, J.Y. (2002) Immune privilege in the anterior chamber of the eye. Critical reviews in immunology, 22, 13-46.

[50] Nistor, G.I., Totoiu, M.O., Haque, N., Carpenter, M.K. & Keirstead, H.S. (2005) Human embryonic stem cells differentiate into oligodendrocytes in high purity and myelinate after spinal cord transplantation. Glia, 49, 385-396.

[51] Noggle, S., Fung, H.L., Gore, A., Martinez, H., Satriani, K.C., Prosser, R., Oum, K., Paull, D., Druckenmiller, S., Freeby, M., Greenberg, E., Zhang, K., Goland, R., Sauer, M.V., Leibel, R.L. & Egli, D. (2011) Human oocytes reprogram somatic cells to a pluripotent state. Nature, 478, 70-75.

[52] Osakada, F., Ikeda, H., Mandai, M., Wataya, T., Watanabe, K., Yoshimura, N., Akaike, A., Sasai, Y. & Takahashi, M. (2008) Toward the generation of rod and cone photoreceptors from mouse, monkey and human embryonic stem cells. Nature biotechnology, 26, 215-224.

[53] Prokhorova, T.A., Harkness, L.M., Frandsen, U., Ditzel, N., Schroder, H.D., Burns, J.S. & Kassem, M. (2009) Teratoma formation by human embryonic stem cells is site dependent and enhanced by the presence of Matrigel. Stem cells and development, 18, 47-54.

[54] Radtke, N.D., Aramant, R.B., Petry, H.M., Green, P.T., Pidwell, D.J. & Seiler, M.J. (2008) Vision improvement in retinal degeneration patients by implantation of retina together with retinal pigment epithelium. American journal of ophthalmology, 146, 172-182.

[55] Reubinoff, B.E., Itsykson, P., Turetsky, T., Pera, M.F., Reinhartz, E., Itzik, A. & Ben-Hur, T. (2001) Neural progenitors from human embryonic stem cells. Nature biotechnology, 19, 1134-1140.

[56] Richards, M., Fong, C.Y., Chan, W.K., Wong, P.C. & Bongso, A. (2002) Human feeders support prolonged undifferentiated growth of human inner cell masses and embryonic stem cells. Nature biotechnology, 20, 933-936.

[57] Rowland, T.J., Buchholz, D.E. & Clegg, D.O. (2012) Pluripotent human stem cells for the treatment of retinal disease. Journal of cellular physiology, 227, 457-466.

[58] Roy, N.S., Cleren, C., Singh, S.K., Yang, L., Beal, M.F. & Goldman, S.A. (2006) Functional engraftment of human ES cell-derived dopaminergic neurons enriched by coculture with telomerase-immortalized midbrain astrocytes. Nature medicine, 12, 1259-1268.

[59] Saha, K. & Jaenisch, R. (2009) Technical challenges in using human induced pluripotent scells to model disease. Cell stem cell, 5, 584-595.

[60] Schuldiner, M., Eiges, R., Eden, A., Yanuka, O., Itskovitz-Eldor, J., Goldstein, R.S. & Benvenisty, N. (2001) Induced neuronal differentiation of human embryonic stem cells. Brain research, 913, 201-205.

[61] Schuldiner, M., Yanuka, O., Itskovitz-Eldor, J., Melton, D.A. & Benvenisty, N. (2000) Effects of eight growth factors on the differentiation of cells derived from human embryonic stem cells. Proceedings of the National Academy of Sciences of the United States of America, 97, 11307-11312.

[62] Schwartz, S.D., Hubschman, J.P., Heilwell, G., Franco-Cardenas, V., Pan, C.K., Ostrick, R.M., Mickunas, E., Gay, R., Klimanskaya, I. & Lanza, R. (2012) Embryonic stem cell trials for macular degeneration: a preliminary report. Lancet, 379, 713-720.

[63] Sharp, J., Frame, J., Siegenthaler, M., Nistor, G. & Keirstead, H.S. (2010) Human embryonic stem cell-derived oligodendrocyte progenitor cell transplants improve recovery after cervical spinal cord injury. Stem Cells, 28, 152-163.

[64] Shih, C.C., Forman, S.J., Chu, P. & Slovak, M. (2007) Human embryonic stem cells are prone to generate primitive, undifferentiated tumors in engrafted human fetal tissues in severe combined immunodeficient mice. Stem cells and development, 16, 893-902.

[65] Stevens, L.C. (1966) Development of resistance to teratocarcinogenesis by primordial germ cells in mice. Journal of the National Cancer Institute, 37, 859-867.

[66] Stewart, C.L., Gadi, I. & Bhatt, H. (1994) Stem cells from primordial germ cells can reenter the germ line. Developmental biology, 161, 626-628.

[67] Strauss, O. (2005) The retinal pigment epithelium in visual function. Physiological reviews, 85, 845-881.

[68] Takahashi, K., Tanabe, K., Ohnuki, M., Narita, M., Ichisaka, T., Tomoda, K. & Yamanaka, S. (2007) Induction of pluripotent stem cells from adult human fibroblasts by defined factors. Cell, 131, 861-872.

[69] Takahashi, K. & Yamanaka, S. (2006) Induction of pluripotent stem cells from mouse embryonic and adult fibroblast cultures by defined factors. Cell, 126, 663-676.

[70] Thomson, J.A., Itskovitz-Eldor, J., Shapiro, S.S., Waknitz, M.A., Swiergiel, J.J., Marshall, V.S. & Jones, J.M. (1998) Embryonic stem cell lines derived from human blastocysts. Science, 282, 1145-1147.

[71] Turovets, N., Semechkin, A., Kuzmichev, L., Janus, J., Agapova, L. & Revazova, E. (2011) Derivation of human parthenogenetic stem cell lines. Methods Mol Biol, 767, 37-54.

[72] Watson, R.A. & Yeung, T.M. (2011) What is the potential of oligodendrocyte progenitor cells to successfully treat human spinal cord injury? BMC neurology, 11, 113.

[73] Wernig, M., Zhao, J.P., Pruszak, J., Hedlund, E., Fu, D., Soldner, F., Broccoli, V., Constantine-Paton, M., Isacson, O. & Jaenisch, R. (2008) Neurons derived from reprogrammed fibroblasts functionally integrate into the fetal brain and improve symptoms of rats with Parkinson's disease. Proceedings of the National Academy of Sciences of the United States of America, 105, 5856-5861.

[74] Wilmut, I., Schnieke, A.E., McWhir, J., Kind, A.J. & Campbell, K.H. (1997) Viable offspring derived from fetal and adult mammalian cells. Nature, 385, 810-813.

[75] Zhang, S.C., Wernig, M., Duncan, I.D., Brustle, O. & Thomson, J.A. (2001) In vitro differentiation of transplantable neural precursors from human embryonic stem cells. Nature biotechnology, 19, 1129-1133.

Chapter 19

Disease Models for the Genetic Cardiac Diseases

Mari Pekkanen-Mattila, Kristiina Rajala and
Katriina Aalto-Setälä

Additional information is available at the end of the chapter

http://dx.doi.org/10.5772/55773

1. Introduction

The ability to reprogram somatic cells into pluripotent stem cells has presented a significant advancement in stem cell research. This technique enables derivation of induced pluripotent stem (iPS) cells from any individual having a unique genotype. iPS cells can be derived from human somatic cells such as fibroblasts, keratinocytes or blood cells. Since, the production of iPS cell lines does not require the destruction of human embryos as in the production of the human embryonic stem cells (hESCs), the legal and ethical issues associated with hESCs can be at least partly avoided. The characteristics of iPS cells are very similar to those of pluripotent hESCs in many respects, including cell morphology, immortal growth characteristics in culture, expression of pluripotent markers, and differentiation potential. The iPS cells combined with the various differentiation protocols developed enable the production of genotype specific cell types. This feature enables also to produce disease-specific iPS cell lines from patients bearing defined genetic mutations. Traditionally, it has been challenging to study genetic cardiac diseases because cardiomyocytes from the heart biopsies of patients are difficult to obtain and the procedure carries a high risk. Additionally these cardiomyocytes do not survive long in culture. Animal models, mostly developed in rodent, have aided in elucidating the basic mechanisms of several genetic cardiac diseases. The disadvantages of small animal models are marked differences in anatomy and physiology of the cardiovascular system in comparison to humans. Rodent models are far from ideal when used in the identification of contractile deficits and signals that initiate pathological growth [1]. Furthermore, the results obtained from neonatal rat cell experiments can be problematic because these cells possess different relative receptor subtypes and cell-signaling mechanisms. It will thus be especially important to investigate functional consequences of genetic cardiac diseases in human cardiomyocytes in which the functional effects of specific proteins have been adjusted to optimize electrical properties, contractile efficiency and power output of larger hearts [2].

Genetic cardiac diseases, such as long QT syndrome, belong to a severe class of diseases which are unpredictable, have variable clinical picture ranging from asymptomatic to sudden cardiac death and lack specific medication. These inherited arrhythmic diseases are caused by single mutations which are relatively common in population. Earlier we did not have *in vitro* models for these diseases, but with the aid of iPS cell derived cardiomyocytes genetic cardiac diseases can now be modeled in cell culture. The patient specific iPS cell derived cardiomyocytes have been demonstrated to manifest the disease-associated electrophysiological abnormalities in a dish [3-6]. Therefore, these cells allow researchers to study and understand disease mechanisms more readily as well as to investigate the effects of different chemical compounds on the electrophysiology of the cardiomyocytes. In addition to basic research, iPS cell derived cardiomyocytes would provide an effective tool for novel drug or treatment discovery. However, before iPS cell derived cardiomyocytes are ready to be considered for use as disease models, the cells produced need to be confirmed to exhibit the essential functional characteristics of human cardiomyocytes.

In this chapter, the production and the characterization of patient specific iPS cell derived cardiomyocytes is described. In addition, we discuss the genetic cardiac disease models so far developed based on iPS technique, their demands, advantages and disadvantages. Furthermore, the future applications for iPS cell derived cardiomyocytes are discussed.

2. Production of disease specific iPS cell lines

The discovery of cellular reprogramming as a technology to generate iPS cells offers a potential solution to the challenge when studying genetic cardiac diseases. In this approach, human adult somatic cells are reprogrammed into stem cells offering comparable function to human pluripotent ESCs in their ability to develop differentiated progeny from all developmental lineages of the human being. When somatic cells are reprogrammed to iPS cells, they shut down the expression of genes specific for that somatic cell type and activate genes that maintain pluripotency. Once reprogramming has occurred, endogenous counterparts of the exogenously supplied reprogramming factors are activated, indicating that exogenous factors are only required for the induction, not for the maintenance of pluripotency [7]. Up to date, various human somatic cell types, including fibroblasts, keratinocytes, and different blood cells have been reprogrammed to iPS cells [7-11].

The initial methods used to generate iPS cells involved the retroviral overexpression of four transcription factors Oct4, Sox2, Klf4, and c-myc observed to be essential in maintaining pluripotency of hESCs [7, 12]. Another set of four transcription factors Oct4, Sox2, Nanog, and Lin-28 was also found to induce pluripotency [9]. Efficient retro- and lentiviral vector systems that have been most widely used to generate iPS cells have several drawbacks including the possibility of proviral genomic integration, which may cause both the reactivation of silenced exogenous genes and the alteration of genomic integrity, thereby increasing the risk for tumorigenesis [12, 13]. Since the seminal discovery the development in this field has been rapid and numerous alternative strategies have been applied to improve the reprogramming safety,

http://dx.doi.org/10.5772/55773

efficiency and kinetics as well as to generate iPS cells without viral integration in the genome (Table 1). The nonintegrating reprogramming methods developed thus far include adeno- and sendai viruses, plasmid- and episomal vector-based approaches, excision systems of integrated transgenes such as Cre/loxP recombination or PiggyBac transposition, and delivery of reprogramming factors directly as RNAs, proteins and chemicals. However, most of these nonintegrating approaches are still highly inefficient when compared to the original retro- or lentiviral reprogramming systems with the exception of nowadays widely used sendai virus reprogramming method.

Methods	Efficiency %	Details	References
Retroviral vectors	Medium, 0.01-0.5	Multiple integration, incomplete silencing, tumorigenicity possible	[7]
Lentiviral vectors	Medium, 0.1-1	Multiple integration, incomplete silencing, tumorigenicity possible	[9]
Adenoviral vectors	Low, 0.001	Non-integrating, however integrated vector-fragment possible	[14]
Sendaiviral vectors	Medium, 0.01-1	Non-integrating, integrated vector-fragment possible, T sensitive Sendai vector allowing removal of the virus	[15]
Plasmids	Low, 0.001	Occasional integration, simple transfection	[16]
OriP/EBNA-1 episomal vectors	Low, 0.0003	Non-integrating, long-term persistent transcription	[17]
Minicircle DNA episomal vectors	Low, 0.005	Non-integrating, multiple transductions needed	[18, 19]
Cre/loxP system	Medium, 0.01-1	Integration but excisable, inefficient loxP site excision, screening needed, tumorigenicity possible	[20]
PiggyBac system	Medium, 0.1	Precise excision possible, screening needed	[21, 22]
RNAs	High, 1	Non-integrating, DNA-free, multiple transfection needed	[23, 24]
Protein	Low, 0.001	Non-integrating, DNA-free, long-term treatment required, genetic abnormality possible	[25]
Factors + small molecules	High, >1	Non-integrating, DNA-free, long-term treatment required, abnormal signaling pathway possible, virus used	[26]

Table 1. Overview of the reprogramming methods for the generation of iPS cells.

3. Cardiomyocyte differentiation

Cardiomyocytes have been differentiated from the hESCs over a decade [27, 28] and multiple cardiac differentiation methods have been developed. The differentiation methods developed for hESC derived cardiomyocytes have been proven to be applicable also for cardiac differentiation of iPS cells.

Overall the differentiation event of hESC and iPS cell derived cardiomyocytes is quite rapid, 10-20 days regardless of the differentiation method used. However, all the differentiation methods share common problems, including uncontrolled differentiation and low differentiation rates. With common differentiation methods the cardiomyocyte yield is between ~1-25 % of the total cell number [28-30]. In addition, the cardiomyocyte differentiation efficiency has been shown to vary markedly between different stem cell lines [31].

All differentiation methods end up with a heterogeneous cell population. In addition to the other cell types, the differentiated population includes all cardiomyocyte subtypes; ventricular, atrial and nodal –like cells [32]. The ventricular cells form usually the majority of differentiated cells (60-80%), atrial cells form usually 10-40 % of the population and only <5% of cells are nodal-like cells [32, 33]. However, these numbers can differ depending on the cell line used [34].

The cardiac differentiation methods are lately reviewed [35] and described in Figure 1.

4. Transdifferentiation of fibroblasts into cardiac cells

Murine fibroblasts can be reprogrammed directly into cardiomyocytes by overexpression of Gata4, Mef2c and Tbx5 (GMT) [36]. This combination of factors has been reported to convert murine cardiac fibroblasts and tail tip fibroblasts into spontaneous beating cells having cardiomyocyte expression profiles. In addition, epigenetic status is typical for cardiomyocytes in these cells. However, Chen and co-workers have shown this method to be inefficient [37]. Overexpression of GMT factors resulted in an increase in cardiac troponin expression but spontaneous action potentials were lacking even though 22% of the cells exhibited voltage-dependent calcium currents.

A lot of effort has been done to transdifferentiate human fibroblasts into cardiomyocytes. So far spontaneously beating human cells have not been obtained. However, with transcription factors mesoderm posterior (MESP) homolog and mammalian v-ets erythoblastosis virus E26 oncogene homolog ETS2 cardiomyocyte progenitors expressing cardiac mesoderm marker KDR have been obtained [38]. It seems that the GMT method alone is not robust enough for direct reprogramming of human cardiomyocytes. Therefore it has been suggested that combination of GMT with other transcription factors, mRNAs or small molecules could provide more efficient reprogramming procedure [39]. In addition, based on animal experiments it can be concluded that cardiac microenvironment has also important role in reprogramming [40].

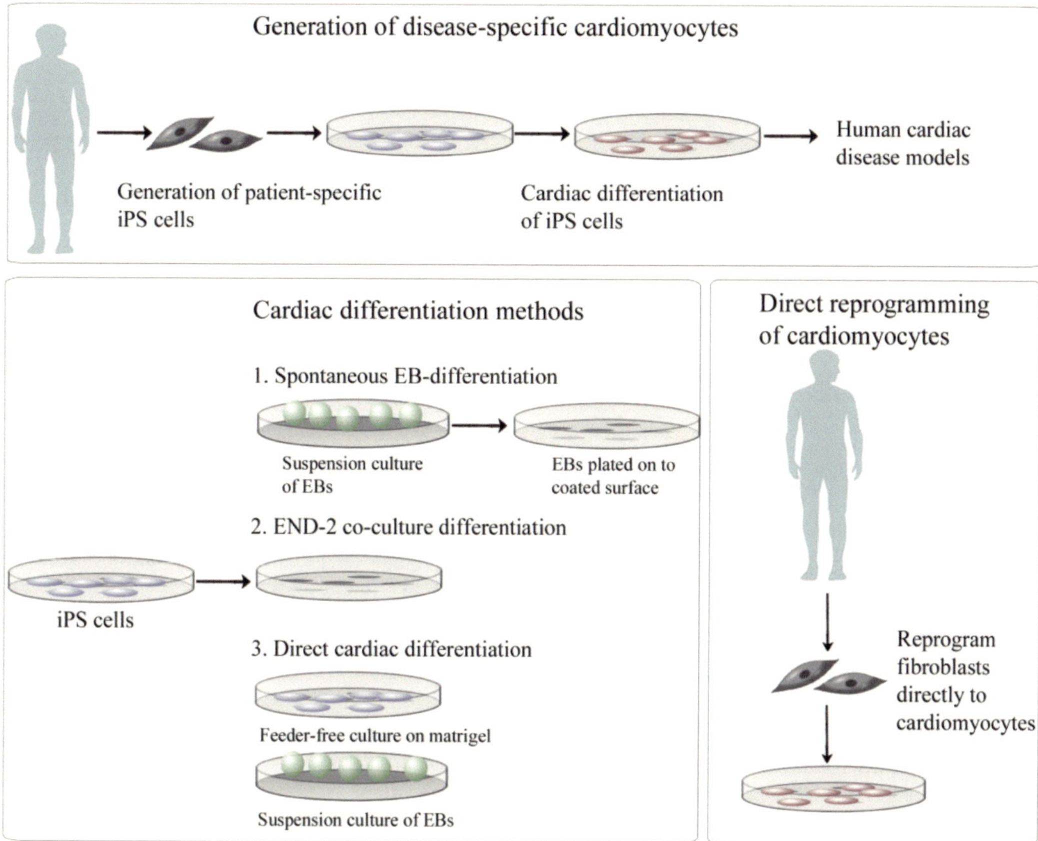

Figure 1. Generation of disease –specific iPS cell lines and cardiomyocytes. Cardiac differentiation methods can be divided into three classes; (1) Embryoid body (EB) based, (2) END-2 coculture based or (3) directed differentiation methods. Traditionally EB method has been based on spontaneous aggregation of EBs and spontaneous differentiation [28]. However, lately multiple methods controlling the EB formation has been developed [41] enhancing the reproducibility and productivity of the cardiac differentiation. END-2 method can be performed in two ways, either co-culturing the hESC or iPS cells in contact with END-2 cells [42] or by using END-2 conditioned media [43]. A lot of effort has been made in enhancing and defining the cardiac differentiation and this has led to the development of directed differentiation methods with growth factors or small molecules. Activin A and BMP-4 has been used in combination with monolayer cultures on matrigel to differentiate cardiomyocytes [44]. A temporal modulation of Wnt signaling by using small molecules has been proven to an even more robust and, in addition, rather inexpensive method for cardiomyocyte differentiation [45]. Directed reprogramming of fibroblasts to cardiomyocytes has been successful with mouse cells. However, this method has not yet been proven to work with human cells.

5. The assesment of the cardiomyocyte functionality

5.1. Cardiomyocyte characterization at gene and protein level

The first characterization step for the differentiated hESC or iPS cell derived cardiomyocytes is the observation of spontaneously beating cells. In addition, cardiomyocyte phenotype can be assessed at the gene or protein level with cardiomyocyte specific markers such as structural proteins troponin, alpha-actinin or myosins. The commonly used markers in monitoring the cardiac differentiation are listed in Table 2.

Cell stage	Markers
Pluripotent cells	OCT4 Nanog SOX2 Tra-1-60 SSEA-4
Precardiac/cardiac mesodermal cells	Brachyury T FoxC1 Dkk-1 Mesp1 Flk-1
Cardiac precursor cells	KDR Nkx2.5 GATA4 Tbx5 Isl-1 Mef2c Hand1/2
Cardiac cells	Troponin I and T Sarcomeric α-actinin Myosin heavy- and light-chain (MHC and MLC)

Table 2. Markers used in monitoring the cardiac differentiation.

5.2. Electrophysiological methods

5.2.1. Patch clamp

Traditional way to study the functionality and the electrical activity of the cardiomyocytes is the patch clamp technique [46]. Originally patch clamp method has been developed to study ion channels in excitable membranes [47]. In this technique micropipette is attached to the cell membrane by a giga seal and this can be exploited to measure current changes and voltage across the membrane. Due to the unique nature of the cardiomyocyte action potential curve, the ion channel composition and the maturation stage of the cardiomyocyte can be assessed and therefore the method has been widely used with stem cell derived cardiomyocyte studies.

Key cardiac ion channels (and respective current) involved in the human action potential are NaV1.5 (INa), KV4.3 (Ito), CaV1.2 (ICa,L) KV11.1 (IKr), KV7.1 (IKs), and Kir2.X (IK1) [48]. The cardiac action potential is composed of co-operation of these channels and the action potential curve can be divided into five different phases (Figure 2). Phase 0 of the action potential is the depolarization phase of the cardiomyocytes from the negative membrane potential to positive, called the upstroke. This is followed by phase 1, the short transient repolarization that is followed by the plateau phase 2. Phase 2 is followed by phase 3, which is the repolarization back to the resting membrane potential. The resting state of the membrane potential is called as the phase 4 [49].

As mentioned, cardiac action potential results from the chain reaction of multiple ion channels. Therefore a malfunction of a single ion channel can be observed from the action potential curve. Figure 2 presents the parameters which are used in analyzing the action potential. In regard to analyzing cardiac disease specific cells, the action potential duration plays an important role because the lengthening of the action potential may lead to severe arrhythmias.

As a method, patch clamp is very informative and provides invaluable data for example for pharmacological and safety pharmacological studies. However, it is very laborious, needs highly specialized machinery and, most importantly, dedicated and specialized users. For these reasons, semi-automated and automated patch clamp machinery are being developed and would be valuable for cardiomyocyte applications [50, 51].

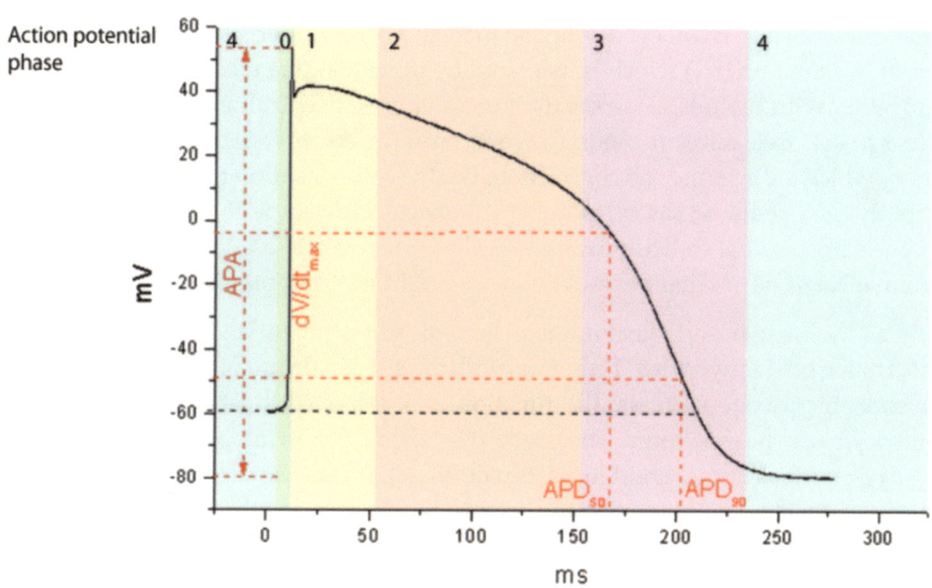

Figure 2. The phases of the cardiac action potential. ADP_{50} and ADP_{90} represent the action potential duration at 50% and 90% of the repolarization and these parameters are used in determining the duration of the action potential. The dV/dt_{max} represents the maximal upstroke velocity and can be used in assessing the electrophysiological phenotype and maturity stage of the cardiomyocytes.

5.2.2. *Micro electrode array*

In addition to the traditional patch clamp technique [46] the micro electrode array (MEA) – platform [52] offers practical, relative easy and non-invasive technique to assess the electrical properties of the differentiated cardiomyocytes [53]. Contrary to the patch clamp, the MEA system measures the electrical activity of a cell population. Therefore the signal resembles electrocardiogram (ECG) and is called field potential instead of action potential. Even though the ion channel function cannot be studied in the similar accuracy as with patch clamp, it allows examination field potential properties, such as cardiac repolarization, and therefore enables drug effect investigation [53]. During the last years, MEA has been widely used in characterization of hESC- and iPS cell derived cardiomyocytes [31, 54]. MEA has been become a basic electrophysiological tool and in addition to cardiomyocytes, it has been successfully used also with other cell types, such as neurons [55].

The MEA system is also applicable in studying cardiac cell responses to pharmaceutical agents [54]. It also enables cells to be measured repeatedly for longer periods of times e.g. multiple days or weeks. However, the analysis of MEA measurement data is laborious. Therefore, semi-automated and automated systems for data analysis have been developed, which makes MEA system more reliable and efficient tool in research [56].

5.2.3. *The assessment of calcium homeostasis*

In addition to the unique co-operation of cardiac ion channels, the interaction of calcium-ions with cardiac structure proteins is another crucial feature in cardiomyocytes that is essential for the proper function of the heart. In human cardiomyocytes, calcium ion (Ca2+) influx through L-type calcium channels during the plateau phase triggers the Ca2+-release from the sarcoplasmic reticulum (SR) which is mediated by the ryanodine receptors (RyR2). The Ca2+ influx together with the release raises the free calcium concentration inside the cardiomyocytes. In sytosol, free calcium binds to troponin C in the myofilaments and triggers the machinery which induces the cell contraction. For the cell relaxation to occur, the calcium has to be rapidly removed from the cytosol. The removal is efficient with the aid of four separate pathways; sarcoplasmic reticulum Ca2+-ATPase (Serca2a), sarcolemmal Ca2+-ATPase, sarcolemmal Na+/Ca2+ exchanger and mitochondrial Ca2+ uniport [57].

Similarly as the regular and synchronous chain of action potentials, calcium concentration fluctuates in the cardiomyocytes. Therefore, with the aid of calcium binding dyes and modern fluorescence microscope systems, the function and response to pharmaceutical agents of cardiomyocytes can be monitored. This method is called calcium imaging [58, 59]. The calcium binding dyes, such as Fura-2 and Fluo-4, can be loaded inside the cardiomyocyte cytosol and when the calcium ions are released to the cytosol, the ions bind to the dyes and a fluorescence signal can be detected. When the fluorescence intensity is measured from the single cell, the calcium handling of the single cardiomyocyte can be monitored and analyzed. From the calcium imaging data, the beating rate and the function of the calcium handling machinery in the cardiomyocytes can be assessed. If the calcium is not released or withdrawn from the cardiomyocyte cytosol in a proper way, irregularity or multiple peaks can be seen in the calcium imaging curve.

5.2.4. Force measurement

Recently a lot of effort has been applied to develop measuring systems to understand the mechanobiology of cardiomyocytes. Force measurement technique can be applied to measure isometric cardiomyocyte force contraction. A number of parameters can be determined by using the cardiomyocyte force measurement such as determination of Ca-sensitivity, cooperativity of force production and maximal Ca-activated force. Kinetics of the contractile responses can also be measured such as the actin-myosin turnover kinetics. These parameters can be useful in the characterization of myofibrillar pathologies of various origin and drug effects. Most of the currently existing systems are only suitable for the study of cardiac tissue slices and therefore inappropriate to be used for iPS cell derived cardiomyocytes. Recently, however, cardiomyocyte force measurement system based on atomic force microscopy (AFM) was developed which can also be used to study single cardiomyocytes and small clusters of cardiomyocytes [60]. With the AFM system they were able to measure contractile forces, beat frequencies and durations of single cardiomyocytes and small cardiomyocyte clusters. The AFM-based method is also applicable for the screening of cardiac-active pharmacological agents. Cardiac microtissues have also been constructed using human pluripotent stem cell derived cardiomyocytes and the contraction force of the beating tissues has been analyzed with custom made platforms [61, 62].

6. Diseases modeled with iPS cell technique

Since the revolutionary discovery of iPS cells, multiple genetic diseases including cardiac and neuronal diseases have been modeled with patient specific iPS cell derived cells. Since primary human cardiomyocytes are not available for research in vitro, iPS cell derived cardiomyocytes are invaluable tool to study the pathophysiology of severe cardiac diseases and will undoubtedly provide groundbreaking innovations in the future.

6.1. Long QT-syndrome

Long QT-syndrome (LQTS) appears as a genetic or a drug-induced form. It is characterized by a prolonged cardiac repolarization phase resulting in a prolonged QT interval in the surface electrocardiogram (ECG). The clinical symptoms of LQTS include palpitations, syncope and seizures and even sudden cardiac death.

More than 700 mutations in 12 different genes (LQT1–12) have been found to affect genetic forms of LQTS [63]. However, two of these subtypes account the majority (>90%) of all the genetically identified LQTSs. Both of these mutations affect potassium channels altering their proper function. LQTS type 1 (LQT1) is the most common LQTS subtype, resulting from mutations in the KCNQ1 gene. This gene encodes the α-subunit of the slow component of the delayed rectifier potassium current (IKs) channel [64]. Individuals with LQT1 typically have symptoms when the heart rate is elevated e.g. during exercise [65, 66].

LQTS type 2 (LQT2) is due to non-proper functioning of the α-subunit of the rapid delayed potassium channel (IKr), which is encoded by the human ether-a-go-go-related gene (HERG), also known as KCNH2-gene [67]. Contrary to type 1, individuals with LQT2 have clinical symptoms when the heart rate is slow [65, 66] and syptoms can be triggered e.g. by an alarm clock during sleep. The drug induced form of LQTS is due to altered function of the HERG-channel by the drug, therefore this channel has a significant importance during drug development and in safety studies.

The prevalence of the genetic form of LQTS is 1:2,000 in the general population [63]. However, the penetrance of the clinical symptoms of LQTS is low and there is considerable variation in phenotypic expression even within families carrying the same mutation [68]. It has also been suggested, that the population prevalence of milder LQTS mutations might be higher. Therefore the prevalence of latent or concealed LQTS, i.e. relatively asymptomatic individuals, would be higher than currently anticipated [69]. Due to this challenging and complex nature of LQTS, in addition, to the great interest of pharmaceutical industry towards this disease, multiple reports of iPS cell- based LQTS cell models have been published since 2007 when the iPS technology was invented.

Moretti and co-workers produced iPS cell derived cardiomycotyes from two patients carrying a KCNQ1 (R190Q) mutation [6]. In this study, the cardiomyocytes possessed the LQT1 genotype and exhibited prolonged action potential duration. The action potential prolongation was determined to be caused by the ion-channel trafficking defect resulting in a 70-80% IKs current density reduction. A β-adrenergic agonist isoproterenol altered the activation and deactivation kinetics of the IKs and this effect was rescued by the β-blockade [6]. Egashira and co-workers also produced a disease model for LQTS type 1 [70]. In their study, the iPS cells were derived from a sporadic patient who did not have a family history of significant QT interval abnormality. The mutation of the patient in the KCNQ1 was novel (1893delC) and the cells exhibited prolonged action potential duration in addition to arrhythmogenity.

Similarly results were found with iPS-CM derived from a patient suffering from the severe LQT type 2 syndrome. The patient had hERG (A614V) mutation and previously presented episodes of torsade de pointes (TdP), a special type of polymorphic ventricular tachycardia which is associated with LQTS [4]. The LQT2-cardiomyocytes derived from the patient's iPS cells demonstrated increased arrhythmogenicity associated with early after depolarizations (EADs) [5]. In addition, significant APD prolongation due to a reduced IKr current density was observed [4]. Arrhytmia and EADs were also induced by a specific HERG-channel blocker E-4031 to iPS cell derived CM having a hERG (G1681A) mutation. In addition, these cells exhibited EADs caused by the isoproterenol treatment and these EADs were rescued by β-blockade [5].

All the aforementioned studies were done with iPS cells derived from the symptomatic LQTS patients. Nevertheless, similar results have been obtained from patients without severe symptoms. In the study made in our institute, iPS cell lines were derived from a patient having a KCNH2 (R176W) mutation and a family history of LQTS. However, this individual was asymptomatic except for occasional palpitations. iPS cell derived cardiomyocytes from this patient manifest the phenotype characteristics to LQT2, such as a prolonged repolarization time and increased arrhythmogenicity [3].

A human cell model for LQT3 has also been produced and its function and characteristics were compared with a mouse models which were based on both mouse ESCs and mouse iPS cells affected with the same disease specific mutation [71]. LQT3 syndrome is due to mutations in the SCN5A gene, which encodes for the α-subunit of the cardiac sodium (Na+) channel. These mutations disrupt the inactivation of the Na+ channel during the action potential plateu phase and this irruption leads to the delay in repolarization and further prolonged QT interval [72]. In addition to LQT3, another kind of cardiac arrhythmia syndromes such as Brugada syndrome and cardiac conduction disease are associated with mutations in the SCN5A gene. In these syndromes the mutations are loss-of-function-type whereas LQT3 syndrome they are gain-of-function-type mutations [72, 73]. The comparison of multiple types of pluripotent stem cell derived cardiomyocytes showed that all of these models manifest the symptoms of the disease and, furthermore, the characteristics are similar within both species [71]. iPS cell models for these loss-of-function diseases have not yet been described.

6.1.1. *Timothy syndrome*

Timothy syndrome is caused by a single mutation in the CACNA1C-gene. This gene encodes the main L-type calcium channel, $Ca_v1.2$, in the mammalian heart which is essential for the cardiac action potential and also for cardiomyocyte contraction [74-76]. Timothy sydrome characterized by LQTS, syndactyly (webbing of fingers and toes), immune deficiency and autism [77] iPS cell derived cardiomyocytes originating from Timothy syndrome patients exhibited irregular functional properties typical for the disease [78]. Interestingly, these irregularities were restored by roscovitine, a compound which increases the voltage-dependent inactivation of $Ca_v1.2$ [78].

6.1.2. *Catecholaminergic Polymorphic Ventricular Tachycardia*

Catecholaminergic polymorphic ventricular tachycardia (CPVT) is an inherited cardiac disorder characterized by stress-induced polymorphic ventricular tachycardia in a structurally normal heart. CPVT is a very severe disease and 30-35% of mutation carriers have had symptoms (stress-related syncope, seizures or sudden death) by the age of 30. This disease is caused by mutations in the genes of RyR2 or calsequestrin (CASQ2) which is a regulatory calcium-buffering protein associated with RyR2 in the SR [79-82].

Multiple iPS-based CPVT disease models have been published, most of them having the disease specific mutation in the RyR2 gene while [83-85] and one having the mutation in the CASQ2 gene [86]. The congruent result from these CPVT model studies was the occurrence of delayed after depolarizations (DADs) and arrhythmias which are caused by the aberrant diastolic Ca^{2+} from the SR. Notably the model with RyR2-P2328S mutation also exhibited early after depolarizations (EADs) in addition to DADs suggesting suggesting another pathophysiological mechanism for CPVT [85]. Intriguing finding was also the effect of dantrolene in rescuing the arrhythmogenic phenotype [84].

6.1.3. Cardiomyopathies

Mutations in the genes expressed in the cardiomyocytes can cause heart diseases known as cardiomyopathies. Cardiomyopathies are currently categorized into the following four classes: arrythmogenic right ventricular cardiomyopathy, dilated cardiomyopathy, hypertrophic cardiomyopathy, and restrictive cardiomyopathy [87]. Cardiomyopathies that are associated with mutations in genes encoding for sarcomeric proteins are a frequent cause of heart failure.

iPS cells have been used to generate cardiomyocytes from patients in a family with inherited dilated cardiomyopathy (DCM) [88]. The researchers generated a large number of individual-specific cardiomyocytes from a family carrying a deleterious point mutation (R173W) in TNNT2, a gene encoding for a sarcomeric protein cardiac troponin T, which regulates cardiomyocyte contraction. When compared to cardiomyocytes derived from iPS cells of healthy controls within the same family, the researchers showed that cardiomyocytes derived from iPS cells of DCM patients exhibited an increased heterogenous myofilament organization due to abnormal distribution of α-actinin, compromised ability to regulate calcium flux, and decreased contraction force. When DCM specific cardiomyocytes were stimulated with a β-adrenergic agonist, the cells showed characteristics of cellular stress such as reduced beating rates, compromised contraction, and a greater number of cells with abnormal sarcomeric α-actinin distribution. The authors also showed that the function of DCM-specific cardiomyocytes was improved with the treatment with β-adrenergic blockers or overexpression of Serca2a.

Arrhythmogenic right ventricular cardiomyopathy (ARVC) is another genetic cardiomyopathy characterized by replacement of cardiomyocytes by adipose and fibrous tissue leading to right ventricular failure, arrhythmias and even sudden death [89]. Twelve different genes have been linked to ARVC and all these encode cardiac cell adhesion proteins resulting in dysfunctional cardiac desmosomes. Cell adhesion proteins resulting in ARVC include plakoglobin (JUP), desmoplakin (DSP) and plakophilin 2 (PKP2). Patient specific iPS cells have been generated from an ARCV patient carrying a PKP2 mutation and having clinical manifestations of the disease [90]. ARVC specific cardiomyocytes revealed reduced amount of desmosomal proteins and more lipid droplets in the cardiomyocytes compared to control cardiomyocytes thus presenting the abnormalities observed in ARCV patients.

The third form of cardiomyopathy, hypertrophic cardiomyopathy (HCM), is a complex autosomal-dominant disease and the affected individuals acquire cardiac hypertrophy without external stimuli. Cardiac hypertrophy can be induced by different exogenous factors such as hypertension and valvular disease and even by severe exercise [91]. Affecting in 1 in 500 individuals within the general population, genetic HCM is the most common inherited cardiovascular disorder and the leading cause of sudden cardiac death in adolescents and young adults, especially in atheletes [92-94]. The majority of gene mutations associated with HCM occur in 13 sarcomere-related genes where several hundred mutations have been identified [94-97]. Typically cardiac hypertrophy affects the left ventricle and the interventricular septum and may eventually lead to left ventricular outflow tract obstruction, arrythmias, diastolic dysfunction, and sudden death. Other hallmark features are myocyte disarray and fibrosis [94-97]. The hypertrophic process in cardiomyocytes is characterized by morphological changes including increase in protein synthesis, enhanced sarcomere reorganization as well as activation of specific cardiac genes [98-100].

iPS cell technology has not yet been reported to model HCM. However, iPS cells were used to generate cardiomyocytes from two LEOPARD syndrome patients carrying mutation in the PTPN11 gene encoding for the SHP2 phosphatase [101]. LEOPARD syndrome is an autosomal-dominant developmental disorder belonging to inherited RAS-mitogen-activated protein kinase signalling diseases. A major disease phenotype of the LEOPARD syndrome patients is HCM [102]. The iPS cell derived cardiomyocytes from LEOPARD syndrome were larger, had a higher degree of sarcomeric organization as well as preferential localization of NFATC4 in the nucleus when compared to iPS cell derived cardiomyocytes from healthy sibling of the LEOPARD syndrome patient thus presenting some indications of hypertrophy in patient specific cardiomyocytes.

7. Challenges with iPS cell technology and disease modeling

There are still several challenges that need to be carefully considered when designing disease modeling studies with specialized cell types derived from iPS cells. One potential challenge relates to the reactivation of silenced exogenous transgenes in the iPS cells or in their differentiated derivatives leading to the altered genomic integrity which may have unknown effects on the differentiation potential and characteristics of differentiated cell types. Efforts to improve the reprogramming methods have led to the technical development of nonintegrating approaches for iPS cell generation which will eliminate this risk in the future iPS cell lines and their differentiated derivatives. The nonintegrating sendai virus tehnique is already widely used in the generation of iPS cells. Regular monitoring of exogenous genes in iPS cells lines generated by using the integrating techniques is advisable.

Many genetic cardiac diseases are complex demonstrating huge clinical heterogeneity even within families and patients having the same mutation. In addition, reprogrammed cells carry genetic alterations that have accumulated through life, thus there is a risk that the variance overwhelms the ability to detect the authentic mechanisms in the pathophysiology of the disease. Thus, it will be essential to investigate adequate number of iPSC lines and patients to be able to demonstrate the common features of the cardiac disease phenotype. Further, it may be advantageous to initially compare the characteristics of cardiomyocytes from patients having severe symptoms.

Most likely in many genetic cardiac diseases various cell types in the heart contribute to the pathophysiological responses of the disease, thus there is a risk that it is impossible to recapitulate the features of the disorder by using solely cardiomyocytes. A 3D human heart tissue model with proper composition of cardiomyocytes, endothelial cells, fibroblasts, smooth muscle cells as well as neurons has not been developed but in recent years the advancement in this field of research has been rapid and hopefully in future we have besides cell models authentic tissue models to study genetic cardiac diseases.

The current cardiomyocyte differentiation protocols generate cells lacking full maturity when compared to human adult cardiomyocytes. This may lead to a situation where it is impossible to detect some molecular or functional basis of the cardiac disease. To reduce this risk it will be advisable to use control cells to compare diseased cardiomyocytes to healthy cardiomyocytes. For reliable and reproducible modeling of cardiac diseases it is necessary to have preferable multiple iPS lines from healthy controls. For monogenic diseases the use of iPS cells derived from the healthy family members would be favorable for minimizing the effect of genetic variation. However, iPS cells from family members are not always available. On possibility to overcome this challenge is to use genome editing techniques such as zinc finger nuclease technology and transcription activator–like effectors (TALEs) in modifying the iPS cells [103, 104]. With these methods, it is possible to correct a targeted point mutation in human iPS cells and produce control cells for disease specific iPS cells.

8. Conclusions

The most relevant human disease model uses cells of human origin, of the appropriate cell type, and with the identical genetic background as the patients. Traditionally, this approach in cardiac diseases has been out of reach as human cardiomyocytes are not easily procured and their propagation in vitro is extremely problematic. The revolutionary discovery of cellular reprogramming as a technology to generate iPS cells enables the production of patient specific cell types such as cardiomyocytes which can be used as authentic and relevant human cell models to study the pathophysiology of genetic cardiac diseases as well as in drug discovery and safety assays. The most relevant aspects in disease modeling are to show that the produced disease specific cell type bears the disease causing mutation and further to present the functional consequences of the mutant protein. Here we have reviewed the genetic cardiac diseases modeled thus far by using the iPS cell technology. Worthwhile of noticing is that the era of iPS cells in disease modeling is just in the very beginning. As the production of iPS cells and cardiomyocytes with more mature phenotype and the methods available for the functional characterization of cardiomyocytes continue to develop the future looks bright for modeling genetic cardiac diseases. Importantly these models will be extremely valuable for drug discovery and toxicology in the future.

Author details

Mari Pekkanen-Mattila[1,2], Kristiina Rajala[1,2] and Katriina Aalto-Setälä[1,2,3]

1 Institute of Biomedical Technology, University of Tampere, Tampere, Finland

2 BioMediTech, University of Tampere, Tampere, Finland

3 Heart Center, Tampere University Hospital, Tampere, Finland

References

[1] Dixon, J. A, & Spinale, F. G. Large animal models of heart failure: a critical link in the translation of basic science to clinical practice. Circ Heart Fail (2009). , 2(3), 262-71.

[2] Cingolani, H. E, & Ennis, I. L. Sodium-hydrogen exchanger, cardiac overload, and myocardial hypertrophy. Circulation (2007). , 115(9), 1090-100.

[3] Lahti, A. L, Kujala, V. J, Chapman, H, Koivisto, A. P, Pekkanen-mattila, M, Kerkela, E, et al. Model for long QT syndrome type 2 using human iPS cells demonstrates arrhythmogenic characteristics in cell culture. Dis Model Mech (2012). , 5(2), 220-30.

[4] Itzhaki, I, Maizels, L, Huber, I, Zwi-dantsis, L, Caspi, O, Winterstern, A, et al. Modelling the long QT syndrome with induced pluripotent stem cells. Nature (2011). , 471(7337), 225-9.

[5] Matsa, E, Rajamohan, D, Dick, E, Young, L, Mellor, I, Staniforth, A, et al. Drug evaluation in cardiomyocytes derived from human induced pluripotent stem cells carrying a long QT syndrome type 2 mutation. Eur Heart J (2011). , 32(8), 952-62.

[6] Moretti, A, Bellin, M, Welling, A, Jung, C. B, Lam, J. T, Bott-flugel, L, et al. Patient-specific induced pluripotent stem-cell models for long-QT syndrome. N Engl J Med (2010). , 363(15), 1397-409.

[7] Takahashi, K, Tanabe, K, Ohnuki, M, Narita, M, Ichisaka, T, Tomoda, K, et al. Induction of pluripotent stem cells from adult human fibroblasts by defined factors. Cell (2007). , 131(5), 861-72.

[8] Haase, A, Olmer, R, Schwanke, K, Wunderlich, S, Merkert, S, Hess, C, et al. Generation of induced pluripotent stem cells from human cord blood. Cell Stem Cell (2009). , 5(4), 434-41.

[9] Yu, J, Vodyanik, M. A, Smuga-otto, K, Antosiewicz-bourget, J, Frane, J. L, Tian, S, et al. Induced pluripotent stem cell lines derived from human somatic cells. Science (2007). , 318(5858), 1917-20.

[10] Aasen, T, Raya, A, Barrero, M. J, Garreta, E, Consiglio, A, Gonzalez, F, et al. Efficient and rapid generation of induced pluripotent stem cells from human keratinocytes. Nat Biotechnol (2008). , 26(11), 1276-84.

[11] Giorgetti, A, Montserrat, N, Aasen, T, Gonzalez, F, Rodriguez-piza, I, Vassena, R, et al. Generation of induced pluripotent stem cells from human cord blood using OCT4 and SOX2. Cell Stem Cell (2009). , 5(4), 353-7.

[12] Takahashi, K, & Yamanaka, S. Induction of pluripotent stem cells from mouse embryonic and adult fibroblast cultures by defined factors. Cell (2006). , 126(4), 663-76.

[13] Hussein, S. M, Batada, N. N, Vuoristo, S, Ching, R. W, Autio, R, Narva, E, et al. Copy number variation and selection during reprogramming to pluripotency. Nature (2011)., 471(7336), 58-62.

[14] Stadtfeld, M, Nagaya, M, Utikal, J, Weir, G, & Hochedlinger, K. Induced pluripotent stem cells generated without viral integration. Science (2008)., 322(5903), 945-9.

[15] Fusaki, N, Ban, H, Nishiyama, A, Saeki, K, & Hasegawa, M. Efficient induction of transgene-free human pluripotent stem cells using a vector based on Sendai virus, an RNA virus that does not integrate into the host genome. Proc Jpn Acad Ser B Phys Biol Sci (2009)., 85(8), 348-62.

[16] Okita, K, Nakagawa, M, Hyenjong, H, Ichisaka, T, & Yamanaka, S. Generation of mouse induced pluripotent stem cells without viral vectors. Science (2008)., 322(5903), 949-53.

[17] Yu, J, Hu, K, Smuga-otto, K, Tian, S, Stewart, R, Slukvin, I. I, et al. Human induced pluripotent stem cells free of vector and transgene sequences. Science (2009)., 324(5928), 797-801.

[18] Ebert, A. D, Yu, J, & Rose, F. F. Jr., Mattis VB, Lorson CL, Thomson JA, et al. Induced pluripotent stem cells from a spinal muscular atrophy patient. Nature (2009)., 457(7227), 277-80.

[19] Jia, F, Wilson, K. D, Sun, N, Gupta, D. M, Huang, M, Li, Z, et al. A nonviral minicircle vector for deriving human iPS cells. Nat Methods (2010)., 7(3), 197-9.

[20] Lee, C. H, Kim, J. H, Lee, H. J, Jeon, K, Lim, H, Choi, H, et al. The generation of iPS cells using non-viral magnetic nanoparticle based transfection. Biomaterials (2011)., 32(28), 6683-91.

[21] Shao, L, Feng, W, Sun, Y, Bai, H, Liu, J, Currie, C, et al. Generation of iPS cells using defined factors linked via the self-cleaving 2A sequences in a single open reading frame. Cell Res (2009)., 19(3), 296-306.

[22] Yusa, K, Rad, R, Takeda, J, & Bradley, A. Generation of transgene-free induced pluripotent mouse stem cells by the piggyBac transposon. Nat Methods (2009)., 6(5), 363-9.

[23] Yakubov, E, Rechavi, G, Rozenblatt, S, & Givol, D. Reprogramming of human fibroblasts to pluripotent stem cells using mRNA of four transcription factors. Biochem Biophys Res Commun (2010)., 394(1), 189-93.

[24] Warren, L, Manos, P. D, Ahfeldt, T, Loh, Y. H, Li, H, Lau, F, et al. Highly efficient reprogramming to pluripotency and directed differentiation of human cells with synthetic modified mRNA. Cell Stem Cell (2010)., 7(5), 618-30.

[25] Kim, D, Kim, C. H, Moon, J. I, Chung, Y. G, Chang, M. Y, Han, B. S, et al. Generation of human induced pluripotent stem cells by direct delivery of reprogramming proteins. Cell Stem Cell (2009). , 4(6), 472-6.

[26] Lin, T, Ambasudhan, R, Yuan, X, Li, W, Hilcove, S, Abujarour, R, et al. A chemical platform for improved induction of human iPSCs. Nat Methods (2009). , 6(11), 805-8.

[27] Mummery, C, & Ward, D. van den Brink CE, Bird SD, Doevendans PA, Opthof T, et al. Cardiomyocyte differentiation of mouse and human embryonic stem cells. J Anat (2002). Pt 3):233-42.

[28] Kehat, I, Kenyagin-karsenti, D, Snir, M, Segev, H, Amit, M, Gepstein, A, et al. Human embryonic stem cells can differentiate into myocytes with structural and functional properties of cardiomyocytes. J Clin Invest (2001). , 108(3), 407-14.

[29] Passier, R, Oostwaard, D. W, Snapper, J, Kloots, J, Hassink, R. J, Kuijk, E, et al. Increased cardiomyocyte differentiation from human embryonic stem cells in serum-free cultures. Stem Cells (2005). , 23(6), 772-80.

[30] Zhang, J, Wilson, G. F, Soerens, A. G, Koonce, C. H, Yu, J, Palecek, S. P, et al. Functional cardiomyocytes derived from human induced pluripotent stem cells. Circ Res (2009). e, 30-41.

[31] Pekkanen-mattila, M, Kerkela, E, Tanskanen, J. M, Pietila, M, Pelto-huikko, M, Hyttinen, J, et al. Substantial variation in the cardiac differentiation of human embryonic stem cell lines derived and propagated under the same conditions--a comparison of multiple cell lines. Ann Med (2009). , 41(5), 360-70.

[32] He, J-Q, Ma, Y, Lee, Y, Thomson, J. A, & Kamp, T. J. Human Embryonic Stem Cells Develop Into Multiple Types of Cardiac Myocytes: Action Potential Characterization. Circ Res (2003). , 93(1), 32-39.

[33] Pekkanen-mattila, M, Chapman, H, Kerkela, E, Suuronen, R, Skottman, H, Koivisto, A. P, et al. Human embryonic stem cell-derived cardiomyocytes: demonstration of a portion of cardiac cells with fairly mature electrical phenotype. Exp Biol Med (Maywood) (2010). , 235(4), 522-30.

[34] Moore, J. C, Fu, J, Chan, Y. C, Lin, D, Tran, H, Tse, H. F, et al. Distinct cardiogenic preferences of two human embryonic stem cell (hESC) lines are imprinted in their proteomes in the pluripotent state. Biochem Biophys Res Commun (2008). , 372(4), 553-8.

[35] Mummery, C. L, Zhang, J, Ng, E. S, Elliott, D. A, Elefanty, A. G, & Kamp, T. J. Differentiation of human embryonic stem cells and induced pluripotent stem cells to cardiomyocytes: a methods overview. Circ Res (2012). , 111(3), 344-58.

[36] Ieda, M, Fu, J. D, Delgado-olguin, P, Vedantham, V, Hayashi, Y, Bruneau, B. G, et al. Direct reprogramming of fibroblasts into functional cardiomyocytes by defined factors. Cell (2010). , 142(3), 375-86.

[37] Chen, J. X, Krane, M, Deutsch, M. A, Wang, L, Rav-acha, M, Gregoire, S, et al. Inefficient reprogramming of fibroblasts into cardiomyocytes using Gata4, Mef2c, and Tbx5. Circ Res (2012)., 111(1), 50-5.

[38] Islas, J. F, Liu, Y, Weng, K. C, Robertson, M. J, Zhang, S, Prejusa, A, et al. Transcription factors ETS2 and MESP1 transdifferentiate human dermal fibroblasts into cardiac progenitors. Proc Natl Acad Sci U S A (2012)., 109(32), 13016-21.

[39] Yoshida, Y, & Yamanaka, S. Labor pains of new technology: direct cardiac reprogramming. Circ Res (2012)., 111(1), 3-4.

[40] Qian, L, Huang, Y, Spencer, C. I, Foley, A, Vedantham, V, Liu, L, et al. In vivo reprogramming of murine cardiac fibroblasts into induced cardiomyocytes. Nature (2012)., 485(7400), 593-8.

[41] Burridge, P. W, Thompson, S, Millrod, M. A, Weinberg, S, Yuan, X, Peters, A, et al. A universal system for highly efficient cardiac differentiation of human induced pluripotent stem cells that eliminates interline variability. PLoS One (2011). e18293.

[42] Mummery, C. Ward-van Oostwaard D, Doevendans P, Spijker R, van den Brink S, Hassink R, et al. Differentiation of human embryonic stem cells to cardiomyocytes: role of coculture with visceral endoderm-like cells. Circulation (2003). , 107(21), 2733-40.

[43] Graichen, R, Xu, X, Braam, S. R, Balakrishnan, T, Norfiza, S, Sieh, S, et al. Enhanced cardiomyogenesis of human embryonic stem cells by a small molecular inhibitor of MAPK. Differentiation (2008)., 38.

[44] Laflamme, M. A, Chen, K. Y, Naumova, A. V, Muskheli, V, Fugate, J. A, Dupras, S. K, et al. Cardiomyocytes derived from human embryonic stem cells in pro-survival factors enhance function of infarcted rat hearts. Nat Biotechnol (2007)., 25(9), 1015-24.

[45] Lian, X, Hsiao, C, Wilson, G, Zhu, K, Hazeltine, L. B, Azarin, S. M, et al. Robust cardiomyocyte differentiation from human pluripotent stem cells via temporal modulation of canonical Wnt signaling. Proc Natl Acad Sci U S A (2012). E, 1848-57.

[46] Hamill, O. P, Marty, A, Neher, E, Sakmann, B, & Sigworth, F. J. Improved patch-clamp techniques for high-resolution current recording from cells and cell-free membrane patches. Pflugers Arch (1981)., 391(2), 85-100.

[47] Sakmann, B, & Neher, E. Patch clamp techniques for studying ionic channels in excitable membranes. Annu Rev Physiol (1984)., 46, 455-72.

[48] Pollard, C. E. Abi Gerges N, Bridgland-Taylor MH, Easter A, Hammond TG, Valentin JP. An introduction to QT interval prolongation and non-clinical approaches to assessing and reducing risk. Br J Pharmacol (2010)., 159(1), 12-21.

[49] Nerbonne, J. M, & Kass, R. S. Molecular physiology of cardiac repolarization. Physiol Rev (2005)., 85(4), 1205-53.

[50] Rajamohan, D, Matsa, E, Kalra, S, Crutchley, J, Patel, A, George, V, et al. Current status of drug screening and disease modelling in human pluripotent stem cells. Bioessays (2012).

[51] Ma, J, Guo, L, Fiene, S. J, Anson, B. D, Thomson, J. A, Kamp, T. J, et al. High purity human-induced pluripotent stem cell-derived cardiomyocytes: electrophysiological properties of action potentials and ionic currents. Am J Physiol Heart Circ Physiol (2011). H, 2006-17.

[52] Reppel, M, Pillekamp, F, Lu, Z. J, Halbach, M, Brockmeier, K, Fleischmann, B. K, et al. Microelectrode arrays: a new tool to measure embryonic heart activity. J Electrocardiol (2004). Suppl:, 104-9.

[53] Reppel, M, Pillekamp, F, Brockmeier, K, Matzkies, M, Bekcioglu, A, Lipke, T, et al. The electrocardiogram of human embryonic stem cell-derived cardiomyocytes. J Electrocardiol (2005). Suppl):, 166-70.

[54] Caspi, O, Itzhaki, I, Kehat, I, Gepstein, A, Arbel, G, Huber, I, et al. In vitro electrophysiological drug testing using human embryonic stem cell derived cardiomyocytes. Stem Cells Dev (2009). , 18(1), 161-72.

[55] Heikkila, T. J, Yla-outinen, L, Tanskanen, J. M, Lappalainen, R. S, Skottman, H, Suuronen, R, et al. Human embryonic stem cell-derived neuronal cells form spontaneously active neuronal networks in vitro. Exp Neurol (2009). , 218(1), 109-16.

[56] Kujala, V. J, Jimenez, Z. C, Vaisanen, J, Tanskanen, J. M, Kerkela, E, Hyttinen, J, et al. Averaging in vitro cardiac field potential recordings obtained with microelectrode arrays. Comput Methods Programs Biomed (2011). , 104(2), 199-205.

[57] Bers, D. M. Cardiac excitation-contraction coupling. Nature (2002). , 415(6868), 198-205.

[58] Wier, W. G, Beuckelmann, D. J, & Barcenas-ruiz, L. Ca2+]i in single isolated cardiac cells: a review of recent results obtained with digital imaging microscopy and fura-2. Can J Physiol Pharmacol (1988). , 66(9), 1224-31.

[59] Herron, T. J, Lee, P, & Jalife, J. Optical imaging of voltage and calcium in cardiac cells & tissues. Circ Res;, 110(4), 609-23.

[60] Liu, J, Sun, N, Bruce, M. A, Wu, J. C, & Butte, M. J. Atomic force mechanobiology of pluripotent stem cell-derived cardiomyocytes. PLoS One (2012). e37559.

[61] Boudou, T, Legant, W. R, Mu, A, Borochin, M. A, Thavandiran, N, Radisic, M, et al. A microfabricated platform to measure and manipulate the mechanics of engineered cardiac microtissues. Tissue Eng Part A (2012).

[62] Schaaf, S, Shibamiya, A, Mewe, M, Eder, A, Stohr, A, Hirt, M. N, et al. Human engineered heart tissue as a versatile tool in basic research and preclinical toxicology. PLoS One (2011). e26397.

[63] Hedley, P. L, Jorgensen, P, Schlamowitz, S, Wangari, R, Moolman-smook, J, Brink, P. A, et al. The genetic basis of long QT and short QT syndromes: a mutation update. Hum Mutat (2009). , 30(11), 1486-511.

[64] Chiang, C. E, & Roden, D. M. The long QT syndromes: genetic basis and clinical implications. J Am Coll Cardiol (2000). , 36(1), 1-12.

[65] Roden, D. M. Clinical practice. Long-QT syndrome. N Engl J Med (2008). , 358(2), 169-76.

[66] Schwartz, P. J, Priori, S. G, Spazzolini, C, Moss, A. J, Vincent, G. M, Napolitano, C, et al. Genotype-phenotype correlation in the long-QT syndrome: gene-specific triggers for life-threatening arrhythmias. Circulation (2001). , 103(1), 89-95.

[67] Curran, M. E, Splawski, I, Timothy, K. W, Vincent, G. M, Green, E. D, & Keating, M. T. A molecular basis for cardiac arrhythmia: HERG mutations cause long QT syndrome. Cell (1995). , 80(5), 795-803.

[68] Priori, S. G, Napolitano, C, & Schwartz, P. J. Low penetrance in the long-QT syndrome: clinical impact. Circulation (1999). , 99(4), 529-33.

[69] Marjamaa, A, Salomaa, V, Newton-cheh, C, Porthan, K, Reunanen, A, Karanko, H, et al. High prevalence of four long QT syndrome founder mutations in the Finnish population. Ann Med (2009). , 41(3), 234-40.

[70] Egashira, T, Yuasa, S, Suzuki, T, Aizawa, Y, Yamakawa, H, Matsuhashi, T, et al. Disease characterization using LQTS-specific induced pluripotent stem cells. Cardiovasc Res (2012). , 95(4), 419-29.

[71] Davis, R. P, & Casini, S. van den Berg CW, Hoekstra M, Remme CA, Dambrot C, et al. Cardiomyocytes derived from pluripotent stem cells recapitulate electrophysiological characteristics of an overlap syndrome of cardiac sodium channel disease. Circulation (2012). , 125(25), 3079-91.

[72] Bennett, P. B, Yazawa, K, Makita, N, & George, A. L. Jr. Molecular mechanism for an inherited cardiac arrhythmia. Nature (1995). , 376(6542), 683-5.

[73] Remme, C. A, & Bezzina, C. R. Sodium channel (dys)function and cardiac arrhythmias. Cardiovasc Ther (2010). , 28(5), 287-94.

[74] Reuter, H. Ion channels in cardiac cell membranes. Annu Rev Physiol (1984). , 46, 473-84.

[75] Flucher, B. E, & Franzini-armstrong, C. Formation of junctions involved in excitation-contraction coupling in skeletal and cardiac muscle. Proc Natl Acad Sci U S A (1996). , 93(15), 8101-6.

[76] Seisenberger, C, Specht, V, Welling, A, Platzer, J, Pfeifer, A, Kuhbandner, S, et al. Functional embryonic cardiomyocytes after disruption of the L-type alpha1C (Cav1.2) calcium channel gene in the mouse. J Biol Chem (2000). , 275(50), 39193-9.

[77] Splawski, I, Timothy, K. W, Sharpe, L. M, Decher, N, Kumar, P, Bloise, R, et al. Ca(calcium channel dysfunction causes a multisystem disorder including arrhythmia and autism. Cell (2004). , 1

[78] Yazawa, M, Hsueh, B, Jia, X, Pasca, A. M, Bernstein, J. A, Hallmayer, J, et al. Using induced pluripotent stem cells to investigate cardiac phenotypes in Timothy syndrome. Nature (2011). , 471(7337), 230-4.

[79] Hayashi, M, Denjoy, I, Extramiana, F, Maltret, A, Buisson, N. R, Lupoglazoff, J. M, et al. Incidence and risk factors of arrhythmic events in catecholaminergic polymorphic ventricular tachycardia. Circulation (2009). , 119(18), 2426-34.

[80] Leenhardt, A, Lucet, V, Denjoy, I, Grau, F, Ngoc, D. D, & Coumel, P. Catecholaminergic polymorphic ventricular tachycardia in children. A 7-year follow-up of 21 patients. Circulation (1995). , 91(5), 1512-9.

[81] Laitinen, P. J, Brown, K. M, Piippo, K, Swan, H, Devaney, J. M, Brahmbhatt, B, et al. Mutations of the cardiac ryanodine receptor (RyR2) gene in familial polymorphic ventricular tachycardia. Circulation (2001). , 103(4), 485-90.

[82] Priori, S. G, Napolitano, C, Tiso, N, Memmi, M, Vignati, G, Bloise, R, et al. Mutations in the cardiac ryanodine receptor gene (hRyR2) underlie catecholaminergic polymorphic ventricular tachycardia. Circulation (2001). , 103(2), 196-200.

[83] Itzhaki, I, Maizels, L, Huber, I, Gepstein, A, Arbel, G, Caspi, O, et al. Modeling of catecholaminergic polymorphic ventricular tachycardia with patient-specific human-induced pluripotent stem cells. J Am Coll Cardiol (2012). , 60(11), 990-1000.

[84] Jung, C. B, & Moretti, A. Mederos y Schnitzler M, Iop L, Storch U, Bellin M, et al. Dantrolene rescues arrhythmogenic RYR2 defect in a patient-specific stem cell model of catecholaminergic polymorphic ventricular tachycardia. EMBO Mol Med (2012). , 4(3), 180-91.

[85] Kujala, K, Paavola, J, Lahti, A, Larsson, K, Pekkanen-mattila, M, Viitasalo, M, et al. Cell model of catecholaminergic polymorphic ventricular tachycardia reveals early and delayed afterdepolarizations. PLoS One (2012). e44660.

[86] Novak, A, Barad, L, Zeevi-levin, N, Shick, R, Shtrichman, R, Lorber, A, et al. Cardiomyocytes generated from CPVTD307H patients are arrhythmogenic in response to beta-adrenergic stimulation. J Cell Mol Med (2012). , 16(3), 468-82.

[87] Richardson, P, Mckenna, W, Bristow, M, Maisch, B, Mautner, B, & Connell, O. J, et al. Report of the 1995 World Health Organization/International Society and Federation of Cardiology Task Force on the Definition and Classification of cardiomyopathies. Circulation (1996). , 93(5), 841-2.

[88] Sun, R. H, Hu, B. C, & Li, Q. Stress-induced cardiomyopathy complicated by multiple organ failure following cephalosporin-induced anaphylaxis. Intern Med (2012). , 51(8), 895-9.

[89] Azaouagh, A, Churzidse, S, Konorza, T, & Erbel, R. Arrhythmogenic right ventricular cardiomyopathy/dysplasia: a review and update. Clin Res Cardiol (2012). , 100(5), 383-94.

[90] Ma, D, Wei, H, Lu, J, Ho, S, Zhang, G, Sun, X, et al. Generation of patient-specific induced pluripotent stem cell-derived cardiomyocytes as a cellular model of arrhythmogenic right ventricular cardiomyopathy. Eur Heart J (2012).

[91] Rohini, A, Agrawal, N, Koyani, C. N, & Singh, R. Molecular targets and regulators of cardiac hypertrophy. Pharmacol Res (2010). , 61(4), 269-80.

[92] Soor, G. S, Luk, A, Ahn, E, Abraham, J. R, Woo, A, Ralph-edwards, A, et al. Hypertrophic cardiomyopathy: current understanding and treatment objectives. J Clin Pathol (2009). , 62(3), 226-35.

[93] Maron, B. J, Gardin, J. M, Flack, J. M, Gidding, S. S, Kurosaki, T. T, & Bild, D. E. Prevalence of hypertrophic cardiomyopathy in a general population of young adults. Echocardiographic analysis of 4111 subjects in the CARDIA Study. Coronary Artery Risk Development in (Young) Adults. Circulation (1995). , 92(4), 785-9.

[94] Poliac, L. C, Barron, M. E, & Maron, B. J. Hypertrophic cardiomyopathy. Anesthesiology (2006). , 104(1), 183-92.

[95] Richard, P, Charron, P, Carrier, L, Ledeuil, C, Cheav, T, Pichereau, C, et al. Hypertrophic cardiomyopathy: distribution of disease genes, spectrum of mutations, and implications for a molecular diagnosis strategy. Circulation (2003). , 107(17), 2227-32.

[96] Lind, J. M, Chiu, C, & Semsarian, C. Genetic basis of hypertrophic cardiomyopathy. Expert Rev Cardiovasc Ther (2006). , 4(6), 927-34.

[97] Maron, M. S, Olivotto, I, Betocchi, S, Casey, S. A, Lesser, J. R, Losi, M. A, et al. Effect of left ventricular outflow tract obstruction on clinical outcome in hypertrophic cardiomyopathy. N Engl J Med (2003). , 348(4), 295-303.

[98] Lorell, B. H, & Carabello, B. A. Left ventricular hypertrophy: pathogenesis, detection, and prognosis. Circulation (2000). , 102(4), 470-9.

[99] Frey, N, Katus, H. A, Olson, E. N, & Hill, J. A. Hypertrophy of the heart: a new therapeutic target? Circulation (2004). , 109(13), 1580-9.

[100] Wolf, C. M, Moskowitz, I. P, Arno, S, Branco, D. M, Semsarian, C, Bernstein, S. A, et al. Somatic events modify hypertrophic cardiomyopathy pathology and link hypertrophy to arrhythmia. Proc Natl Acad Sci U S A (2005). , 102(50), 18123-8.

[101] Carvajal-vergara, X, Sevilla, A, Souza, D, Ang, S. L, Schaniel, Y. S, & Lee, C. DF, et al. Patient-specific induced pluripotent stem-cell-derived models of LEOPARD syndrome. Nature (2010). , 465(7299), 808-12.

[102] Sarkozy, A, Digilio, M. C, & Dallapiccola, B. Leopard syndrome. Orphanet J Rare Dis (2008).

http://dx.doi.org/10.5772/55773

[103] Soldner, F, Laganiere, J, Cheng, A. W, Hockemeyer, D, Gao, Q, Alagappan, R, et al. Generation of isogenic pluripotent stem cells differing exclusively at two early onset Parkinson point mutations. Cell (2011). , 146(2), 318-31.

[104] Hockemeyer, D, Wang, H, Kiani, S, Lai, C. S, Gao, Q, Cassady, J. P, et al. Genetic engineering of human pluripotent cells using TALE nucleases. Nat Biotechnol (2011). , 29(8), 731-4.

Chapter 20

Pluripotent Stem Cells to Model Human Cardiac Diseases

Calvin C. Sheng and Charles C. Hong

Additional information is available at the end of the chapter

http://dx.doi.org/10.5772/54373

1. Introduction

For past several decades, laboratory animal models have been the prevailing paradigm for studying human diseases. A classic approach is to study the impact of specific genes through the use of gain- or loss-of-function mutant animals. While the animal models have greatly contributed to our understanding of the etiology and mechanisms of disease, they often fall short of fully recapitulating human pathophysiology and translating to clinical applications due to interspecies physiologic differences. In a review of preclinical studies of animal models published in high-impact scientific journals, approximately one-third translated to the level of human randomized trials and only one-tenth were subsequently approved clinically for patient use [1]. This attrition rate would have been even higher if less frequently cited animal research had been included. These unresolved issues with animal models have set the stage for the emergence of human embryonic stem cell (hESC) and human induced pluripotent stem cell (hiPSC) for modeling human diseases.

Laid out in this chapter, we will discuss the development of various stem cell paradigms including mESC, hESC, and hiPSC (Figure 1); examine the utilization of these models via studies of cardiac diseases; assess the current limitations and future challenges; and finally conclude with the prospective outlook and viability of the field holistically in the scope of disease modeling.

2. Human cardiovascular diseases

According to the American Heart Association, cardiovascular diseases (CVD) remains the leading cause of deaths in United States, accounting for 32.8% of all deaths or roughly one of every three deaths [2]. To put into perspective, that is an average of 1 death every 39 seconds. CVD is a generic term that encompasses conditions that affect the circulatory system, including myocardial infarction, angina pectoris, heart failure, stroke, and congenital cardiovascular defects. Both genetic and environmental factors are implicated in the pathogenesis of CVDs. While some risk factors such as lifestyle habits and family history have been identified for CVDs, much more remains to be learned about the pathophysiology, optimal management, and proper prevention. Moreover, genetic predispositions like abnormalities in specific ion channels and sarcomere proteins pose special diagnostic and therapeutic challenges. In fact, for most heritable forms of heart diseases, current treatment options leave much to be desired.

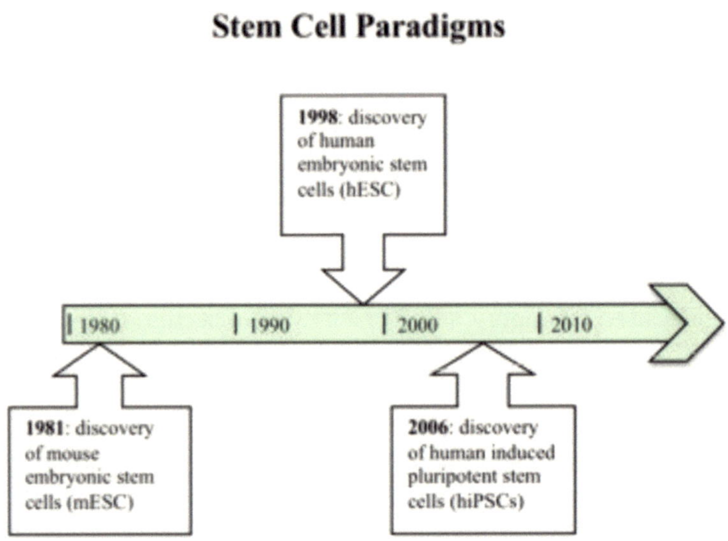

Figure 1. Timeline of stem cell modeling progress. Stem cell platforms are a new technology that was only introduced within the last two decades. The most recent breakthrough in hiPSC occurred just six years ago.

3. Stem cell disease modeling

Despite much progress in the past couple decades in the discovery of the molecular and genetic causes of many heart diseases, a detailed mechanic understanding of failing heart at the cellular level remains rudimentary. The main reason for this situation is the lack of access to live human tissues and unproven human cardiomyocyte cell culture models. Postpartum, cardiomyocytes become terminally differentiated and cease to proliferate, thus making isolation and culture of human myocardial cells extremely challenging. One surrogate for human cardiomyocyte culture is the use of rat neonatal cardiomyocytes, which has been shown to yield 8.4×10^6 cells per heart [3]. However, with both human and rat neonatal cardiomyocytes, the inability to continuously passage cells and scarcity of resource make them unsustainable candidates for disease modeling.

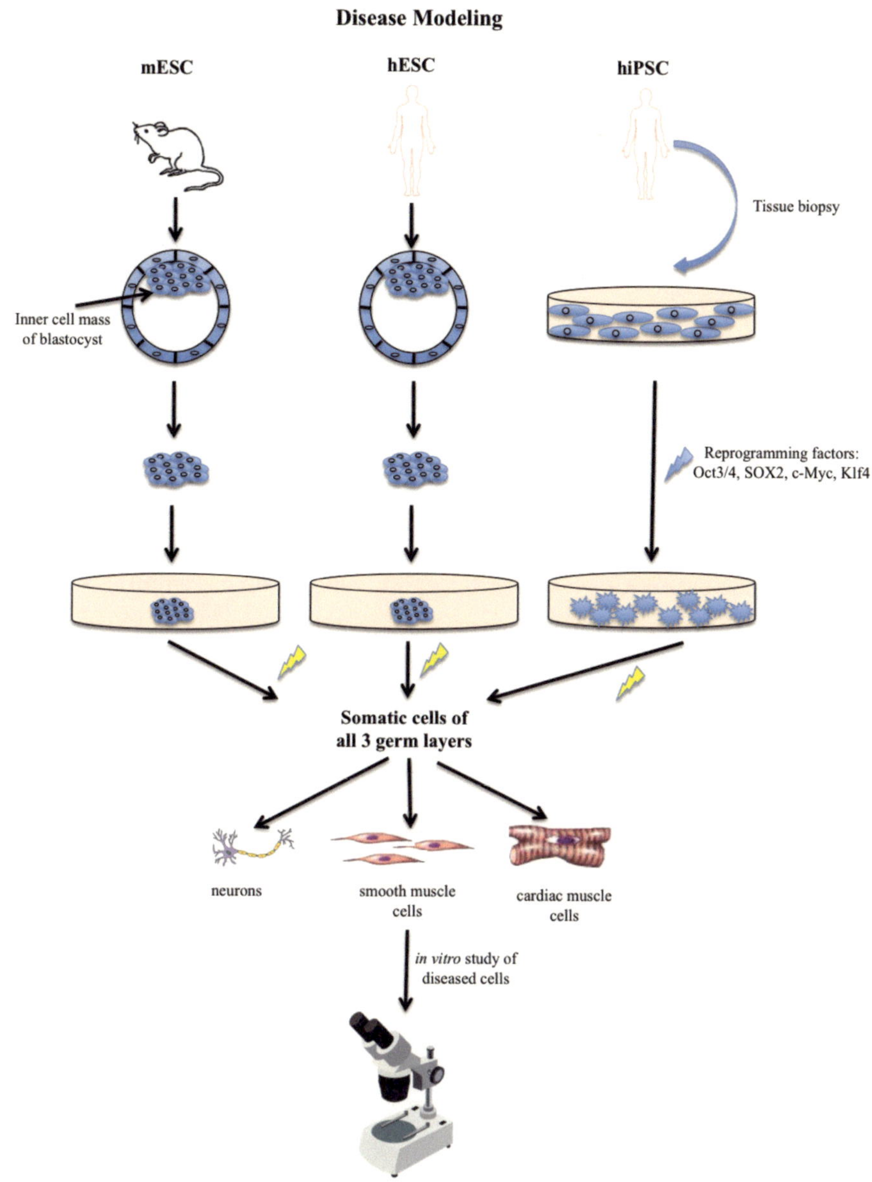

Figure 2. Overview of the stem cell disease modeling process. The blue and yellow lightning bolts indicate the addition of reprogramming and directed differentiation factors, respectively.

Furthermore, special considerations must be taken into account for critical differences between animal and human cardiomyocytes, in terms of cell biological, mechanical and electrophysiological properties. The lack of appropriate human heart disease models have hindered development of rational therapies, and the prospects for new therapies to treat heart diseases remain dim despite tremendous advances in various animal models. An alternative human biology based approach for heart disease modeling is to use human stem cells as a renewable source of cells for cardiomyocytes. In the following section, we will discuss the various stem cell platforms (mESC, hESC, & hiPSC) for disease modeling, with specific focus on cardiovascular diseases (Figure 2).

3.1. mESC paradigm

In 1981, the first pluripotent mouse embryonic stem cells (mESCs) were isolated *in vitro* by culturing the inner cell mass of pre-implantation mouse blastocysts [4, 5]. These cells were capable of self-renewal and pluripotent differentiation into all three germ layers (ectoderm, mesoderm, and endoderm) [6]. The initial studies demonstrated a proof of concept, showing the feasibility of isolating pluripotent cells directly from early embryos. The unique capability of culturing pluripotent cells *in vitro* provided the means for genetic manipulation via selection or transformation of specific DNA fragments, and importantly to develop genetic mouse models of human disease. This platform allowed researchers to begin exploring pathways in cardiac development to dissect underlying molecular and cellular mechanisms causing congenital defects and other abnormalities.

While the general use of mESCs was promising, inherent problems with using animal models remained in the context of studying disease pathogenesis and pathophysiology. One of the crucial points of divergence is the shear difference in size and complexity between humans and mice both macroscopically and genomically [7]. Consequently, disease susceptibility may vary drastically. For instance, a mouse heart is ten thousand times smaller but beats roughly seven times faster than that of a human. The two organisms also differ in their expression of myosin heavy chain (MHC) isoforms. βMHC is the predominant isoform in fetal mouse hearts, whereas mainly αMHC is expressed in adults; conversely, the vice versa is true for humans [8]. Furthermore, mice are resistant to the development of coronary atherosclerosis even on a high-fat, high-cholesterol diet, because they lack cholesteryl ester transfer protein (CETP), an enzyme responsible for the transfer of cholesterol from high-density lipoprotein to low-density lipoprotein [9].

3.2. hESC paradigm

Building on the initial discovery of mESC technologies, increased research focus has been directed towards developing a human-based stem cell approach in anticipation of creating a more accurate disease model. It would be another seventeen years before human embryonic stem cells (hESCs) derived from the inner cell mass of the human blastocyst (stage 4-5 days post-fertilization) were isolated by Thomson *et al.* in 1998 [10]. Many factors hindered the transition from mESC to hESC, such as the limited availability of surplus human embryos, stringent growth requirements for culturing hESC, and the shroud of ethical controversies. Generating hESCs require the destruction of the donor embryo that is considered a potential human life by some ethical and religious groups. The debate revolving around hESC has deterred many researchers, mainly in the United States, from pursuit of this technology. In August of 2001, President Bush became the first President to provide federal funding for embryonic stem cell research, albeit limited to experimenting with only the 15 existing stem cell lines [11]. Nonetheless, this discovery paved the way for modeling diseases directly on a human-based paradigm.

In a study in 2009, Lu et al. evaluated long-term safety and function of retinal pigment epithelium (RPE) as preclinical models of macular degeneration using hESCs [10]. When hESC-induced RPE were subsequently transplanted into mutant mice, they demonstrated long-term functional rescue, though progressively deteriorating function was noted due to the immunogenic response elicited by the xenografts. The initial data showed promise for future elucidation of macular degenerative disease pathophysiology. However, there were important obstacles to widespread clinical translation. First, transplantation of hESC requires immunosuppression, since the cells are allogeneic. In addition, a well known risk of this technology is the formation of teratomas, tumor-like formations containing tissues belonging to all three germ layers, if some undifferentiated pluripotent cells are transplanted [12]. Finally, perhaps the biggest obstacle to a widespread acceptance of human ESC transplantation is ethical and religious, as derivation of human ESCs typically involves the consumption of a human embryo.

3.3. hiPSC paradigm

Given these obstacles to a widespread use of the human ESCs, a new stem cell technology, human induced pluripotent stem cells (hiPSCs), has rapidly overtaken hESC research. Introduced in 2006 by Takahashi and Yamanaka, hiPSCs have been hailed as "the molecular equivalent of the discovery of antibiotics or vaccines in the last century [13]." The technology revolutionized the stem cell field, and for his achievement, Yamanaka received the 2012 Nobel Prize in Medicine." In a span of just six years, the field has rapidly expanded the repertoire of reprogrammable terminally differentiated tissue into hiPSC (keratinocytes [14, 15], hepatocytes [16], adipose-derived stem cells [17, 18], neural stem cells [19], astrocytes [20], cord blood [21, 22], amniotic cells [23], peripheral blood [24, 25], mesenchymal stromal cells [26], oral mucosa fibroblasts [27], and T-cells [28]). Most recently, the ability to generate hiPSC from Epstein-Barr virus (EBV)-immortalized B cell lines (lymphoblastoid B-cell lines) provides the opportunity to obtain samples from disease cohort repositories such as the Coriell Institute for Medical Research or the UK Biobank [29, 30].

In parallel, tremendous progress has been made towards the directing differentiation of these hiPSCs into various cell fates (neural progenitors [31], [32] motor neurons [33] [34], dopaminergic neurons [35], retinal cells [36], hepatocytes [37], blood cells [25, 38], adipocytes [39], endothelial cells [37, 38], fibroblasts [40, 41], and cardiomyocytes [42]). In theory, these patient-derived hiPSCs should be capable of differentiating into all of the >210 adult cell lineages. Nonetheless, our current growing repertoire sets the stage for studying various disease mechanisms in the laboratory, with the caveat that monogenic diseases such as long-QT syndrome will be much easier to model than complex diseases like Parkinson's.

As alluded to above, the somatic cell reprogramming offers several distinct advantages over embryonic stem cells. In the U.S. particularly, funding may be scarce at times due to the government's political stance regarding stem cell research. Importantly, somatic cells can be obtained from individual patients, enabling the development of truly personalized diagnostics and therapeutics.

4. Modeling cardiovascular diseases

While there is a wide array of cardiovascular diseases, we chose to focus on several with well-defined clinical presentation, strong genetic component, and significant research progress (Long QT syndrome types 1 and 2, Timothy syndrome, LEOPARD syndrome, & dilated cardiomyopathy; see Table 1). As discussed below, the paradigm of using stem cells to model inherited cardiovascular diseases is rapidly being established and validated. Moreover, these advances with the rare inherited conditions may lead to new paradigms to study the much more prevalent acquired heart and vascular diseases at the cellular and molecular levels.

Genetic Disorder	Mutation	Main findings
Long QT syndrome Type 1		
Moretti et al. (2010) [48]	*KCNQ1* R190Q	marked prolonged action potentials; dominant negative trafficking defect associated with a 70 to 80% reduction in I_{ks} current; altered channel activation and deactivation properties; increased susceptibility to catecholamine-induced tachyarrhythmia attenuated by β-blockage
Long QT syndrome Type 2		
Lahti et al. (2012) [83]	*KCNH2* R176W	prolonged action potential; reduced I_{kr} density; more sensitive to potentially arrhythmogenic drugs; more pronounced inverse correlation between the beating rate and repolarization time
Itzhaki et al. (2011) [50]	*KCNH2* A614V	significant reduction of potassium current I_{Kr}; marked arrhythmogenecity; evaluated potency of existing & novel pharmacological agents
Matsa et al. (2011) [84]	*KCNH2* G1681A	prolonged field/action potential duration; I_{kr} blocker & isoprenaline induced arrhythmias presenting as early after depolarizations; attenuated by β-blockers propranolol & nadolol
Timothy syndrome		
Yazawa et al. (2011) [52]	*CACNA1C* G1216A	irregular contraction; excess Ca^{2+} influx; prolonged action potentials; irregular electrical activity; abnormal calcium transients in ventricular-like cells; roscovitine restored electrical and Ca^{2+} signaling properties
LEOPARD syndrome		
Carvajal-Vergara et al. (2010) [57]	*PTPN11* T468M	hypertrophic cardiomyopathy; higher degree of sarcomeric organization; preferential localization of NFATC4 in the nucleus
Dilated cardiomyopathy		
Sun et al. (2012) [67]	*TNNT2* R173W	altered regulation of Ca^{2+}; decreased contractility; abnormal distribution of sarcomeric α-actinin; β-drenergic agonist induced cellular stress; β-adrenergic blockers or overexpression of Serca2a improved function

Table 1. hiPSC studies modeling cardiovascular diseases.

4.1. Long QT syndrome

Long-QT syndrome (LQTS) is a rare congenital channelopathy disease that is characterized by an abnormally prolonged ventricular repolarization phase, inherited primarily in an autosomal dominant manner but sometimes autosomal recessively. It was first described in 1957 in a family with normal parents and two healthy children but also in which three children experienced recurrent syncope and sudden death [43]. Electrocardiography (EKG) studies showed prolonged QT interval due to increased ventricular action potential, hence the name of the disease (Figure 3). The prevalence of LQTS in the U.S. is approximately 1 in 7,000 individuals, causing 2,000 to 3,000 sudden deaths annually in children or adolescents [44]. This abnormality can lead to an increased risk of such reported incidence of sudden death, usually triggered by the resulting ventricular fibrillation or torsade de pointes (polymorphic ventricular tachycardia). Depending on the specific gene mutation, long-QT syndrome can be classified into 12 genetic subtypes [45]. Together, LQT1, LQT2, and LQT3 genotypes account for 97% of the mutations identified to date [46].

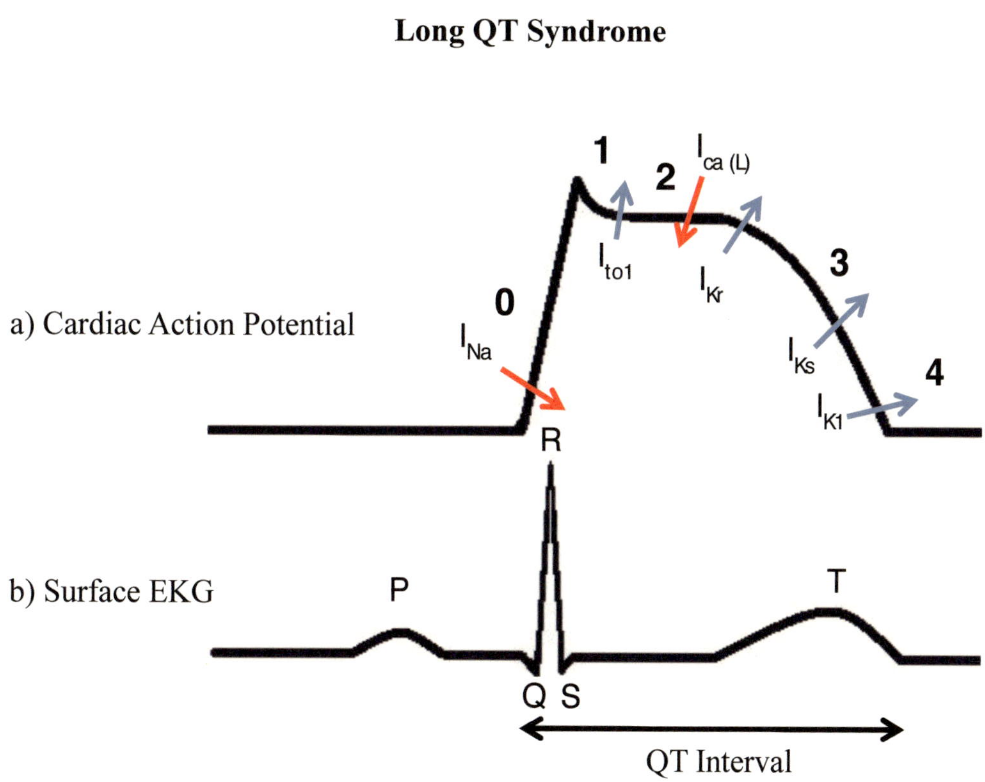

Figure 3. Long QT Syndrome. a) a visual representation of the cardiac action potential during depolarization and repolarization of the cell. There are 4 phases of the cycle in which various ion channels open and close, causing the flux of charged ions (red: into the cell & blue: out of the cell) and reflecting the change in overall action potential. **b)** an illustration of a normal surface EKG plot, highlighting the QT interval in particular. In long QT syndrome, a clear indication is the prolongation of that interval on an EKG.

Our current understanding of how mutations in ion channels cause disease can only be extrapolated from, at best, mammalian cell lines such as immortalized human embryonic kidney 293 cells or *Xenopus* oocytes using heterologous expression systems designed with the mutant channel of interest [47]. Commonly used mouse models are not apt for studying LQTS because the I_{Kr} current is essentially absent in the mouse heart. With the advent of patient-derived iPSC technology, cardiac induction of these cell lines may recapitulate their respective disease pathophysiology *in vitro*, providing a unique platform for studying cellular and molecular mechanisms and assessing the efficacy of various therapies.

4.1.1. Long QT syndrome type 1

The most common type LQT1, accounting for roughly 45% of genotyped patients, results from mutations of the alpha subunit of the slow delayed rectifier potassium channel KvLQT1, encoded by gene KCNQ1 on chromosome 11 [48]. In a recent study aimed at recapitulating disease phenotype using patient-derived iPSCs, Moretti et al. initially screened a family affected by LQTS type 1 through genotyping and electrophysiology studies, identifying an autosomal dominant missense mutation R190Q of KCNQ1 [48]. Then, they reprogrammed skin fibroblast from two affected family members into iPSCs and directed cardiac induction to yield spontaneously beating cardiomyocytes. Finally, they characterized these heart cells through whole-cell patch clamp, observing reduced I_{ks}, a slow delayed rectifier potassium current, by 70-80%, altered I_{ks} activation and deactivation properties, and an abnormal response to catecholamine stimulation.

Not only were Moretti et al. able to capture characteristics of LQTS type 1 *in vitro*, they were also able to demonstrate physiologically how beta-blockers, clinically administered as a prophylactic therapy for asymptomatic LQTS type 1 patients, had protective effects against catecholamine-induced tachyarrhythmia by reducing early afterdepolarizations [49]. This ability to mimic LQTS type 1 in an *in vitro* human model paved way for similar studies involving other genetic diseases.

4.1.2. Long QT syndrome type 2

Similar to LQTS type 1, LQTS type 2 is another mutation arising from the alpha subunit of a potassium channel, but one with different properties: a KCNH2 (also known as hERG)-encoded rapid delayed rectifier potassium channel [50]. A diagnostic finding in patients is the onset of clinical symptoms such as syncope triggered by sudden loud noises [45].

In a study by Itzhaki et al., A614V missense mutation was identified in the KCNH2 gene in a 28 year old patient with clinically diagnosed type 2 LQTS [50]. Dermal fibroblast samples were obtained, reprogrammed to generate patient-specific human iPSCs, and through retroviral transduction, differentiated into embryoid bodies of spontaneously beating cardiomyocytes. Through the use of these iPSC-generated heart cells, they were able to conduct electrophysiology studies and test the effects of pharmacological intervention. Itzhaki et al. found marked prolonged action potential duration and significantly reduced peak ampli-

http://dx.doi.org/10.5772/54373

tudes of I_{Kr} activation and tail currents in the cells derived from the LQTS patient compared to those generated from a healthy individual, both hallmark signatures of LQTS. They also reported observing early-after depolarizations in 66% of the iPSC-CMs on both cellular and multicellular levels, a key finding suggestive of arrhythmogenicity that explains sudden death in LQTS patients clinically. With the amount of clinical evidence extracted from these patient-derived cardiomocytes, this novel technology can serve as an excellent *in vitro* disease model for understanding cellular & molecular pathogenesis and becomes a very viable option for personalized medicine in the future.

4.2. Timothy syndrome

In contrast to the previously detailed potassium channel defects that lead to LQTS, Timothy syndrome is a form of LQTS caused by a missense mutation in the L-type calcium channel $Ca_v1.2$, encoded by the CACNA1C gene. This is the predominant L-type channel in the mammalian heart, which is essential for normal heart development and excitation-contraction coupling [51]. Mutations in this Ca^{2+} channel cause delayed channel closing and consequently, increased cellular excitability.

Concurrent with Itzhaki et al.'s publication LQTS type 2, Yazawa et al. reported their findings on Timothy syndrome using a patient-derived iPSC-CM disease model [52]. To summarize, using a similar cardiac induction protocol, they successfully reproduced *in vitro* cardiomyocytes exhibiting clinical Timothy syndrome phenotypes. Electrophysiology and calcium imaging studies showed irregular contraction, excess Ca^{2+} influx, prolonged action potentials, irregular electrical activity, and abnormal calcium transients in ventricular-like cells.

One of the key findings in their study was the functional difference between Timothy syndrome and LQTS type 1 cardiomyocytes. Unlike the latter where both ventricular- and atrial-like cells had prolonged action potentials, only ventricular Timothy syndrome cardiomyocytes exhibited this phenotype. Additionally, drug-induced triggering of arrhythmias and delayed depolarizations in LQTS type 1 cells were not necessary, because they were observed spontaneously in Timothy syndrome cells. While a direct correlation has yet to be established to the clinical outcomes (i.e. torsades de points and ventricular fibrillation), this study is another proof-of-concept that iPSC-CMs are invaluable for examining detailed pathogenesis of human diseases.

4.3. LEOPARD syndrome

LEOPARD syndrome is an autosomal-dominant developmental disorder with clinical manifestations described by its acronym: lentigines, electrocardiographic abnormalities, ocular hypertelorism, pulmonary valve stenosis, abnormal genitalia, retardation of growth, and deafness [53]. It is caused by a mutation in the PTPN11 gene, which impairs the catalytic region of the encoded SHP2 phosphatase [54]. Currently, drosophila [55] and zebrafish [56] models of LEOPARD syndrome have been described in literature, but the molecular basis of pathogenesis remains to be addressed.

In 2010, Carvajal-Vergara et al. successfully demonstrated the use of iPSC technology to characterize LEOPARD syndrome *in vitro* [57]. One of the clinical hallmarks of the disease is hypertrophic cardiomyopathy. In this study, iPSC-CMs derived from a 25-year old female patient with the condition were compared to those differentiated a healthy brother as a control. Carvajal-Vergara et al. showed, by comparison to the wild-type, larger patient-derived iPSC-CMs with a higher degree of sarcomeric organization and preferential localization of NFATC4 (calcineurin-NFAT pathway is an important regulator of cardiac hypertrophy [58]) in the nucleus [57]. Using antibody microarrays on patient-specific iPSCs, they also noted increased phosphorylation of certain proteins such as MEK1 leading to perturbations in the RAS-MAPK signaling cascade, which can begin to provide some preliminary understanding and elucidation of LEOPARD syndrome's pathogenesis on a molecular level [57].

4.4. Dilated cardiomyopathy

As previously mentioned, cardiovascular disease is the leading cause of morbidity and mortality worldwide, projected to represent 30% of all deaths in 2015 [59]. In the United States alone, heart disease accounts for roughly one-third of all deaths [60]. Of those, dilated cardiomyopathy (DCM) is one of the leading causes of heart failure and is associated with substantial mortality [61]. It leads to progressive cardiac remodeling, characterized by ventricular dilatation, hypertrophy, and systolic dysfunction [62]. In an estimated 20% to 48% of cases depending on the study, DCM is identified as a familial disorder with strong heritability [63]. Mutations in over 30 genes have been shown to be disease causing or disease associated [64].

One of the more common genetic defects causing DCM is a mutation in the cardiac troponin T gene (TNNT2) [65]. Mouse models have already provided invaluable insight to the disease mechanism. For instance, mice still displayed normal phenotype after knockout of one TNNT2 allele, because they only lead to a mild deficit in transcript but not protein [66]. Furthermore, the severity of DCM depends on the ratio of mutant to wild-type TNNT2 transcript, since mutant protein is associated with cardiomyocyte Ca^{2+} desensitization [66]. However, given the differences in electrophysiological and developmental properties, *in vitro* human models of DCM would conceivably provide a more precise platform for understanding molecular basis of pathogenesis.

In Sun et al.'s study published in 2012, they characterized iPSC-CMs from a family carrying a point mutation (R173W) in the TNNT2 gene by comparing to healthy individuals in the same cohort [67]. These patient-specific cardiomyocytes from diseased individuals exhibited dysregulated calcium handling, decreased contractility, and abnormal heterogenous distribution of sarcomeric α-actinin. The overexpression of Serca2a, a gene therapy treatment for heart failure currently in clinical trials [68], significantly improved the contractility force generated by iPSC-CMs derived from DCM patients [67]. Much like the use of hiPSC technology for other cardiovascular diseases discussed previously, it appears to be a robust system for describing pathogenesis of disease that has yielded preliminary positive results.

5. Stem cell disease modeling challenges

In the framework of disease modeling, both hESC and hiPSC technologies still have unresolved issues to address. For instance, hESCs display chromosomal instability with long-term *in vitro* culture [69], and iPSCs undergo dynamic changes in copy number variations during reprogramming, especially in the early passages [70]. In the U.S., research funding for hESC often fluctuates, subjecting to restrictions imposed by Congress and its current stance on the destruction of fertilized human embryos. The advantage of hiPSC over hESC is that it bypasses this controversy and generates autologous cells while maintaining key characteristics: morphology, feeder dependency, surface markers, gene expression, promoter methylation status, telomerase activities, *in vitro* differentiation potential, and *in vivo* teratoma forming capacity [71]. These features heavily favor hiPSC technology as the predominant approach for disease modeling over hESCs.

In the near future, the hiPSC model faces several main challenges. One of the concerns is developing a robust and efficient methodology for yielding large quantities of differentiated and functional cells of a designated lineage. Depending on the protocol and cell lines used, efficiencies can range anywhere from <0.0001% to >50%. Specifically in the case of cardiac induction, the hiPSC-induced cardiomyocytes resemble immature fetal cardiomyocytes in their gene expression profile (key marker is β-tubulin) as well as electrophysiologic and structural properties [72]. Resolving this hindrance will also have great impact on facilitating *in vivo* studies and widespread applications in drug discovery and development.

The practicality of studying disease pathogenesis *in vitro*, especially those with systemic involvement, raises another question. This intrinsic lack of an *in vivo* environment prevents a global understanding of how a disease may impact the body and simplifies interactions of basic signaling pathways. For more complex diseases, it may also be difficult to replicate conditions in a petri dish with a single lineage cell type, even if done via co-cultures. Furthermore, the current designation of a control line is arbitrary since it is mainly a criterion of exclusions. In diseases such as Alzheimer's or Parkinson's, there is a long latency period, which would be hard to mimic *in vitro* due to the dynamics of real-time disease progression. Studies are currently underway to assess the possibility of accelerating disease progression *in vitro* via exposure to environmental factors contributing to the disease such as oxidative stress [73].

Finally, not all diseases can be readily modeled using hiPSCs. For example, patients with Fanconi anemia have a defective DNA repair mechanism, and therefore cannot be reprogrammed without antecedent gene correction [74]. For other conditions, some may exhibit low penetrance or do not follow a simple Mendelian form of inheritance and are affected by a multitude of factors ranging from the environment to epigenetics. The latter in diseased state may become an inevitable confounding factor working with iPSCs, because of its contribution to the low efficiency of reprogramming and its stochastic nature. In a study by Meissner et al., sub-clone lineages transfected with an Oct4-GFP reporter were obtained from early appearing iPSC colonies and displayed temporally different expression patterns of GFP, some never expressing it at all [75]. Because of the sensitivity to epigenetic events, the use of histone deacetylase (HDAC) inhibitors may help promote self-renewal and/or directed differentiation of stem cells [76].

6. Future outlook & research direction

The intent of stem cell technology was to recapitulate, as closely as possible, disease phenotype in the human body for three primary outcomes: disease modeling, drug discovery & development, and regenerative medicine. The first of which will provide the initial platform from which drugs and therapeutic applications can be derived. In some cases, a treatment could be discovered before the underlying disease mechanism is understood, because patient-derived hiPSCs can be differentiated without any genetic modifications *in vitro* into the desired cell type and characterized in drug screenings.

In the context of patient-derived cardiomyocytes, while not a perfect *in vivo* surrogate, they will still be one of the better models currently available due to their identical genomes and phenotype. The complex interactions of normal human physiology is incredibly difficult to mimic outside the host, let alone recapitulating a diseased phenotype. The mouse model is currently the most common mammalian system used to study human physiology for several reasons: 90% genetic homology with comparable genomic sizes, relatively easy maintenance, rapid cost-effective breeding under laboratory settings, and capability for genetic manipulation. It is great for initial studies and insight into basic understanding and elucidation of the mechanisms underlying the disease.

Building on the gradual advancement from mESC to hESC to the current hiPSC technology, one of the technical goals remains to be removing all extrinsic factors with the goal of mimicking *in vivo* conditions. Most established mESC and hESC protocols relied on a fibroblast feeder-cell layer for culture and proliferation, which secrete undefined substrates into the medium and cause batch-to-batch variation [77]. Similarly, initial hiPSC protocols used a mouse embryonic fibroblast (MEF) feeder-cell layer that had similar problems [78]. In 2011, Yu et al. developed a feeder-free system with chemically defined medium and also replaced conventional transfection of somatic cells with footprint-less episomal reprogramming using small molecules to generate hiPSCs [79].

Furthermore, mESC and hESC-directed differentiation formed embryoid bodies (EBs), which are spheroids with an inner layer of ectoderm and a single outer layer of endoderm. These EBs differentiate to derivatives of all 3 primary germ layers, leaving a very low yield of spontaneously contracting cardiomyocytes. While this was sufficient for initial studies, larger quantities of pure cardiomyocytes are necessary to establish a scalable system for disease modeling and drug development. In 2007, Laflamme et al. reported the use of a monolayer cardiac induction system based on activin A and BMP4 with a 30-fold higher yield of pure cardiomyocytes than through the formation of EBs [80]. Most recently, Lian et al. of the Wisconsin stem cell group identified temporal modulation of canonical Wnt signaling as a key step for robust cardiomyocyte differentiation reporting efficient yields of up to 98% [81].

Further studies are needed to evaluate the optimal cardiac induction protocol. Once a robust, universal, and scalable system for directed differentiation of iPSCs into cardiomyocytes is established, we can provide an inexhaustible supply of patient-derived cells for research and therapeutic purposes.

6.1. Zinc finger nucleases

With some host-specific modifications, currently available technologies such as zinc finger nucleases can be applied as the next step in disease modeling after understanding the pathogenesis, developing a cure. Zinc finger nucleases are enzymes that manipulate specific sites of the host genome, generating transgenic lines via knocking-in and knocking-out of genes. The homologous recombination pathway, naturally occurring at DNA replication forks and repairing double stranded breaks, can be exploited to selectively target a locus for modification while leaving the rest of the genome in tact [82]. Through this method, we have been able to identify new gene function in mouse and other homologous mammalian models. The same concept can be applied to gene therapy for humans. For example, with patient-specific cardiomyocytes, constructs can be created and tested *in vitro* to restore wild-type function.

6.2. High-throughput screening

High-throughput screening is another means of advancing disease therapy, but it hinges on its scalability; in other words, whether or not cells of the disease model can be mass-produced. With current protocols for directed cardiac differentiation, every round of experiments would take at least 2 weeks [81]. If hiPSC-derived cardiomyocytes could be consistently generated in 96-well plates, then these high-throughput screenings that could propel translational research from a cellular and molecular level of disease directly to therapeutic applications.

7. Conclusion

In the new era of personalized medicine, the stem cell platform for disease modeling appears very viable, especially given the rapid advancements in the field over the past several decades. We have thoroughly discussed the advantages and disadvantages of using mESC, hESC, and hiPSC, all of which have the common end goal of best recapitulating disease phenotype *in vitro*. Of those, we strongly believe that hiPSC-derived cells can eventually be the gold standard for personalized medicine. Using a heritable cardiovascular class of diseases as an example, we endeavored to convey the potential benefits of harnessing iPSC technology to study the pathogenesis of various disorders. One of the most difficult challenges currently is establishing a robust, universal, and scalable cardiac induction protocol. Combined with the genetic tools available, we will be able to break the barriers to disease modeling with the limitless supply of human cells *in vitro*.

Acknowledgements

CCH was funded by the Veterans Affairs VA Merit Award BX000771, NIH grants 5U01HL100398 and 1R01HL104040. CCS was funded by NIH NRSA 5T35HL090555.

Author details

Calvin C. Sheng[1] and Charles C. Hong[1,2*]

*Address all correspondence to: charles.c.hong@vanderbilt.edu

1 Division of Cardiovascular Medicine, Center for Inherited Heart Disease, Department of Cell and Developmental Biology, Department of Pharmacology, Vanderbilt University School of Medicine, Nashville, USA

2 Research Medicine, Veterans Affairs TVHS, Nashville, USA

References

[1] Hackam, D.G. and D.A. Redelmeier, *Translation of research evidence from animals to humans.* JAMA: the journal of the American Medical Association, 2006. 296(14): p. 1731-1732.

[2] Members, W.G., et al., *Heart Disease and Stroke Statistics,Äî2012 Update.* Circulation, 2012. 125(1): p. e2-e220.

[3] Fu, J., et al., *An optimized protocol for culture of cardiomyocyte from neonatal rat.* Cytotechnology, 2005. 49(2): p. 109-116.

[4] Martin, G.R., *Isolation of a pluripotent cell line from early mouse embryos cultured in medium conditioned by teratocarcinoma stem cells.* Proceedings of the National Academy of Sciences, 1981. 78(12): p. 7634.

[5] Evans, M.J. and M.H. Kaufman, *Establishment in culture of pluripotential cells from mouse embryos.* Nature, 1981. 292(5819): p. 154-156.

[6] Urbach, A., M. Schuldiner, and N. Benvenisty, *Modeling for Lesch,ÄêNyhan Disease by Gene Targeting in Human Embryonic Stem Cells.* Stem Cells, 2004. 22(4): p. 635-641.

[7] Musunuru, K., I.J. Domian, and K.R. Chien, *Stem cell models of cardiac development and disease.* Annual review of cell and developmental biology, 2010. 26: p. 667-687.

[8] Krenz, M. and J. Robbins, *Impact of beta-myosin heavy chain expression on cardiac function during stress.* Journal of the American College of Cardiology, 2004. 44(12): p. 2390-2397.

[9] Thomson, J.A., et al., *Embryonic stem cell lines derived from human blastocysts.* science, 1998. 282(5391): p. 1145-1147.

[10] Lu, B., et al., *Long,ÄêTerm Safety and Function of RPE from Human Embryonic Stem Cells in Preclinical Models of Macular Degeneration.* Stem Cells, 2009. 27(9): p. 2126-2135.

[11] Walters, L.R., *Human embryonic stem cell research: an intercultural perspective.* Kennedy Institute of Ethics Journal, 2004. 14(1): p. 3-38.

[12] Zhang, Q., et al., *Stem cells and cardiovascular tissue repair: Mechanism, methods, and clinical applications.* Journal of Cardiothoracic Renal Research, 2006. 1(1): p. 3-14.

[13] Wu, S.M. and K. Hochedlinger, *Harnessing the potential of induced pluripotent stem cells for regenerative medicine.* Nature cell biology, 2011. 13(5): p. 497-505.

[14] Aasen, T., et al., *Efficient and rapid generation of induced pluripotent stem cells from human keratinocytes.* Nature biotechnology, 2008. 26(11): p. 1276-1284.

[15] Carey, B.W., et al., *Reprogramming of murine and human somatic cells using a single polycistronic vector.* Proceedings of the National Academy of Sciences, 2009. 106(1): p. 157-162.

[16] Liu, H., et al., *Generation of endoderm,Äêderived human induced pluripotent stem cells from primary hepatocytes.* Hepatology, 2010. 51(5): p. 1810-1819.

[17] Sun, N., et al., *Feeder-free derivation of induced pluripotent stem cells from adult human adipose stem cells.* Proceedings of the National Academy of Sciences, 2009. 106(37): p. 15720-15725.

[18] Aoki, T., et al., *Generation of induced pluripotent stem cells from human adipose-derived stem cells without c-MYC.* Tissue Engineering Part A, 2010. 16(7): p. 2197-2206.

[19] Kim, J.B., et al., *Direct reprogramming of human neural stem cells by OCT4.* Nature, 2009. 461(7264): p. 649-643.

[20] Ruiz, S., et al., *High-efficient generation of induced pluripotent stem cells from human astrocytes.* PLoS One, 2010. 5(12): p. e15526.

[21] Haase, A., et al., *Generation of induced pluripotent stem cells from human cord blood.* Cell Stem Cell, 2009. 5(4): p. 434-441.

[22] Giorgetti, A., et al., *Generation of induced pluripotent stem cells from human cord blood using OCT4 and SOX2.* Cell Stem Cell, 2009. 5(4): p. 353.

[23] Li, C., et al., *Pluripotency can be rapidly and efficiently induced in human amniotic fluid-derived cells.* Human molecular genetics, 2009. 18(22): p. 4340-4349.

[24] Loh, Y.H., et al., *Generation of induced pluripotent stem cells from human blood.* Blood, 2009. 113(22): p. 5476-5479.

[25] Ye, Z., et al., *Human-induced pluripotent stem cells from blood cells of healthy donors and patients with acquired blood disorders.* Blood, 2009. 114(27): p. 5473-5480.

[26] Oda, Y., et al., *Induction of pluripotent stem cells from human third molar mesenchymal stromal cells.* Journal of Biological Chemistry, 2010. 285(38): p. 29270-29278.

[27] Miyoshi, K., et al., *Generation of human induced pluripotent stem cells from oral mucosa.* Journal of bioscience and bioengineering, 2010. 110(3): p. 345-350.

[28] Seki, T., et al., *Generation of induced pluripotent stem cells from human terminally differentiated circulating T cells.* Cell Stem Cell, 2010. 7(1): p. 11.

[29] Rajesh, D., et al., *Human lymphoblastoid B-cell lines reprogrammed to EBV-free induced pluripotent stem cells.* Blood, 2011. 118(7): p. 1797-1800.

[30] Choi, S.M., et al., *Reprogramming of EBV-immortalized B-lymphocyte cell lines into induced pluripotent stem cells.* Blood, 2011. 118(7): p. 1801-1805.

[31] Chambers, S.M., et al., *Highly efficient neural conversion of human ES and iPS cells by dual inhibition of SMAD signaling.* Nature biotechnology, 2009. 27(3): p. 275-280.

[32] Neely, M.D., et al., *DMH1, a highly selective small molecule BMP inhibitor promotes neurogenesis of hiPSCs: Comparison of PAX6 and SOX1 expression during neural induction.*

[33] Dimos, J.T., et al., *Induced pluripotent stem cells generated from patients with ALS can be differentiated into motor neurons.* science, 2008. 321(5893): p. 1218-1221.

[34] Ebert, A.D., et al., *Induced pluripotent stem cells from a spinal muscular atrophy patient.* Nature, 2008. 457(7227): p. 277-280.

[35] Soldner, F., et al., *Parkinson's disease patient-derived induced pluripotent stem cells free of viral reprogramming factors.* Cell, 2009. 136(5): p. 964-977.

[36] Osakada, F., et al., *In vitro differentiation of retinal cells from human pluripotent stem cells by small-molecule induction.* Journal of cell science, 2009. 122(17): p. 3169-3179.

[37] Sullivan, G.J., et al., *Generation of functional human hepatic endoderm from human induced pluripotent stem cells.* Hepatology, 2010. 51(1): p. 329-335.

[38] Choi, K.D., et al., *Hematopoietic and endothelial differentiation of human induced pluripotent stem cells.* Stem Cells, 2009. 27(3): p. 559-567.

[39] Taura, D., et al., *Adipogenic differentiation of human induced pluripotent stem cells: comparison with that of human embryonic stem cells.* FEBS letters, 2009. 583(6): p. 1029-1033.

[40] Hockemeyer, D., et al., *A drug-inducible system for direct reprogramming of human somatic cells to pluripotency.* Cell Stem Cell, 2008. 3(3): p. 346-353.

[41] Maherali, N., et al., *A high-efficiency system for the generation and study of human induced pluripotent stem cells.* Cell Stem Cell, 2008. 3(3): p. 340-345.

[42] Uosaki, H., et al., *Efficient and scalable purification of cardiomyocytes from human embryonic and induced pluripotent stem cells by VCAM1 surface expression.* PLoS One, 2011. 6(8): p. e23657.

[43] Jervell, A. and F. Lange-Nielsen, *Congenital deaf-mutism, functional heart disease with prolongation of the QT interval, and sudden death.* American heart journal, 1957. 54(1): p. 59-68.

[44] Vincent, G.M., *The long QT syndrome.* Indian pacing and electrophysiology journal, 2002. 2(4): p. 127.

[45] Roden, D.M., *Long-QT Syndrome.* New England Journal of Medicine, 2008. 358(2): p. 169-176.

[46] Goldenberg, I., W. Zareba, and A.J. Moss, *Long QT syndrome.* Current problems in cardiology, 2008. 33(11): p. 629-694.

[47] Kamp, T.J., *An electrifying iPSC disease model: long QT syndrome type 2 and heart cells in a dish.* Cell Stem Cell, 2011. 8(2): p. 130-131.

[48] Moretti, A., et al., *Patient-specific induced pluripotent stem-cell models for long-QT syndrome.* New England Journal of Medicine, 2010. 363(15): p. 1397-1409.

[49] Vincent, G.M., et al., *High Efficacy of Œ≤-Blockers in Long-QT Syndrome Type 1.* Circulation, 2009. 119(2): p. 215-221.

[50] Itzhaki, I., et al., *Modelling the long QT syndrome with induced pluripotent stem cells.* Nature, 2011. 471(7337): p. 225-229.

[51] Xu, M., et al., *Enhanced expression of L-type Cav1. 3 calcium channels in murine embryonic hearts from Cav1. 2-deficient mice.* Journal of Biological Chemistry, 2003. 278(42): p. 40837-40841.

[52] Yazawa, M., et al., *Using induced pluripotent stem cells to investigate cardiac phenotypes in Timothy syndrome.* Nature, 2011. 471(7337): p. 230-234.

[53] Legius, E., et al., *PTPN11 mutations in LEOPARD syndrome.* Journal of medical genetics, 2002. 39(8): p. 571-574.

[54] Kontaridis, M.I., et al., *PTPN11 (Shp2) mutations in LEOPARD syndrome have dominant negative, not activating, effects.* Journal of Biological Chemistry, 2006. 281(10): p. 6785-6792.

[55] Oishi, K., et al., *Phosphatase-defective LEOPARD syndrome mutations in PTPN11 gene have gain-of-function effects during Drosophila development.* Human molecular genetics, 2009. 18(1): p. 193-201.

[56] Jopling, C., D. Van Geemen, and J. Den Hertog, *Shp2 knockdown and Noonan/LEOPARD mutant Shp2,Äìinduced gastrulation defects.* PLoS genetics, 2007. 3(12): p. e225.

[57] Carvajal-Vergara, X., et al., *Patient-specific induced pluripotent stem-cell-derived models of LEOPARD syndrome.* Nature, 2010. 465(7299): p. 808-812.

[58] Heineke, J. and J.D. Molkentin, *Regulation of cardiac hypertrophy by intracellular signalling pathways.* Nature Reviews Molecular Cell Biology, 2006. 7(8): p. 589-600.

[59] Fuster, V. and B.B. Kelly, *Epidemiology of Cardiovascular Disease.* 2010.

[60] Roger, V.L., et al., *Heart Disease and Stroke Statistics--2012 Update: A Report From the American Heart Association.* Circulation, 2011.

[61] Refaat, M.M., et al., *Genetic Variation in the Alternative Splicing Regulator, RBM20, is associated with Dilated Cardiomyopathy.* Heart Rhythm, 2011.

[62] Maron, B.J., et al., *Contemporary definitions and classification of the cardiomyopathies an American heart association scientific statement from the council on clinical cardiology, heart failure and transplantation committee; quality of care and outcomes research and functional genomics and translational biology interdisciplinary working groups; and council on epidemiology and prevention.* Circulation, 2006. 113(14): p. 1807-1816.

[63] Brauch, K.M., et al., *Mutations in ribonucleic acid binding protein gene cause familial dilated cardiomyopathy.* Journal of the American College of Cardiology, 2009. 54(10): p. 930-941.

[64] Li, D., et al., *Identification of novel mutations in RBM20 in patients with dilated cardiomyopathy.* Clinical and translational science, 2010. 3(3): p. 90-97.

[65] Willott, R.H., et al., *Mutations in Troponin that cause HCM, DCM AND RCM: What can we learn about thin filament function?* Journal of molecular and cellular cardiology, 2010. 48(5): p. 882-892.

[66] Ahmad, F., et al., *The role of cardiac troponin T quantity and function in cardiac development and dilated cardiomyopathy.* PLoS One, 2008. 3(7): p. e2642.

[67] Sun, N., et al., *Patient-Specific Induced Pluripotent Stem Cells as a Model for Familial Dilated Cardiomyopathy.* Science Translational Medicine, 2012. 4(130): p. 130ra47-130ra47.

[68] Jessup, M., et al., *Calcium Upregulation by Percutaneous Administration of Gene Therapy in Cardiac Disease (CUPID) Clinical Perspective A Phase 2 Trial of Intracoronary Gene Therapy of Sarcoplasmic Reticulum Ca2+-ATPase in Patients With Advanced Heart Failure.* Circulation, 2011. 124(3): p. 304-313.

[69] Hanson, C. and G. Caisander, *Human embryonic stem cells and chromosome stability.* Apmis, 2005. 113(11,Äê12): p. 751-755.

[70] Hussein, S.M., et al., *Copy number variation and selection during reprogramming to pluripotency.* Nature, 2011. 471(7336): p. 58-62.

[71] Takahashi, K., et al., *Induction of pluripotent stem cells from adult human fibroblasts by defined factors.* Cell, 2007. 131(5): p. 861-872.

[72] Josowitz, R., et al., *Induced pluripotent stem cell-derived cardiomyocytes as models for genetic cardiovascular disorders.* Current Opinion in Cardiology, 2011. 26(3): p. 223.

[73] Saha, K. and R. Jaenisch, *Technical challenges in using human induced pluripotent stem cells to model disease.* Cell Stem Cell, 2009. 5(6): p. 584.

[74] Raya, Å., et al., *Disease-corrected haematopoietic progenitors from Fanconi anaemia induced pluripotent stem cells.* Nature, 2009. 460(7251): p. 53-59.

[75] Meissner, A., M. Wernig, and R. Jaenisch, *Direct reprogramming of genetically unmodified fibroblasts into pluripotent stem cells.* Nature biotechnology, 2007. 25(10): p. 1177-1181.

[76] Kretsovali, A., C. Hadjimichael, and N. Charmpilas, *Histone deacetylase inhibitors in cell pluripotency, differentiation, and reprogramming.* Stem Cells International, 2012. 2012.

[77] Boheler, K.R., et al., *Differentiation of pluripotent embryonic stem cells into cardiomyocytes.* Circulation research, 2002. 91(3): p. 189-201.

[78] Zhang, J., et al., *Functional cardiomyocytes derived from human induced pluripotent stem cells.* Circulation research, 2009. 104(4): p. e30-e41.

[79] Yu, J., et al., *Efficient feeder-free episomal reprogramming with small molecules.* PLoS One, 2011. 6(3): p. e17557.

[80] Laflamme, M.A., et al., *Cardiomyocytes derived from human embryonic stem cells in pro-survival factors enhance function of infarcted rat hearts.* Nature biotechnology, 2007. 25(9): p. 1015-1024.

[81] Lian, X., et al., *Robust cardiomyocyte differentiation from human pluripotent stem cells via temporal modulation of canonical Wnt signaling.* Proceedings of the National Academy of Sciences, 2012. 109(27): p. E1848-E1857.

[82] Lombardo, A., et al., *Gene editing in human stem cells using zinc finger nucleases and integrase-defective lentiviral vector delivery.* Nature biotechnology, 2007. 25(11): p. 1298-1306.

[83] Lahti, A.L., et al., *Model for long QT syndrome type 2 using human iPS cells demonstrates arrhythmogenic characteristics in cell culture.* Disease Models & Mechanisms, 2012. 5(2): p. 220.

[84] Matsa, E., et al., *Drug evaluation in cardiomyocytes derived from human induced pluripotent stem cells carrying a long QT syndrome type 2 mutation.* European heart journal, 2011. 32(8): p. 952-962.

Chapter 21

Pluripotent Stem Cells for Cardiac Cell Therapy: The Application of Cell Sheet Technology

Hidetoshi Masumoto and Jun K. Yamashita

Additional information is available at the end of the chapter

http://dx.doi.org/10.5772/56326

1. Introduction

Cardiovascular disease remains the leading cause of death worldwide despite many years of declining mortality rates in the Western world [1,2]. Myocardial infarction carries a short term mortality rate of about 7% even with aggressive therapy, and congestive heart failure with even more distressing 20% one-year mortality [3]. Despite significant advances in therapeutic modalities and risk-reduction strategies, the substantial burden remains. This continued health problem has prompted research into new therapeutic strategies including cardiac regenerative therapy as a new approach for severe cardiac diseases resistant to conventional therapies [4,5].

Acute ischemic injury and chronic cardiomyopathies lead to permanent loss of cardiac tissue, leading to heart failure. For pathologic situations, cell transplantation is thought to be an ideal therapeutic method for supplying *de novo* myocardium [6]. Of the available cell sources for cardiac cell therapy, stem cells (e.g. pluripotent stem cells, bone-marrow derived stem cells, skeletal myoblasts and cardiac stem cells) are now being prioritized for basic research and clinical trials [4,7]. The discoveries of various stem cell populations possessing cardiogenic potential and the advance of methods to isolate and expand these cells have shaped the notion of cell-based restorative therapy [8-11]. Despite much knowledge gained through numerous basic researches, significant challenges for true cardiac regeneration remain, and the field lacks sufficient results conclusive to support full-scale implementation of such treatments. Furthermore, results of clinical researches in cardiac stem cell therapy with a relatively small cohort scale were marginal, thus only showing little clinical advantages so far [12].

Among the stem cell types, pluripotent stem cells (PSCs) [Embryonic stem cells (ESCs) / induced pluripotent stem cells (iPSCs)] possess great capacity for cardiac regeneration mainly due to the prominent potential to expand and differentiate into most somatic cell lineages [13,14]. To date, no human trials using PSCs for cardiac repair have been attempted. Intensive translational researches, including the demonstration of effectiveness and safety, are needed to realize clinical application of PSCs.

Another concern is the actual phenomena which are taking place in the niche of transplanted site: does cardiac stem cell therapy bring *de novo* functional myocardium, or some indirect mechanisms mediate cardiac repair? It is reported that very few of the transplanted tissue stem cells seem to differentiate into mature cardiovascular cell types, suggesting that transplanted cells exert indirect paracrine effects by which humoral factors induce or support favorable processes, including angiogenesis, prevention of apoptosis, and promotion of healing, in the injured myocardium rather than differentiating into *de novo* myocardium [4,15]. PSCs might possess advantages in this context; defined cell populations differentiated from PSCs might be effective to elucidate underlying paracrine mechanisms in cardiac restoration compared to bulk cell mixture derived from somatic stem cells with various cell lineages and differentiation stages [16].

Concerning stem cell transplantation, as well as the transplanted cell type, the method for transplantation is also important to overcome the poor efficiency of engraftment with needle injection. A promising approach is the creation of cell sheets that better support effective engraftment of the transplants. We have shown the effectiveness of temperature-responsive cell sheet technology in basic studies [16].

In this chapter, we introduce the clarification for the progress and drawbacks of current cardiac stem cell therapy, and finally indicate the future directions of cardiac cell therapy through our recent researches combining PSCs and cell sheet technology.

2. Various somatic stem cell populations for cardiac stem cell therapy

To date, various somatic stem cells have been investigated for their feasibility to cardiac regenerative therapy with many basic studies.

Bone marrow hematopoietic stem cells (or circulating peripheral-blood progenitor cells) are an abundant and well characterized source of progenitor cells. A number of studies have shown that direct transplantation of bone marrow-derived cells or mobilization from endogenous reservoirs of the cell population significantly improves cardiac function [17,18]. However, other investigations found limited differentiation of bone marrow cells into cardiovascular cell types [19]. This suggests that beneficial results were mainly due to indirect paracrine effects such as neovascularization, independent of direct tissue regeneration.

Mesenchymal stem cells (MSCs) are a subset of stem cells found in the stroma of the bone marrow, adipose tissue, fetal membrane and many other tissues that can differentiate into osteoblasts, chondrocytes, and adipocytes [20,21] and also into small numbers of cardiomyocytes [8]. MSCs are thought to be either less immunogenic than other stem cell populations or inherently immunomodulatory [22], alleviating the need for immunosuppression prior to transplantation. Transplantation of MSCs into infarct animal models demonstrated improved left ventricular function, reduced infarct size, and increased survival rate [8,22,23]. The major disadvantage of MSCs for this clinical application is the broad differentiation capacity; MSC populations remain highly heterogeneous and are less predictable after transplantation. Some studies have shown that MSCs differentiated into osteoblasts inside ventricular tissue after transplantation [24].

Endothelial progenitor cells (EPCs) are another promising stem cell subset which accumulate to vascular injury sites from bone marrow and incorporate into the microvasculature (vasculogenesis) [9]. EPCs can be identified by the ability to acquire the expression of endothelial cell surface makers, such as cluster of differentiation molecule 133 (CD133), CD34 and so on, both in vitro and in vivo [25]. The research into their therapeutic use began with attempts to enhance their mobilization or incorporate EPCs directly into the vasculature of injured sites [26]. Preclinical studies of the injection of EPCs to infarct myocardium improved left ventricular function [15]. Although EPCs remain promising as a potential therapeutic material, they have several disadvantages for cell therapy: 1) Their heterogeneity. EPCs circulating in the peripheral blood span the full range of differentiation from angioblasts to mature endothelial cells. 2) Limited stem cell pool. Ex vivo expansion would be the only way to obtain a sufficient amount of EPCs for the treatment of an ischemic injury [27]. 3) The pool of EPCs is reduced in patients with common comorbidities of cardiac ischemia (e.g. diabetes mellitus, hypertension, and hyperlipidemia) [28].

Skeletal myoblasts (SMs) are a stem cell population derived from the satellite cells which exists beneath the basal membrane of adult skeletal muscle tissue [29]. SMs have been considered as an attractive source for cardiac restoration because of the small potential for teratoma formation, availability for autologous transplantation, resistance to ischemic condition and so on [10]. Most transplantations in animal disease models improved left ventricular function and decreased ventricular remodeling [10,30]. There are however, two main limitations; the first is the arrhythmogenic potential of the engrafted SMs. It is reported that only a fraction of skeletal myoblasts differentiate into cardiomyocytes after transplantation, and the generated myotubules may not synchronically work with the native myocardium [31]. A large scale clinical trial, Myoblast Autologous Grafting in Ischemic Cardiomyopathy (MAGIC) trial, showed a higher number of arrhythmic events in myoblast-treated patients [32]. The second limitation is the relatively poor engraftment of the transplanted cells into the host myocardium. It is reported that less than 10% of transplanted cells could survive within the first few days after injection in mice [33].

Several populations of cardiac progenitor / stem cells derived from mature cardiac tissue have been reported, which may hold the natural and endogenous cardiac regenerative mechanisms. Traditionally, the heart has been considered to be a post-mitotic organ, and withdrawn from the proliferative cell cycle. However, some contradictory data have reported, as cardiomyocyte proliferation and cell cycling have been observed under pathological conditions (e.g. hypertension or myocardial infarction) [34, 35] and even in the healthy heart [36]. These evidences prompted further research for such resident cardiac cells. The first cell population with stem cell properties is called the side population (SP) cells. Isolated cardiac SP cells represent cardiac and vascular progenitor cells and can differentiate into cardiomyocytes, endothelial cells, or smooth muscle cells [37]. The second progenitor population is the cells expressing the stem cell factor receptor c-Kit (also designed as CD117), which are located in small clusters within the adult cardiac tissue. c-Kit$^+$ cells hold regenerative potential after transplantation and give rise to cardiomyocytes, endothelial cells, and smooth muscle cells [38]. The third cell type expresses stem cell antigen 1 (Sca-1). Sca-1$^+$ cells migrate to infarcted myocardium and differentiate into cardiomyocytes around the injured area [39]. Finally, enzymatic digestion of heart tissue obtained via endomyocardial biopsy or during cardiac surgery yields cardiac progenitor cells that form what is called cardiospheres. Cardiosphere derived cells (CDCs) can also differentiate into cardiomyocytes, endothelial cells, and smooth muscle cells, exhibiting prominent capacities for proliferation and differentiation [11]. This population can be differentiated into aggregates of cardiomyocytes that when transplanted into injured myocardium produced functional improvement in preclinical studies [40]. It is unclear whether the various cardiac stem cells shown here are different populations, or represent various stages of a single cell lineage. A major limitation of cardiac progenitor / stem cell populations is that the cardiac stem cell pool appears to diminish along with age, which may limit the efficacy of regeneration in elderly people [41]. Considering that it is mostly the elderly who suffer increased mortality from cardiac ischemia, intensive research aiming to rejuvenate this senescent stem cell population is required.

Many clinical studies have been conducted using these somatic stem cells so far: TOPCARE-AMI [42], BOOST [43], REPAIR-AMI [44], LateTime [45] (Bone marrow hematopoietic stem cells), REGENT [46] (EPCs), MAGIC [32], CAuSMIC [47] (SMs), CADUCEUS [48], SCIPIO [49] (cardiac progenitor / stem cells) and so on. However, most of these clinical studies have shown relatively limited clinical benefits in general. These marginal results indicate that more efficient approaches for stem cell therapy are needed to realize full-scale stem cell-based therapy.

3. Advantages of pluripotent stem cells in cardiac regeneration

Embryonic stem cells (ESCs) are one of the stem cell populations which can be removed from the inner cell mass of the blastocyst and expanded in vitro with practically no limitations [13]. Yamanaka and colleagues have discovered that reprogramming of adult somatic cells with transcription factor genes that confer pluripotency generates ESC-like cells, called induced pluripotent stem cells (iPSCs) [14,50]. Among the stem cell types, these pluripotent stem cells (PSCs) [ESCs / iPSCs] possess great capacity especially for cardiac regeneration due to several reasons.

The first reason is that PSCs can be expanded practically indefinitely in vitro remaining pluripotent in an undifferentiated state in culture, and can give rise to most somatic cell lineages once allowed to differentiate. In this regard, the regenerative capacity is theoretically limitless [51]. The merit of PSCs is larger especially for the heart compared to other organs, such as endocrine or sensory organs, as the heart functions as an assembly of a large number of cells including cardiomyocytes and other cell types (e.g. vascular cells, cardiac fibroblasts), and numerous ($>10^8$) heart-composing cells might be required to fully compensate for the damaged human heart [5].

The second reason is that the capacity for the differentiation towards a desired cell type, such as cardiomyocytes or other vascular cell types is the highest among various stem cell populations known to possess cardiogenic potential. The differentiation of PSCs can be driven towards cardiomyocytes or others by culture conditions as monolayers or embryoid bodies in various growth media [52–55]. Previously, we have developed a novel monolayer culture-based ESC / iPSC differentiation system that recapitulates early cardiovascular developmental processes using Flk1 (also designed as vascular endothelial cell growth factor [VEGF] receptor-2)-positive cells as common cardiovascular progenitors. Cardiovascular cell types, namely cardiomyocytes [53], endothelial cells, and vascular mural cells [52], can be systematically induced and purified with this system (Figure 1A). In fact, of the various stem cell populations studied so far, PSCs have demonstrated probably the greatest capacity for cardiac cell differentiation and long-term cell survival [56].

The third reason is that PSCs might be advantageous for further elucidation of regenerative mechanisms. In the field of cardiac restoration with stem cell therapy, it has been widely believed that transplanted cells act as an inducer of indirect paracrine effects such as angiogenesis, prevention of apoptosis, and so on rather than regeneration of *de novo* myocardium [4,15]. Considering this point, the transplantation of somatic stem cells, which are largely performed thus far as mentioned above, may raise a question, "which cells are really effective?", because the transplanted cells from somatic stem cells might consist of heterogeneous cell populations. In this regard, the transplantation of defined cardiovascular cell populations systematically derived from PSCs might be much more superior to that of somatic stem cell-derived populations for the sake of the elucidation of regenerative mechanisms (Figure 1B).

The final reason is the discovery of iPSCs. The generation of iPSCs by reprogramming autologous somatic cells with genes regulating pluripotency may resolve the ethical and immunogenic issues associated with the use of ESCs. Furthermore, we have reported that cardiovascular cell types can be differentiated respectively from mouse iPSCs almost identically with those from mouse ESCs [57]. This indicates that iPSCs possess almost the same regenerative capacity as that of ESCs. A potent differentiation protocol based on high-density monolayer culture and chemically defined factors, and modifications thereof, have been reported to induce cardiomyocytes from human iPSCs with a robust efficiency of 40–70 % [54, 58]. The application of this method would strongly promote cardiac regeneration using human iPSCs.

The transplantation of cardiac cells derived from PSCs has been tested in animal studies with encouraging results [16, 54]. However, no human studies using PSCs for cardiac repair have been attempted so far. A major concern regarding iPSC transplantation as a treatment modality is related to the potential tumor formation. The differentiating cells from PSCs contain derivatives from three germ layers (ectoderm, mesoderm and endoderm), possessing the capacity to differentiate along any or all of these three lineages. This increases the risk of teratoma formation at the transplantation site. Although such teratomas are believed to be largely benign, some teratoma cells have been reported to express markers similar to those seen in malignant tumors [59]. Recently, protocols for generating human iPSCs without genomic integration by utilizing episomal vectors [60] or human artificial chromosome vectors

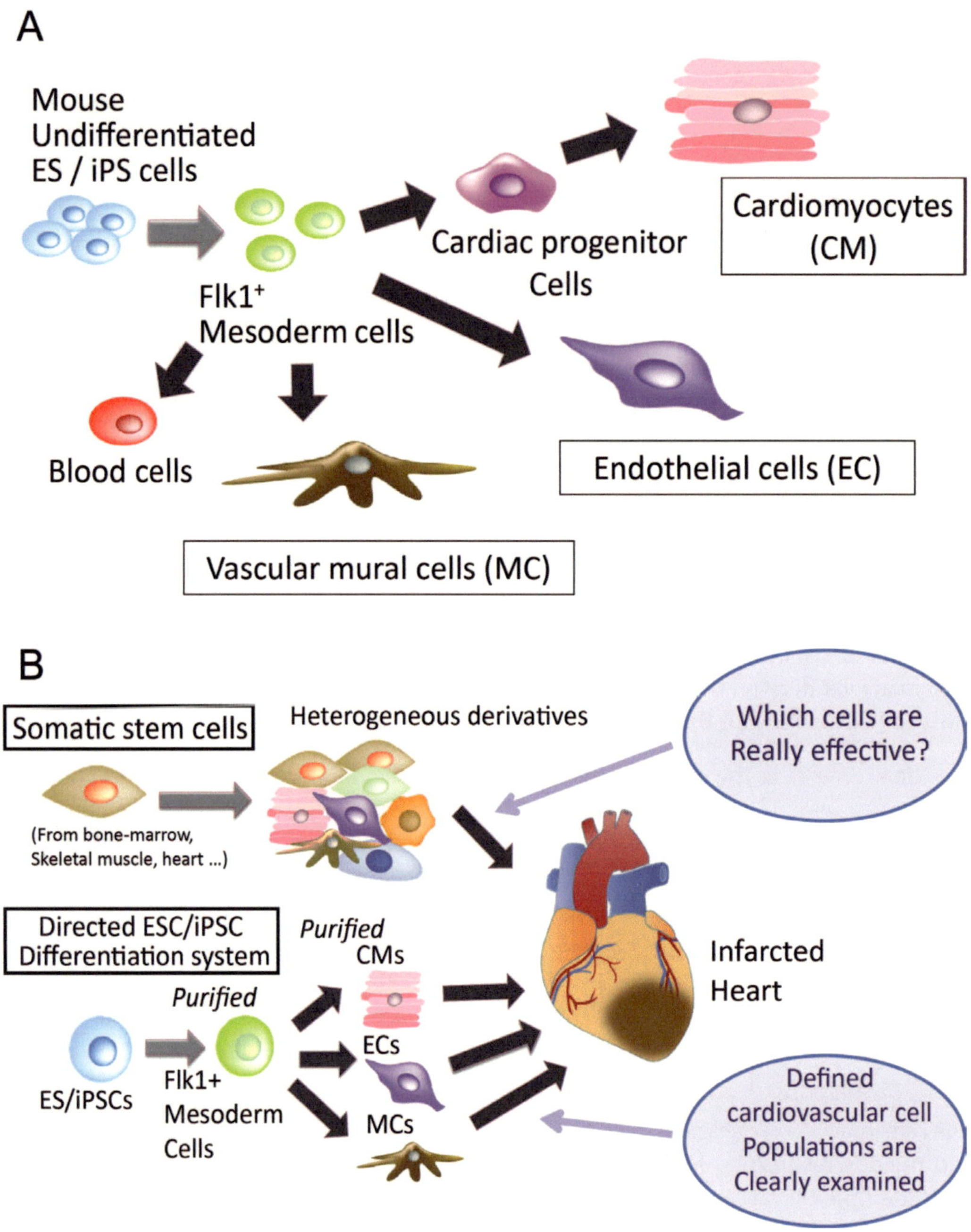

Figure 1. The advantages of PSCs for cardiac regeneration. (A): The capacity for the differentiation towards a desired cardiac cell type. The scheme of directed mouse PSC differentiation system from Flk1+ mesoderm cells as a common progenitor is shown. (B): Effectiveness for further elucidation of regenerative mechanisms. The usage of somatic stem cells (upper) may lead to the transplantation of heterogeneous derivatives in lineage and differentiation stage. On the other hand, the usage of directed PSC differentiation system with purifying processes (lower) clarify which cell populations are actually transplanted. ES cell, embryonic stem cell; iPS cell, induced pluripotent stem cell.

[61] have been reported. These may reduce tumorigenesis due to mutations, which could otherwise limit the clinical application of iPSCs.

Considering the results of these basic studies, PSCs (especially iPSCs) are currently recognized to be one of the most promising cell sources for cardiac regeneration. However, further careful exploration for the feasibility of this new modality will be needed to realize the clinical application.

4. Cell sheet technology as a novel method for PSC-derived cell transplantation

In addition to the transplanted cell type, the transplantation method is also important to overcome poor efficiency of engraftment associated with needle injection. The low level of grafted cell survival and engraftment diminishes their potential for paracrine effects, besides regeneration of *de novo* myocardium, and is a major technical limitation for stem cell therapy [62]. It is reported that >70% of injected cells die during the first 48 hours after needle injection, progressively diminishing during the following days possibly due to the hypoxic, inflammatory, and/or fibrotic environment [63]. Another report shows that only 5.4 to 8.8% of microspheres remain just after direct injection into the beating myocardium due to massive mechanical loss [64]. To overcome this problem, a combination of bioengineering techniques have been developed and investigated for their efficacy, suggesting that these new strategies may improve the efficiency of stem cell therapies [65].

Initial experiments were performed by combining the cells with injectable biomaterials such as collagen, fibrin, gelatin or matrigel as a sccafold. In general, early results showed an increased survival of the transplanted cells, and a greater improvement in cardiac function of the treated hearts [66]. However, these approaches did not assure complete cell retention or an adequate distribution of the transplanted cells within the host heart.

The creation of cell sheets without scaffold support would be a more promising approach. The advantages of this method are as follows: 1) Potent increase of the efficiency of transplantation compared to that of needle injection. 2) Potential for construction of three-dimentional tissue-like structure as a graft. 3) Avoidance of inflammatory reactions against the biomaterials constituting the scaffolds. 4) Larger scalability and accessibility due to two-dimensional cell culture.

Several methods have been reported for cell sheet formation [67-69]. Among them, we have utilized temperature-responsive culture surface-based method [16]. This technique was made possible by using a culture dish covalently grafted with temperature-responsive polymer poly (N-isopropylacrylamide) (PIPAAm) which enables the generation of cell sheets without enzymatic digestion, retaining intact extracellular matrices or adhesion molecules [67]. The benefits of this technique have been demonstrated by many experiments of stem cell therapy such as the transplantation of monolayer adipose tissue-derived MSCs to the infarcted rat heart

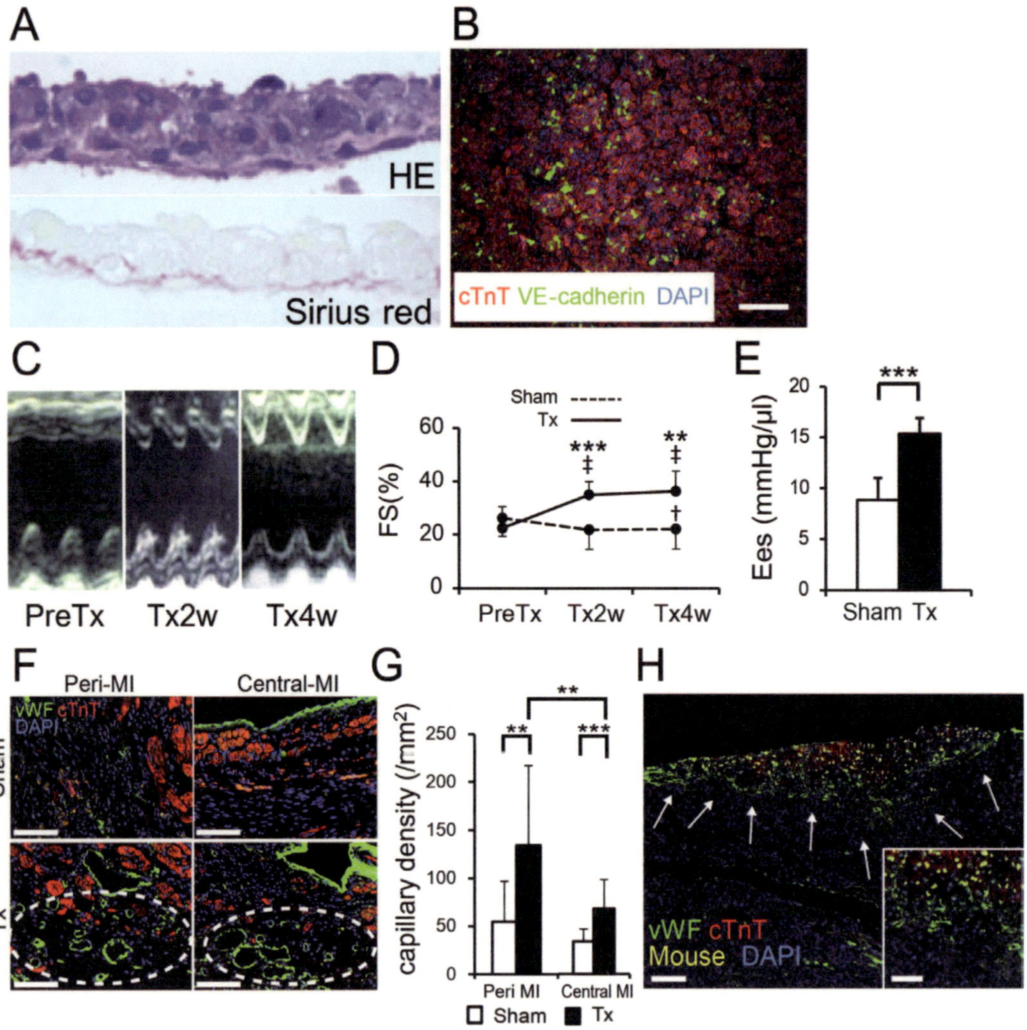

Figure 2. The improvement of infarcted heart function after transplantation of cardiac tissue sheets bioengineered with mouse ES cell-derived defined cardiac cell populations. (A): Cross-sections of the sheet. Upper panel: H&E staining showing cell appearance of the sheet. Lower panel: Sirius red staining showing intact extracellular matrix. (B): Immunostaining of sheets for cTnT (red), VE-cadherin (green), and DAPI. (C,D): Echocardiogram (n=9). (C): Representative M-mode image. Note that infarct anterior wall started to move 2–4 weeks after transplantation (Tx). (D): Fractional shortening (FS). (E): LV pressure-volume loop study 4 weeks after Tx (n=8). Ees: End-systolic elastance. (F, G): Capillary formation at Tx-d28. (F): Double staining for vWF (ECs, green) and cTnT (cardiomyocytes [CMs], red) at peri-MI and central-MI areas. Note that newly formed capillaries are clearly observed in transplantation group (dotted circles). (G): Quantification of capillary density (capillary number per square millimeter). Peri-MI area (left panel) and central-MI area (right panel) (15 views each). (H): Triple staining for vWF, cTnT, and species-specific fluorescent in situ hybridization (mouse nuclei, yellow) (Tx-d3). Most of the accumulated vWF-positive cells are negative for mouse nuclear staining (arrows). Inset: higher magnification view. **, $p < .01$; and ***, $p < .001$ (unpaired t test), †, $p < .05$ and ‡, $p < .01$ (vs. PreTx, paired t test). PreTx; Pretransplantation, Tx2w, Tx4w; 2 and 4 weeks after transplantation, respectively. Scale bars: 200 μm in (B), 100 μm in (F) and (H) (main panel), 50 μm in (H) (inset). HE, Hematoxylin and Eosin; cTnT, cardiac troponin-T; DAPI, 4,6-diamidino-2-phenylindole; vWF, von Willebrandfactor.; MI, myocardial infarction. (quote from ref. 16 with revision)

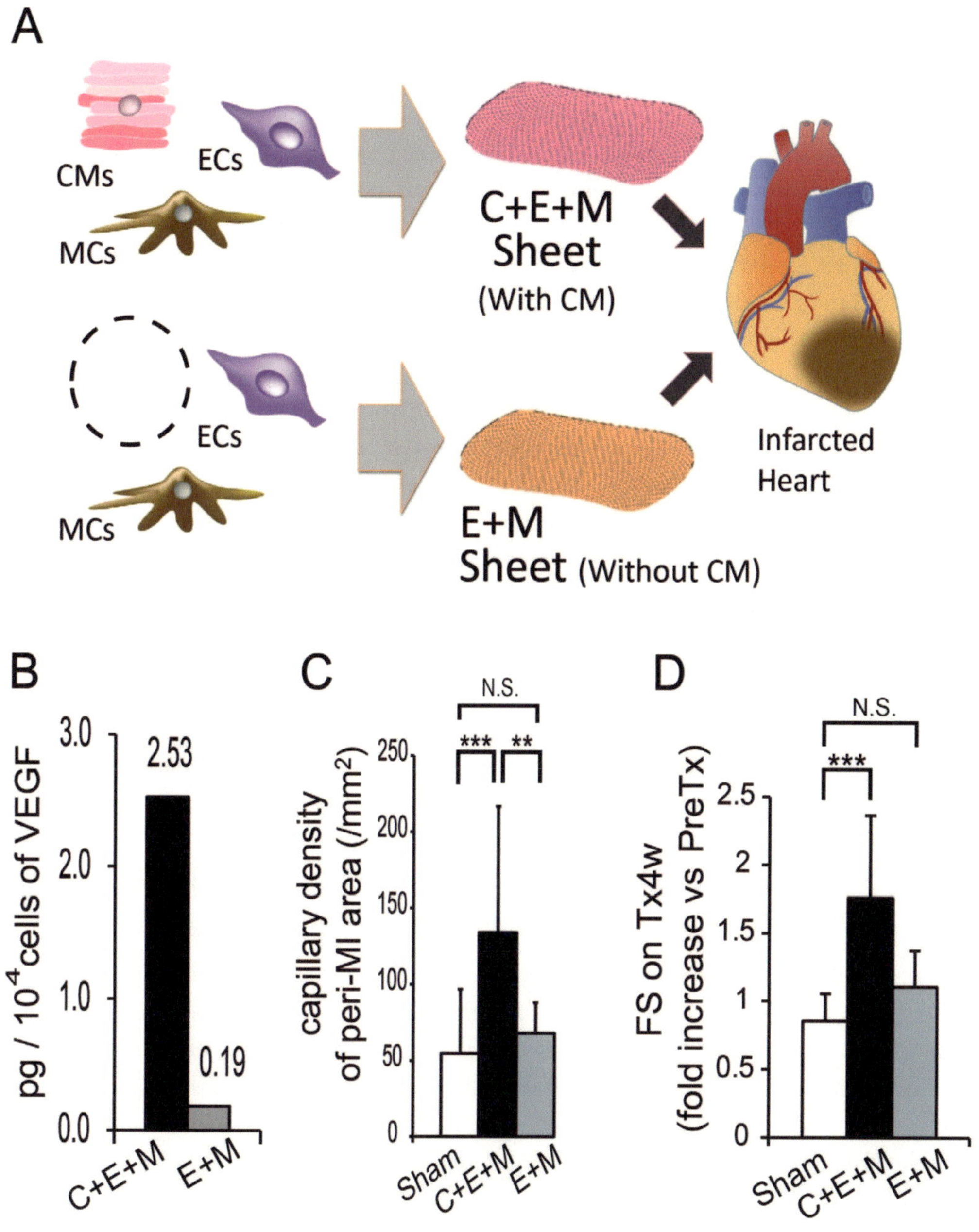

Figure 3. Cell type-controlled sheet analyses. (A): The Scheme of cell sheets with CMs (C+E+M) or without CMs (E+M). (B): ELISA for VEGF secretion (picogram per10^4 cells) in culture supernatants of C+E+M and E+M sheets. (C,D): Transplantation of sham operation (n=9) versus C+E+M sheets (n=9) versus E+M sheets (n=3) (Tx-d28). (C): Capillary density in peri-MI area (capillary number per square millimeter). (15 views each). (D): Fractional shortening (FS) on echocardiogram (fold increase vs. PreTx). **, p <.01, and ***, p <.001 (unpaired t test). C: cardiomyocytes, E: endothelial cells, M: vascular mural cells. N.S., not significant; VEGF, vascular endothelial cell growth factor. (quote from ref. 16 with revision).

[23]. Recently, we have reported transplantation of a three-layered cardiac tissue sheet bioengineered with mouse ESC-derived defined cardiac cell populations in the infarcted heart (Figure 2) [16]. In both cases, increased tissue neovascularization together with a prominent attenuation of cardiac remodeling responsible for the improvement in cardiac function were demonstrated. Furthermore, our research indicated the potential for cell sheet-based prospective elucidation of the cellular mechanisms of cardiac restoration. The combinations of cell populations composing the transplanted cell sheets enabled us to elucidate the contributions of each cell type (for example, the comparison of cell sheets with or without cardiomyocytes is useful for the elucidation of the cellular function of cardiomyocytes). This cell-type controlled analysis led us to identify one of the important cellular mechanisms of cardiac restoration following cell therapy, that is, cardiomyocytes are essential for the functional improvement of ischemic heart through neovascularization (Figure 3). These results show that the tissue-like cell sheet system is advantageous for the elucidation of cardiac regenerative mechanism, as well as for therapeutic purposes.

5. Future directions

One future direction of this PSC-derived cell sheet technology is its utilization as a novel experimental tool for elucidation of regenerative mechanisms. Although the present results of clinical trials using stem cell therapy are marginal, further elucidation of the actual mechanisms of cardiac repair following cell therapy would enhance the potential of stem cell therapy to be a full-scale therapy. It would be a breakthrough for further improvement of cardiac cell therapy to understand the role of each cell population as well as the various cellular interactions in the chaos of heterogeneity.

Another direction is a more efficient survival of transplanted sheets to realize regeneration of functioning *de novo* myocardium. Considering that more cells that survived were observed in peri-infarction than central infarction region in our study [16], it would be possible that the severe ischemic condition may not be suitable for sheet survival. Novel techniques increasing blood supply in the graft should be applied, such as prevascularization in 3-dimensional tissue formation [70,71] or vascularized flap grafts.

6. Conclusion

In this chapter, we have reviewed the status quo of current cardiac stem cell therapy, and shown the promising potential of PSC-derived cardiac tissue-like sheets. The knowledge yielded from this cell sheet-based study would provide a hallmark for cell therapy with PSCs and a strategic principle for future cardiac restoration therapy.

http://dx.doi.org/10.5772/56326

Author details

Hidetoshi Masumoto[1,2*] and Jun K. Yamashita[1]

*Address all correspondence to: masumoto@kuhp.kyoto-u.ac.jp

1 Laboratory of Stem Cell Differentiation, Department of Cell Growth and Differentiation, Center for iPS Cell Research and Application (CiRA), Kyoto University, Kyoto, Japan

2 Department of Cardiovascular Surgery, Kyoto University Graduate School of Medicine, Kyoto, Japan

References

[1] World Health Organization (WHO)The global burden of disease: 2004 update. 2008. http://www.who.int/healthinfo/global_burden_disease/2004_report_update/en/ index.html accessed 31 August (2012).

[2] Ford, E. S, & Capewell, S. (2011). Proportion of the Decline in Cardiovascular Mortality Disease due to Prevention Versus Treatment: Public Health Versus Clinical Care. Annu Rev Public Health , 32, 5-22.

[3] Assessment of the Safety and Efficacy of a New Thrombolytic Regimen (ASSENT)-3 Investigators(2001). Efficacy and safety of tenecteplase in combination with enoxaparin, abciximab, or unfractionated heparin: the ASSENT-3 randomised trial in acute myocardial infarction. Lancet , 358, 605-613.

[4] Joggerst, S. J, & Hatzopoulos, A. K. (2009). Stem cell therapy for cardiac repair: benefits and barriers. Expert Rev Mol Med 11,e20.

[5] Masumoto, H, & Sakata, R. Cardiovascular surgery for realization of regenerative medicine. Gen Thorac Cardiovasc Surg. In press.

[6] Sakakibara, Y, Tambara, K, Lu, F, Nishina, T, Sakaguchi, G, Nagaya, N, Nishimura, K, Li, R. K, Weisel, R. D, & Komeda, M. (2002). Combined procedure of surgical repair and cell transplantation for left ventricular aneurysm: an experimental study. Circulation 106,II197., 193.

[7] Chien, K. R. (2008). Regenerative medicine and human models of human disease. Nature , 453, 302-305.

[8] Tomita, S, Li, R. K, Weisel, R. D, Mickle, D. A, Kim, E. J, Sakai, T, & Jia, Z. Q. (1999). Autologous transplantation of bone marrow cells improves damaged heart function. Circulation 100,IIII256., 247.

[9] Asahara, T, Murohara, T, Sullivan, A, Silver, M, Van Der Zee, R, Li, T, Witzenbichler, B, Schatteman, G, & Isner, J. M. (1997). Isolation of putative progenitor endothelial cells for angiogenesis. Science , 275, 964-967.

[10] Murry, C. E, Wiseman, R. W, Schwartz, S. M, & Hauschka, S. D. (1996). Skeletal myoblast transplantation for repair of myocardial necrosis. J Clin Invest , 98, 2512-2523.

[11] Messina, E, De Angelis, L, Frati, G, Morrone, S, Chimenti, S, Fiordaliso, F, Salio, M, Battaglia, M, Latronico, M. V, Coletta, M, et al. (2004). Isolation and expansion of adult cardiac stem cells from human and murine heart. Circ Res , 95, 911-921.

[12] Rosenzweig, A. (2006). Cardiac cell therapy--mixed results from mixed cells. N Engl J Med , 355, 1274-1277.

[13] Thomson, J. A, Itskovitz-eldor, J, Shapiro, S. S, Waknitz, M. A, Swiergiel, J. J, Marshall, V. S, & Jones, J. M. (1998). Embryonic stem cell lines derived from human blastocysts. Science , 282, 1145-1147.

[14] Takahashi, K, Tanabe, K, Ohnuki, M, Narita, M, Ichisaka, T, Tomoda, K, & Yamanaka, S. (2007). Induction of pluripotent stem cells from adult human fibroblasts by defined factors. Cell , 131, 861-872.

[15] Kocher, A. A, Schuster, M. D, Szabolcs, M. J, Takuma, S, Burkhoff, D, Wang, J, Homma, S, Edwards, N. M, & Itescu, S. (2001). Neovascularization of ischemic myocardium by human bone-marrow derived angioblasts prevents cardiomyocyte apoptosis, reduces remodeling and improves cardiac function. Nat Med , 7, 430-436.

[16] Masumoto, H, Matsuo, T, Yamamizu, K, Uosaki, H, Narazaki, G, Katayama, S, Marui, A, Shimizu, T, Ikeda, T, Okano, T, et al. (2012). Pluripotent stem cell-engineered cell sheets reassembled with defined cardiovascular populations ameliorate reduction in infarct heart function through cardiomyocyte-mediated neovascularization. Stem Cells , 30, 1196-1205.

[17] Orlic, D, Kajstura, J, Chimenti, S, Jakoniuk, I, Anderson, S. M, Li, B, Pickel, J, Mckay, R, Nadal-ginard, B, Bodine, D. M, et al. (2001). Bone marrow cells regenerate infarcted myocardium. Nature , 410, 701-705.

[18] Orlic, D, Kajstura, J, Chimenti, S, Limana, F, Jakoniuk, I, Quaini, F, Nadal-ginard, B, Bodine, D. M, Leri, A, & Anversa, P. (2001). Mobilized bone marrow cells repair the infarcted heart, improving function and survival. Proc Natl Acad Sci USA. , 98, 10344-10349.

[19] Murry, C. E, Soonpaa, M. H, Reinecke, H, Nakajima, H, Nakajima, H. O, Rubart, M, Pasumarthi, K. B, Virag, J. I, Bartelmez, S. H, Poppa, V, et al. (2004). Haematopoietic stem cells do not transdifferentiate into cardiac myocytes in myocardial infarcts. Nature , 428, 664-668.

[20] Jiang, Y, Jahagirdar, B. N, Reinhardt, R. L, Schwartz, R. E, Keene, C. D, Ortiz-gonzalez, X. R, Reyes, M, Lenvik, T, Lund, T, Blackstad, M, et al. (2002). Pluripotency of mesenchymal stem cells derived from adult marrow. Nature , 418, 41-49.

[21] Ishikane, S, Yamahara, K, Sada, M, Harada, K, Kodama, M, Ishibashi-ueda, H, Hayakawa, K, Mishima, K, Iwasaki, K, Fujiwara, M, et al. (2010). Allogeneic administration of fetal membrane-derived mesenchymal stem cells attenuates acute myocarditis in rats. J Mol Cell Cardiol. , 49, 753-761.

[22] Dai, W, Hale, S. L, Martin, B. J, Kuang, J. Q, Dow, J. S, Wold, L. E, & Kloner, R. A. (2005). Allogeneic mesenchymal stem cell transplantation in postinfarcted rat myocardium: short- and long-term effects. Circulation , 112, 214-223.

[23] Miyahara, Y, Nagaya, N, Kataoka, M, Yanagawa, B, Tanaka, K, Hao, H, Ishino, K, Ishida, H, Shimizu, T, Kangawa, K, et al. (2006). Monolayered mesenchymal stem cells repair scarred myocardium after myocardial infarction. Nat Med , 12, 459-465.

[24] Yoon, Y. S, Park, J. S, Tkebuchava, T, Luedeman, C, & Losordo, D. W. (2004). Unexpected severe calcification after transplantation of bone marrow cells in acute myocardial infarction. Circulation , 109, 3154-3157.

[25] Hristov, M, & Weber, C. (2008). Endothelial progenitor cells in vascular repair and remodeling. Pharmacol Res , 58, 148-151.

[26] Llevadot, J, Murasawa, S, Kureishi, Y, Uchida, S, Masuda, H, Kawamoto, A, Walsh, K, Isner, J. M, & Asahara, T. (2001). HMG-CoA reductase inhibitor mobilizes bone marrow-derived endothelial progenitor cells. J Clin Invest , 108, 399-405.

[27] Jujo, K, Ii, M, & Losordo, D. W. (2008). Endothelial progenitor cells in neovascularization of infarcted myocardium. J Mol Cell Cardiol , 45, 530-544.

[28] Vasa, M, Fichtlscherer, S, Aicher, A, Adler, K, Urbich, C, Martin, H, Zeiher, A. M, & Dimmeler, S. (2001). Number and migratory activity of circulating endothelial progenitor cells inversely correlate with risk factors for coronary artery disease. Circ Res 89,EE7., 1.

[29] Buckingham, M, & Montarras, D. Skeletal muscle stem cells. ((2008). Curr Opin Genet Dev. , 18, 330-336.

[30] Taylor, D. A, Atkins, B. Z, Hungspreugs, P, Jones, T. R, Reedy, M. C, Hutcheson, K. A, Glower, D. D, & Kraus, W. E. (1998). Regenerating functional myocardium: improved performance after skeletal myoblast transplantation. Nat Med , 4, 929-933.

[31] Farahmand, P, Lai, T. Y, Weisel, R. D, Fazel, S, Yau, T, Menasche, P, & Li, R. K. (2008). Skeletal myoblasts preserve remote matrix architecture and global function when implanted early or late after coronary ligation into infarcted or remote myocardium. Circulation 118,SS137., 130.

[32] Menasche, P, Alfieri, O, Janssens, S, Mckenna, W, Reichenspurner, H, Trinquart, L, Vilquin, J. T, Marolleau, J. P, Seymour, B, Larghero, J, et al. (2008). The Myoblast Au-

tologous Grafting in Ischemic Cardiomyopathy (MAGIC) trial: first randomized placebo-controlled study of myoblast transplantation. Circulation , 117, 1189-1200.

[33] Suzuki, K, Murtuza, B, Beauchamp, J. R, Smolenski, R. T, Varela-carver, A, Fukushima, S, Coppen, S. R, Partridge, T. A, & Yacoub, M. H. (2004). Dynamics and mediators of acute graft attrition after myoblast transplantation to the heart. FASEB J , 18, 1153-1155.

[34] Anversa, P, Palackal, T, Sonnenblick, E. H, Olivetti, G, & Capasso, J. M. (1990). Hypertensive cardiomyopathy. Myocyte nuclei hyperplasia in the mammalian rat heart. J Clin Invest , 85, 994-997.

[35] Beltrami, A. P, Urbanek, K, Kajstura, J, Yan, S. M, Finato, N, Bussani, R, Nadal-ginard, B, Silvestri, F, Leri, A, Beltrami, C. A, et al. (2001). Evidence that human cardiac myocytes divide after myocardial infarction. N Engl J Med , 344, 1750-1757.

[36] Bergmann, O, Bhardwaj, R. D, Bernard, S, & Zdunek, S. Barnabe´-Heider F, Walsh S, Zupicich J, Alkass K, Buchholz BA, Druid H et al. ((2009). Evidence for cardiomyocyte renewal in humans. Science , 324, 98-102.

[37] Oyama, T, Nagai, T, Wada, H, Naito, A. T, Matsuura, K, Iwanaga, K, Takahashi, T, Goto, M, Mikami, Y, Yasuda, N, et al. (2007). Cardiac side population cells have a potential to migrate and differentiate into cardiomyocytes in vitro and in vivo. J Cell Biol , 176, 329-341.

[38] Beltrami, A. P, Barlucchi, L, Torella, D, Baker, M, Limana, F, Chimenti, S, Kasahara, H, Rota, M, Musso, E, Urbanek, K, et al. (2003). Adult cardiac stem cells are multipotent and support myocardial regeneration. Cell , 114, 763-776.

[39] Oh, H, Bradfute, S. B, Gallardo, T. D, Nakamura, T, Gaussin, V, Mishina, Y, Pocius, J, Michael, L. H, Behringer, R. R, Garry, D. J, et al. (2003). Cardiac progenitor cells from adult myocardium: homing, differentiation, and fusion after infarction. Proc Natl Acad Sci USA , 100, 12313-12318.

[40] Takehara, N, Tsutsumi, Y, Tateishi, K, Ogata, T, Tanaka, H, Ueyama, T, Takahashi, T, Takamatsu, T, Fukushima, M, Komeda, M, et al. (2008). Controlled delivery of basic fibroblast growth factor promotes human cardiosphere-derived cell engraftment to enhance cardiac repair for chronic myocardial infarction. J Am Coll Cardiol , 52, 1858-65.

[41] Torella, D, & Ellison, G. M. Me´ndez-Ferrer S, Ibanez B and Nadal-Ginard B. ((2006). Resident human cardiac stem cells: role in cardiac cellular homeostasis and potential for myocardial regeneration. Nat Clin Pract Cardiovasc Med 3,SS13., 8.

[42] Leistner, D. M, Fischer-rasokat, U, Honold, J, & Seeger, F. H. Scha°chinger V, Lehmann R, Martin H, Burck I, Urbich C, Dimmeler S et al. ((2011). Transplantation of progenitor cells and regeneration enhancement in acute myocardial infarction (TOP-

CARE-AMI): final 5-year results suggest longterm safety and efficacy. Clin Res Cardiol , 100, 925-934.

[43] Meyer, G. P, Wollert, K. C, Lotz, J, Steffens, J, Lippolt, P, Fichtner, S, Hecker, H, Schaefer, A, Arseniev, L, Hertenstein, B, et al. (2006). Intracoronary bone marrow cell transfer after myocardial infarction: eighteen months' follow-up data from the randomized, controlled BOOST (BOne marrOw transfer to enhance ST-elevation infarct regeneration) trial. Circulation , 113, 1287-1294.

[44] Assmus, B, Rolf, A, & Erbs, S. Elsa°sser A, Haberbosch W, Hambrecht R, Tillmanns H, Yu J, Corti R, Mathey DG et al. ((2010). Clinical outcome 2 years after intracoronary administration of bone marrow-derived progenitor cells in acute myocardial infarction. Circ Heart Fail , 3, 89-96.

[45] Traverse, J. H, Henry, T. D, Ellis, S. G, Pepine, C. J, Willerson, J. T, Zhao, D. X, Forder, J. R, Byrne, B. J, Hatzopoulos, A. K, Penn, M. S, et al. (2011). Effect of intracoronary delivery of autologous bone marrow mononuclear cells 2 to 3 weeks following acute myocardial infarction on left ventricular function: the LateTIME randomized trial. JAMA , 306, 2110-2119.

[46] Tendera, M, Wojakowski, W, Ruzyllo, W, Chojnowska, L, Kepka, C, Tracz, W, Musialek, P, Piwowarska, W, Nessler, J, Buszman, P, et al. (2009). Intracoronary infusion of bone marrowderived selected CD34+CXCR4+ cells and non-selected mononuclear cells in patients with acute STEMI and reduced left ventricular ejection fraction: results of randomized, multicenter Myocardial Regeneration by Intracoronary Infusion of Selected Population of Stem Cells in Acute Myocardial Infarction (REGENT) Trial. Eur Heart J , 30, 1313-1321.

[47] Dib, N, Dinsmore, J, Lababidi, Z, White, B, Moravec, S, Campbell, A, Rosenbaum, A, Seyedmadani, K, Jaber, W. A, Rizenhour, C. S, et al. (2009). One-year follow-up of feasibility and safety of the first US, randomized, controlled study using 3-dimensional guided catheter-based delivery of autologous skeletal myoblasts for ischemic cardiomyopathy (CAuSMIC study). JACC Cardiovasc Interv , 2, 9-16.

[48] Makkar, R. R, Smith, R. R, Cheng, K, Malliaras, K, Thomson, L. E, Berman, D, Czer, L. S, Marbán, L, Mendizabal, A, Johnston, P. V, et al. (2012). Intracoronary cardiosphere-derived cells for heart regeneration after myocardial infarction (CADUCEUS): a prospective, randomised phase 1 trial. Lancet , 379, 895-904.

[49] Bolli, R, Chugh, A. R, Amario, D, Loughran, D, Stoddard, J. H, Ikram, M. F, Beache, S, Wagner, G. M, Leri, S. G, & Hosoda, A. T et al. ((2011). Cardiac stem cells in patients with ischaemic cardiomyopathy (SCIPIO): initial results of a randomised phase 1 trial. Lancet , 378, 1847-1857.

[50] Takahashi, K, & Yamanaka, S. (2006). Induction of pluripotent stem cells from mouse embryonic and adult fibroblast cultures by defined factors. Cell , 126, 663-676.

[51] Murry, C. E, & Keller, G. (2008). Differentiation of embryonic stem cells to clinically relevant populations: lessons from embryonic development. Cell , 132, 661-680.

[52] Yamashita, J, Itoh, H, Hirashima, M, Ogawa, M, Nishikawa, S, Yurugi, T, Naito, M, Nakao, K, & Nishikawa, S. (2000). Flk1-positive cells derived from embryonic stem cells serve as vascular progenitors. Nature , 408, 92-96.

[53] Yamashita, J. K, Takano, M, Hiraoka-kanie, M, Shimazu, C, Yan, P, Yanagi, K, Nakano, A, Inoue, E, Kita, F, & Nishikawa, S. (2005). Prospective identification of cardiac progenitors by a novel single cell-based cardiomyocyte induction. FASEB J , 19, 1534-1536.

[54] Laflamme, M. A, Chen, K. Y, Naumova, A. V, Muskheli, V, Fugate, J. A, Dupras, S. K, Reinecke, H, Xu, C, Hassanipour, M, Police, S, et al. (2007). Cardiomyocytes derived from human embryonic stem cells in pro-survival factors enhance function of infarcted rat hearts. Nat Biotechnol , 25, 1015-1024.

[55] Sone, M, Itoh, H, Yamahara, K, Yamashita, J. K, Yurugi-kobayashi, T, Nonoguchi, A, Suzuki, Y, Chao, T. H, Sawada, N, Fukunaga, Y, et al. (2007). Pathway for differentiation of human embryonic stem cells to vascular cell components and their potential for vascular regeneration. Arterioscler Thromb Vasc Biol , 27, 2127-2134.

[56] Van Laake, L. W, Passier, R, Monshouwer-kloots, J, Verkleij, A. J, Lips, D. J, & Freund, C. den Ouden K, Ward-van Oostwaard D, Korving J, Tertoolen LG et al. ((2007). Human embryonic stem cell-derived cardiomyocytes survive and mature in the mouse heart and transiently improve function after myocardial infarction. Stem Cell Res. , 1, 9-24.

[57] Narazaki, G, Uosaki, H, Teranishi, M, Okita, K, Kim, B, Matsuoka, S, Yamanaka, S, & Yamashita, J. K. (2008). Directed and systematic differentiation of cardiovascular cells from mouse induced pluripotent stem cells. Circulation , 118, 498-506.

[58] Uosaki, H, Fukushima, H, Takeuchi, A, Matsuoka, S, Nakatsuji, N, Yamanaka, S, & Yamashita, J. K. (2011). Efficient and scalable purification of cardiomyocytes from human embryonic and induced pluripotent stem cells by VCAM1 surface expression. PLoS One 6,e23657.

[59] Blum, B, & Benvenisty, N. (2008). The tumorigenicity of human embryonic stem cells. Adv Cancer Res , 100, 133-158.

[60] Okita, K, Matsumura, Y, Sato, Y, Okada, A, Morizane, A, Okamoto, S, Hong, H, Nakagawa, M, Tanabe, K, Tezuka, K, et al. (2011). A more efficient method to generate integration-free human iPS cells. Nat Method , 8, 409-412.

[61] Hiratsuka, M, Uno, N, Ueda, K, Kurosaki, H, Imaoka, N, Kazuki, K, Ueno, E, Akakura, Y, Katoh, M, Osaki, M, et al. (2011). Integration-free iPS cells engineered using human artificial chromosome vectors. PLoS One 6,e25961.

[62] Pelacho, B, Mazo, M, Gavira, J. J, & Prósper, F. (2011). Adult stem cells: from new cell sources to changes in methodology. J Cardiovasc Transl Res , 4, 154-160.

[63] Müller-ehmsen, J, Whittaker, P, Kloner, R. A, Dow, J. S, Sakoda, T, Long, T. I, Laird, P. W, & Kedes, L. (2002). Survival and development of neonatal rat cardiomyocytes transplanted into adult myocardium. J Mol Cell Cardiol , 34, 107-116.

[64] Teng, C. J, Luo, J, Chiu, R. C, & Shum-tim, D. (2006). Massive mechanical loss of microspheres with direct intramyocardial injection in the beating heart: implications for cellular cardiomyoplasty. J Thorac Cardiovasc Surg , 132, 628-632.

[65] Masumoto, H, & Yamashita, J. K. Strategies in cell therapy for cardiac regeneration. Inflammation and Regeneration. In press.

[66] Cortes-morichetti, M, Frati, G, & Schussler, O. Duong Van Huyen JP, Lauret E and Genovese JA. ((2007). Association between a cell-seeded collagen matrix and cellular cardiomyoplasty for myocardial support and regeneration. Tissue Eng , 13, 2681-2687.

[67] Okano, T, Yamada, N, Sakai, H, & Sakurai, Y. (1993). A novel recovery system for cultured cells using plasma-treated polystyrene dishes grafted with poly (N-isopropylacrylamide). J Biomed Mater Res , 27, 1243-1251.

[68] Furuta, A, Miyoshi, S, Itabashi, Y, Shimizu, T, Kira, S, Hayakawa, K, Nishiyama, N, Tanimoto, K, Hagiwara, Y, Satoh, T, et al. (2006). Pulsatile cardiac tissue grafts using a novel three-dimensional cell sheet manipulation technique functionally integrates with the host heart, in vivo. Circ Res , 98, 705-712.

[69] Ishii, M, Shibata, R, Numaguchi, Y, Kito, T, Suzuki, H, Shimizu, K, Ito, A, Honda, H, & Murohara, T. (2011). Enhanced angiogenesis by transplantation of mesenchymal stem cell sheet created by a novel magnetic tissue engineering method. Arterioscler Thromb Vasc Biol , 31, 2210-2215.

[70] Sekine, H, Shimizu, T, Hobo, K, Sekiya, S, Yang, J, Yamato, M, Kurosawa, H, Kobayashi, E, & Okano, T. (2008). Endothelial cell coculture within tissue-engineered cardiomyocyte sheets enhances neovascularization and improves cardiac function of ischemic hearts. Circulation 118,SS152., 145.

[71] Stevens, K. R, Kreutziger, K. L, Dupras, S. K, Korte, F. S, Regnier, M, Muskheli, V, Nourse, M. B, Bendixen, K, Reinecke, H, & Murry, C. E. (2009). Physiological function and transplantation of scaffold-free and vascularized human cardiac muscle tissue. Proc Natl Acad Sci USA , 106, 16568-16573.

Chapter 22

Human Pluripotent Stem Cells Modeling Neurodegenerative Diseases

Roxana Nat, Andreas Eigentler and Georg Dechant

Additional information is available at the end of the chapter

http://dx.doi.org/10.5772/55772

1. Introduction

Modeling of human neurodegenerative diseases in animals has led to important advances in the understanding of pathogenic mechanisms and has opened avenues for curative approaches. However, inherent genetic, developmental and anatomical species differences between humans and animals frequently resulted in imperfect phenotypic correlations between animal models and human diseases. This might account for the observed hampered translation of promising preclinical treatment studies in animal models towards clinics.

Pluripotent stem (PS) cells hold considerable promise as a novel tool for modeling human diseases. Human PS cells include human embryonic stem (hES) cells and induced PS (IPS) cells. IPS cells are generated *via* reprogramming of somatic cells through the forced expression of key transcription factors and share salient characteristics of ES cells, which are derived from the preimplantation blastocyst.

Both types of PS cells show the capacity to self-renew and to differentiate *in vitro* and *in vivo* into the cell types that make up the human body. This includes the various types of mature neurons affected by neurodegenerative diseases. The combination of the key advantages of PS cells allows for the first time to generate large numbers of postmitotic human neurons for preclinical research in cell culture. In particular, the IPS cell technology opens doors for intensified research on human PS-derived neurons because, in comparison to hES cells, ethical concerns can be dispelled. Furthermore, the isolation of patient-derived IPS cell lines from skin biopsies enables the study of pathogenic mechanisms in human cells carrying relevant pathogenic allelic constellations.

During recent years the generation of IPS cell lines from human material has become routine. However, for neurological research a remaining major challenge is to guide *in vitro* differentiation of IPS cells into defined and homogeneous neuronal populations that are required for modeling neurodegenerative diseases. A hallmark of human neurodegenerative diseases is the chronic and progressive loss of specific types of neurons: cerebral cortex glutamatergic and basal forebrain cholinergic neurons in Alzheimer's disease, midbrain dopaminergic neurons in Parkinson's disease, striatal GABAergic neurons in Huntington's disease, motor neurons in amyotrophic lateral sclerosis and spinal muscular atrophy, cerebellar and peripheral sensory neurons in ataxias and others. To fully tap into the potential of the IPS technology and to progress towards a fundamental understanding of the causes of disease selectivity in the loss of neuron subtypes it will be necessary to establish reproducible and tailored protocols for differentiation of IPS cells specifically into these neuronal subtypes *in vitro.*

To date, reprogramming of patient somatic cells into IPS cell-based models has been achieved for several neurodegenerative diseases. The results show that IPS cells or their derivatives can display at least some of the cellular and/or molecular characteristics of the respective diseases. These findings provide first proof for etiological validity of these models. Here, we review the existing reports demonstrating the generation of human PS cell-based models for neurodegenerative diseases, including also the studies showing the differentiation of human PS cells, both ES and IPS cells, toward telencephalic neurons (glutamatergic, GABAergic and cholinergic), midbrain dopaminergic neurons, cerebellar neurons, spinal motor neurons and peripheral neurons. We further discuss the perspectives of these cellular models.

2. Generation of human IPS cells

It was in 2006 when the first IPS cells were generated by Takahashi and Yamanaka *via* reprogramming of mouse somatic fibroblasts through retroviral transduction with a specific set of factors [1]. A screen of pluripotency-associated genes yielded a successful combination of transcription factors, comprising Oct4, Sox2, Klf4 and c-Myc (OSKM), which are commonly referred to as the 'Yamanaka factors'. Shortly afterwards, the same group [2], concurrently with other groups that used different combinations of transcription factors, for example substituting c-Myc and Klf4 by Lin28 or Nanog [3-5], were able to demonstrate that also fibroblasts obtained from adult human beings can be induced to undergo the transformation into PS cells.

Since these first descriptions of IPS cell derivation significant improvements in efficiency of the protocols, in the quality of the resulting IPS lines and in the depth of their analysis have been achieved. So far, fibroblasts remain the most popular donor cell type, and were used in more than 80% of all published reprogramming experiments. Figure 1 illustrates the steps in generating human IPS cell from skin fibroblasts, as well as cell morphology transition in culture.

However, other cell sources for inducing pluripotency have been used, amongst them keratinocytes [6], cord blood cells [7] and mesenchymal stem cells [8] with sometimes higher efficiency compared to fibroblasts. Furthermore, different combinations of reprogramming

http://dx.doi.org/10.5772/55772

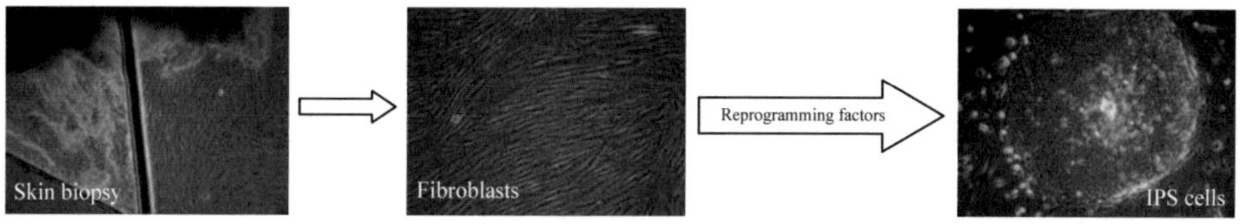

Figure 1. Generation of human IPS cells from a skin biopsy.

factors have been developed, ranging in number between two to six [3;4;9]. Each of these reprogramming factors contributes to the kinetics and efficiency of IPS induction.

Genetic material coding for these reprogramming factors has been introduced into cells via a variety of methods, comprising genome integrating as well as non-integrating techniques [10]. The most commonly used method for factor delivery is the transduction using retroviruses, originally with Moloney murine leukemia virus (MMLV), vectors, later on with modified lentiviral vectors. The efficiency of IPS cell generation using sets of four MMLV-derived retroviruses expressing single genes from the OSKM set separately is ~0.01% in human fibroblasts.

Silencing of the permanently integrated transgenes is important because only an IPS cell that has up regulated the endogenous pluripotency gene network but down regulated the expression of the transgenes can be considered fully reprogrammed [11]. Although the use of retroviruses is efficient and yields reproducible results, random insertional mutagenesis, permanent alteration of gene expression as well as reactivation of silenced transgenes during differentiation cannot be excluded. The use of Cre-deletable or dox-inducible lentiviruses has overcome some of these problems and allows factor expression in a more controlled manner [12;13]. Other attempts to generate integration-free IPS cells focused on replication-defective adenoviral vectors, or Sendai viral vectors [14;15] which efficiently deliver foreign genes into a multitude of cell types.

To avoid the use of viral vectors, direct delivery of episomal vectors (plasmids) as well as standard DNA transfections using liposomes or electroporation have also been used, but with low transfection efficiency [16-18]. A polycistronic expression cassette flanked by loxP sites enabled the excision of the reprogramming cassette after expressing Cre recombinase also in the non-viral system [19].

Alternatively, Warren et al. [20] developed a novel mRNAs-based system and achieved an efficient conversion of different human somatic donor cells into IPS cells using a direct delivery of high dosages of modified mRNAs encoding OSKM and Lin28 packaged in a cationic vehicle. The efficiency reached with this approach was much higher when compared with other non-integrative protocols [20].

Recently, a potential role of specific microRNAs (miRNAs) for pluripotency has been elucidated. The miRNAs from the miR-302 cluster contribute to unique ES cells features such as cell cycle and pluripotency maintenance [21;22]. Based on these findings protocols for highly efficient miRNA-mediated reprogramming of mouse and human somatic cells to pluripotency were reported [23;24]. The resulting miR-IPS cells are subject to a reduced risk of mutations and tumorigenesis relative to most other protocols because mature miRNAs function without genomic integration [23;24].

Finally, another promising possibility of inducing pluripotency is to deliver the reprogramming factors directly as proteins. To this end Zhou et al. generated recombinant OSKM proteins fused with a poly-arginine transduction domain [25]. However, this protein-based strategy induced pluripotency with extremely slow kinetics and poor efficiencies [25].

Apart from the delivery methods of reprogramming factors, other parameters, including culture conditions and the application of small pharmacological compounds, exert an influence on reprogramming efficiency. For example, it has been demonstrated that culturing IPS cells under hypoxic conditions mimicking the *in vivo* environment, enhances the efficiency rate [26]. The addition of small molecules, that either modifies epigenetic states like DNA methylation or histone acetylation, or influences specific receptor mediated signaling pathways, enhances the generation of IPS cells [27-31].

Eventually, the reactivation of endogenous pluripotency genes leads to establishment of cell lines with pluripotent characteristics. However, even though IPS lines share many characteristics with hES cells with regard to morphology and pluripotent gene expression, further research is required to establish more precisely communalities and differences between hES and IPS cells. Differences in epigenetic status and *in vitro* and *in vivo* differentiation potential have been reported [32-34].

3. Neuronal differentiation of human PS cells

The *in vitro* production of neurons from PS cells, following similar mechanism as in vivo development, involves several sequential steps precisely orchestrated by signaling events (reviewed in [35;36]).

In vivo, during embryonic development, the initial step is neural induction, the specification of neuroepithelia from ectoderm cells [37]. When the neuroectodermal fate is determined, the neural plate folds to form the neural tube, from which cells differentiate into various neurons and glia [38;39]. The neural tube is patterned along its anteroposterior (A/P) and dorsoventral (D/V) axes to establish a set of positional cues. The neural plate acquires an anterior character, and is subsequently posteriorized by exposure to Wingless/Int proteins (Wnt), fibroblast growth factors (FGF), bone morphogenic proteins (BMP) and retinoic acid (RA) signals to establish the main subdivisions of the central nervous system (CNS): forebrain, midbrain, hindbrain, and spinal cord, as well as the neural crest from which the peripheral neurons derive [40-42]. Therefore, the precursor cells in each subdivision along the A/P axis are fated to subtypes of neurons and glia depending on its exposure to unique sets of morphogens at specific concentrations (Figure 2).

As reviewed in Petros et al. [35], specific PS cell-bases protocols, following the principles of nervous system development, can generate neuronal types with markers consistent with telencephalic, midbrain, hindbrain spinal cord and peripheral neurons (Table 1).

http://dx.doi.org/10.5772/55772

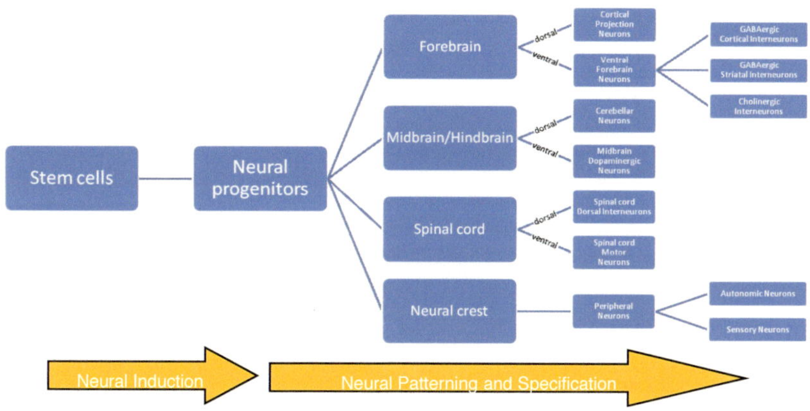

Figure 2. Stem cell fates aligned to nervous system development

Differentiated neural subtype	PS cell type	Key patterning differentiation factors	References
General telencephalic neurons	mES cells	DKK, LeftyA, Wnt3a, Shh	Watanabe et al. (2005)[43], Li et al. (2009)[44]
Cortical pyramidal neurons	mES cells hES cells	Cyclopamine, Fgf2, RA	Eiraku et al. (2008)[45], Gaspard et al. (2008)[46], Gaspard et al. (2009)[47], Ideguchi et al. (2010)[48], Nat et al 2012[36]
Cortical interneurons	mES cells hES cells	Shh, Fgf2, IGF, Activin	Maroof et al. (2010)[49], Danjo et al. (2011)[50], Goulburn et al. (2011, 2012) [51;52], Cambray et al. (2012)[53], Nat et al 2012[36]
Basal forebrain cholinergic neurons	hES cells	RA, bFGF, FGF8, Shh, BMP9	Wicklund et al (2010)[54], Bissonnette et al. (2011)[55]
Striatal medium spiny neurons	mES cells, hES cells	Shh, BDNF, DKK1, cAMP, valproic acid	Aubry et al. (2008)[56], Zhang et al. (2010)[57], Danjo et al. (2011)[50]
Floor plate cells	hES cells	Shh, dual SMAD inhibition	Fasano et al. (2010)[58]
Midbrain dopaminergic neurons	mES cells hES cells, hIPS cells	Shh, AA, FGF8, bFGF	Kawasaki et al. (2000)[59], Lee et al. (2000)[60], Perrier et al. (2004)[61], Yan et al. (2005)[62], Chambers et al. (2009)[63], Sánchez-Danés et al. (2012)[64]

Differentiated neural subtype	PS cell type	Key patterning differentiation factors	References
Cerebellar granule cells	mES cells, hES cells	Wnt1, Fgf8, RA, BMP 6/7, GDF7, Shh, JAG1	Salero and Hatten (2007)[65], Erceg et al. (2010)[66]
Cerebellar Purkinje cells	mES cells	BMP4, Fgf8	Su et al. (2006)[67], Tao et al. (2010)[68]
Spinal cord motor neurons	mES cells, hES cells, IPS cells, hMS cells, hADS cells	Shh, RA, SB431542, Olig2, HB9	Wichterle et al. (2002)[69], Li et al. (2005)[70], Soundararajan et al. (2006)[71], Lee et al. (2007)[72], Dimos et al. (2008)[73], Peljto et al. (2010)[74], Patani et al. (2011)[75], Park et al. (2012)[76], Liqing et al. (2011)[77]
Spinal cord interneurons	mES cells	Wnt3A, Shh, RA, BMP2	Murashov et al. (2005)[78]
Neural crest	hES cells, hIPS cells	SB431542, noggin, BDNF, NGF, AA, dbcAMP	Lee et al. (2010)[79], Menendez et al. (2011)[80], Goldstein et al. (2010)[81]

Table 1. Neural cell types derived from PS cells to date (modified from Petros et al. 2011[35])

Recognizing that all resulting cell populations, although enriched in specific neurons, remain heterogeneous, there is a need for additional selection methods to further purify neuronal subtype lineages. Whilst a key aim of positionally specifying human neurons is to work towards the generation of cell-based therapies for diseases that target a sub-population of cells, this system will be particularly powerful in attempting to understand disease specificity when applied to patient-derived IPS cells.

4. Neurodegenerative diseases and related models

Neurodegenerative diseases are characterized by the chronic and progressive loss of neuronal functions in selected neurons. Classical neurodegenerative diseases are Alzheimer's disease, Parkinson's disease, Huntington's disease, amyotrophic lateral sclerosis, spinal muscular atrophy and ataxias. Other rare diseases such as familial dysautonomia and Fragile-X syndrome contain neurodegenerative aspects as well.

Here we aim to present the main characteristics of these diseases, focusing on their pathogenesis and its reflection into the disease models, including the recent cellular models derived via IPS cell technology. The most important publications and aspects regarding the IPS cell-related models for neurological diseases are reviewed in Han et al. 2011[82] and updated for the neurodegenerative diseases in Table 2.

Neurodegenerative disease	Types of affected neurons	Histopathology	Gene (Mutation)	Donor cell	Reprogramming method	Reported disease-related phenotype	References
Alzheimer's disease	Basal forebrain cholinergic neurons, cortical neurons	Neurofibrillary tangles, Amyloid plaque, Loss of neurons and synapses	PS1, PS2 mutations; Sporadic and APP duplication	SF	LV: OSKLN RV:OSKM	Increased amyloid β42 secretion increased β(1-40) and phospho-τ levels	Yagi et al. (2011)[83], Israel et al. (2012)[84]
Parkinson's disease	Midbrain nigro-striatal dopaminergic neurons	Lewy-bodies, loss of dopaminergic neurons	idiopathic	SF	LV: Cre-excisable, DOX-inducible; OSK or OSKM	NA	Soldner et al. (2009)[12], Hargus et al. (2010)[85]
			LRRK2 (G2019S)	SF	RV: OSK	Increased caspase-3 activation and DA neuron death with various cell stress conditions	Nguyen et al. (2011)[86]
			PINK1 (Q456X; V170G)	SF	RV: OSKM	Impaired stress-induced mitochondrial translocation of Parkin in DA neurons	Seibler et al. (2011)[87]
			SCNA triplication	SF	RV: OSKM	Increased neural α-Synuclein protein levels sensitivity to oxidative stress	Devine et al. (2011)[88], Byers et al. (2011)[89]
Huntington's disease	Striatal GABAergic medium spiny neurons, cortical neurons	Neural inclusion bodies, loss of striatal/cortical neurons	HTT (CAG repeats)	SF	RV: OSKM	Increase in lysosomal activity	Park et al. (2008)[4], Camnasio et al. (2012)[90]
Amyotrophic lateral sclerosis	Upper and lower motor neurons	Ubiquitinated inclusion bodies, loss of motoneurons	SOD1	SF	RV: OSKM, OSK	NA	Dimos et al. (2008)[73], Boulting et al. (2011)[33]
Spinal muscular atrophy, type I	Spinal motor neurons	Loss of anterior horn cells	SMN1 deletion	SF	RV: OSNL Episomal plasmid OSKMNL combinations	Reduced number of motor neurons, decreased soma size, synaptic defects	Ebert et al. (2009)[91] Sareen et al. (2012)[92]
Friedrich Ataxia	Dorsal root ganglia (DRG) peripheral	Reduced size of DRG-neurons, iron	FXN (GAA expansion)	SF	RV: OSKM	GAA repeat instability	Ku et al. (2010)[93], Liu

Neurodegenerative disease	Types of affected neurons	Histopathology	Gene (Mutation)	Donor cell	Reprogramming method	Reported disease-related phenotype	References
	neurons, cerebellar neurons	misdistribution, decreased myelination					et al. (2010) [94]
Spinocerebellar Ataxia Type 3 (Machado-Joseph Disease)	Cerebellar neurons, striatal and cortical neurons	Intranuclear inclusion bodies, neuronal loss	ATAXIN 3(CAG expansion)	SF	RV: OSKM	NA	Koch et al. (2011)[95]
Familial dysautonomia	Sensory and autonomic neurons	Reduced size of DRG neurons, reduced number of non-myelinated small fibers and intermediolateral column neurons	IKBKAP	SF	RV: OSKM	Defects in neurogenesis and migration	Lee et al. (2009)[96]
Fragile-X syndrome	Hippocampal, cerebellar neurons	Dendritic spine abnormalities, neuronal loss	FMR1 (CGG repeat)	SF	RV: OSKM	NA	Urbach et al. (2010)[97]

SF-skin fibroblasts, RV retroviruses, LV-lentiviruses and NA- not assessed

Table 2. Overview of the iPS the cell-related models for neurodegenerative diseases (modified from Han et al. 2011 [82])

4.1. Alzheimer's disease

Alzheimer's disease (AD) is the most common neurodegenerative disease, affecting 35 million patients worldwide. Clinically, it is characterized by progressive loss of short-term memory and other cognitive functions toward a state of profound dementia.

AD is histopathologically characterized by neuronal and synapse loss and the appearance of extracellular amyloid plaques (AP) and intracellular neurofibrillary tangles (NFTs) in affected brain regions, especially cerebral cortex, hippocampus and basal forebrain [98;99]. The AP and NFTs form by aggregation of two proteins, beta amyloid (Aβ) and hyperphosphorylated tau (pTau), respectively [100]. Aβ is formed from the cleavage of the amyloid precursor protein (APP) into soluble monomers that then aggregate into fibrils and are eventually deposited in the extracellular space [101]. Tau is a microtubule-associated protein that undergoes hyperphosphorylation and accumulates as intraneuronal inclusions or tangles in the brains of individuals with AD [100;102].

Degeneration of basal forebrain cholinergic neurons is a principal feature of AD and the reduction in the level of acetylcholine and choline acetyltransferase activity in the hippocampus and cerebral cortex has been reported in the brains of AD patients [103;104].

The majority of AD cases are sporadic; in these cases the major genetic risk factor disease is the *APOE* gene. ApoE is synthesized in astrocytes and acts as a ligand for the receptor-mediated endocytosis of cholesterol-containing lipoprotein particles. Whether ApoE affects Aβ clearance or operates through its function in lipid metabolism is not yet fully established [105].

Few familial AD (FAD) cases are an early-onset autosomal dominant disorder. Three genes have been identified that account for FAD: the first mutations causing Mendelian AD were identified in the *APP* gene [106], although mutations in two other genes, *presenilin 1* and *2* (*PSEN1* and *PSEN2*), that form the γ-secretase complex components, are more commonly found. The mutations cause different clinical phenotypes, but for all the aberrant processing of Aβ led to its aggregation [107].

By classical transgene and knockout approaches, there were established mouse models that reflect different aspects of AD [108]. Representative models are APP mutant strains (such as PDAPP, J20, APP23 or Tg2576) with a robust APP/Aβ pathology and tau mutant strains with NFT formation such as (JNPL3 or pR5). The histopathology in these strains is associated with behavioral impairment [109].

The modeling of AD via IPS cell technology was recently reported [83;84]. The first study used AD patient fibroblasts carrying mutations in PS1 and PS2. The IPS cells kept the mutations and differentiated into neural cells, showing increased amyloid β42 secretion as compared to the healthy controls [83].

In the second study, IPS cells were generated from both patients with sporadic AD or caring APP duplication. Interestingly, increased level of both Aβ (1-40) and pTau were detected in neural cells cultures after neural progenitor expansion of about five weeks, followed by differentiation of about four weeks [84].

4.2. Parkinson's disease

Parkinson's Disease (PD) is the second most common neurodegenerative disorder, afflicting over 6 million people worldwide. Clinically, there are progressive motor dysfunctions comprising bradykinesia, rigidity and tremor, as well as non-motor features.

Pathologically, PD is identified by intracellular inclusions known as Lewy bodies and dopaminergic neuronal loss that initiates in the substantia nigra.

PD is largely a late onset sporadic neurodegenerative condition. However, 5–10% cases are familial, transmitted in either an autosomal-dominant or autosomal recessive fashion [110]. A number of genes have been linked to our understanding of pathogenesis. The gene α-synuclein (*SNCA)* product is the major component of the Lewy body in sporadic and in some cases of autosomal dominant types and therefore appears to be central to PD pathophysiology [111;112]. The most common mutation related to autosomal-dominant PD occurs in the gene encoding leucine-rich repeat kinase-2 (LRRK2) [113]. One missense mutation, the G2019S mutation, occurs in 5% of familial cases and 1–2% of sporadic cases of PD. Mutations in *PARK2*, *PINK1* and *PARK7* (also known as *DJ1*) cause autosomal-recessive, early onset PD [114-116]. These genetic discoveries have highlighted the importance of the ubiquitin proteasome system, mitochondrial dysfunction and oxidative stress in PD pathogenesis.

The most common genetic risk factor for PD appears to be heterozygous mutations in the *glucocerebrosidase* gene (*GBA*) [117]. The frequency of heterozygous mutations in *GBA* reaches ~4% in sporadic PD populations.

Because PD results from the loss of dopaminergic neurons, the prospect of utilizing cell replacement therapies has attracted substantial interest. Several methods are able to improve the effectiveness of midbrain dopamine neuron generation and/or retrieval from fetal tissue and stem cells.

The ability of deriving large quantities of correctly differentiated dopamine neurons makes stem cells promising cell sources for transplantation in PD; having the transplantation as a main goal, many studies improved the directed differentiation of PS cells toward dopaminergic neurons, opening the doors to IPS cell-derived models.

Soldner *et al.* induced pluripotency in fibroblasts derived from idiopathic PD patients and controls and subsequently differentiated both into dopaminergic neurons. As they did not find significant differences between the expression of *SNCA* or *LRKK2* between patients and controls, they went on to suggest that it might still be necessary to further accelerate PD-pathology related phenotypes *in vitro* with neurotoxins such as MPTP, or the overexpression of PD-related genes such as *SNCA* or *LRKK2* in order to obtain a valid PD model [12].

Hargus et al. [85] used a similar protocol of inducing PS cells for idiopathic PD patients and controls, and further differentiated them into dopaminergic neurons. Additionally, they performed intrastriatal transplantation studies into 6-OHDA lesioned rats, demonstrating improvements in motor symptoms.

Regarding familiar PD, Nguyen *et al.* [86] used a classical protocol for IPS cells generation and differentiation and found that IPS cell-derived dopaminergic neurons from patients carrying a LRRK2 mutation had increased expression of oxidative stress response genes and α-synuclein protein. The mutant neurons were also more sensitive to caspase-3 activation and cell death caused by exposure to hydrogen peroxide, MG-132 (a proteasome inhibitor), and 6-hydroxydopamine than control neurons. The finding of increased susceptibility to stress in patient-derived neurons provides insights into the pathogenesis of PD and a potential basis for a cellular screen.

Seibler *et al.* [87] generated IPS cells form PD patients carrying mutation in PINK1 gene (Q456X; V170G). They compared the mitochondrial translocation of Parkin in DA neurons under mitochondrial stress conditions and found a difference between patients and controls, making a step forward into PD pathogenesis *in vitro*.

Two recent studies focused on the IPS cell-derived models of PD carrying a triplication in SNCA genes. Devine *et al.* showed that the levels of α-Synuclein protein were increased in the dopaminergic population derived from patients, compared to the healthy controls [88], while

http://dx.doi.org/10.5772/55772

Byers *et al.* focused on the differences in sensitivity to oxidative stress in correlation with this mutation [89].

4.3. Huntington's disease

Huntington's disease (HD) is an autosomal dominant neurodegenerative disorder resulting from an expanded CAG triplet repeat in the Huntingtin gene (*HTT*) on chromosome 4 [118]. This expansion accounts for an attachment of a polyglutamine strand of variable length at the N-terminus of the protein leading to a toxic gain of function [119]. HD together with eight other CAG triplet repeat expansion disorders forms the group of PolyQ diseases which share some specific pathophysiological features [120].

Although the protein huntingtin is ubiquitously expressed in mammalian cells, mainly striatal GABAergic medium spiny neurons with a dopamine- and cyclic AMP-regulated phosphoprotein (DARPP-32)-positive phenotype are the most susceptible to neurodegeneration in HD [121]. As a consequence a prominent cell loss and atrophy in the caudate nucleus and putamen can be observed. Other brain regions and neuronal subtypes involved in HD comprise the substantia nigra, hippocampus, cerebellar Purkinje cells and thalamic nuclei [119;122].

One of the histopathological hallmarks of Huntington's disease, as in other PolyQ disorders too, is the appearance of nuclear and cytoplasmic inclusion bodies containing the mutant huntingtin and polyglutamine [123;124]. Much debate regarding the meaning and function of these inclusions is going on, and although indicative of pathological mutant protein processing they do not correlate with cellular dysfunction and might even confer a protective role [125;126].

Numerous studies indicated that wild-type huntingtin might be involved in a variety of intracellular functions such as in protein trafficking, vesicle and axonal transport, mitochondrial function, postsynaptic signaling; transcriptional regulation, as wells as in anti-apoptotic pathways [127;128]. Therefore a disruption and detrimental impairment of these various intracellular pathways is supposed to be the consequence of accumulation of mutant huntingtin, finally leading to neuronal death.

Over the years, several different HD models had been introduced, ranging from invertebrate models like Drosophila and C. elegans to various rodent models [129;130]. Genetically modified animals (especially mouse) models such as transgenic, knock-in and conditional ones recapitulated some features of HD like neuronal polyglutamine inclusions [131].

The intrastriatal injection of excitotoxic glutamic acid analogues like kainic acid, quinolinic acid and 3-nitropropionic acid into animals resulted in neuronal cell death similar to the pathology observed in HD patients [132-134]. They proved to be useful in studying pathogenetic processes involved in the progressive disease course although some limitations regarding the selective neuronal cell loss as well as aggregate formation properties and variable phenotypes have to be kept in mind.

Transplantation studies in animal HD models aimed at providing neuroprotective support or intended to replace damaged and lost neuronal subtypes. Successful application of stem cell-based therapy in animal models of HD with functional recovery has been reported [135;136]. Different cell types ranging from neural stem/progenitor cells from mouse and rat or human fetal brain tissue to bone marrow and mesenchymal stem cells have been transplanted into excitotoxic animal HD models[137].

In order to facilitate research in the HD field with human material, Bradley *et al.* [138] derived four hES cell lines containing more than 40 CAG repeats from donated embryos obtained through an informed consent. Those hES cells were able to differentiate in to neuronal cells expressing the mutant huntingtin protein.

The first HD-IPS cell lines were successfully generated by Park *et al.* from patient with a 72 CAG repeat tract using the classical lentiviral vectors [4]. In a subsequent study Zhang et al used these patient-specific IPS cells in order to generate HD specific neural stem cells that were then differentiated into striatal neurons. Besides a stable CAG repeat expansion in all patient-derived cells, an enhanced caspase 3/7 activity was found.

A second group successfully generated HD-specific IPS cells via lentiviral transduction of transcription factors and was able to demonstrate a stable CAG triple repeat length in all IPS cell clones as well as in IPS cell-derived neurons. Interestingly, they observed an enhanced lysosomal activity in IPS cells and their derived neuronal populations [139].

4.4. Motor neuron diseases

Motor neurons (MNs) are essential effector cells for the control of motor function. Degenerative MN diseases, such Spinal Muscular Atrophy (SMA) and Amyotrophic Lateral Sclerosis (ALS) are devastating disorders due to a selective loss of MNs, which in turn leads to progressive muscle atrophy and weakness.

SMA is the most common form of degenerative motor neuron disease in children and young adults, characterized by the selective degeneration of lower MNs in the brainstem and spinal cord [140]. SMA is a classical autosomal recessive disorder with the vast majority of SMA cases caused by homozygous mutations in the gene named Survival of Motor Neuron-1 (SMN1) [141;142].

Interestingly, in humans the SMN exists in a telomeric copy, SMN1, and several centromeric highly homologous copies, SMN2, with both genes being transcribed [143]. Due to the fact that the vast majority of SMN2 transcripts lack an exon due to a splicing defect, it is only partially and poorly able to compensate for reduced SMN1 levels [144;145].

SMN is a ubiquitously expressed gene involved in the biogenesis of small nuclear ribonucleo-proteins important for pre-mRNA splicing, but might also have a specific role in RNA transport in neurons [146]. However, it remains to be elucidated how a deficiency in SMN is responsible for the selective degeneration of lower motor neurons [147].

Several experimental models have been used to study the putative cellular and molecular processes involved in SMA. Mouse models have become the most often used, albeit lacking the duplication of the SMN gene in humans. As a consequence, homologous recombination technology of the *Smn* locus in mice leads to complete depletion of the SMN protein, causing early embryonic lethality [148], which has necessitated generating transgenic mice that harbor human SMN2 [149;150] on a SMN-/- background. Although this model provided invaluable protein and disease information, reflecting a gene dosage–dependent phenotype similar to severe forms of SMA, these mice normally die shortly after birth.

It was in 2009 when patient-derived IPS cells were used for the first time to model SMA [91]. Therefore, skin fibroblasts from a three-year old child with SMA as well as from the unaffected mother were successfully reprogrammed via transduction with lentiviral vectors comprising OCT4, SOX2, NANOG, and LIN28. Characterization of the obtained IPS cells demonstrated lack of SMN1 expression and reduced levels of the full-length protein compensated by SMN2. Patient and control IPS cells were further differentiated into neurons. Within these neural cultures, significant differences regarding the number of motoneurons as well as their soma size and synapse formation ability could be observed between patient-specific and control cells, therefore reflecting disease-specific phenotypes. Furthermore, valproic acid and tobramycin, two drugs known to increase full-length SMN mRNA levels from the SMN2 locus, were tested on this human cellular SMA model. A 2-3-fold increase in SMN protein expression in SMA-IPS cells and an increased nuclear punctuate localization of SMN protein were found.

In a continuative experiment, Sareen et al [92] generated SMA-specific IPS cells using a virus-free plasmid-based approach with subsequent differentiation of IPS cells into NSCs and further MN differentiation. Besides the already described SMA-specific phenotypes, increased apoptosis was detected in SMA-specific cells, which might be another potential target for therapeutic intervention.

ALS, also known as Lou Gehrig's disease is the most common form of MN disease; it as a rapidly progressing, fatal disorder, usually as a result of respiratory failure. In contrast to SMA, it is characterized by a progressive loss of both upper and lower motor neurons in the cerebral cortex, brainstem and spinal cord. Two forms of ALS can be distinguished, the more frequent sporadic form accounts for about 90% of cases and the less common familial form (FALS) for the remaining 10% [151].

Mutations in the Cu/Zn superoxide dismutase 1 (SOD1) gene are responsible for about 20% of the familial cases [152]. Recently, several other gene mutations were identified as important causing typical FALS, such as the gene encoding the TAR DNA-binding protein 43 (TDP-43) [153]. The role of TDP-43 was first suspected when it was identified as one of the major constituents of the intra-neuronal inclusions characteristically observed in ALS and in frontotemporal lobar degeneration–ubiquitin (FTLD-U); [154]. Subsequently, mutations in the TARDBP gene encoding TDP-43 were identified in some FALS [153].

On macroscopic and microscopic examination of the nervous system in ALS variable neuronal inclusion bodies in lower motor neurons of the spinal cord and brain stem can be detected [155]. Morphologically these inclusions are reliably demonstrated only by their immunoreactivity to ubiquitin, and have been reported in both sporadic and familial cases and are present in transgenic models of ALS. It is now well established that ALS is typically characterized by the presence of these inclusion bodies.

Furthermore, it has been reported that an ALS genotype in glial cells (astrocytes) has an effect on the survival of motor neurons and contributes a crucial role in motor neuron degeneration [156].

ALS research has focused mainly on models of the familial SOD1-mediated form, although all forms of ALS share striking similarities in pathology and clinical symptoms. A toxic gain of function of this enzyme with the exact mechanism still unclear is thought to be responsible which subsequently results in mitochondrial dysfunction, oxidative damage, glutamate excitotoxicity, protein aggregation, proteasome dysfunction, cytoskeletal and axonal transport defects and inflammation [151;157].

Transgenic mice or rats overexpressing mutant SOD1 develop MN degeneration with progressive muscle weakness, muscle wasting and reduced life span [158]. Furthermore, mutant SOD1 as well as TDP-43 models have been generated in zebrafish and C. elegans, mimicking at least some of the pathological hallmarks (e.g. selective vulnerability of MN and MN dysfunction) and therefore making them suitable for genetic and small compound screening [157;159].

Transgenic ALS models have also already been utilized for stem cell therapies by transplanting different types of cells comprising human as well as rodent fetal neural stem and progenitor cells, umbilical cord blood stem cells, mesenchymal stem cells and bone marrow. In some of the studies, a moderate improvement of motor function and a delayed disease progression could be observed [160]. However, the translation of stem cell transplantation therapies into clinical trials did not show any therapeutic benefit in ALS patients.

In order to get more insight into human pathophysiology, Dimos *et al.* [161] were the first to generate ALS-patient specific IPS cells using retroviral transduction of the classical Yamanaka factors OSKM. They successfully obtained IPS cells from an 82-year old sibling suffering from a familial form of ALS with a mutation in the SOD1 gene. Subsequently, patient-specific IPS cells were forced to differentiate into MN and glia. Due to the fact that more than 90% of ALS cases are sporadic, patient-specific IPS cell models from sporadic ALS might overcome this drawback through the integration of the genetic as well as environmental individual background.

4.5. Ataxias

The degenerative ataxias are a group of hereditary or idiopathic diseases that are clinically characterized by progressive ataxia resulting from degeneration of cerebellar-brainstem structures and spinal pathways [162].

Autosomal recessive cerebellar ataxias are heterogeneous, complex, disabling inherited neurodegenerative diseases that become manifest usually during childhood and adolescence.

Friedreich Ataxia (FRDA), an autosomal-recessive ataxia, is the most common inherited ataxic disorder in the white Caucasian population with a prevalence of 2-4/100,000 and with an age of onset in the teenage years. Clinical characteristics include progressive ataxia of gait and limbs, dysarthria, muscle weakness, spasticity in the legs, scoliosis, bladder dysfunction, and loss of position and vibration sense [163;164]. Cardiomyopathy and diabetes mellitus are systemic complications in some patients [165].

FRDA is caused in 96% of individuals by a GAA triplet expansion in the first intron of the Frataxin (*FXN*) gene on chromosome 9q13 [166]. The mutation leads to transcriptional silencing as a result of heterochromatin formation, adoption of an abnormal DNA-RNA hybrid structure, or triplex DNA formation [167] with reduced Frataxin protein expression. About 4% of the individuals affected with FRDA are compound heterozygous. Disease-causing expanded alleles present with 66 to 1700 GAA repeats with the majority ranging from 600 up to 1200 GAA repeats [166;168]. Major neuropathologic findings comprise a degeneration of dorsal root ganglia (DRG), with loss of large sensory neurons, followed by degeneration of posterior columns, corticospinal tracts and spinocerebellar tracts, and the deep nuclei in the cerebellum [165;169].

The gene product Frataxin is a ubiquitously expressed and evolutionary conserved mitochondrial protein that has been proposed to exhibit roles in mitochondrial iron metabolism and the production of iron-sulfur (Fe-S) clusters.

Several FRDA disease models, from yeast, C. elegans and Drosophila to mice have been used to get more insight into the disease [170;171]. Viable transgenic mouse models were generated through conditional gene targeting [172] which have been crucial in the development as models for FRDA, although some with ambiguous results. The complete knock-out of Frataxin resulted in embryonic lethality [173], whereas conditional mouse models under the control of different promoters were capable to recapitulate some of the disease phenotypes [174]. In order to circumvent the non-physiologic complete loss of Frataxin at a specific time point in conditional models, GAA based mouse models were introduced [175;176], shedding more light on tissue-dependent GAA dynamics and putative pathophysiologic pathways.

Despite a general genotype-phenotype correlation it is not possible to predict the specific clinical outcome in any individual based on GAA repeat length. The inherent variability in FRDA may be caused by genetic background, somatic heterogeneity of the GAA expansion [177;178], and yet other unidentified factors.

Therefore, FRDA-IPS cell lines have already been established by Ku et al. [93] and Liu et al. [94]. Data showed that, although a specific disease-related phenotype was not reported, these FRDA IPS cells were able to recapitulate some of the molecular genetic aspects of FRDA, including the phenomenon of repeat-length instability, epigenetic silencing of the FXN locus and low levels of Frataxin expression [93].

With regard to GAA repeat instability, IPS cells showed repeat expansions whereas parental fibroblasts did not [93]. Instability was specific to the abnormally expanded FXN as GAA expansions in normal FXN alleles or at two unrelated loci with short GAA repeats remained unchanged. To understand the mechanism of instability in this IPS cell system, analysis of differences in mRNA expression showed that MSH2, a critical component of the DNA mismatch repair (MMR) machinery and important for mediating repeat-length instability, was highly expressed in FRDA-IPS cells relative to donor fibroblasts. ShRNA-mediated silencing of MSH2 resulted in shorter repeat lengths suggesting that FRDA IPS cells could be a useful system to evaluate the mechanisms of repeat expansions and contractions in disease.

GAA repeat mutations are unstable and progressive and postnatal instability occurs in various tissues throughout life. For example, large GAA repeat expansions are especially prominent in the dorsal root ganglia of FRDA patients, which harbor cell bodies of sensory neurons, a neuronal subtype especially affected in FRDA [179].

Given FRDA-IPS cells can be directed to differentiate into sensory neurons, as well as cardiomyocytes [94], the presence and mechanisms of tissue-specific expansion should be testable. The major focus of FRDA IPS cell differentiation research is currently focused on generating appropriate disease-relevant cell types. For example, sensory neurons of the DRG are crucially affected in individuals with FRDA.

The autosomal dominant Spinocerebellar Ataxias (SCAs) comprise a genetically and clinically heterogeneous group of inherited neurodegenerative progressive disorders affecting various parts of the CNS. The number of known SCAs continues to grow and comprises meantime over 30 entities.

Spinocerebellar ataxia type 3 (SCA3), also known as Machado-Joseph disease (MJD), is the most frequent entity among the autosomal dominantly inherited cerebellar ataxias in Europe, Japan, and the United States [180].

Genetically, SCA3 belongs to the group of CAG-triple repeat disorders, also known as PolyQ-disorders due to abnormally long polyglutamine tracts within the corresponding protein. The majority of patients suffering from SCA3 carry one allele of the ataxin3 (*ATXN3*) gene with 60–82 CAG repeats and a second allele containing the normal number of repeats, which is usually between 13 and 41 [181].

As in most of these polyglutamine diseases, patients with a repeat expansion above a critical threshold form neuronal intranuclear inclusion bodies, one important hallmark of polyQ diseases [182]. Further neuropathological features include a depigmentation of the substantia nigra as well as a pronounced atrophy of the cerebellum, pons and medulla oblongata, altogether culminating in an overall reduced brain weight compared to healthy individuals [183].

As most of the PolyQ disease proteins are ubiquitously expressed it still remains unclear why only specific neuronal cell populations are prone to neurodegeneration. Many animal models, like rodents, C.elegans and Drosophila, overexpressing specific forms of ATXN3 are available to study the molecular and phenotypic aspects of MJD involving aggregation, proteolysis and toxicity of expanded ATXN3, as well as the apparent neuroprotective role of wild-type ATXN3 [184].

Kakizuka's group was the first to demonstrate neurodegeneration and a neurological phenotype in mice transgenic for the CAG repeat expansion [185]. Mouse models further provided evidence for the subcellular site of pathogenesis, the processing and trafficking of the mutant protein in order to cause cellular dysfunction and neuronal cell loss.

While some of the transgenic mouse models expressing the full-length ATXN3 under control of various exogenous promoters were able to mimic some aspects of the disease, they all overexpress only a single isoform of ATXN3. Taken this into account, a YAC MJD transgenic model was established which more closely recapitulates the human disease as all elements, including regulatory regions of the gene, are present [186]. Research in animal models of SCA has now begun to focus on therapeutic strategies to prevent protein misfolding and aggregation in polyglutamine diseases by overexpressing chaperones.

Koch *et al* [95], investigated the formation of early aggregates and their behavior in time by making use of patient- specific IPS cell-derived neurons. They demonstrated that MJD-IPS cell derived neurons constitute an appropriate cellular model in the study of aberrant human protein processing. Moreover, they concluded that neurons are able to cope, at least in the beginning, with the aggregated mutant material and cytotoxicity evolved over time. Besides, a key role for the protease calpain in ATXN3-aggregation formation was found which could further display a putative benefit of calpain inhibitors.

4.6. Familial dysautonomia and fragile X syndrome

Familial Dysautonomia (FD), also known as Riley-Day Syndrome or Hereditary Sensory Autonomic Neuropathy (HSAN) Type III, is a rare autosomal recessive disease mostly occurring in persons of Ashkenazi Jewish descent [187]. The disease is characterized by degeneration of sensory and autonomic neurons, leading to severe and often lethal central and peripheral autonomic perturbations, as well as small-fiber sensory dysfunction. The underlying mutation induces a splicing defect in the IkB kinase complex-associated protein (*IKBKAP*) gene, which results in tissue-specific loss of function or reduced levels of the IKAP protein [188]. Individuals affected with FD suffer from incomplete neuronal development as well as progressive neuronal degeneration with the sensory and autonomic neurons mainly affected [189].

Although the exact function of the IKAP protein is not clearly understood, researchers have identified IKAP as the scaffold protein required for the assembly of a holo-elongator complex [190]. As a consequence, an impaired transcriptional elongation of genes responsible for cell motility is thought to be the cause for the observed cell migration deficiency in FD neurons [191]. Besides, the IKAP protein is also thought to be involved in other cellular processes, including tRNA and epigenetic modifications and exocytosis [192].

To better understand the function of IKAP, Dietrich *et al.* [193] created a mouse model with two distinct alleles that result in either loss of Ikbkap expression, or expression of the mutated truncated protein. Besides, a humanized IKBKAP transgenic mouse model for FD had been created that recapitulated the tissue-specific splicing defect, i.e. skipping of exon 20, in nervous tissues [194].

In order to untangle the tissue-specific pattern of IKBKAP mRNA splicing in FD, Boone et al. [195] created a human olfactory ecto-mesenchymal stem cell (hOEMSC) model of FD. It has been shown that these multipotent hOE-MSCs exhibit the potential to differentiate in vitro into neurons, astrocytes, and oligodendrocytes as well as other cell types [196]. Classical features of the FD phenotype, like the expression of the mutant IKBKAP transcript, notably lower IKBKAP levels as well as an impaired migration, were observed. Besides, drug testing experiments with kinetin, which had been shown effective in previous studies [197], had the potential to correct the splicing in a dose-dependent manner in FD hOE-MSCs.

Furthermore, IPS cells were generated from a patient with FD using the classical Yamanaka factors and subsequently differentiated into neural crest derivatives [96]. This was one of the earliest reports of a phenotype for a neurological disease to be modeled with IPS cells. FD-IPS cell derived neural precursors showed particularly low levels of IKBKAP, mis-splicing of IKBKAP, and defects in neurogenic differentiation and migration behavior. Again, the plant hormone kinetin was tested as a candidate and showed a reduction of mutant IKBKAP splice forms, an improvement in neuronal differentiation, but not in cell migration.

Fragile-X (FX) syndrome belongs to the autism spectrum disorders, and is the most common cause of inherited mental retardation with a prevalence of 1/3600 [198]. In the vast majority of cases, the disease is caused by a silencing of the FMR1 gene due to a CGG repeat expansion (>200 repeats) in the 5-UTR of the *FMR1* gene [199]. The FMR1 gene codes for the cytoplasmic protein FMRP, which has RNA-binding properties and is thought to play a role in synaptic plasticity and dendrite maturation. This could be demonstrated in histopathological studies of FX where dendritic spine abnormalities were found [200].

Several animal models revealed important insights into the role of the FMR protein. A Drosophila model showed a role of FRMP in the regulation of the microtubule network [201].

The first fmr1 KO mouse model was generated shortly after the discovery of the disease-causing gene and showed classical clinical features of FXS like macroorchidism, learning deficits, and hyperactivity[202].

Although current mouse models for FX syndrome are useful for studying the clinical phenotype, they do not recapitulate the hallmark, i.e. silencing of the FMR1 gene due to the triplet repeat expansion [203]. Loss of function studies using morpholino antisense oligonucleotides in zebrafish revealed a function of FMRP in terms of normal axonal branching.

Primary and transformed cell cultures obtained with an unmethylated full mutation in the FMR1 showed that the CGG expansion per se does not block transcription [204]. In undifferentiated human FX embryonic stem cells (FX-ES cells) derived from affected blastocyst-stage embryos, FMR1 is expressed and gene silencing occurs only upon differentiation [205] indicating a developmentally dependent process.

Recently, Urbach et al. [97] generated FX-IPS cell lines from three patients. In contrast to FX-ES cells, FX-IPS cells presented with a transcriptionally silent FMR1 gene, both in the pluripotent and differentiated states. This was reflected by corresponding epigenetic heterochromatin modifications in the gene promoter. IPS cells were further differentiated into neural derivatives and different potential epigenetic modifiers were tested. Amongst those, 5-azacytidin showed an upregulation of FMR1 transcripts both in pluripotent as well as neuronal FX-cells.

http://dx.doi.org/10.5772/55772

5. Conclusions and perspectives

In this chapter we have described the first successful attempts to harness the IPS technology for the generation of models for neurodegenerative diseases of the human nervous system. The key advantage of IPS based models over animal models is that they offer researchers for the first time a realistic chance to work in cell culture with large numbers of primary human cells that closely resemble the postmitotic neurons affected by neurodegeneration.

The first studies in which patient-derived disease-susceptible cellular phenotypes were compared with those of cells derived from healthy individuals, provide strong indications that such cellular models reflect key pathological molecular and cellular aspects of the neurological diseases. Therefore a future concept for patient-derived cellular models will be to correct neuronal malfunctions diseases by *in vitro* treatment of affected cells. A first such attempts aspect has been in the SMA models [91].

These *in vitro* treatments will include hypothesis driven approaches based on knowledge about pathophysiological mechanisms. Equally important patient derived lines will be used as *in vitro* assays for the screening of compound libraries. Drug safety screens with IPS cell-derived neurons will help to reduce the animal dependency of the current drug development pipeline. Finally, IPS cell technology will be an important driver of personalized medicine. Prior to patient treatment drug types and doses can be tested on patient-derived IPS cells or differentiated progenies in order to tailor a personalized curative approach according to the individual genetic and cellular profile.

There is even hope that the novel approach bypasses the laborious, time-consuming and expensive IPS cell generation by direct reprogramming of mouse and human somatic cells into functional neurons, called induced neurons (INs) [206;207], will come to fruition. Several groups have already generated dopaminergic INs [208;209] and motor INs [210]. Patient-specific INs could be generated to enhance the study of developmental disorders and other neurological diseases [211]. The significant decrease in time and resources to derive neurons directly from somatic cells justifies further investigation into this strategy.

But despite the enormous potential of IPS cell derived neurons for studies involving cell biological, physiological and pharmacological methods important question remain to be solved. One major drawback is that we still know very little about the specific cell biology of IPS cells and even less of their neuronal derivatives. This includes for example changes in chromatin structure and epigenetic signatures that accompany the reprogramming process. And there is exceedingly little information about membrane physiology of the IPS cell- derived neurons. Electrophysiological recordings and parallel studies of synaptic proteins and ion specific channel composition should be a focus of future research.

We have already pointed out the difficulties to design specific differentiation protocols for specific neuronal populations from IPS cells. The underlying hypothesis for all existing protocols is that cells should be guided through a shortcut version of embryonic development. A hindrance for progress in this regard is the lack of specific information of human embryonic development since most of our knowledge about vertebrate brain development derives from work with rodents. Recent reports about surprising differences between rodent and human developmental processes emphasize the demand for further comparative studies of human and rodent brain development [36;44].

The biggest limitation of IPS cell models is that they do not offer straightforward possibilities to study functions of neurons *in vivo*, as parts of the brain circuitries that regulate higher brain functions and organismic behavior. Obviously, cellular models alone will never be able to produce clinically important read-outs, such as memory dysfunction and behavioral changes in AD, tremor, bradykinesia, and rigidity in PD, or reduced forced vital capacity, swallowing dysfunction, dysarthria, or limb motor impairment in ALS. Therefore, in the foreseeable future research on neurodegenerative diseases will combine *in vitro* and *in vivo* approaches. *In vivo* transplantation of stem cell derivatives in relevant animal models could bring additional information regarding the potential of hIPS cells for *in vivo* differentiation and their survival in a pathological brain environment. This is first exemplified in a study of directed differentiation of IPS cells to midbrain neurons and their transplantation into a rat model of PD, which led to functional recovery [64].

This result and many others that we summarized in this chapter raise hopes that IPS cells derived from affected and healthy human individuals will provide a unique opportunity to gain insights into the human pathophysiology and pharmacologic responses in yet incurable neurodegenerative diseases.

Acknowledgements

This work was supported by SPIN FWF W1206-B05, Austria

Author details

Roxana Nat[1], Andreas Eigentler[2] and Georg Dechant[1]

1 Institute for Neuroscience, Innsbruck Medical University, Innsbruck, Austria

2 Department of Neurology, Innsbruck Medical University, Innsbruck, Austria

References

[1] K. Takahashi and S. Yamanaka, "Induction of pluripotent stem cells from mouse embryonic and adult fibroblast cultures by defined factors," *Cell*, vol. 126, no. 4, pp. 663-676, 2006.

[2] K. Takahashi, K. Tanabe, M. Ohnuki, M. Narita, T. Ichisaka, K. Tomoda, and S. Yamanaka, "Induction of pluripotent stem cells from adult human fibroblasts by defined factors," *Cell*, vol. 131, no. 5, pp. 861-872, 2007.

[3] J. Yu, M. A. Vodyanik, K. Smuga-Otto, J. Antosiewicz-Bourget, J. L. Frane, S. Tian, J. Nie, G. A. Jonsdottir, V. Ruotti, R. Stewart, I. I. Slukvin, and J. A. Thomson, "Induced pluripotent stem cell lines derived from human somatic cells," *Science*, vol. 318, no. 5858, pp. 1917-1920, Dec.2007.

[4] I. H. Park, N. Arora, H. Huo, N. Maherali, T. Ahfeldt, A. Shimamura, M. W. Lensch, C. Cowan, K. Hochedlinger, and G. Q. Daley, "Disease-specific induced pluripotent stem cells," *Cell*, vol. 134, no. 5, pp. 877-886, Sept.2008.

[5] M. Nakagawa, M. Koyanagi, K. Tanabe, K. Takahashi, T. Ichisaka, T. Aoi, K. Okita, Y. Mochiduki, N. Takizawa, and S. Yamanaka, "Generation of induced pluripotent stem cells without Myc from mouse and human fibroblasts," *Nat. Biotechnol.*, vol. 26, no. 1, pp. 101-106, Jan.2008.

[6] T. Aasen, A. Raya, M. J. Barrero, E. Garreta, A. Consiglio, F. Gonzalez, R. Vassena, J. Bilic, V. Pekarik, G. Tiscornia, M. Edel, S. Boue, and J. C. Izpisua Belmonte, "Efficient and rapid generation of induced pluripotent stem cells from human keratinocytes," *Nat. Biotechnol.*, vol. 26, no. 11, pp. 1276-1284, Nov.2008.

[7] A. Giorgetti, N. Montserrat, T. Aasen, F. Gonzalez, I. Rodriguez-Piza, R. Vassena, A. Raya, S. Boue, M. J. Barrero, B. A. Corbella, M. Torrabadella, A. Veiga, and J. C. Izpisua Belmonte, "Generation of induced pluripotent stem cells from human cord blood using OCT4 and SOX2," *Cell Stem Cell*, vol. 5, no. 4, pp. 353-357, Oct.2009.

[8] J. Cai, W. Li, H. Su, D. Qin, J. Yang, F. Zhu, J. Xu, W. He, X. Guo, K. Labuda, A. Peterbauer, S. Wolbank, M. Zhong, Z. Li, W. Wu, K. F. So, H. Redl, L. Zeng, M. A. Esteban, and D. Pei, "Generation of human induced pluripotent stem cells from umbilical cord matrix and amniotic membrane mesenchymal cells," *J. Biol. Chem.*, vol. 285, no. 15, pp. 11227-11234, Apr.2010.

[9] E. P. Papapetrou, M. J. Tomishima, S. M. Chambers, Y. Mica, E. Reed, J. Menon, V. Tabar, Q. Mo, L. Studer, and M. Sadelain, "Stoichiometric and temporal requirements of Oct4, Sox2, Klf4, and c-Myc expression for efficient human iPSC induction and differentiation," *Proc Natl Acad Sci U S A*, vol. 106, no. 31, pp. 12759-12764, 2009.

[10] F. González, S. Boué, and J. C. I. Belmonte, "Methods for making induced pluripotent stem cells: reprogramming á la carte," *Nat Rev Genet*, vol. 12, no. 4, pp. 231-242, Apr. 2011.

[11] A. Hotta and J. Ellis, "Retroviral vector silencing during iPS cell induction: an epigenetic beacon that signals distinct pluripotent states," *J. Cell Biochem.*, vol. 105, no. 4, pp. 940-948, Nov.2008.

[12] F. Soldner, D. Hockemeyer, C. Beard, Q. Gao, G. W. Bell, E. G. Cook, G. Hargus, A. Blak, O. Cooper, M. Mitalipova, O. Isacson, and R. Jaenisch, "Parkinson's disease pa-

tient-derived induced pluripotent stem cells free of viral reprogramming factors," *Cell*, vol. 136, no. 5, pp. 964-977, Mar.2009.

[13] D. Hockemeyer, F. Soldner, E. G. Cook, Q. Gao, M. Mitalipova, and R. Jaenisch, "A drug-inducible system for direct reprogramming of human somatic cells to pluripotency," *Cell Stem Cell*, vol. 3, no. 3, pp. 346-353, 2008.

[14] W. Zhou and C. R. Freed, "Adenoviral gene delivery can reprogram human fibroblasts to induced pluripotent stem cells," *Stem Cells*, vol. 27, no. 11, pp. 2667-2674, Nov.2009.

[15] T. Seki, S. Yuasa, M. Oda, T. Egashira, K. Yae, D. Kusumoto, H. Nakata, S. Tohyama, H. Hashimoto, M. Kodaira, Y. Okada, H. Seimiya, N. Fusaki, M. Hasegawa, and K. Fukuda, "Generation of Induced Pluripotent Stem Cells from Human Terminally Differentiated Circulating T Cells," *Cell Stem Cell*, vol. 7, no. 1, pp. 11-14, July2010.

[16] J. Yu, K. Hu, K. Smuga-Otto, S. Tian, R. Stewart, I. I. Slukvin, and J. A. Thomson, "Human Induced Pluripotent Stem Cells Free of Vector and Transgene Sequences," *Science*, vol. 324, no. 5928, pp. 797-801, May2009.

[17] K. Okita, M. Nakagawa, H. Hyenjong, T. Ichisaka, and S. Yamanaka, "Generation of mouse induced pluripotent stem cells without viral vectors," *Science*, vol. 322, no. 5903, pp. 949-953, Nov.2008.

[18] K. Woltjen, I. P. Michael, P. Mohseni, R. Desai, M. Mileikovsky, R. Hamalainen, R. Cowling, W. Wang, P. Liu, M. Gertsenstein, K. Kaji, H. K. Sung, and A. Nagy, "piggyBac transposition reprograms fibroblasts to induced pluripotent stem cells," *Nature*, vol. 458, no. 7239, pp. 766-770, Apr.2009.

[19] C. A. Sommer, A. G. Sommer, T. A. Longmire, C. Christodoulou, D. D. Thomas, M. Gostissa, F. W. Alt, G. J. Murphy, D. N. Kotton, and G. Mostoslavsky, "Excision of Reprogramming Transgenes Improves the Differentiation Potential of iPS Cells Generated with a Single Excisable Vector," *Stem Cells*, vol. 28, no. 1, pp. 64-74, 2010.

[20] L. Warren, P. D. Manos, T. Ahfeldt, Y. H. Loh, H. Li, F. Lau, W. Ebina, P. K. Mandal, Z. D. Smith, A. Meissner, G. Q. Daley, A. S. Brack, J. J. Collins, C. Cowan, T. M. Schlaeger, and D. J. Rossi, "Highly efficient reprogramming to pluripotency and directed differentiation of human cells with synthetic modified mRNA," *Cell Stem Cell*, vol. 7, no. 5, pp. 618-630, Nov.2010.

[21] S. L. Lin, D. C. Chang, C. H. Lin, S. Y. Ying, D. Leu, and D. T. Wu, "Regulation of somatic cell reprogramming through inducible mir-302 expression," *Nucleic Acids Res.*, vol. 39, no. 3, pp. 1054-1065, Feb.2011.

[22] K. N. Ivey, A. Muth, J. Arnold, F. W. King, R. F. Yeh, J. E. Fish, E. C. Hsiao, R. J. Schwartz, B. R. Conklin, H. S. Bernstein, and D. Srivastava, "MicroRNA regulation of cell lineages in mouse and human embryonic stem cells," *Cell Stem Cell*, vol. 2, no. 3, pp. 219-229, Mar.2008.

[23] F. Anokye-Danso, C. Trivedi, D. Juhr, M. Gupta, Z. Cui, Y. Tian, Y. Zhang, W. Yang, P. Gruber, J. Epstein, and E. Morrisey, "Highly Efficient miRNA-Mediated Reprogramming of Mouse and Human Somatic Cells to Pluripotency," *Cell Stem Cell*, vol. 8, no. 4, pp. 376-388, Apr.2011.

[24] N. Miyoshi, H. Ishii, H. Nagano, N. Haraguchi, D. Dewi, Y. Kano, S. Nishikawa, M. Tanemura, K. Mimori, F. Tanaka, T. Saito, J. Nishimura, I. Takemasa, T. Mizushima, M. Ikeda, H. Yamamoto, M. Sekimoto, Y. Doki, and M. Mori, "Reprogramming of Mouse and Human Cells to Pluripotency Using Mature MicroRNAs," *Cell Stem Cell*, vol. 8, no. 6, pp. 633-638, June2011.

[25] H. Zhou, S. Wu, J. Y. Joo, S. Zhu, D. W. Han, T. Lin, S. Trauger, G. Bien, S. Yao, Y. Zhu, G. Siuzdak, H. R. Scholer, L. Duan, and S. Ding, "Generation of induced pluripotent stem cells using recombinant proteins," *Cell Stem Cell*, vol. 4, no. 5, pp. 381-384, May2009.

[26] H. Shimada, Y. Hashimoto, A. Nakada, K. Shigeno, and T. Nakamura, "Accelerated generation of human induced pluripotent stem cells with retroviral transduction and chemical inhibitors under physiological hypoxia," *Biochem. Biophys. Res. Commun.*, vol. 417, no. 2, pp. 659-664, Jan.2012.

[27] D. Huangfu, R. Maehr, W. Guo, A. Eijkelenboom, M. Snitow, A. E. Chen, and D. A. Melton, "Induction of pluripotent stem cells by defined factors is greatly improved by small-molecule compounds," *Nat Biotech*, vol. 26, no. 7, pp. 795-797, July2008.

[28] Y. Shi, J. T. Do, C. Desponts, H. S. Hahm, H. R. Scholer, and S. Ding, "A combined chemical and genetic approach for the generation of induced pluripotent stem cells," *Cell Stem Cell*, vol. 2, no. 6, pp. 525-528, June2008.

[29] W. Li, W. Wei, S. Zhu, J. Zhu, Y. Shi, T. Lin, E. Hao, A. Hayek, H. Deng, and S. Ding, "Generation of Rat and Human Induced Pluripotent Stem Cells by Combining Genetic Reprogramming and Chemical Inhibitors," *Cell Stem Cell*, vol. 4, no. 1, pp. 16-19, Jan.2009.

[30] J. Silva, O. Barrandon, J. Nichols, J. Kawaguchi, T. W. Theunissen, and A. Smith, "Promotion of Reprogramming to Ground State Pluripotency by Signal Inhibition," *PLoS Biol*, vol. 6, no. 10, p. e253, Oct.2008.

[31] T. Lin, R. Ambasudhan, X. Yuan, W. Li, S. Hilcove, R. Abujarour, X. Lin, H. S. Hahm, E. Hao, A. Hayek, and S. Ding, "A chemical platform for improved induction of human iPSCs," *Nat. Methods*, vol. 6, no. 11, pp. 805-808, Nov.2009.

[32] N. Maherali and K. Hochedlinger, "Guidelines and techniques for the generation of induced pluripotent stem cells," *Cell Stem Cell*, vol. 3, no. 6, pp. 595-605, 2008.

[33] G. L. Boulting, E. Kiskinis, G. F. Croft, M. W. Amoroso, D. H. Oakley, B. J. Wainger, D. J. Williams, D. J. Kahler, M. Yamaki, L. Davidow, C. T. Rodolfa, J. T. Dimos, S. Mikkilineni, A. B. MacDermott, C. J. Woolf, C. E. Henderson, H. Wichterle, and K.

Eggan, "A functionally characterized test set of human induced pluripotent stem cells," *Nat Biotech*, vol. 29, no. 3, pp. 279-286, Mar.2011.

[34] M. Wernig, A. Meissner, R. Foreman, T. Brambrink, M. Ku, K. Hochedlinger, B. E. Bernstein, and R. Jaenisch, "In vitro reprogramming of fibroblasts into a pluripotent ES-cell-like state," *Nature*, vol. 448, no. 7151, pp. 318-324, July2007.

[35] T. J. Petros, J. A. Tyson, and S. A. Anderson, "Pluripotent stem cells for the study of CNS development," *Frontiers in Molecular Neuroscience*, vol. 4 2011.

[36] R. Nat, A. Salti, L. Suciu, S. Strom, and G. Dechant, "Pharmacological modulation of the Hedgehog pathway differentially affects dorsal/ventral patterning in mouse and human embryonic stem cell models of telencephalic development," *Stem Cells Dev.*, vol. 21, no. 7, pp. 1016-1046, May2012.

[37] A. J. Levine and A. H. Brivanlou, "Proposal of a model of mammalian neural induction," *Dev. Biol.*, vol. 308, no. 2, pp. 247-256, Aug.2007.

[38] M. Gotz and W. B. Huttner, "The cell biology of neurogenesis," *Nat. Rev. Mol. Cell Biol.*, vol. 6, no. 10, pp. 777-788, Oct.2005.

[39] R. Nat, M. Nilbratt, S. Narkilahti, B. Winblad, O. Hovatta, and A. Nordberg, "Neurogenic neuroepithelial and radial glial cells generated from six human embryonic stem cell lines in serum-free suspension and adherent cultures," *Glia*, vol. 55, no. 4, pp. 385-399, 2007.

[40] C. D. Stern, "Initial patterning of the central nervous system: how many organizers?," *Nat. Rev. Neurosci.*, vol. 2, no. 2, pp. 92-98, Feb.2001.

[41] T. Kudoh, S. W. Wilson, and I. B. Dawid, "Distinct roles for Fgf, Wnt and retinoic acid in posteriorizing the neural ectoderm," *Development*, vol. 129, no. 18, pp. 4335-4346, Sept.2002.

[42] C. Patthey, L. Gunhaga, and T. Edlund, "Early development of the central and peripheral nervous systems is coordinated by Wnt and BMP signals," *PLoS. One.*, vol. 3, no. 2, p. e1625, 2008.

[43] K. Watanabe, D. Kamiya, A. Nishiyama, T. Katayama, S. Nozaki, H. Kawasaki, Y. Watanabe, K. Mizuseki, and Y. Sasai, "Directed differentiation of telencephalic precursors from embryonic stem cells," *Nat Neurosci*, vol. 8, no. 3, pp. 288-296, Mar.2005.

[44] X. J. Li, X. Zhang, M. A. Johnson, Z. B. Wang, T. Lavaute, and S. C. Zhang, "Coordination of sonic hedgehog and Wnt signaling determines ventral and dorsal telencephalic neuron types from human embryonic stem cells," *Development*, vol. 136, no. 23, pp. 4055-4063, Dec.2009.

[45] M. Eiraku, K. Watanabe, M. Matsuo-Takasaki, M. Kawada, S. Yonemura, M. Matsumura, T. Wataya, A. Nishiyama, K. Muguruma, and Y. Sasai, "Self-organized forma-

tion of polarized cortical tissues from ESCs and its active manipulation by extrinsic signals," *Cell Stem Cell*, vol. 3, no. 5, pp. 519-532, Nov.2008.

[46] N. Gaspard, T. Bouschet, R. Hourez, J. Dimidschstein, G. Naeije, J. van den Ameele, I. Espuny-Camacho, A. Herpoel, L. Passante, S. N. Schiffmann, A. Gaillard, and P. Vanderhaeghen, "An intrinsic mechanism of corticogenesis from embryonic stem cells," *Nature*, vol. 455, no. 7211, pp. 351-357, Sept.2008.

[47] N. Gaspard, T. Bouschet, A. Herpoel, G. Naeije, J. van den Ameele, and P. Vanderhaeghen, "Generation of cortical neurons from mouse embryonic stem cells," *Nat. Protoc.*, vol. 4, no. 10, pp. 1454-1463, 2009.

[48] M. Ideguchi, T. D. Palmer, L. D. Recht, and J. M. Weimann, "Murine embryonic stem cell-derived pyramidal neurons integrate into the cerebral cortex and appropriately project axons to subcortical targets," *J. Neurosci.*, vol. 30, no. 3, pp. 894-904, Jan.2010.

[49] A. M. Maroof, K. Brown, S. H. Shi, L. Studer, and S. A. Anderson, "Prospective isolation of cortical interneuron precursors from mouse embryonic stem cells," *J. Neurosci.*, vol. 30, no. 13, pp. 4667-4675, Mar.2010.

[50] T. Danjo, M. Eiraku, K. Muguruma, K. Watanabe, M. Kawada, Y. Yanagawa, J. L. Rubenstein, and Y. Sasai, "Subregional specification of embryonic stem cell-derived ventral telencephalic tissues by timed and combinatory treatment with extrinsic signals," *J. Neurosci.*, vol. 31, no. 5, pp. 1919-1933, Feb.2011.

[51] A. L. Goulburn, D. Alden, R. P. Davis, S. J. Micallef, E. S. Ng, Q. C. Yu, S. M. Lim, C. L. Soh, D. A. Elliott, T. Hatzistavrou, J. Bourke, B. Watmuff, R. J. Lang, J. M. Haynes, C. W. Pouton, A. Giudice, A. O. Trounson, S. A. Anderson, E. G. Stanley, and A. G. Elefanty, "A targeted NKX2.1 human embryonic stem cell reporter line enables identification of human basal forebrain derivatives," *Stem Cells*, vol. 29, no. 3, pp. 462-473, Mar.2011.

[52] A. L. Goulburn, E. G. Stanley, A. G. Elefanty, and S. A. Anderson, "Generating GABAergic cerebral cortical interneurons from mouse and human embryonic stem cells," *Stem Cell Res.*, vol. 8, no. 3, pp. 416-426, May2012.

[53] S. Cambray, C. Arber, G. Little, A. G. Dougalis, P. de, V, M. A. Ungless, M. Li, and T. A. Rodriguez, "Activin induces cortical interneuron identity and differentiation in embryonic stem cell-derived telencephalic neural precursors," *Nat. Commun.*, vol. 3, p. 841, 2012.

[54] L. Wicklund, R. N. Leao, A. M. Stromberg, M. Mousavi, O. Hovatta, A. Nordberg, and A. Marutle, "Beta-amyloid 1-42 oligomers impair function of human embryonic stem cell-derived forebrain cholinergic neurons," *PLoS. One.*, vol. 5, no. 12, p. e15600, 2010.

[55] C. J. Bissonnette, L. Lyass, B. J. Bhattacharyya, A. Belmadani, R. J. Miller, and J. A. Kessler, "The controlled generation of functional basal forebrain cholinergic neurons from human embryonic stem cells," *Stem Cells*, vol. 29, no. 5, pp. 802-811, May2011.

[56] L. Aubry, A. Bugi, N. Lefort, F. Rousseau, M. Peschanski, and A. L. Perrier, "Striatal progenitors derived from human ES cells mature into DARPP32 neurons in vitro and in quinolinic acid-lesioned rats," *Proc. Natl. Acad. Sci. U. S. A*, vol. 105, no. 43, pp. 16707-16712, Oct.2008.

[57] N. Zhang, M. C. An, D. Montoro, and L. M. Ellerby, "Characterization of Human Huntington's Disease Cell Model from Induced Pluripotent Stem Cells," *PLoS. Curr.*, vol. 2, p. RRN1193, 2010.

[58] C. A. Fasano, S. M. Chambers, G. Lee, M. J. Tomishima, and L. Studer, "Efficient derivation of functional floor plate tissue from human embryonic stem cells," *Cell Stem Cell*, vol. 6, no. 4, pp. 336-347, Apr.2010.

[59] H. Kawasaki, K. Mizuseki, S. Nishikawa, S. Kaneko, Y. Kuwana, S. Nakanishi, S. I. Nishikawa, and Y. Sasai, "Induction of midbrain dopaminergic neurons from ES cells by stromal cell-derived inducing activity," *Neuron*, vol. 28, no. 1, pp. 31-40, Oct.2000.

[60] S. H. Lee, N. Lumelsky, L. Studer, J. M. Auerbach, and R. D. McKay, "Efficient generation of midbrain and hindbrain neurons from mouse embryonic stem cells," *Nat. Biotechnol.*, vol. 18, no. 6, pp. 675-679, June2000.

[61] A. L. Perrier, V. Tabar, T. Barberi, M. E. Rubio, J. Bruses, N. Topf, N. L. Harrison, and L. Studer, "Derivation of midbrain dopamine neurons from human embryonic stem cells," *Proc. Natl. Acad. Sci. U. S. A*, vol. 101, no. 34, pp. 12543-12548, Aug.2004.

[62] Y. Yan, D. Yang, E. D. Zarnowska, Z. Du, B. Werbel, C. Valliere, R. A. Pearce, J. A. Thomson, and S. C. Zhang, "Directed differentiation of dopaminergic neuronal subtypes from human embryonic stem cells," *Stem Cells*, vol. 23, no. 6, pp. 781-790, June2005.

[63] S. M. Chambers, C. A. Fasano, E. P. Papapetrou, M. Tomishima, M. Sadelain, and L. Studer, "Highly efficient neural conversion of human ES and iPS cells by dual inhibition of SMAD signaling," *Nat. Biotechnol.*, vol. 27, no. 3, pp. 275-280, Mar.2009.

[64] A. Sanchez-Danes, A. Consiglio, Y. Richaud, I. Rodriguez-Piza, B. Dehay, M. Edel, J. Bove, M. Memo, M. Vila, A. Raya, and J. C. Izpisua Belmonte, "Efficient generation of A9 midbrain dopaminergic neurons by lentiviral delivery of LMX1A in human embryonic stem cells and induced pluripotent stem cells," *Hum. Gene Ther.*, vol. 23, no. 1, pp. 56-69, Jan.2012.

[65] E. Salero and M. E. Hatten, "Differentiation of ES cells into cerebellar neurons," *Proc. Natl. Acad. Sci. U. S. A*, vol. 104, no. 8, pp. 2997-3002, Feb.2007.

[66] S. Erceg, M. Ronaghi, I. Zipancic, S. Lainez, M. G. Rosello, C. Xiong, V. Moreno-Manzano, F. J. Rodriguez-Jimenez, R. Planells, M. Alvarez-Dolado, S. S. Bhattacharya,

and M. Stojkovic, "Efficient differentiation of human embryonic stem cells into functional cerebellar-like cells," *Stem Cells Dev.*, vol. 19, no. 11, pp. 1745-1756, Nov.2010.

[67] H. L. Su, K. Muguruma, M. Matsuo-Takasaki, M. Kengaku, K. Watanabe, and Y. Sasai, "Generation of cerebellar neuron precursors from embryonic stem cells," *Dev. Biol.*, vol. 290, no. 2, pp. 287-296, Feb.2006.

[68] O. Tao, T. Shimazaki, Y. Okada, H. Naka, K. Kohda, M. Yuzaki, H. Mizusawa, and H. Okano, "Efficient generation of mature cerebellar Purkinje cells from mouse embryonic stem cells," *J. Neurosci. Res.*, vol. 88, no. 2, pp. 234-247, Feb.2010.

[69] H. Wichterle, I. Lieberam, J. A. Porter, and T. M. Jessell, "Directed differentiation of embryonic stem cells into motor neurons," *Cell*, vol. 110, no. 3, pp. 385-397, Aug.2002.

[70] X. J. Li, Z. W. Du, E. D. Zarnowska, M. Pankratz, L. O. Hansen, R. A. Pearce, and S. C. Zhang, "Specification of motoneurons from human embryonic stem cells," *Nat Biotech*, vol. 23, no. 2, pp. 215-221, Feb.2005.

[71] P. Soundararajan, G. B. Miles, L. L. Rubin, R. M. Brownstone, and V. F. Rafuse, "Motoneurons derived from embryonic stem cells express transcription factors and develop phenotypes characteristic of medial motor column neurons," *J. Neurosci.*, vol. 26, no. 12, pp. 3256-3268, Mar.2006.

[72] H. Lee, G. A. Shamy, Y. Elkabetz, C. M. Schofield, N. L. Harrsion, G. Panagiotakos, N. D. Socci, V. Tabar, and L. Studer, "Directed differentiation and transplantation of human embryonic stem cell-derived motoneurons," *Stem Cells*, vol. 25, no. 8, pp. 1931-1939, Aug.2007.

[73] J. T. Dimos, K. T. Rodolfa, K. K. Niakan, L. M. Weisenthal, H. Mitsumoto, W. Chung, G. F. Croft, G. Saphier, R. Leibel, R. Goland, H. Wichterle, C. E. Henderson, and K. Eggan, "Induced pluripotent stem cells generated from patients with ALS can be differentiated into motor neurons," *Science*, vol. 321, no. 5893, pp. 1218-1221, Aug.2008.

[74] M. Peljto, J. S. Dasen, E. O. Mazzoni, T. M. Jessell, and H. Wichterle, "Functional diversity of ESC-derived motor neuron subtypes revealed through intraspinal transplantation," *Cell Stem Cell*, vol. 7, no. 3, pp. 355-366, Sept.2010.

[75] R. Patani, A. J. Hollins, T. M. Wishart, C. A. Puddifoot, S. Alvarez, A. R. de Lera, D. J. Wyllie, D. A. Compston, R. A. Pedersen, T. H. Gillingwater, G. E. Hardingham, N. D. Allen, and S. Chandran, "Retinoid-independent motor neurogenesis from human embryonic stem cells reveals a medial columnar ground state," *Nat Commun.*, vol. 2, p. 214, 2011.

[76] H. W. Park, J. S. Cho, C. K. Park, S. J. Jung, C. H. Park, S. J. Lee, S. B. Oh, Y. S. Park, and M. S. Chang, "Directed Induction of Functional Motor Neuron-Like Cells from Genetically Engineered Human Mesenchymal Stem Cells," *PLoS One*, vol. 7, no. 4, p. e35244, Apr.2012.

[77] Y. Liqing, G. Jia, C. Jiqing, G. Ran, C. Fei, K. Jie, W. Yanyun, and Z. Cheng, "Directed differentiation of motor neuron cell-like cells from human adipose-derived stem cells in vitro," *Neuroreport*, vol. 22, no. 8, pp. 370-373, June2011.

[78] A. K. Murashov, E. S. Pak, W. A. Hendricks, J. P. Owensby, P. L. Sierpinski, L. M. Tatko, and P. L. Fletcher, "Directed differentiation of embryonic stem cells into dorsal interneurons," *The FASEB Journal*, Nov.2004.

[79] G. Lee, S. M. Chambers, M. J. Tomishima, and L. Studer, "Derivation of neural crest cells from human pluripotent stem cells," *Nat Protoc.*, vol. 5, no. 4, pp. 688-701, Apr. 2010.

[80] L. Menendez, T. A. Yatskievych, P. B. Antin, and S. Dalton, "Wnt signaling and a Smad pathway blockade direct the differentiation of human pluripotent stem cells to multipotent neural crest cells," *Proceedings of the National Academy of Sciences*, vol. 108, no. 48, pp. 19240-19245, Nov.2011.

[81] R. S. Goldstein, O. Pomp, I. Brokhman, and L. Ziegler, "Generation of neural crest cells and peripheral sensory neurons from human embryonic stem cells," *Methods Mol. Biol.*, vol. 584, pp. 283-300, 2010.

[82] S. Han, L. Williams, and K. Eggan, "Constructing and Deconstructing Stem Cell Models of Neurological Disease," *Neuron*, vol. 70, no. 4, pp. 626-644, May2011.

[83] T. Yagi, D. Ito, Y. Okada, W. Akamatsu, Y. Nihei, T. Yoshizaki, S. Yamanaka, H. Okano, and N. Suzuki, "Modeling familial Alzheimer's disease with induced pluripotent stem cells," *Human Molecular Genetics*, vol. 20, no. 23, pp. 4530-4539, Dec.2011.

[84] M. A. Israel, S. H. Yuan, C. Bardy, S. M. Reyna, Y. Mu, C. Herrera, M. P. Hefferan, S. Van Gorp, K. L. Nazor, F. S. Boscolo, C. T. Carson, L. C. Laurent, M. Marsala, F. H. Gage, A. M. Remes, E. H. Koo, and L. S. B. Goldstein, "Probing sporadic and familial Alzheimer/'s disease using induced pluripotent stem cells," *Nature*, vol. advance online publication Jan.2012.

[85] G. Hargus, O. Cooper, M. Deleidi, A. Levy, K. Lee, E. Marlow, A. Yow, F. Soldner, D. Hockemeyer, P. J. Hallett, T. Osborn, R. Jaenisch, and O. Isacson, "Differentiated Parkinson patient-derived induced pluripotent stem cells grow in the adult rodent brain and reduce motor asymmetry in Parkinsonian rats," *Proc. Natl. Acad. Sci. U. S. A*, vol. 107, no. 36, pp. 15921-15926, Sept.2010.

[86] H. Nguyen, B. Byers, B. Cord, A. Shcheglovitov, J. Byrne, P. Gujar, K. Kee, B. Schüle, R. Dolmetsch, W. Langston, T. Palmer, and R. Pera, "LRRK2 Mutant iPSC-Derived DA Neurons Demonstrate Increased Susceptibility to Oxidative Stress," *Cell Stem Cell*, vol. 8, no. 3, pp. 267-280, Mar.2011.

[87] P. Seibler, J. Graziotto, H. Jeong, F. Simunovic, C. Klein, and D. Krainc, "Mitochondrial Parkin Recruitment Is Impaired in Neurons Derived from Mutant PINK1 In-

duced Pluripotent Stem Cells," *The Journal of Neuroscience*, vol. 31, no. 16, pp. 5970-5976, Apr.2011.

[88] M. J. Devine, M. Ryten, P. Vodicka, A. J. Thomson, T. Burdon, H. Houlden, F. Cavaleri, M. Nagano, N. J. Drummond, J. W. Taanman, A. H. Schapira, K. Gwinn, J. Hardy, P. A. Lewis, and T. Kunath, "Parkinson's disease induced pluripotent stem cells with triplication of the α-synuclein locus," *Nat Commun*, vol. 2, p. 440, Aug.2011.

[89] B. Byers, B. Cord, H. N. Nguyen, B. Schüle, L. Fenno, P. C. Lee, K. Deisseroth, J. W. Langston, R. R. Pera, and T. D. Palmer, "SNCA Triplication Parkinson's Patient's iPSC-derived DA Neurons Accumulate α-Synuclein and Are Susceptible to Oxidative Stress," *PLoS ONE*, vol. 6, no. 11, p. e26159, Nov.2011.

[90] S. Camnasio, A. D. Carri, A. Lombardo, I. Grad, C. Mariotti, A. Castucci, B. Rozell, P. L. Riso, V. Castiglioni, C. Zuccato, C. Rochon, Y. Takashima, G. Diaferia, I. Biunno, C. Gellera, M. Jaconi, A. Smith, O. Hovatta, L. Naldini, D. S. Di, A. Feki, and E. Cattaneo, "The first reported generation of several induced pluripotent stem cell lines from homozygous and heterozygous Huntington's disease patients demonstrates mutation related enhanced lysosomal activity," *Neurobiol. Dis.*, vol. 46, no. 1, pp. 41-51, Apr.2012.

[91] A. D. Ebert, J. Yu, F. F. Rose, V. B. Mattis, C. L. Lorson, J. A. Thomson, and C. N. Svendsen, "Induced pluripotent stem cells from a spinal muscular atrophy patient," *Nature*, vol. 457, no. 7227, pp. 277-280, Jan.2009.

[92] D. Sareen, A. D. Ebert, B. M. Heins, J. V. McGivern, L. Ornelas, and C. N. Svendsen, "Inhibition of Apoptosis Blocks Human Motor Neuron Cell Death in a Stem Cell Model of Spinal Muscular Atrophy," *PLoS ONE*, vol. 7, no. 6, p. e39113, June2012.

[93] S. Ku, E. Soragni, E. Campau, E. A. Thomas, G. Altun, L. C. Laurent, J. F. Loring, M. Napierala, and J. M. Gottesfeld, "Friedreich's ataxia induced pluripotent stem cells model intergenerational GAATTC triplet repeat instability," *Cell Stem Cell*, vol. 7, no. 5, pp. 631-637, Nov.2010.

[94] J. Liu, P. J. Verma, M. V. Evans-Galea, M. B. Delatycki, A. Michalska, J. Leung, D. Crombie, J. P. Sarsero, R. Williamson, M. Dottori, and A. Pebay, "Generation of induced pluripotent stem cell lines from Friedreich ataxia patients," *Stem Cell Rev.*, vol. 7, no. 3, pp. 703-713, Sept.2011.

[95] P. Koch, P. Breuer, M. Peitz, J. Jungverdorben, J. Kesavan, D. Poppe, J. Doerr, J. Ladewig, J. Mertens, T. Tuting, P. Hoffmann, T. Klockgether, B. O. Evert, U. Wullner, and O. Brustle, "Excitation-induced ataxin-3 aggregation in neurons from patients with Machado-Joseph disease," *Nature*, vol. 480, no. 7378, pp. 543-546, Dec.2011.

[96] G. Lee, E. P. Papapetrou, H. Kim, S. M. Chambers, M. J. Tomishima, C. A. Fasano, Y. M. Ganat, J. Menon, F. Shimizu, A. Viale, V. Tabar, M. Sadelain, and L. Studer, "Modelling pathogenesis and treatment of familial dysautonomia using patient-specific iPSCs," *Nature*, vol. 461, no. 7262, pp. 402-406, 2009.

[97] A. Urbach, O. Bar-Nur, G. Q. Daley, and N. Benvenisty, "Differential Modeling of Fragile X Syndrome by Human Embryonic Stem Cells and Induced Pluripotent Stem Cells," *Cell Stem Cell*, vol. 6, no. 5, pp. 407-411, May2010.

[98] H. Braak and E. Braak, "Frequency of Stages of Alzheimer-Related Lesions in Different Age Categories," *Neurobiology of Aging*, vol. 18, no. 4, pp. 351-357, July1997.

[99] D. J. Selkoe, "Alzheimer's Disease Is a Synaptic Failure," *Science*, vol. 298, no. 5594, pp. 789-791, Oct.2002.

[100] L. M. Ittner and J. Götz, "Amyloid-β and tau - a toxic pas de deux in Alzheimer's disease," *Nat Rev Neurosci*, vol. 12, no. 2, pp. 67-72, Feb.2011.

[101] T. Iwatsubo, T. C. Saido, D. M. Mann, V. M. Lee, and J. Q. Trojanowski, "Full-length amyloid-beta (1-42(43)) and amino-terminally modified and truncated amyloid-beta 42(43) deposit in diffuse plaques," *Am. J. Pathol.*, vol. 149, no. 6, pp. 1823-1830, Dec. 1996.

[102] A. Alonso, I. Grundke-Iqbal, H. Barra, and K. Iqbal, "Abnormal phosphorylation of tau and the mechanism of Alzheimer neurofibrillary degeneration: Sequestration of microtubule-associated proteins 1 and 2 and the disassembly of microtubules by the abnormalGÇëtau," *Proceedings of the National Academy of Sciences*, vol. 94, no. 1, pp. 298-303, Jan.1997.

[103] E. Giacobini, "Cholinergic function and Alzheimer's disease," *Int. J. Geriat. Psychiatry*, vol. 18, no. S1, p. S1-S5, 2003.

[104] J. T. Coyle, D. L. Price, and M. R. DeLong, "Alzheimer's disease: a disorder of cortical cholinergic innervation," *Science*, vol. 219, no. 4589, pp. 1184-1190, Mar.1983.

[105] W. J. Strittmatter and A. D. Roses, "Apolipoprotein E and Alzheimer disease," *Proceedings of the National Academy of Sciences*, vol. 92, no. 11, pp. 4725-4727, May1995.

[106] M. C. Chartier-Harlin, F. Crawford, H. Houlden, A. Warren, D. Hughes, L. Fidani, A. Goate, M. Rossor, P. Roques, J. Hardy, and M. Mullan, "Early-onset Alzheimer's disease caused by mutations at codon 717 of the [beta]-amyloid precursor protein gene," *Nature*, vol. 353, no. 6347, pp. 844-846, Oct.1991.

[107] R. E. Tanzi, D. M. Kovacs, T. W. Kim, R. D. Moir, S. Y. Guenette, and W. Wasco, "The gene defects responsible for familial Alzheimer's disease," *Neurobiol. Dis.*, vol. 3, no. 3, pp. 159-168, 1996.

[108] J. Gotz and L. M. Ittner, "Animal models of Alzheimer's disease and frontotemporal dementia," *Nat Rev Neurosci*, vol. 9, no. 7, pp. 532-544, July2008.

[109] D. Van Dam and P. P. De Deyn, "Drug discovery in dementia: the role of rodent models," *Nat Rev Drug Discov*, vol. 5, no. 11, pp. 956-970, Nov.2006.

[110] S. Lesage and A. Brice, "Parkinson's disease: from monogenic forms to genetic susceptibility factors," *Hum. Mol. Genet.*, vol. 18, no. R1, p. R48-R59, Apr.2009.

[111] M. H. Polymeropoulos, C. Lavedan, E. Leroy, S. E. Ide, A. Dehejia, A. Dutra, B. Pike, H. Root, J. Rubenstein, R. Boyer, E. S. Stenroos, S. Chandrasekharappa, A. Athanassiadou, T. Papapetropoulos, W. G. Johnson, A. M. Lazzarini, R. C. Duvoisin, G. Di Iorio, L. I. Golbe, and R. L. Nussbaum, "Mutation in the +¦-Synuclein Gene Identified in Families with Parkinson's Disease," *Science*, vol. 276, no. 5321, pp. 2045-2047, June1997.

[112] A. B. Singleton, M. Farrer, J. Johnson, A. Singleton, S. Hague, J. Kachergus, M. Hulihan, T. Peuralinna, A. Dutra, R. Nussbaum, S. Lincoln, A. Crawley, M. Hanson, D. Maraganore, C. Adler, M. R. Cookson, M. Muenter, M. Baptista, D. Miller, J. Blancato, J. Hardy, and K. Gwinn-Hardy, "+¦-Synuclein Locus Triplication Causes Parkinson's Disease," *Science*, vol. 302, no. 5646, p. 841, Oct.2003.

[113] W. P. Gilks, P. M. Abou-Sleiman, S. Gandhi, S. Jain, A. Singleton, A. J. Lees, K. Shaw, K. P. Bhatia, V. Bonifati, N. P. Quinn, J. Lynch, D. G. Healy, J. L. Holton, T. Revesz, and N. W. Wood, "A common LRRK2 mutation in idiopathic Parkinson's disease," *The Lancet*, vol. 365, no. 9457, pp. 415-416, Jan.2005.

[114] T. Kitada, S. Asakawa, N. Hattori, H. Matsumine, Y. Yamamura, S. Minoshima, M. Yokochi, Y. Mizuno, and N. Shimizu, "Mutations in the parkin gene cause autosomal recessive juvenile parkinsonism," *Nature*, vol. 392, no. 6676, pp. 605-608, Apr.1998.

[115] E. M. Valente, P. M. Abou-Sleiman, V. Caputo, M. M. K. Muqit, K. Harvey, S. Gispert, Z. Ali, D. Del Turco, A. R. Bentivoglio, D. G. Healy, A. Albanese, R. Nussbaum, R. González-Maldonado, T. Deller, S. Salvi, P. Cortelli, W. P. Gilks, D. S. Latchman, R. J. Harvey, B. Dallapiccola, G. Auburger, and N. W. Wood, "Hereditary Early-Onset Parkinson's Disease Caused by Mutations in PINK1," *Science*, vol. 304, no. 5674, pp. 1158-1160, May2004.

[116] V. Bonifati, P. Rizzu, M. J. van Baren, O. Schaap, G. J. Breedveld, E. Krieger, M. C. J. Dekker, F. Squitieri, P. Ibanez, M. Joosse, J. W. van Dongen, N. Vanacore, J. C. van Swieten, A. Brice, G. Meco, C. M. van Duijn, B. A. Oostra, and P. Heutink, "Mutations in the DJ-1 Gene Associated with Autosomal Recessive Early-Onset Parkinsonism," *Science*, vol. 299, no. 5604, pp. 256-259, Jan.2003.

[117] J. Neumann, J. Bras, E. Deas, S. S. O'Sullivan, L. Parkkinen, R. H. Lachmann, A. Li, J. Holton, R. Guerreiro, R. Paudel, B. Segarane, A. Singleton, A. Lees, J. Hardy, H. Houlden, T. Revesz, and N. W. Wood, "Glucocerebrosidase mutations in clinical and pathologically proven Parkinson's disease," *Brain*, vol. 132, no. 7, pp. 1783-1794, July2009.

[118] "A novel gene containing a trinucleotide repeat that is expanded and unstable on Huntington's disease chromosomes. The Huntington's Disease Collaborative Research Group," *Cell*, vol. 72, no. 6, pp. 971-983, Mar.1993.

[119] Rubinsztein DC, "Molecular biology of Huntington's disease (HD) and HD-like disorders.," in *Genetics of Movement Disorders*. Pulst S, Ed. San Diego: Academic Press, 2003, pp. 365-377.

[120] C. M. Everett and N. W. Wood, "Trinucleotide repeats and neurodegenerative disease," *Brain*, vol. 127, no. Pt 11, pp. 2385-2405, Nov.2004.

[121] I. J. Mitchell, A. J. Cooper, and M. R. Griffiths, "The selective vulnerability of striatopallidal neurons," *Prog. Neurobiol.*, vol. 59, no. 6, pp. 691-719, Dec.1999.

[122] N. F. H. S. Gutekunst C, "The neuropathology of Huntington's disease," in *Huntington's disease*. H. P. J. L. e. Bates G, Ed. New York: Oxford University Press, 2002, pp. 251-275.

[123] M. DiFiglia, E. Sapp, K. O. Chase, S. W. Davies, G. P. Bates, J. P. Vonsattel, and N. Aronin, "Aggregation of huntingtin in neuronal intranuclear inclusions and dystrophic neurites in brain," *Science*, vol. 277, no. 5334, pp. 1990-1993, Sept.1997.

[124] S. W. Davies, M. Turmaine, B. A. Cozens, M. DiFiglia, A. H. Sharp, C. A. Ross, E. Scherzinger, E. E. Wanker, L. Mangiarini, and G. P. Bates, "Formation of neuronal intranuclear inclusions underlies the neurological dysfunction in mice transgenic for the HD mutation," *Cell*, vol. 90, no. 3, pp. 537-548, Aug.1997.

[125] M. Arrasate, S. Mitra, E. S. Schweitzer, M. R. Segal, and S. Finkbeiner, "Inclusion body formation reduces levels of mutant huntingtin and the risk of neuronal death," *Nature*, vol. 431, no. 7010, pp. 805-810, Oct.2004.

[126] H. Mukai, T. Isagawa, E. Goyama, S. Tanaka, N. F. Bence, A. Tamura, Y. Ono, and R. R. Kopito, "Formation of morphologically similar globular aggregates from diverse aggregation-prone proteins in mammalian cells," *Proc. Natl. Acad. Sci. U. S. A*, vol. 102, no. 31, pp. 10887-10892, Aug.2005.

[127] J. M. Gil and A. C. Rego, "Mechanisms of neurodegeneration in Huntington's disease," *Eur. J. Neurosci.*, vol. 27, no. 11, pp. 2803-2820, June2008.

[128] C. A. Ross and S. J. Tabrizi, "Huntington's disease: from molecular pathogenesis to clinical treatment," *Lancet Neurol.*, vol. 10, no. 1, pp. 83-98, Jan.2011.

[129] P. Kumar, H. Kalonia, and A. Kumar, "Huntington's disease: pathogenesis to animal models," *Pharmacol. Rep.*, vol. 62, no. 1, pp. 1-14, Jan.2010.

[130] F. O. Walker, "Huntington's Disease," *Semin. Neurol.*, vol. 27, no. 2, pp. 143-150, Apr. 2007.

[131] L. Mangiarini, K. Sathasivam, M. Seller, B. Cozens, A. Harper, C. Hetherington, M. Lawton, Y. Trottier, H. Lehrach, S. W. Davies, and G. P. Bates, "Exon 1 of the HD gene with an expanded CAG repeat is sufficient to cause a progressive neurological phenotype in transgenic mice," *Cell*, vol. 87, no. 3, pp. 493-506, Nov.1996.

[132] M. F. Beal, R. J. Ferrante, K. J. Swartz, and N. W. Kowall, "Chronic quinolinic acid lesions in rats closely resemble Huntington's disease," *J. Neurosci.*, vol. 11, no. 6, pp. 1649-1659, June1991.

[133] M. F. Beal, E. Brouillet, B. G. Jenkins, R. J. Ferrante, N. W. Kowall, J. M. Miller, E. Storey, R. Srivastava, B. R. Rosen, and B. T. Hyman, "Neurochemical and histologic characterization of striatal excitotoxic lesions produced by the mitochondrial toxin 3-nitropropionic acid," *J. Neurosci.*, vol. 13, no. 10, pp. 4181-4192, Oct.1993.

[134] J. T. Coyle and R. Schwarcz, "Lesion of striatal neurones with kainic acid provides a model for Huntington's chorea," *Nature*, vol. 263, no. 5574, pp. 244-246, Sept.1976.

[135] S. T. Lee, K. Chu, J. E. Park, K. Lee, L. Kang, S. U. Kim, and M. Kim, "Intravenous administration of human neural stem cells induces functional recovery in Huntington's disease rat model," *Neurosci. Res.*, vol. 52, no. 3, pp. 243-249, July2005.

[136] J. K. Ryu, J. Kim, S. J. Cho, K. Hatori, A. Nagai, H. B. Choi, M. C. Lee, J. G. McLarnon, and S. U. Kim, "Proactive transplantation of human neural stem cells prevents degeneration of striatal neurons in a rat model of Huntington disease," *Neurobiol. Dis.*, vol. 16, no. 1, pp. 68-77, June2004.

[137] M. Kim, S. T. Lee, K. Chu, and S. U. Kim, "Stem cell-based cell therapy for Huntington disease: A review," *Neuropathology*, vol. 28, no. 1, pp. 1-9, 2008.

[138] C. K. Bradley, H. A. Scott, O. Chami, T. T. Peura, B. Dumevska, U. Schmidt, and T. Stojanov, "Derivation of Huntington's disease-affected human embryonic stem cell lines," *Stem Cells Dev.*, vol. 20, no. 3, pp. 495-502, Mar.2011.

[139] S. Camnasio, A. D. Carri, A. Lombardo, I. Grad, C. Mariotti, A. Castucci, B. Rozell, P. L. Riso, V. Castiglioni, C. Zuccato, C. Rochon, Y. Takashima, G. Diaferia, I. Biunno, C. Gellera, M. Jaconi, A. Smith, O. Hovatta, L. Naldini, S. Di Donato, A. Feki, and E. Cattaneo, "The first reported generation of several induced pluripotent stem cell lines from homozygous and heterozygous Huntington's disease patients demonstrates mutation related enhanced lysosomal activity," *Neurobiology of Disease*, vol. 46, no. 1, pp. 41-51, Apr.2012.

[140] S. Lefebvre, L. Burglen, S. Reboullet, O. Clermont, P. Burlet, L. Viollet, B. Benichou, C. Cruaud, P. Millasseau, M. Zeviani, and ., "Identification and characterization of a spinal muscular atrophy-determining gene," *Cell*, vol. 80, no. 1, pp. 155-165, Jan.1995.

[141] S. Lefebvre, L. Burglen, J. Frezal, A. Munnich, and J. Melki, "The role of the SMN gene in proximal spinal muscular atrophy," *Hum. Mol. Genet.*, vol. 7, no. 10, pp. 1531-1536, 1998.

[142] D. D. Coovert, T. T. Le, P. E. McAndrew, J. Strasswimmer, T. O. Crawford, J. R. Mendell, S. E. Coulson, E. J. Androphy, T. W. Prior, and A. H. Burghes, "The survival motor neuron protein in spinal muscular atrophy," *Hum. Mol. Genet.*, vol. 6, no. 8, pp. 1205-1214, Aug.1997.

[143] U. R. Monani, C. L. Lorson, D. W. Parsons, T. W. Prior, E. J. Androphy, A. H. Burghes, and J. D. McPherson, "A single nucleotide difference that alters splicing patterns distinguishes the SMA gene SMN1 from the copy gene SMN2," *Hum. Mol. Genet.*, vol. 8, no. 7, pp. 1177-1183, July1999.

[144] C. L. Lorson, E. Hahnen, E. J. Androphy, and B. Wirth, "A single nucleotide in the SMN gene regulates splicing and is responsible for spinal muscular atrophy," *Proc. Natl. Acad. Sci. U. S. A*, vol. 96, no. 11, pp. 6307-6311, May1999.

[145] U. R. Monani, M. Sendtner, D. D. Coovert, D. W. Parsons, C. Andreassi, T. T. Le, S. Jablonka, B. Schrank, W. Rossoll, T. W. Prior, G. E. Morris, and A. H. Burghes, "The human centromeric survival motor neuron gene (SMN2) rescues embryonic lethality in Smn(-/-) mice and results in a mouse with spinal muscular atrophy," *Hum. Mol. Genet.*, vol. 9, no. 3, pp. 333-339, Feb.2000.

[146] A. H. Burghes and C. E. Beattie, "Spinal muscular atrophy: why do low levels of survival motor neuron protein make motor neurons sick?," *Nat. Rev. Neurosci.*, vol. 10, no. 8, pp. 597-609, Aug.2009.

[147] U. R. Monani, "Spinal muscular atrophy: a deficiency in a ubiquitous protein; a motor neuron-specific disease," *Neuron*, vol. 48, no. 6, pp. 885-896, Dec.2005.

[148] B. Schrank, R. Gotz, J. M. Gunnersen, J. M. Ure, K. V. Toyka, A. G. Smith, and M. Sendtner, "Inactivation of the survival motor neuron gene, a candidate gene for human spinal muscular atrophy, leads to massive cell death in early mouse embryos," *Proc. Natl. Acad. Sci. U. S. A*, vol. 94, no. 18, pp. 9920-9925, Sept.1997.

[149] H. M. Hsieh-Li, J. G. Chang, Y. J. Jong, M. H. Wu, N. M. Wang, C. H. Tsai, and H. Li, "A mouse model for spinal muscular atrophy," *Nat. Genet.*, vol. 24, no. 1, pp. 66-70, Jan.2000.

[150] U. R. Monani, M. Sendtner, D. D. Coovert, D. W. Parsons, C. Andreassi, T. T. Le, S. Jablonka, B. Schrank, W. Rossoll, T. W. Prior, G. E. Morris, and A. H. Burghes, "The human centromeric survival motor neuron gene (SMN2) rescues embryonic lethality in Smn(-/-) mice and results in a mouse with spinal muscular atrophy," *Hum. Mol. Genet.*, vol. 9, no. 3, pp. 333-339, Feb.2000.

[151] L. I. Bruijn, T. M. Miller, and D. W. Cleveland, "Unraveling the mechanisms involved in motor neuron degeneration in ALS," *Annu. Rev. Neurosci.*, vol. 27, pp. 723-749, 2004.

[152] D. R. Rosen, T. Siddique, D. Patterson, D. A. Figlewicz, P. Sapp, A. Hentati, D. Donaldson, J. Goto, J. P. O'Regan, H. X. Deng, and ., "Mutations in Cu/Zn superoxide dismutase gene are associated with familial amyotrophic lateral sclerosis," *Nature*, vol. 362, no. 6415, pp. 59-62, Mar.1993.

[153] J. Sreedharan, I. P. Blair, V. B. Tripathi, X. Hu, C. Vance, B. Rogelj, S. Ackerley, J. C. Durnall, K. L. Williams, E. Buratti, F. Baralle, B. J. de, J. D. Mitchell, P. N. Leigh, A. Al-Chalabi, C. C. Miller, G. Nicholson, and C. E. Shaw, "TDP-43 mutations in familial

and sporadic amyotrophic lateral sclerosis," *Science*, vol. 319, no. 5870, pp. 1668-1672, Mar.2008.

[154] M. Neumann, D. M. Sampathu, L. K. Kwong, A. C. Truax, M. C. Micsenyi, T. T. Chou, J. Bruce, T. Schuck, M. Grossman, C. M. Clark, L. F. McCluskey, B. L. Miller, E. Masliah, I. R. Mackenzie, H. Feldman, W. Feiden, H. A. Kretzschmar, J. Q. Trojanowski, and V. M. Lee, "Ubiquitinated TDP-43 in frontotemporal lobar degeneration and amyotrophic lateral sclerosis," *Science*, vol. 314, no. 5796, pp. 130-133, Oct.2006.

[155] J. Lowe, "New pathological findings in amyotrophic lateral sclerosis," *J. Neurol. Sci.*, vol. 124 Suppl, pp. 38-51, July1994.

[156] M. Nagai, D. B. Re, T. Nagata, A. Chalazonitis, T. M. Jessell, H. Wichterle, and S. Przedborski, "Astrocytes expressing ALS-linked mutated SOD1 release factors selectively toxic to motor neurons," *Nat Neurosci*, vol. 10, no. 5, pp. 615-622, May2007.

[157] A. Bento-Abreu, D. P. Van, L. Van Den Bosch, and W. Robberecht, "The neurobiology of amyotrophic lateral sclerosis," *Eur. J. Neurosci*, vol. 31, no. 12, pp. 2247-2265, June2010.

[158] L. Van Den Bosch, "Genetic rodent models of amyotrophic lateral sclerosis," *J. Biomed. Biotechnol.*, vol. 2011, p. 348765, 2011.

[159] M. A. Gama Sosa, G. R. De, and G. A. Elder, "Modeling human neurodegenerative diseases in transgenic systems," *Hum. Genet.*, vol. 131, no. 4, pp. 535-563, Apr.2012.

[160] J. R. Thonhoff, L. Ojeda, and P. Wu, "Stem cell-derived motor neurons: applications and challenges in amyotrophic lateral sclerosis," *Curr. Stem Cell Res. Ther.*, vol. 4, no. 3, pp. 178-199, Sept.2009.

[161] J. T. Dimos, K. T. Rodolfa, K. K. Niakan, L. M. Weisenthal, H. Mitsumoto, W. Chung, G. F. Croft, G. Saphier, R. Leibel, R. Goland, H. Wichterle, C. E. Henderson, and K. Eggan, "Induced Pluripotent Stem Cells Generated from Patients with ALS Can Be Differentiated into Motor Neurons," *Science*, vol. 321, no. 5893, pp. 1218-1221, Aug. 2008.

[162] A. E. Harding, "Classification of the hereditary ataxias and paraplegias," *Lancet*, vol. 1, no. 8334, pp. 1151-1155, May1983.

[163] A. E. Harding, "Friedreich's ataxia: a clinical and genetic study of 90 families with an analysis of early diagnostic criteria and intrafamilial clustering of clinical features," *Brain*, vol. 104, no. 3, pp. 589-620, Sept.1981.

[164] J. B. Schulz, S. Boesch, K. Bük, A. Dürr, P. Giunti, C. Mariotti, F. Pousset, L. Schöls, P. Vankan, and M. Pandolfo, "Diagnosis and treatment of Friedreich ataxia: a European perspective," *Nat Rev Neurol*, vol. 5, no. 4, pp. 222-234, 2009.

[165] M. Pandolfo, "Friedreich ataxia: the clinical picture," *J. Neurol.*, vol. 256 Suppl 1, pp. 3-8, Mar.2009.

[166] V. Campuzano, L. Montermini, M. D. Molt•, L. Pianese, M. Coss,e, F. Cavalcanti, E. Monros, F. Rodius, F. Duclos, A. Monticelli, F. Zara, J. Ca¤izares, H. Koutnikova, S. I. Bidichandani, C. Gellera, A. Brice, P. Trouillas, G. De Michele, A. Filla, R. De Frutos, F. Palau, P. I. Patel, S. Di Donato, J. L. Mandel, S. Cocozza, M. Koenig, and M. Pandolfo, "Friedreich's ataxia: autosomal recessive disease caused by an intronic GAA triplet repeat expansion," *Science*, vol. 271, no. 5254, pp. 1423-1427, 1996.

[167] S. I. Bidichandani, T. Ashizawa, and P. I. Patel, "The GAA triplet-repeat expansion in Friedreich ataxia interferes with transcription and may be associated with an unusual DNA structure," *Am. J. Hum. Genet.*, vol. 62, no. 1, pp. 111-121, Jan.1998.

[168] A. Dürr, M. Cossee, Y. Agid, V. Campuzano, C. Mignard, C. Penet, J. L. Mandel, A. Brice, and M. Koenig, "Clinical and genetic abnormalities in patients with Friedreich's ataxia," *N Engl J Med*, vol. 335, no. 16, pp. 1169-1175, 1996.

[169] A. H. Koeppen, "Friedreich's ataxia: pathology, pathogenesis, and molecular genetics," *J. Neurol. Sci.*, vol. 303, no. 1-2, pp. 1-12, Apr.2011.

[170] H. Puccio, "Multicellular models of Friedreich ataxia," *J. Neurol.*, vol. 256 Suppl 1, pp. 18-24, Mar.2009.

[171] A. Martelli, M. Napierala, and H. Puccio, "Understanding the genetic and molecular pathogenesis of Friedreich's ataxia through animal and cellular models," *Dis. Model. Mech.*, vol. 5, no. 2, pp. 165-176, Mar.2012.

[172] H. Puccio, "Conditional mouse models for Friedreich ataxia, a neurodegenerative disorder associating cardiomyopathy," *Handb. Exp. Pharmacol.*, no. 178, pp. 365-375, 2007.

[173] M. Cossee, H. Puccio, A. Gansmuller, H. Koutnikova, A. Dierich, M. LeMeur, K. Fischbeck, P. Dolle, and M. Koenig, "Inactivation of the Friedreich ataxia mouse gene leads to early embryonic lethality without iron accumulation," *Hum. Mol. Genet.*, vol. 9, no. 8, pp. 1219-1226, May2000.

[174] D. Simon, H. Seznec, A. Gansmuller, N. Carelle, P. Weber, D. Metzger, P. Rustin, M. Koenig, and H. Puccio, "Friedreich ataxia mouse models with progressive cerebellar and sensory ataxia reveal autophagic neurodegeneration in dorsal root ganglia," *J. Neurosci.*, vol. 24, no. 8, pp. 1987-1995, Feb.2004.

[175] C. J. Miranda, M. M. Santos, K. Ohshima, J. Smith, L. Li, M. Bunting, M. Cossee, M. Koenig, J. Sequeiros, J. Kaplan, and M. Pandolfo, "Frataxin knockin mouse," *FEBS Lett.*, vol. 512, no. 1-3, pp. 291-297, Feb.2002.

[176] S. Al-Mahdawi, R. M. Pinto, D. Varshney, L. Lawrence, M. B. Lowrie, S. Hughes, Z. Webster, J. Blake, J. M. Cooper, R. King, and M. A. Pook, "GAA repeat expansion mutation mouse models of Friedreich ataxia exhibit oxidative stress leading to progressive neuronal and cardiac pathology," *Genomics*, vol. 88, no. 5, pp. 580-590, Nov. 2006.

[177] L. Montermini, E. Andermann, M. Labuda, A. Richter, M. Pandolfo, F. Cavalcanti, L. Pianese, L. Iodice, G. Farina, A. Monticelli, M. Turano, A. Filla, M. G. De, and S. Cocozza, "The Friedreich ataxia GAA triplet repeat: premutation and normal alleles," *Hum. Mol. Genet.*, vol. 6, no. 8, pp. 1261-1266, Aug.1997.

[178] R. Sharma, S. Bhatti, M. Gomez, R. M. Clark, C. Murray, T. Ashizawa, and S. I. Bidichandani, "The GAA triplet-repeat sequence in Friedreich ataxia shows a high level of somatic instability in vivo, with a significant predilection for large contractions," *Hum. Mol. Genet.*, vol. 11, no. 18, pp. 2175-2187, Sept.2002.

[179] I. De Biase, A. Rasmussen, D. Endres, S. Al-Mahdawi, A. Monticelli, S. Cocozza, M. Pook, and S. I. Bidichandani, "Progressive GAA expansions in dorsal root ganglia of Friedreich's ataxia patients," *Ann Neurol*, vol. 61, no. 1, pp. 55-60, 2007.

[180] L. Schols, P. Bauer, T. Schmidt, T. Schulte, and O. Riess, "Autosomal dominant cerebellar ataxias: clinical features, genetics, and pathogenesis," *Lancet Neurol.*, vol. 3, no. 5, pp. 291-304, May2004.

[181] Y. Kawaguchi, T. Okamoto, M. Taniwaki, M. Aizawa, M. Inoue, S. Katayama, H. Kawakami, S. Nakamura, M. Nishimura, I. Akiguchi, and ., "CAG expansions in a novel gene for Machado-Joseph disease at chromosome 14q32.1," *Nat. Genet.*, vol. 8, no. 3, pp. 221-228, Nov.1994.

[182] H. L. Paulson, M. K. Perez, Y. Trottier, J. Q. Trojanowski, S. H. Subramony, S. S. Das, P. Vig, J. L. Mandel, K. H. Fischbeck, and R. N. Pittman, "Intranuclear inclusions of expanded polyglutamine protein in spinocerebellar ataxia type 3," *Neuron*, vol. 19, no. 2, pp. 333-344, Aug.1997.

[183] U. Rub, E. R. Brunt, and T. Deller, "New insights into the pathoanatomy of spinocerebellar ataxia type 3 (Machado-Joseph disease)," *Curr. Opin. Neurol.*, vol. 21, no. 2, pp. 111-116, Apr.2008.

[184] O. Riess, U. Rüb, A. Pastore, P. Bauer, and L. Schöls, "SCA3: neurological features, pathogenesis and animal models," *Cerebellum*, vol. 7, no. 2, pp. 125-137, 2008.

[185] H. Ikeda, M. Yamaguchi, S. Sugai, Y. Aze, S. Narumiya, and A. Kakizuka, "Expanded polyglutamine in the Machado-Joseph disease protein induces cell death in vitro and in vivo," *Nat. Genet.*, vol. 13, no. 2, pp. 196-202, June1996.

[186] C. K. Cemal, C. J. Carroll, L. Lawrence, M. B. Lowrie, P. Ruddle, S. Al-Mahdawi, R. H. King, M. A. Pook, C. Huxley, and S. Chamberlain, "YAC transgenic mice carrying pathological alleles of the MJD1 locus exhibit a mild and slowly progressive cerebellar deficit," *Hum. Mol. Genet.*, vol. 11, no. 9, pp. 1075-1094, May2002.

[187] C. Maayan, E. Kaplan, S. Shachar, O. Peleg, and S. Godfrey, "Incidence of familial dysautonomia in Israel 1977-1981," *Clin. Genet.*, vol. 32, no. 2, pp. 106-108, Aug.1987.

[188] S. A. Slaugenhaupt, A. Blumenfeld, S. P. Gill, M. Leyne, J. Mull, M. P. Cuajungco, C. B. Liebert, B. Chadwick, M. Idelson, L. Reznik, C. Robbins, I. Makalowska, M.

Brownstein, D. Krappmann, C. Scheidereit, C. Maayan, F. B. Axelrod, and J. F. Gusella, "Tissue-specific expression of a splicing mutation in the IKBKAP gene causes familial dysautonomia," *Am. J. Hum. Genet.*, vol. 68, no. 3, pp. 598-605, Mar.2001.

[189] J. Pearson and B. A. Pytel, "Quantitative studies of sympathetic ganglia and spinal cord intermedio-lateral gray columns in familial dysautonomia," *J. Neurol. Sci.*, vol. 39, no. 1, pp. 47-59, Nov.1978.

[190] N. A. Hawkes, G. Otero, G. S. Winkler, N. Marshall, M. E. Dahmus, D. Krappmann, C. Scheidereit, C. L. Thomas, G. Schiavo, H. Erdjument-Bromage, P. Tempst, and J. Q. Svejstrup, "Purification and characterization of the human elongator complex," *J. Biol. Chem.*, vol. 277, no. 4, pp. 3047-3052, Jan.2002.

[191] P. Close, N. Hawkes, I. Cornez, C. Creppe, C. A. Lambert, B. Rogister, U. Siebenlist, M. P. Merville, S. A. Slaugenhaupt, V. Bours, J. Q. Svejstrup, and A. Chariot, "Transcription impairment and cell migration defects in elongator-depleted cells: implication for familial dysautonomia," *Mol. Cell*, vol. 22, no. 4, pp. 521-531, May2006.

[192] G. Lee and L. Studer, "Modelling familial dysautonomia in human induced pluripotent stem cells," *Philosophical Transactions of the Royal Society B: Biological Sciences*, vol. 366, no. 1575, pp. 2286-2296, Aug.2011.

[193] P. Dietrich, J. Yue, S. E., and I. Dragatsis, "Deletion of Exon 20 of the Familial Dysautonomia Gene Ikbkap in Mice Causes Developmental Delay, Cardiovascular Defects, and Early Embryonic Lethality," *PLoS ONE*, vol. 6, no. 10, p. e27015, Oct.2011.

[194] M. M. Hims, R. S. Shetty, J. Pickel, J. Mull, M. Leyne, L. Liu, J. F. Gusella, and S. A. Slaugenhaupt, "A humanized IKBKAP transgenic mouse models a tissue-specific human splicing defect," *Genomics*, vol. 90, no. 3, pp. 389-396, Sept.2007.

[195] N. Boone, B. Loriod, A. Bergon, O. Sbai, C. Formisano-Treziny, J. Gabert, M. Khrestchatisky, C. Nguyen, F. Feron, F. B. Axelrod, and E. C. Ibrahim, "Olfactory stem cells, a new cellular model for studying molecular mechanisms underlying familial dysautonomia," *PLoS. One.*, vol. 5, no. 12, p. e15590, 2010.

[196] W. Murrell, F. Feron, A. Wetzig, N. Cameron, K. Splatt, B. Bellette, J. Bianco, C. Perry, G. Lee, and A. Mackay-Sim, "Multipotent stem cells from adult olfactory mucosa," *Dev. Dyn.*, vol. 233, no. 2, pp. 496-515, June2005.

[197] S. A. Slaugenhaupt, J. Mull, M. Leyne, M. P. Cuajungco, S. P. Gill, M. M. Hims, F. Quintero, F. B. Axelrod, and J. F. Gusella, "Rescue of a human mRNA splicing defect by the plant cytokinin kinetin," *Hum. Mol. Genet.*, vol. 13, no. 4, pp. 429-436, Feb.2004.

[198] D. C. Crawford, J. M. Acuna, and S. L. Sherman, "FMR1 and the fragile X syndrome: human genome epidemiology review," *Genet. Med.*, vol. 3, no. 5, pp. 359-371, Sept. 2001.

[199] A. J. Verkerk, M. Pieretti, J. S. Sutcliffe, Y. H. Fu, D. P. Kuhl, A. Pizzuti, O. Reiner, S. Richards, M. F. Victoria, F. P. Zhang, and ., "Identification of a gene (FMR-1) contain-

ing a CGG repeat coincident with a breakpoint cluster region exhibiting length variation in fragile X syndrome," *Cell*, vol. 65, no. 5, pp. 905-914, May1991.

[200] S. A. Irwin, R. Galvez, and W. T. Greenough, "Dendritic spine structural anomalies in fragile-X mental retardation syndrome," *Cereb. Cortex*, vol. 10, no. 10, pp. 1038-1044, Oct.2000.

[201] A. Yao, S. Jin, X. Li, Z. Liu, X. Ma, J. Tang, and Y. Q. Zhang, "Drosophila FMRP regulates microtubule network formation and axonal transport of mitochondria," *Hum. Mol. Genet.*, vol. 20, no. 1, pp. 51-63, Jan.2011.

[202] "Fmr1 knockout mice: a model to study fragile X mental retardation. The Dutch-Belgian Fragile X Consortium," *Cell*, vol. 78, no. 1, pp. 23-33, July1994.

[203] J. R. Brouwer, E. J. Mientjes, C. E. Bakker, I. M. Nieuwenhuizen, L. A. Severijnen, H. C. Van der Linde, D. L. Nelson, B. A. Oostra, and R. Willemsen, "Elevated Fmr1 mRNA levels and reduced protein expression in a mouse model with an unmethylated Fragile X full mutation," *Experimental Cell Research*, vol. 313, no. 2, pp. 244-253, Jan. 2007.

[204] R. Pietrobono, E. Tabolacci, F. Zalfa, I. Zito, A. Terracciano, U. Moscato, C. Bagni, B. Oostra, P. Chiurazzi, and G. Neri, "Molecular dissection of the events leading to inactivation of the FMR1 gene," *Hum. Mol. Genet.*, vol. 14, no. 2, pp. 267-277, Jan.2005.

[205] R. Eiges, A. Urbach, M. Malcov, T. Frumkin, T. Schwartz, A. Amit, Y. Yaron, A. Eden, O. Yanuka, N. Benvenisty, and D. Ben-Yosef, "Developmental Study of Fragile X Syndrome Using Human Embryonic Stem Cells Derived from Preimplantation Genetically Diagnosed Embryos," *Cell Stem Cell*, vol. 1, no. 5, pp. 568-577, Nov.2007.

[206] T. Vierbuchen, A. Ostermeier, Z. P. Pang, Y. Kokubu, T. C. Sudhof, and M. Wernig, "Direct conversion of fibroblasts to functional neurons by defined factors," *Nature*, vol. 463, no. 7284, pp. 1035-1041, Feb.2010.

[207] Z. P. Pang, N. Yang, T. Vierbuchen, A. Ostermeier, D. R. Fuentes, T. Q. Yang, A. Citri, V. Sebastiano, S. Marro, T. C. Sudhof, and M. Wernig, "Induction of human neuronal cells by defined transcription factors," *Nature*, vol. 476, no. 7359, pp. 220-223, Aug. 2011.

[208] M. Caiazzo, M. T. 'Anno, E. Dvoretskova, D. Lazarevic, S. Taverna, D. Leo, T. D. Sotnikova, A. Menegon, P. Roncaglia, G. Colciago, G. Russo, P. Carninci, G. Pezzoli, R. R. Gainetdinov, S. Gustincich, A. Dityatev, and V. Broccoli, "Direct generation of functional dopaminergic neurons from mouse and human fibroblasts," *Nature*, vol. 476, no. 7359, pp. 224-227, Aug.2011.

[209] U. Pfisterer, A. Kirkeby, O. Torper, J. Wood, J. Nelander, A. Dufour, A. Björklund, O. Lindvall, J. Jakobsson, and M. Parmar, "Direct conversion of human fibroblasts to dopaminergic neurons," *Proceedings of the National Academy of Sciences*, vol. 108, no. 25, pp. 10343-10348, June2011.

[210] E. Y. Son, J. K. Ichida, B. J. Wainger, J. S. Toma, V. F. Rafuse, C. J. Woolf, and K. Eggan, "Conversion of mouse and human fibroblasts into functional spinal motor neurons," *Cell Stem Cell*, vol. 9, no. 3, pp. 205-218, Sept.2011.

[211] L. Qiang, R. Fujita, T. Yamashita, S. Angulo, H. Rhinn, D. Rhee, C. Doege, L. Chau, L. Aubry, W. B. Vanti, H. Moreno, and A. Abeliovich, "Directed conversion of Alzheimer's disease patient skin fibroblasts into functional neurons," *Cell*, vol. 146, no. 3, pp. 359-371, Aug.2011.

Chapter 23

Induced Pluripotent Stem Cells as a Source of Hepatocytes

Minoru Tomizawa, Fuminobu Shinozaki,
Takao Sugiyama, Shigenori Yamamoto,
Makoto Sueishi and Takanobu Yoshida

Additional information is available at the end of the chapter

http://dx.doi.org/10.5772/54552

1. Introduction

Human induced pluripotent stem (hiPS) cells are generated with cellular reprogramming factors [1], and they have the potential to differentiate into a variety of cells. Ethical issues and graft-versus-host disease may be avoided with hiPS cells because they can be established in each patient individually. hiPS cells may therefore be an ideal cell source for patients.

The liver is a single large organ, the cells of which are 70–80% hepatocytes. These liver-specific cells play a major role in protein synthesis, glucose metabolism, and detoxification. Methods of producing hepatocytes from hiPS cells have been under development for some time. In this chapter, we will cover the following topics:

- Hepatocyte culture
- Applications of hepatocyte culture
- Production of hepatocytes from human embryonic stem (hES) cells
- Protocols for differentiation of hiPS cells into hepatocytes
- Current applications of hepatocytes differentiated from hiPS cells
- Limitations of differentiation
- Future directions

First we will discuss primary hepatocyte culture. The knowledge on primary hepatocyte culture is applicable to maintenance of hepatocytes differentiated from hiPS cells. Next, application of hepatocyte culture will be discussed because the application would provide potential usage of hiPS cells. Then production of hepatocytes from ES cells will be presented. Methods presented in this section are prototypes of differentiation protocols of hiPS cell into hepatocytes. Sequentially, current protocols of differentiation of hiPS cells into hepatocytes will be summarized. Applications of hepatotyes from hiPS cells will be presented specific to human diseases such as hepatitis C virus. Even with the protocols above mentioned, differentiation of hiPS cells to functioning hepatocytes is difficult. Limitations of differentiation will be discussed. Finally, potential new approaches will be presented in the last section.

2. Primary hepatocyte culture

Before the era of ES cells or iPS cells, primary hepatocyte culture had been the only method to investigate differentiation and function of hepatocytes. The accumulated knowledge on hepatocytes would be applicable to maintain hepatocytes differentiated from hiPS cells. Hepatocyte culture is useful for developing drugs, cell therapies, and disease models. Primary hepatocyte culture is an ideal in vitro model of drug metabolism and toxicology, and primary hepatocytes can be transplanted into patients with liver failure [2]. Hepatocytes from patients with metabolic diseases can be used to investigate disease mechanisms. However, primary hepatocyte culture remains technically difficult. Hepatocytes are isolated from a fragment of resected donor liver with a 2-step collagenase perfusion [2]. Fetal hepatocytes (10^7 cells) have been transplanted into patients with hepatic encephalopathy [3], and while the disease improved, there was no increase in survival time. The speculated reason is that not enough cells were transplanted [4]. Isolated hepatocytes are prone to apoptosis and damage [5] and have difficulty proliferating once cultured [6]. Primary hepatocyte culture also presents ethical issues when cells are harvested from humans. Now hiPS cells could overcome problems that primary hepatocyte culture encounters.

3. Application of hepatocytes differentiated from iPS cells

If hiPS cells could differentiate into hepatocytes, they would be useful for medical practice and biological study. Potential applications would be as follows:

- Transplantation into patients with hepatic insufficiency
- A method to support patients with hepatic insufficiency such as hemodialysis
- Drug screening
- Toxicology
- In vitro model of hepatitis C virus infection
- In vitro model of hepatocyte differentiation

http://dx.doi.org/10.5772/54552

- In vitro model of liver diseases

One of the most important applications of hepatocytes from hiPS cells would be transplantation into patients with hepatic insufficiency caused by fulminant hepatitis. The disease could be treated perfectly with transplanted hepatocytes because it is caused by significant loss of functioning hepatocytes. Hepatic progenitor cells have potential to differentiate into mature hepatocytes and bile duct epithelial cells. Hepatic progenitor cells would be expected to construct normal liver structure such as hepatic lobule and bile ducts. Hepatic progenitor cells derived from mouse embryonic stem (ES) cells engraft in host liver tissue and differentiate into hepatocytes when transplanted into partially hepatectomized mice [7]. Hepatocytes will also engraft in mice with acute liver failure caused by carbon tetrachloride intoxication [8]. This is a promising finding that suggests that hepatocytes from pluripotent cells are transplantable. Hepatocytes have indeed been differentiated from human ES cells and transplanted [9]. One disadvantage of the use of human ES cells is that they may provoke graft-versus-host disease. This could be overcome if hepatocytes are derived from iPS cells established from the individual patient. Patients with acute liver failure could be successfully treated in this manner.

Another application of hepatocytes from hiPS cells would be metabolic diseases. The disease could be cured with transplantation of functioning hepatocytes because they play pivotal roles in metabolism. High levels of low-density lipoprotein cholesterol (LDL-Chol) in the plasma is known to cause cardiovascular disease. Successful reduction of LDL-Chol may lead to prevention of cardiovascular disease. Mutations in the LDL receptor gene result in familial hypercholesterolemia (FH); iPS cells derived from patients with FH provide a good model for analyzing the mechanism of this condition [10].

4. Differentiation of ES cells into hepatocytes

Cultured primary hepatocytes do not proliferate but disappear and lose their function quickly. Pluripotent stem cells have been focused as a cell source of hepatocytes. Before the advent of iPS cells, ES cells had been the center of investigation of differentiation methods into hepatocytes. The topics of the investigation have been growth factors, transcription factors, extracellular matrix, and three-dimensional (3D) culture

Mouse ES cells start differentiation into the hepatocyte lineage once leukemia inhibitory factor (LIF) is deprived and embryoid bodies are formed [11-13]. Hepatocyte-like cells derived from mouse ES cells take up indocyanine green, express albumin, and form bile canaliculi [14]. The induced cells express specific live genes such as α-1-antitrypsin and phosphoenolpyruvate carboxykinase (PEPCK). Withdrawal of LIF is not an appropriate method for inducing hiPS cell differentiation because these cells are not LIF dependent [15]. Human ES cells differentiate into mesoderm, endoderm, and ectoderm after withdrawal of the LIF and basic fibroblast growth factor (bFGF) [16], but they do not necessarily differentiate into hepatocytes. Therefore, growth factors are expected to be needed for hepatocyte differentiation from human ES cells. Nerve growth factor (NGF) and hepatocyte growth factor (HGF) induce differentiation into endoderm and eventually liver cells [17]. Transcription factors also play an important role in hepatocyte differentiation. Transcription factor forkhead box protein (Fox) A2 promotes differentiation of mouse ES cell into the hepatocyte lineage [18], and these hepatocye-like cells express phosphoenolpyruvate (PEPCK) and albumin.

To search for more efficient protocols to promote differentiation of ES cells into hepatoctyes, combinations of growth factors and extracellular matrices have been investigated [19]. Shirahashi et al. reported that a mixture of Iscove's modified Dulbecco's medium with 20% fetal bovine serum, human insulin, dexamethasone, and type 1 collagen is optimum for mouse and human ES cell differentiation into the hepatocyte lineage. Bovine serum should not be used because xeno-proteins are not suitable for human application. This study suggests that extracellular matrix is important in hepatocyte differentiation.

Hepatic progenitor cells differentiate into hepatocytes in 3D structure in liver. It is expected that 3D culture is more suitable environment for ES cells to differentiate into hepatocytes. Indeed, 3D cultures of mouse ES cells have been shown to differentiate into hepatocytes [20]. Embryoid bodies (EB) were inserted into a collagen scaffold 3D culture system and stimulated with exogenous growth factors and hormones to produce hepatic differentiation.

Hepatocytes should be isolated from the other cells because ES cells could be among hepatocytes. Undifferentiated cells have been shown to form teratoma when transplanted into recipient cells mixed with hepatocytes [21]. A practical method to avoid this is to enrich the hepatocytes and eliminate the undifferentiated cells by Percoll discontinuous gradient centrifugation [22, 23].

Rambhatla et al. [24] reported that the addition of sodium butylate leads to significant cell death and induction of hepatocyte differentiation in human ES cells. Cells cultured with sodium butylate express albumin, α-1-antitrypsin, and cytochrome P450 and also accumulate glycogen. However, the induced cells do not express alpha-fetoprotein (AFP). Sodium butylate is a possible candidate for a small molecule to eliminate undifferentiated cells and induce hepatic differentiation.

5. Protocols for differentiation of hiPS cells into hepatocytes

Protocols for differentiation of hiPS cells into hepatocytes follow those for mouse ES cells as mentioned above. Stepwise protocols are currently used to promote the differentiation [25-28] (Table 1). These protocols consist of sequential application of growth factors and introduction of transcription factors to mimic hepatocyte differentiation during liver development. The progression is endodermal cell, immature hepatocyte (often referred as hepatoblast), and finally mature hepatocyte.

http://dx.doi.org/10.5772/54552

DeLaForest [25]	D0-5		D5-10	D10-15	D15-20
	Activin A, LY294002		BMP4, FGF2	HGF	OncoM
S-Tayeb [26]	D0-5	D5-10		D10-15	D15-20
	O2: 20%	O2: 4%		O2: 4%	O2: 20%
	Activin A	BMP4, FGF2		HGF	OncoM
Song [27]	D0-3	D4-7	D8-13	D14-18	D19-21
	Activin A	FGF4, BMP3	HGF, KGF	OncoM	OncoM, Dex

D: day; BMP4: bone morphogenic protein 4; FGF: fibroblast growth factor 2; HGF: hepatocyte growth factor; OncoM: oncostatin M; KGF: keratinocyte growth factor; Dex: dexamethasone.

Table 1. Protocols for hepatocyte differentiation from human induced pluripotent stem cells.

6. Endodermal differentiation

All differentiation protocols apply activin A (a member of the tumor growth factor β superfamily) at a high concentration of 100-ng/mL. LY294002 (a specific inhibitor of phosphatidylinositol 3 phosphatase), B27 supplement, or bFGF are added, depending on the purpose of the research. After 3–5 days of culture, iPS cells differentiate into endodermal cells. From days 5–10, a combination of bone morphogens 2 or 4 and fibroblast growth factors 2 or 4 is applied. Takayama et al. [28] introduced sex-determining region Y box 17 to promote differentiation at this stage after incubation with activin A. Sekine et al. [29] used LY294002 in addition to 100-ng/mL activin A. In their study, FoxA2 and Sox17 expressions appeared but AFP and albumin were not analyzed. Phosphatidyl inositol (PI) 3 kinase may control differentiation of iPS cells into endodemal cells, but other factors are still needed.

7. Differentiation into immature hepatocytes

Hepatocyte growth factor (HGF) or keratinocyte growth factor (KGF) is applied from days 10–14. Inamura et al. introduced hematopoietically expressed homeobox (HEX) to promote differentiation into hepatoblasts [30].

8. Differentiation into mature hepatocytes

HGF or oncostatin M is added to promote differentiation of hepatoblasts into mature hepatocytes. Takayama et al. [28] introduced hepatocyte nuclear factor-4 to provide the terminal differentiation of hepatoblasts into hepatocytes. Mature hepatocytes appeared at approximately 20 days after the initiation of the differentiation process. Si-Tayeb et al. [26] cultured cells under 4% oxygen from days 5 to 15.

In another study, Nakamura et al. [31] derived hepatocytes from human ES and iPS cells under feeder- and serum-free conditions. They succeeded in producing cholangiocytes and proliferating progenitors. The cells produced with their protocol were confirmed to function as mature hepatocytes. Indocyanine green was taken up by 30% of the hepatocytes, and 80% stored glycogen. They also maintained the metabolic activity of CYP3A4.

Chen et al. [32] proposed another multistep protocol. They do not apply any transcription factors, but growth factors. They have succeeded in differentiation of hiPS cells into mature hepatocytes within only 12 days. The period is significantly shorter than the other researchers. With their method, activin A (100 ng/mL) and HGF (10 ng/mL) were added from days 1 to 3, and prior to that, HGF had been added at the last step of hepatocyte maturation. They also added HGF at the first step of differentiation and successfully derived hepatocyte-like cells. Sox17 and FoxA2, induced by activin A, are important markers of endodermal differentiation. HGF and activin A may have synergistic effects on the differentiating cells.

Transcription factors play an important role in liver development and hepatocyte differentiation [33]. Generally, pluripotent stem cells are hard to transfect plasmids. Adenovirus vectors provide highly efficient transduction to hiPS cells [34]. Inamura et al. [30] transduced HEX into hES and hiPS cells to efficiently produce hepatoblasts (Table 2). After differentiation into hepatoblasts, transduction of HNF4α finally produces mature hepatocytes [28].

Inamura [34]	D0-6	D6-8		D9-18
	Activin A	BMP4, FGF4		FGF4, HGF, OncoM, Dex
	D5: passage, D6:Ad-Hex			D9: passage
Takayama [28]	D0-3	D3-6	D7-9	D10-20
	Activin A	Activin A	BMP4, FGF4	HGF, OncoM, Dex
	D3: Ad-Sox17	D5: passage, D6: Ad-Hex	D9: Ad-HNF4A	

BMP4: bone morphogen protein 4; FGF4: fibroblast growth factor 4; HGF: hepatocyte growth factor; OncoM: oncostatin M; Dex: dexamethasone; Ad-Hex, Sox17, HNF4A: adenovirus vector transducing Hex, Sox17, and HNF4A, respectively; Hex: hematopoietically expressed homeobox; Sox17: sex determining region Y box 7; HNF4A: hepatocyte nuclear factor 4 α.

Table 2. Protocols for hepatocyte differentiation from human induced pluripotent stem cells with adenovirus vectors.

9. Current applications of hepatocytes differentiated from hiPS cells

Hepatocytes from hiPS cells are perfect for in vitro model of human diseases because human primary hepatocytes have both ethical and technical issues. Hepatitis C virus (HCV) causes liver cirrhosis and hepatocellular carcinoma (HCC). Primary human hepatocyte culture is a relevant in vitro model for HCV infection, but it presents some ethical issues. Human iPS cells are not permissive to HCV. Interestingly, hepatocyte-like cells derived from hiPS cells recapitulate permissiveness and are infected with HCV [35, 36]. Hepatocyte-like cells derived from hiPS cells exert an inflammatory response to infection [37] and may provide a suitable in vitro model to study the mechanism of HCV infection. Such a model may potentially lead to innovative methods to inhibit HCV and prevent liver cirrhosis and HCC.

Hepatocyte-like cells derived from mouse iPS cells have been shown to improve acute liver failure caused by carbon tetrachloride [38]. These cells were transplanted through peritoneal injection and significantly reduced the extent of necrotic liver. The authors concluded that the hepatoprotective effects were based on antioxidant activity.

10. Limitations of hepatocytes differentiated from hiPS cells

Cells cultured under the protocols mentioned above are referred to as hepatocyte-like cells. In these cells, detoxification activity is lower than in primary hepatocyte culture [26, 28]. Hepatocytes differentiated from hiPS cells have lower expression levels of FoxA1, FoxA2, FoxA3, and HNF1α, and Takayama et al. [28] speculated that other factors are still needed. iPS cells retain their donor cell gene expressions. Lee et al. [35] generated mouse iPS cells from hepatoblasts and adult hepatocytes. Hepatocytes differentiated from hiPS cells express mRNA that is normally not found in fetal or adult liver [25]. An interesting finding is that hepatocytes are differentiated more efficiently from hepatoblast-derived iPS cells than from adult hepatocytes. This suggests that the efficiency of hepatocyte differentiation may depend on the origin of the iPS cells. Protocols need to be further developed given that the mentioned liver-specific genes are important for clinical and pharmacological applications.

11. Future directions

To overcome these limitations, novel approaches are under investigation. Current research efforts can be categorized into extracellular matrix, 3D culture, and cell sheet approaches.

An extracellular matrix (ECM) provides conditions suitable for cultured cells to differentiate to hepatocytes. M15, a mesonephric cell line, induces differentiation of mouse ES cells into the hepatocyte lineage [39]. Eighty percent of mouse ES cells cultured with M15 express AFP, and 9% express albumin. It is interesting that even the fixed M15 cells can promote mouse ES cell differentiation. Shiraki et al. reported a synthesized basement membrane composed of human recombinant laminin 511 [40] that induced differentiation of mouse ES cells into hepatocyte lineages.

A 3D culture system is composed of gelatin and extracellular matrix from Swiss 3T3 cells [41]. This system preserves the functions of hepatocyte-like cells differentiated from hiPS cells. The most important component of the ECM has been determined to be type 1 collagen.

Cells are 3D cultured in hollow fibers similar to embryoid bodies. Hollow fibers are useful because the efficiency of embryoid body formation is low compared with mouse ES cells, which also differentiate into hepatocytes in hollow fibers [42]. The organoid culture system efficiently allows mouse ES cells to form cellular aggregates in their lumen. Liver-specific functions of mouse ES cells are comparable with those of primary hepatocytes.

Primary rat hepatocytes have been successfully cultured for 200 days on temperature-responsive sheets [43]. These sheets attach on the bottoms of culture dishes at 37ºC and detach at 25ºC.

They provide easy culturing and handling of cells. Primary rat hepatocytes have preserved liver-specific functions for 28 days in hybrid sheets with endothelial cells [44]. This system enables easy manipulation of iPS cells and may promote differentiation into hepatocytes.

12. Conclusion

Human iPS cells are a promising source for hepatocytes and may be used for drug screening, for cell transplantation, and as a model for studying human diseases. Protocols have been presented for the differentiation of human iPS cells into hepatocytes; however, the differentiated cells have limited hepatocyte characteristics. In the future, as more sophisticated methods are expected to be developed, new applications of these cells will be realized.

Acknowledgements

This work was supported in part by a Grant-in-Aid for Scientific Research (C) (grant No. 23591002) from the Japan Society for the Promotion of Science (JSPS).

Author details

Minoru Tomizawa[1*], Fuminobu Shinozaki[2], Takao Sugiyama[3], Shigenori Yamamoto[4], Makoto Sueishi[3] and Takanobu Yoshida[5]

*Address all correspondence to: nihminor-cib@umin.ac.jp

1 Department of Gastroenterology, National Hospital Organization Shimoshizu Hospital, Yotsukaido City, Japan

2 Department of Radiology, National Hospital Organization Shimoshizu Hospital, Yotsukaido City, Japan

3 Department of Rheumatology, National Hospital Organization Shimoshizu Hospital, Yotsukaido City, Japan

4 Department of Pediatrics, National Hospital Organization Shimoshizu Hospital, Yotsukaido City, Japan

5 Department of Internal Medicine, National Hospital Organization Shimoshizu Hospital, Yotsukaido City, Japan

References

[1] Takahashi K, Tanabe K, Ohnuki M, Narita M, Ichisaka T, Tomoda K, Yamanaka S. Induction of pluripotent stem cells from adult human fibroblasts by defined factors. Cell 2007;131(5):861-872.

[2] Strom SC, Chowdhury JR, Fox IJ. Hepatocyte transplantation for the treatment of human disease. Semin Liver Dis 1999;19(1):39-48.

[3] Habibullah CM, Syed IH, Qamar A, Taher-Uz Z. Human fetal hepatocyte transplantation in patients with fulminant hepatic failure. Transplantation 1994;58(8):951-952.

[4] Riehle KJ, Dan YY, Campbell JS, Fausto N. New concepts in liver regeneration. J Gastroenterol Hepatol 2011;26 Suppl 1(203-212.

[5] Fisher RA, Bu D, Thompson M, Wolfe L, Ritter JK. Optimization of conditions for clinical human hepatocyte infusion. Cell Transplant 2004;13(6):677-689.

[6] Mitaka T, Sattler CA, Sattler GL, Sargent LM, Pitot HC. Multiple cell cycles occur in rat hepatocytes cultured in the presence of nicotinamide and epidermal growth factor. Hepatology 1991;13(1):21-30.

[7] Yin Y, Lim YK, Salto-Tellez M, Ng SC, Lin CS, Lim SK. AFP(+), ESC-derived cells engraft and differentiate into hepatocytes in vivo. Stem Cells 2002;20(4):338-346.

[8] Yamamoto H, Quinn G, Asari A, Yamanokuchi H, Teratani T, Terada M, Ochiya T. Differentiation of embryonic stem cells into hepatocytes: biological functions and therapeutic application. Hepatology 2003;37(5):983-993.

[9] Basma H, Soto-Gutierrez A, Yannam GR, Liu L, Ito R, Yamamoto T, Ellis E, et al. Differentiation and transplantation of human embryonic stem cell-derived hepatocytes. Gastroenterology 2009;136(3):990-999.

[10] Cayo MA, Cai J, Delaforest A, Noto FK, Nagaoka M, Clark BS, Collery RF, et al. 'JD' iPS cell-derived hepatocytes faithfully recapitulate the pathophysiology of familial hypercholesterolemia. Hepatology 2012.

[11] Chinzei R, Tanaka Y, Shimizu-Saito K, Hara Y, Kakinuma S, Watanabe M, Teramoto K, et al. Embryoid-body cells derived from a mouse embryonic stem cell line show differentiation into functional hepatocytes. Hepatology 2002;36(1):22-29.

[12] Abe K, Niwa H, Iwase K, Takiguchi M, Mori M, Abe SI, Abe K, et al. Endoderm-specific gene expression in embryonic stem cells differentiated to embryoid bodies. Exp Cell Res 1996;229(1):27-34.

[13] Jones EA, Tosh D, Wilson DI, Lindsay S, Forrester LM. Hepatic differentiation of murine embryonic stem cells. Exp Cell Res 2002;272(1):15-22.

[14] Yamada T, Yoshikawa M, Kanda S, Kato Y, Nakajima Y, Ishizaka S, Tsunoda Y. In vitro differentiation of embryonic stem cells into hepatocyte-like cells identified by cellular uptake of indocyanine green. Stem Cells 2002;20(2):146-154.

[15] Hirai H, Firpo M, Kikyo N. Establishment of LIF-Dependent Human iPS Cells Closely Related to Basic FGF-Dependent Authentic iPS Cells. PLoS One 2012;7(6):e39022.

[16] Itskovitz-Eldor J, Schuldiner M, Karsenti D, Eden A, Yanuka O, Amit M, Soreq H, et al. Differentiation of human embryonic stem cells into embryoid bodies compromising the three embryonic germ layers. Mol Med 2000;6(2):88-95.

[17] Schuldiner M, Yanuka O, Itskovitz-Eldor J, Melton DA, Benvenisty N. Effects of eight growth factors on the differentiation of cells derived from human embryonic stem cells. Proc Natl Acad Sci U S A 2000;97(21):11307-11312.

[18] Ishizaka S, Shiroi A, Kanda S, Yoshikawa M, Tsujinoue H, Kuriyama S, Hasuma T, et al. Development of hepatocytes from ES cells after transfection with the HNF-3beta gene. Faseb J 2002;16(11):1444-1446.

[19] Shirahashi H, Wu J, Yamamoto N, Catana A, Wege H, Wager B, Okita K, et al. Differentiation of human and mouse embryonic stem cells along a hepatocyte lineage. Cell Transplant 2004;13(3):197-211.

[20] Imamura T, Cui L, Teng R, Johkura K, Okouchi Y, Asanuma K, Ogiwara N, et al. Embryonic stem cell-derived embryoid bodies in three-dimensional culture system form hepatocyte-like cells in vitro and in vivo. Tissue Eng 2004;10(11-12):1716-1724.

[21] Teramoto K, Hara Y, Kumashiro Y, Chinzei R, Tanaka Y, Shimizu-Saito K, Asahina K, et al. Teratoma formation and hepatocyte differentiation in mouse liver transplanted with mouse embryonic stem cell-derived embryoid bodies. Transplant Proc 2005;37(1):285-286.

[22] Kumashiro Y, Asahina K, Ozeki R, Shimizu-Saito K, Tanaka Y, Kida Y, Inoue K, et al. Enrichment of hepatocytes differentiated from mouse embryonic stem cells as a transplantable source. Transplantation 2005;79(5):550-557.

[23] Kumashiro Y, Teramoto K, Shimizu-Saito K, Asahina K, Teraoka H, Arii S. Isolation of hepatocyte-like cells from mouse embryoid body cells. Transplant Proc 2005;37(1): 299-300.

[24] Rambhatla L, Chiu CP, Kundu P, Peng Y, Carpenter MK. Generation of hepatocyte-like cells from human embryonic stem cells. Cell Transplant 2003;12(1):1-11.

[25] DeLaForest A, Nagaoka M, Si-Tayeb K, Noto FK, Konopka G, Battle MA, Duncan SA. HNF4A is essential for specification of hepatic progenitors from human pluripotent stem cells. Development 2011;138(19):4143-4153.

[26] Si-Tayeb K, Noto FK, Nagaoka M, Li J, Battle MA, Duris C, North PE, et al. Highly efficient generation of human hepatocyte-like cells from induced pluripotent stem cells. Hepatology 2010;51(1):297-305.

[27] Song Z, Cai J, Liu Y, Zhao D, Yong J, Duo S, Song X, et al. Efficient generation of hepatocyte-like cells from human induced pluripotent stem cells. Cell Res 2009;19(11):1233-1242.

[28] Takayama K, Inamura M, Kawabata K, Katayama K, Higuchi M, Tashiro K, Nonaka A, et al. Efficient Generation of Functional Hepatocytes From Human Embryonic Stem Cells and Induced Pluripotent Stem Cells by HNF4alpha Transduction. Mol Ther 2012;20(1):127-137.

[29] Sekine K, Takebe T, Suzuki Y, Kamiya A, Nakauchi H, Taniguchi H. Highly efficient generation of definitive endoderm lineage from human induced pluripotent stem cells. Transplant Proc 2012;44(4):1127-1129.

[30] Inamura M, Kawabata K, Takayama K, Tashiro K, Sakurai F, Katayama K, Toyoda M, et al. Efficient Generation of Hepatoblasts From Human ES Cells and iPS Cells by Transient Overexpression of Homeobox Gene HEX. Mol Ther 2010.

[31] Nakamura N, Saeki K, Mitsumoto M, Matsuyama S, Nishio M, Hasegawa M, Miyagawa Y, et al. Feeder-free and serum-free production of hepatocytes, cholangiocytes, and their proliferating progenitors from human pluripotent stem cells: application to liver-specific functional and cytotoxic assays. Cell Reprogram 2012;14(2):171-185.

[32] Chen YF, Tseng CY, Wang HW, Kuo HC, Yang VW, Lee OK. Rapid generation of mature hepatocyte-like cells from human induced pluripotent stem cells by an efficient three-step protocol. Hepatology 2012;55(4):1193-1203.

[33] Zaret KS, Watts J, Xu J, Wandzioch E, Smale ST, Sekiya T. Pioneer factors, genetic competence, and inductive signaling: programming liver and pancreas progenitors from the endoderm. Cold Spring Harb Symp Quant Biol 2008;73(119-126.

[34] Xu ZL, Mizuguchi H, Sakurai F, Koizumi N, Hosono T, Kawabata K, Watanabe Y, et al. Approaches to improving the kinetics of adenovirus-delivered genes and gene products. Adv Drug Deliv Rev 2005;57(5):781-802.

[35] Lee SB, Seo D, Choi D, Park KY, Holczbauer A, Marquardt JU, Conner EA, et al. Contribution of hepatic lineage stage-specific donor memory to the differential potential of induced mouse pluripotent stem cells. Stem Cells 2012;30(5):997-1007.

[36] Wu X, Robotham JM, Lee E, Dalton S, Kneteman NM, Gilbert DM, Tang H. Productive hepatitis C virus infection of stem cell-derived hepatocytes reveals a critical transition to viral permissiveness during differentiation. PLoS Pathog 2012;8(4):e1002617.

[37] Schwartz RE, Trehan K, Andrus L, Sheahan TP, Ploss A, Duncan SA, Rice CM, et al. Modeling hepatitis C virus infection using human induced pluripotent stem cells. Proc Natl Acad Sci U S A 2012;109(7):2544-2548.

[38] Chang HM, Liao YW, Chiang CH, Chen YJ, Lai YH, Chang YL, Chen HL, et al. Improvement of Carbon Tetrachloride-Induced Acute Hepatic Failure by Transplantation of Induced Pluripotent Stem Cells without Reprogramming Factor c-Myc. Int J Mol Sci 2012;13(3):3598-3617.

[39] Shiraki N, Umeda K, Sakashita N, Takeya M, Kume K, Kume S. Differentiation of mouse and human embryonic stem cells into hepatic lineages. Genes Cells 2008;13(7): 731-746.

[40] Shiraki N, Yamazoe T, Qin Z, Ohgomori K, Mochitate K, Kume K, Kume S. Efficient differentiation of embryonic stem cells into hepatic cells in vitro using a feeder-free basement membrane substratum. PLoS One 2011;6(8):e24228.

[41] Nagamoto Y, Tashiro K, Takayama K, Ohashi K, Kawabata K, Sakurai F, Tachibana M, et al. The promotion of hepatic maturation of human pluripotent stem cells in 3D co-culture using type I collagen and Swiss 3T3 cell sheets. Biomaterials 2012;33(18): 4526-4534.

[42] Amimoto N, Mizumoto H, Nakazawa K, Ijima H, Funatsu K, Kajiwara T. Hepatic differentiation of mouse embryonic stem cells and induced pluripotent stem cells during organoid formation in hollow fibers. Tissue Eng Part A 2011;17(15-16): 2071-2078.

[43] Ohashi K, Yokoyama T, Yamato M, Kuge H, Kanehiro H, Tsutsumi M, Amanuma T, et al. Engineering functional two- and three-dimensional liver systems in vivo using hepatic tissue sheets. Nat Med 2007;13(7):880-885.

[44] Kim K, Ohashi K, Utoh R, Kano K, Okano T. Preserved liver-specific functions of hepatocytes in 3D co-culture with endothelial cell sheets. Biomaterials 2012;33(5): 1406-1413.

Chapter 24

Induced Pluripotent Stem Cells: Therapeutic Applications in Monogenic and Metabolic Diseases, and Regulatory and Bioethical Considerations

Antonio Liras, Cristina Segovia and Aline S. Gabán

Additional information is available at the end of the chapter

http://dx.doi.org/10.5772/55816

1. Introduction

The potential use of stem cells in advanced therapies such as tissue engineering, regenerative medicine, cell therapy and gene therapy by virtue of their significant therapeutic potential and clinical applications has aroused keen interest among scientists [1,2]. Cell therapy is based on the transplantation of living cells into an organism with a view to repairing tissue or restoring a lost or deficient function. Stem cells are the most frequently used cells for such purposes given their ability to differentiate into other more specialized cells [3].

The chief defining feature of stem cells is their capacity for self-renewal and their ability to differentiate into cells of various lineages. Stem cells can be classified on the basis of their potency and their source into (i) Totipotent stem cells (zygote and 2-4 cell embryo), since these cells are capable of giving rise to the entire organism (both embryonic and extra-embryonic tissues); (ii) Pluripotent stem cells (embryonic stem and embryonic germ cells), which can give rise to derivatives of all three germ layers (embryonic tissues only, but not the extra-embryonic ones); (iii) Multipotent stem cells (adult stem cells) [4,5].

Adult stem cells are undifferentiated cells that provide a natural reservoir that is available to replace damaged or ageing cells throughout the lifetime of the individual. They can be found in virtually any kind of tissue including bone marrow, trabecular bone, periosteum, synovium, muscle, adipose tissue, breast gland, gastrointestinal tract, central nervous system, lung, peripheral blood, dermis, hair follicle, corneal limbus, etc. [6]. The clinical application of this type of cell is associated with potentially better prospects than that of embryonic stem cells since use of adult stem cells does not raise any ethical conflicts nor does it involve immune rejection problems in the event of autologous implantation.

The possibility to generate induced pluripotent stem cells (iPSCs) by reprogramming somatic stem cells through the introduction of certain transcription factors [7-12] is radically transforming received scientific wisdom. The pluripotency of these cells, which enables them to differentiate into cells of all three germ layers (endoderm, mesoderm, and ectoderm), makes them an extremely valuable tool for the potential design of cell therapy protocols. iPSC technology can indeed allow the development of patient-specific cell therapy protocols [13] as the use of cells like iPSCs, which are genetically identical to the donor, may protect the individual from immune rejection. Furthermore, unlike embryonic stem cells, iPSCs are not associated with bioethical problems and are considered a "consensus" alternative that does not require use of human oocytes or embryos and is therefore not subject to any specific regulations. Lastly, iPSCs are very similar to embryonic stem cells as far as their molecular and functional characteristics are concerned [14-15].

Although research into iPSCs is still at an early stage, interesting results have already been obtained in a number of monogenic and polygenic diseases of different etiologies: cardiovascular and liver diseases, immunologic, infectious, metabolic diseases, rare diseases and cancer [16-19]. Researchers have also looked into the application of iPSCs to toxigological and pharmacological screening for the presence of toxic and teratogenic substances [20].

Stem cell therapy is emerging as a new concept of medical application in pharmacology. For all practical purposes, human embryonic stem cells are used in 13% of treatments, whereas fetal stem cells are used in 2%, umbilical cord stem cells in 10%, and adult stem cells in 75% of cases. The most significant treatment indications for gene and cell therapy have so far been cardiovascular and ischemic diseases, diabetes, hematopoietic diseases, liver diseases and, more recently, orthopaedics [21]. For example, over 25,000 transplants of hematopoietic stem cells are performed every year for treatment of lymphoma, leukemia, immunodeficiency disorders, congenital metabolic defects, hemoglobinopathies, and myelodysplastic and myeloproliferative syndromes [22].

Each type of stem cell has its own advantages and disadvantages, which vary depending on the different treatment protocols and the requirements of each clinical condition. Thus, embryonic stem cells have the advantages of being pluripotent, easy to isolate and highly productive in culture, in addition to showing a high capacity to integrate into fetal tissue during development. By contrast, their disadvantages include immune rejection and the possibility that they may spontaneously and uncontrollably differentiate into inadequate cell types or even induce tumors. Adult stem cells have a high differentiation potential, are less likely to induce an undesirable immune response and may be stimulated by drugs. Their disadvantages include that they are scarce and difficult to harvest, grown slowly, differentiate poorly in culture and are difficult to handle and produce in adequate amounts for transplantation. In addition, they behave differently depending on the source tissue, show telomere shortening, and may carry the genetic abnormalities inherited or acquired by the donor.

At least three different strategies are available for proper use of stem cells. The first one is stimulation of endogenous stem cells by growth factors, cytokines, and second messengers, which must be able to induce tissue self-repair. The second alternative is direct administration of the cells so that they differentiate at the damaged or non-functional tissue sites. The third

http://dx.doi.org/10.5772/55816

possibility is transplantation of cells, tissues, or organs taken from cultures of stem cell-derived differentiated cells. The US Food and Drug Administration defines somatic cell therapy as the administration to humans of autologous, allogeneic or xenogeneic living non-germline cells, other than transfusion blood products, which have been manipulated, processed, propagated or expanded *ex vivo*, or are drug-treated.

The most significant applications of cell therapy as a whole are expected to be related to the treatment of organ-specific conditions such as diabetes —a typically metabolic disease—, liver and cardiovascular conditions, immunological disorders and hereditary monogenic diseases such as haemophilia. As one of the key advanced therapies —together with gene therapy and tissue engineering— cell therapy will require a new legal framework that affords generalized patient accessibility to these products and that allows governments to discharge their regulatory and control duties. In this respect, the main advantage of iPSCs lies in the fact that their use does not raise bioethical questions, which means that regulatory provisions governing their use need not be overly stringent.

2. Induced pluripotent stem cells technology and general clinical applications

iPSCs are obtained through the reprogramming of an individual's somatic stem cells by the introduction of certain transcription factors. Their chief value is based on their pluripotency to differentiate into cells of all three germ layers, which makes them an useful tool for the discovery of new drugs and the establishment of cell therapy programs.

iPSC technology makes it possible to develop patient-specific cell therapy protocols as they are genetically identical to the donor and thus prevent the occurrence of an immune rejection in autologous transplantations. Moreover, unlike embryonic stem cells, they are not associated with any ethical controversies and therefore regulatory conditions governing their use are much less stringent.

Induced pluripotent stem cells were generated for the first time by Shinya Yamanaka's team [8] from murine and human fibroblasts by transfecting certain transcription factors (Oct4, Sox2, c-Myc, and Klf4) by means of retroviral vectors. (Figure 1). Thomson *et al.* replicated Yamanaka's experiments with human cells and two additional factors: Nanog and Lin28, which rendered the reprogramming process more efficient [9].

The same group developed an alternative reprogramming method using non-integrating episomal vectors derived from the Epstein-Barr virus (oriP/EBNA1), which may be maintained in a stable form in transfected cells by pharmacological selection [23]. Nonetheless, it was later reported that only two transcription factors (Oct4 and Klf4) are needed for generating the iPSCs from neural stem cells that endogenously express high Sox2 concentrations [24].

All of these strategies require transfection through retroviral vectors and integration for *in vitro* and *in vivo* modeling, which precludes their clinical use because of the potential risks involved. This is the reason why several research teams have looked into the reprogramming

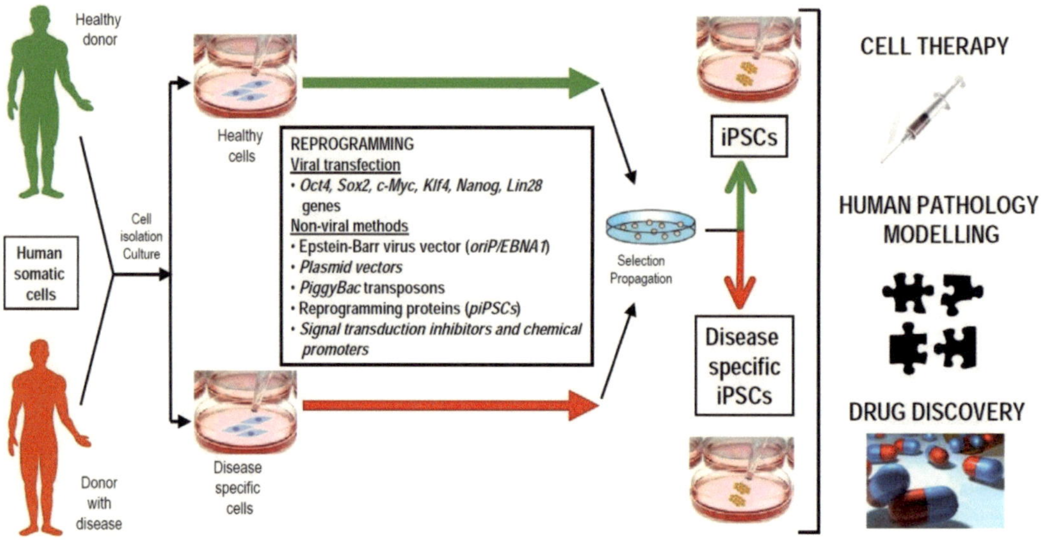

Figure 1. Generation of human induced pluripotent stem cells for use in cell therapy, in vitro human pathology modelling and in drug discovery. Reprogramming of human somatic cells can be induced by: Viral transfection of Oct4, Sox2, c-Myc, Klf4, Nanog and Lin28 genes; non-viral methods using a nonintegrating episomal vector derived from Epstein-Barr virus (oriP/EBNA1), plasmid vectors or piggyBac transposon/transposase systems; direct delivery of the reprogramming proteins (piPSCs) and signal transduction inhibitors and chemical promoters cell survival.

of cells using plasmid vector rather than viral vector transfection [10-12]. Although reprogramming efficacy with plasmid vectors is lower —as is also the case with non viral gene therapy— this method significantly increases the safety of the procedure, which makes it clinically applicable and also constitutes a source of valuable cell material that can be used for research into reprogramming and pluripotency.

Another promising strategy consists in the direct release of reprogramming proteins through modified versions of reprogramming factors in some of their molecular domains. These protein-induced pluripotent stem cells (piPSCs) bind to the membrane of cells reaching their nucleus [25]. Ding *et al.*, have also shown that the addition of two signal transduction inhibitors and certain cell-survival promoting chemicals (e.g. thiazovivin) can induce a 200-fold increase in reprogramming efficacy [26].

As explained above, iPSCs technology makes it possible to establish patient-specific cell therapy protocols [13]. On the one hand, this reduces the risk of immune rejection in autologous transplantations by virtue of gene identity. On the other, it provides treatment that is customized to the specific characteristics of each patient and takes into account the etiology and severity of the condition. Moreover, induction of pluripotency has been developed for a great variety of tissue types [9,24,27] as it is a relatively straightforward procedure and —as mentioned above— subject to fewer regulatory constraints [28].

Important as these advantages are, there are still a few uncertainties that need to be resolved. One of the most pressing ones is related to determining the likelihood that these iPSCs may undergo genetic aberrations further to the reprogramming process [29].

In order for the clinical application of these cells to become a reality both for diagnostic purposes and for the design of cell therapy protocols, a few methodological hurdles must still be resolved in connection, as is often the case with pharmacological products, with their safety profile [30]. This means basically that efforts must be directed at removing the genome in the integrating viral vectors, eliminating the risk of tumor formation and establishing more efficient reprogramming and differentiation protocols. Clearly our knowledge on the reprogramming mechanisms leading to pluripotency are still insufficient to understand and more importantly control the adverse events that could potentially occur. Therefore the most important goal for research in this field will be to study genetic modifications in animal models by means of large-scale genome sequencing programs. This task will require sharing cell lines with other researchers, with appropriate confidentiality protections and, eventually, patenting scientific discoveries and developing commercial tests and therapies. It will also be necessary to fully ascertain and confirm that pluripotency confers iPSCs with functions similar to those of embryonic stem cells regardless of the initial source of somatic cells used [14,15].

Undoubtedly, the most attractive application of this type of strategy is the production of patient-specific or healthy individual-specific iPSCs for replacement of damaged non-functional tissue. Thus for example skin fibroblast-derived iPSCs have been shown to possess a high potential to differentiate into islet-like clusters and to release insulin, which is highly relevant for diabetes [16]. Such developments are also relevant for amyotrophic lateral sclerosis (Lou Gehrig's disease) [17]; adenosine deaminase deficiency-related severe combined immunodeficiency, Shwachman-Bodian-Diamond syndrome, Gaucher disease type III, Duchenne and Becker muscular dystrophy, Parkinson disease, Huntington disease, juvenile-onset, type 1 diabetes mellitus and Down syndrome (trisomy 21) [31]; spinal muscular atrophy [19]; and in toxicology and pharmacology for screening toxics for embryo and/or teratogenic substances [20].

The great promise of iPSCs (Figure 1) is associated to their role in the investigation of the phyisiological mechanisms related with the biology of stem cells themselves; in the modeling of different pathologies; and, fundamentally, in the development of therapies for human diseases and in drug screening. In fact, since they were discovered in 2008, almost one-hundred-and-fifty iPSCs have been established from nearly thirty fibroblast cell lines related to over a dozen conditions, including some complex diseases such as schizophrenia and autism and other genetic or acquired disorders such as cardiovascular or infectious diseases. Numerous types of functional cells have already been derived from iPSCs including neurons [17,32], hematopoietic cells [33], and cardiomyocytes [34,35].

Taking into account the far-from-trivial fact that iPSCs can be obtained from individuals affected by a disease and that they are indefinitely self-renewable and fully of human origin, it could well be that these cells, obtained from several individuals suffering from the same disease and presenting with similar clinical manifestations, may provide highly valuable information about certain predisposing genes —as in the case of diabetes mellitus— and therefore allow physicians to provide well-grounded genetic guidance.

Human iPSCs have the potential to be used in regenerative medicine for the design of individualized therapies and also in the field of research and development. However, it is still necessary to optimize iPSC protocols, particularly with respect to the possible modifications to their genome, and to increase the efficacy of the transfection process leading to iPSC reprogramming [36,37]. The present state of the art of reprogramming mechanisms —viral transfection of Oct4, Sox2, c-Myc, Klf4, Nanog and Lin28 genes; non-viral transfection using a non-integrating episomal vector derived from the Epstein-Barr virus (oriP/EBNA1), plasmid vectors or piggyback transposon/transposase systems; direct delivery of the reprogramming proteins (piPSCs); and signal transduction inhibitors and chemical promoter cell survival— will allow safe integration and the removal of ectopic transgenes, improving the efficiency of iPSC production using a minimally invasive strategy.

3. Advanced therapies for monogenic and metabolic diseases

The progression of the different areas of biology, biotechnology and medicine leads to the development of highly innovative new treatments and pharmacological products. In this regard, advanced therapies based on the by-products of gene therapy, cell therapy and nanomedicine/tissue engineering are of great importance for their potential to radically improve treatment of a large number of conditions. The different schools of thought that advocate the emerging concept of advanced therapies agree that the latter must be used for the treatment of diseases (both hereditary and non-transmissible) caused by the anomalous behavior, or complete lack of function, of a single gene (also called monogenic hereditary diseases) or by an anomaly in several genes (polygenic diseases).

Metabolic diseases, or congenital metabolic errors, are conditions highly amenable to be treated by the new advanced therapies as such treatments have been shown to restore mutation-induced alterations of gene products. Proteins are the most commonly affected gene products, although messenger RNA is also a usual victim. Alterations affect gene products, i.e. proteins, most of which are enzymes but there is also a group of other proteins fulfilling all kinds of different functions (structural proteins, transport proteins and signal cascade activation proteins). Of particular interest are the proteins that participate in homeostasis and exert their functions outside the cells that synthesize them. This is the case of coagulation factors VIII and IX (FVIII and FIX), whose deficiency results in the development of haemophilia A or B, respectively. Another member of this class of proteins is antitrypsin, also of hepatic origin and secreted into the bloodstream, whose function is to prevent the digestion of pulmonary alveoli by proteolytic enzymes. Lastly, mention should be made of proteins with such diverse functions as transcription factors, oncogenes, tumor-suppressing genes and even some hormones and their receptors, the latter being specifically related with diabetes mellitus, a typically metabolic disease.

The nature of the monogenic or metabolic disease is the main factor that determines whether a treatment that can eradicate or at least mitigate its clinical consequences is possible or not. Before the concept of advanced therapies came to be applied to these (wide ranging) conditions, many of them were treated using both conventional/classical and more advanced approaches.

http://dx.doi.org/10.5772/55816

Advanced therapies are applied following three basic approaches: replacement of a deficient gene by a healthy gene so that it generates a certain functional, structural or transport protein (gene therapy); incorporation of a full array of healthy genes and proteins through perfusion or transplantation of healthy cells (cell therapy); or tissue transplantation and formation of healthy organs (tissue engineering). In this context, induced pluripotent stem cells can play a very significant role and hold an enormous therapeutic potential in the fields of cell therapy and tissue engineering.

4. Advanced therapies and induced pluripotent stem cells in the treatment of haemophilia

Haemophilia is a recessive X-linked hereditary disorder caused by a deficiency of coagulation factor VIII (haemophilia A) or IX (haemophilia B). The disease is considered to be severe when factor levels are below 1% of normal values, moderate when they are between 1 and 5% and mild when levels range between 5% and 40%. Haemophilia A is four times more common than haemophilia B and, in terms of severity for both types, 35% of patients have the severe form, 15% the moderate form and 55% have mild haemophilia. Incidence of the disease is 1:6,000 males born alive for haemophilia A and 1:30,000 for haemophilia B [38].

The etiopathogenesis of the disease is related to different kinds of mutations (large deletions and insertions, inversions and point mutations) that occur in the gene expressing the deficient coagulation factor. The clinical characteristics of both types of haemophilia are very similar: spontaneous or traumatic hemorrhages, muscle hematomas, haemophilic arthropathy resulting from the articular damage caused by repetitive bleeding episodes in the target joints, or hemorrhages in the central nervous system. In the absence of appropriate replacement treatment with exogenous coagulation factors, these manifestations of the disease can have disabling or even fatal consequences thus negatively impacting patients' quality of life and reducing their life expectancy [39].

At present, patients with haemophilia benefit from optimized treatment schedules based on the intravenous systemic delivery of exogenous coagulation factors, either prophylactically or on demand. The current policy in developed countries is in general to administer a prophylactic treatment (2 or 3 times a week) from early childhood into adulthood [39]. Such prophylactic protocols result in a clear improvement in patients' quality of life on account of the prevention of haemophilic arthropaty and other fatal manifestations of the disease as well as a reduction in the long-term costs of treatment because of a decrease in the need of surgical procedures such as arthrodesis, arthroplasty or synovectomy [40].

Conventional treatment of haemophilia [41,42] is currently based on the use of plasma-derived or recombinant high-purity coagulation factor concentrates. The former are duly treated with heat and detergent to inactivate lipid-coated viruses [43], and the latter are a recently developed product that does not contain proteins of human or animal origin [44,45]. Both kinds of factor boast high efficacy and safety profiles, at least for the inactivation-susceptible pathogens known to date. The choice of one product over the other is usually based on the clinical characteristics of the patient and on cost and availability considerations [46,47].

Now that infections by pathogenic viruses (HIV, HCV) that were common a few decades ago have been eradicated, the most distressing adverse effect observed when using either product is the development of antibodies (inhibitors) against the perfused exogenous factors [48,49]. The appearance of inhibitors renders current treatment with factor concentrates inefficient, increasing morbidity and mortality, leading to the early onset of haemophilic arthropathy and disability and to a consequent reduction in patients' quality of life. Lastly, inhibitors result in higher costs as treatment must be provided both for bleeding episodes and inhibitor eradication (immune tolerance induction). The incidence of inhibitors is around 30% in haemophilia A and 6% in haemophilia B.

The immunologic mechanism whereby these neutralizing antibodies are generated is highly complex and involves several messenger molecules (tumor necrosis factor, interleukins…), and cells (T-lymphocytes B-lymphocytes, macrophages...). They are directed at certain regions in the factor molecule that interact with other components of the coagulation cascade and, depending on their titre level and on whether they are transient or persistent, will bring about greater or lesser alterations in the said cascade. The causes that influence inhibitor development may be genetic, i.e. inherent in the patients themselves [48], such as ethnicity, familial history, type of mutation or certain changes in some of the genes involved in the immune response; or non-genetic, i.e. environmental [50], such as age at first factor infusion, breastfeeding, stimulation of the immune system by other antigens or the treatment regimen used (prophylactic vs. on demand). Whether the factor concentrate used is plasma-derived or recombinant does not have a significant influence on the inhibitor incidence rate [51].

Short and medium-term perspectives for the treatment of haemophilia strongly rely on the current research efforts directed at increasing the safety levels of (especially) plasma-derived factors. Such research focuses on the detection and subsequent inactivation of emerging bloodborne pathogens in donors such as the prions causing variant Creutzfeldt-Jakob disease, or other potential emerging agents [52-54]. It is also important to increase the efficiency of recombinant factors increasing their half-life (by PEGylating the factor molecule or using fusion proteins [55-58] and attenuating their immunogenic capacity to produce inhibitors, by chemically modifying them [59] or by developing recombinant factors of human origin [60].

In the long term, efforts must be directed at the development of advanced therapies, particularly strategies in the field of gene therapy (using of adeno-associated viral vectors) and cell therapy (using of adult stem cells or induced pluripotent stem cells). The chief goal of these new strategies will be to address some of the shortcomings associated with current treatment options such as the short *in vivo* half-life of administered factors, the impending risk of a pathogen-induced infection and the development of inhibitors. Another goal of the advanced therapies (cell therapy) will be palliative treatment of the articular consequences derived from haemophilic arthropathy [40].

Haemophilia is optimally suited for advanced therapies as it is a monogenic condition and does not require very high expression levels of a coagulation factor to reach moderate disease

http://dx.doi.org/10.5772/55816

status (Figure 2). For this reason, significant progress has been possible with respect to these kinds of therapies: cell therapy has broken new ground with the use of several types of target cells and gene therapy has shown particular promise with the use of viral and non-viral vectors. In fact, haemophilia is now recognized as a condition amenable to gene therapy [61-64]. Strategies available include use of lentiviral (LVV) [65] and adeno-associated (AAV) [66] vectors in adult stem cells and autologous fibroblasts, in platelets and in hematopoietic stem cells; transfer by means of non-viral vectors; and repair of mutations with chimeric oligonucleotides. The studies published so far have, in the most part, not reported any severe adverse effect resulting from the application of such strategies in the clinical trials performed.

Specifically, gene therapy trials in haemophilic patients have shown adeno-associated vectors to represent the most promising treatment option given their excellent safety profile, even if on occasion they may create immune response problems. Efforts are currently centered on minimizing the incidence of immune rejection and increasing efficacy and expression time. In this connection, several studies have been published with a view to optimizing the use of this type of viral vectors. Among them, in a landmark study on patients with severe haemophilia B (<1% FIX), Nathwani *et al.* infused their subjects with a dose of a serotype-8-pseudotyped, self-complementary AAV vector that expresses factor IX and can efficiently transduce hepatocytes [66]. Their results showed that factor IX expression ranged between 3 and 11% of normal values. Significant as they may seem, these results must be considered with caution as the expression levels achieved rather than normalize the patient's phenotype convert it to a mild-to-moderate form. Also, concomitant treatment with glucocorticoids is needed to prevent immune rejection and elevation of liver transaminase levels. Due account must also be taken of the fact that the adeno-associated vector has the potential to induce hepatotoxicity. For all these reasons, these undoubtedly encouraging results can only be considered a first step in the development of safe and effective advanced therapies for the treatment of haemophilia.

Non-viral strategies also have a role to play in the treatment of haemophilia as they could in the long term provide a safer alternative than viral vectors which, as we have seen, are fraught with significant biosafety and efficacy-related problems, which have so far limited their clinical application. Sivalingam *et al*. [67] evaluated the genotoxic potential of phiC31 bacteriophage integrase-mediated transgene integration in cord-lining epithelial cells cultured from the human umbilical cord. This non-viral strategy has made it possible to obtain stable factor VIII secretion *in vitro*. Xenoimplantation of these protein-secreting cell lines into immunocompetent haemophilic mice corrects the severe form of the disease. Such implantation could prove extremely useful as a bioimplant in the context of monogenic diseases such as haemophilia.

Our laboratory has advanced the use of nucleofection as a non-viral transfection method to obtain factor IX expression and secretion in adult adipose tissue-derived mesenchymal stem cells [68]. Although it is certainly true that expression efficacy with these types of protocols is lower than when viral vectors are used, it must be underscored that these protocols do offer much higher safety levels, with the additional advantage that increasing factor activity to above 5% of normal values already places the patient in the mild phenotype group.

The use of cell therapy in the treatment of haemophilia has to date consisted mainly in the transplantation of healthy cells in an attempt to repair or replace a coagulation factor defi-

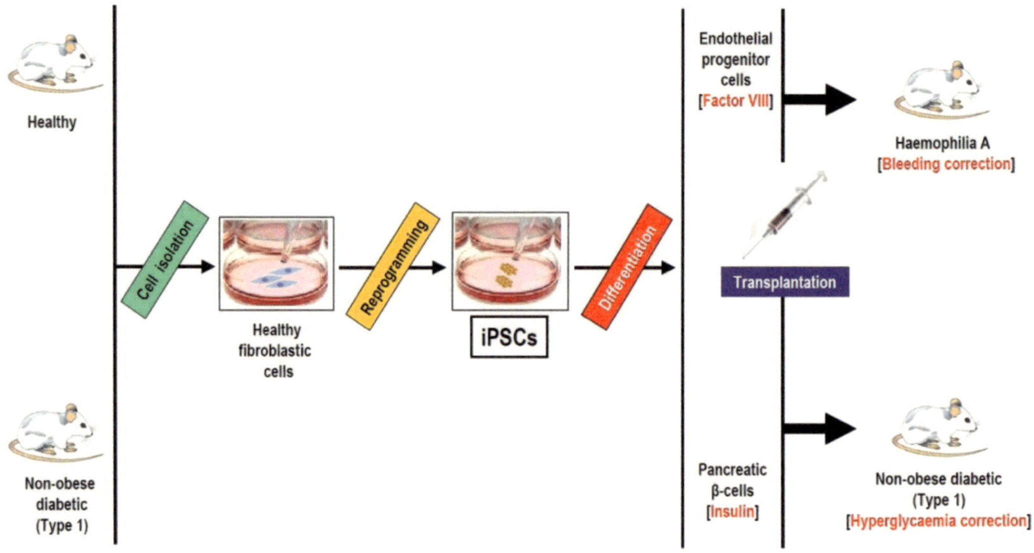

Figure 2. Induced pluripotent stem cells application to the treatment of haemophilia and diabetes mellitus. Autologous transplantation of healthy differentiated cells, obtained from iPSCs, into an animal model with haemophilia or diabetes mellitus type 1, normalizes the corresponding altered function by in vivo production of the deficient protein or hormone.

ciency. These procedures have been conducted mainly with adult stem cells and, more recently, with progenitor cells partially differentiated from iPSCs, albeit in most cases the mechanisms by which transplanted cells (to a greater or lesser extent) engraft and go on to proliferate and function remain unknown.

Aronovich *et al.* [69], have shown that transplantation of embryonic spleen tissue (embryonic day 42 spleen tissue) in immunocompetent mice with haemophilia A attenuates the severity of the disease in the 2-3 months after the procedure. These results would seem to indicate that transplantation of a fetal spleen (obtained from a developmental stage prior to the appearance of T-cells) may potentially be used to treat some genetic disorders. For their part, Follenzi *et al.* [70] reported that once liver sinusoidal endothelial cells were transplanted and successfully engrafted into mice with haemophilia A, they were seen to proliferate and partially replace some areas of the hepatic endothelium. This resulted in a restoration of factor VIII plasma levels and in the correction of the bleeding phenotype. More recently, this same team [71] demonstrated that transplantation of bone marrow cells (healthy mouse Kupffer cells —liver macrophage/mononuclear cells— and healthy bone marrow derived mesenchymal stromal cells) can correct the phenotype of haemophilic mice and restore factor VIII levels.

As far as the use of iPSCs is concerned, the first paper came from Xu *et al.* [72], who reported on the generation of murine iPSCs from tail-tip fibroblasts and their differentiation into endothelial cells and their precursors. These iPSC-derived cells express specific

http://dx.doi.org/10.5772/55816

membrane markers for these cells such as CD31, CD34 and Flk1, as well as factor VIII. Following transplantation of these cells into mice with haemophilia A, the latter survived the tail-clip bleeding assay by over 3 months and their factor VIII plasma levels increased to 8%-12%. Yadav *et al.* [73] studied transdifferentiation of iPSC-derived endothelial progenitor cells into hepatocytes (primary cells of FVIII synthesis). These transplanted cells were injected into the liver parenchyma where they integrated functionally and made correction of the haemophilic phenotype. High levels of FVIII mRNA were detected in the spleen, heart, and kidney tissues of injected animals with no indication of tumor formation or any other adverse events in the long-term. Alipio *et al.* [74] for their part also reported on the generation of factor VIII in a haemophilic murine model one year after transplantation of iPSC-derived endothelial cells.

5. Induced pluripotent stem cells in the treatment of diabetes mellitus

Diseases caused by the destruction or loss of function of a limited number of cells are good candidates for cell therapy. Such is the case of diabetes mellitus (Figure 2).

Diabetes mellitus (DM) is classified into two broad categories: type 1 DM, which is a genetic disease, and type 2 DM, a more generalized variety related with insulin resistance. DM, especially the type 1 form, is associated with microvascular complications, such as retinopathy, neuropathy or nephropathy, as well as cardiovascular problems. Type 1 DM is a T-cell mediated autoimmune disease specifically aimed against pancreatic beta cells, which results in insulin deficiency [75,76].

Symptoms of DM include episodes of lethargy and fatigue, polyuria, enuresis, noctu-ria, polydipsia, polyphagia, weight loss and abdominal pain. The disorder has a strong genetic component related with the susceptibility to inherit and develop the disease through the HLA complex (HLA-DR and HLA-DQ genotypes) and other loci involved in immunologic recognition and cell-to-cell signaling in the immune system (graft compatibility) [77,78].

Abnormal T-cell activation in susceptible individuals results in both an inflammatory response within the Langerhans islets and a humoral immune response involving the production of antibodies against insulin-specific beta cell antigens, decarboxylase glutamic acid or the protein tyrosine phosphatase [79]. The presence of one or more types of antibodies may precede the appearance of type 1 diabetes and its subsequent development [80,81]. In any case, the final result is the destruction of beta cells and progressive impairment of the blood glucose metabolism [82]. Some patients with type 1 diabetes may show a higher susceptibility to other conditions such as thyroiditis, Graves disease, Adisson disease, celiac disease, myasthenia gravis or to degenerative skin conditions such as vitíligo [83-85].

The greatest incidence of type 1 DM occurs during childhood and in the early years of adulthood with significant variations across different geographies. Diagnosis is usually made before the age of 20 (between 16 and 18 in 50-60% of cases) [75]. The factors involved in the development of type 1 DM include the so-called familial predisposing factors, gestational status, age and other iatrogenic causes.

Type 2 DM is characterized by a functional deficiency of insulin per se or by a resistance to the hormone resulting from an alteration of the function or structure of the insulin receptor at the level of the membrane or of any of the molecules involved in the intracytoplasmic signal transduction cascade [86]. The metaboilic effects of insulin vary depending on the action of the molecules that participate in signaling pathways to regulate gene expression in striated muscle cells, adipocytes, hepatocytes and in pancreatic beta cells [87-90]. Thus, for example, insulin resistance caused by the impairment of glucose transporter GLUT4 initially results in a metabolic syndrome, type 2 diabetes, lipodystrophy, hypertension, polycystic ovary syndrome or atherosclerosis.

In general, the morbidity and mortality of DM is related with the different long-term cardiovascular complications associated with the disease, also taking into account other proactivating factors such as smoking, obesity, a sedentary lifestyle, hypertension, early onset and prolonged duration of type 1 DM, genetic predisposition and hyperglycemia.

Nephropathy, retinopathy and diabetic neuropathy are the most common microvascular complications of DM. As regards diabetic neuropathy, this can be a focal complication associated with diabetic amyotrophy or with cranial nerve III oculomotor palsy, or a more generalized occurrence that can take the form of a sensorimotor polyneuropathy affecting the autonomic nervous system, gastric motility and cardiac function. Peripheral neuropathy together with peripheral vascular disease may lead to a diabetic foot syndrome, characterized by ulcerations and poor healing in the lower limbs [91]. As a macrovascular complication, cardiovascular disease accounts for 70% of mortality in individuals with type 2 DM, with the incidence of coronary artery disease being higher in women than in men suffering from type 1 DM [92]. Atherosclerotic processes are in turn more common in patients with type 1 DM [93].

Although treatment and diagnosis of diabetes is well-established, there is a constant quest for new drugs that may be more effective at lowering blood glucose levels, controlling their therapeutical management —especially in younger patients—, and preserving patients' long-term quality of life by reducing the incidence of complications resulting from the disease. Current research is centered on unveiling the structure and function of glucose transporters, which may offer significant therapeutic advantages [86], as well as on the development of new fast-acting insulin analogs and more accurate subcutaneous pumps [94-98]. Commendable as these initiatives are, it is difficult to anticipate and control factors that exert a variable influence upon glucose levels such as nutrition, physical activity or stress. These factors alter the glycemic environment and consequently the amount of insulin required at each point in time, which reinforces the need to establish sophisticated artificial pumping systems that may simulate the natural endocrine pancreas.

The continuous advancement of our understanding of the mechanisms that govern the physiopathology of diabetes and gene susceptibility together with the multiple possibilities currently offered by biotechnology have fuelled the researchers' interest in the development of all three types of advanced therapies: gene therapy, cell therapy and tissue engineering. In this regard, although we are still at a very incipient stage [99,100], procedures based on transplantation of insulin-secreting cells or islets obtained from stem cell differentiation may hold valuable hope for the future.

The need to justify the human and financial investment made in the development of new advanced therapies is as strong in diabetes as it is in haemophilia. However, in the case of the former justification is even more compelling taking into account that an optimal and efficient treatment is already available for the disease. The discovery of insulin as a therapeutic tool for DM constituted an important milestone in the history of medicine even if administration of this hormone does not fully compensate for the function lost. This is also the case with factor replacement in haemophilia. Moreover, both coagulation factor and insulin treatment are only palliative, never curative, which is the basic idea underlying treatment of DM and haemophilia. Moreover, it is also important to take into account the potential adverse effects of these therapies, and particularly the complications associated with DM, which derive from the fact that it is a long-term disease.

In addition, advances in terms of the clinical transplantation of Langerhans islets have not met with the expected success as a result of the inadequate number of donors available and the incidence of immune rejection of the newly transplanted beta cells [101]. This has intensified efforts aimed at developing insulin-producing cells from stem cells. iPSC technology could turn the tide in this respect as such cells may be induced to form endodermal structures, pancreatic and endocrine progenitors and, naturally, differentiated insulin-producing cells [102-104].

Built upon the knowledge gained from studies on embryonic cells about the differentiation process, the first studies on iPSCs, whereby human cells were reprogrammed to become *in vitro* differentiated insulin-producing cells, showed great promise [105,106]. However, as only partial cell differentiation was achieved, those studies failed in their attempt to enrich insulin-producing cell lines or assess their function.

Drawing on current knowledge on the embryonic development of the pancreas, Zhu *et al.* [107] recently reported on the generation of insulin-producing pancreatic cells from iPSCs obtained from a rhesus monkey [108]. These authors established a quantitative cytometric method to evaluate the efficacy of cell differentiation. In addition, they increased the level of precision in the assessment of the competence and function of the iPSCs from a rhesus monkey by means of transplantation into immunodeficient mice. These cells were induced to form endodermal structures, pancreatic and endocrine progenitors and insulin-producing cells. By means of a TGF-β inhibitor, generation of endocrine precursor cells capable of generating insulin-producing cells that respond to glucose stimulation *in vitro* was undertaken. Transplantation of these cells into a type 1 DM murine model decreased blood glucose levels in 50% of the mice. These results show the high efficacy that can be achieved by obtaining iPSCs from a superior animal model as well as the capacity of iPSCs to be transformed into insulin-producing cells, which opens up the possibility for carrying out autologous transplantations in the future.

Along the same lines, Jeon *et al.* [109] studied the functionality of iPSC-derived insulin-producing cells generated from pancreas-derived epithelial cells in non-obese diabetic mice. The insulin-producing cells obtained in this way express different pancreatic β cell markers and secrete insulin in response to glucose stimulation. Transplantation of these cells into non-obese diabetic mice (a model of autoimmune type 1 DM very similar to the human form) results in a kidney graft with a functional response to glucose stimulation and a consequent normalization of blood glucose levels (Figure 2).

Until recently, iPSC generation from patients with type 2 DM had not been reported in the literature. However, Ohmine *et al.* [110] described not long ago the generation of iPSCs from keratinocytes of elderly patients with type 2 DM. These cells were reprogrammed by lentiviral transduction with human transcription factors OCT4, SOX2, KLF4 and cMYC, telomere elongation, and down-regulation of senescence and apoptosis-related genes, and were subsequently differentiated into insulin-producing islet-like cells. Reprogramming of keratinocytes from elderly type 2 DM patients produces efficient iPSCs with a "privileged" senescence status that allows them to transform into insulin-producing islet-like cells, which may lead to the development of a versatile strategy for modeling the disease as well as an advanced therapy for treating it.

Generally speaking, several problems must yet be resolved before iPSCs can be applied clinically, specifically to the treatment of haemophilia or diabetes. In the first place, it is essential to optimize the reprogramming process so that it provides maximum safety assurances against the potential risks derived from undesirable genetic changes in iPSCs [111]. Recent studies have revealed significant chromosomal changes that take place during the long-term culture of iPSCs as well as variations in the number of copies of certain genes and point mutations, which could clearly be related with the reprogramming of somatic cells and result in damage to the DNA [112-115].

The second hurdle that must be overcome is the high variability that exists between the different cell lines in the context of differentiation into pancreatic lineages [16]. The epigenetic and functional trials that should be performed in this respect are complicated by the fact that iPSCs have a high epigenetic content [116]. The third obstacle has to do with the purification of iPSC-derived β cells to prevent the transplantation of undifferentiated cells, which could result in the formation of teratomas. Moreover, it is necessary to develop new reagents to make direct differentiation of pancreatic progenitors into functional β cells more efficient and to design highly specific surface markers for these cells so that a more precise fluorescence analysis can be performed in order to isolate homogeneous populations of this kind of cell so that their function can be rigorously controlled.

6. General regulatory and bioethical issues

Cell therapy, as one of the bedrocks of the advanced therapies —together with gene therapy and tissue engineering—, requires a new legislative framework in order to guarantee that patients can avail themselves of the products they need and provide governments with a robust

http://dx.doi.org/10.5772/55816

protection, control and regulation mechanism. The existing framework regulating advanced therapies will have to be adapted fast in order to keep pace with the proliferation of new knowledge in this rapidly developing field. However, desirable that this may be, the pace of legislative reform is unfortunately slow and inevitably lags behind the development of new science.

The aspects to be regulated include mainly those related with controlling the development, manufacturing and quality of release and stability testing programs; non-clinical aspects such as promoting research on biodistribution, cell viability and proliferation levels and ratios, and the persistence of in vivo function; clinical aspects such as dose-specific characteristics, risk stratification; and aspects specifically connected to pharmacovigilance and traceability.

The guidelines for therapeutic products based on human cells must be drawn up by the drug agencies of the different countries [117,118] both as regards the development of clinical and preclinical trials and with respect to pharmacovigilance, taking in all cases a multidisciplinary perspective.

For any product based on cells or on tissue, it should be made compulsory to verify that the desired physiological functions are preserved after the preparation process, both in isolation and in combination with other non-biological components, as many of these products will be used with a metabolic purpose [119,120]. Nevertheless, many things remain to be learned about the procedures that should be followed to guarantee the safety and efficacy of cell therapy products, especially with respect to the biology of stem cells, their self-renewal and differentiation potential and, above all, the evaluation and prediction of potential risks.

Most cell therapy products are not controversial from a bioethical point of view. The exception to this is therapy with human embryonic stem cells, which raises moral and bioethical problems [121,122]. Such consideration refer to the donor's informed consent and to problems associated with the harvesting of oocytes and the destruction of human embryos. In this regard, the guidelines used by the different countries range from total prohibition to regulated authorization. In general, there is an international consensus that the results obtained in stem cell research should be applied to humans without prior bioethical scrutiny, with the understanding that scientific research and the use of scientific knowledge must respect human rights and the dignity of the individual in accordance with the Universal Declaration of Human Rights and the Universal Declaration of the Human Genome [123].

The main advantage of induced pluripotent stem cells is that their use, unlike that of embryonic stem cells, does not raise moral or bioethical issues as the scientific community, as well as society at large, consider it a valid alternative for the generation of pluripotent stem cells without the need to use human oocytes or embryos. Furthermore, these cells have shown themselves to be functionally and molecularly similar to embryonic cells, but without their bioethical problems, which means that their use in humans will not require an overly stringent regulatory framework. The importance of this cannot be overstated as, in many instances, and in some countries more than in others, legislation can hinder the development of science and, consequently, the application of new knowledge and new therapeutic strategies.

7. Concluding remarks

iPSCs offer an unprecedented alternative for basic, clinical and applied biomedical research. The most significant applications of these cells to the field of cell therapy are related to the treatment of such organ-specific conditions as diabetes —a typically metabolic disease—, hepatic and cardiovascular diseases, immunological disorders and monogenic hereditary conditions in general such as haemophilia.

However, many aspects remain to be unveiled about the safety of iPSCs and about their reprogramming mechanisms, although no-one denies that this technology offers new, until-recently-unimaginable possibilities for correcting alterations in a large number of conditions, particularly in monogenic and metabolic diseases [124]. Also, some technical problems will also have to be resolved such as finding a way to produce these cells using risk-free viral vector transfection as well as safer alternative methods such as viral vector-mediated reprogramming.

Other more general, though no less important, issues that remain to be addressed include optimal extrapolation to humans of the high levels of safety and expression obtained in animal models and finding out whether it is adult mesenchymal stem cells or iPSCs that constitute the best and most easily applicable alternative for the administration of combined cell therapy/gene therapy.

For the reasons mentioned it is imperative not to create false expectations in patients suffering from a disease that is amenable to advanced therapies, specifically cell therapy, as these strategies are still in their "infancy". In the longer term, once the challenges mentioned above have been overcome, both cell and gene therapy will become plausible alternatives. Optimism is in order, but fantasy is best avoided.

As far as haemophilia is concerned, the first article discussing the benefits of gene therapy for the treatment of the disease was published a decade ago. At that time, experts in the field anticipated that a cure for haemophilia would be found by the first decade of the 21st century [125], a prediction that did not come true because of multiple problems related to biosafety. Although many steps have been taken in the right direction with respect to gene therapy, cellular reprogramming of iPSCs and the safety of transfer vectors, efforts must continue in order to resolve problems related to immune response, insertional mutagenesis, efficacy and expression time, the collateral (particularly hepatotoxic) damage caused by viral vectors and the risk of teratoma and neoplasia derived from the application of certain cell types. Sight should not be lost of the difficulties inherent in recruiting patients for clinical trials and in the large-scale production of vectors and cell lines, needed to facilitate optimal and efficient implementation in the clinical setting.

One of the first things that must be addressed when doing research into advanced therapies is whether the expected benefits of such therapies will be able to offset the investment needed. In the case of haemophilia, the answer is clearly in the affirmative as it is a chronic disease that requires high-frequency life-long treatment, very costly in patients on prophylaxis, and which poses a potential risk of infection by emerging pathogens. The second question is whether advanced therapies are at all feasible. In this regard, haemophilia is considered an optimal

http://dx.doi.org/10.5772/55816

candidate for such treatments for several reasons: it is a monogenic disease; the expression of low levels of coagulation factor (1-5%) can result in a moderate phenotype; a large variety target cells can be applied; there is no need to regulate factor expression, and a large amount of animal models are available for experimentation. In this regard, application of strategies that are less demanding in terms of efficacy, i.e. level of protein expression, but that afford much greater safety, may be an alternative for this condition, taking into account that both physicians and patients are highly sensitive to the special immunologic situation of the haemophilic population and that viral infections (HIV/HCV) have had lethal consequences for these individuals in the past [76].

As regards diabetes as a typically metabolic disease, advances in the understanding of its physio- and etiopathology, together with the greater biotechnological possibilities available, have made new alternatives possible as a result of the development of advanced therapies to treat it. Transplantation of insulin-secreting cells or of islets obtained a from differentiation of stem cells could hold some hope in the long term.

As in haemophilia, in diabetes it is also necessary to justify the investment of human and financial resources required for the development of new advanced therapeutical strategies, taking account of the fact that patients with this condition also benefit from an optimal and efficient treatment at present. The justification for the said investment is that diabetes gives rise to vascular and neurological complications in the long term and that transplantation of Langerhans islets has not achieved the success that scientists hoped for because of the dearth of donors and the high rate of immune rejection that characterizes diabetic patients.

In a nutshell, iPSCs technology has the potential to produce an about-face in the way we conceive cell behavior as iPSCs can be induced to form hormone-producing differentiated cells. In this regard, several authors have reported on the generation of insulin-producing pancreatic cells from iPSCs from rhesus monkey and murine models which, after transplantation, are capable of producing insulin *in vivo* in response to glucose stimulation. Nonetheless, some general issues affecting iPSCs remain to be resolved before these cells can be used clinically in the treatment of diabetes. Prominent among these are optimizing the reprogramming process as well as their genetic safety, controlling the high differentiation variability of the different pancreatic lines by means of epigenetic trials and enhancing the purification, isolation and characterization of homogeneous populations of iPSC-derived insulin-producing β cells.

Acknowledgements

Antonio Liras is Principal Researcher in a preclinical project —not clinical trial— with adipose mesenchymal stem cells and gene/cell therapy protocols for treatment of haemophilia. This project is supported by funding from a grant from the Royal Foundation Victoria Eugenia of Haemophilia.

Author details

Antonio Liras[1,2*], Cristina Segovia[1] and Aline S. Gabán[1,3]

*Address all correspondence to: aliras@hotmail.com

1 Department of Physiology, School of Biology, Complutense University of Madrid, and Cell Therapy and Regenerative Medicine Unit, La Paz University Hospital Health Research Institute-IdiPAZ. Madrid, Spain

2 Royal Foundation Victoria Eugenia of Haemophilia. Madrid, Spain

3 University for the Development of State and the Pantanal Region. Campo Grande, Brazil

References

[1] Chagastelles PC, Nardi NB, Camassola M. Biology and applications of mesenchymal stem cells. *Sci Prog* 2010;93:113-27.

[2] Thiede MA. Stem Cell: applications and opportunities in drug discovery. *Drug Discov World* 2009;10:9-16.

[3] Ahrlund-Richter L, De Luca M, Marshak DR, et al. Isolation and production of cells suitable for human therapy: challenges ahead. *Cell Stem Cell* 2009;4:20-6.

[4] Thomson JA, Itskovitz-Eldor J, Shapiro SS, et al. Embryonic Stem Cell Lines Derived from Human Blastocysts. *Science* 1998;282:1145-7.

[5] Aflatoonian B, Moore H. Human primordial germ cells and embryonic germ cells, and their use in cell therapy. *Curr Opin Biotechnol* 2005;16:530-5.

[6] Alison MR, Islam S. Attributes of adult stem cells. *J Pathol* 2009;217:144-60.

[7] Sommer CA, Mostoslavsky G. Experimental approaches for the generation of induced pluripotent stem cells. *Stem Cell Res Ther* 2010;1:26.

[8] Takahashi K, Yamanaka S. Induction of pluripotent stem cells from mouse embryonic and adult fibroblast cultures by defined factors. *Cell* 2006;126:663-76.

[9] Yu J, Vodyanik MA, Smuga-Otto K, et al. Induced pluripotent stem cell lines derived from human somatic cells. *Science* 2007;318:1917-20.

[10] Okita K, Hong H, Takahashi K, et al. Generation of mouse-induced pluripotent stem cells with plasmid vectors. *Nat Protoc* 2010;5:418-28.

[11] Kaji K, Norrby K, Paca A, et al. Virus-free induction of pluripotency and subsequent excision of reprogramming factors. *Nature* 2009;458:771-5.

[12] Woltjen K, Michael IP, Mohseni P, et al. PiggyBac transposition reprograms fibroblasts to induced pluripotent stem cells. *Nature* 2009;458:766-70.

[13] Yamanaka S. Strategies and new developments in the generation of patient-specific pluripotent stem cells. *Cell Stem Cell* 2007;1:39-49.

[14] Amabile G, Meissner A. Induced pluripotent stem cells: current progress and potential for regenerative medicine. *Trends Mol Med* 2009;15:59-68.

[15] Hochedlinger K, Plath K. Epigenetic reprogramming and induced pluripotency. *Development* 2009;136:509-23.

[16] Tateishi K, He J, Taranova O, et al. Generation of insulin-secreting islet-like clusters from human skin fibroblasts. *J Biol Chem* 2008;283:31601-7.

[17] Dimos JT, Rodolfa KT, Niakan KK, et al. Induced pluripotent stem cells generated from patients with ALS can be differentiated into motor neurons. *Science* 2008;321:1218-21.

[18] Park IH, Arora N, Huo H, et al. Disease-specific induced pluripotent stem cells. *Cell* 2008;134:877-86.

[19] Ebert AD, Yu J, Rose FF Jr, et al. Induced pluripotent stem cells from a spinal muscular atrophy patient. *Nature* 2009;457:277-80.

[20] Caspi O, Itzhaki I, Kehat I, et al. In vitro electrophysiological drug testing using human embryonic stem cell derived cardiomyocytes. *Stem Cells Dev* 2009;18:161-72.

[21] Razvi ES, Oosta GM. Stem Cells for cellular therapy space. *Drug Discov Today* 2010;11:37-40.

[22] Hatzimichael E, Tuthill M. Hematopoietic stem cell transplantation. *Stem Cells Cloning: Advances and Applications* 2010;3:105-17.

[23] Yu J, Hu K, Smuga-Otto K, et al. Human induced pluripotent stem cells free of vector and transgene sequences. *Science* 2009;324:797-801.

[24] Park IH, Zhao R, West JA, et al. Reprogramming of human somatic cells to pluripotency with defined factors. *Nature* 2008;451:141-6.

[25] Zhou H, Wu S, Joo JY, et al. Generation of induced pluripotent stem cells using recombinant proteins. *Cell Stem Cell* 2009;4:381-4.

[26] Lin T, Ambasudhan R, Yuan X, et al. A chemical platform for improved induction of human iPSCs. *Nat Methods* 2009;6:805-8.

[27] Lowry WE, Richter L, Yachechko R, et al. Generation of human induced pluripotent stem cells from dermal fibroblasts. *Proc Natl Acad Sci USA* 2008;105:2883-8.

[28] Alternative sources of human pluripotent stem cells. Available at White paper: The President's Council on Bioethics. Washington, D.C; 2005. Available: http://bioeth-

ics.georgetown.edu/pcbe/reports/white_paper/alternative_sources_white_paper.pdf. (Accessed 2012 July 30).

[29] Martins-Taylor K, Xu RH. Concise Review: Genomic Stability of Human Induced Pluripotent Stem Cells. *Stem Cells* 2012;30:22–7.

[30] Rolletschek A, Wobus AM. Induced human pluripotent stem cells: promises and open questions. *Biol Chem* 2009;390:845-9.

[31] Jang J, Yoo JE, Lee JA, et al. Disease-specific induced pluripotent stem cells: a platform for human disease modeling and drug discovery. *Exp Mol Med* 2012;44:202-13.

[32] Wernig M, Zhao JP, Pruszak J, et al. Neurons derived from reprogrammed fibroblasts functionally integrate into the fetal brain and improve symptoms of rats with Parkinson's disease. *Proc Natl Acad Sci USA* 2008;105:5856-61.

[33] Hanna J, Wernig M, Markoulaki S, et al. Treatment of sickle cell anemia mouse model with iPS cells generated from autologous skin. *Science* 2007;318:1920-3.

[34] Mauritz C, Schwanke K, Reppel M, et al. Generation of functional murine cardiac myocytes from induced pluripotent stem cells. *Circulation* 2008;118:507-17.

[35] Zhang J, Wilson GF, Soerens AG, et al. Functional cardiomyocytes derived from human induced pluripotent stem cells. *Circ Res* 2009;104:e30-41.

[36] Mason C, Manzotti E. Induced pluripotent stem cells: an emerging technology platform and the Gartner hype cycle. *Regen Med* 2009;4:329-31.

[37] Nelson TJ, Terzic A. Induced pluripotent stem cells: reprogrammed without a trace. *Regen Med* 2009;4:333-55.

[38] Bolton-Maggs PHB, Pasi KJ. Haemophilias A and B. *Lancet* 2003;361:1801-9.

[39] Berntorp E, Shapiro AD. Modern haemophilia care. *Lancet* 2012;379:1447-56.

[40] Liras A, Gaban AS, Rodriguez-Merchan EC. Cartilage restoration in haemophilia: advanced therapies. *Haemophilia* 2012;1-8.

[41] Schaub RG. Recent advances in the development of coagulation factors and procoagulants for the treatment of hemophilia. *Biochemical Pharmacology* 2011;82:91-8.

[42] Key NS, Negrier C. Coagulation factor concentrates: past, present, and future. *Lancet* 2007;370:439-48.

[43] Farrugia A. Plasma fractionation issues. *Biologicals* 2009;37:88-93.

[44] Hermans C, Brackmann HH, Schinco P, et al. The case for wider use of recombinant factor VIII concentrates. *Critical Rev Oncology/Hematology* 2012;83:11-20.

[45] Liras A. Recombinant proteins in therapeutics: Haemophilia treatment as an example. *Int Arch Med* 2008;1:4.

http://dx.doi.org/10.5772/55816

[46] Batlle J, Villar A, Liras A, et al. Consensus opinion for the selection and use of therapeutic products for the treatment of haemophilia in Spain. *Blood Coagul Fibrinolysis* 2008;19:333-40.

[47] Keeling D, Tait C, Makris M. Guideline on the selection and use of therapeutic products to treat haemophilia and other hereditary bleeding disorders. A United Kingdom Haemophilia Center Doctors' Organisation (UKHCDO) Guideline. *Haemophilia* 2008;14:671-84.

[48] Astermark J. Inhibitor development: patient-determined risk factors. *Haemophilia* 2010;16:66-70.

[49] Green D. Factor VIII inhibitors: a 50-year perspective. *Haemophilia* 2011;17:831-8.

[50] Astermark J, Altisent C, Batorova A, et al. Non-genetic risk factors and the development of inhibitors in haemophilia: a comprehensive review and consensus report. *Haemophilia* 2010;16:747-66.

[51] Franchini M, Tagliaferri A, Mengoli C, et al. Cumulative inhibitor incidence in previously untreated patients with severe hemophilia A treated with plasma-derived versus recombinant factor VIII concentrates: A critical systematic review. *Critical Rev Oncology/Hematology* 2012;81:82-93.

[52] Andréoletti O, Litaise C, Simmons H, et al. Highly Efficient Prion Transmission by Blood Transfusion. *PLoS Pathog* 2012;8:e1002782.

[53] Peden A, McCardle L, Head MW, et al. Variant CJD infection in the spleen of a neurologically asymptomatic UK adult patient with haemophilia. *Haemophilia* 2010;16:296-304.

[54] Zaman SM, Hill FG, Palmer B, et al. The risk of variant Creutzfeldt-Jakob disease among UK patients with bleeding disorders, known to have received potentially contaminated plasma products. *Haemophilia* 2011;17:931-7.

[55] Mei B, Pan C, Jiang H, et al. Rational design of a fully active, long-acting PEGylated factor VIII for hemophilia A treatment. *Blood* 2010;116:270-9.

[56] Negrier C, Knobe K, Tiede A, et al. Enhanced pharmacokinetic properties of a glycoPEGylated recombinant factor IX: a first human dose trial in patients with hemophilia B. *Blood* 2011;118:2695-701.

[57] Dumont JA, Liu T, Low SC, et al. Prolonged activity of a recombinant factor VIII-Fc fusion protein in hemophilia A mice and dogs. *Blood* 2012;119:3024-30.

[58] Shapiro AD, Ragni MV, Valentino LA, et al. Recombinant factor IX-Fc fusion protein (rFIXFc) demonstrates safety and prolonged activity in a phase 1/2a study in hemophilia B patients. *Blood* 2012;119:666-72.

[59] Peng A, Kosloski MP, Nakamura G, et al. PEGylation of a factor VIII-phosphatidylinositol complex: pharmacokinetics and immunogenicity in hemophilia A mice. *AAPS J* 2012;14:35-42.

[60] Casademunt E, Martinelle K, Jernberg M, et al. The first recombinant human coagulation factor VIII of human origin: human cell line and manufacturing characteristics. *Eur J Haematol* 2012;89:165-76.

[61] Nichols TC, Dillow AM, Franck HW, et al. Protein replacement therapy and gene transfer in canine models of haemophilia A, haemophilia B, von Willebrand disease, and factor VII deficiency. *ILAR J* 2009;50:144-67.

[62] Nienhuis AW. Development of gene therapy for blood disorders. *Blood* 2009;111:4431-44.

[63] Liras A. Gene therapy for haemophilia: The end of a "royal pathology" in the third millennium? *Haemophilia* 2001;7:441-5.

[64] Liras A, Olmedillas S. Gene therapy for haemophilia…yes, but…with non-viral vectors? *Haemophilia* 2009;15:811-6.

[65] Jeon HJ, Oh TK, Kim OH, et al. Delivery of factor VIII gene into skeletal muscle cells using lentiviral vector. *Yonsei Med J* 2010;51:52-7.

[66] Nathwani AC, Tuddenham EG, Rangarajan S, et al. Adenovirus-associated virus vector mediated gene transfer in hemophilia B. *N Engl J Med* 2011;365:2357-65.

[67] Sivalingam J, Krishnan S, Ng WH, et al. Biosafety assessment of site-directed transgene integration in human umbilical cord-lining cells. *Mol Ther* 2010;18:1346-56.

[68] Liras A, García-Arranz M, García-Gómez I, et al. Factor IX secretion in human adipose-derived stem cells by non-viral gene transfer. *Haemophilia* 2012;18(Suppl 3):A65.

[69] Aronovich A, Tchorsh D, Katchman H, et al. Correction of hemophilia as a proof of concept for treatment of monogenic diseases by fetal spleen transplantation. *Proc Natl Acad Sci USA* 2006;103:19075-80.

[70] Follenzi A, Benten D, Novikoff P, et al. Transplanted endothelial cells repopulate the liver endothelium and correct the phenotype of hemophilia A mice. *J Clin Invest* 2008;118:935-45.

[71] Follenzi A, Raut S, Merlin S, et al. Role of bone marrow transplantation for correcting hemophilia A in mice. *Blood* 2012;119:5532-42.

[72] Xu D, Alipio Z, Fink LM, et al. Phenotypic correction of murine hemophilia A using an iPSCs cell-based therapy. *Proc Natl Acad Sci USA* 2009;106:808-13.

[73] Yadav N, Kanjirakkuzhiyil S, Kumar S, et al. The therapeutic effect of bone marrow-derived liver cells in the phenotypic correction of murine hemophilia A. *Blood* 2009;114:4552-61.

[74] Alipio Z, Adcock DM, Waner M, et al. Sustained factor VIII production in hemophiliac mice 1 year after engraftment with induced pluripotent stem cell-derived factor VIII producing endothelial cells. *Blood Coagul Fibrinolysis* 2010;21:502-4.

[75] Daneman D. Type 1 diabetes. *Lancet* 2006;367:847-58.

[76] Retnakaran R, Zinman B. Type 1 diabetes, hyperglycaemia, and the heart. *Lancet* 2008;371:1790-9.

[77] Sayad A, Akbari MT, Pajouhi M, et al. The influence of the HLA-DRB, HLA-DQB and polymorphic positions of the HLA-DRβ1 and HLA-DQβ1 molecules on risk of Iranian type 1 diabetes mellitus patients. *Int J Immunogenet* 2012 (In press).

[78] Steck AK, Wong R, Wagner B, et al. Effects of non-HLA gene polymorphisms on development of islet autoimmunity and type 1 diabetes in a population with high-risk HLA-DR,DQ genotypes. Diabetes 2012;61:753-8.

[79] Krischer JP, Cuthbertson DD, Yu L, et al. Screening strategies for the identification of multiple antibody-positive relatives of individuals with type 1 diabetes. *J Clin Endocrinol Metab* 2003;88:103-8.

[80] Maclaren N, Lan M, Coutant R, et al. Only multiple autoantibodies to islet cells (ICA), insulin, GAD65, IA-2 and IA-2beta predict immune-mediated (Type 1) diabetes in relatives. *J Autoimmun* 1999;12:279-87.

[81] Barker JM, Barriga KJ, Yu L, et al. Prediction of autoantibody positivity and progression to type 1 diabetes: diabetes autoimmunity study in the young (DAISY). *J Clin Endocrinol Metab* 2004;89:3896-902.

[82] Barker JM, Yu J, Yu L, et al. Autoantibody "subspecificity" in type 1 diabetes: risk for organ-specific autoimmunity clusters in distinct groups. *Diabetes Care* 2005;28:850-5.

[83] Kordonouri O, Hartmann R, Deiss D, et al. Natural course of autoimmune thyroiditis in type 1 diabetes: association with gender, age, diabetes duration, and puberty. *Arch Dis Child* 2005;90:411-4.

[84] Skovbjerg H, Tarnow L, Locht H, et al. The prevalence of coeliac disease in adult Danish patients with type 1 diabetes with and without nephropathy. *Diabetologia* 2005;48:1416-7.

[85] Norris JM, Barriga K, Hoffenberg EJ, et al. Risk of celiac disease autoimmunity and timing of gluten introduction in the diet of infants at increased risk of disease. *JAMA* 2005;293:2343-51.

[86] Taton J, Czech A, Piątkiewicz P. Insulin as the main regulator of cellular glucose utilization-etiological aspects of insulin resistance. *Pol J Endocrinol* 2010;61:388-94.

[87] Azpiazu I, Manchester J, Skurat AV et al. Control of glycogen synthesis is shared between glucose transport and glycogen synthase in skeletal muscle fibers. *Am J Physiol Endocrinol Metab* 2000;278:E234-E43.

[88] Abel ED, Peroni O, Kim JK, et al. Adipose-selective targeting of the GLUT4 gene impairs insulin action in muscle and liver. *Nature* 2001;409:729-33.

[89] Boden G, Shulman GI. Free fatty acids in obesity and type 2 diabetes: defining their role in the development of insulin resistance and β-cell dysfunction. *Eur J Clin Invest* 2002;32(Suppl 3):14-23.

[90] Bruning JC, Gautam D, Burks DJ, et al. Role of brain insulin receptor in control of body weight and reproduction. *Science* 2000;289:2122-5.

[91] Callaghan BC, Cheng HT, Stables CL, et al. Diabetic neuropathy: clinical manifestations and current treatments. *Lancet Neurol* 2012;11:521-34.

[92] Laing SP, Swerdlow AJ, Slater SD, et al. Mortality from heart disease in a cohort of 23,000 patients with insulin-treated diabetes. *Diabetologia* 2003;46:760-5.

[93] Pajunen P, Taskinen MR, Nieminen MS, et al. Angiographic severity and extent of coronary artery disease in patients with type 1 diabetes mellitus. *Am J Cardiol* 2000;86:1080-5.

[94] Danne T, Bolinder J. New insulins and insulin therapy. *Int J Clin Pract Suppl* 2011;170:26-30.

[95] Nicholson G, Hall GM. Diabetes mellitus: new drugs for a new epidemic. *Br J Anaesth* 2011;107:65-73.

[96] Boyle ME, Seifert KM, Beer KA, et al. Guidelines for application of continuous subcutaneous insulin infusion (insulin pump) therapy in the perioperative period. *J* Diabetes *Sci Technol* 2012;6:184-90.

[97] Hanaire H. External insulin pump treatment in the day-to-day management of diabetes: benefits and future prospectives. Diabetes *Metab* 2011;37(Suppl 4):S40-7.

[98] Schaepelynck P, Darmon P, Molines L, et al. Advances in pump technology: insulin patch pumps, combined pumps and glucose sensors, and implanted pumps. Diabetes *Metab* 2011;37(Suppl 4):S85-93.

[99] Hansson M, Tonning A, Frandsen U, et al. Artifactual insulin release from differentiated embryonic stem cells. *Diabetes* 2004;53:2603-9.

[100] Dor Y, Brown J, Martinez OI, et al. Adult pancreatic beta cells are formed by self-duplication rather than stem-cell differentiation. *Nature* 2004;429:41-6.

[101] Azzi J, Geara AS, El-Sayegh S, et al. Immunological aspects of pancreatic islet cell transplantation. *Expert Rev Clin Immunol* 2010;6:111-24.

[102] Maehr R. iPS Cells in Type 1 Diabetes Research and Treatment. *Clin Pharmacol Ther* 2011;89:750-3.

[103] Kao DI, Chen S. Pluripotent stem cell-derived pancreatic β-cells: potential for regenerative medicine in diabetes. *Regen Med* 2012;7:583-93.

[104] Schroeder IS. Potential of Pluripotent Stem Cells for Diabetes Therapy. *Curr Diab Rep* 2012 (In press).

[105] Maehr R, Chen S, Snitow M, et al. Generation of pluripotent stem cells from patients with type 1 diabetes. *Proc Natl Acad Sci USA* 2009;106:15768-73.

[106] Zhang D, Jiang W, Liu M, et al. Highly efficient differentiation of human ES cells and iPS cells into mature pancreatic insulin-producing cells. *Cell Res* 2009;19:429-38.

[107] Zhu FF, Zhang PB, Zhang DH, et al. Generation of pancreatic insulin-producing cells from rhesus monkey induced pluripotent stem cells. *Diabetologia* 2011;54:2325-36.

[108] Pan FC, Wright C. Pancreas Organogenesis: From Bud to Plexus to Gland. *Dev Dynamics* 2011;240:530–65.

[109] Jeon K, Lim H, Kim JH, al. Differentiation and transplantation of functional pancreatic beta cells generated from induced pluripotent stem cells derived from a type 1 diabetes mouse model. *Stem Cells Dev* 2012 (In press).

[110] Ohmine S, Squillace KA, Hartjes KA, et al. Reprogrammed keratinocytes from elderly type 2 diabetes patients suppress senescence genes to acquire induced pluripotency. *Aging (Albany NY)* 2012;4:60-73.

[111] Mummery C. Induced Pluripotent Stem Cells - A Cautionary Note. *N Engl J Med* 2011;364:2160-2.

[112] Hussein SM, Batada NN, Vuoristo S, et al. Copy number variation and selection during reprogramming to pluripotency. *Nature* 2011;471:58-62.

[113] Gore A, Li Z, Fung HL, et al. Somatic coding mutations in human induced pluripotent stem cells. *Nature* 2011;471:63-7.

[114] Lister R, Pelizzola M, Kida YS, et al. Hotspots of aberrant epigenomic reprogramming in human induced pluripotent stem cells. *Nature* 2011;471:68-73.

[115] Pera MF. Stem cells: the dark side of induced pluripotency. *Nature* 2011;471:46-7.

[116] Kim K, Doi A, Wen B, et al. Epigenetic memory in induced pluripotent stem cells. *Nature* 2010;467:285–90.

[117] European Medicines Agency. Available: http://www.ema.europa.eu/ema/index.jsp?curl=/pages/home/Home_Page.jsp&jsenabled=true. (Accessed 2012 July 30).

[118] Halme DG, Kessler DA. FDA Regulation of Stem-Cell-Based Therapies. *N Engl J Med* 2006;355:1730-5.

[119] CFR - Code of Federal Regulations Title 21. Part 1271-Human Cells, Tissues, and Cellular and Tissue-Based Products. Available: http://www.accessdata.fda.gov/scripts/cdrh/cfdocs/cfcfr/CFRSearch.cfm?CFRPart=1271. (Accessed 2012 July 30).

[120] 120. Proposed approach to regulation of cellular and tissue-based products. The Food and Drug Administration. Available: http://www.fda.gov/downloads/Biologi-

csBloodVaccines/GuidanceComplianceRegulatoryInformation/Guidances/Tissue/UCM062601.pdf. (Accessed 2012 July 30).

[121] Rao M, Condic ML. Alternative sources of pluripotent stem cells: scientific solutions to an ethical dilemma. *Stem Cells Dev* 2008;17:1-10.

[122] Hyun I. The bioethics of stem cell research and therapy. *J Clin Invest* 2010;120:71-5.

[123] Science for the Twunty-First Century: a New Commitment. UNESCO. http://www.unesco.org/science/wcs/eng/declaration_e.htm.

[124] Wong GK, Chiu AT. Gene therapy, gene targeting and induced pluripotent stem cells: Applications in monogenic disease treatment. *Biotechnol Adv* 2011;29:1-10.

[125] Mannucci PM, Tuddenham EG. The hemophilias – from royal genes to gene therapy. *N Engl J Med* 2001;344:1773-9.

Chapter 25

Safety Assessment of Reprogrammed Cells Prior to Clinical Applications: Potential Approaches to Eliminate Teratoma Formation

Juan Carlos Polanco and Andrew L. Laslett

Additional information is available at the end of the chapter

http://dx.doi.org/10.5772/54400

1. Introduction

Human pluripotent stem cells (hPSC) include human embryonic stem cells (hESC) and human induced pluripotent stem cells (hIPSC). Due to their inherent ability to self-renew indefinitely in vitro and to give rise to essentially all cell lineages, both cell types have enormous potential for applications in regenerative medicine, but differ in their origin. HESC are derived from early pre-implantation stage embryos and have the capacity, known as *pluripotency*, to generate any other cell type of the human body. HESC can be differentiated in the laboratory, a procedure aimed at the generation of healthy somatic cells that eventually could be used in a large variety of applications including therapeutic options. However, work with hESC raises ethical concerns regarding the use of human early pre-implantation embryos, as well as concerns regarding the future use of hESC-derived cells in non-autologous cell transplantation therapies due to immune rejection of hESC-derived tissues, given that hESC are non-self. These concerns appeared to be overcome when it was demonstrated that pluripotency could be induced in differentiated somatic (adult) cells of the body by introduction of a cocktail of pluripotency-associated transcription factors, usually *OCT4*, *SOX2*, *KLF4* and *c-MYC* [1]. This process is known as reprogramming, and generates human induced pluripotent stem cells (hIPSC), which show an embryonic-like state similar to hESC (for review see [2]). Human iPSC are considered to have immense potential for regenerative medicine, do not require the use of donated human embryos for their generation and may provide an alternative and suitable resource for autologous cell-based therapies, in which cells obtained from the patient could be used to generate self-hIPSC followed by differentiation to relevant lineages required for therapeutic intervention. However, disturbingly, mouse experiments have shown that autologous mouse iPSC can induce unexpected T-cell-dependent immune response in syngeneic recipients [3], suggesting that hIPSC-derived cell types should also be evaluated for immunogenicity before any clinical application.

Given that: (i) the generation of human iPSC does not require destruction of embryos, (ii) that many iPSC lines can be established from a single patient, (iii) hIPSC are predicted to lead to patient specific therapies and (iv) that hIPSC could be used as a source of somatic cells for toxicology and drug screening studies, many research programs have shifted their focus from solely hESC-based research to also include work on hIPSC. However, despite the phenotypic similarities with hESC, recent reports described the worrying phenomena of elevated genetic [4-6] and epigenetic abnormalities [7-9] in hIPSC, raising concern about the suitability of hiPSC-derived cell types for future clinical applications. Nevertheless, it appears that these abnormalities are not present in all iPSC cell lines and that at least in mouse studies the current reprogramming methods can produce pluripotent mouse IPSC lines that lack identifiable genomic alterations [10], a result that calls for additional experiments to explain the discrepancies with respect to hIPSC [4-6]. It is becoming increasingly obvious, based on the studies described, that it is extremely important for hIPSC-derived therapies to become a reality in the clinic, that researchers develop diagnostic tools to definitively recognise clinically "safe" and "unsafe" hIPSC lines. This is likely to be a complex and cumbersome task due to the large number of methodological approaches used. To date hIPSC lines have been generated (for review see [2]); using a large number of different vectors to introduce the transgenes, with variations in the combinations of genes used to induce pluripotency, with significant modifications in culture conditions aimed at improving reprogramming efficiency, and from many of the more than 200 cell types in the human body. It will be a challenging undertaking to develop individual safety profiles for the multitude of hIPSC lines developed to date. Additionally, hIPSC-derived cells/tissues intended for clinical applications will need to comply with the following conditions: (i) adequate numbers of cells for transplantation therapy, (ii) hIPSC differentiated progeny need to be tolerated (not immunorejected) by a patient's immune system and (iii) hIPSC-derived cells should not generate teratoma-like tumours at any time after transplantation. In vitro and pre-clinical optimisations for these parameters are essential before hIPSC-derived technologies reach the clinic.

In this Chapter, we discuss the prospects for clinical applications using pluripotent cells, focusing on an evaluation of hIPSC cell potential and on the development of methods for the identification and removal of unwanted residual tumorigenic pluripotent cells from hIPSC-derived cell populations following differentiation.

2. The risk of tumour formation from residual pluripotent cells

In vivo, pluripotent stem cells reside only during a short time in embryonic development. Conversely, in vitro, hESC and hIPSC lines can be propagated indefinitely in the embryonic-like state and remain pluripotent, or with the appropriate cues they can give rise to a range of body cell types. For human cells, the most accepted *in vivo* assay to prove pluripotency is the generation of *teratomas* in immuno-deficient mice (ie: NOD-SCID and NOD/SCID $IL2R\gamma^{-/-}$ mice), by injection of putative pluripotent hPSC into organs like testis, kidney or muscle. Teratomas are benign solid tumours that contain a mixture of differentiated tissues such as nerve cells, muscle cells or cartilage. If a human cell line generates teratomas, it is considered pluripotent, because teratomas emulate differentiation in the developing embryo, albeit in a disorganised fashion, by generation of tissues resembling different parts of the embryo known as embryonic germ layers (i.e.: Ectoderm, Mesoderm and Endoderm).

In the clinical context, pluripotent stem cells will not be transplanted, rather the progenitors and/or specialised somatic cell types that are derived from hPSC will be used. It is the hope of researchers working in the expanding field of regenerative medicine that hPSC-derived cell populations will integrate into tissues and receive appropriate cues to functionally correct diseased or injured tissue, (i.e.: Parkinson's disease, Huntington's disease, cardiac failure, multiple sclerosis or macular degeneration). Therefore, differentiated somatic cell types are the final product for transplantation and therapeutic applications, and pluripotent stem cells are the stable source to generate those somatic cells or their progenitors (depending upon disease context) in the laboratory. In this context, the presence of even low frequency residual undifferentiated stem cells capable of teratoma formation becomes a highly undesirable feature when considering hPSC-derived somatic cells for transplantation into patients. Differentiated cells will not be deemed safe for use in regenerative medicine if they generate tumours at any time after transplantation. To comply with this requirement, we consider that researchers should aim at the generation of pluripotent stem cell-free samples. Therefore, it will be essential to be able to monitor if any undifferentiated pluripotent cells remain after differentiation protocols, and if so, remove them without damaging the potentially therapeutic differentiated cells. Evidence supporting this statement is that it is known that the numbers of pluripotent cells injected experimentally have a directly proportional effect on how fast the teratomas develop and the size of the tumour [11-13]. It has also been reported that at doses of 1,000 pluripotent cells, teratomas developed with 40% efficiency but with 10,000 cells the efficiency increased to 100% [12]. However, as few as two pluripotent cells have been reported to induce teratoma formation in immuno-deficient mice, although with lower efficiency [11]. Taken together, this might mean that one remaining pluripotent stem cell in a patient bound cell preparation could lead to teratoma formation. There is some limited evidence that potentially refutes the tumorgenic potential of low doses of pluripotent cells. This evidence is demonstrated by experiments showing that two pluripotent cells transplanted into syngeneic immunocompetent mice practically abolished tumour formation [11], most likely because those stem cells were cleared by the immune system. This could be taken to imply that in the clinical context of immuno-competent patients, low contamination with human pluripotent stem cells may be safe, but nevertheless for hPSC-derived cell populations to be approved for use in clinical trials their stringent elimination will be a requirement. Furthermore, the site of transplantation needs to be taken into account as not all places in the body are equally permissive for teratoma growth and development and contaminating hPSC may also migrate to alternative and possibly more permissive sites for teratoma growth post transplantation. For instance, it has been reported that similar number of pluripotent stem cells injected into immuno-deprived mice induced teratomas with 12.5% efficiency in intramuscular injections, 33% in subcutaneous injections, 60% in intratesticular, and approximately 100% under the kidney capsule [14]. Although many variables can potentially affect teratoma formation, we consider that the most ethical and safest cell population for transplantation into patients should be classified as pluripotent stem cell-free.

3. How to purge residual tumorigenic pluripotent stem cells from differentiated cell types?

To guarantee that no undifferentiated pluripotent stem cells are present in a hESC or hiPSC-differentiated progeny intended for transplantation into patients, researchers need assays to detect those residual pluripotent cells and efficient methods to purge stem cells from the differentiated cell populations. A good strategy to detect pluripotent cells is using antibodies that detect surface markers on live hPSC that are not present on differentiated cell types. After antibody-mediated detection of stem cells, other technologies could be coupled to the antibodies in order to eliminate residual pluripotent stem cells from the transplantation sample. For instance, Fluorescent or Magnetic Activated Cell Sorting (FACS and MACS) could be used with antibody detection for elimination of the targeted cells.

There are only a few available antibodies that detect cell surface markers on live human pluripotent stem cells (See table 1). Researchers, utilising the available antibodies, have described methods to eliminate residual pluripotent cells from samples of differentiated cell types. For instance the SSEA-4 antibody first demonstrated its utility in purging pluripotent stem cells from simian ESC-derived hematopoietic precursors used for transplantations into monkeys [15]. In this study, researchers used SSEA-4 antibody to detect residual pluripotent cells that persisted despite rigorous and extended differentiation protocols for hematopoietic precursors. SSEA-4 negative cells obtained by fluorescence activated cell sorting (FACS) did not develop teratomas, whereas teratomas were consistently observed in hematopoietic precursors showing presence of SSEA-4 positive cells [15]. The SSEA-4 and Tra-1-60 antibodies have also been compared for their efficiency in detecting and removing residual hPSC, by FACS or magnetic-activated cell sorting MACS [16]. This comparison revealed that MACS technology was not efficient for complete depletion of hESCs, with an average of 82% retention of hESCs, and highlighted that negative selection via FACS may be a preferred approach to eliminate undesirable hESCs from differentiated populations [16]. However, a note of caution against the use of single antibodies to detect hESCs emerged from data showing that 47% of SSEA-4 low-expressing hESCs exhibited a high level of expression for TRA-1-60. Therefore, detection of a single cell-surface marker may not be sufficient to eliminate all pluripotent stem cells, and methods that use multiple antibodies detecting different epitopes expressed by hESCs are more likely to be successful [16].

http://dx.doi.org/10.5772/54400

Antibody	Isotype	Cell-surface antigen	Source/Supplier	Literature reference
GCTM-2	IgM	Keratan sulphate proteoglycan (KSPG)-protein core	Kindly donated by Prof. Martin Pera	Laslett *et al.*, 2003 [27]; Pera *et al.*,2003 [28].
mAB 84	IgM	Podocalyxin (PODXL); CD34 family member.	Millipore MAB4414 http://www.millipore.com	Choo *et al.*, 2008 [17].
PHM-5	IgG1	Podocalyxin (PODXL); CD34 family member.	Millipore MAB430 http://www.millipore.com	Kerjaschki *et al.*, 1986 [29].
SSEA-3	IgM	Globoseries glycolipid	Millipore MAB4303 http://www.millipore.com	Kannagi *et al.*, 1983 [30].
SSEA-4	IgG3	Globoseries glycolipid	Millipore MAB4304 http://www.millipore.com	Kannagi *et al.*, 1983 [31].
TG30 (CD9)	IgG2a	25kDa tetraspannin protein CD9	Millipore MAB4427 http://www.millipore.com	Laslett *et al.*, 2003 [27]; Pera *et al.*, 2003 [28].
TG343	IgM	KSPG-protein core (detects the same antigen as the GCTM-2 antibody).	Millipore MAB4346 http://www.millipore.com	Cooper *et al.*, 2002 [32].
TRA-1-60	IgM	KSPG-carbohydrate side chain	Millipore MAB4360 http://www.millipore.com	Andrews *et al.*, 1984 [33].
TRA-1-81	IgM	KSPG-carbohydrate side chain	Millipore MAB4381 http://www.millipore.com	Andrews *et al.*, 1984 [33].

Table 1. Antibodies that are reactive with cell surface markers expressed on human pluripotent stem cells

The studies described above point to FACS technology coupled to antibody detection of surface markers as a good strategy to eliminate residual undifferentiated pluripotent cells and recover differentiated live cells for further applications such as re-culture or transplantation.

However, as the viability of hPSC-derived lineage progenitors or more mature cell types can be compromised post-FACS, caused by shearing forces, laser damage or osmotic stress, other technologies such as MACS may be better suited in these instances. Although MACS does not completely remove all hESCs in a single pass [16], this technology exhibits higher cell viability than FACS and it is possible that subsequent positive selections by MACS using multiple antibodies for different hESC cell surface markers could completely remove all hESCs. An alternative approach to MACS could be to use cytotoxic antibodies directed against hESC surface antigens or chemicals that could selectively eliminate hESCs without affecting their derivatives. An example of a cytotoxic antibody that detects and removes hESCs is the monoclonal antibody mAB-84 [17], which binds to PODXL (Podocalyxin-like protein 1) on hESCs and initiates a sequence of events that leads to hESC-membrane damage by formation of leaking pores [18]. It has been proposed that using the monoclonal antibody mAB-84 in a two-step cell-cell separation approach can eliminate teratoma-forming hESC from differentiated cell types [19]. In this strategy, an initial depletion of hESCs was achieved via MACS using a panel of commonly used hESC cell-surface markers, which was followed by selective elimination of residual undifferentiated stem cells post-MACS using the cytotoxic antibody mAB-84, an approach that appears to increase the safety of cell transplantation [19].

Selective elimination of residual human pluripotent stem cells after differentiation can also be achieved by targeting apoptosis-meditating receptors that are differentially expressed in undifferentiated stem cells and absent in hESC derivatives. Therefore, stimulation of these specific hESC apoptotic receptors induce programmed cell death only in the residual stem cells without affecting their differentiated progeny. One example of this kind of receptor is the prostate apoptosis response-4 (PAR-4), which mediates ceramide or ceramide-analogue-induced apoptosis in proliferating stem cells [20]. The apoptotic response appears to be specific for PAR-4(+) stem cells, and given that ESC-differentiated progenies such as neuro-progenitors express very low levels of PAR-4, they are less sensitive to ceramide induced apoptosis [20]. Using this approach, ceramide treatment appears to prevent teratoma formation when transplanting neural progenitors derived from ES cells [20] although it is likely that regulatory assays will require a more stringent method. Although PAR-4 induced apoptosis by ceramides appears an effective way to eliminate residual pluripotent stem cells following differentiation, this approach has not been broadly tested.

4. Antibodies against cell surface markers of human stem cells

The scarcity of antibodies directed against cell surface markers that recognize live human pluripotent stem cells (See table 1) is compounded by the fact that most of these antibodies lack identification of their encoding gene. Indeed, some cell surface antibodies do not recognize proteins, but complex carbohydrate and lipid moieties for which the corresponding gene is not yet identified. Despite this, these complex moieties are strong

http://dx.doi.org/10.5772/54400

antigens that elicit highly sensitive antibodies that recognize human pluripotent stem cells. Furthermore, a caveat is that stem-cell antibodies could also be immunoreactive with some embryonic tissues, or some mature cell types, becoming problematic with some hESC differentiation protocols. Therefore, depending on the phenotype of the target somatic cells, selected antibodies used to detect human pluripotent cells should be selected that do not react with the differentiated cells intended for transplantation. For instance, if working with hESC-derived renal tissues for treatment of kidney disorders, PODXL antibodies should not be used alone to detect stem cells because Podocalyxin protein is also expressed in glomerular podocytes.

The information in the previous section demonstrates that FACS and MACS technologies are potential methods for the elimination of residual pluripotent cells following *in vitro* differentiation (Figure 1). Both methodological approaches use cell surface antibodies for the labelling and detection of undifferentiated live hPSC. The advantage of live cell detection using either FACS or MACS is the ability to retrieve live hESC or hIPSC-derivatives that could be used for *in vitro* re-culture and expansion, or, ultimately, transplantation. However, FACS and MACS studies have also revealed the immunological complexity of *in-vitro* hESC cultures. HESC cultures contain a continuum of different subpopulations, where some hESC subpopulations express low levels of one surface marker and at the same time high levels of another [16, 21-23]. These findings imply strongly that a single cell-surface marker is not sufficient to eliminate all pluripotent stem cells [16, 21-23]. Therefore, any attempt to eliminate all hESC pluripotent subpopulations should rely on methods that use multiple antibodies detecting different epitopes expressed by hESCs. For instance, SSEA-4-coupled MACS showed an average 82% retention of hESCs [16], but when a panel of cell surface antibodies directed to different epitopes was used with MACS, the removal of undifferentiated hESCs raised to 98% on average [19].

In our laboratory, we have been working on the development of monoclonal antibody panels against extracellular markers that allow efficient human pluripotent cell separation from mixed populations of cultured cells, an essential requirement for safe hESC or hIPSC-based therapeutics [21-24]. Towards this end, we have reported a FACS-based immuno-transcriptional profiling system based on the detection of two pluripotency-associated cell surface antigens TG30 (CD9) and GCTM-2, [25-26]. This method is useful to characterise multiple human pluripotent stem cell lines, and to identify the subpopulations that are found in hESC *in-vitro* continuous culture [21-22]. Ongoing unpublished observations indicate that this double staining of human stem cells using two cell-surface markers is a better way to eliminate residual and persistent undifferentiated pluripotent cells using FACS in both hESC and hIPSC lines. Nevertheless, we are aware that there will be differentiation contexts in which TG30 (CD9) and GCTM-2 might not be appropriate or sufficient to purge pluripotent cells from particular differentiated hPSC-derivatives. Therefore there is a real need for new monoclonal antibodies that detect cell surface proteins on live hPSC.

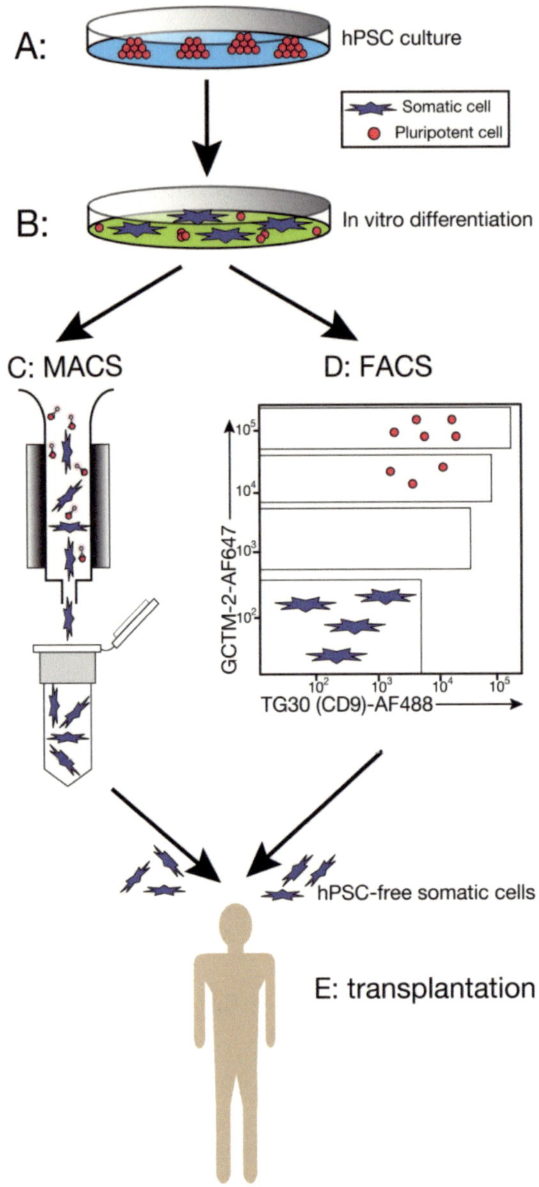

Figure 1. Potential approaches to eliminate residual pluripotent stem cells after in vitro differentiation. Shown are two potential methods that could be used to purge residual tumorigenic pluripotent stem cells from differentiated cell types. (A): Human pluripotent stem cells (hPSC) are able to self-renew indefinitely *in vitro*. (B): These pluripotent cells can be induced to differentiate *in vitro* to generate healthy progenitors and/or specialised somatic cell types that could potentially be used for transplantation and therapeutic applications. However, it is essential to monitor if any residual undifferentiated pluripotent cells remain after differentiation protocols. If undifferentiated stem cells remain, these cells should be removed without damaging the potentially therapeutic differentiated cells. Two good strategies for elimination of residual pluripotent cells are Magnetic Activated Cell Sorting (C: MACS) and Fluorescence Activated Cell Sorting (D: FACS). Both technologies are coupled to antibody detection of cell surface markers and allow retrieval of live hPSC-derivatives that could be used for further *in vitro* re-culture and expansion, or in due course transplantation (E).

http://dx.doi.org/10.5772/54400

5. Conclusions

Human pluripotent stem cells, namely hESC and hIPSC lines, may be the future mainstay of medicine, providing a plethora of medical applications and transplantation therapies aimed at the correction of an important number of pathological disorders. However, reaching clinical applications based on hPSC-therapies has not been as fast as expected. The ability to generate hIPSC lines from a variety of tissue sources has brought hIPSC research clearly into the spotlight, but reports on their epigenetic instability and genetic variability suggest that these cells are not yet clinic-ready. In addition, the concern of tumorigenesis or teratoma formation is an unsolved problem for both hESC and hIPSC research. If differentiation protocols are not 100% efficient and yield a mixture of differentiated and undifferentiated cells, this presents a significant risk of teratoma formation after transplantation. It is clear that adequate safety assays for hESC or hIPSC-derived technologies are of the utmost importance to aid in the safe translation from the bench to the clinic. This includes the essential monitoring of any residual undifferentiated pluripotent cells after differentiation protocols, an unavoidable methodological step in any sample to be used in the clinic. A variety of approaches have been discussed in this chapter to help to eliminate the undesirable residual pluripotent stem cells from samples intended for transplantation. However, there is an ongoing need to improve these separation methods in order to achieve hPSC free samples in a rapid, easy, safe, cost effective, scalable and clinically applicable way. We expect that novel cell-surface antibodies recognizing live pluripotent stem cells will strongly contribute to this ongoing search.

Acknowledgements

Work in ALL's laboratory contributes to Stem Cells Australia and was supported by the Australian Stem Cell Centre, the National Health and Medical Research Council (Australia) and the Victoria-California Stem Cell Alliance (CIRM grant TR1-01250). Comments and suggestions on draft versions of the manuscript by Drs. Carmel O'Brien and Tung-Liang Chung are gratefully acknowledged.

Author details

Juan Carlos Polanco[1,2] and Andrew L. Laslett[1,2*]

*Address all correspondence to: Andrew.Laslett@csiro.au

1 CSIRO, Materials Science and Engineering, Clayton, Victoria, Australia

2 Department of Anatomy and Developmental Biology, Monash University, Victoria, Australia

References

[1] Takahashi K, Tanabe K, Ohnuki M, Narita M, Ichisaka T, *et al.* Induction of pluripotent stem cells from adult human fibroblasts by defined factors. Cell. 2007;131(5) 861-872.

[2] Stadtfeld M, Hochedlinger K. Induced pluripotency: history, mechanisms, and applications. Genes and Development. 2010;24(20) 2239-2263.

[3] Zhao T, Zhang ZN, Rong Z, Xu Y. Immunogenicity of induced pluripotent stem cells. Nature. 2011;474(7350) 212-215.

[4] Gore A, Li Z, Fung HL, Young JE, Agarwal S, *et al.* Somatic coding mutations in human induced pluripotent stem cells. Nature. 2011;471(7336) 63-67.

[5] Hussein SM, Batada NN, Vuoristo S, Ching RW, Autio R, *et al.* Copy number variation and selection during reprogramming to pluripotency. Nature. 2011;471(7336) 58-62.

[6] Laurent LC, Ulitsky I, Slavin I, Tran H, Schork A, *et al.* Dynamic Changes in the Copy Number of Pluripotency and Cell Proliferation Genes in Human ESCs and iPSCs during Reprogramming and Time in Culture. Cell Stem Cell. 2011;8(1) 106-118.

[7] Kim K, Zhao R, Doi A, Ng K, Unternaehrer J, *et al.* Donor cell type can influence the epigenome and differentiation potential of human induced pluripotent stem cells. Nature Biotechnology. 2011;29(12) 1117-1119.

[8] Lister R, Pelizzola M, Kida YS, Hawkins RD, Nery JR, *et al.* Hotspots of aberrant epigenomic reprogramming in human induced pluripotent stem cells. Nature. 2011;471(7336) 68-73.

[9] Ohi Y, Qin H, Hong C, Blouin L, Polo JM, *et al.* Incomplete DNA methylation underlies a transcriptional memory of somatic cells in human iPS cells. Nature Cell Biology. 2011;13(5) 541-549.

[10] Quinlan A, Murat D, Vali H, Komeili A. The HtrA/DegP family protease MamE is a bifunctional protein with roles in magnetosome protein localization and magnetite biomineralization. Molecular Microbiology. 2011;80(4) 1075-1087.

[11] Lawrenz B, Schiller H, Willbold E, Ruediger M, Muhs A, Esser S. Highly sensitive biosafety model for stem-cell-derived grafts. Cytotherapy. 2004;6(3) 212-222.

[12] Shih CC, Forman SJ, Chu P, Slovak M. Human embryonic stem cells are prone to generate primitive, undifferentiated tumors in engrafted human fetal tissues in severe combined immunodeficient mice. Stem Cells and Development. 2007;16(6) 893-902.

[13] Reubinoff BE, Pera MF, Fong CY, Trounson A, Bongso A. Embryonic stem cell lines from human blastocysts: somatic differentiation in vitro. Nature Biotechnology. 2000;18(4) 399-404.

[14] Prokhorova TA, Harkness LM, Frandsen U, Ditzel N, Schroder HD, *et al.* Teratoma Formation by Human Embryonic Stem Cells Is Site Dependent and Enhanced by the Presence of Matrigel. Stem Cells and Development. 2009;18(1) 47-54.

[15] Shibata H, Ageyama N, Tanaka Y, Kishi Y, Sasaki K, *et al.* Improved safety of hematopoietic transplantation with monkey embryonic stem cells in the allogeneic setting. Stem Cells. 2006;24(6) 1450-1457.

[16] Fong CY, Peh GS, Gauthaman K, Bongso A. Separation of SSEA-4 and TRA-1-60 labelled undifferentiated human embryonic stem cells from a heterogeneous cell population using magnetic-activated cell sorting (MACS) and fluorescence-activated cell sorting (FACS). Stem Cell Reviews. 2009;5(1) 72-80.

[17] Choo AB, Tan HL, Ang SN, Fong WJ, Chin A, *et al.* Selection against undifferentiated human embryonic stem cells by a cytotoxic antibody recognizing podocalyxin-like protein-1. Stem Cells. 2008;26(6) 1454-1463.

[18] Tan HL, Fong WJ, Lee EH, Yap M, Choo A. mAb 84, a Cytotoxic Antibody that Kills Undifferentiated Human Embryonic Stem Cells via Oncosis. Stem Cells. 2009;27(8) 1792-1801.

[19] Schriebl K, Satianegara G, Hwang A, Tan HL, Fong WJ, *et al.* Selective removal of undifferentiated human embryonic stem cells using magnetic activated cell sorting followed by a cytotoxic antibody. Tissue Engineering Part A. 2012;18(9-10) 899-909.

[20] Bieberich E, Silva J, Wang G, Krishnamurthy K, Condie BG. Selective apoptosis of pluripotent mouse and human stem cells by novel ceramide analogues prevents teratoma formation and enriches for neural precursors in ES cell-derived neural transplants. Journal of Cell Biology. 2004;167(4) 723-734.

[21] Kolle G, Ho M, Zhou Q, Chy HS, Krishnan K, *et al.* Identification of human embryonic stem cell surface markers by combined membrane-polysome translation state array analysis and immunotranscriptional profiling. Stem Cells. 2009;27(10) 2446-2456.

[22] Laslett AL, Grimmond S, Gardiner B, Stamp L, Lin A, *et al.* Transcriptional analysis of early lineage commitment in human embryonic stem cells. BMC Developmental Biology. 2007;7(1) 12.

[23] Hough SR, Laslett AL, Grimmond SB, Kolle G, Pera MF. A continuum of cell states spans pluripotency and lineage commitment in human embryonic stem cells. PLoS One. 2009;4(11) e7708.

[24] Wada N, Wang B, Lin NH, Laslett AL, Gronthos S, Bartold PM. Induced pluripotent stem cell lines derived from human gingival fibroblasts and periodontal ligament fibroblasts. Journal of Periodontal Research. 2011;46(4) 438-447.

[25] Zhou Q, Chy H, Laslett AL. Preparation of defined human embryonic stem cell populations for transcriptional profiling. Current Protocols in Stem Cell Biology. 2010;Chapter 1 Unit 1B7.

[26] Ho MS, Fryga A, Laslett AL. Flow cytometric analysis of human pluripotent stem cells. Methods in Molecular Biology. 2011;767221-230.

[27] Laslett AL, Filipczyk AA, Pera MF. Characterization and culture of human embryonic stem cells. Trends in Cardiovascular Medicine. 2003;13(7) 295-301.

[28] Pera MF, Filipczyk AA, Hawes SM, Laslett AL. Isolation, characterization, and differentiation of human embryonic stem cells. Methods in Enzymology. 2003;365429-446.

[29] Kerjaschki D, Poczewski H, Dekan G, Horvat R, Balzar E, *et al.* Identification of a major sialoprotein in the glycocalyx of human visceral glomerular epithelial cells. Journal of Clinical Investigation. 1986;78(5) 1142-1149.

[30] Kannagi R, Levery SB, Ishigami F, Hakomori S, Shevinsky LH, *et al.* New globoseries glycosphingolipids in human teratocarcinoma reactive with the monoclonal antibody directed to a developmentally regulated antigen, stage-specific embryonic antigen 3. Journal of Biological Chemistry. 1983;258(14) 8934-8942.

[31] Kannagi R, Cochran NA, Ishigami F, Hakomori S, Andrews PW, *et al.* Stage-specific embryonic antigens (SSEA-3 and -4) are epitopes of a unique globo-series ganglioside isolated from human teratocarcinoma cells. Embo Journal. 1983;2(12) 2355-2361.

[32] Cooper S, Bennett W, Andrade J, Reubinoff BE, Thomson J, Pera MF. Biochemical properties of a keratan sulphate/chondroitin sulphate proteoglycan expressed in primate pluripotent stem cells. Journal of Anatomy. 2002;200(Pt 3) 259-265.

[33] Andrews PW, Banting G, Damjanov I, Arnaud D, Avner P. Three monoclonal antibodies defining distinct differentiation antigens associated with different high molecular weight polypeptides on the surface of human embryonal carcinoma cells. Hybridoma. 1984;3(4) 347-361.

Chapter 26

Stem Cells in Tissue Engineering

Shohreh Mashayekhan, Maryam Hajiabbas and
Ali Fallah

Additional information is available at the end of the chapter

http://dx.doi.org/10.5772/54371

1. Introduction

With the increasing number of patients suffering from damaged or diseased organs and the shortage of organ donors, the need for methods to construct human tissues outside the body has arisen. Tissue engineering is a newly emerging biomedical technology and methodology which combines the disciplines of both the materials and life sciences to replace a diseased or damaged tissue or organ with a living, functional engineered substitute [1, 2]. The so-called triad in tissue engineering encompasses three basic components called scaffold, cell and signaling biomolecule.

Whatever the approach being used in tissue engineering, the critical issues to optimize any tissue engineering strategy toward producing a functional equivalent tissue are the source of the cells and substrate biomaterial to deliver the cells in particular anatomical sites where a regenerative process is required. Due to their unique properties, stem cells and polymeric biomaterials are key design options. Briefly, stem cells have the ability to self-renew and commit to specific cell lineages in response to appropriate stimuli, providing excellent regenerative potential that will most likely lead to functionality of the engineered tissue. Polymeric materials are biocompatible, degradable, and flexible in processing and property design. A major focus of tissue engineering, therefore, is to utilize functional polymers with appropriate characteristics, as a means of controlling stem cell function. Based on their differentiation potential, stem cells used for tissue engineering can be divided into two categories: pluripotent stem cells and multipotent stem cells. Pluripotent stem cells include embryonic stem cells (ESCs) as well as induced pluripotent stem cells (iPSCs). Because ESCs are isolated from the inner cell mass of the blastocyst during embryological development, their use in tissue engineering is controversial and more limited while more attention has been paid to adult stem cells, which are multipotent and have a larger capacity to differentiate into a limited number of cell types [3]. Adult stem cells can be found in many adult tissue types including bone marrow, peripheral blood, adipose tissues, nervous tissues, muscles, dermis, etc. For instance, mesenchymal stem cells (MSCs) which reside in the bone marrow can differentiate into bone (osteoblasts) [4], muscle (myoblasts) [5], fat (adipocytes) [6] and cartilage (chrondocytes) [3] cells, while neural stem cells (NSCs) either give rise to support cells in the nervous system of vertebrates (astrocytes and oligodendrocytes) or neurons [7]. *In vivo*, differentiation and self-renewal of stem cells are dominated by signals from their surrounding microenvironment [8]. This microenvironment or "niche" is composed of other cell types as well as numerous chemical, mechanical and topographical cues at micro- and nano-scales, which are believed to serve as signaling mechanisms to determine cell-specific recruitment, migration, proliferation, differentiation as well as the production of numerous proteins required for hierarchical tissue organization [9].

In vivo, the cells are surrounded by a biological matrix comprising of tissue-specific combinations of insoluble proteins (e.g. collagens, laminins, and fibronectins), glycosaminoglycans (e.g. hyaluronan) and inorganic hydroxyapatite crystals (in bone) that are collectively referred to as the extracellular matrix (ECM). The varied composition of the ECM components not only contains a reservoir of cell-signaling motifs (ligands) and growth factors that guide cellular anchorage and behavior, but also provides physical architecture and mechanical strength to the tissue. The spatial distribution and concentration of ECM ligands, together with the tissue-specific topography and mechanical properties (in addition to signals from adjacent cells—juxtacrine signalling—and the surrounding fluid), provide signaling gradients that direct cell migration and cellular production of ECM constituents. In this dynamic environment, the bidirectional flow of information between the ECM and the cells mediates gene expression, ECM remodeling and ultimately tissue/organ function.

Native ECM exhibits macro- to nano-scale patterns of chemistry and topography [10]. Tissue stiffness is also known to vary depending on the organ type, disease state and aging process [11-13]. In tissue culture, stem cell differentiation has traditionally been controlled by the addition of soluble factors to the growth media [14]. However, most stem cell differentiation protocols yield heterogeneous cell types [15, 16]. Moreover, cells encounter very different, unfamiliar surfaces and environments when cultured *in vitro* or when materials are implanted into the body. Therefore, it is desirable to use more biomimetic *in vitro* culture conditions to regulate stem cell fate so as to advance clinical translation of stem cells through better expansion techniques and scaffolding for the regeneration of many tissues. Recent advances have facilitated further the creation of substrates with precise micro- and nano-cues, variable stiffness and chemical composition to better mimic the *in vivo* microenvironment [2, 17, and 18]. By employing various novel approaches, tissue engineers aim to incorporate topographical, mechanical and chemical cues into biomaterials to control stem cell fate decisions [2, 18, and 19].

This chapter will present various biomaterial designing considerations and strategies for stem cell-based tissue engineering for development as carriers for stem cells facilitating the *in vivo* use of stem cells in tissue engineering. This part first presents some biomimetic approaches to designing novel polymeric biomaterials with appropriate physical,

http://dx.doi.org/10.5772/54371

chemical, mechanical, and biological cues mimicking the natural stem cell niche in order to direct the desired stem cell behavior to facilitate the regeneration of desired tissues with particular emphasis on using adult stem cells including MSCs and NSCs. The next part will introduce some new trends emerging in the field of tissue engineering in terms of both cellular biology and biomaterial point of view in order to improve the overall efficiency of tissue regeneration for effectively controlling the cell fate and translating the stem cell research into much needed clinical applications in a not-too-distant future. The topics discussed in the latter part include 2D polysaccharide-based hydrogel scaffolds designed in the authors' studies for muscle tissue engineering applications. Hydrogel scaffolds made of natural polymers with proper handling for surgery and mechanical properties similar to muscle tissue, which could promote the desired muscle-derived stem cell behavior on the surface were developed in this study.

2. Biomimetic microenvironment design strategies

Damaged tissues often lose deeper layers which contain stem cell niches. In such cases, biomaterials could be useful tools for reestablishing the niches' functionality [20]. Artificial niches would need to incorporate appropriate 'homing' signals able to either localize endogenous stem cells or direct the desired incorporated exogenous stem cell behavior by means of developing various microenvironment design parameters including the dynamic control of soluble and surface-bound cytokines, ECM, cell-cell interactions, mechanical forces and physicochemical cues [21, 22].

The use of biomaterials as scaffolds is a fundamental component of tissue engineering since these materials serve as templates for tissue formation and are engineered depending on the tissue of interest. These scaffolds provide structural and mechanical support for the cells as well as present cues inducing tissue repair. The structure, morphology, degradation and presentation of bioactive sites are all important parameters in material design for these applications and may signal the differentiation of stem cells. Beside all the parameters related to the biomaterials scaffold, there are some other factors such as chemical cues (e.g. soluble reagents in terms of both concentration and their gradient, medium pH), mechanical cues (e.g. fluid shear stress) and other types of cues (electric and magnetic field) which are believed to have significant effect on stem cell behavior. These factors are reviewed extensively elsewhere [23, 24].

Figure 1 summarizes the biomimetic microenvironment design strategies for controlling stem cell behavior including chemical/biochemical (e.g. growth/differentiation factor presentation, density and gradient), structural, mechanical and some other types of cues.

Engineering these design parameters will effectively yield materials that create an architecture resembling the native environment for stem cells, and have controlled mechanical properties enabling adhesion and thus enhancing contractility in the cellular cytoskeleton, and present ligands directing intracellular signaling and gene expression. This section provides an overview of biomimetic microenvironment design strategies to direct the stem cell behavior for tissue engineering applications.

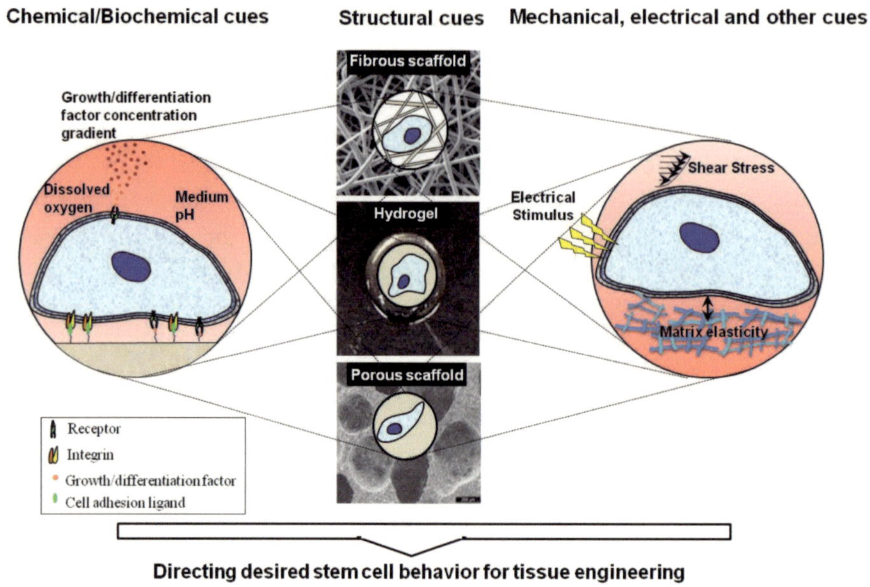

Figure 1. The biomimetic microenvironment design strategies for controlling stem cell fate

2.1. Chemical and biochemical cues

Biochemical cues are generally provided by soluble ligands, which may be either secreted by paracrinal cells or supplied by a capillary network in the human body. Insoluble ligands, which are adhesion proteins or molecules such as collagen, laminin and carbohydrates, are also present. Biochemical factors typically influence the cell microenvironment in a concentration or gradient-dependent manner.

Chemical and biochemical means are the first choice for stem cell differentiation. Small ions, growth factors, and cytokines can exert potent, long-range effects over stem cell microenvironments. Owing to their relative ease of study, *soluble biochemical cues* and their downstream signal transduction pathways are the best characterized determinants of stem cell fate and have been extensively used in *ex vivo* stem cell culture systems, as extensively discussed elsewhere [24-26]. Therefore, the following section will mainly focus upon the application of other types of *soluble* signals such as dissolved oxygen as well as insoluble chemical and biochemical cues (e.g., immobilized growth factor, extracellular matrix material, etc) to engineered niches.

In vivo, numerous growth factors and morphogens are immobilized by binding to the ECM through specific heparin-binding domains or by direct binding to ECM molecules such as collagen, or direct anchoring to cell membranes [27]. Immobilization of growth factors in

http://dx.doi.org/10.5772/54371

this manner can serve to increase local concentration of the protein by hindering diffusion and receptor-mediated endocytosis. For example, the morphogen Sonic hedgehog (Shh) is modified at its termini by lipids that link it to the cell membrane and thereby limit its mobility. Removing the lipids dilutes the factor to a lower concentration and thereby shrinks its effectiveness [28, 29]. Accordingly, mimicking the natural immobilization of cytokines is one approach utilized by engineers to concentrate factors in proximity to the cell surface in a manner that activates target signaling pathways effectively, and reduces, as well, the levels of growth factor necessary to elicit a potent cellular response.

An early study exploring this design concept focused on epidermal growth factor (EGF) [30] which is beneficial in repairing the damaged tissues, but is often difficult to deliver at sufficiently high concentrations to mediate downstream signaling events as it does not contain a matrix-binding domain and rapidly undergoes receptor-mediated endocytosis [31]. In a recent example involving human and porcine MSCs, amine-targeting chemistry was used to tether EGF to the surface of poly (methyl methacrylate)-graft-poly (ethylene oxide) comb polymers [32]. The tethered EGF led to sustained EGF receptor signaling and subsequent cellular responses including cell spreading and protection from apoptosis, whereas saturating levels of soluble EGF did not. Sakiyama-Elbert et al. incorporated heparin into biomaterial scaffolds to allow for immobilization of basic fibroblast growth factor (bFGF) [33]. bFGF was released either passively by diffusion, or actively via heparinases secreted by neighboring cells, thereby allowing for a controlled release and presentation of signal which was not possible with soluble growth factor delivery. The same delivery system has been used for differentiation of murine ESCs into mature neural cell types, including neurons and oligodendrocytes, indicating that biomaterials scaffolds functionalized with immobilized growth factors may be a potential strategy for generation of engineered tissue for treatment of spinal cord injury [34]. Finally, in a recent study, polymer substrates functionalized with the signaling domain of Shh supported enhanced osteogeneic differentiation of bone marrow-derived MSCs, as compared to cells cultured on the same surfaces with soluble Shh at the same concentration [35]. This example further demonstrates how growth factor or morphogen immobilization serves as an effective means to achieve sustained activation of downstream signaling pathways due in part to the finding that the local concentration in the scaffold was greater for immobilized growth factor than for soluble form.

There is a significant scope in the application of surface modifications, despite the use of protein biomolecules to provide more cues for cell adhesion, proliferation and differentiation. Arg-Gly-Asp (RGD) sequence and several natural proteins like collagen, laminin and fibronectin were shown to be essential for cell attachment to polymeric material surfaces devoid of any cell recognition sites [36, 37]. The immobilization of these proteins to polymers not only promotes cell adhesion and proliferation but also increases hydrophilicity of the polymers such as aliphatic polyesters. One such surface functionalization for biopolymer substrate surfaces is attachment of RGD peptides that is the most effective and often employed peptide sequence for stimulating cell adhesion on synthetic polymer surfaces. This peptide sequence can interact with integrin receptors at the focal adhesion points. Once the RGD sequence is recognized by the integrins, it will initiate an integrin-mediated cell attachment pathway and activate signal transduction between the cell and ECM, thus influencing various cell behaviors on the substrate including proliferation, differentiation, survival and migration [38]. Roeker et al. showed that the composite materials modified by immobilizing poly-L-lysine and BMP-2 as bioactive ligands on the ceramic surface had promising potential to enhance the adhesion of hMSCs and directing cell differentiation into osteoblasts [39]. In another study, it was demonstrated that hMSCs encapsulated in poly (ethylene glycol) (PEG)/ RGD hydrogels undergo chondrogenic differentiation in the presence of TGF-β3. More importantly, this effect has been found to be RGD-dose dependent and there is an optimal concentration of RGD present in PEG hydrogels, which improves cell viability and promotes chondrogenesis [40].

In spite of the addition of differentiation factors in the culture media, the matrix materials which support the cells affect the differentiation of stem cells as well. Mauney et al. found that the matrix-denatured collagen type I is more capable in retaining the osteogenic differentiation potential *in vitro* and even bone-forming capacity *in vivo* of hMSCs than the conventional tissue culture plastic [41]. Mwale et al. discerned that bi-axially oriented polypropylene plasma treated in ammonia reduced upregulation of the expression of osteogenic marker genes, such as alkaline phosphatase (ALP), bone sialoprotein and osteocalcin significantly [42]. According to a report presented by Ager et al. [43], collagen I/III and PLLA porous scaffolds showed certain osteoinductive properties without Dex, ascorbic acid, and βGP (DAG) stimulation, verified by immunocytochemical staining against osteoblast-typical markers and completed by calcified matrix detection. Wang et al. demonstrated that ascorbic acid-functionalized poly (methyl methacrylate) can modulate the proliferation and osteogenic differentiation of early and late-passage bone marrow-derived hMSCs [44].

More recently, Xu et al. showed that hMSCs attached, and subsequently proliferated and differentiated toward the osteogenic lineage on the biomimetic bioglass-collagen-hyaluronic acid-phosphatidylserine (BG-COL-HYA-PS) composites to a significantly higher degree compared to those cells on the BG-COL, BG-COL-HYA composites, suggesting the BG-COL-HYA-PS composite porous scaffolds have high potential for bone tissue engineering [45]. In another study, it was shown that the incorporation of gelatin in the poly [(L-lactide)-co-(e-caprolactone)] (PLCL) nano-fibers stimulated the adhesion and osteogenic differentiation of hMSCs, suggesting that the chemical composition of the underlying scaffolds play a key role in regulating the osteogenic differentiation of hMSCs [46].

Regarding chondrogenic differentiation, investigating the effect of cartilage-tissue chondroitin-sulfate (CS) in a fibrin scaffold on the differentiation of adipose-derived adult stem cells into chondrocytes revealed the significant effect of CS on the differentiation efficiency. It can be concluded that the fibrin–CS matrices mimicking native cartilage extracellular matrix could act as a three-dimensional scaffold for cartilage tissue engineering and have the potential for promoting the differentiation of adipose-derived adult stem cells into chondrocytes [47].

Since the chemical properties of substrates (e.g., hydrophobicity) play an important role in the kinetics of protein adsorption and folding, which in turn influence cellular activities, direct the stem cells' fate can be controlled by chemical modification of the sub-

http://dx.doi.org/10.5772/54371

strate. Surface modification techniques such as plasma treatment, ion sputtering, oxidation and corona discharge affect the chemical and physical properties of the polymer surface without significantly changing the bulk material properties. For example, plasma processes makes it possible to change the chemical composition and properties of the polymer system such as hydrophobicity, surface energy, refractive index, hardness, chemical inertness and biocompatibility [48]. Plasma techniques can easily be used to induce the desired groups or chains onto the surface of a polymer [49, 50]. Appropriate selection of the plasma source facilitates the introduction of diverse functional groups on the polymer surface to improve biocompatibility or to allow subsequent covalent immobilization of various bioactive cues. For instance, plasma treatments with oxygen, ammonia, or air can generate carboxyl groups or amine groups on the polymer surface [51, 52]. A variety of ECM protein components such as gelatin, collagen, laminin, and fibronectin could be immobilized onto the plasma-treated surface to enhance cellular functions [53]. Curran et al. show that stem cell differentiation is guided by surface chemistry and energy, independent of inductive media [54]. Although all the surfaces tested maintained cell viability, silanized hydrophobic surfaces with CH_3 end groups (with low surface energy) maintain MSC phenotype, while increasing the surface energy by adding NH_2- or SH- terminal groups promotes osteogenesis. Further increase of surface energy by addition of OH or COOH moieties promotes chondrogenesis. However, there are reports indicating that both hydrophobicity and surface energy play a role in cell adhesion, but only in the short term until cells themselves modulate their extracellular environment [55, 56].

Probably one of the best known soluble reagents is dissolved oxygen. Typical oxygen concentrations *in vivo* vary from 12.5 to 5%, whilst the oxygen concentration in cell culture incubators is the same as that in the air, which is 20%. Several reports show that lowered oxygen concentrations (5%) increase stem cell proliferation [57-59]. Grayson et al. [60] have shown that even lower oxygen concentrations of about 2% increase MSC proliferation whilst maintaining an undifferentiated state, thus suggesting that hypoxic conditions are the characteristic of the niche environment. Some authors have observed an induction of adipose-like phenotype in MSCs in severe hypoxia (1%) [61], whilst others showed that adipogenesis is suppressed at 6% oxygen compared to 20% oxygen [62]. Lennon et al. reported that rat MSCs exposed to 5% oxygen during amplification show enhanced osteogenesis after implantation, compared with cells amplified in 20% which may probably be due to increased proliferation as suggested above [63]. Buckley et al. showed the beneficial response of chondrocyte cells to a low oxygen environment in the absence of TGF-β, suggesting that hypoxia can be used as an alternative to growth factor stimulation to engineer cartilage from culture-expanded chondrocyte [64].

2.2. Structural cues

Biomaterial scaffolds take on a variety of structures based on their material composition and processing for forming 3D environments. These materials consist of natural polymers such as collagen, hyaluronic acid, fibrin, alginate, or synthetic polymers such as polyethylene glycol (PEG), dextran, or polyvinyl alcohol and can be formed into hydrogels, fibrous structures, and microporous scaffolds [65,66]. Figure 1 illustrates examples of the structure of each of these scaffold types. The biomaterial structure controls how a cell interacts with the material and is important in stem cell fate decisions as the presentation of cues and cellular morphology are dependent on this structure.

Hydrogels are comprised of insoluble networks of cross-linked polymers with high water contents [67]. Hydrogels with the ability to encapsulate stem cells have been used for applications such as cartilage [68, 69] and cardiac [70, 71] tissue regeneration. In order to achieve tissue formation, stem cells must either be encapsulated within or recruited to the hydrogel. Some recently reported applications of hydrogel in tissue engineering are presented the following part.

Hydrogels such as those derived from alginate, collagen and hyaluronic acid have been found to be quite promising – they provide a homogeneous, structureless soft 3D environment which is probably ideal for stem cell proliferation and maintenance, as well as for differentiation into softer tissues such as neural or hepatic [72, 73]. Pranga et al. showed the promotion of oriented axonal regrowth in the injured spinal cord by alginate-based anisotropic capillary hydrogels [74]. In a recent study, Nguyen et al. demonstrated that a three-layer polyethylene glycol-based hydrogel creates native-like articular cartilage with spatially-varying mechanical and biochemical properties that can direct a single MSC population to differentiate into the superficial, transitional, or deep zones of articular cartilage. They concluded that spatially-varying biomaterial compositions within single 3D scaffolds can stimulate efficient regeneration of multi-layered complex tissues from a single stem cell population. The ability to generate such zone-specific tissue could eventually allow tissue-engineering of more native-like articular cartilage substitutes with spatially varying ECM composition and mechanical properties [75, 76]. Moreover, injectable hydrogels have been extensively explored as cell delivery systems with the advantage that cells and biomolecules can be readily integrated into the gelling matrix [77, 78]. The injectable nature of the hydrogels provides the attractive feature of facile and homogenous cell distribution within any defect size or shape prior to gelation. In addition, injectable hydrogels allow good physical integration into the defect and facilitating the use of minimally invasive approaches for material delivery [79, 80]. Tan et al. demonstrated the usefulness of the aminated hyaluronic acid-g-poly (N isopropylacrylamide) copolymer as an injectable hydrogel for adipose tissue engineering [81]. Recently, Tan et al. demonstrated that the thermo-sensitive alginate-based injectable hydrogel has attractive properties that make it suitable as cell or pharmaceutical delivery vehicles for a variety of tissue engineering applications [82].

Although hydrogels provide a highly controlled 3D microenvironment for cells, the nature of this scaffold does not entirely mimic the structure of native ECM. Generally the cells encounter and respond to basement membrane topography in the *in vivo* environment mainly composed of networks of pores, ridges, and fibers made by ECM molecules such as collagen, fibronectin and laminin at length scales ranging from nano- to micro-scale [83]. It is therefore important to incorporate features at such length scales into the development of biomaterials suitable for stem cell therapies.

One of the most widely used biomaterial structures for tissue engineering involves microporous scaffolds, which can form interconnected porous networks that allow for cellular infiltration and tissue formation. These scaffolds are often formed with leachable components around which the desired polymer forms a scaffold [84]. Upon removal of the leachable components, a 3D structure can be obtained with varying parameters such as pore size, porosity, and interconnectivity. Aronin et al. created poly-(*e*-caprolactone) scaffolds with varied pore sizes and interconnectivity to monitor osteogenesis of dura mater stem cells [85]. High porosity and adequate pore-size are key requisites to increase the surface area available for cell attachment and tissue in-growth in order to facilitate the uniform distribution of cells and the adequate transport of nutrients. Murphy et al. has investigated the effect of mean pore size on cell behavior in collagen–glycosaminoglycan scaffolds for bone tissue engineering application [86]. The results show that cell number was highest in scaffolds with the largest pore size of 325 μm. While the increased surface area provided by scaffolds with small pores may have a beneficial effect on initial cell adhesion but ultimately the improved cellular infiltration provided by scaffolds with larger pores outweighs this effect and suggests these scaffolds might be optimal for bone tissue repair. Kasten et al. also showed that porosity, distribution and size of the pores of beta-tricalcium phosphate ceramic scaffold can influence protein production and osteogenic differentiation of hMSCs [87]. Tayton et al. have compared the porous and non-porous versions of poly (DL-lactide) for potential clinical use as alternatives to allografts in impaction bone grafting [88]. The results showed that the skeletal stem cells differentiated along the osteoblastic lineage in porous samples compared to the non-porous versions. This feature may result from the fact that the 3D microarchitecture could distribute cellular binding sites in a variety of specific spatial locations rather than on only the single plane of rigid substrate, as in traditional two-dimensional 2D architecture of cell culture plastic or the surface of the non-porous polymers. Cells, therefore, may have cytoskeletal adaptor proteins on a 3D matrix in addition to proteins present in 2D focal adhesions [89, 90]. Such differences in cell adhesion on the porous and non-porous polymers may therefore lead to different signal transduction and subsequent alteration in cellular rearrangement.

Natural ECM consists of various protein fibrils and fibers interwoven within a hydrated network of glycosaminoglycan chains [91]. The nano-scale structure of the ECM offers a natural network of intricate nano-fibers to support cells and present an instructive background to guide their behavior [92-94]. Each nano-fiber provides the way for cells to form tissues as complex as bone, liver, heart, and kidney. Researchers try to fabricate fibers to mimic the natural ECM as a support for cell growth. The proliferation and osteogenic differentiation of MSCs was investigated in 3D non-woven fabrics prepared from polyethylene terephthalate (PET) microfiber by Takahashi et al. They showed that the attachment, proliferation and bone differentiation of MSCs were influenced by the fiber diameter and porosity of non-woven fabrics in the scaffolds [95]. Several reports have demonstrated that nano-fibers are more favorable than micro-fibers, suggesting that cell activities can further be regulated by the size of the fiber [96-98] in terms of the biological response of chondrocytes, NSCs and endothelial cells cultured on nanofibrous and microfibrous scaffolds. Although the mechanisms by which a nano-fibrous scaffold acts as a selective substrate are not known yet, it is clear that the enhanced adsorption of cell adhesion matrix molecules enhances cell adhesion. Xin et al. also confirmed that PLGA nano-fibers accommodate the survival and proliferation of human MSCs. hMSCs, as well as hMSC-derived chondrogenic and osteogenic cells, apparently attach to PLGA nano-fibers, and yet assume different morphological features [99]. These results demonstrate the full support of multi-lineage differentiation of MSCs within nano-fibrous scaffolds and the feasibility of multi-phasic tissue engineering constructs using a single cell source, which is of particular relevance to the development of multi-phasic tissue constructs. However, there are very few in-depth studies on nano-fiber topographical effects on stem cell differentiation. Other nano-scaled topographical features such as steps, grooves, pillars and pits also modulate cell behavior, as reviewed elsewhere [100].

Currently, there are three techniques available for the synthesis of nano-fibers: electrospinning, self-assembly, and phase separation. In particular, electrospinning technique is the most widely studied technique which has attracted wide attention due to its applicability for a variety of synthetic and natural polymers, exhibiting the most promising results for tissue engineering applications. Electrospinning is a spinning method to generate submicron to nanometer scale fibers from polymer melts or solutions. It is a physical process to obtain fibers from a bulk polymer of interest under the applied electric field. The most commonly used polymers for nano-fiber fabrication using electrospinning are the aliphatic polyesters [101]. There are several reports describing the potential of nanofibers fabricated by electrospinning method for neural [102-104], bone [105-108] and cartilage [109, 110] tissue engineering which mimic the native tissue environment and support the cell adhesion, proliferation and differentiation.

Nano-fibers hold great promise as potential scaffolds owing to their high porosity and high surface area-to-volume ratio, which are favorable parameters for cell attachment, growth, and proliferation in addition to possessing favorable mechanical properties [111]. Furthermore the effect of nano-fibers for stem cells' differentiation is promising further applications of nano-fibers for tissue engineering. Stem cells can be induced to differentiate into different cell types by growth/differentiation factors in the media, and we can incorporate such biomolecules into the nano-fibers to direct differentiation to a desired cell type. The biomimetic morphology of nano-fibers with different patterns may also help to direct the stem cells' differentiation, which is particularly attractive given differentiation induction by some of medium supplements, although successful, is not physiologically relevant and offers the possibility for development of improved clinical prostheses with topographies that can directly modulate stem cell fate.

2.3. Mechanical cues

Importantly, the various tissues of the body exhibit a range of matrix stiffness, and such differences in substrate stiffness have long been known to influence cell fate decisions in differentiated cell types [112]. An emerging area of study in stem cell biology and engineering is investigation of the role of these mechanical cues in stem cell fate decisions. Because MSCs can differentiate *in vitro* into cell types from tissues ranging from muscle, bone, and potentially brain, it can be hypothesized that the mechanical cues provided by the ECM are particularly instructive in lineage specification. The study carried out by Engler et al. revealed that matrix elasticity influences differentiation of hMSCs into osteogenic, myogenic, and neurogenic cells [113]. Softer gels (0.1–1 kPa) were neurogenic, the hardest (24–40 kPa) were osteogenic, and the gels with intermediate elastic moduli (8–17 kPa) were myogenic. In all three cases, the elastic modulus matches that of the corresponding native tissue. It has recently been found out that substrate stiffness collaborates with soluble medium conditions to regulate the proliferation and differentiation of adult NSCs [114]. Cells exhibit optimum proliferation (in FGF-2) and optimum neuronal differentiation (in retinoic acid) at an intermediate stiffness that is characteristic of brain tissue. Furthermore, under conditions that induce nonspecific cell differentiation, stiff substrates support the differentiation of GFAP-expressing astrocytes, whereas soft substrates preferentially support the differentiation of β-tubulin III expressing neurons. This research demonstrates how the mechanical and biochemical properties of an adult NSCs microenvironment can be tuned to regulate the self-renewal and differentiation of adult NSCs. In another study, Leipzig et al. demonstrated that an optimal stiffness exists for both proliferation (3.5 kPa) as well as differentiation of neural stem/progenitor cell to neurons (<1 kPa) [115].

The study conducted by Banerjee et al. [116] provided insights into the influence of the mechanical properties of 3D alginate hydrogel scaffolds on the proliferation and differentiation of NSCs, where varying the concentrations of alginate and calcium chloride provided facile control over the elastic modulus of the hydrogels. They demonstrated that the properties of the 3D scaffolds significantly impacted both the proliferation and the neuronal differentiation of encapsulated NSCs. In addition, they observed the greatest enhancement in expression of the neuronal marker β-tubulin III within hydrogels having an elastic modulus comparable to that of brain tissues. They noted that the optimal value of the elastic modulus might depend on the stem cell type and the lineage to which differentiation is being directed. Wang et al. reported an injectable hydrogel scaffold composed of gelatin-hydroxyphenylpropionic acid conjugate system with tunable stiffness for controlling the proliferation rate and differentiation of hMSCs in a 3D context in normal growth media. The rate of hMSC proliferation increased with the decrease in the stiffness of the hydrogel. Also, the neurogenesis of hMSCs was controlled by the hydrogel stiffness in a 3D context without the use of any additional biochemical signal. These cells which were cultured for 3 weeks in hydrogels with lower stiffness expressed much more neuronal protein markers compared to those cultured in stiffer hydrogels for the same period of time [117]. In another study, lower cross-linked matrix of hydrogel system comprising hyaluronic acid-tyramine conjugates enhanced chondrogenesis with increases in the percentage of cells with chondrocytic morphology, biosynthetic rates of glycosaminoglycan and type II collagen, and hyaline cartilage tissue formation. By increasing cross-linking degree and matrix stiffness, a shift in MSC differentiation toward fibrous phenotypes with the formation of fibrocartilage and fibrous tissues was observed [118]. In general, the ability to control stem cell fate – possibly without the use of chemical inducers – would be broadly useful for applications in regenerative medicine and tissue engineering [116].

Except mechanical properties of the matrix, the external mechanical stimulus can also induce stem cell differentiation. Bioreactors provide various active environments for stem cell growth under specific mechanical conditions. Flow perfusion culture of scaffold/cell constructs has been witnessed to enhance the osteoblastic differentiation of rat MSCs over static culture in the presence of osteogenic supplements such as Dex. Although Dex is known to be a powerful induction agent of osteogenic differentiation in MSCs, Holtorf et al. showed that the mechanical shear force caused by fluid flow in a flow perfusion bioreactor would be sufficient to induce osteoblast differentiation in the absence of Dex [119]. Flow perfusion also accelerates the proliferation and differentiation of rat MSCs seeded on non-woven PLLA microfibrous scaffolds toward the osteoblastic phenotype, and improves the distribution of the calcified extracellular matrix generated *in vitro* [120]. Li et al. reported that MSCs are also mechano-sensitive and that Ca^{2+} may play a role in the signaling pathway since MSCs subjected to oscillatory fluid flow exhibited increased intracellular Ca^{2+} mobilization [121]. More recently, studies have shown that *shear stress* can induce *differentiation* of *stem cells* toward both endothelial and bone-producing cell phenotypes. The current data supporting the role of shear stress in stem cell fate and potential mechanisms and signaling cascades for transducing shear stress into a biological signal are reviewed elsewhere [122].

In another study, it was shown that the cyclic compressive loading alone will induce chondrogenic differentiation as effectively as the TGF-β alone or TGF-β plus loading in short term culture. Regarding MSCs angiogenesis, DNA microarray experiments [123] showed that uniaxial strain increased smooth muscle cell (SMC) markers. But cyclic equiaxial strain downregulated SM α-actin and SM-22α in MSCs on collagen- or elastin-coated membranes after 1 day, and decreased α-actin in stress fibers. This result suggests that uniaxial strain, which better mimics the type of mechanical strain experienced by SMCs, may promote MSCs differentiation into SMCs if cell orientation can be controlled. Solvig Diederichs et al. applied singular and repetitive cyclic strain of short- and long-time strains [124]. Additionally, a gradually increasing strain scheme commencing with short-time strain and continuing elongated strain periods was applied. Adipose tissue–derived MSCs on planar silicone and a three-dimensionally structured collagen I mesh were exposed to these strain regimes. The results revealed that even short-time strain can enhance osteogenic differentiation. Elongation and repetition of strain, however, resulted in a decline of the observed short-time strain effects, which was interpreted as positively induced cellular adaptation to the mechanically active surroundings. With regard to cellular adaptation, the gradually increasing strain scheme was especially advantageous.

Taken together, these results suggest that the design of *ex vivo* stem cell culture systems should consider all types of mechanical cues in the microenvironment including matrix stiffness, compressive loading and shear stress as factors in guiding proper lineage specification.

2.4. Electrical stimulus and other cues

Several studies have recently shown the response of NSCs to electric fields. The studies reported by Matos et al. showed the response of murine NSCs encapsulated in alginate hydrogel beads to alternating current electric fields [125]. They found an enhanced propensity for astrocyte differentiation over neuronal differentiation in the 1 Hz cultures. In another study, Park et al. discovered the enhanced neuronal differentiation of hNSCs on graphene, which had a good electrical coupling with the differentiated neurons for electrical stimulation [126]. The application of an electrical stimulus causes fibroblasts to change cell shape and reorient in the 3D collagen scaffold perpendicularly to the direction of electrical stimulus, while the same electrical stimulus applied to MSCs induces much less significant reorientation. A stimulus as strong as 10 V/cm is needed to induce a δV of 50 mV or greater, which would be sufficient to activate voltage-gated Ca^{2+} channels and regulate Ca^{2+}-dependent sub-cellular processes, including cytoskeletal reorganization that is likely to cause changes in the cell morphology and reorientation signaling pathways [127]. It needs to be identified as to whether the differentiation of stem cell following adhesion will change under electrical stimulus. Endothelial progenitor cells and muscle precursor cells can also be stimulated by electromagnetic fields to promote myocyte differentiation [128,129]. Interestingly, electrical stimulation (10–40 V, 5 ms, 0.5 Hz pulses) of human embryonic fibroblasts was found to cause loss of cell proliferation and cell number but also led to differentiation of fibroblasts into multinucleated myotube-like structures [130].

Ultrasound has also been shown to induce differentiation. In low-intensity ultrasound field studies, MSCs differentiate towards a chondrocytic phenotype [131]. In one study, Abramovitch-Gottlib L et al. have illustrated that the use of low level laser irradiation (~0.5 mW/cm^2) applied to a MSC/coralline construct stimulates the proliferation and differentiation of MSC into an osteoblastic phenotype during the initial culture period and significantly induced *in vitro* osteogenesis over time [132]. Thus, low level laser irradiation quickens the differentiation of MSC into an osteoblastic phenotype during bone formation processes in early culture periods.

Numerous recent papers have sprouted showing how even minor experimental modifications can change cell phenotype. Indeed, stem cells are so sensitive and unstable that even cell seeding density and seeding protocol have been observed to influence cell shape and gene expression [133].

3. Some novel trends emerging in the field of tissue engineering

In the following part we will introduce some novel trends emerging in the field of tissue engineering in terms of both cellular biology (cell reprogramming) and biomaterial (multifactorial design strategies) point of view in order to improve the overall efficiency of tissue regeneration.

3.1. Cell reprogramming

Though all somatic cells of the human body have the same genome structure, differences in chromatin organization and expression pattern of genes lead to the formation of various types of cells with different physiology, function and morphology [134,135]. Therefore, one could speculate that by changing chromatin structure and pattern of gene expression, all cells can be converted to other cell types [136]. The first cell reprogramming report has been presented in an earlier report [137] in which fibroblast cells converted into myocyte through the overexpression of MyoD gene. In a later study, the nucleus of the fibroblast cell has been transferred to the enucleated oocytes which finally led to the birth of Dolly sheep [135]. Yamanaka (2006) shed some light on the biology underlying cell differentiation and cell fate by converting the mouse fibroblast to iPS cells in his study; one year later, Yamanaka and Thompson [138-140] reported the generation of human iPS cells from fibroblast cells.

The possibility of directing lineage specific reprogramming of cells opens a window to a vast range of new possibilities in tissue engineering and regenerative medicines [141]. Herein, generation of iPS cell lines is an important issue in the way to derive pluripotent cells from somatic cells. Instability of the genome, high cost of culture, lack of an efficient protocol for differentiation as well as the presence of tumorigenic potential upon transplantation are among the main reasons for the slow progress of its clinical application [142].

Differentiation of stem cells into different types of tissue or organ is still a major limiting factor in the area of tissue engineering mainly due to the complexity and multicellular structure of the tissues and organs. To overcome such a limitation, it is highly demanded to have different types of cells for tissue engineering which is considered to be as important as mimicking the physiological condition *in vivo*. Self-renewing and pluripotency are unique properties of pluripotent stem cells that make the embryonic developmental process possible for the complex and integrated tissue-engineered systems. Accordingly, to make complex and integrated tissues, intrinsic developmental programs of inner cell mass of blastocysts such as those of post gastrulation events can be followed. Eiraku et al. [143] in a recent study managed to recreate the 3D structure of an organ for the first time in the world. They succeeded in growing a structure like the optic cup with the six cell types present in normal retina tissue. They mimicked aggregation and self-induction of mESCs as embryoid body and neurosphere formation to make optic cup that can be the source of retinal neurons like embryonic process of eye formation. For this, they used genetic engineered mES with tissue specific reporter RX-venues DNA construct for capturing the early stages of optic cup-cell mass formation and their separation for more maturation. Scientists hope to begin applying the same technologies used for retinal tissue to make 3D structure of other organs such as the brain, lung and kidney. However, despite advances like these, it is quick to note that we can determine as to whether pluripotent stem cells can be used for regenerative therapy. The best idea is not always to uprise the cells to the tip of potency pyramid and then downrise it to a low level with differentiation, whereas one can directly convert one cell type to another [143]. It has been shown that the fibroblast cells can be converted to myocyte, neuron, hepatocyte, cardiomyocyte simply with direct reprogramming [137, 144-146]. This provides us good tools for having wide ranges of cells for regenerative medicines [147]. New approaches to cell reprogramming such as direct reprogramming of somatic cells to tissue-specific stem cells and conversion of fibroblast to neural stem cells have been proposed [148]. Providing three types of cells, namely astrocyte, oligodendrocyte and neuron, which are required in neural systems, is the advantage of cell reprogramming [148]. Another advantage of using direct reprogramming to tissue specific stem cells instead of reprogramming to full matured cells is that all types of cells which are necessary for the regenerating of that specific tissue will be provided in the former approach. For instance, it has recently been well demonstrated that convection of fibroblast cells to NSC is more promising than the conversion of the same cells to the neuron [149]. Moreover, adult stem cell generation through direct reprogramming has more capacity for self-renewal, which can be expanded and stored for different clinical applications. Tissue specific adult stem cells are natural stem cells of any tissue and match the normal homing tissue [149] and can respond to niche messages under both stress and damage condition.

Human body is a complex system that works with many regulatory and check points in coordination with many flexible programs. Using direct reprogramming, progression in regenerative therapy will be possible if all demanding material such as adult stem cells, ES, iPS are well prepared in a suitable place and appropriate manner.

3.2. Multifactorial design strategies

In contrast to elements of living systems' ECM, the designed scaffolds are very poor in information, which make them suboptimal for many tissue engineering applications. These passive biomaterials are unlikely to guide cell migration and differentiation or controlled matrix deposition, a problem that becomes even more evident in complex tissues with more than one cell type. Furthermore, they also cannot induce tissue neo-formation while preventing other undesirable tissue repair processes such as scarring; they are also unable to promote functional tissue integrations, such as vascular and/or nervous connectivity, in the host. Finally, these passive scaffolds largely lack the capacity to induce cell differentiation, thus resulting in a major limitation for their use together with current stem cell-based therapies [150]. A promising strategy to overcome these limitations is to consider the *multi-factorial design strategies* by combining various external cues with one another for efficient and controlled formation of complex tissues.

3.2.1. Combining structural and biological cues for scaffold bioactivation

While combining the structural and biological cues, a bioactive scaffold can be constructed in which biological functionality has been integrated to provide an information-rich support material for tissue engineering. Bioactive scaffolds are designed to control cell and tissue responses, and to provide a more efficient integration with the host. Indeed, bioactive scaffolds can also be prepared from synthetic materials by physical adsorption or chemical immobilization of biomolecules or oligopeptides on the scaffold surface, or by physical entrapment of bioactive molecules alone or incorporated in a drug delivery system into the scaffold. These strategies can also be applied to enhance the bioactivity of scaffolds made from ECM-native materials.

Engineered tissues need not only to remedy a defect and to integrate into a host tissue, but they also need to meet the demands of a constantly changing tissue. It was hypothesized that those tissues capable of growing with time could be engineered by supplying growth stimulus signals to cells from the biomaterials used for cell transplantation [151]. Smart drug delivery system is able to transmit multiple signals to the cells in a timely controlled release pattern. This release may be controlled through properties of the drug delivery system itself such as biodegradation-controlled release devices or stimulisensitive systems. Polymeric materials can be used as tissue-engineering *scaffolds and drug release* carriers, a strategy that has been mainly used for soluble signaling molecules such as growth factors. Cell recruitment and migration to the site of injury may be promoted through various signaling molecules. Many of these factors, e.g. TGF-βs, BMPs and IGF-1, are not only involved in cell attraction but also affect stem cell proliferation and differentiation [152-155].

Drug delivery strategies are designed to provide a platform for the localized delivery of the growth factors at the site of implantation. This is to protect the bioactivity of the molecule, to provide a controlled release pattern of the drug over a desired time frame, and *deliver angiogenic factors so as to promote angiogenesis*.

Two approaches have been mainly used for scaffold bioactivation: growth factors can be encapsulated in a selected drug delivery system such as a microsphere or nanoparticle formulation, and these can be incorporated into the scaffolds. Otherwise, growth factors can be incorporated directly into the scaffold itself [156-158]. For example, IGF-1 has been directly incorporated into porous 3D silk fibroin scaffolds [159]. Silk scaffolds incorporating IGF-1 were able to preserve growth factor bioactivity, and prompted chondrogenic stimuli to seeded MSCs *in vitro*. By definition, implantation of growth factor-loaded scaffolds results in the localized delivery of the signaling molecule. Still, a certain fraction of the incorporated drug can reach the lymphatics or the circulation, and then distribute to non-target tissues. Therefore, even for these localized therapies, potential adverse effects of growth factor need to be carefully monitored.

Silk fibroin nano-fibrous scaffolds containing BMP-2 and/or nanoparticles of hydroxyapatite which were prepared via electrospinning were selected as matrix for *in vitro* bone formation from human bone marrow derived hMSCs. Li et al. [160] reported that silk fibroin nano-fibrous scaffolds with BMP-2 supported higher calcium deposition and enhanced transcript levels of bone-specific markers in comparison with controls without BMP-2, suggesting that nano-fibrous electrospun silk scaffolds can be an efficient delivery system for BMP-2. The mild aqueous process required for electrospinning, offers an important option for delivery of labile cytokines and other biomolecules. Lee et al. reported that calcium phosphate cement (CPC(combined with alginate solution to form a porous scaffold showed the capability to safely load biological proteins (BSA and lysozyme) during preparation and to release them *in vitro* for over a month [161]. CPC–alginate scaffolds can further be developed into tissue engineered constructs which deliver biological molecules for bone regeneration stimulation.

In case of building biofunctionality into electrospun nano-fibers for neural tissue engineering, the challenge to produce nano-fibers with more bioactive surfaces, significantly improving specific targeting of cell substrate interactions and consequently creating a more biomimetic microenvironment for implanted cells remains. There are several methods, such as polymer blending and surface biofunctionalization, for improvement of nano-fibrous scaffolds bioactivity for nerve tissue engineering which are reviewed elsewhere [162]. It is possible to fabricate electrospun scaffolds from blends of synthetic and natural polymers,

http://dx.doi.org/10.5772/54371

which will then have improved cell substrate interactions. The orientation of neurites from chick embryonic dorsal root ganglia is enhanced on aligned blended polycaprolactone/collagen (PCL/collagen) (72:25) nano-fibers compared with that on aligned, pure PCL [163]. The migration and proliferation of Schwann cells is also significantly improved on aligned PCL/ collagen nano-fibers, indicating more specific biomolecular interactions between cells and the collagen polymers on the nano-fiber surface [164].

Instead of direct electrospinning the naturally derived polymers such as collagen together with synthetic polymers to provide biomemitic nano-fibrous scaffolds, one can immobilize some specific peptide motifs derived from ECM protein, which have been discerned to play an important role in tissue regeneration to the synthetic nano-fiber surface, which provides an alternative method to render the fibers bioactive. For instance, immobilization of molecules, such as specific peptide motifs derived from fibronectin and collagen VI, to the synthetic nano-fiber surface provides an alternative method to render the fibers bioactive. Therefore, surface immobilization of these small molecules that are neuroactive can provide a great advantage for neural tissue engineering. In addition, immobilized growth factors such as brain-derived neurotrophic factor [165] and basic fibroblast growth factor [166] can also promote cell survival and neurite outgrowth.

3.2.2. Combining structural and mechanical cues for engineering large-scale and/or complex tissues

The successful replacement of large-scale defects using tissue-engineering approaches will likely require composite biomaterial scaffolds that have biomimetic structural and mechanical properties and can provide cell-instructive cues to control the growth and differentiation of embedded stem or progenitor cells.

The depth-dependent composition and structure of articular cartilage gives rise to its complex, non-homogeneous mechanical properties. Articular cartilage is generally composed of chondrocytes and a dense ECM, which mainly includes type II collagen and proteoglycans [167]. Articular cartilage is structurally comprised of four different layers that can be distinguished from one another by collagen fiber alignment and proteoglycan composition. The depth-dependent alignment of collagen leads to important tensile and shear properties, whereas the depth-dependent proteoglycan content contributes more to the compressive properties of each zone [168, 169]. Nguyen et al. demonstrated in a recent study that layer-by-layer organization of specific biomaterial compositions creates 3D niches that allow a single MSC population to differentiate into zone-specific chondrocytes and organize into a complex tissue structure [75]. The results indicated that a three-layer polyethylene glycol (PEG)-based hydrogel with chondroitin sulfate (CS) and matrix metalloproteinase-sensitive peptides (MMP-pep) incorporated into the top layer (superficial zone, PEG:CS:MMP-pep), CS incorporated into the middle layer (transitional zone, PEG:CS) and hyaluronic acid incorporated into the bottom layer (deep zone, PEG:HA) which ultimately created native-like articular cartilage with spatially-varying mechanical and biochemical properties. They concluded that spatially-varying biomaterial compositions within single 3D scaffolds can stimulate efficient regeneration of multi-layered complex tissues from a single stem cell population.

In another study, the potency of scaffold stiffness and topology in driving cardiac stem cell differentiation in a 3D culture context was confirmed by Forte et al. [170]. Cardiac stem cells adopted the cardiomyocytic phenotype only when cultured in strictly controlled conditions characterized by a critical combination of chemical, biochemical, structural and mechanical factors, and emulation of the inner myocardial environment. In these studies, the mimicry of myocardial environment was achieved by fine-tuning the array of growth factors dissolved in the culture medium and the chemistry, topology and stiffness of three-dimensional supports on which stem cells were seeded. Scaffold stiffness was modulated in this study by changing the topology of the structure using a rapid prototyping technique. The optimal stiffness to induce cardiomyocyte differentiation was around 300 kPa on the scaffolds with square pores of about 150 μm.

4. 2D Polysaccharide-based hydrogel scaffolds for muscle tissue engineering

Hydrogels have been used for a variety of biomedical applications [171-175], and because of their viscoelastic characteristics [176], similarities with ECM, excellent biological performance, inherent cellular interaction capability [177], ability to allow transfer of gases and nutrients [177], and their amiability of fabrication into specific shapes, they have recently been explored as scaffolding materials for tissue engineering applications [178-180]. On the other hand, in the recent decade, researchers realized that the mechanical properties of the used hydrogel material had to be adapted to the elastic properties of the damaged tissue [181]. Hydrogels such as alginate, chitosan, collagen and hyaluronic acid, which are derived from natural polymers, have been proved to be quite promising for stem cell proliferation, maintenance and differentiation for tissue engineering applications.

The authors of this paper tried to prepare hydrogels made of natural polymers (chitosan (CS) and gelatin (G)) with proper handling for surgery, and with mechanical properties similar to those of muscle tissues as well as good cell adhesion properties. In the current study, we investigated the effect of CS and G concentration in blend scaffolds on mechanical properties of the CS-G hydrogel sheets as well as the seeded muscle-derived stem cells (MDSCs) and smooth muscle cells' (SMCs) behavior on the CS-G hydrogel sheets. MDSCs and SMCs were isolated, expanded in culture and characterized with respect to the expression of surface markers with flow cytometry analysis. After crosslinking of CS and G, the CS-G blend hydrogel sheets were prepared by a casting method and used for 2D cell culture.

While the elasticity of the CS-G hydrogel sheets increased by increasing the CS concentration, the gelatin concentration did not have any notable effect on the hydrogel mechanical properties.

The MDSCs attachment on the surface with elastic modulus of 25 kPa stiffness and proliferation on different CS-G hydrogel sheet surfaces having varying modulus of elasticity is shown in Figure 2. The cell observation result on day 1 showed that by increasing the

http://dx.doi.org/10.5772/54371

elasticity of hydrogel sheets, most of the cells on the hydrogel surfaces with high elasticity (E=100 kPa, CS=4.5% w/v) didn't fully expand on the hydrogel surface, while the cells on the hydrogel surfaces with low and intermediate elasticity (E=15 kPa, CS=1.5 % w/v; E=25 kPa, CS=3% w/v) had more spindle shape (data not presented). Gelatin concentration was fixed (18% w/v) for all the samples. The greatest proliferation of the cells was found on the hydrogels with intermediate elasticity (25 kPa) and the number of cells increased over time during the 7-day culture (Figure2). Hydrogel blends with lower or higher gelatin concentration showed significantly lower attached cell numbers (data not presented). Recent studies have illustrated the profound *dependence of cellular behavior* on *the stiffness* of 2D hydrogel sheets. Boontheekul et al. demonstrated that alginate gel with higher mechanical strength (increasing from 13 kPa to 45 kPa) increased myoblast adhesion, proliferation, and differentiation in a 2D cell culture model [182]. They also showed that primary mouse myoblasts were more highly responsive to this cue than the C2C12 myoblast cell line.

An innovative approach has recently been described by Gilbert et al. as well. Using a bioengineered substrate in conjunction with a highly automated single-cell tracking algorithm, the authors showed that substrate elasticity is a potent regulator of muscle stem cells' fate in culture. In fact, muscle stem cells cultured on soft hydrogel substrates that mimic the elasticity of muscle self-renew *in vitro*, and contribute extensively to muscle regeneration when subsequently transplanted into mice. This study has provided novel evidence showing that recapitulating physiological tissue rigidity allows the propagation of adult muscle stem cells [183].

In the current study, the authors investigated the behavior of MDSCs and SMCs cultured on the prepared hydrogel surfaces. The results indicate that increasing the hydrogel mechanical strength from E=15 kPa to E=25 kPa, increases MDSCs adhesion and proliferation. The authors further found that MDSCs were more responsive to mechanical properties of the hydrogel sheets compared to SMCs, due to their higher ability and relatively smaller size (Data not presented). In contrast, for engineering central nervous system tissue, Leipzig et al. demonstrated that gels with lower mechanical properties of methacrylamide chitosan hydrogel sheet (E $\leq$3.5 kPa) were more appropriate for neural stem progenitor cell differentiation and proliferation [115]. As mentioned above, mechanical properties of hydrogel can regulate the cell adhesion, proliferation, and differentiation. However, the response and sensitivity to this variable is highly dependent on the cell source. In the current work, MDSCs exhibited maximal proliferation on hydrogel surface with 25 kPa elasticity. The same hydrogel sheet showed also the best handling qualities for surgery, with elasticity in the range of elastic modulus for muscle tissues [184], showing its potential for being used in muscle tissue engineering applications.

The strategy applied in the current study provides an opportunity to independently control mechanical and bioadhesive properties of the hydrogels so as to probe stem cell behavior. By changing both material mechanical and biochemical properties of the hydrogel blend, we could find the optimum condition for MDSCs attachment and proliferation in contact with CS-G hydrogel sheets.

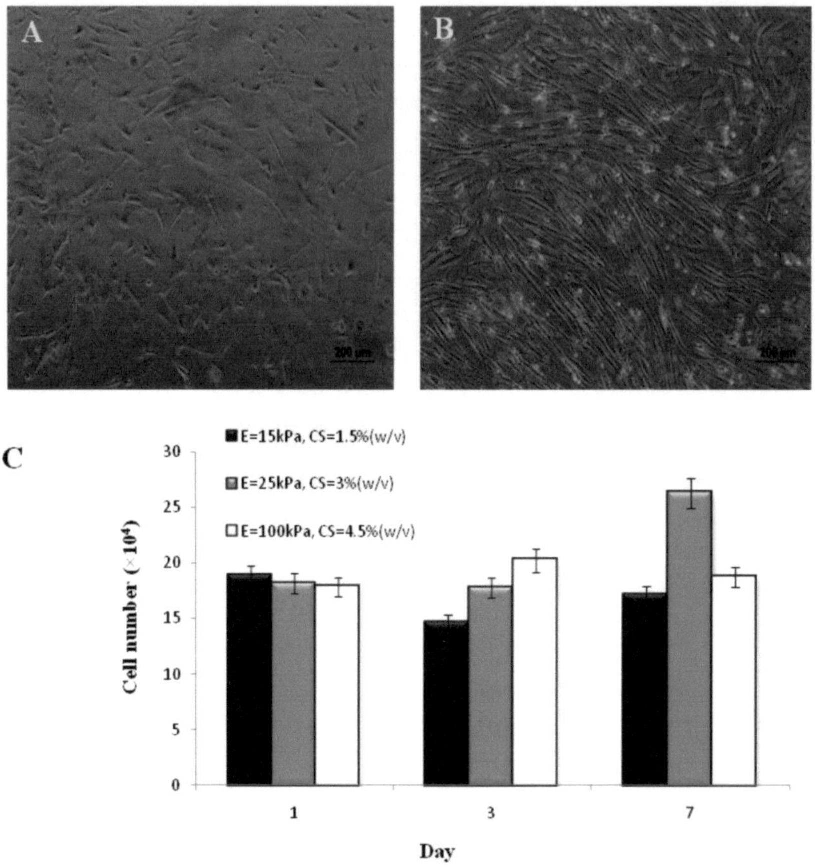

Figure 2. MDSCs adhesion and proliferation on CS-G hydrogel surfaces. Photomicrographs of MDSCs attachment on the surface with intermediate elasticity (25kPa, CS=3 % w/v) at: (A) day 1, (B) day 7 and (C) cell proliferation on CS-G hydrogel surfaces with different mechanical strength. CS-G hydrogel sheets prepared at different chitosan concentration (4.5, 3 & 1.5 % w/v) with constant gelatin (18 %w/v). MDSCs were seeded onto all hydrogel surfaces at the density of 7500cells/cm^2.

5. Conclusion and outlook for the future

In tissue engineering, directing the cells to differentiate at the right time, in the right place, and into the right phenotype, requires an environment providing the same factors that govern cellular processes *in vivo*. The current chapter described various biomaterials and external cues designing considerations mimicking the natural stem cell microenvironment in order to direct the desired stem cell fate, facilitating the regeneration of desired tissues. In addition we introduced our approach to designing a 2D polysaccharide-based hydrogel scaffolds as a potential and suitable biomaterial for muscle tissue engineering applications.

Overall, this chapter provides an overview of recent progresses made by application of novel engineering strategies that have been developed to emulate the stem cell niche for effectively controlling the cell fate and translating the stem cell research into much needed clinical applications in the not-too-distant future.

Future directions in tissue engineering will involve elucidation of molecular mechanisms by which all types of external cues influence stem cells' behavior, followed by translation of these scientific data to clinical applications. Further advances in controlling stem cell fate can be achieved by combining the above mentioned parameters in a more scalable and combinatorial manner to address the complexity of the natural stem cell niche. To this end, collaborative efforts between cell biologists and materials scientists are critical for answering the key biological questions and promoting interdisciplinary stem-cell researches in the direction of clinical relevance.

Author details

Shohreh Mashayekhan[1*], Maryam Hajiabbas[1] and Ali Fallah[2,3]

*Address all correspondence to: mashayekhan@sharif.edu

1 Department of Chemical & Petroleum Engineering, Sharif University of Technology, Tehran, Iran

2 Molecular Medicine Group, Faculty of Medicine, Shahid Beheshti University of Medical Sciences, Tehran, Iran

3 Maad Systems Biomedicine, Tehran, Iran

References

[1] Griffith LG, Naughton G. Tissue engineering-current challenges and expanding opportunities. Science 2002;295(5557):1009-1014.

[2] Khademhosseini A, Langer R, Borenstein J, Vacanti JP. Microscale technologies for tissue engineering and biology. Proc. Natl. Acad. Sci. USA 2006;103:2480-2487.

[3] Zandstra PW, Nagy A. Stem cell bioengineering. Annu. Rev. Biomed. Eng. 2001;3;275-305.

[4] Oh S, Brammer KS, Li JYS, Teng D, Engler AJ, Chien S, Jin S. Stem cell fate dictated solely by altered nanotube dimension. Proc. Natl. Acad. Sci. USA 2009;106:2130-2135.

[5] Engler AJ, Sen S, Sweeney HL, Discher DE. Matrix elasticity directs stem cell lineage specification. Cell 2006;126:677-689.

[6] Kilian KA, Bugarija B, Lahn BT, Mrksich M. Geometric cues for directing the differentiation of mesenchymal stem cells. Proc. Natl. Acad. Sci.USA 2010;107:4872-7487.

[7] Solanki A, Shah S, Memoli KA, Park SY, Hong S, Lee K-B. Controlling differentiation of neural stem cells using extracellular matrix protein patterns. Small 2010;6:2509-2513.

[8] Moore KA, Lemischka IR. Stem cells and their niches. Science 2006;311:1880-1885.

[9] Place ES, Evans ND, Stevens MM. Complexity in biomaterials for tissue engineering. Nat. Mater. 2009;8:457-470.

[10] Stevens MM, George JH. Exploring and engineering the cell surface interface. Science 2005;310: 1135-8.

[11] Gefen A, Gefen N, Zhu Q, Raghupathi R, Margulies SS. Age-dependent changes in material properties of the brain and braincase of the rat. J. Neurotrauma 2003;20:1163-1177.

[12] Engler AJ, Griffin MA, Sen S, Bönnemann CG, Sweeney HL, Discher DE. Myotubes differentiate optimally on substrates with tissue-like stiffness: Pathological implications for soft or stiff microenvironments. J. Cell Biol. 2004;166:877-887.

[13] Mitchell GF, Parise H, Benjamin EJ, Larson MG, Keyes MJ, Vita JA, Vasan RS, Levy D. Changes in arterial stiffness and wave reflection with advancing age in healthy men and women: The Framingham Heart Study. Hypertension 2004;43:1239-1245.

[14] Pittenger MF, Mackay AM, Beck SC, Jaiswal RK, Douglas R, Mosca JD, Moorman MA, Simonetti DW, Craig S, Marshak DR. Multilineage potential of adult human mesenchymal stem cells. Science 1999;284:143-147.

[15] Ding S, Schultz PG. A role for chemistry in stem cell biology. Nat. Biotechnol. 2004;22:833-840.

[16] Hwang NS, Varghese S, Elisseeff J. Controlled differentiation of stem cells. Adv. Drug Delivery Rev. 2008;60:199-214.

[17] Park TH, Shuler ML. Integration of cell culture and microfabrication technology. Biotechnol. Prog. 2003;19:243-253.

[18] Bettinger CJ, Langer R, Borenstein JT. Engineering substrate topography at the micro- and nanoscale to control cell function. Angew. Chem. Int. Ed. 2009;48:5406-5415.

[19] Pollard TD, Earnshaw WC, Lippincott-Schwartz J. Cell Biology. 2nd ed. Philadelphia,PA,USA: Elsevier Inc.; 2008.

[20] Lutolf MP, Doyonnas R, Havenstrite K, Koleckar K, Blau HM. Perturbation of single-hematopoietic stem cell fates in artificial niches. Integr Biol (Camb) 2009;1:59–69.

[21] Dickinson LE, Kusuma S, Gerecht S. Reconstructing the differentiation niche ofembryonic stem cells using biomaterials. Macromol Biosci 2011;1:36–49.

[22] Peerani R, Zandstra PW. Enabling stem cell therapies through synthetic stem cell-nicheengineering. J Clin Invest 2010;120:60–70.

[23] Nardo PD, Minieri M, Ahluwalia A. Engineering the Stem Cell Niche and the Differentiative Micro- and Macroenvironment: Technologies and Tools for Applying Biochemical, Physical and Structural Stimuli and Their Effects on Stem Cells. Stem Cell Engineering 2011;41-59.

[24] Liao S, Chan CK, Ramakrishna S. Stem cells and biomimetic materials strategies for tissue engineering, Materials Science and Engineering C 2008;28(8):1189–1202.

[25] Lowry WE, Richter L. Signaling in adult stem cells. Front Biosci. 2007;12:3911–3927.

[26] Molofsky AV, Pardal R, Morrison SJ. Diverse mechanisms regulate stem cell self-renewal. Curr Op in Cell Biol 2004;16:700–707.

[27] Rider CC. Heparin/heparan sulphate binding in the TGF-beta cytokine superfamily. Biochem Soc Trans 2006;34:458–460.

[28] Li Y, Zhang H, Litingtung Y, Chiang C. Cholesterol modification restricts the spread of Shh gradient in the limb bud. Proc Natl Acad Sci U S A 2006;103:6548–6553.

[29] Saha K, Schaffer DV. Signal dynamics in Sonic hedgehog tissue patterning. Development 2006;133:889–900.

[30] Kuhl PR, Griffith-Cima LG. Tethered epidermal growth factor as a paradigm for growth factor-induced stimulation from the solid phase. Nat Med 1996;2:1022–1027.

[31] Reddy CC, Niyogi SK, Wells A, Wiley HS, Lauffenburger DA. Engineering epidermal growth factor for enhanced mitogenic potency. Nat Biotechnol 1996;14:1696–1699.

[32] Fan VH, Tamama K, Au A, Littrell R, Richardson LB, Wright JW, Wells A, Griffith LG. Tethered epidermal growth factor provides a survival advantage to mesenchymal stem cells. Stem Cells 2007;25:1241–1251.

[33] Sakiyama-Elbert SE, Hubbell JA. Development of fibrin derivatives for controlled release of heparin binding growth factors. J Control Release 2000;65:389–402.

[34] Willerth SM, Rader AR, Sakiyama-Elbert SE. The Effect of Controlled Growth Factor Delivery on Embryonic Stem Cell Differentiation Inside of Fibrin Scaffolds. Stem Cell Res. 2008;1(3):205-218.

[35] Ho JE, Chung EH, Wall S, Schaffer DV, Healy KE. Immobilized sonic hedgehog N-terminal signaling domain enhances differentiation of bone marrow-derived mesenchymal stem cells. J Biomed Mater Res A 2007;83:1200–1208.

[36] Anselme K. Osteoblast adhesion on biomaterials. Biomaterials 2000;21(7):667-81.

[37] Mrksich M. What can surface chemistry do for cell biology? Curr. Opin. Chem. Biol. 2002;6: 794.

[38] Hersel U, Dahmen C, Kessler H. RGD modified polymers: biomaterials for stimulated cell adhesion and beyond. Biomaterials 2003;24(24):4385-415.

[39] Roeker S, Böhm S, Diederichs S, Bode F, Quade A, Korzhikov V, van Griensven M, Tennikova TB, Kasper C. A study on the influence of biocompatible composites with bioactive ligands toward their effect on cell adhesion and growth for the application in bone tissue engineering. J Biomed Mater Res B Appl Biomater. 2009;91(1):153-62.

[40] Liu SQ, Tian Q, Wang L, Hedrick JL, Hui JHP, Yang YY, Ee PLR. Injectable Biodegradable Poly(ethylene glycol)/ RGD Peptide Hybrid Hydrogels for in vitro Chondrogenesis of Human Mesenchymal Stem Cells. Macromolecular Rapid Communications 2010;31(13): 1148-54

[41] Mauney JR, Kirker-Head C, Abrahamson L, Gronowicz G, Volloch V, Kaplan DL. Matrix-mediated retention of in vitro osteogenic differentiation potential and in vivo bone-forming capacity by human adult bone marrow-derived mesenchymal stem cells during ex vivo expansion. J Biomed Mater Res A. 2006;79(3):464-75.

[42] Mwale F, Wang HT, Nelea V, Li L, John A, Wertheimer MR. The effect of glow discharge plasma surface modification of polymers on osteogenic differentiation of committed human mesenchymal stem cell. Biomaterials 2006;27(10):2258-2264.

[43] Ager MJ, Feser T, Denck H, Krauspe R. Proliferation and Osteogenic Differentiation of Mesenchymal Stem Cells Cultured onto Three Different Polymers In Vitro. Annals of Biomedical Engineering 2005;33(10):1319-1332.

[44] Wang YZ, Singh A, Xu P, Pindrus MA, Blasioli DJ, Kaplan DL. Expansion and osteogenic differentiation of bone marrow-derived mesenchymal stem cells on a vitamin C functionalized polymer. Biomaterials 2006; 27(17) 3265-3273.

[45] Xu C, Wang Y, Yu X, Chen X, Li X, Yang X, Li S, Zhan X, Xiang AP. Evaluation of human mesenchymal stem cells response to biomimetic bioglass-collagen-hyaluronic acid-phosphatidylserine composite scaffolds for bone tissue engineering. Materials Research Part A 2009;88(1):264-73.

[46] Rim NG, Lee JH, Jeong SI, Lee BK, Kim CH, Shin H. Modulation of Osteogenic Differentiation of Human Mesenchymal Stem Cells by Poly[(Llactide)-co-(e caprolactone)]/Gelatin Nanofibers. Macromolecular Bioscience 2009;9(8):795–804.

[47] Wei Y, Hu Y, Hao W, Han Y, Meng G, Zhang D, Wu Z, Wang H. A Novel Injectable Scaffold for Cartilage Tissue Engineering Using Adipose-Derived Adult Stem Cells. Orthopaedic Research 2008;26(1):27–33.

[48] Wan Y, Qu X, Lu J, Zhu C, Wan L, Yang J. Characterization of surface property of poly(lactide-co-glycolide) after oxygen plasma treatment. Biomaterials 2004;25:4777-4783.

[49] Favia P, d'Agostino R. Plasma treatments and plasma deposition of polymers for biomedical applications. Surf. Coat. Technol. 1998;98:1102-1106.

[50] Hsu SH, Chen WC. Improved cell adhesion by plasma-induced grafting of l-lactide onto polyurethane surface.Biomaterials 2000;21:359-367.

[51] Hegemann D, Brunner H, Oehr C, Instrum N. Plasma treatment of polymers for surface and adhesion improvement.Methods Phys. Res. B Beam Interact. Mater. Atoms 2003;208:281-286.

[52] Park H, Lee KY, Lee SJ, Park KE, Park WH. Plasma-treated poly (lacticco-glycolic acid) nanofibers for tissue engineering.Macromol. Res. 2007;15:238-243.

[53] Baek HS, Park YH, Ki CS, Park JC, Rah DK. Enhanced chondrogenic responses of articular chondrocytes onto porous silk fibroin scaffolds treated with microwave-induced argon plasma.Surf. Coat. Technol. 2008;202:5794-5797.

[54] Curran JM, Chen R, Hunt JA. The guidance of human mesenchymal stem cell differentiation in vitro by controlled modifications to the cell substrate. Biomaterials 2006;27(27):4783–4793.

[55] Bianchi F, Vozzi G, Pescia C, Domenici C, Ahluwalia A. A comparative study of chemical derivatisation methods for spatially differentiated cell adhesion on 2-dimensional microfabricated polymeric matrices. J Biomater Sci Polymer Edn. 2003;14:1077–1096.

[56] Liu X, Lim JY, Donahue HJ, Dhurjati R, Mastro AM, Vogler EM. Influence of substratum surface chemistry/energy and topography on the human fetal osteoblastic cell line hFOB 1.19: phenotypic and genotypic responses observed in vitro. Biomaterials 2007;28:4535–4550.

[57] Csete M. Oxygen in the cultivation of stem cells. Ann NY Acad Sci. 2005;1049:1–8.

[58] Studet L, Csete M, Lee SH. Enhanced proliferation, survival, and dopaminergic differentiation of CNS precursors in lowered oxygen. J Neurosci. 2000;20:7377–7383.

[59] Moussavi-Harami F, Duwayri Y, Martin JA, Moussavi-Harami F, Buckwalter JA. Oxygen effects on senescence in chondrocytes and mesenchymal stem cells: consequences for tissue engineering. Iowa Orthop J. 2004;24:15–20.

[60] Grayson WL, Zhao F, Izadpanah R, Bunnell B, Ma T. Effects of hypoxia on human mesenchymal stem cell expansion and plasticity in 3D constructs. J Cell Physiol. 2006 May;207(2):331–339.

[61] Fink T, Abildtrup L, Fogd K, Abdallah BM, Kassem M, Ebbesen P, Zachara V. Induction of adipocyte-like phenotype in human mesenchymal stem cells by hypoxia. Stem Cells. 2004; 22:1346–1355.

[62] Csete M, Walkikonis J, Slawany N, Wei Y, Korsnes S, Doyle JC, Wold B. Oxygen-mediated regulation of skeletal muscle satellite cell proliferation and adipogenesis in culture. J Cell Physiol. 2001;189:189–196.

[63] Lennon DP, Edminson JM, Caplan AI. Cultivation of rat marrow-derived mesenchymal stem cells in reduced oxygen tension: effects on in vitro and in vivo osteochondrogenesis. J Cell Physiol. 2001;187:345–355.

[64] Buckley CT, Vinardell T, Kelly DJ. Oxygen tension differentially regulates the functional properties of cartilaginous tissues engineered from infrapatellar fat pad derived MSCs and articular chondrocytes. Osteoarthritis and Cartilage 2010;18(10): 1345-54.

[65] Burdick JA, Vunjak-Novakovic G. Engineered Microenvironments for Controlled Stem Cell Differentiation. Tissue Eng. Part A 2008;15(2):205-219.

[66] Ifkovits JL, Burdick JA. Photopolymerizable and Degradable Biomaterials for Tissue Engineering Applications. Tissue Eng. 2007;13:2369-2385.

[67] Lutolf M, Hubbell JA. Synthetic biomaterials as instructive extracellular microenvironments for morphogenesis in tissue engineering. Nature Biotechnology 2005;23(1): 47-55.

[68] Chung C, Burdick JA. Influence of three-dimensional hyaluronic acid microenvironments on mesenchymal stem cell chondrogenesis. Tissue Engineering. Part A 2009;15:243-254.

[69] Xu JA, Wang W, Ludeman M, Cheng K, Hayami T, Lotz JA, Kapila S. Chondrogenic differentiation of human mesenchymal stem cells in three-dimensional alginate gels. Tissue Eng. Part A 2008;14:667-680.

[70] Kraehenbuehl TP, Zammaretti P, Van der Vlies AJA, Schoenmakers RG, Lutolf MP, Jaconi ME, Hubbell JA. Three-dimensional extracellular matrix-directed cardioprogenitor differentiation: systematic modulation of a synthetic cell-responsive PEG-hydrogel. Biomaterials 2008;29(18):2757-2766.

[71] Yeo Y, Geng W, Ito T, Kohane D, Burdick JA, Radisic M. A photocrosslinkable hydrogel for myocyte cell culture and injection. J. Biomed. Mater. Res. 2007;81B:312-322.

[72] Baharvand H, Hashemi SM, Ashtiani SM, Farrokhi A. Differentiation of human embryonic stem cells into hepatocytes in 2D and 3D culture systems in vitro. Int J Dev Biol. 2006;50:645–652.

[73] Brännvall K, Bergman K, Wallenquist U, Svahn S, Bowden T, Hilborn J, Forsberg-Nilsson K. Enhanced neuronal differentiation in a three-dimensional collagen–hyaluronan matrix. J Neurosci Res. 2007 Aug 1;85(10):2138–2146.

[74] Pranga P, Müller R, Eljaouharib A, Heckmannb K, Kunzb W,Weberc T, Faberc C, Vroemena M, Bogdahna U, Weidner N. Thepromotion of oriented axonal regrowth in the injured spinal cord by alginate-based anisotropic capillary hydrogels. Biomaterials 2006;27:3560–3569.

[75] Nguyen LH, Kudva AK, Saxena NS, Roy K. Engineering articular cartilage with spatially-varying matrix composition and mechanical properties from a single stem cell population using a multi-layered hydrogel. Biomaterials 2011;32:6946-6952.

[76] Nguyen LH, Kudva AK, Guckert NL, Linse KD, Roy K. Unique biomaterial compositions direct bone marrow stem cells into specific chondrocytic phenotypes corresponding to the various zones of articular cartilage. Biomaterials 2011;32:1327-1338.

[77] Nuttelman CR, Rice MA, Rydholm AE, Salinas CN, Shah DN, Anseth KS. Macromolecular monomers for the synthesis of hydrogel niches and their application in cell encapsulation and tissue engineering. Prog Polym Sci 2008;33:167–79.

[78] Hou QP, De Bank PA, Shakesheff KM. Injectable scaffolds for tissue regeneration. J Mater Chem 2004;14:1915–23.

[79] Pratt AB, Weber FE, Schmoekel HG, Müller R, Hubbell JA. Synthetic extracellular matrices for in situ tissue engineering. Biotechnol Bioeng 2004;86:27–36.

[80] Hennink WE, van Nostrum CF. Novel crosslinking methods to design hydrogels. Adv Drug Delivery Rev 2002;54:13–36.

[81] Tan H, Ramirez CM, Miljkovic N, Li H, Rubin JP, Marra KG. Thermosensitive injectable hyaluronic acid hydrogel for adipose tissue engineering. Biomaterials 2009;30:6844–6853.

[82] Tana R, She Zh, Wang M, Fang Zh, Liu Y, Feng Q. Thermo-sensitive alginate-based injectable hydrogel for tissue engineering. Carbohydrate Polymers 2012;87:1515–1521.

[83] Stevens MM, George JH. Exploring and engineering the cell surface interface. Science2005;310: 1135-1138.

[84] Liu X, Ma P, Ann. Polymeric scaffolds for bone tissue engineering. Biomed. Eng. 2004;32:477-86.

[85] Petrie Aronin CE, Cooper JA, Sefcik LS, Tholpady SS, Ogle RC, Botchwey EA. Osteogenic differentiation of dura mater stem cells cultured in vitro on three-dimensional porous scaffolds of poly(ε-caprolactone) fabricated via co-extrusion and gas foaming. Acta Biomater. 2008;4:1187-1197.

[86] Murphy CM, Haugh MG, O'Brien FJ. The effect of mean pore size on cell attachment, proliferation and migration in collagen–glycosaminoglycan scaffolds for bone tissue engineering. Biomaterials 2010;31(3):461–6.

[87] Kasten P, Beyen I, Niemeyer P, Luginbühl R, Bohner M, Richter W. Porosity and pore size of b-tricalcium phosphate scaffold can influence protein production and osteogenic differentiation of human mesenchymal stem cells: An in vitro and in vivo study. Acta Biomater. 2008;4(6):1904-15.

[88] Tayton E, Purcell M, Aarvold A, Smith JO, Kalra S, Briscoe A, Shakesheff K, Howdle SM, Dunlop DG, Oreffo ROC. Supercritical CO2 fluid-foaming of polymers to in-

crease porosity: A method to improve the mechanical and biocompatibility characteristics for use as a potential alternative to allografts in impaction bone grafting? Acta Biomaterialia 2012;8:1918–1927.

[89] Chang Z, Meyer K, Rapraeger AC, Friedl A. Differential ability of heparan sulfate proteoglycans to assemble the fibroblast growth factor receptor complex in situ. FASEB J 2000;14:137–44.

[90] Cukierman E, Pankov R, Stevens DR, Yamada KM. Taking cell-matrix adhesions to the third dimension. Science 2001;294:1708–12.

[91] Lutolf MP, Hubbell JA. Synthetic biomaterials as instructive extracellular microenvironments for morphogenesis in tissue engineering. Nature biotechnology 2005;23(1): 47-55.

[92] Stevens MM, George JH. Exploring and Engineering the Cell Surface Interface. Science 2005;310:1135-1138.

[93] Griffith LG, Naughton G. Tissue Engineering--Current Challenges and Expanding Opportunities. Science 2002;295:1009-1014.

[94] Langer R, Tirrell DA. Designing materials for biology and medicine. Nature 2004;428:487-92.

[95] Takahashi Y, Tabata Y. Effect of the fiber diameter and porosity of non-woven PET fabrics on the osteogenic differentiation of mesenchymal stem cells. Journal of Biomaterials Science Polymer Ed 2004;15(1):41-57.

[96] Li WJ, Jiang YJ, Tuan RS. Chondrocyte phenotype in engineered fibrous matrix is regulated by fiber size. Tissue Engineering 2006;12(7):1775-1785.

[97] Yang F, Murugan R, Wang S, Ramakrishna S. Electrospinning of nano/micro scale poly (L-lactic acid) aligned fibers and their potential in neural tissue engineering. Biomaterials 2005;26:2603-2610.

[98] Xu CY, Yang F, Wang S, Ramakrishna S. in vitro study of human vascular endothelial cell function on materials with various surface roughness. Journal of Biomedical Materials Research Part A 2004;71A: 154-161.

[99] Xin XJ, Hussain M, Mao JJ. Continuing differentiation of human mesenchymal stem cells and induced chondrogenic and osteogenic lineages in electrospun PLGA nanofiber scaffold. Biomaterials 2007;28(2):316-325.

[100] Liao S, Chan CK, Ramakrishna S. Stem cells and biomimetic materials strategies for tissue engineering. Materials Science and Engineering C 2008;28:1189–1202.

[101] Boland ED, Wnek GE, Simpson DG, Pawlowski KJ, Bowlin GL, Tailoring tissue engineering scaffolds using electrostatic processing techniques: A study of poly(glycolic acid) electrospinning. J. Macromol. Sci. A, Pure Appl. Chem., 2001;38(12):1231–1243.

[102] Eberli D. Tissue Engineering.Bio-nanotechnology Approaches to Neural Tissue Engineering. Published by In-The; 2010,Chap-23.

[103] Yang F, Xu CY, Kotaki M, Wang S, Ramakrishna S. Characterization of neural stem cells on electrospun poly(L-lactic acid) nanofibrous scaffold. J. Biomater. Sci. Polym. Ed. 2004; 15(12):1483–1497.

[104] Christopherson GT, Song H, Mao H-Q. The influence of fiber diameter of electrospun substrates on neural stem cell differentiation and proliferation. Biomaterials 2009;30(4):556–564.

[105] Zilberman M. Active implants and scaffolds for tissue regeneration(Studies in Mechanobiology, Tissue Engineering and Biomaterials). Springer; 2011.

[106] Shin M, Yoshimoto H, Vacanti JP. In vivo bone tissue engineering using mesenchymal stem cells on a novel electrospun nanofibrous scaffold. Tissue Eng. 2004;10(1–2): 33–41.

[107] Yoshimoto H, Shin YM, Terai H, Vacanti JP. A biodegradable nanofiber scaffold by electrospinning and its potential for bone tissue engineering. Biomaterials 2003;24(12):2077–2082.

[108] Li WJ, Tuli R, Huang X, Laquerriere P, Tuan RS. Multilineage differentiation of human mesenchymal stem cells in a three-dimensional nanofibrous scaffold. Biomaterials 2005;26(25):5158–5166.

[109] Li W-J, Tuli R, Okafor C, Derfoul A, Danielson KG, Hall DJ, Tuan RS. A three-dimensional nanofibrous scaffold for cartilage tissue engineering using human mesenchymal stem cells. Biomaterials 2005;26:599–609.

[110] Xin X, Hussain M, Mao JJ. Continuing differentiation of human mesenchymal stem cells and induced chondrogenic and osteogenic lineages in electrsopun PLGA nanofiber scaffold. Biomaterials 2007;28:316–325.

[111] Li W-J, Laurencin CT, Caterson EJ, Tuan RS, Ko FK, Electrospun nanofibrous structure: A novel scaffold for tissue engineering. J. Biomed. Mater. Res. 2002;60(4):613–621.

[112] Discher DE, JanmeyP, WangYL. Tissue cells feel and respond to the stiffness of their substrate. Science 2005;310:1139–1143.

[113] Engler AJ, Sen S, Sweeney HL, DischerDE. Matrix elasticity directs stem cell lineage specification. Cell 2006;126:677–689.

[114] Saha K, Keung AJ, Irwin EF, Li Y, Little L, Schaffer DV, Healy KE.Substrate modulus directs neural stem cell behaviour. Biophys J. 2008;95(9):4426-4438.

[115] Leipzig ND, Shoichet MS. The effect of substrate stiffness on adult neural stem cell behaviour. Biomaterials 2009;30:6867–6878.

[116] Banerjee A, Arha M, Choudhary S, Ashton RS, Bhatia SR, Schaffer DV, Kane RS. The influence of hydrogel modulus on the proliferation and differentiation of encapsulated neural stem cells. Biomaterials 2009;30:4695–4699.

[117] Wang LS, Chung JE, Chan PP, Kurisawa M. Injectable biodegradable hydrogels with tunable mechanical properties for the stimulation of neurogenesic differentiation of human mesenchymal stem cells in 3D culture. Biomaterials 2010;31(6):1148-57.

[118] Toh WS, Lim TCh, Kurisawa M, Spector M. Modulation of mesenchymal stem cell chondrogenesis in a tunable hyaluronic acid hydrogel microenvironment. Biomaterials 2012;33: 3835-3845.

[119] Holtorf HL, Jansen JA, Mikos AG. Flow perfusion culture induces the osteoblastic differentiation of marrow stromal cell-scaffold constructs in the absence of dexamethasone. Journal of Biomedical Materials Research 2005;72(A): 326-334.

[120] Sikavitsas VI, Bancroft GN, Lemoine JJ, Liebschner MAK, Dauner M, Mikos AG. Flow perfusion enhances the calcified matrix deposition of marrow stromal cells in biodegradable nonwoven fiber mesh scaffolds. Annals of Biomedical Engineering 2005;33:63-70.

[121] Li YJ, Batra NN, You LD, Meier SC, Coe IA, Yellowley CE, Jacobs CR. Oscillatory fluid flow affects human marrow stromal cell proliferation and differentiation. Journal of Orthopaedic Research 2004;22:1283-1289.

[122] Stolberg S, McCloskey KE. Can shear stress direct stem cell fate? Biotechnol Prog. 2009; 25(1):10-19.

[123] Park JS, Chu JSF, Cheng C, Chen F, Chen D, Li S. Differential effects of equiaxial and uniaxial strain on mesenchymal stem cells. Biotechnology and Bioengineering 2004;88:359-368.

[124] Diederichs S, Bo°hm S, Peterbauer A, Kasper C, Scheper T, van Griensven M.Application of different strain regimes in two-dimensional and three-dimensional adipose tissue–derived stem cell cultures induces osteogenesis: Implications for bone tissue engineering. Journal of Biomedical Materials Research Part A 2010;94(A):927-936.

[125] Matos MA, Cicerone MT. Alternating current electric field effects on neural stem cell viability and differentiation. Biotechnology Progress 2010; 26(3):664–670.

[126] Park SY, Park J, Sim SH, Sung MG, Kim KS, Hong BH, Hong S. Enhanced Differentiation of Human Neural Stem Cells into Neurons on Graphene. Adv. Mater. 2011;23: 263–267.

[127] Schienbein M, Gruler H. Physical Review. E- Statistical Physics, Plasmas, Fluids, and Related Interdisciplinary Topics. The American Physical Society 1995;52: 4183-4197.

[128] Abilez O, Benharash P, Miyamoto E, Gale A, Xu C, Zarins CK. P19 progenitor cells progress to organized contracting myocytes after chemical and electrical stimulation: implications for vascular tissue engineering. Endovasc Ther 2006;13(3):377–388.

[129] Serena E, Flaibani M, Carnio S, Boldrin L, Vitiello L, De Coppi P, Elvassore N. Electrophysiologic stimulation improves myogenic potential of muscle precursor cells grown in a 3D collagen scaffold. Neurol Res. 2008;30(2):207–214.

[130] Genovese JA, Spadaccio C, Langer J, Habe J, Jackson J, Patel AN. Electrostimulation induces cardiomyocyte predifferentiation of fibroblasts. Biochem Biophys Res Commun. 2008;370(3):450–455.

[131] Ebisawa K, Hata K, Okada K, Kimata K, Ueda M, Torii S, Watanabe H. Ultrasound enhances transforming growth factor beta-mediated chondrocyte differentiation of human mesenchymal stem cells. Tissue Eng. 2004; 10(5–6):921–929.

[132] Abramovitch-Gottlib L, Naveh TGD, Geresh S, Rosenwaks S, Bar I, Vago R. Low level laser irradiation stimulates osteogenic phenotype of mesenchymal stem cells seeded on a three-dimensional biomatrix. Lasers in Medical Science

[133] McBride SH, Knothe Tate ML. Modulation of stem cell shape and fate A: the role of density and seeding protocol on nucleus shape and gene expression. Tissue Eng Part A. 2008;14(9):1561–1572.

[134] Wilmut I, Schnieke AE, McWhir J, Kind AJ, Campbell KH. Viable offspring derived from fetal and adult mammalian cells. Nature 1997;385(6619): 810–3.

[135] MacArthur BD, Ma'ayan A, Lemischka IR. Systems biology of stem cell fate and cellular reprogramming. Nature Reviews Molecular Cell Biology 2009;10: 672-681.

[136] Takahashi K, Tanabe K, Ohnuki M, Narita M, Ichisaka T, Tomoda K, Yamanaka S. Induction of pluripotent stem cells from adult human fibroblasts by defined factors. Cell 2007;131(5):861-872.

[137] Choi J, Costa ML, Mermelstein CS, Chagas C, Holtzer S, Holtzer H. MyoD converts primary dermal fibroblasts, chondroblasts, smooth muscle, and retinal pigmented epithelial cells into striated mononucleated myoblasts and multinucleated myotubes. Proc Natl Acad Sci USA 1990;87(20):7988–7992.

[138] Takahashi K, Yamanaka S. Induction of pluripotent stem cells from mouse embryonic and adult fibroblast cultures bydefined factors. Cell 2006;126(4):663-76.

[139] Takahashi K, Tanabe K, Ohnuki M, Narita M, Ichisaka T, Tomoda K, Yamanaka S. Induction of pluripotent stem cells from adult human fibroblasts by defined factors. Cell 2007;131(5):861-72.

[140] Yu J, Vodyanik MA, Smuqa-Otta K, Antosiewica-Bourqet J, Frane JL, Tian S, Nie J, Jonsdottir GA, Ruotti V, Stewart R, Slukvin II, Thomson JA. Induced Pluripotent Stem Cell Lines Derived from Human Somatic Cells. Science 2007;318(5858):1917–1920.

[141] Blanpain C, Daley GQ, Hochedlinger K, Passegué E, Rossant J, Yamanaka S. Stem cells assessed. Nature Reviews Molecular Cell Biology 2012;13:471-476

[142] Okita K, Yamanaka S. Induced pluripotent stem cells: opportunities and challenges. Philos Trans R Soc Lond B Biol Sci. 2011;366(1575):2198-207.

[143] Eiraku M, Takata N, Ishibashi H, Kawada M, Sakakura E, Okuda S, Sekiguchi K, Adachi T, Sasai Y. Self-organizing optic-cup morphogenesis in three-dimensional culture. Nature 2011;472:51–56.

[144] Selvaraj V, Plane JM, Williams AJ, Deng W. Switching cell fate: the remarkable rise of induced pluripotent stem cells and lineage reprogramming technologies. Trends in Biotechnology 2010 Feb 9;28(4):214-223.

[145] Vierbuchen T, Ostermeier A, Pang ZP, Kokubu Y, Sudhof TC, Wernig M. Direct conversion of fibroblasts to functional neurons by defined factors. Nature 2010 Feb 25;463(7284):1035-41.

[146] Huang P, He Z, Ji S, Sun H, Xiang D, Liu C, Hu Y, Wang X, Hui L. Induction of functional hepatocyte-like cells from mouse fibroblasts by defined factors. Nature 2011 May 11;475(7356):386-9.

[147] Ieda M, Fu JD, Delgado-Olguin P, Vedantham V, Hayashi Y, Bruneau BG, Srivastava D. Direct Reprogramming of Fibroblasts into Functional Cardiomyocytes by Defined Factors. Cell 2010;142(3):375-386.

[148] Nicholas CR, Kriegstein AR. Regenerative medicine: Cell reprogramming gets direct. Nature 2010;463:1031-1032.

[149] Ring KL, Tong LM, Balestra ME, Javier R, Andrews-Zwilling Y, Li G, Walker D, Zhang WR, Kreitzer AC, Huang Y. Direct reprogramming of mouse and human fibroblasts into multipotent neural stem cells with a single factor. Cell Stem Cell 2012;11(1):100-9.

[150] Huebsch N, Mooney DJ. Inspiration and Application in the Evolution of Biomaterials. Nature 2009;462(7272):426-432.

[151] Alsberg E, Anderson KW, Albeiruti A, Rowley JA, Mooney DJ. Engineering Growing Tissues. Proceedings of the National Academy of Sciences of the United States of America 2002;99(19):12025.

[152] Lieberman JR, Daluiski A, Einhorn TA. The Role of Growth Factors in the Repair of Bone: Biology and Clinical Applications. The Journal of Bone and Joint Surgery 2002;84(6):1032-1044.

[153] Reddi AH. Bone Morphogenetic Proteins: From Basic Science to Clinical Applications. The Journal of Bone and Joint Surgery 2001;83(S1):S1-S6.

[154] Reddi AH. Interplay Between Bone Morphogenetic Proteins and Cognate Binding Proteins in Bone and Cartilage Development: Noggin, Chordin and DAN. Arthritis-Research 2001;3(1):1-5.

[155] Sundelacruz S, Kaplan DL. Stem Cell-And Scaffold-Based Tissue Engineering Approaches to Osteochondral Regenerative Medicine. Seminars in Cell &Developmental Biology 2009;20(6):646-655.

[156] Holland TA, Mikos AG. Biodegradable Polymeric Scaffolds. Improvements in bone tissue engineering through controlled drug delivery. Advances inBiochemical Engineering/Biotechnology 2006;102:161-185.

[157] Holland TA, Bodde EWH, Cuijpers V, Baggett LS, Tabata Y, Mikos AG, Jansen JA. Degradable Hydrogel Scaffolds for in Vivo Delivery of Single and Dual Growth Factors in Cartilage Repair. Osteoarthritis and Cartilage 2007;15(2):187-197.

[158] Liu H, Zhang L, Shi P, Zou Q, Zuo Y, Li Y. Hydroxyapatite/Polyurethane Scaffold Incorporated with Drug-Loaded Ethyl Cellulose Microspheres for Bone Regeneration. Journal of Biomedical Materials Research Part B: Applied Biomaterials 2010;95(1):36-46.

[159] Uebersax L, Merkle HP, Meinel L. Insulin-Like Growth Factor I Releasing Silk Fibroin Scaffolds Induce Chondrogenic Differentiation of Human Mesenchymal Stem Cells. Journal of Controlled Release 2008;127(1):12-21.

[160] Li C, Vepari C, Jina HJ, Kim HJ, Kaplan DL. Electrospun silk-BMP-2 scaffolds for bone tissue engineering. Biomaterials 2006;27(16): 3115–3124.

[161] Lee GS, Park JH, Shin US, Kim HW. Direct deposited porous scaffolds of calcium phosphate cement with alginatefor drug delivery and bone tissue engineering. Acta Biomaterialia 2011;7(8):3178–3186.

[162] Zhou K, Nisbet DR, Thouas G, Bernard C, Forsythe JS., editor. Eberli. Bio-nanotechnological approaches to neural tissue engineering in Tissue Engineering. In-Tech; 2010. p459-483.

[163] Schnell E, Klinkhammer K, Balzer S, Brook G, Klee D, Datton P, Mey J. Guidance of glial cell migration and axonal growth on electrospun nanofibers of poly-e-caprolactone and a collagen/poly-ecaprolactone blend. Biomaterials 2007;28(19): 3012-3025.

[164] Geiger B, Bershadsky A, Pankov R, Yamada KM. Transmembrane crosstalk between the extracellular matrix and the cytoskeleton. Nat Rev Mol Cell Biol 2001;2:793-805.

[165] Horne MK, NisbetDR,Forsythe JS, Parish CL. Three dimensional nanofibrous scaffolds incorporating immobilized BDNF promote proliferation and differentiation of cortical neural stem cells. Stem Cells and Development 2010;19(6):843-852.

[166] Patel S, Kurpinski K, Quigley R, Gao H, Hsiao BS, Poo MM, Li S. Bioactive Nanofibers: Synergistic Effects of Nanotopography and Chemical Signaling on Cell Guidance. Nano Letters 2007;7(7): 2122-2128.

[167] Bobick BE, Chen FH, Le AM, Tuan RS. Regulation of the chondrogenic phenotype in culture. Birth Defects Res C Embryo Today 2009;87:351-71.

[168] Schinagl RM, Ting MK, Price JH, Sah RL. Video microscopy to quantitate the inhomogeneous equilibrium strain within articular cartilage during confined compression. Ann Biomed Eng 1996;24:500-512.

[169] Wang CC, Hung CT, Mow VC. An analysis of the effects of depth-dependent aggregate modulus on articular cartilage stress-relaxation behavior in compression. J Biomech 2001;34:75-84.

[170] Forte G, Carotenuto F, Pagliari F, Pagliari S, Cossa P, Fiaccavento R, Ahluwalia A, Vozzi G, Vinci B, Serafino A, Rinaldi A, Traversa E, Carosella L, Minieri M, Di Nardo P. Criticality of the biological and physical stimuli array inducing resident cardiac stem cell determination. Stem Cells 2008;26(8):2093–2103.

[171] Lowman AM, Morishita M, Kajita M, Nagai T, Peppas NA. Oral delivery of insulin using pH-responsive complexation gels. Journal Pharmacy Science 1999;88(9): 933-937.

[172] Elisseeff J, Anseth K, Sims D, McIntosh W, Randolph M, Langer R. Transdermal photopolymerization for minimally invasive implantation. Proc Natl Academic Science USA 1999; 96(6):3104-3107.

[173] Mongia NK, Anseth KS, Peppas NA. Mucoadhesive poly (vinyl alcohol) hydrogels produced by freezing/thawing processes: applications in the development of wound healing systems. Journal of Biomaterial Science Polymer Edition 1996; 7(12): 1055-1064.

[174] Lu S, Anseth KS. Photopolymerization of multilaminated poly (HEMA) hydrogels for controlled release. Journal of Control Release 1999; 57(3):291-300.

[175] Babensee JE, Cornelius RM, Brash JL, Sefton MV. Immunoblot analysis of proteins associated with HEMAMMA microcapsules: human serum proteins in vitro and rat proteins following implantation. Biomaterials 1998; 19(7-9):839-849.

[176] Ahearne M, Yang Y, El Haj AJ, Then KY, Liu KK.Characterizing the viscoelastic properties of thin hydrogel-based constructs for tissue engineering applications. J. R. Soc. Interface 2005; 2:455-463.

[177] Annabi N, Nichol JW, Zhong X, Ji Ch, Koshy S, Khademhosseini A, Dehghani F. Controlling the Porosity and Microarchitecture of Hydrogels for Tissue Engineering. Tissue engineering Part B 2010 August;16(4):371–383.

[178] Suggs LJ, Payne RG, Yaszemski MJ, Alemany LB, Mikos AG. Preparation and characterization of poly (propylene fumarate-co-ethylene glycol) hydrogels. Journal of Biomaterial Science Polymer Edition 1998; 9(7): 653-666.

[179] Hadlock TA, Elisseeff J, Langer R, Vacanti J, Cheney MA. Tissue engineered conduit for peripheral nerve repair. Arch Otolaryngol Head Neck Surg 1998; 124(10):1081-6.

[180] Atala A, Kim W, Paige KT, Vacanti CA, Retik AB. Endoscopic treatment of vesicoureteral reflux with a chondrocyte-alginate suspension. Journal of Urology 1994; 152(2 Pt 2) 641-643; discussion 644.

[181] Rosellini E, Cristallini C, Barbani N, Vozzi G, Giusti P. Preparation and characterization of alginate/gelatin blend films for cardiac tissue engineering. Journal of Biomedical Material Research A 2009 Nov;91(2):447-53.

[182] Boontheekul T, Hill EE, Kong HU, Mooney DJ. Regulating myoblast phenotype through controlled gel stiffness and degradation. Tissue engineering 2007; 13 (7): 1431-1442.

[183] Gilbert PM, Havenstrite KL, Magnusson KEG, Sacco A, Leonardi NA, Kraft P, Nguyen NK, Thrun S, Lutolf MP, Blau HM. Substrate elasticity regulates skeletal muscle stem cell self-renewal in culture. Science 2010;329:1078–81.

[184] Fisher OZ, Khademhosseini A, Langer R, Peppas NA. Bioinspired Materials for Controlling Stem Cell Fate. Accounts of chemical research 2010; 43 (3):419-428.

Chapter 27

Ethical Considerations on Stem Cell Research

Andreas M. Weiss, Michael Breitenbach,
Mark Rinnerthaler and Günter Virt

Additional information is available at the end of the chapter

http://dx.doi.org/10.5772/54375

1. Introduction

Definitions: First we have to clearly define what we are talking about in the field of stem cells. The zygote (fertilized egg cell) and the cells of the very young embryo up until the eigth-cell stage are totipotent. This expression means that in the appropriate environment (the uterus) these cells can form a complete and normal individual.

In contrast to this notion, the embryonal stem cells of mammals are derived from the inner cell mass of the blastocyst, a slightly later stage of embryonal development. These cells are no longer totipotent, but pluripotent. This means that those cells, if artificially inserted into a heterologous young embryo, survive and give rise to all tissues and cell types in this embryo including cells of the germ line, thus creating a chimeric embryo, which consists of two types of cells that are genetically different form each other. Embryonal stem cells (ES cells) display a few properties that make them highly interesting for regenerative medicine: they can be grown and multiplied indefinitely in the presence of the appropriate "factors" (proteins, growth factors, small molecules) without major genetic changes and without loss of pluripotency, and they can be modified by genetic engineering without major chromosomal changes and without using viral vehicles [1]. The latter property is essential for the future application of those cells for gene therapy. Mammalian ES cell technology was first developed in the mouse model system beginning with the landmark paper of Martin [2]. Human ES cells (hESC) were first isolated by Thomson [3]. The patenting of the isolation of hESC (the so-called WARF patents) led to a huge public discussion regarding the moral and legal implications of those patents [4]. Ultimately the US supreme court acknowledged those patents as being legal, while the Court of Justice of the European Union ruled that no procedures can be patented, which use embryo research, i.e. the destruction of human embryos [81]. However, human induced pluripotent stem cells (hiPSC) can now be created from differentiated adult cells, like dermal fibroblasts (see below), which according to biochemical criteria (transcriptome, proteome), are very near identical to hESC [5]. It has been shown, in the mouse, that not only by biochemical criteria, but also in terms of the developmental potential, mouse iPSC are identical with mouse ESC [6].

In contrast to the pluripotent ES cells, somatic stem cells are multipotent, meaning that their developmental potential is rather limited to a number of related cell types. For instance, the well-known hematopoietic stem cells of the red bone marrow can generate *in vivo* all cells that are found in the blood of humans. Until recently it was believed that this commitment to a number of related developmental fates is absolute, however it is now known that even in normal individuals *in vivo,* a low percentage of bone marrow stem cells can become quite different cells [7], and, to give just one example, fibroblasts can be induced, by expression of two to three transcription factors, to become *bona fide* heart muscle cells [8].

Currently, an ever increasing number of papers on hiPSC (human induced pluripotent stem cells) are being published as documented by indexing services such as PubMed. In vitro methods of creating hiPSC from the easily available dermal fibroblasts were first described in 2006 and 2007 [9, 10]. Due to longer experience with the stem cells of the mouse and due to ethical and legal considerations, there is still a technical gap between procedures applicable to mouse iPSC and hiPSC. Since 2008, a nearly exponential increase in papers dealing with hiPSC is appearing and well over 1000 papers are now being published every year. Many of those papers mention that hiPSCs in contrast to hESCs (human embryonal stem cells) are considered to be ethically acceptable while an intensive debate was and is going on concerning the ethical implications of hESCs [4, 11]; (see below in the next part of this chapter).

Another unsolved probem in stem cell therapy is "homing" of the repaired cells to the "niche" in the body where they are needed and can function. Only in exceptional cases does homing occur automatically (bone marrow stem cells in the mouse), but in other cases (brain) the cells must be directly injected into the relevant area. Modern nanotechnological methods may be helpful for this immense task in the future [12].

What we would like to do in the current paper is (paragraph 2) to give a very short overview of the present and the anticipated future status of hiPSCs and their use in biomedicine including the new topic of differentiated cell plasticity [7]; (paragraph 3) to explain the ethical arguments that were brought forward concerning hESCs; and (paragraph 4) to discuss some remaining ethical arguments concerning hiPSCs with special emphasis on the argument of complicity [13].

2. Short overview of the present and the anticipated future status of hiPSCs

Stem cell and related techniques, such as direct reprogramming of differentiated cells, offer an immense promise for the future of regenerative medicine using stem cell therapy and/or a combination of stem cell and gene therapy. This promise is, as we now know, a realistic one, but the enormous technical difficulties and the requirements imposed by clinical safety (for instance concerning the cancer risk) are not easily overcome and we estimate that many years will pass before these methods become clinical routine for many diseases. Presently, very few clinical examples exist that successfully show the efficacy of stem cell and gene therapy [14].

The theoretical and biological basis for the techniques to be discussed here are, among others, the fact that somatic cells of animals (and of the human animal, of course) contain the same genetic complement as the fertilized egg cell (the zygote). This means that every gene needed for the complete development of an individual is present in every somatic cell of a mature individual. The direct and undisputable proof for this is shown by the cloning of animals [15]. However a similar result was obtained decades before "Dolly the sheep" by John Gurdon [16], working with frogs. Therefore, the phenotypic differences between different somatic cells of an adult individual must depend on differences in gene expression, or to use a modern term on the "epigenome" of those cells. At present Bio-medicine is, at an increasing speed, discovering methods to change this differentiated state from one well defined cell type (say fibroblasts) to another (say, for example, a specific subtype of neurons needed for an individual patient) [7]. Previously, the differentiated state of somatic cells was believed to be immutable, at least *in vivo*, but this paradigm clearly is no longer true. Why are such procedures needed in regenerative medicine? This question leads us to the genetic differences between human individuals and the immunological incompatibility between humans who are not monozygotic twins. For reasons that are not entirely clear to scientists who study the evolutionary history of mankind, it appears that differences in the antigens of the HLA type (human lymphocyte antigen; displayed on cell surfaces) occur between any two humans and are large enough to lead to immunological attack (host versus graft disease) after the transplantation of cells and organs. Therefore, it is desirable to use autologous (HLA-compatible) cells for therapy, which raise no immune response and make immune suppression of the patient superfluous. In organ transplantation, this problem is generally overcome (although, perhaps, insufficiently) by the pharmacological immune suppression of the patient who receives a transplant. For the combination of gene and cell therapy, the idea is to use autologous cells which, however, must conform to strict safety standards before a clinical trial is granted by the authorities and can be started. There are also a number of unresolved problems if the autologous cells to be transplanted need a genetic "repair" because the patient to be treated suffers from a genetic disease whose underlying mutation is known and will be corrected by sophisticated genetic engineering as is applicable to human cells.

Genome editing: For several reasons which have to do with differences that exist between mouse and human iPSCs, as well as with the low success rate of current methods for genome editing [17], the originally developed ingenious method of selection and counter-selection in mouse ESCs [1] seems not to be suitable for a safe repair of known mutations in genes of a patient suffering from a particular and genetically well-known inherited disease. Ideally, the presence of the mutation in question should be known by DNA sequencing of the relevant part (or the whole genome) of the patient. Instead, the scientific community is now seeking to improve the efficiency of point-directed genome editing to clinically acceptable levels [17]. The cells to be used for these procedures should be as close as possible to the original patient-derived cells, avoiding prolonged proliferation of hiPSCs. The tools that must be developed to achieve this are the so-called ZNF-nucleases (zinc finger nucleases) based on a concept by Kim [18] which can produce a double strand break at a precisely defined point in the whole human genome [17]. This double strand break is then recombinogenic enough to lead to homologous recombination with a co-transformed plasmid that carries the corrected DNA sequence [19]. Alternatively, the TALEN strategy can be used [20]. One problem that must be overcome here in the future, is the limited capacity for proliferation of differentiated cells and their general reluctance to be transformed by plasmids, which is true for instance for dermal fibroblasts.

Cancer risk: One of the greatest obstacles that must be overcome before stem cell therapy can become clinical routine is the inherent cancer risk conferred by both ESCs and iPSCs. In one of the very few and frequently-quoted clinical trials for gene therapy of X-SCID, some of the affected and essentially cured children came down with leukemia. The reason for the cancer incidence in this case was the lack of control of the point of integration of the viral vector used to introduce the genetically corrected gene sequence, which was inserted at locations in the genome where it caused leukemia [21, 22]. However, even ESCs or iPSCs which are not genetically manipulated, by their "stemness" alone can cause cancer. It must not be forgotten that embryonal stem cells were first discovered during the study of teratocarcinomas and one of the most important decisive traits was the ability to form teratocarcinomas in nude mice [2]. Therefore, for some time, the idea was to re-differentiate the hiPSCs to the needed cells after genetic manipulation and then purify these cells until they were essentially free of remaining stem cells [23]. This proved to be a difficult job. The other solution to this problem is to directly produce the desired cell type using the action of transcription factors and small molecule signalling substances without ever going through a stage of stem cells [7]. This way is very promising but also not yet matured enough for clinical practice.

In summary, we may say that it is still too early to decide in which direction future cell and gene therapy will go. For some time, hESCs, and even more importantly, hiPSCs will be needed for biomedical research. This is not restricted to gene therapy and cell therapy without genetic corrections (as in the case of acquired diseases), but equally is needed for the establishment of disease models and for drug testing, which is, however, not the topic of this chapter. For all of those reasons, we think it is timely to discuss the ethical implications of stem cell research.

3. Ethical arguments brought forward concerning hESCs

The central ethical concern that is raised by production and use of hESC is the question concerning the moral status of human embryos. The derivation of hESCs from early embryos (blastocysts) is, in practice, necessarily connected with their destruction. Because of that, we have to ask, if a human embryo is recognized as a being endowed with human dignity and a right to life comparable to that of born human beings. Destruction for research purposes raises the serious ethical issues of exploitation, instrumentalisation and killing of human beings. Concerning both ethical issues, human dignity and the prohibition of killing, in regards to human embryos in spite of the long discussions an ethical consensus is nowhere in sight. In the following passage some explanations will be given regarding the fundamental question of the moral status of embryos [12, 24-31].

Further intensively discussed issues in hESC research are research cloning (the procurement of embryos for research purposes by nuclear transfer in enucleated egg cells) and the donation of egg-cells. For a long time, the development of therapeutic applications seemed to involve research cloning (also called "therapeutic cloning"). Research cloning of humans would represent a clear instance of exploiting humans solely for the benefit and interests of others. Establishing this technique in humans requires further destructive embryo research and is feared to prepare a slippery slope for reproductive cloning of humans, which is generally considered as ethically unacceptable [32-37].

If this way to therapeutic applications had succeeded, the demand for a high number of donated egg-cells would have been a consequence. For women, egg donation causes health risks and the danger of commercial exploitation. The alternative to produce hybrids of humans and animals is also seen as offending human dignity [38]. These ethical problems have lost some urgency, since this strategy doesn't seem to be succeeding. The fundamental question of the moral status of human embryos is still a matter of open discussion in ethics.

3.1. The discussion about the moral status of early human embryos

hESCs needed for research are obtained from different sources that entail a different ethical evaluation. While extraction of stem cells from adults, from umbilical cord blood or from aborted foetuses, is considered to be ethically acceptable under certain conditions, the procurement of hESCs is confronted with ethical objections, since it is necessarily connected with the destruction of human embryos. It is a kind of consuming embryo research. The possible sources are already established embryonic stem cell lines, supernumerary embryos from IVF-treatment, embryos produced specifically for research purposes or even embryos cloned by nuclear transfer as a logical consequence in case of successful therapeutic applications.

Different regulations worldwide and in the EU, as well as an on-going discussion about the funding of research projects are taking place [12, 39]. As a minimal consensus, creation of embryos solely for research purposes is forbidden in the European Council's Convention for the Protection of Human Rights and Dignity of the Human Being with regard to the Application of Biology and Medicine [32].

By obtaining hESCs from the inner cell mass of a blastocyst for research purposes beginning human life is destroyed. The embryo is obviously a human being, a member of the human species, has an individual genome, neither identical with that of the mother nor that of the father, in contrast to other human tissue, can develop into the full shape of a human being (totipotent) and has a small, but realistic chance to be born and live its own life.

Since hESC research, on the one hand, gives hope in terms of therapeutic applications for severe diseases and, on the other hand, is connected with the destruction of embryos as necessary means to this end, two ethically high standing aims are opposed. Basic research (freedom of research) and the hopes connected with therapeutic application (principle of beneficence, value of health and life of patients) are confronted with the respect for human dignity and the right to life of human embryos. The question is: May human embryos be produced and destroyed as biological material for research and therapy or even for industrial applications?

In relation to already born humans we would never accept such destruction or killing no matter how great the benefit for research or therapy could be. For born humans there is a strong agreement: They have moral status and equal human dignity independent of their actual abilities or disabilities. The statement about the moral status is a value judgement. At first it means that humans have intrinsic value. If the moral status of humans is determined in the tradition of the German philosopher, Immanuel Kant, with the term "dignity", an unconditional value is proclaimed, which goes beyond the intrinsic value of non-human beings and can't be balanced with the benefit of others. Kant makes this clear in a well-known quote regarding his categorical imperative: “Act in such a way that you treat humanity, whether in your own person or in the person of any other, never merely as a means to an end, but always at the same time as an end.” [40, 41].

The central consequences of the recognition of equal human dignity are the fundamental equality of all humans with regard to this dignity, the same right to welfare and the prohibition of arbitrary instrumentalisation and exploitation for the purposes of others. Killing for research purposes definitely falls under this prohibition. Whether and to which degree these moral demands are already valid in the early stages of development, is a matter of the controversy concerning the ontological, moral, and legal status of human embryos [36, 42-47].

It is therefore clear why this discussion is unavoidable. Before discussing freedom of research, hopes for therapeutic applications, and different possibilities of regulations, the question, of whether or not embryos, in an ethical respect, belong to the community of beings deserving equal and impartial consideration, must be answered. Is impartiality (the “golden rule”), to be applied even to embryos, or not at all, or merely in a gradually weaker sense?

These issues were discussed extensively in the last decades and, regrettably, have not achieved a consensus. Here we will shortly explain the general lines of reasoning. Summarized in a simplified overview there are three types of answers: (a) Personalistic positions maintain human dignity and a right to life of human embryos. (b) Non-personalistic positions deny that and impute to embryos a status similar to human tissues or cadavers. A third group proposes to find a kind of middle position by giving several types of (c) relative or gradualistic answers.

http://dx.doi.org/10.5772/54375

a. **Personalistic positions** claim that already the embryo must be respected as a person and, therefore, has a right to life in the earliest stage and also outside the mother [42, 48-52].

The reason for personal positions is a certain view of the embryonic development. The development from fertilization up to birth is understood as a continuous development (*argument of continuity*) of something that is, basically, already present, and under natural conditions, has the inner capability of further development into a fully evolved person (*argument of potentiality*) and remains the same being (*argument of identity*). The embryo is not a preliminary stage of a human but a human in the earliest stage. Although it doesn't have the actual abilities of a person (self-consciousness, reason, freedom), the embryo must be treated as a person because of its inner potential to develop these qualities and, under normal circumstances, become such a person.

This reasoning can be combined with two additional arguments. The *species argument* points out that the embryo's membership in the human species is a biological fact. Biological facts alone are not sufficient reasonings for moral judgement. However, the argument may serve as a determination of the scope, the application area of dignity: All members of the human species are included. Being a member of the human species and being endowed with dignity and certain rights is actually coinciding with each other. Therefore, the species-membership suffices to claim the corresponding rights. If this argument also applies to embryos, then this is controversial and presupposes the first three arguments. The four previously mentioned arguments are often described as a "SKIP-quadrology": species, continuity, identity, potentiality [44].

Sometimes another argument is added in respect to the remaining uncertainties of empirical knowledge, as well as philosophical interpretation of early embryonic development. The *precautionary principle* generally calls for a careful proceeding in small steps and imposes the burden of proof on those, who want to change existing attitudes and moral norms. They have to offer evidence, not those who defend them. According to this position, doubts about being a person may not lead to an arbitrary restriction of human dignity. No man is subject to the constraint of having to justify his existence. This corresponds to the basic structure of the human dignity argument, which should, primarily, serve as protection of the weak against any kind of discrimination. Everyone is basically interested in safe conditions, in which she/he need not fear being excluded from the common protection area due to some actual lack of abilities or characteristics [45, 53].

The consequences of the personal position are unambiguous: Destroying embryos for research purposes and research cloning is forbidden. Freedom of research is subjected to moral limits. Therapies, which cost the lives of other humans, are not acceptable. Even the hope for therapy for serious diseases is no adequate reason for the specific production and destruction of human embryos. Nevertheless, each mentioned argument is subjected to criticism and the personal position hasn't turned out to provide a consensus [42, 43].

b. **Non-personalistic positions** deny what personalistic positions proclaim.

A far-reaching objection to the personal position is, for example, represented by the Australian moral philosopher, Peter Singer. He denies the human dignity of embryos, foetuses and even newborn children due to a very narrow concept of personality based solely on actual abilities: "My suggestion, then, is that we accord the life of a fetus no greater value than the life of a nonhuman animal at a similar level of rationality, self-consciousness, awareness, capacity to feel, etc. Since no fetus is a person, no fetus has the same claim to life as a person." [54]. For these positions there is, in principle, no objection to hESC research as long as the rights of the donors of gametes or embryos are respected.

c. **Gradualistic positions** try to find a way of maintaining special respect for human embryos and restrictions of research purposes and, at the same time, allowing research for high standing objectives. They are quite frequently supported [29, 31, 55-58].

According to this kind of reasoning full protection of human embryos starts at a later stage of development. The time before the moral status is gradually weakened, but not reduced to that of some other human tissue. Most frequently nidation, or the end of the possibility of twin formation, is seen as the relevant moment. When nidation is complete, the embryos' chance of survival increases significantly. Sometimes other stages of development are argued as being relevant e.g. the beginning of the first nerve cells in the fifth or sixth week. This is seen as relevant, if the ability to feel pain is seen as a decisive ethical quality.

Finally, there are suggestions in which the moral status of embryos isn't differentiated depending on the stage of development, but according to the context and target of its creation. In such an "extrinsic" determination of the moral status surplus embryos from IVF-treatment and research embryos don't have any dignity, because they lack the necessary conditions for further development, or according to their creators' intentions, never should be born at all, while embryos produced for IVF-treatment already have this dignity in a very early stage, since the intention and hope is that they be born [24, 59, 60]. In this way of reasoning, dignity and the right to life are conferred or awarded by society. Dignity depends on the allocation to the research department or the IVF department. Some authors turn this reasoning into the field of metaethics and proclaim, that human dignity is always invented and awarded by society and not based on an objective moral reality [58].

If the protection of some early stage or research embryos or surplus embryos is weakened, the interests and well-being of embryos and patients can be balanced against each other and destruction of embryos can be justified for high standing objectives. Strict embryo protection is argued to be valid for later stages and a clear limit seems possible for the time being. Nevertheless this reasoning is not free of some arbitrariness and, if the restrictions are sustained, one can fear for the time, when interests for research with later stages of embryonic development will emerge. In principle, everything seems justifiable, if dignity depends on society or the intentions of the embryo's creators.

Some authors try to justify hESC research without weakening the moral status of embryos through a special reasoning within the prohibition of killing [61, 62]. In an opinion of the

Austrian Bioethics Commission these attempts are summarized as follows: "The first argument chooses the comparison with the removal of organs from brain dead patients. This does not violate the prohibition of killing nor the prohibition of the complete instrumentalisation of a human life that is derived from the concept of human dignity. Even less should the use of fertilised egg cells at a stage in which one cannot speak of either an organ or brain development be rejected as such on ethical reasons. The second argument compares the obtaining of embryonic stem cells from surplus embryos with the medical use of tissue from aborted foetuses, which can be ethically justified in so far as the abortion was not performed for the purpose of obtaining foetal tissue. Both lines of argument imply that at the moment it is no longer used for reproduction, the embryo created in vitro undergoes a change of status that is equivalent to that of a person's transition from life into death. Even if one wishes to accord the fertilised egg cell personhood, this does not mean that there is an irresolvable conflict of values between the protection of life for the embryo and the freedom of research in the service of present and future patients"[31]. These arguments cannot be discussed here [63, 64]. The intention to escape the endless discussion about the moral status of embryos is clever, the hope to prevent the weakening of the human dignity argument may be honourable, but as a matter of fact, the relevant embryos are not dead prior to the destruction for research. One might wonder, what results this kind of reasoning could have, when applied to disabled persons or patients at the end of life, which could also be said to have no chance for further development (a logical version of the slippery slope-argument).

3.2. Results of the status debate

Each modification of ethical reasoning and central moral attitudes must be paid attention to in terms of consistency, rationality and possible side effects for other areas of life. This examination of the arguments is sometimes more important than the solution itself. Bad arguments are counterproductive, promote distrust against ethical reasoning and science in the long run, and weaken their aptitude to give orientation. The first task of ethics is the effort to obtain good reasons, not fast answers [65]. The personalistic positions are consistent in the protection of the right to life, but have trouble convincing society and researchers. The non-personalistic positions will not find approval because of the openly declared consequences for new-born children, disabled, or dying persons. The middle positions try to release research from some ethical boarders, without damaging the conviction of equal human dignity. But their methods of reasoning don't really convince and, in the long run, leave open too many options.

Nevertheless two fundamental considerations seem to support maintaining a rejection of the destruction of human embryos for research purposes:

a. If someone wants to justify hESC research, either within a limited extent or up to research cloning, she/he must be able to give convincing arguments, why embryos might be treated in a different way than born humans. This seems to be impossible without weakening or denying the moral status of early human embryos. This method of reasoning possesses danger of weakening the protection of the human dignity in general. If the coincidence of human species and human dignity is given up and exchanged for a dignity awarded by society, corresponding to actual research interests, serious doubts may arise, whether the desired protection standard can be maintained in other areas of life, e.g. for coma patients, disabled people or new born children.

b. hESC research including destruction of human embryos is not without alternatives. The promised therapeutic applications of hESC research are still lacking, while research in adult stem cells and hiPSC research seem very promising and are reducing the ethical objections. When opposition to hESC research is still accused of impeding research and preventing necessary new therapies, this could also be seen as a clever policy of small steps to deceive moral convictions. Also other objectives are highly relevant, for example industrial applications in toxicity testing with human embryos as a substitute for animal experiments: "These cell lines may provide more clinically relevant biological systems than animal models for drug testing and are therefore expected to contribute to the development of safer and more effective drugs for human diseases and ultimately to reduce the use of animals. They also offer the possibility to develop better in vitro models to enhance the hazard identification of chemicals. It is possible that these applications will turn out to be the major medical impact of human ES cell research..." [66].

4. Remaining ethical arguments concerning hiPSCs with special emphasis on the argument of complicity in another's wrongdoing and double effect reasoning

If it is true that successful therapeutic applications are more likely to result from hiPSC research than from hESC research, ethical problems would be reduced significantly [4, 26, 67, 68]. Research cloning could be avoided. It would never be necessary for therapeutic application. hESC-research would, at least, be reduced to the domain of basic research and control experiments. For this remaining need it seems realistic that already existing cell-lines will be sufficient [12]. In this case, the destruction of human embryos for research is completely avoidable in the future and even the destruction of surplus human embryos may be unnecessary.

Nevertheless, even in hiPSC-research, some ethical issues remain and are in need of intensive consideration:

Can the distinction between hESC and hiPSC be explained in a consistent and convincing way? Is it possible to find a reliable delimitation between pluripotent and totipotent stem cells? Is it possible to prevent the production of germ cells out of hiPSCs, as well as their use to create new research-embryos [26]?

Is the assumption that hESC-research is completely dispensable, or will be after a period of time, justified, or is it only a means of sedating the conscience? Some scientist say, that is too early to decide [11]. Even a temporal limited "exception", or a limited number cannot be seen as an exception of ethical principles but must be justified. If further destruction of a limited number of human embryos for research purposes would be necessary during a transition period, some ethicists argue for the use of surplus embryos from IVF-treatment [31]. The ethical objections were indicated above. This way is surely not acceptable, if, according to our appraisal, existing cell lines are sufficient. If not, the use of surplus embryos needs to be justified in a consistent way without denying the human dignity of embryos and without opening the way to the creation of research embryos on demand and even for non-therapeutic applications.

How can the cell donors' right to voluntary and informed consent, as well as the protection of personal data, especially in the case of application of hiPSCs as disease models, be guaranteed? How can the relevant questions of property rights and patent law be solved [69]?

Even hiPSC-research is, in several ways, confronted with the ethical problem of "complicity in others' wrongdoing": How can someone consistently reject the destruction of human embryos and, at the same time, use the result of former destructive research [13, 47, 70, 71]? Katrien Devolder draws attention to this problem of complicity. She contradicts the opinion, that hiPSC research is ethically correct, while hESC research is wrong because it involves destruction of human embryos: "Many who object to human embryonic stem cell (hESC) research because they believe it involves complicity in embryo destruction have welcomed induced pluripotent stem cell (iPSC) research as an ethical alternative. This opinion article aims to show that complicity arguments against hESC research are *prima facie* inconsistent with accepting iPSC research as it is currently done." [13].

In this passage we would like to scrutinize her theses and her suggestions for a solution. We are convinced that the problem of complicity is no obstacle for hiPSC research, if certain requirements are met.

4.1. Double effect reasoning

In theological and philosophical ethics, problems like this (cooperation with another's sin, "cooperatio in malo") can be discussed in relation to the so-called "principle of an action with double effect", in brief "principle of double effect", or "double-effect reasoning" [72-74]. In this principle, a distinction is drawn between direct consequences of an action and side effects, which are only indirectly wanted or accepted as unavoidable. The principle wasn't interpreted and used uniformly and has undergone some changes. In philosophical and theological ethics, it is relevant in two different contexts. The first and original context is the question of cooperating with the sin of another person. In these cases, the wrongness of the action is presupposed and the question concerns only the legitimacy, or culpability of the cooperation. Furthermore the principle of double effect is relevant in the context of some specific moral norms, such as the prohibition of killing to determine moral rightness or wrongness. In these cases it is a principle of restrictive interpretation of deontological moral norms [75]. This is an issue of high complexity and not necessary for the question of complicity. In the first context, the principle draws one's attention to several relevant aspects that may be helpful for our question of complicity in hESC and hiPSC research.

The basis of the argument of complicity with another's wrongdoing is the estimation that somebody, who cooperates in, or profits from the morally reprehensable actions of other persons, makes himself responsible in a certain way as an accomplice. "Complicity" means a culpable cooperation in the ethically wrong action of another person. The conviction that we are responsible not only for the immediate results of our behaviour, but also for the influence we exert by our behaviour on convictions and behaviour of others in the long run, as far as this is foreseeable, is fundamental.

Just as the demands of morality are aimed at the inner attitude as well as the outer actions of man, accusations of complicity are not only aimed at a voluntary and deliberate cooperation in the wrong actions of others, but also at inadequate attitudes towards the wrong actions of others. Our inner disposition, our fundamental attitude, our character is the central content of our moral obligation. Morality primarily consists in the fundamental attitude of impartial benevolence, in the respect for the equal dignity of all humans. Motives cannot be recognized directly but only inferred from our behaviour. Sometimes adequate symbolic actions can help to express the inner attitudes and prevent misunderstandings. Symbolic actions partly get credibility by the costs they cause and by the disadvantages somebody is ready to accept [76].

This effort especially is necessary if somebody profits from the wrong actions of others and thus, gives the impression of approval or inner consent of these actions. This can even be the case, if one wasn't involved in the wrong actions at all. The use of research results from morally reprehensible experiments in the past [77] without an explicit dissociation can give the impression of lacking sensibility and missing respect for the victims or even the impression of an inner consent, of condoning or justifying these actions. If there are scientific reasons to use the results, the rejection of these crimes must be articulated by explicitly remembering the victims and condemning the crimes.

Complicity with another's wrongdoing can happen in different constellations. In the tradition of moral theology, different types of cooperation with the sin of another one were distinguished and relevant distinctions were made for the degree of guilt [74, 78, 79].

In any case, the rejection of a sin, a willingly performed wrong action of another person, is required. Complicity, as an inner consent when another one's sin "is wanted as such", is called "*formal*" cooperation and is always wrong. Even an implicit inner consent is seen as a formal cooperation, especially in the case of serious offenses. If the inner consent is missing because the cooperation happens involuntarily or without knowledge, this is called a "*material*" cooperation. However, this kind of cooperation requires a justification, but, in contrast to a formal cooperation, this is possible. According to traditional arguments a material cooperation is permitted, if the other's sin is "wanted only indirectly" and the action corresponds to the rules of the "principle of double effect".

Within the principle of double effect, a distinction is drawn between direct consequences of an action and side effects, which are only indirectly wanted, or accepted as unavoidable. While direct cooperation is regarded as forbidden, the indirect one can be justified by adequately important, so-called *proportionate reasons* for accepting the others' sin. In this way, teleological reasoning, on the basis of balancing good and bad consequences, is made possible for the indirect causation of the others' sin. Nevertheless, this remains excluded for a di-

http://dx.doi.org/10.5772/54375

rect causation or a direct intention, in which the wrong action is intended itself (per se), or as a means to an end [80]. In these cases the sin must be seen as directly intended. As a minimum for speaking of an indirect causation of an evil, it was demanded that good and bad consequences must result from the action "at least equal immediately" [73, 74, 78].

In casuistry, further types of a "material" cooperation were distinguished: A *positive* cooperation by an active action is more serious than a *negative* cooperation by omission of an action. An *immediate* cooperation is more serious than *a mediate.* A *near* cooperation is more serious than a *remote* one. Necessary cooperation, without which the wrong action of another one wouldn't have happened at all, is worse than cooperation, when it would have been performed anyway. A direct intention could be suspected, the more immediate and more near one's own action is connected with another one's sin and the more probably the other one wouldn't sin without this cooperation. Here the principle includes a difficult question: Does the indirectness and justifiability of complicity primarily depend on the causal proximity, or on the probability of another person's wrong action? Is it really less problematic to promote a wrong action with high probability, if the number of mediating instances is increased? In the theological tradition there was no agreement on this matter. According to a teleological method, responsibility refers to all foreseeable consequences that can be influenced by one's actions. In this point of view, probability is more important than proximity. For the credibility of the inner consent, proximity may be the greater problem.

These distinctions show the difficulties in dissociating oneself consistently from another's wrongdoing while cooperating or profiting from it. While the distinction between formal and material cooperation is a clear alternative, the distinctions of types of material cooperation seems in real life often to be a matter of degree. Al least one could say, that the effort to make one's own inner rejection of anothers' wrongdoing credible to other people is greater, the more a cooperation is near, immediately and necessarily.

The principle of double effect includes at least three relevant aspects that may help to evaluate the problem of complicity in hESC and iPSC research: (a) In any case, the rejection of another one's action, which one determines as ethically wrong, is required as matter of inner consistency. (b) A material cooperation can, nevertheless, be ethically justified, if intention and causal relation can be seen as indirect, which is sometimes clearly identifiable, but is often a matter of degree. (c) In any case, a proportionate reason for accepting the others' sin must be given. Additionally sometimes symbolic actions will be necessary to maintain one's credibility.

4.2. Complicity according to Devolder

Devolder's statements to complicity partly correspond with these arguments. She introduces the following variants [13]:

1. "Causally contributing": "When I induce or encourage you, or provide you with the means to commit a murder, and as a result you commit it, I am complicit in that murder." In these cases, the other's wrong action is also the result of one's own action.

2. "Promoting wrongdoing through increasing demand for embryonic stem cell lines": "One can be complicit in wrongdoing by increasing the likelihood of that wrongdoing (or future instances of it) in certain ways, even if one does not in fact cause it."

3. "Promoting wrongdoing through altering attitudes to embryo destruction": Further ways of promoting wrongdoing "include condoning a wrong or fostering more permissive social attitudes towards it." Profiting from the use of the results of a wrong action can awake the assumption that one excuses this action. This can in the long run weaken social attitudes and promote wrong behaviour.

4. "Implicitly condoning wrongdoing and disrespecting its victims": Complicity can also be supposed, independent of the consequences, if an implicit excuse of a wrong action, or disrespect towards the victims seems to be expressed.

In the terminology of theological ethics, paradigms 1-3 refer to different forms of material cooperation. The first includes examples of direct and indirect cooperation specified as near forms of cooperation. Category 2 and 3 are examples of mediate cooperation of a more remote type, the acceptance of a wrong action as a side effect. One's own action is not sufficient for the realization of this side effect, but increases its probability in connection with others. In contrast to Devolder, this can also be seen as a kind of causation, but an indirect one. In Example 3, the side effect is a problematic change of social attitudes. This effect is even more remote. The connection is a very complex one. It is unquestionable that research often changes social attitudes. Researchers should think about such consequences, which occur as a result of their work. But they aren't alone responsible for it and their actions are seldom a sufficient condition for a change of social attitudes. Category 4 refers to the appearance of an inner consent, which is called an implicit formal cooperation. Either the actual inner attitude or the publicly noticeable expression is not adequate.

4.2.1. Devolder's criticism of hESC research

According to Devolder hESC research is confronted with the problem of complicity even if researchers use already existing cell lines and don't themselves destroy human embryos. Even if there is no direct causal contribution, they contribute to an "increasing demand for embryonic cell lines" [13, p 2176] and, in this way, promote the likelihood of "further embryo destruction" [13, p 2176]. At least at a collective level, this mediate and remote effect is a reality. Presupposition for this criticism is that destroying human embryos is determined as ethically not justified.

A strategy to prevent this contribution is "separating the use of hESCs from their derivation by instituting a cut-off date" [13, p 2176]. This method was used by the jurisdiction in Germany when trying to deal with the problem in 2002. When the cut-off date was moved in 2007, the credibility of the proclaimed objection to the destruction of embryos was damaged. If the shift of a cut-off date can be anticipated, contribution to an increasing demand is not prevented any more. Devolder emphasizes, that even when using hESCs produced before a cut-off date successful research may promote the destruction of embryos in less restrictive countries. As a counter-argument, she points out that hESC lines are mostly derived from discarded IVF embryos. Since they are available in a large number, hESC research will not increase the likelihood of embryo-destruction in any way. Of course this objection presupposes the acceptability of the destruction of surplus IVF embryos, which is an open discussion. In addition to this, the question arises, of whether or not research interests truly have no effect on the production of surplus IVF embryos [71].

Furthermore, Devolder indicates complicity by contributing to altering attitudes in society, changing moral beliefs, legislation or incentives. In this way, the potential benefits of hESC research for many people and the good reputation of biomedical research in general may weaken efforts to reduce the number of embryos discarded in IVF.

Finally, hESC research is accused of "implicitly condoning wrongdoing and disrespecting its victims". If the destruction of embryos is evaluated as a kind of wrongdoing, it is inconsistent and not credible, when researchers, who benefit from it, would regret or try to distance themselves from the practice of destruction of embryos. By using the stem cell lines, they seem to condone the way, they were obtained.

4.2.2. Devolder's Criticism of hiPSC research

hiPSC research enables the development of illness specific or patient specific pluripotent stem cells without supply of oocytes and without the creation and destruction of embryos. Thus, the central ethical objections seem to be removed. Contrary to widespread opinion, Devolders thesis is that, regarding complicity with the destruction of human embryos, hiPSC research is in a similar situation as hESC research. hiPSC research wouldn't be a solution for the ethical problems connected to hESC research. She "aims to show that complicity arguments against hESC research are *prima facie* inconsistent with accepting iPSC research as it is currently done." [13]. She suggests that, in a consistent way, both should be accepted or rejected.

Devolder accuses hiPSC research of „promoting and condoning embryonic stem cell research". The connections between hiPSC and hESC research seem to be similar to the connections between hESC research and embryo destruction: "Research on hESCs arguably promotes embryo destruction through increasing demand; similarly iPSC research arguably promotes hESC research in the same way. Engaging in hESC research arguably also implicitly condones embryo destruction, in part because it involves significant interaction with those who destroy embryos. Engaging in iPSC research involves even more significant interaction with hESC researchers and thus, even more plausibly, implicitly condones hESC research.... Consistency requires that considerations of complicity are invoked in both cases." [13]. To a great extent, hiPSC research uses results of hESC research and therefore cannot dissociate itself in a credible way from it. It seems to be contributing at least implicitly to weakening the rejection of the destruction of embryos. If hESC research is opposed because of complicity, according to Devolder, even hiPSC research must be seen as highly problematic, unless several modifications are implemented [13].

4.3. Application of double effect reasoning

The argument of complicity legitimately asks for justification of the involvement of hESC research and in a more remote way hiPSC research in the destruction of human embryos, even if researchers don't perform it themselves. Double effect reasoning can give some general guidance for performing research with including benefits from objected research in the past and unintended side-effects in the future. Researchers must look back and consider, how they think about the way cell lines, were obtained via the destruction of human embryos in the past. Their research should be in consistency with this judgement. They should also think about their contribution to further destruction of human embryos in the future. They should pay attention to the way their research changes the attitudes of society. Both kinds of consequences are part of the responsibility of researches to the extent they can be foreseen as being in some direct or indirect, close or remote way connected to their scientific work.

The possible indirect and more remote consequences of hiPSC research on the destruction of embryos cannot be denied. Who opposes the destruction of embryos for ethical reasons and nevertheless participates in hiPSC research, can be justified in the line of double effect reasoning only, if the rejection of the destruction of embryos and of possible problematic research in other countries is honest and proven by the attempt to minimize the effect of one's own research on promoting further embryo destruction. This objection should also be made public in some clear and unambiguous way and should be accompanied by institutional or legal precautions to avoid further embryo destruction and weakening of social attitudes. The remaining indirect or remote contributing can be justified, if the benefit of the research is adequately high.

4.4. Consistent solutions?

Devolder suggests 5 possible solutions [13]:

1. Rejection of hESC research, as well as hiPSC research.
2. Radical separation of the two research areas and "a change in the ways iPSC research is done so that it would no longer involve complicity in hESC research."
3. One could argue that hiPSC research is considerably more remote from the destruction of human embryos and is, in this respect, less contributing to a weakening of the social sensibility for the victims. In this respect, the "moral costs" could be justified more easily.
4. Complicity arguments could be rejected or limited to cases "when one actually and significantly causally contributes to more embryo deaths", which is not the case for research with stem cells obtained by others.
5. The wrongness of the destruction of human embryos for important research areas could be denied. In this case, the discussed complicity arguments would no longer be pertinent to both ways of research.

http://dx.doi.org/10.5772/54375

Rejection or radical separation of the two research areas are regarded as unappealing by Devolder, because this would be connected with considerable disadvantages for research. A complete renunciation would retard important research projects and be a disadvantage for potential patients hoping for new therapies. The renunciation would be a credible sign, but a burden for others is a problematic proof of one's own integrity.

A possible solution might be seen in a combination of Devolder's suggestions 2 and 3. The change in the ways hiPSC research is performed could be a radical constraint on the already existing stem cell lines and a credible renunciation of obtaining new stem cell lines, or using new ones from other countries, such as e.g. the European Group on Ethics proposes in its opinion 22: "The derivation of new toti-potent cells or pluri-potent stem cell lines from donated pre-implantation human embryos or embryonic cells, or via nuclear reprogramming, is not funded by the EU Research Programme." [12]. If existing cell lines are sufficient for the necessary comparison studies, research for therapeutic applications will not be hampered or retarded any way and no direct or near contribution to further destruction of embryos is remaining. If applicable regulations were found on a broad basis, protected in a credible way and maintained in the long run, complicity arguments pertaining to embryo destruction in the future wouldn't be applicable anymore to hiPSC research. If, according to the latest reports, the stage of pluripotency were dispensable for therapeutic applications and adult stem cells could be developed into desired cell types without this step [7], even the control studies with hESCs would become less important.

An important step in the direction of a limitation of research to existing hESC lines is the European registry of existing hESC lines: "The European Commission has therefore decided to establish and fund a European registry for human embryonic stem cell lines in order to help researchers to optimise the hESC resources available, avoid duplication of work and/or the creation of new cell lines where possible." [12]. This kind of policy helps to avoid the new destruction of embryos and enables transparency and credibility. Regulated in such a way hiPSC has a good chance, not to contribute to a weakening of the social sensibility for the victims of research and to changing attitudes to the dignity of human embryos. More likely it is a step towards the opposite direction of more respect for human dignity.

Devolder's suggestion 3 and 4 refer to the distinction of causally direct and indirect action. The argument, "that the complicity arguments for rejecting hESC research are stronger than the complicity arguments for rejecting iPSC research" [13] seems appropriate to us. Conforming to the principle of double effect, the distinction between immediate consequences and side effects, which are only wanted or accepted indirectly, opens a way to justify these kind of consequences by proportionate reasons like the high benefit of research for fighting diseases in the future. The remaining indirect and remote contribution to the destruction of embryos can be estimated as balanced as long as it is not actively supported and possible usage of results out of this kind of rejected research is not secretly hoped for.

Of course clarification is needed, which research objectives are regarded as adequately high for the use of hESC lines. Therapeutical applications for humans can be regarded as adequate, also necessary control experiments for research with adult stem cells or hiPSCs. But serious doubts appear in relation to non-therapeutic industrial applications like toxicity testing to replace or reduce animal experimentation. Here the opinions are divided and depending on the ethical background, using hESCs for applications like these are seen as a welcome improvement by the one side [12, 66], or as a disproportionate means and a way of damaging human dignity that is not acceptable by the other side. The European Group on Ethics stated clearly: "Although the Group is aware of the importance of respecting animal welfare, it is concerned that respect for human dignity may not be maintained when hESCs are used in toxicity testing of industrial or other commercially produced chemicals not related to drugs, such as cosmetics, or for replacement of animal testing. Therefore, particular attention is to be drawn to this issue." [12, 38, 69, 81]. The demand for further destruction of embryos would be increasing enormously and one can suppose that social attitudes would really change in the long run, if cell lines derived from human embryos are used as commodity, as raw material in industrial dimensions.

Devolder's fourth solution, narrowing "complicity" to cases "when one actually and significantly causally contributes to more embryo deaths" [13], is no convenient way. It tends to reduce researchers responsibility too much. Mediate and remote consequences of research are part of the researchers' moral responsibility. Abuse of discoveries and inventions, the promotion of personally rejected methods and applications and even a problematic modification of social attitudes are relevant objects of responsibility, as far as they can be foreseen and are enabled or promoted by one's own activity. Taking responsibility of course doesn't mean being accused for every effect, but being willing to give a justification for accepting unwanted side effects or long term consequences. If appropriate reasons are given, research is justifiable despite these problems. Thus, the principle of double effect opens a way of dealing with negative and unwanted side effects in a responsible way. Research does not justify everything. But complicity is reduced to cases of voluntary and deliberate cooperation in the actions of others, which one claims to evaluate as morally wrong, (1) when there is formal inner consent, even an implicit one, which is inconsistent, (2) when the cooperation is so near and direct, that an inner rejection is not credible any more, or (3) when the damage and harm caused by the wrong action is not balanced by a proportionate high benefit.

Devolder's fifth solution shows the necessary precondition for this discussion about complicity of hiPSC research, the determination of the destruction of human embryos for research purposes as morally wrong. This judgement mostly corresponds to a personalistic position regarding the moral status of human embryos. Non-personalistic and gradualistic positions don't determine destruction of embryos as morally wrong generally or under specific conditions. Of course they don't have a problem with the discussed type of complicity. As indicated in section 2 of this chapter, the ways of justifying the destruction of human embryos haven't been able to obtain an agreement until now: Denying or weakening of the moral status and dignity of early human embryos, of research embryos or at least of surplus IVF-embryos, always contains the risk of weakening this basic ethical argument of equal human dignity in general and causing bad effects for humans in other stages of life. The second way, a justification of their destruction, as a legitimate way of killing without denying dignity of human embryos, is not convincing and may cause similar side-effects.

5. Conclusion

A consensus conferring the moral status of human embryos and the ethical evaluation of creating and destructing human embryos hasn't been achieved in the past and doesn't seem probable in the near future. Attempts to justify the destruction of human embryos for research have not succeeded in answering the ethical objections in a sufficient and convincing way. Since fundamental moral attitudes and convictions are concerned, it is adequate to impose the burden of proof on those, who advocate these ways of research. Liberty of research finds its limits where the basic moral convictions of a society are violated.

In areas of close scientific cooperation the search for agreement in fundamental ethical questions remains an urgent challenge. In a pluralistic society, despite all efforts for an ethical basic consensus, it is possible that over a longer period of time, a consensus on a certain moral question cannot be found. In such cases, the principle of tolerance is applicable only if both positions, at least, share a common basis that allows to include the contradicting positions as rational and consistent lines of reasoning. The problem is that the positions regarding the moral status of human embryos don't seem to be reconcilable within a shared basic consensus.

In this situation, the only rational way seems to be the renunciation of any further destruction of human embryos, a concentration on research with adult stem cells, iPSCs, and, where necessary, with existing hESC-lines. According to the newest developments in stem cell research, this position doesn't retard research for therapeutic objectives. It has a chance to serve as a minimal consensus and, in the long run, possibly will prove to be the better way, scientifically, ethically, in relation to social acceptability and maybe even economically.

The concern for common and strong ethical standards is part of the external responsibility of science. Science itself is dependent on social agreement and legal certainty and would suffer from a distrust and hostility towards science. In the end, there should be no difference between ethical requirements and a science that is striving for an improvement of human living conditions in a sustainable and comprehensive way: "An ethics turned towards the future and a politics of comprehensive ecological, social and humane sustainability are guided by the insight, that there cannot be a double truth. Both, ethics and politics, should be guided by the conviction that in a humane society the moral right in the long run will also be the really beneficial for humans. Though one must realistically anticipate that single groups and perhaps even societies will try to provide themselves with short-term advantages by overriding ethical boundaries, this won't be to the advantage of most people and the world of future generations" [82].

Acknowledgements

This work was supported by Austrian Science Fund (FWF) Grant S9302-B05 (to M.B.).

Author details

Andreas M. Weiss[1], Michael Breitenbach[2], Mark Rinnerthaler[2] and Günter Virt[3]

1 Department of Practical Theology, University of Salzburg, Austria

2 Department of Cell Biology, University of Salzburg, Austria

3 Department of Moral Theology, University of Vienna, Austria

References

[1] Capecchi MR. Altering the genome by homologous recombination. Science 1989; 244:1288-92.

[2] Martin GR. Isolation of a pluripotent cell line from early mouse embryos cultured in medium conditioned by teratocarcinoma stem cells. Proc Natl Acad Sci U S A 1981; 78:7634-8.

[3] Thomson JA, Itskovitz-Eldor J, Shapiro SS, Waknitz MA, Swiergiel JJ, Marshall VS, Jones JM. Embryonic stem cell lines derived from human blastocysts. Science 1998; 282:1145-7.

[4] Condic ML, Rao M. Alternative sources of pluripotent stem cells: ethical and scientific issues revisited. Stem Cells Dev 2010; 19:1121-9.

[5] Guenther MG, Frampton GM, Soldner F, Hockemeyer D, Mitalipova M, Jaenisch R, Young RA. Chromatin structure and gene expression programs of human embryonic and induced pluripotent stem cells. Cell Stem Cell 7:249-57.

[6] Boland MJ, Hazen JL, Nazor KL, Rodriguez AR, Gifford W, Martin G, Kupriyanov S, Baldwin KK. Adult mice generated from induced pluripotent stem cells. Nature 2009; 461:91-4.

[7] Bonfanti P, Barrandon Y, Cossu G. 'Hearts and bones': the ups and downs of 'plasticity' in stem cell biology. EMBO Mol Med 2012; 4:353-61.

[8] Efe JA, Hilcove S, Kim J, Zhou H, Ouyang K, Wang G, Chen J, Ding S. Conversion of mouse fibroblasts into cardiomyocytes using a direct reprogramming strategy. Nat Cell Biol 2011; 13:215-22.

[9] Takahashi K, Tanabe K, Ohnuki M, Narita M, Ichisaka T, Tomoda K, Yamanaka S. Induction of pluripotent stem cells from adult human fibroblasts by defined factors. Cell 2007; 131:861-72.

[10] Takahashi K, Yamanaka S. Induction of pluripotent stem cells from mouse embryonic and adult fibroblast cultures by defined factors. Cell 2006; 126:663-76.

[11] Hug K, Hermeren G. Do we still need human embryonic stem cells for stem cell-based therapies? Epistemic and ethical aspects. Stem Cell Rev 2011; 7:761-74.

[12] Recommendations on the ethical review of hESC FP7 research projects. Opinion No 22.: EGE (The European Group on Ethics in Science and New Technologies to the European Commission); 2007.

[13] Devolder K. Complicity in stem cell research: the case of induced pluripotent stem cells. Hum Reprod 2010; 25:2175-80.

[14] Yoshida Y, Yamanaka S. Recent stem cell advances: induced pluripotent stem cells for disease modeling and stem cell-based regeneration. Circulation 2010; 122:80-7.

[15] Wilmut I, Schnieke AE, McWhir J, Kind AJ, Campbell KH. Viable offspring derived from fetal and adult mammalian cells. Nature 1997; 385:810-3.

[16] Gurdon JB, Elsdale TR, Fischberg M. Sexually mature individuals of Xenopus laevis from the transplantation of single somatic nuclei. Nature 1958; 182:64-5.

[17] Cheng LT, Sun LT, Tada T. Genome editing in induced pluripotent stem cells. Genes Cells 2012; 17:431-8.

[18] Kim YG, Cha J, Chandrasegaran S. Hybrid restriction enzymes: zinc finger fusions to Fok I cleavage domain. Proc Natl Acad Sci U S A 1996; 93:1156-60.

[19] Sebastiano V, Maeder ML, Angstman JF, Haddad B, Khayter C, Yeo DT, Goodwin MJ, Hawkins JS, Ramirez CL, Batista LF, Artandi SE, Wernig M, Joung JK. In situ genetic correction of the sickle cell anemia mutation in human induced pluripotent stem cells using engineered zinc finger nucleases. Stem Cells 2011; 29:1717-26.

[20] Miller JC, Tan S, Qiao G, Barlow KA, Wang J, Xia DF, Meng X, Paschon DE, Leung E, Hinkley SJ, Dulay GP, Hua KL, Ankoudinova I, Cost GJ, Urnov FD, Zhang HS, Holmes MC, Zhang L, Gregory PD, Rebar EJ. A TALE nuclease architecture for efficient genome editing. Nat Biotechnol 2011; 29:143-8.

[21] Hacein-Bey-Abina S, Hauer J, Lim A, Picard C, Wang GP, Berry CC, Martinache C, Rieux-Laucat F, Latour S, Belohradsky BH, Leiva L, Sorensen R, Debre M, Casanova JL, Blanche S, Durandy A, Bushman FD, Fischer A, Cavazzana-Calvo M. Efficacy of gene therapy for X-linked severe combined immunodeficiency. N Engl J Med 363:355-64.

[22] Kaiser J. Gene therapy. Seeking the cause of induced leukemias in X-SCID trial. Science 2003; 299:495.

[23] Chen HF, Chuang CY, Lee WC, Huang HP, Wu HC, Ho HN, Chen YJ, Kuo HC. Surface marker epithelial cell adhesion molecule and E-cadherin facilitate the identification and selection of induced pluripotent stem cells. Stem Cell Rev 2011; 7:722-35.

[24] Maio G, Hilt A. Der Status des extrakorporalen Embryos im interdisziplinären Zugang – Grundlagen, Herausforderungen, Ergebnisse. In: Maio G (ed.) Der Status des

extrakorporalen Embryos. Perspektiven eines interdisziplinären Zugangs (Medizin und Philosophie 9). Stuttgart/Bad Cannstadt: frommann-holzboog; 2007. p 11-44.

[25] Maio G. Der Status des extrakorporalen Embryos. Perspektiven eines interdisziplinären Zugangs (Medizin und Philosophie 9). Stuttgart/Bad Cannstadt: frommann-holzboog; 2007.

[26] Kummer C. Induzierte pluripotente Stammzellen und Totipotenz. Die Bedeutung der Reprogrammierbarkeit von Körperzellen für die Potentialitätsproblematik in der Stammzellforschung. In: Hilpert K (ed.) Forschung contra Lebensschutz. Der Streit um die Stammzellforschung (Quaestiones disputatae 233). Freiburg i. Br./Basel/Wien: Herder; 2009. p 322-338.

[27] Holland S, Lebacqz K, Zoloth L editors. The Human Embryonic Stem Cell Debate. Science Ethics and Public Policy. Cambridge Mass: Massachusetts Institute of Technology; 2001.

[28] Lenzen W (ed.) Wie bestimmt man den "moralischen Status" von Embryonen? Paderborn: 2004.

[29] Wagner-Westerhausen K. Die Statusfrage in der Bioethik (Ethik in der Praxis, Kontroversen 26). Münster: LIT-Verlag; 2008.

[30] Ostnor L. Stem Cells, Human Embryos and Ethics. Interdisciplinary Perspectives. Oslo: Springer; 2008.

[31] Research on Human Embryonic Stem Cells. Opinion of the Austrian Bioethics Commission. Wien: Austrian Bioethics Commission 2009

[32] Convention for the Protection of Human Rights and Dignity of the Human Being with regard to the Application of Biology and Medicine (Convention on Human Rights and Biomedicine). Oviedo: Council of Europe; 1997.

[33] Additional Protocol to the Convention for the Protection of Human Rights and Dignity of the Human Being with regard to the Application of Biology and Medicine, on the Prohibition of Cloning Human Beings. Paris: Council of Europe 1998.

[34] Cloning in Biomedical Research and Reproduction. Scientific Aspects – Ethical, Legal and Social Limits. Bonn: Bonn University Press; 2003.

[35] Virt G. Zur ethischen Debatte um das Klonen. In: Marschütz G, Prüller-Jagenteufel G (eds.) Damit Menschsein Zukunft hat. Theologische Ethik im Einsatz für eine humane Gesellschaft. Würzburg: Echter Verlag; 2007. p 196-203.

[36] Stammzellenforschung und therapeutisches Klonen. Göttingen: Vandenhoeck und Ruprecht; 2002.

[37] Human Cloning and Human Dignity. With a foreword by Leon R. Kass and M.D. Chairman. New York: Public Affairs: The Report of the President's Council on Bioethics 2002.

[38] Directive 98/44/EC OF THE on the legal protection of biotechnological inventions. European Parliament and the Council; 1998.

[39] Ethical Aspects of Human Stem Cell Research and Use. Opinion No 15. EGE (The European Group on Ethics in Science and New Technologies to the European Commission); 2000.

[40] Kant I. Groundwork of the Metaphysic of Morals, translated by James W. Ellington. Hackett; 1785/1993.

[41] Wolbert W. Der Mensch als Mittel und Zweck. Die Idee der Menschenwürde in normativer Ethik und Metaethik. Münster: Aschendorff; 1987.

[42] Wolbert W. Du sollst nicht töten. Systematische Überlegungen zum Tötungsverbot. Freiburg i.Ue./Freiburg i. Br.: Herder/Academic Press Fribourg; 2009.

[43] Knoepffler N. Der Beginn der menschlichen Person und bioethische Konfliktfälle. Anfragen an das Lehramt. Freiburg i. Br.: Herder; 2012.

[44] Damschen G, Schönecker D (eds.) Der moralische Status menschlicher Embryonen. Pro und contra Spezies-, Kontinuums-, Identitäts- und Potentialitätsargument. In: Berlin/New York: Walter de Gruyter; 2003.

[45] Düwell M. Der moralische Status von Embryonen und Feten. In: Düwell M, Steigleder K (eds.) Bioethik. Eine Einführung. Frankfurt 2003. p 221-229.

[46] Oduncu FS. Moralischer Status von Embryonen. In: Düwell M, Steigleder K (eds.) Bioethik. Eine Einführung. Frankfurt: Suhrkamp; 2003. p 213-220.

[47] Maio G. Mittelpunkt Mensch. Ethik in der Medizin. Ein Lehrbuch. Stuttgart: Schattauer; 2012. p 201-220.

[48] Congregation for the Doctrine of the Faith, Instruction on respect for human life in its origin and on the dignity of procreation – Donum vitae (February 22, 1987). Donum vitae AAS 80; 1987. p 70-102.

[49] Congregation for the Doctrine of the Faith, Instruction Dignitas Personae on Certain Bioethical Questions (September 8, 2008). Dignitas personae AAS 100; 2008. p 858-887.

[50] Bormann FJ. Embryonen, Menschen und die Stammzellforschung. Plädoyer für eine differenzierte Identitätsthese in der Statusfrage. In: Theologie und Philosophie. 2002. p 216-232.

[51] Virt G. Verantwortung für das Menschenleben an seinem Beginn. In: Marschütz G, Prüller-Jagenteufel G (eds.) Damit Menschsein Zukunft hat. Theologische Ethik im Einsatz für eine humane Gesellschaft. Würzburg: Echter Verlag; 2007. p 170-186.

[52] Schockenhoff E. Ethik des Lebens. Grundlagen und neue Herausforderungen. Freiburg/Basel/Wien: Herder; 2009.

[53] Virt G. Die Spaltung des menschlichen im Horizont der modernen Lebenswissenschaften. In: Oduncu FS, Schroth U, Vossenkuhl W (eds.) Stammzellenforschung und therapeutisches Klonen. Göttingen: Vandenhoeck und Ruprecht; 2002. p 201-210.

[54] Singer P. Practical Ethics. Cambridge: Cambridge University Press; 1993.

[55] Körtner HJ, Bünker M. Verantwortung für das Leben. Eine evangelische Denkschrift zu Fragen der Biomedizin. Im Auftrag des Evangelischen Oberkirchenrats A. und H.B. der Evangelischen Kirche A. und H.B. in Österreich erarbeitet von Ulrich H. J. Körtner in Zusammenarbeit mit Michael Bünker. Wien: 2001.

[56] Fischer J. Vom Etwas zum Jemand. Zeitzeichen 2002; 3(1) 13. 2002.

[57] Vieth A. Einführung in die Angewandte Ethik. Darmstadt Wissenschaftliche Buchgesellschaft; 2006.

[58] Wiesemann C. Von der Verantwortung, ein Kind zu bekommen. Eine Ethik der Elternschaft. München: C.H. Beck; 2006.

[59] Beckmann JP. Ontologische Status- oder pragmatische Umgangsanalyse? Zur Ergänzungsbedürftigkeit des Fragens nach dem Seinsstatus des extrakorporalen frühen menschlichen Embryos in ethischen Analysen. In: Maio G (ed.) Der Status des extrakorporalen Embryos. Perspektiven eines interdisziplinären Zugangs (Medizin und Philosophie 9). Stuttgart/Bad Cannstadt: frommann-holzboog; 2007. p 275-304.

[60] Koch HG. Disziplinspezifische Vorannahmen: Intrinsische und extrinsische Statusbestimmung des extrakorporalen Embryos – Vermittlung und Kombinatorik der verschiedenen Kriterien. In: Maio G (ed.) Der Status des extrakorporalen Embryos. Perspektiven eines interdisziplinären Zugangs (Medizin und Philosophie 9). Stuttgart/Bad Cannstadt: frommann-holzboog; 2007.

[61] Körtner UHJ. Embryonenschutz und Embryonenforschung aus der Sicht evangelischer Theologie. In: Körtner UHJ, Kopetzki C (eds.) Embryonenschutz – Hemmschuh für die Biomedizin? Wien: Springer; 2003. p 84-111.

[62] Dabrock P, Klinnert P. Verbrauchende Embryonenforschung. Kommt allen Embryonen Menschenwürde zu? In: Dabrock P, Klinnert P, Schardien S (eds.) Menschenwürde und Lebensschutz. Herausforderungen theologischer Bioethik. Gütersloh: Gütersloher Verlagshaus; 2004. p 173-210.

[63] Weiss AM. Moralischer Status von Embryonen – Hindernis für die Forschung? Anmerkungen zu Thesen von Ulrich Körtner. In: Fischer M, Zänker KS (eds.) Medizin- und Bioethik (Ethik transdisziplinär 1). Frankfurt: Peter Lang; 2006. p 117-138.

[64] Weiss AM. Abwägung im Tötungsverbot? Zur Kontroverse um die Forschung an überzähligen Embryonen. In: Haering S, Hirnsberger J, Katzinger G, Rees W (eds.) In mandatis meditari. FS f. Hans Paarhammer zum 65. Geburtstag. Berlin: Duncker & Humblot; 2012. p 387-402.

[65] Morscher E. Why is it morally wrong to clone a human being? How to evaluate arguments of biopolitics, biomorality, and bioethics. In: Thiele F, Ashcroft RE (eds.) Bioethics in a Small World. Berlin-Heidelberg Springer; 2005. p 121-128.

[66] Commission Staff Working Paper: Report on Human Embryonic Stem Cell Research. ("Matthiessen report"). Commission of the European Union; 2003.

[67] Müller AM, Obier N, Choi SW, Li X, Dinger TC, Brousos N. Möglichkeiten und Chancen der Stammzellenforschung: Stammzellen für Alle? In: Hilpert K (ed.) Forschung contra Lebensschutz. Der Streit um die Stammzellforschung (Quaestiones disputatae 233). Freiburg i. Br. / Basel / Wien: Herder; 2009. p 30-44.

[68] Breitenbach M, Laun P. Einige biologische Grundlagen der modernen Reproduktionsmedizin und der Stammzell- bzw. Gentherapie. In: Fischer M, Zänker KS (eds.) Medizin- und Bioethik (Ethik transdisziplinär 1). Frankfurt: Peter Lang; 2006. p 29-50.

[69] Ethical Aspects of Patenting Inventions involving Human Stem Cells. Opinion No 16. EGE (The European Group on Ethics in Science and New Technologies to the European Commission); 2002

[70] Green RM. The human embryo research debates. Oxford: Oxford University Press; 2001.

[71] Wolbert W. Zum Vorwurf der Doppelmoral in der Diskussion um die embryonale Stammzellforschung. Deutsche Medizinische Wochenschrift 2003; 128:453-456.

[72] Cavanaugh TA. Double Effect reasoning. A Critique and Defense. Dissertation. University of Notre Dame: 1995.

[73] Kaczor C. Proportionalism and the Natural Law Tradition. Washington: The Catholic University of America Press; 2002.

[74] Weiss AM. Sittlicher Wert und nichtsittliche Werte. Zur Relevanz der Unterscheidung in der moraltheologischen Diskussion um deontologische Normen (Studien zur theologischen Ethik 73). Freiburg i. Ue./Freiburg i. Br.: Universitätsverlag Freiburg/Herder; 1996.

[75] Schüller B. Die Begründung sittlicher Urteile. Typen ethischer Argumentation in der Moraltheologie. Düsseldorf: Patmos Verlag 1987.

[76] Ginters R. Die Ausdruckshandlung. Eine Untersuchung ihrer sittlichen Bedeutsamkeit (Moraltheologische Studien. Systematische Abteilung 4). Düsseldorf: Patmos Verlag; 1976.

[77] Fragwürdige Medizin. Unmoralische Forschung in Deutschland, Japan und den USA im 20. Jahrhundert. Frankfurt a. M.: Campus Verlag; 2008.

[78] Mausbach J. Katholische Moraltheologie, überarbeitet von G. Ermecke. Münster: Aschendorff; 1959.

[79] Göpfert FA. Moraltheologie. Paderborn: Schöningh; 1905.

[80] Schüller B. Wholly Human. Essays on the Theory and Language of Morality. Washington D.C.: Georgetown University Press; 1986.

[81] Judgement in case C-34/10 Oliver Brüstle vs. Greenpeace. Court of Justice of the European Union 2011.

[82] Virt G. Zukunftswelten – Lebenswelten. In: Marschütz G, Prüller-Jagenteufel G (eds.) Damit Menschsein Zukunft hat. Theologische Ethik im Einsatz für eine humane Gesellschaft. Würzburg: Echter Verlag; 2007. p 106-114.